Contaminated Soils

Volume 2

Contaminated Soils

Volume 2

Site Assessment
Chemical Analysis and Environmental Fate
Risk Assessment, Remediation
Bioremediation, State Regulatory
Federal and Military Considerations
Manufactured Gas Plant Sites
Radioactivity, MTBE

edited by
Paul T. Kostecki
Edward J. Calabrese
Marc Bonazountas

Amherst Scientific Publishers   Amherst, Massachusetts

Amherst Scientific Publishers, 150 Fearing Street, Amherst, Massachusetts

The material contained in the document was obtained from independent and highly respected sources. Every attempt has been made to ensure accurate, reliable information. However, the publisher cannot be held responsible for the information or how the information is applied. Opinions expressed in this book are those of the authors and/or contributors and do not reflect those of the publisher.

ASP recognizes that its business necessarily entails the use of a great deal of paper. We acknowledge a certain irony in working hard to protect the environment through the extensive use of one of the environment's most precious resources. ASP therefore participates in the U.S. Forest Service's Plant-A-Tree Program. This program arranges for the planting of trees in our 156 national forests. Trees are planted for multiple purposes, including recreation area development, wildlife habitat and commercial timber production. A portion of our proceeds is earmarked for this worthy program. Thank you for joining us in our efforts on behalf of the environment.

Printed in the United States of America

ISBN 1-884940-10-2

Contents

Foreword

Each year, additional sites are found to be contaminated with petroleum hydrocarbons, volatile organic compounds, and other substances, and yet the financial resources to remediate these sites have been stretched to their limits. Because expenditures for environmental cleanup are driven by regulatory compliance requirements rather than profit generation, responsible parties in the industrial and commercial sectors often are reluctant to commit resources to restoration of sites with low regulatory priority. Today, expensive remediation projects can only be justified at sites with an immediate or impending risk to human health or the environment. In recognition of this fact, many regulatory agencies have issued more flexible cleanup standards which promote a risk-based approach to remediation. The American Society for Testing and Materials recently released a Risk-Based Corrective Action (RBCA) standard that is gaining increasing acceptance, particularly among state agencies responsible for leaking underground storage tank cleanups. The success of this approach depends on careful site characterization, scientifically documented estimates of natural attenuation processes, and proper selection and monitoring of effective low-cost remediation technologies.

Contaminated Soils, Volume 2 provides 52 technical papers that were presented at the 11th Annual Contaminated Soils Conference held at the University of Massachusetts at Amherst in October 1996. This volume provides unique solutions and technical information on a wide spectrum of environmental issues. Discussed are site assessment (Part I), chemical analysis and environmental fate (Part II), and risk assessment (Part III). Part IV discusses various technologies for remediating contaminated sites. Applications of bioremediation are presented in Part V. Current perspectives from state regulatory programs and for federal/military sites are presented in Parts VI and VII, respectively. A special session was held at the conference to address the investigation, management, and remediation of manufactured gas plant site (Part VIII). Parts IX and X present papers on treatment of readionuclide-contaminated soils and MTBE in groundwater, respectively.

Papers representative of both oral and poster sessions are included as part of this volume. The majority of the authors are from the consulting industry, although representatives from academia, industry, and the regulatory community have also contributed. Consulting firms, federal, state and local government employees, industrial representatives, engineers, researchers, scientist, educators and other professionals involved in hazardous waste management and treatment will find that this volume provides useful information on the current state of the science of environmental cleanup.

Peter Guest
Senior Associate
Parsons Engineering Science, Inc.
Denver, CO

Acknowledgments

We wish to thank all the agencies, organizations and companies that sponsored the conference, without their generosity and assistance the conference and this book would not have been possible.

Benefactors

> *American Petroleum Institute*
> *Association of American Railroads*
> *Gas Research Institute*
> *Massachusetts Department of Environmental Protection*
> *Maxxus Energy Corporation*
> *Naval Facilities Engineering Command*
> *Naval Facilities Engineering Service Center*

Sponsors

> *Camp Dresser & McKee*
> *ENSR Consulting and Engineering*
> *Geomatrix*
> *McLaren/Hart*
> *Parsons Engineering Science, Inc.*

Supporters

> *3M*
> *Association for the Environmental Health of Soils (AEHS)*
> *Earth Tech*
> *New York Department of Environmental Conservation*
> *Northeast Utilities Service Company*
> *Ogden Environmental and Energy Services*
> *Remediation Technologies, Inc.*
> *Roy F. Weston, Inc.*
> *Texaco, Inc.*
> *United Retek Corporation*

In addition we express our deepest appreciation to the members of the Advisory Committees who volunteered their time to provide guidance and encouragement.

Scientific Advisory Boards

General

Ralph Baker	*ENSR Consulting and Engineering*
Christopher P.L. Barkan	*Association of American Railroads*
Bruce Bauman	*American Petroleum Institute*
Barbara Beck	*Gradient Corporation*
Robert Block	*Remediation Technologies, Inc.*
Marco Boscardin	*GEI Consultants, Inc.*
Clifford Bruell	*University of Massachusetts Lowell*
David Burmaster	*Alceon Corporation*
John Button	*CER Corporation*
Leo Carden	*Roy F. Weston*
Mark Chrisos	*Earth Tech*

Rob Clemens	*Ogden Environmental and Energy Services*
James Colman	*Massachusetts DEP*
Michael Conway	*United Retek Corporation*
Jeffrey Dagdigian	*McLaren Hart Environmental Engineering*
John Del Pup	*Texaco, Inc.*
James Dragun	*The Dragun Corporation*
Mohamed Elnabarawy	*3M Environmental Technology and Services.*
Brent Finley	*McLaren Hart/ChemRisk*
Gary Foote	*Geomatrix*
George Furst	*Massachusetts Turnpike Authority*
Peter Guest	*Parsons Engineering Science*
Annette Guiseppi-Elie	*Exxon Biomedical Sciences*
Haim Gunner	*Biosolutions International*
John Gustafson	*Shell Development Company*
Tom Hayes	*Gas Research Institute*
Robert Hazen	*New Jersey DEP*
Wally Hise	*Radian Corporation*
Michael Huesemann	*Battelle*
Don Jacobs	*ENSR*
Paul Jacobsen	*Northeast Utilities*
Evan Johnson	*Tighe & Bond Consulting Engineers*
Sara Kimball	*Consultant*
Dick Kraybill	*Roy F. Weston, Inc.*
Bill Kucharski	*GECCO Inc.*
Guy Lanza	*University of Massachusetts, Amherst*
Ralph Ludwig	*Wastewater Technology Centre*
John Lynch	*RETEC*
Thomas McMahon	*Southern California Gas Co.*
David Nakles	*RETEC*
Tom Pederson	*Camp Dresser & McKee, Inc.*
Frank Peduto	*New York DEC*
Tom Potter	*University of Massachusetts, Amherst*
Lynn Preslo	*Earth Tech*
Mark Richardson	*O'Connor Associates*
Eugene Rutz	*University of Cincinnati*
Martina Schlauch	*Radian Corporation*
Corinne Schultz	*Roy F. Weston, Inc.*
Lee Shull	*Foster Wheeler*
Baoshan Xing	*University of Massachusetts, Amherst*

Federal Advisory Board

John Cullinane	*US Army Engineer Waterways Exp. Station*
Joseph Graf	*Naval Facilities Engineering Command*
Leslie Karr	*Naval Facilities Engineering Service Center*
Lynn S. Kucharski	*McLaren Hart*
Mark Nickelson	*Martin Marietta Energy Systems, Inc.*
William Quade	*Naval Facilities Engineering Command*
Paul Rakowski	*Atlantic Division Naval Facilities Engineering Command*
Fred Ramsey	*ATSDR*
Debby Tremblay	*US EPA*
Wade Weisman	*Wright-Patterson AFB*

PART I—SITE ASSESMENT

CHAPTER 1

Management of Environmental Data Generated During Large-Scale Site Investigations - A GIS Database Case Study

Donald J. Bourgeois, Jeffrey A. Hamel, and Eric W. Daley, EMCON, Andover, Massachusetts

INTRODUCTION

Concurrent large-scale Remedial Investigations (RI) are being conducted at a large defense contractor's six facilities located in the eastern United States, to comply with either state and/or federal (RCRA Corrective Action Program) regulations. Due to the extensive amount of environmental data anticipated to be generated during the investigations, a database was selected and implemented to more thoroughly and cost-effectively manage and interpret the data.

Data management objectives of the project team that were evaluated prior to the selection of the specific database included: management of both hydrogeological and analytical data; generation of and compatibility with AutoCAD™ drawings; quick response, immediate access to data, and control of output formats; electronic validation of quality assurance/quality control (QA/QC) data; and a system specifically designed to manage environmental data and be compatible with other software packages.

The database selected was a three-dimensional Geographical Information System (GIS) relational database (GIS\Key™) coupled with statistical and visualization software. GIS\Key™ was installed on a Hewlett-Packard Series 3 Vectra VL personal computer equipped with a pentium processor operating at 75 MHz. Hard drive capacity was 1 gigabyte with 32K bytes of Random Access Memory (RAM). Other software used in conjunction with GIS/Key™ were FoxPro™, AutoCAD™, Site View™, and The Monitor System™.

Field activities for the RI were initiated in June 1993. As of December 1995, the environmental data generated and managed by the database includes: 560 soil borings and monitoring wells; 3,600 portable field gas chromatograph screening results; 1450 soil gas points; greater than 4,500 individual environmental samples for laboratory analysis (soil, groundwater, surface water, sediment, wipe, and QA/QC samples); numerous groundwater level measuring events; and elevation and location survey data.

DATA MANAGEMENT

To satisfy the project's data management objectives, data management procedures were developed and implemented, including:

- sample tracking from collection through analyses and reporting
- base map development
- electronic data delivery
- data coding
- data validation
- database outputs

Discussions of each of these procedures are presented in the following sections.

Sample Tracking from Collection Through Analyses and Reporting

A sample tracking spreadsheet was developed by the database managers which enabled the daily field input of samples collected, their specific identification number, sample matrix, requested analysis, and sample type (i.e., primary or QA/QC). Upon receipt of the samples at the laboratory, a sample confirmation was sent to the field to verify the requested sample analyses, sample numbering (to avoid transcription errors), and the due date for the results.

This program was also used to generate a field and office listing of QA/QC samples and their percentage compared to the number of primary samples (i.e., one of the project QA/QC goals was to collect 10% duplicate samples). An example of this sampling tracking sheet is presented as Table 1.

The program was also accessible to the database manager who compared the input produced in the field with the electronic files provided by the laboratory. This ensured that all sample data (both electronic and hard copy) were received from the laboratory and properly identified.

The primary benefit of this program was that it provided the baseline, which was completed by the sampler in the field on the sampling date, to verify that all samples were received, analyzed, and reported for the appropriate parameters by the laboratory.

Base Map Development

The key link between the analytical data and the specific geographic location was the AutoCAD™ site base map. Because the database is a GIS database, it relates all analytical, geological, and hydrological information in the database to a precise X,Y,Z coordinate on the base map.

The topographic and planimetric features on the site base maps were generated photogrammetrically from aerial photography. All site features and the specific coordinates of each surveyed sampling location (e.g., soil borings or monitoring wells) were electronically downloaded directly onto the site base map.

Table 1. **Sample Tracking Spreadsheet**

Primary Sample Number	Custocy Seal #	Sample Date	Sample Time	Sample Depth	Method	Sample Matrix	Preparation Fraction	Result Type	Duplicate Sample Number
PR02-SS-SS01-1033	30314	23-Aug-95	11:00	2	8260	S	T	PP	
PR02-SS-SS01-1033	30314	23-Aug-95	11:00	2	D3328	S	P	PP	
PR02-SS-SS01-1033	30314	23-Aug-95	11:00	2	D3328	S	T	PP	
PR02-SS-SS50-1034	30314	23-Aug-95	11:00	2	8260	S	T	PD	SH02-SS-SS01-1033
PR02-SS-TB03-1040	30314	23-Aug-95	17:00	0	8260	W	T	BT	
PR02-SS-SS99-1041	30314	23-Aug-95	16:50	0	8260	W	T	BF	
PR05-SS-SS01-1035	30314	23-Aug-95	12:00	3	8270	S	T	PP	
PR05-SS-SS01-1035	30314	23-Aug-95	12:00	3	6010	S	T	PP	
PR05-SS-SS01-1035	30314	23-Aug-95	12:00	3	7470	S	T	PP	
PR05-SS-SS01-1035	30314	23-Aug-95	12:00	3	D3328	S	P	PP	
PR05-SS-SS01-1035	30314	23-Aug-95	12:00	3	6010	S	P	PP	
PR05-SS-SS01-1035	30314	23-Aug-95	12:00	3	7470	S	P	PP	
PR05-SS-SS01-1035	30314	23-Aug-95	12:00	3	D3328	S	T	PP	
PR05-SS-SS01-1035	30314	23-Aug-95	12:00	3	TPH/EPH	S	T	PP	
PR05-SS-SS50-1036	30314	23-Aug-95	12:00	3	8270	S	T	PD	SH05-SS-SS01-1035
PR05-SS-SS50-1036	30314	23-Aug-95	12:00	3	6010	S	T	PD	SH05-SS-SS01-1035
PR05-SS-SS50-1036	30314	23-Aug-95	12:00	3	7470	S	T	PD	SH05-SS-SS01-1035
PR05-SS-SS50-1036	30314	23-Aug-95	12:00	3	D3328	S	P	PD	SH05-SS-SS01-1035
PR05-SS-SS50-1036	30314	23-Aug-95	12:00	3	6010	S	P	PD	SH05-SS-SS01-1035
PR05-SS-SS50-1036	30314	23-Aug-95	12:00	3	7470	S	P	PD	SH05-SS-SS01-1035
PR05-SS-SS50-1036	30314	23-Aug-95	12:00	3	D3328	S	T	PD	SH05-SS-SS01-1035
PR05-SS-SS50-1036	30314	23-Aug-95	12:00	3	TPH/EPH	S	T	PD	SH05-SS-SS01-1035
PR07-SS-MW02-1042	30314	23-Aug-95	11:00	2	8260	S	T	PP	
PR07-SS-MW02-1042	30314	23-Aug-95	11:00	2	8270	S	T	PP	

Preparation Fraction   P = SPLP Extraction   T = Total

Result Type   BF = Field Blank   BT = Trip Blank   PD = Duplicate Sample   PP = Primary Sample

Electronic Data Delivery

All analytical results are provided by the laboratories in an electronic data delivery (EDD) format developed by the database managers to be compatible with the database. The EDD format virtually eliminated potential errors frequently associated with repeated manual translation of data.

Upon receipt from the laboratory, a series of programs (written by the database managers) are run to compare the data with the sample tracking spreadsheet. Differences in data sets are reported on summary tables and corrected (i.e., obvious transcription errors are changed [8 instead of B], adding sample depths from field books, etc.). The database has two primary fields to identify a sample (sample number and site identifier). The sample number, which is recorded on the actual sample bottle and hard copy laboratory results, is never changed. The site identifier, which identifies a specific location, was modified, as necessary, to reflect actual site conditions (e.g., a sample may be labeled as soil boring SB-04 [XY12-SS-SB04-001], but completed as monitoring well MW-03; therefore, the site identifier was modified to XY12-MW03, while the sample number remained at XY12-SS-SB04-001).

The programs, in conjunction with the sample tracking spreadsheet, also confirm that every sample identifier has a corresponding sample location on the site base map, that the analytical parameters in the EDD were requested for the specific samples, and that the sampling information (dates, depths, units, etc.) were in an acceptable format.

Data Coding

As data are entered into the database, a series of codes are assigned that are used to identify and filter the data. Specifically, the data are designated as primary or QA/QC samples and as soil or aqueous samples, and identified with a program code. Program codes used during this investigation included: soil, groundwater, surface water, sediment, free product, pilot test samples, confirmatory soil samples from excavations, primary samples removed during soil excavation, biota samples, wipe samples, and air samples.

The program codes allow the user to specify the types of samples to be included in the database output. For example, by selecting the appropriate program codes, the user can generate a table of soil samples that include only confirmatory soil samples taken from excavations. In addition, the use of filters can further limit the output to only results above the laboratory's detection limit for soil samples collected from specific excavations, from specific depths, or from specific dates.

Data Validation

Once the data is in the database, validation of the data can be performed using available menu-driven queries, in accordance with the United States Environmental Protection Agency's (USEPA) protocol and guidance documents for analytical data validation. Analytical data are validated for numerous parameters including: holding time; compounds detected in blanks; surrogate, spike, and control sample recoveries; and duplicate and split analyses.

Although the actual time saved will vary depending on project-specific factors, the use of electronic data validation in this case study reduced the time spent by more than 60% when compared with a similar project where data validation was performed manually. A significant advantage was the virtual elimination of typical human errors associated with repeated and labor-intensive procedures.

An example of useful data validation output from the database is presented as Table 2. This table is a validation summary per sample matrix and lists the sample number, date, depth, concentration, laboratory detection limit, data validation qualifier, and reason code.

Database Outputs

Outputs from the database can be divided into two main categories: tabular output and graphical output, as discussed below.

Tabular

The database allowed presentation of geologic, hydrogeologic, and analytical results in a variety of report-ready tables. Generation of these tables greatly improved the efficiency with which results were evaluated by a variety of project personnel because the results could be filtered so that only those specific results needed by the user would be included on the table (i.e., report only positive detects over the laboratory's detection limit).

Typical tables generated included: primary results tables (i.e., water and soil results; groundwater elevations; soil boring and monitoring well construction details, etc.), QA/QC tables (i.e., blank hits, MS/MSD and surrogate recoveries, duplicate and split sample relative percent difference, etc.), and tables of samples that exceed a project action level (state or federal standard). A sample action level table, for Massachusetts Soil Standards, is presented as Table 3.

Tabular outputs can also be generated in an ASCII comma-delimited format for importing into other software packages. This format was used during the project for importing data into a Microsoft Excel™ spreadsheet to standardize formats for reports and to download data into other software, such as The Monitor System ™ for additional statistical evaluation. An example of an Excel™ table format is presented as Table 4.

Graphical

The database enabled the generation of several different types of graphs from either the database or directly from AutoCAD™. Typical graphs included concentration versus time, depth, or distance; groundwater elevations versus time; well screen elevations versus depth, etc. Examples of a concentration versus distance

Table 2.   **Data Validation Summary for Soil Samples**

Site	Sample Date	Sample Depth	Compound Name	Conc.		D.L.	Units	Lab Qual.	Expert Qual.	R1	R2
E01B07	05-Oct-94	1	Antimony		<	1.6	mg/Kg		J	BM	
E01B07	05-Oct-94	1	Barium	21		0.54	mg/Kg		J	DU	
E01B07	05-Oct-94	1	Zinc	51		1.2	mg/Kg		J	DU	
E01B07	05-Oct-94	1	Arsenic	2.0		0.29	mg/Kg		J	DU	
E01B07	05-Oct-94	1	Lead	27		2.9	mg/Kg		J	DU	AM
E01B07	05-Oct-94	1	Total petroleum hydrocarbons	4.7	<	11	mg/Kg	J	J	RD	
E01B14	07-Aug-95	0	Pyrene	3800		520	µg/Kg		J	BM	
E01B14	07-Aug-95	0	Benzo(a)pyrene	1200		520	µg/Kg		J	BM	
E01W01	03-Oct-94	3	Tin	4.1		2.4	mg/Kg		J	BM	
E01W01	03-Oct-94	3	Antimony	2.4		1.7	mg/Kg		J	BM	
E01W01	03-Oct-94	3	Lead	16		1.2	mg/Kg		J	AM	
E01W02	04-Oct-94	5	Tin	6.1		2.4	mg/Kg		J	BM	
E01W02	04-Oct-94	5	Antimony		<	1.7	mg/Kg		J	BM	
E01W02	04-Oct-94	5	Lead	8.0		1.4	mg/Kg		J	AM	
E01W03	28-Sep-94	1	Tin	3.3		2.2	mg/Kg		J	BM	
E01W03	28-Sep-94	1	Antimony	2.6		1.5	mg/Kg		J	BM	
E01W03	28-Sep-94	1	Lead	78		11	mg/Kg		J	AM	
E01W03	28-Sep-94	1	Total petroleum hydrocarbons	21		11	mg/Kg		J	RD	
E01W05	11-Oct-94	5	Antimony		<	1.6	mg/Kg		J	BM	
E01W05	11-Oct-94	5	Lead	2.5		0.31	mg/Kg		J	AM	
E01W05	11-Oct-94	5	Tin		<	2.3	mg/Kg		R	BM	
E01W06	06-Oct-94	1	Di-n-octylphthalate		<	360	µg/Kg		J	IS	
E01W06	06-Oct-94	1	Benzo (g,h,i) perylene		<	360	µg/Kg		J	IS	
E01W06	06-Oct-94	1	Indeno(1,2,3-cd)pyrene		<	360	µg/Kg		J	IS	
E01W06	06-Oct-94	1	Benzo(b)fluoranthene		<	360	µg/Kg		J	IS	
E01W06	06-Oct-94	1	Benzo(k)fluoranthene		<	360	µg/Kg		J	IS	
E01W06	06-Oct-94	1	Nickel	5.9		0.69	mg/Kg		J	DU	
E01W06	06-Oct-94	1	Tin	6.4		2.4	mg/Kg		J	BM	
E01W06	06-Oct-94	1	Antimony	2.2		1.7	mg/Kg		J	BM	
E01W06	06-Oct-94	1	Barium	19		0.56	mg/Kg		J	DU	
E01W06	06-Oct-94	1	Cadmium	0.91		0.44	mg/Kg		J	DU	

Qualifier Codes

B: Constituent found in blank as well as sample

J: Estimated concentration

R: Unusable Data

Reason Codes (R1, R2, R3)

AM: Recovery above matrix spike QC limit

BM: Recovery below matrix spike QC limit

DU: Duplicate RPD greater than QC limit

IS: Internal standard area outside required QC limits

RD: Raw data review indicates a problem with result

Table 3. Analytes Exceeding Action Levels in Soil Samples

Sample ID and CAS Number	Sample Depth and Analyte	Action Level Code	Action Conc	Level	Sample Conc	
S06S03						
08/24/95	@ 1.00'					
7439-92-1	Lead	MAS-S1/GW1	300	mg/Kg	1890	mg/Kg
7440-36-0	Antimony	MAS-S1/GW1	10	mg/Kg	55.3	mg/Kg
S35S01						
09/21/95	@ 14.00'					
56-55-3	Benzo(a)anthracene	MAS-S1/GW1	0.7	mg/Kg	1.3	mg/Kg
205-99-2	Benzo(b)fluoranthene	MAS-S1/GW1	0.7	mg/Kg	2.7	mg/Kg
50-32-8	Benzo(a)pyrene	MAS-S1/GW1	0.7	mg/Kg	1.3	mg/Kg
S35S03						
09/21/95	@ 14.00'					
56-55-3	Benzo(a)anthracene	MAS-S1/GW1	0.7	mg/Kg	1.3	mg/Kg
205-99-2	Benzo(b)fluoranthene	MAS-S1/GW1	0.7	mg/Kg	2.3	mg/Kg
50-32-8	Benzo(a)pyrene	MAS-S1/GW1	0.7	mg/Kg	1.6	mg/Kg
193-39-5	Indeno(1,2,3-cd)pyrene	MAS-S1/GW1	0.7	mg/Kg	0.92	mg/Kg
S35S04						
09/21/95	@ 10.00'					
56-55-3	Benzo(a)anthracene	MAS-S1/GW1	0.7	mg/Kg	21	mg/Kg
205-99-2	Benzo(b)fluoranthene	MAS-S1/GW1	0.7	mg/Kg	40	mg/Kg
207-08-9	Benzo(k)fluoranthene	MAS-S1/GW1	7	mg/Kg	7.1	mg/Kg
50-32-8	Benzo(a)pyrene	MAS-S1/GW1	0.7	mg/Kg	23	mg/Kg
218-01-9	Chrysene	MAS-S1/GW1	7	mg/Kg	18	mg/Kg
53-70-3	Dibenz(a,h)anthracene	MAS-S1/GW1	0.7	mg/Kg	2.4	mg/Kg
193-39-5	Indeno(1,2,3-cd)pyrene	MAS-S1/GW1	0.7	mg/Kg	8	mg/Kg
7439-92-1	Lead	MAS-S1/GW1	300	mg/Kg	354	mg/Kg

graph and a groundwater elevation versus time graph are depicted in Figures 1 and 2, respectively.

The database was also used to generate all the boring logs for the project. At one of the facilities, the boring logs were completed in the field by the geologists. An example of a boring log generated by the database is presented as Figure 3.

By combining the hydrogeological and analytical data, the database was able to quickly generate hydrogeological and chemical cross-sections which greatly facilitated the understanding of site conditions. An example of a typical cross-sectional drawing is presented as Figure 4. The quickness and ease with which these drawings were generated enabled the project team to evaluate several different cross-sectional views and select which views were more suitable for report presentation.

Multiple drawings of analytical concentrations for soil and groundwater samples (single parameter, total concentrations, samples that exceed an action level)

Table 4.   Summary of Semivolatile Organic Compounds in Groundwater Samples (µg/L)

Sample Location Identifier	Acenaphthene	Bis(2-ethyl hexyl)phthalate	Fluorene	2-Methyl Naphthalene	Naphthalene	Phenanthrene	Pyrene
MW01R	< 10	< 10	10	46	320 J	< 10	< 10
MW02R	< 50	< 50	86	< 50	< 50	< 50	< 50
MW04	< 10	< 10	20	19	< 10	22	< 10
MW03R	< 10	< 10	10	11	< 10	< 10	< 10
MW06	< 10 R	< 10 R	10 R	12 J	< 10 R	< 10 R	< 10 R
MW09	26 J	40 J	130	530	< 100	190	15 J
MW112	11	16	16	19	< 10	23	< 10
MW14	12	< 10	12	< 10	< 10	42	27
MW21	< 10	< 10	14	130	69	20	< 10
MW33	< 10	< 10	14	73	23	23	< 10

NOTE: In addition to the results indicated above, SVOCs were not detected in 58 samples submitted for analytical testing.

µg/L = micrograms per liter.

Figure 1. Trichloroethylene concentration as a function of distance from the source.

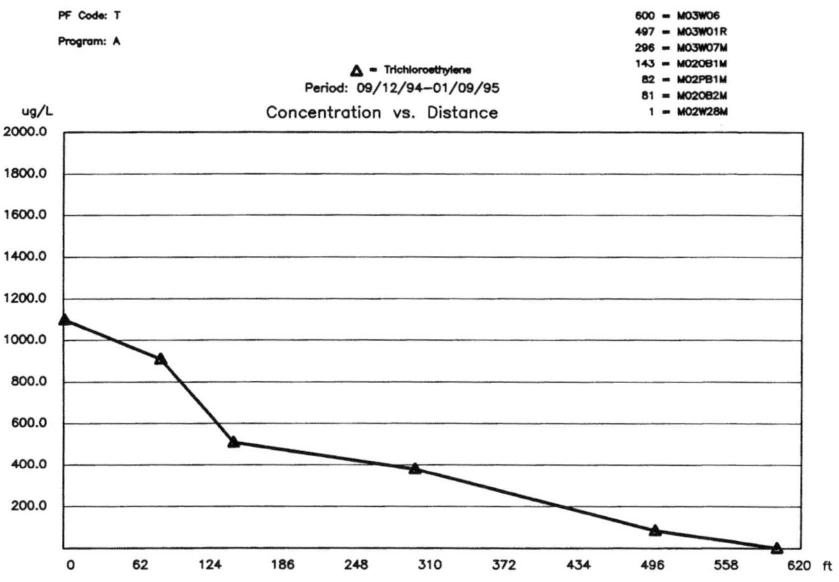

Figure 2. Historic groundwater levels from MW-6.

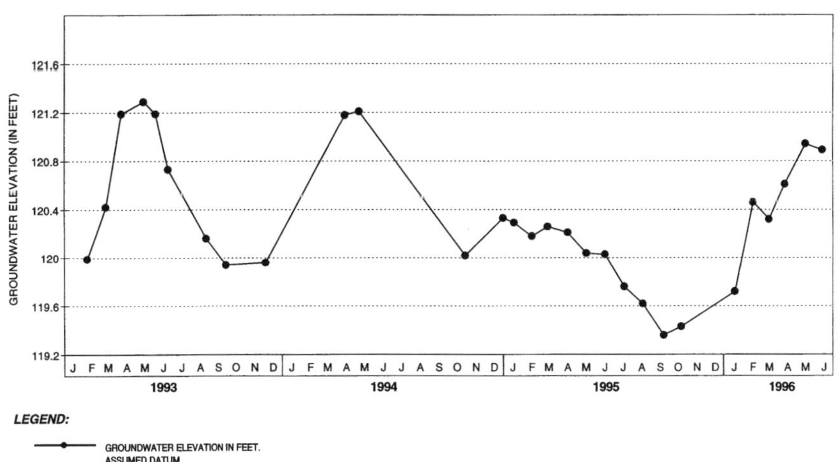

Figure 3. **Example boring log.**

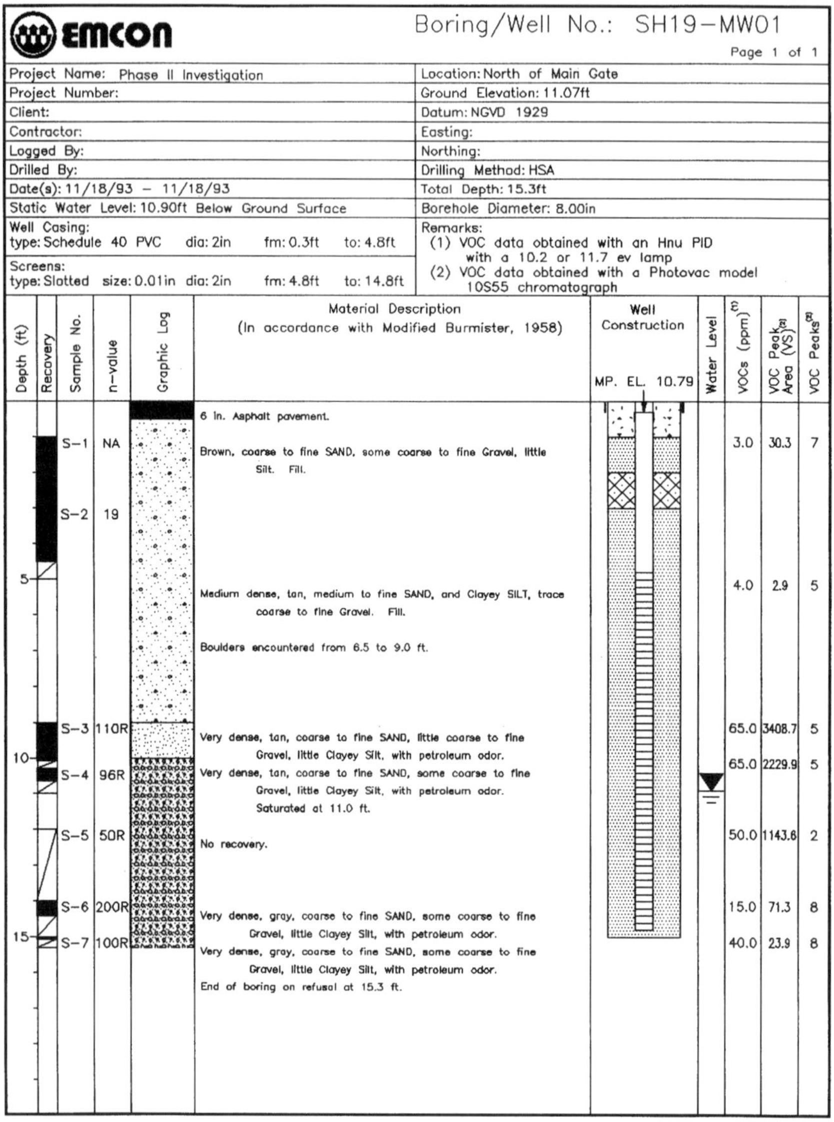

Figure 4. **Hydrogeologic/chemical section.**

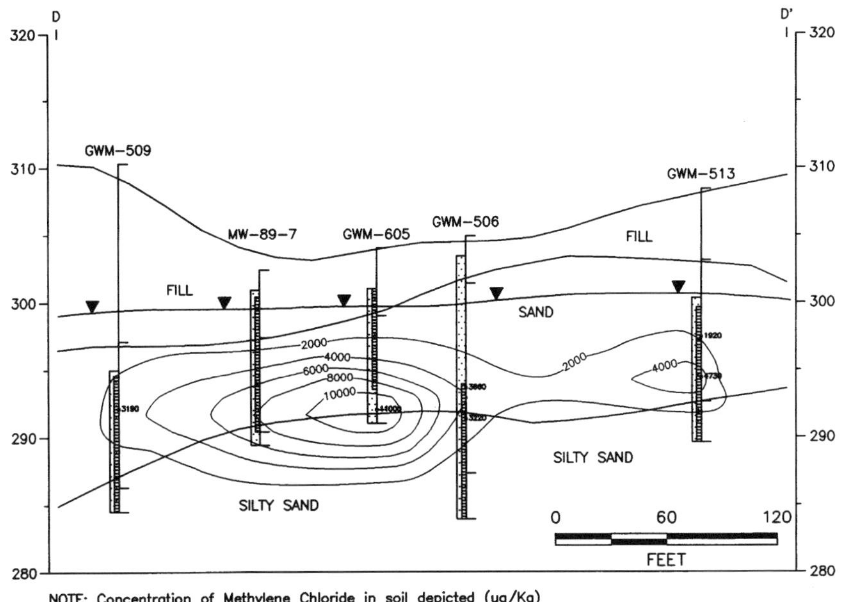

NOTE: Concentration of Methylene Chloride in soil depicted (ug/Kg)

were also generated on the site base map. The generation of isopleth concentration contour maps was an easy process, requiring several minutes versus several hours when done by conventional methods. An example of an isopleth concentration contour map is presented as Figure 5.

OTHER SOFTWARE COMPATIBILITY

The database also allowed exportation of results to numerous other software packages which were used for generating outputs not available from the database. Software packages that were used included: Access™ for custom reporting formats, The Monitor System™ for advanced statistical testing procedures, and Site View™ for generation of three-dimensional drawings of contaminant plumes. An example of a three-dimensional contaminant plume map generated from Site View™ is presented as Figure 6. Without the ability to selectively sort and present the data in a spreadsheet, the generation of such a drawing would have been extremely labor-intensive.

ADVANTAGES

Use of the database on this project greatly improved the overall technical quality of the project by allowing for extensive data interpretation that would not have been practical using conventional data management methods. The database greatly

Figure 5.   **1,1-DCE in groundwater, 1994 samples.**

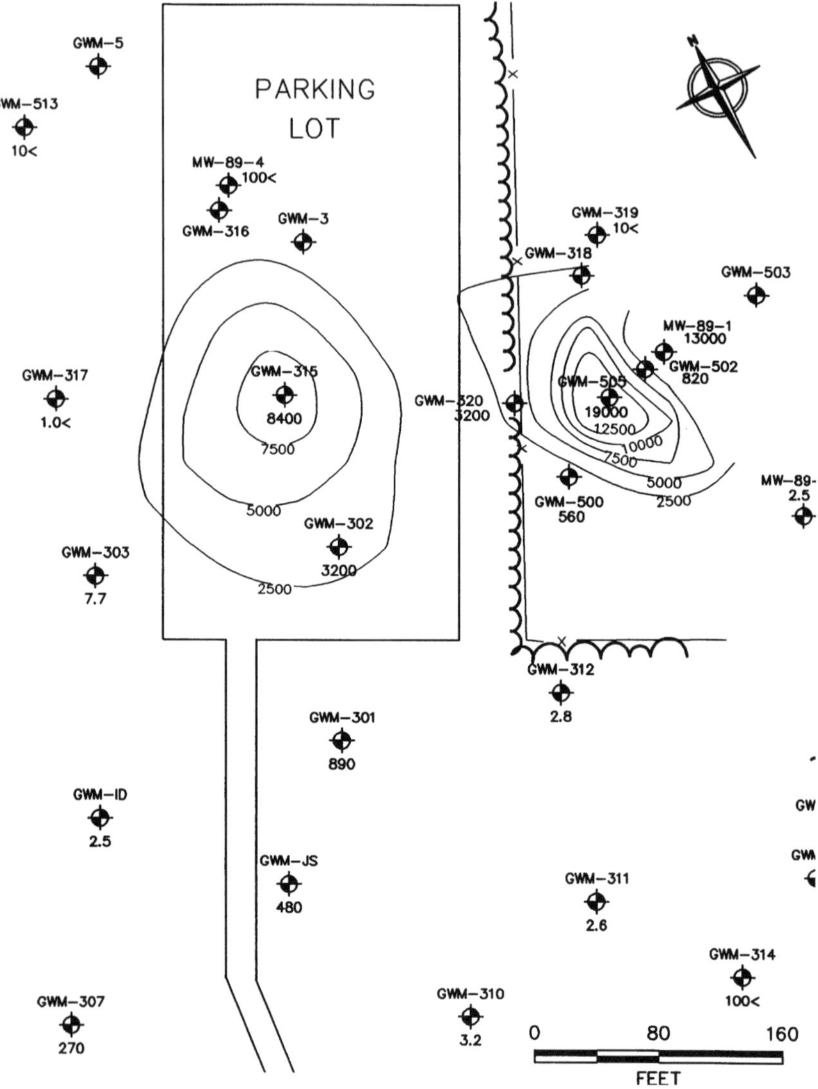

Figure 6.   **Sliced 3-D view of trichloroethylene plume in groundwater.**

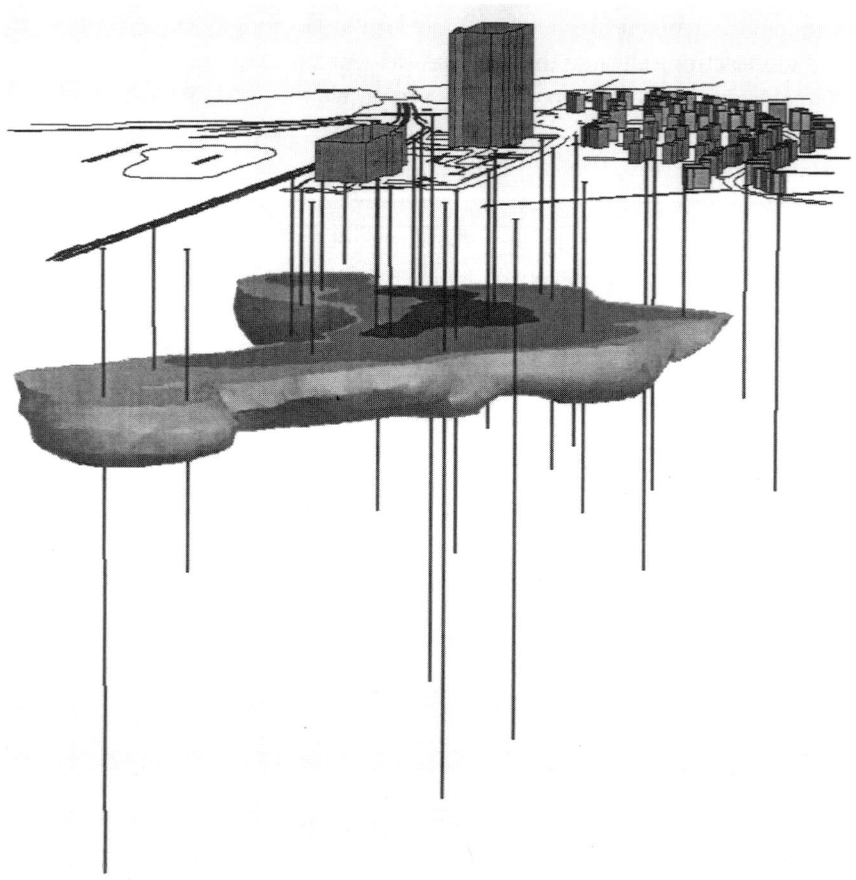

improved the integrity of the data due to a centralized storage location with controlled access and one-time data modification. The ability to easily and accurately generate tables, graphics, cross-sectional drawings, and isopleth maps was critical in the evaluation of the data. Analytical data was validated rapidly and accurately in significantly less time than conventional methods. The database also met the needs of the project team by allowing data to be sent to files which then were customized for the user's specific needs.

LIMITATIONS

Due to the strict data integrity required in managing complex environmental data, manipulation of the data through the database screens was sometimes deemed to be too slow for some users. To circumvent this issue, considerable use was made of FoxPro ™ to enter and edit data, customize reports, and run programs to export data in specific formats. The project team developed numerous programs which were not components of the original database, to track, confirm, and format data into useable/acceptable formats. The EDD format required for importing analytical results electronically into the database was generated with mixed success by the different laboratories participating in this project. Some laboratories needed to go through several iterations to generate an acceptable EDD format.

CONCLUSIONS

The GIS database enabled significant amounts of data to be managed in a much more efficient manner than conventional methods, resulting in an overall savings in project costs and time. Although the database requires an up-front investment in capital and labor, due to the sheer volume of data to be managed on large-scale investigations, the cost savings realized are substantial through the reduction in the amount of labor hours for table generation, data validation, graphs, etc., and typical human errors associated with repeated and labor-intensive procedures. In addition to being cost-effective, the database also increased the technical quality of the project by allowing for extensive data interpretation that would not have been practical using conventional methods.

CHAPTER 2

Using Rapid Sampling and Field Screening Techniques for the Characterization of a Gasoline Spill Site

Jimmy Kao, James Holly, Lynn Sutton, and **Eric Powers**, Geophex, Ltd., Raleigh, North Carolina

Linda Blalock, North Carolina Department of Environmental Management, Ground Water Section, Raleigh, North Carolina

David Ariail, EPA - Region 4, Atlanta, Georgia

INTRODUCTION

Geological heterogeneities and preferential groundwater flow paths create complex hydrogeologic conditions at most contaminated sites. A thorough understanding of the resulting three-dimensional distribution of contaminants is a necessity prior to determining a need for remediation. Therefore, many technologies have been employed in each phase of environmental investigation projects to improve and streamline site investigations. These technologies allow for faster sample collection and sample analysis on-site with less cost. One of the more common technologies in use today is the application of hydraulic and percussion systems (push technologies) to drive sampling tools into the ground for the acquisition of vapor, soil, and groundwater samples. After the sample collection, on-site screening instruments can be applied for sample analysis to maximize the efficiency of site investigations. Application of efficient sample collection and screening techniques are important factors in conducting cost-effective and comprehensive site investigation.

The objectives of this study were to cross-compare several sampling and screening techniques, provide guidance for technique selection and on-site data interpretation, and compare field data to laboratory analyses for the disparity of the results. This report presents the results of applying push technology for conducting preliminary site assessments, and also presents a comparison of several field screening techniques, which includes portable gas chromatography (GC), photoionization detector (PID), flame ionization detector (FID), drager tube, immunoassay, and Petrex survey.

SITE BACKGROUND

An underground storage tank (UST) Federal Trust Fund site (Mayo's Grocery site) was selected for this study. Mayo's Grocery is located in Garysburg, Northampton County, North Carolina, 400 feet southwest of the Northampton County Progressive Community Well #1 (Figure 1). Because benzene (3.9 µg/L) was discovered in groundwater from the well in 1993, the well has been taken off-line. Before this study, we conducted a geophysical survey and located four USTs at this site (T1 to T4) (Figure 1). All USTs were removed after the completion of this study.

Figure 1. **Site map showing the locations of Mayo's Grocery, Community Well #1, USTs, and water table elevation contour.**

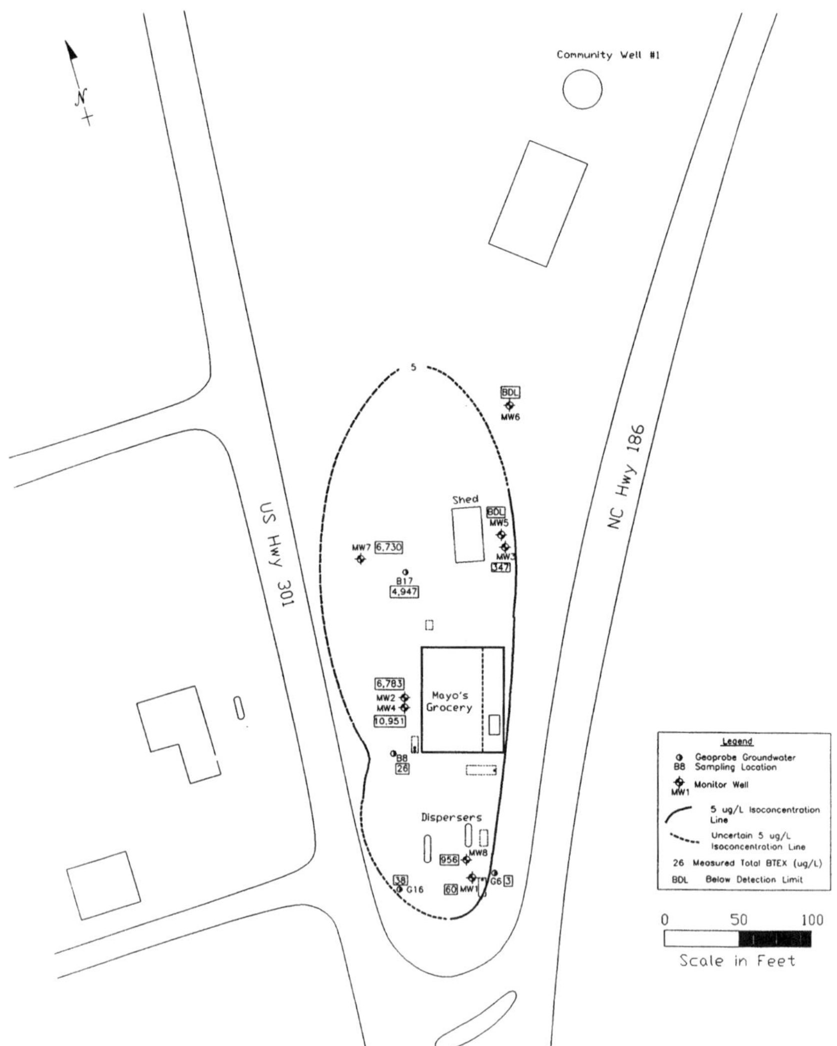

According to the Geologic Map of North Carolina (NCGS, 1985), the site lies in the Coastal Plain Province, just east of the Fall Line. The geology consists of Tertiary sediments of the Yorktown Formation. The soils at this site are primarily classified as Goldsboro sandy loam (USDA, 1994). Site soils consist of yellowish-brown sandy clay loam, gray sandy clay loam with brown and dark reddish-brown mottles, and light gray sandy clay loam with red, reddish-brown, and yellowish-brown mottles. The measured water table was 4 to 7 feet below land surface (bls) at this site.

METHODS AND PROCEDURES

Sampling Technology

Push technology (Geoprobe™ system) was applied in this study for sample collection. This system is a four-wheel drive, van-mounted, hydraulic soil penetration system designed specifically for the collection of subsurface samples and on-site screening studies. The system allows collection of soil, soil gas, and groundwater samples directly and efficiently. Although soil gas is usually the easiest medium to sample with direct push tools, it was not applicable at this site due to the tight clay materials and high soil moisture content in the subsurface.

On-Site Screening Technologies

Petrex Technology

Petrex technology is a shallow soil gas survey method designed to delineate both total areal extent of contamination and areas of heaviest contamination (Vroblesky et al., 1991). It qualitatively screens both soils and groundwater for volatile organic compounds (VOCs) and semi-VOCs (SVOCs). This process begins with the design of a sampling grid, followed by the installation of the Petrex samplers. Petrex samplers consist of activated charcoal adsorption elements (collectors) housed in a resealable glass container in an inert atmosphere. Soil gas sample collection is performed by exposing the collector to the soil gas of the subsurface environment at the base of a shallow borehole. Following a controlled period of time, the sampler is retrieved from the borehole, and analyzed by thermal desorption/mass spectrometry (MS). The Petrex survey consisted of the analysis of soil gas samples from 30 sampling points and collection of headspace vapor from three monitor wells.

Immunoassay

Immunoassay technology is an on-site, semi-quantitative, and colorimetric analytical method for organics contamination in water and soil. This system is comprised of a disposable analyte detector coated with antibodies specific to organic contaminants. A handheld reflectometer is used to quantify the test results. Water samples are analyzed directly; soil contaminants are extracted with methanol prior to analysis. Analysis is performed by adding the sample, an enzyme substrate, and a fixing solution to the surface of the detector. As each solution is absorbed by the detector, it passes through the surface zone of immobilized antibody. The assay device couples to the handheld dual-beam reflectometer, which compares the color intensity of the sample zone to that of the reference zone. The results are displayed as percent inhibition or as actual analyte concentration extrapolated from a preprogrammed curve. Eight soil samples were analyzed in the field for total petroleum hydrocarbons (TPH) using this technology.

Portable GC

Portable GC (or field GC) is a quantitative analytical method that can detect trace levels of contamination by individual compounds in various media. Because of its high sensitivity, and results directly comparable to laboratory analyses, a portable GC eliminates delays in obtaining analytical laboratory-quality information. A photovac 10S+ equipped with PID detector and capillary column was used in this study. For the soil sample analysis, approximately 10 grams of soil were placed into a 40-mL sample vial containing 15 mL of water. A headspace sample was withdrawn from the vial and analyzed with the GC.

The Drager Tube System

The drager tube system is a quick and easy method to screen vapors derived from soil or groundwater samples to obtain preliminary contamination levels on-site. This system includes a simple hand-operated sampling pump and drager detection tube for different compounds. Soil or groundwater vapors can be screened on-site for the presence of specific compounds (e.g., benzene, TPH) without further sample preparation.

PID and FID

PID and FID are currently the most popular types of instruments for organic vapor analysis (Capacci and Wilcove, 1995). The basic premise behind the PID is to pass the sample past a lamp that produces ultraviolet light of a certain energy. If the energy of the lamp is greater than or equal to the ionization potential of the compound in the sample, free electrons are generated that produce a current proportional to the number of ions present. This resulting signal is translated as a total organic vapor (TOV) concentration. The FID draws a sample into a combustion chamber with an internal pump. A hydrogen flame ionizes the sample, and the concentration reading is proportional to the ion flow rate through the detector. The PID used in this site assessment project was a Thermo Environmental Instruments Inc. PID (Model 580B OVM). The FID used was a Century OVA FID (Model OVA-128).

Lab-In-A-Bag

Lab-In-A-Bag (LIAB) employs a three-way ball valve attached to a quart plastic bag, to which a PID or FID can be attached. Twenty-five grams of soil were extruded from a sampler into the bag containing 100 mL of water. The bag was filled with air, and the soil was stirred for five minutes with a magnetic stir bar. After stirring, the headspace gases in the bag were analyzed by PID, FID, and drager tube.

Monitor Well Installation

The Geoprobe^{TM} system can be used to install piezometers or monitor wells for hydrogeologic data collection. However, due to the tight clay materials in the subsurface, this method proved impractical. We installed three deep permanent monitor wells using a conventional drill rig (hollow stem auger, Model Mobil Drill - B57), and five shallow monitor wells using hand augers. The eight monitor wells were constructed of two-inch, flush-jointed, PVC casing, and 0.01-inch slot screen.

RESULTS AND DISCUSSION

Soil Sample Analyses

Prior to this site assessment, there were no data regarding the plume characteristics or the downgradient extent of gasoline contamination. We advanced 18 soil borings, and collected a soil core at a depth of 4 to 6 feet bls at each boring location. We used the LIAB system followed by PID, FID, and drager tube to determine TOV concentrations, and portable GC and immunoassay to analyze BTEX and TPH concentrations. PID, FID, and portable GC results were used to generate contour maps for soil contamination (Figures 2 to 4). Table 1 presents the drager tube and immunoassay results for soil samples.

Although the contaminant concentrations as determined by these technologies are not quantitatively comparable, the areal distributions of contaminated soils determined by each method are similar. All maps indicate that the highest soil contamination occurred at seven different locations (B1, B7, B8, B9, B10, B11, and B16). Contamination at boring locations B1 and B10 is most likely caused by a release at tank T1. Contamination at boring location B7 is possibly due to a release at both tanks T1 and T4 and/or the underground pipelines. Contamination at boring locations B8 and B9 is also most likely due to leakage or spills at tank T4. Concentrations detected at B11 and B16 may have been caused by surface spills or other releases at the dispenser.

Table 1. **Soil sample analyses using drager tube and immunoassay.**

	B1	B2	B3	B4	B5	B6	B7	B8	B9
Drager[a] (ppm)	130	nd[c]	nd	na[d]	30	nd	110	40	50
Immuno[b] (ppm)	250	na	<70	na	na	na	150	500	100

	B10	B11	B12	B13	B14	B15	B16	B17	B18
Drager (ppm)	nd	80	nd	nd	nd	nd	90	nd	nd
Immuno (ppm)	na	500	na	na	<70	na	180	na	na

[a]Drager: drager tube; [b]Immuno: immunoassay; [c]nd: not detectable; [d]na: not available.

Figure 2. **Contour map showing TOV concentrations in soil samples detected by PID.**

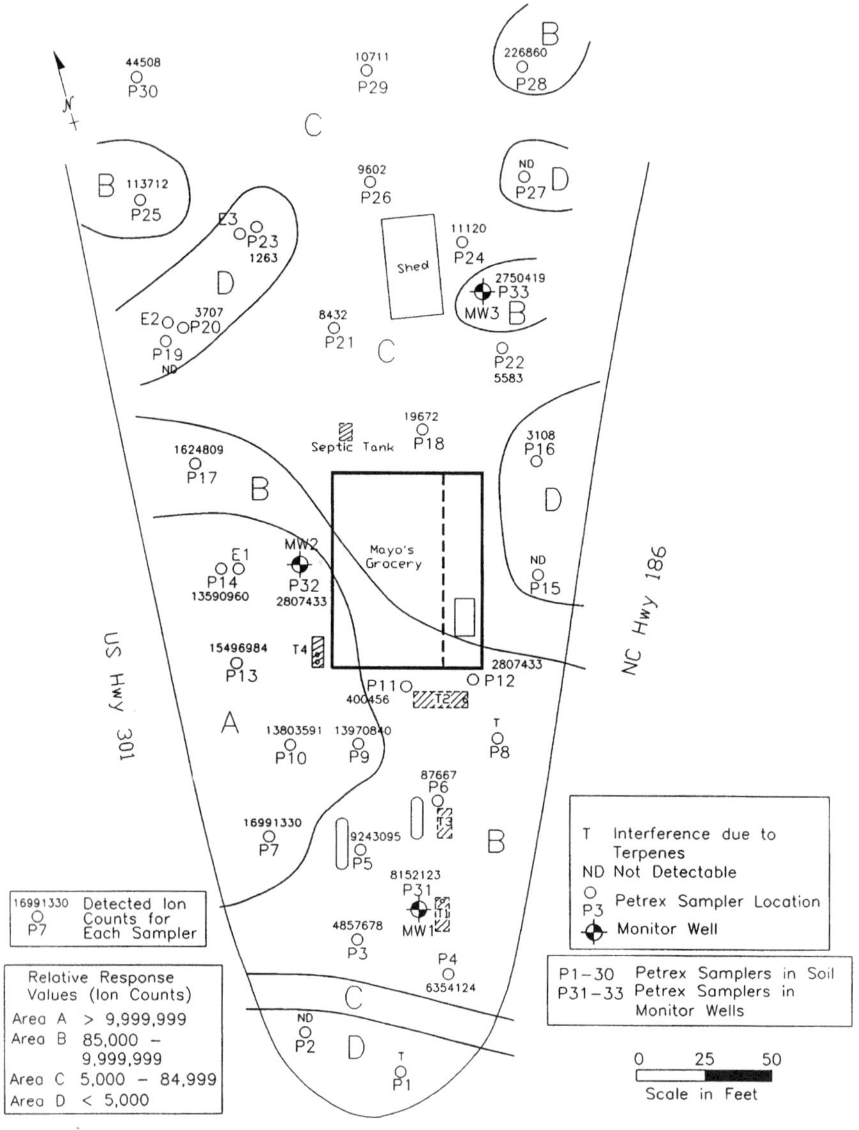

Figure 3. **Contour map showing TOV concentrations in soil samples detected by FID.**

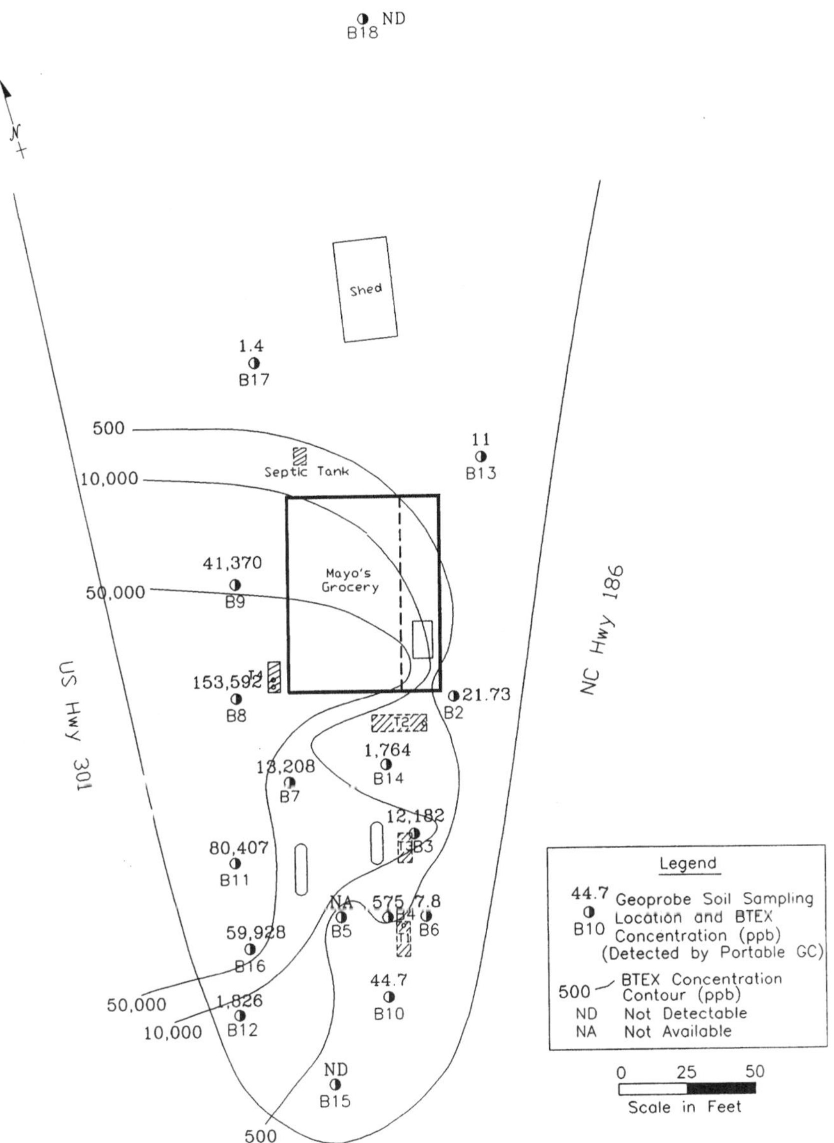

Figure 4. **Contour map showing BTEX concentrations in soil samples detected by portable GC.**

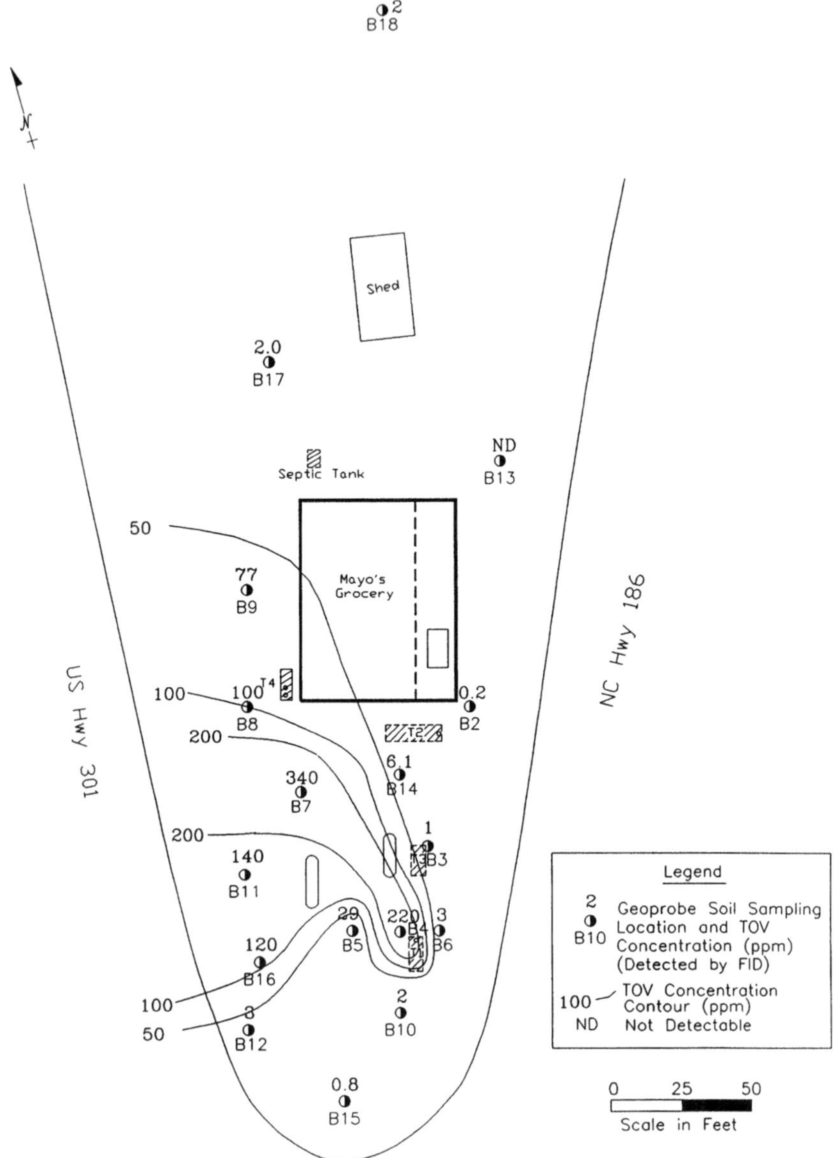

The following equations were developed to describe the correlation of the TOV concentrations in the soil samples as determined by each technique:

$$\text{PID (ppm)} = 7 + 0.44 \times \text{FID (ppm)} \tag{1}$$

$$\text{PID (ppm)} = 6.5 + 1 \times \text{Drager Tube (ppm)} \tag{2}$$

$$\text{PID (ppm)} = 13.3 + 0.22 \times \text{Portable GC (ppm)} \tag{3}$$

Based on the results from the statistical hypothesis F test, we observed evidence of linear association at significance level 0.05 between the measured and predicted PID values for the above three relationships. The correlation coefficients (R^2) of the three regression curves are 0.65, 0.62, and 0.56, respectively. The TOV analyzers seem to be good instruments for the detection of petroleum contamination at higher contamination levels. With low contaminant concentrations, results determined by TOV detectors may be misleading because of their high detection limits (10 ppm for drager tube and 0.1 ppm for PID and FID).

Portable GC results indicate that most of the soil samples contained high concentrations of methyl-tertiary-butyl-ether (MTBE). However, laboratory results (Table 1) indicate that only trace amounts of MTBE were contained in the soil samples. We selected four soil samples (B7, B9, B11, and B14) for laboratory conformation. Table 2 presents the comparison of laboratory analyses and portable GC analyses for the four selected soil samples. Theoretically, TPH in a soil sample as determined by EPA Method 5030 should be higher than the total VOC concentrations (total identified and total unknown compounds) determined by portable GC, because the purge-and-trap technique is used in the laboratory for sample equilibration. However, samples B7, B9, and B11 showed higher portable GC data than laboratory data. The F test suggests that there is no linear association between the measured and predicted portable GC values ($R^2 = 0.37$). The variations may have occurred for the following reasons: (1) Total volatile unknown concentrations in the portable GC method were calculated using average standard response, which is not an accurate way to present the total unknown volatile organics in the soil samples. Method 5030 selects 10 to 15 known major components in gasoline to calibrate for TPH, which can result in more accurate data. (2) The heterogeneity of the contaminant distribution in soil samples will cause variations, because only 10 grams of each sample were analyzed. (3) Organic volatilization may occur during the sample collection, preparation, and transportation. (4) Different sample preparation techniques (Method 5030 analyzes the extracts after sample extraction with solvent, and portable GC analyzes the headspace in the sample vial) may cause some variations, even though each method develops its own standard curves.

TPH concentrations determined by the immunoassay method for samples B7, B9, B11, and B13 (Table 1) were higher compared to the laboratory GC results (Table 2). Again, the heterogeneity of the contaminant distribution in the

Table 2.   Comparison of laboratory and portable GC soil sample analyses.

Compound	Lab[a] B7	Port[b] B7	Lab B9	Port B9	Lab B11	Port B11	Lab B14	Port B14
Benzene (µg/kg)	1.7	179	<5.0	4,009	<10	14,928	338	35
Ethylbenzene (µg/kg)	1.6	577	11.3	292	19	856	1,890	121
Toluene (µg/kg)	3.3	9,308	<5.0	28,167	13	58,165	1,190	1,107
Total Xylenes (µg/kg)	4.8	3,149	149.0	8,903	189	6,458	10,800	501
MTBE (µg/kg)	<1.0	0	<5.0	12,063	<10	85,231	<200	535
Total BTEX (µg/kg)	11.4	13,208	160.3	41,370	221	80,407	14,218	1,746
Total Unknown (mg/kg)	na[c]	32.4	na	35.9	na	133.9	na	1.9
TPH (mg/kg)	0.63	45.6	20.4	89.3	23.4	299.5	12.6	3.9

[a]Drager: drager tube; [b]Immuno: immunoassay; [c]nd: not detectable; [d]na: not available.

samples may cause the analysis variation. The main disadvantage of this technology is its high detection limits. The detection range for gasoline TPH is from 70 to 5,000 ppm; therefore, this technology is not useful if TPH concentrations are below 70 ppm.

Immunoassays rely on highly specific, animal-derived, antibody proteins, and relatively simple apparatus to detect and quantify a wide variety of target materials in a broad range of matrices (Deshpande, 1994). The sensitivity of the antibodies and immunoassay methods can significantly reduce sample preparation times. Additional research, however, is needed to develop and refine immunoassay tests to improve their sensitivity and expand the scope of detection of components in the environment.

Of the techniques evaluated here, the portable GC, by far, gave the most information on the contaminants present. However, portable GCs require a higher degree of operator competency. Furthermore, the investigator must select which analytes are to be targeted prior to analyzing the sample, so that the instrument can be properly calibrated.

Petrex Survey

We used the Petrex samplers to detect the presence of VOCs and SVOCs in the subsurface. We installed the Petrex samplers at 33 locations in the field (three of them were placed in the monitor wells). We retrieved and analyzed the samplers after two weeks of exposure. Results for gasoline components extracted from the activated carbon ranged from undetectable to greater than 16,991,330 ion counts. Figure 5 is the contour map showing the distribution of gasoline components. Areas A, B, C, and D indicate the most, second most, third most, and least contaminated zones, respectively. The observed contaminant distribution matches the results of other field screening methods used for the delineation of soil contamination. Results also show that monitor well MW2 contained higher contaminant concentrations in comparison to monitor wells MW1 and MW3. The Petrex survey results indicate that tank T4 is a major source of contamination in this area.

The Petrex method provides the most benefit when investigating sites where little is known about the chemicals present, because the MS does not need to be precalibrated for the contaminants of concern, as does the GC. This technology also works very effectively in environments where contaminant levels are very low, because the sampling tubes can be left in the ground as long as needed to obtain positive results. One disadvantage of this method is that the results are not quantitative. Rather, the results reflect relative variations in soil gas content and concentrations. A second disadvantage is that it commonly takes longer (usually three weeks) to obtain analytical results compared to other on-site techniques.

Groundwater Sample Analyses

Using the water level elevations in Geoprobe™ borings, we developed a water table contour map (Figure 1). Based on the soil analytical results and water table contour, we installed three deep monitor wells (MW1 to MW3) and five shallow monitor wells (MW4 to MW8) to intercept the probable path of contaminant plume movement (Table 3). Groundwater samples from eight monitor wells and four Geoprobe™ borings (G6, G8, G16, and G17 collected from 14 to 16 feet bls) were analyzed for organics in the laboratory using EPA Method 502.2 (Table 3). Figure 6 presents the extent of the BTEX plume and analytical data from monitor wells and Geoprobe™ borings.

Hydraulic data collected at this site from monitor wells demonstrates variation of hydraulic head with depth. Water levels in proximal shallow and deeper monitor wells indicate a vertical downward flow potential (e.g., MW2 vs. MW4 and MW3 vs. MW5) (Table 3 and Figure 6). This head variation is typical of sites

Table 3.   Laboratory analyses of groundwater samples.

ID	MTBE (µg/L)	Benzene (µg/L)	Toluene (µg/L)	Ethylben (µg/L)	Xylenes (µg/L)	BTEX (µg/L)	WL[a] (ft)	SI[b] (ft -ft)
MW1	<1.0	3.4	7.8	8.3	40.2	59.7	na3	17-22
MW2	<10	5,070	143	54.2	1,516	6,783.2	93.89	15-22
MW3	<5.0	2/4	<2.5	<2.5	73.3	347.3	91.63	19-24
MW4	<100	431	1,680	1,630	7,210	10,951	96.01	3-10
MW5	<1.0	<0.5	<0.5	<0.5	<0.5	<0.5	95.39	3.5-8.5
MW6	<1.0	<0.5	<0.5	<0.5	<0.5	<0.5	94.54	3-8
MW7	<100	605	855	1,200	4,070	6,730	95.38	3.5-8.5
MW8	<1.0	30.3	7.1	25.2	893	955.6	96.78	3-8
B6	<1.0	1.2	0.7	<1.0	0.8	2.7	97.69	na
B8	<1.0	18.5	<1.0	<1.0	7.0	25.5	96.1	na
B16	<1.0	12.6	1.8	2.2	21.5	38.1	na	na
B17	60.4	2,340	73.4	164	2,370	4,947	na	na

[a]WL: water level elevation; [b]SI: screen interval; [c]na: not available.

Figure 5. Contour map showing distribution of gasoline components in the subsurface detected by Petrex technology.

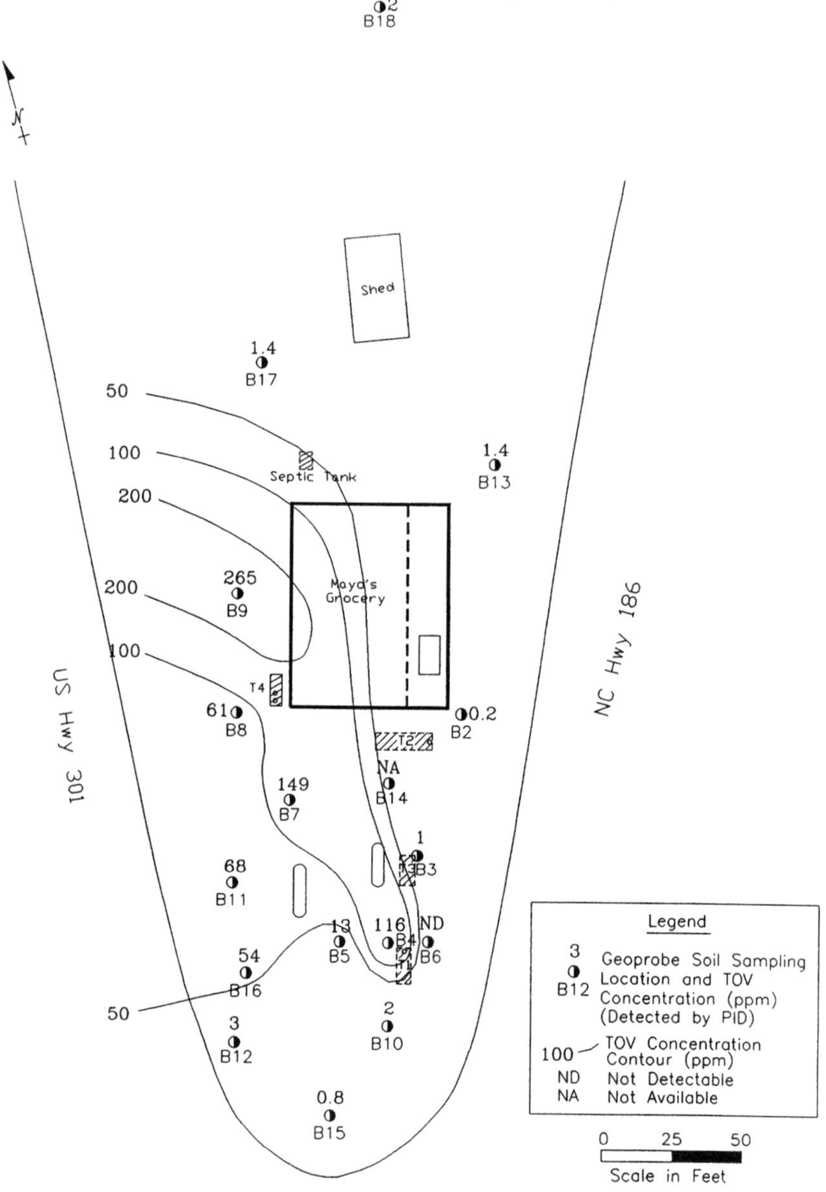

Figure 6.   **Site map showing the BTEX plume and laboratory results from monitor wells and Geoprobe™ borings.**

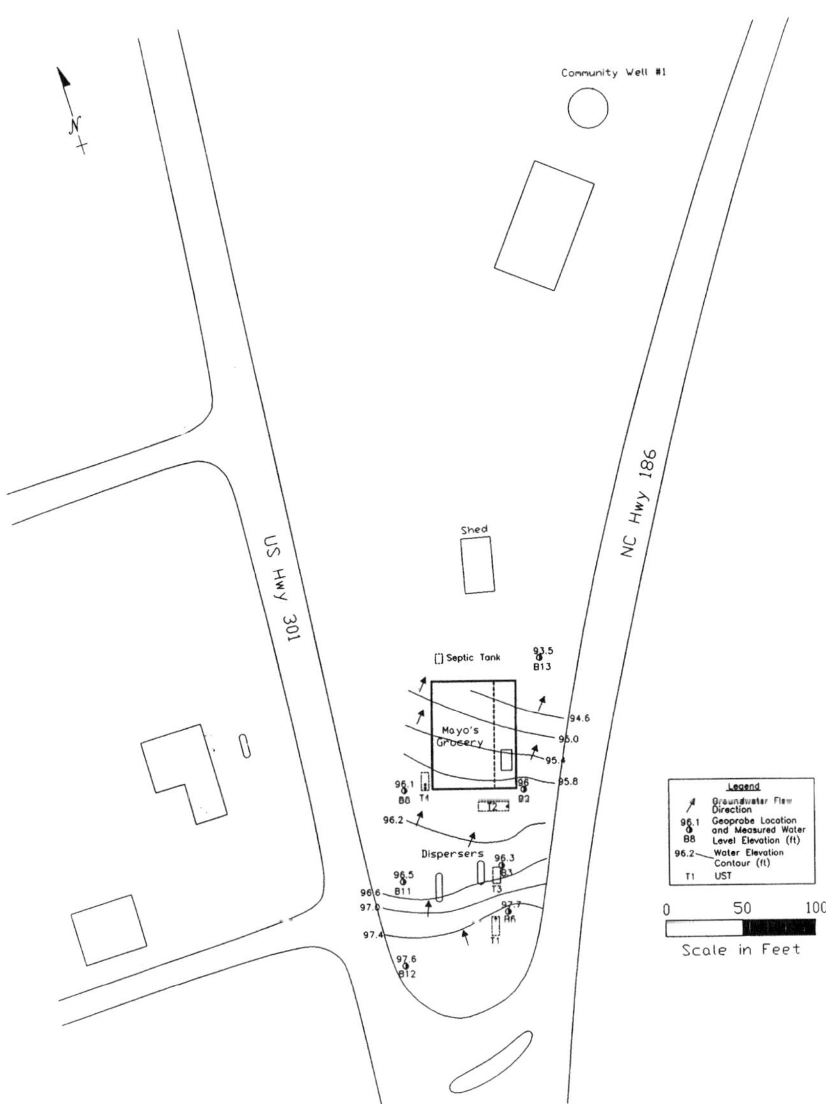

which occur on areas of recharge. Vertical downward groundwater flow carries dissolved contaminants into deeper aquifer zones. High concentrations of BTEX occur in the wells screened into the shallow aquifer zones at locations near the presumed release source areas (USTs). As the contaminants move downgradient from these source areas, they become diluted by uncontaminated recharge water. This enhances the biodegradation rates of contaminants near the water table surface by the introduction of oxygen. The surficial recharge also drives contaminants into progressively deeper aquifer zones. This hydraulic system creates a heterogeneous, three-dimensional distribution of dissolved contaminants within the aquifer. The vertical distribution heterogeneity is demonstrated by comparing samples from wells MW3 and MW5. Well MW5, screened from 3.5 to 8.5 feet bls, is uncontaminated. However, well MW3, which is screened from 19 to 24 feet bls, contains dissolved contaminants. Thus, it appears that the shallow well MW5 is screened above the plume.

The heterogeneous distribution of contaminants within the aquifer demonstrates that proper groundwater assessments should incorporate three-dimensional groundwater sampling approaches. At sites with higher permeability aquifers, such three-dimensional groundwater sampling could be accomplished with the use of the Geoprobe™ sampling techniques. However, the tight clays at the Mayo's Grocery site did not allow for shallow groundwater sampling with the Geoprobe™ system.

BTEX concentrations in MW8 and MW1 were 956 µg/L and 60 µg/L, respectively. This indicates that a smaller plume appeared around monitor well MW1 and MW8 due to a release near the former UST T1 location.

CONCLUSIONS

The following advantages can be summarized for the push technology: (1) The push technology approach is less intrusive than conventional methods. Handling and disposal costs for drill cuttings are minimized. (2) Collected samples are representative of the discrete zone sampled. (3) Under appropriate site conditions, the push technology can be used for monitor well or piezometer installation. (4) Using push technology can accelerate the time-frame for completion of a site assessment.

The disadvantages of push technology can be summarized as follows: (1) Push technology is very useful in soil materials. However, stiff clays, bedrock, or boulders and cobbles will restrict the use of this technology and limit its utility. (2) The soil vapor survey using push technology is not applicable at sites with tight clay materials and high soil moisture content. (3) Push technology is not useful for free-product observation. To collect and determine the appearance and amount of free-product, monitor well installation is needed. (4) For low conductivity aquifers, identification of the water table depth by analyzing the 2" x 4' soil core, collected using the Geoprobe™ soil sampler, may be misleading because the push technology tends to smear clays along the borehole walls and inhibit groundwater entry into the boring.

The advantages of the on-site chemical analyses obtained from this study include: (1) The time spent waiting on laboratory analyses is reduced. (2) The on-site analyses allow rapid and efficient field decisions regarding the best way to proceed to fully delineate the extent of contamination. (3) The increased site information, due to the increase in sample locations, allows easier decisions on where to install monitor wells and how many are needed.

The disadvantages of the field screening include the following: (1) The application of analytical quality assurance and quality control procedures is more difficult in the field. (2) Laboratory analyses are still required to confirm field screening results. (3) Field screening data are not directly comparable to cleanup and health risk standards based on controlled laboratory analyses.

Cost differences between field and laboratory analyses are strongly dependent on the number of samples from a site that must be analyzed, with the cost advantage tending to shift to field analysis as the number of samples increases. Results indicate that if less than 20 cumulative samples are required, laboratory GC analyses are likely to be less expensive than if a portable GC rental is required. Results of this study represent a step forward toward a better understanding of the suitability, efficiency, accuracy, and practicality of utilizing push technology and field screening methods for leaking UST site assessments.

REFERENCES

Capacci, M.J. and Wilcove, M.S. 1995. OVA Readings: Useful Tool or Misleading Information? *Environmental Protection.* November, 31-33.

Deshpande, S.S. 1994. Immunodiagnostics in Agricultural, Food, and Environmental Quality Control. *Food Technol.* June, 136-141.

North Carolina Geologic Survey (NCGS). 1985. Geologic Map of North Carolina: North Carolina Department of Natural Resources and Community Development, Geologic Survey Section, scale 1:500,000.

USDA, United States Department of Agriculture Soil Conservation Service. 1994. Soil Survey of Northampton County, North Carolina.

Vroblesky, D.A., Lorah, M.M., and Trimble, S.P. 1991. Mapping Zones of Contaminated Ground-Water Discharge using Creek-Bottom-Sediment Vapor Samplers, Aberdeen Proving Ground, Maryland. *Ground Water,* 29, 7-11.

CHAPTER 3

Near-Surface Organic Vapor Analyses as an Indicator of Subsurface Hydrocarbon Plume Migration

Douglas G. Mose and **George W. Mushrush**, Chemistry Department, George Mason University, Fairfax, Virginia

INTRODUCTION

Over the past decade, a decreasing number of significant hydrocarbon plumes have reached national attention. State and federal regulations, designed to encourage fuel storage operators to prevent subsurface hydrocarbon losses, plus a growing list of remediated spills, have contributed to a cleaner environment. The more recently discovered contamination events involve small hydrocarbon plumes, or large plumes in remote locations. However, the remediation industry recently watched closely when the largest-ever east coast hydrocarbon plume was discovered under a densely populated affluent suburban neighborhood just outside of Washington, D.C.

In the fall of 1990, the Star Enterprise tank farm in northern Virginia (Fairfax County) was identified as the likely source of a large and relatively rapidly advancing subsurface hydrocarbon plume. The Star Enterprise operation is one of four immediately adjacent tank farms which, through the Colonial Pipe Line Company, receive jet fuel, gasoline and diesel fuels from the Gulf Coast. All four operators receive their fuels through a 24-inch diameter multiproduct transcontinental branch line from a federally regulated pipeline that supplies the eastern seaboard northward to New York. The branch line is paralleled by a 6-inch diameter pipeline from the facility that carries petroleum products to Dulles International Airport, approximately 10 miles to the west.

Late in 1990, when the plume was first discovered, its origin was thought to be a break in the Colonial multiproduct transmission line. Quickly drilled monitor wells revealed that the spill originated in the vicinity of the Star Enterprise tanker truck loading rack, and a portion of the spill had migrated into the gravel-filled trenches which hold the Colonial pipelines.

The Star Enterprise tank farm covers approximately 20 acres on the top and east side of a small hilltop, surrounded by the urbanized part of Fairfax County.

The hilltop was flattened during the construction of this facility in 1964, and on the site were constructed 9 aboveground storage tanks, 11 underground storage tanks, and a loading rack for trucks. The entire bulk storage facility, which includes three other operators, covers about 110 acres. At the present time the total flammable and combustible liquid storage capacity at the entire facility is about 70,000,000 gallons; more than 100 trucks (at ca. 7,500 gallons/truck) per day are filled with diesel and gasoline for distribution throughout Fairfax County.

At the time the Star Enterprise tank farm became operational in 1965, few homes were within one-half mile of this facility. During the following 25 years, the facility became surrounded by several hundred homes, a shopping mall, and roadside commercial buildings. The closest homes were along the east side of this facility, and typically the homes were evaluated at $300,000 to $400,000 prior to the discovery of the plume.

The petroleum storage facility is slightly uphill from an eastward flowing storm sewer system. The storm sewer system was invaded by the subsurface plume, and the storm sewer system emptied into a southeast flowing stream on which floating fuels were discovered in late 1990, flowing through the southeast part of the community of relatively expensive homes. At the same time, but unknown to the home owners, only a small portion of the plume was moving southeastward. Most had already traveled northeastward, having moved down the slope of the water table (which dips toward the east) and along partings in the subsurface, and accumulated under homes not adjacent to the contaminated stream.

The homes to the northeast of the tank farm were not aware of their proximity to the plume until monitor wells were placed in their vicinity in late 1991. Occupants of some of the homes reported hydrocarbon vapors released when, during the course of planting trees or flowers, they noticed a gasoline-like smell in the soil. Local environmental contractors also detected elevated concentrations of hydrocarbon (between 100 and 500 ppm) in the soil, before local monitor wells revealed the arrival of Phase Separated Hydrocarbon (PSH). Neighborhood concern reached its maximum early in 1992 when the plume was discovered to be within a few feet of the basement floor of one home 1500 feet away from the facility, and contaminated water seeping through the gravel below the home exsolved enough explosive vapor to force an emergency evacuation of the occupants.

The volume of petroleum product released into the ground underlying the petroleum distribution facility is not well known. Product management records from the facility, though incomplete and suspect, do not reveal the loss of any fuel. However, it is possible to estimate the volume of fuel by noting the thickness of fuel that accumulated in monitor wells, and by noting the area of the plume as defined by those wells around the plume margin which do not accumulate any fuel. It is necessary to take into consideration the density of the fuel as well as the soil porosity. Based on reasonable assumptions, the loss is calculated to be about 200,000 gallons.

The subsurface plume passed under ca. 15 of the expensive homes. The plume extends about 1800 feet along its northeast-oriented axis. The first ca. 200 feet were between the loading rack area to the northeast corner of the Star Enterprise facility. The next ca. 500 feet were under a four-lane highway and then under the building and parking lot of a privately owned ice skating arena. The plume passed under a densely wooded narrow parkland which separates the ice arena and other commercial buildings from a subdivision of several hundred expensive homes, and finally came to a stop after passing under two parallel streets in the subdivision.

Questions immediately arose about the age of the spill, since a recent spill of this magnitude would incur federal, state, and county penalties. The tank farm began its operation in 1965, and if the plume started to accumulate in 1965 (25 years prior to its discovery in 1990), responsibility for the spill would be shared among several generations of employees. If the plume began in 1965, the estimated loss would average 8000 gallons/year, and the average rate of subsurface plume migration would be about 70 feet per year. However, it was pointed out that the fuels recovered from the plume appear relatively young, in that it is little changed due to subsurface processes of preferential dissolution, volatilization, and degradation. We are impressed by the possibility that the origin of the plume is related to construction at the facility in 1985.

During the construction activities, some of which were at the tanker truck loading rack, subsurface pipes were installed. Some of these pipes were recently found to be not firmly connected, and some pipes were found to be cracked. It is also important that chemical analyses of fuel in the plume revealed that the plume contains no leaded gasoline, and leaded gasoline was distributed until the late 1980s. If the plume began to develop in the late 1980s, and accumulated over 5 years, the estimated loss would average almost 40,000 gallons/year, and the average migration rate would be more than 350 feet/year.

The subsurface plume is about 5 to 10% gasoline, with approximately equal parts of diesel fuel and av-jet fuel. In this report, new chemical analyses are presented for a sample collected along the central axis of the plume, under private property (ice arena parking lot) approximately 300 feet from the northeastern corner of the Star Enterprise facility.

SAMPLE COLLECTION AND INTERPRETATION

Chemistry of Phase Separated Hydrocarbon

Many commercial properties and private properties were adversely affected by the subsurface plume. The largest commercial property is a privately owned ice skating arena. The ice arena is ca. 500 feet northeast of the tank farm, and is located directly over and entirely over the central part of the northeast-flowing subsurface hydrocarbon plume. Figure 1 shows the commercial buildings, and 3779 Pickett Road is the ice arena. Monitor wells placed prior to the vapor study, to determine the thickness of the PSH, are identified in Figure 1 using MW-(...).

Figure 1. Map showing ice arena building (3779 Pickett Road), and showing as contour lines the relative concentrations of hydrocarbon vapor around the ice arena. Notice that while the subsurface hydrocarbon plume underlies the entire map area at a depth of about 35 feet, the vapor is only detected to the right of the ice arena, under an asphalt-covered parking lot which extends from the back (right side) of the ice arena to the right margin of the map.

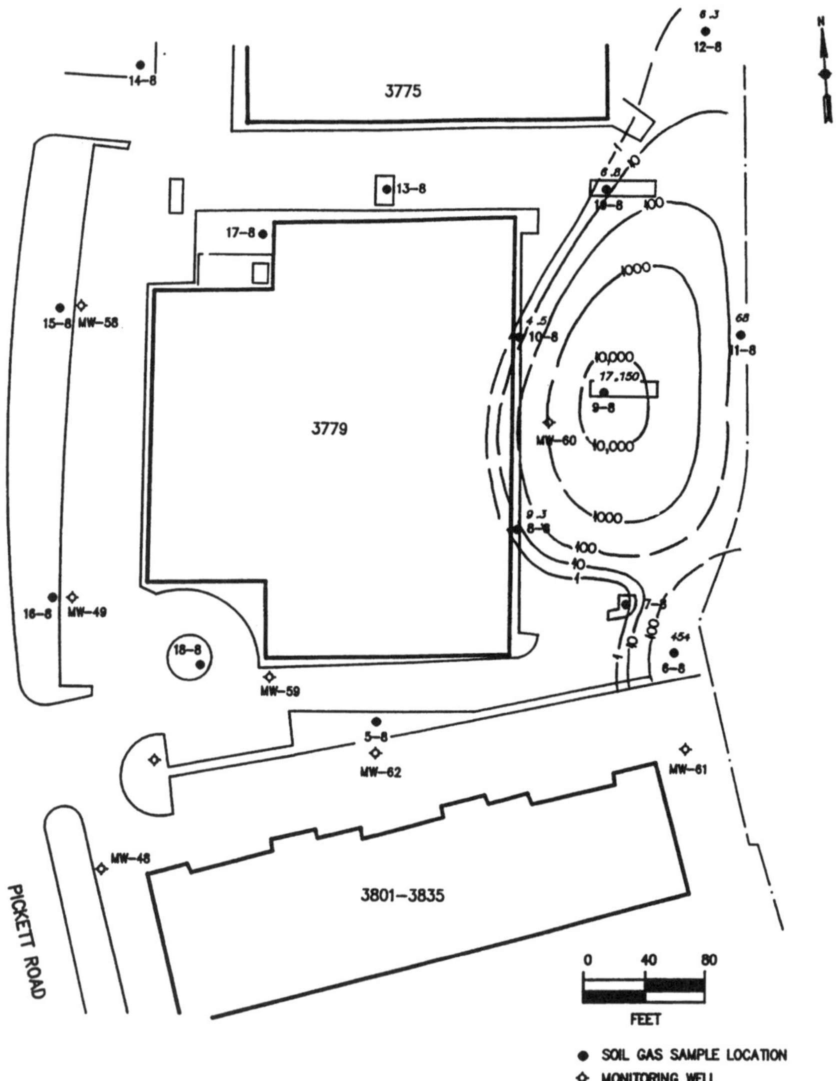

Hydraulic soil probes, to determine the concentration of hydrocarbon vapor, were placed at 15 locations identified as 6-8, 7-8,...., and the concentrations at a depth of 8 feet are shown in Figure 1, in parts per billion. The highest concentration, at site 9-8, was ca. 17,000 ppb.

Samples of phase-separated hydrocarbon mixed with groundwater were obtained from monitor wells located immediately behind the ice arena (monitor well MW-60). The sample was placed in a 2-L separatory funnel, and the aqueous and petroleum fractions were separated. A 100-mL water sample was then adjusted to pH = 9 with Na_2CO_3. The water was subsequently extracted with two 50-mL portions of dichloromethane. The dichloromethane extract was reduced in volume at room temperature to 15-mL by rotary-flash evaporation. The concentrated extract was then dried over anhydrous magnesium sulfate.

The petroleum compounds contained in the dichloromethane extract were identified by using combined capillary column gas chromatography/mass spectrometry. The GC/MS system consisted of a Finnigan INCOS 50B mass spectrometer equipped with a Hewlett-Packard 5890A gas chromatograph. The GC/MS was fitted with a 30 m x 0.25 mm DB-5 (95% dimethyl, 5% diphenyl siloxane; J & W Scientific) fused-silica capillary column that was operated with the following parameters: sample size - 2 µL, injector port -250°C and configured for splitless injection, temperature program - 50°C/5 min, ramped at 4°C/min to 285°C with a 10 min final hold. These parameters gave the necessary resolution to distinctly separate and identify both the hydrocarbon and organonitrogen components. Internal standards were isopropyl benzene for the hydrocarbons and 2-(n-pentyl)pyridine for the nitrogen components.

The mass spectrometer was operated in the electron impact ionization mode (70 eV) with continuous scan acquisition from m/z of 50 to 550 daltons at a cycling rate of 1 scan/s. The mass spectrometer parameters were set up with the electron multiplier at 1050 V, source temperature of 200°C, and transfer line temperatures both at 290°C. The MS was tuned and calibrated with perfluorotributylamine immediately before use. The INCOS 50 data system software was used to process the acquired spectral information.

DISCUSSION

Geological Factors That Determined Plume Migration

This part of Virginia, containing the Fairfax bulk storage facilities and the surrounding neighborhoods, is in the Virginia Piedmont Province characterized by a tree-covered rolling upland surface. The elevation of the hill on which the facilities are constructed is about 410 feet, with a local relief of about 40 feet. During construction of the hilltop facilities and during the development of the surrounding neighborhood, soil removed from some areas was placed in other areas, to depths as much as 30 feet. Filled areas were of particular concern during the investigation of the plume, since fill has very different hydrologic properties than native soils.

Over a geologic interval of time estimated at approximately 300 million years, persistent uplift and erosion has brought deeply buried products of geologic recrystallization (rocks called schists, phyllites, and gneisses) toward the surface (Froelich and Langer, 1983). Thousands of centuries of weathering by rainfall infiltration and groundwater has resulted in the chemical decomposition of most of the minerals in the bedrock, and has formed and maintained the thick soil cover. The upper few feet of soil is a layer of sand, silt, and clay, made homogeneous by the root-growing activity of vegetation and by burrowing animals. Below the surficial homogeneous layer, down to the crystalline rocks, is a 50-100 foot thick layer of chemically weathered rock called saprolite by Appalachian geologists (Froelich and Heironimus, 1977). Because the saprolite forms by chemical weathering in the absence of mechanical weathering, the saprolite retains many features of the parent rock such as near-vertical layering oriented toward the northeast and much of the original mineralogy (e.g., mica and quartz minerals).

The parent rock is relatively hard and formed from layers of sediments and volcanic debris deposited about 500 million years ago. These ancient sediments and volcanics were recrystallized into much harder rocks by intense heat and pressure during the formation of the Appalachian Mountains about 350 million years ago. The ancient sediments and volcanics originally had horizontal layering, but the layering has been rotated and is now vertical. This layering now strikes toward the northeast.

The northeast-oriented layering slowly twists and bends during regional uplift, and as the rocks move up toward the earth's surface and saprolite forms, long and narrow parallel planar partings develop, particularly where the saprolite contains the planar mineral mica. It is along these northeast-oriented partings that fluids can rapidly move. Other pathways of rapid movement develop when some saprolite layers are partially dissolved. In this process, a portion of the saprolite minerals dissolve (e.g., feldspar turns to clay particles, and many iron-rich minerals form soluble metal oxides). The residual saprolite minerals (mostly sand-size quartz grains) are porous and allow rapid fluid movement. In short, the partings and sandy layers are known to produce a groundwater permeability that is 10 times faster than what would be measured across the layering (Sowers and Richardson, 1983).

It is the partings and sandy layers which formed the streams which first carried the stream contamination miles away from the tank farm. Over geologic time, the northeast strike of these partings and sandy layers in the saprolite layers produced areas of more rapidly weathered soils, so most stream tributaries in this area have a northeast (or southwest) orientation. However, more important determinators of surface flow patterns appear to be the southeast-oriented faults and joints, because the local creek and river systems drain toward the southeast, presumably along channels developed along the weakened rocks and saprolite in fault and joint systems. In fact, all the east coast major rivers, including the nearby Potomac River into which the local streams drain, have a southeast orientation.

Subsurface Contamination and Migration

Since the Star Enterprise tank farm is constructed on the top of a small hill, the groundwater recharge over the plume has unfortunately probably been highest in the facility's bermed area around the storage tanks, immediately west of the tanker truck loading rack. In the bermed area, groundwater recharge is enhanced by a thick gravel layer which promotes infiltration after rainfall and snowfall, minimizes subsequent evapotranspiration, and pushes hydrocarbon down toward the water table.

The subsurface plume which escaped the tank farm floats on the local water table at depths of 10-40 feet. The discovery in late 1990 of fuel floating on a stream southeast of the facility was interpreted to mean that the pathway of the subsurface plume was determined by the inclination of the subsurface water table (slopes down toward the southeast) which mirrors the general inclination of the surface water flow. It was subsequently discovered, more than a year later, that most of the subsurface plume moved toward the northeast, diagonally down the water table and along partings and sandy layers in the saprolite.

In this part of the Appalachians, groundwater occurs primarily in alluvial sediments within a few feet of the land surface, in void spaces in the saprolite, and within fractures (faults, joints, and partings) in the underlying bedrock. Potable water comes from bedrock aquifers, typically at less than 50 gallons per minute, at depths of less than 1000 feet. Precipitation is normally about 40 inches per year and snowfall is normally about 20 inches per year. The water table is unconfined, so precipitation determines seasonal changes in the depth.

The volume of sediments that were contaminated by the Star Enterprise plume has been significantly increased by the seasonal changes in the depth of the water table. A normal fluctuation of about 5 feet, which spreads the plume (typically about 0.1 to 1 foot thick) over the intervening saprolite, coupled with an estimated plume area of about 1,000,000 square feet, results in a volume of about 150,000 to 170,000 cubic yards of contaminated sediment.

The advance of the plume was halted in 1993 when the petroleum distribution facility operators constructed a recovery trench, perpendicular to the advance of the plume, and approximately 100 feet downhill from the evacuated home. The trench was constructed in a valley which was selected because the down-to-the-east water table on which the northeastward moving plume traveled reached a down-to-the-west water table originating under the eastern side of the valley. The recovery trench still is used to remove phase separated hydrocarbon and contaminated groundwater, and on the assumption that the source of the plume has been repaired (probably disconnected and/or cracked pipes, now replaced by an aboveground system), the recovery trench will probably continue to remove decreasing amounts of contamination for 10-100 years.

Chemistry of Vapor and Phase Separated Hydrocarbons

Gas chromatography revealed the vapor to be gasoline, jet A, and diesel fuel. Vapor samples from soil probe sites 7 and 19, located within the parking lot, contained the most vapor. From probe site 7, the vapor contained BTEX (Benzene:Toluene:Ethylbenzene:Xylene) ratio of ca. 20:50:10:400. From probe site 19, the BTEX ratio was ca. 30:100:30:100.

The soil probe sites 17 and 18 from the front of the ice arena, and soil probe sites 7 and 19 from within the parking lot, were used to collect vapor samples every 2 feet, from depths of 2 feet to 24 feet. In front of the ice area, at locations known to be underlain by PSH, concentrations of ca. 200-2000 ppb were found; in back of the arena, under the parking lot, concentrations of 20,000 to 60,000 ppb were found.

A contour of the vapor concentrations (Figure 1) reveals that while the entire map area is underlain by a PSH plume at a depth of ca. 30 feet, the hydrocarbon vapor concentrations are highest in the center of the parking lot immediately behind the ice arena. Attempts to correlate variations in the depth of PSH with vapor concentrations in this area were unsuccessful. Similarly, a correlation was not found between thickness of the PSH (in this area it ranged between 0.1 and 1 foot during the study interval) and the vapor concentration. The best correlation was between the geometry of the parking lot and the vapor concentration below the lot.

The petroleum liquid samples analyzed were from a monitoring well that had approximately one foot of hydrocarbon fuel, as free product, floating on the groundwater. Migration of free product into a monitor well causes a depression of the water table, so the monitor well thickness of free product is about 10 times as thick as in the surrounding soil. Free product from the monitor well was analyzed by combined capillary column GC/MS (Keith, 1992). The resulting chromatograms were analyzed first for petroleum product distribution; see Table 1. The hydrocarbon component distribution was assigned by taking the area percent of all compounds (1) from C_4 to toluene as representative of the gasoline fraction, (2) from toluene to C_{14} as representative of the jet A fraction, (3) from C_{14} to C_{20} as representative of diesel fuel, and (4) above C_{20} as representative of fuel oil. The GC/MS analysis of this free product revealed that jet fuel at 48% and diesel at 45% comprised the majority of the hydrocarbon material present, while gasoline and fuel oil were present at only 5 and 2%, respectively. The quantity of each fuel was not as significant as the quantity of the individual components present. The low molecular weight components contribute not only to the groundwater contamination, but more importantly, to the soil gases that emanate from hydrocarbon liquids (Luckner and Schestakow, 1991).

Table 1.   **Composition of the Free Hydrocarbon Product Above the Groundwater**

Hydrocarbon Components[a]	Concentration in Wt %
Jet A	48
Diesel	45
Gasoline	5
Fuel Oil	2

[a]Summation of peaks in the various boiling point ranges for the individual fuels.

CONCLUSIONS

The subsurface hydrocarbon plume discovered east of the Star Enterprise tank farm migrated away from the tank farm at between 70 and 350 feet/year. The plume, estimated to contain ca. 200,000 gallons of fuel, extends ca. 1500 feet beyond the property margin of the tank farm, underlying ca. 1,000,000 square feet, and contaminating in excess of 150,000 cubic yards of soil.

Analyses of hydrocarbon vapor and phase-separated hydrocarbon removed from the plume reveal that the plume is a mixture of diesel fuel, aviation jet fuel, and gasoline. Volatiles from the plume have reached the surface in several areas, and in the study area examined in this report, hydrocarbon vapors were found to have accumulated below an asphalt-covered parking lot immediately adjacent to a public ice arena. Evidently the gravel-based fill under the asphalt surface became a reservoir for upward-moving hydrocarbon vapors, and unlike surrounding grass-covered areas where the vapors escaped, the asphalt-covered parking lot capped the vapor and prevented its escape.

REFERENCES

Froelich, A.J. and Heironimus, T.L. 1977. Map Showing Contours of the Base of Saprolite, Fairfax County, Virginia. U.S. Geological Survey Open File. Report 77-710.

Froelich, A.J. and Langer, W.H. 1983. Map Showing Geological Provinces, Landforms, Drainage Characteristics and Flooding in Fairfax County, Virginia. U.S. Geological Survey Miscellaneous Map Investigation Series. Map I-1534.

Keith, L.H. 1992. Semivolatile Organic Compounds In: *Compilation of E.P.A.'s Sampling and Analysis Methods*, Chapter 2, Chelsea, MI, Lewis Publishers.

Luckner, L. and Schestakow, W.M. 1991. In: *Migration Processes in the Soil and Groundwater Zone*, Chapter 3, Chelsea, MI, Lewis Publishers.

Sowers, G.F. and Richardson, T.L. 1983. Residual Soils of the Piedmont and Blue Ridge. Special Problems in Residual Soils and Rock, National Research Council, Transportation Research Board, Transportation Research Record. 919, 10-16.

PART II — ANALYSIS/FATE

CHAPTER 4

Measurement of Dissolved Petroleum in Groundwater: Pitfalls of TPH Analyses

Dawn A. Zemo, Geomatrix Consultants, Inc., San Francisco, California 94111

INTRODUCTION

This paper discusses the measurement of total petroleum hydrocarbons (TPH) in groundwater samples using EPA Method 8015M or equivalent and its representation of dissolved-phase petroleum. This paper is a continuation and expansion of the work first presented in Zemo and Synowiec (1995) that demonstrated positive interferences to TPH measurements in groundwater. Conclusions about the reliability of TPH measurements are important because many sites remain active for years solely because "TPH" is detected in groundwater samples, even when benzene, toluene, ethylbenzene, and xylenes (BTEX) or polynuclear aromatics (PNAs) are not present.

The following two sections present a brief summary of the theoretical background for the subject, including the general chemistry of petroleum products and their water-soluble fraction, and a discussion of the Method 8015M analysis. Following the theoretical background is a compilation of case studies that demonstrates the prevalence of positive interferences to the measurement of TPH concentrations in groundwater samples.

THE WATER-SOLUBLE FRACTION OF PETROLEUM HYDROCARBON PRODUCTS

Petroleum products are complex mixtures of hundreds to thousands of individual petroleum constituents. The specifications for various products are performance-based rather than constituent-based (Bruya, 1993). Accordingly, the exact composition of a product type varies but generally falls within certain ranges. Gasolines commonly contain C_4 to C_{12} alkanes, C_4 to C_7 alkenes, C_6 to C_{11} monoaromatics (BTEX, substituted benzenes), C_{10} to C_{11} PNAs, and performance additives. Kerosene, jet fuel, diesel, and home heating oil commonly contain C_{10} to C_{24} alkanes, C_6 to C_{11} monoaromatics, and C_{10} to C_{22} PNAs. Diesel Nos. 4 and 6, Bunker C, lube and motor oils, commonly contain C_{20} to C_{78} alkanes, C_{14} and larger PNAs, and metals such as nickel and vanadium (Bruya, 1993; Zemo et al., 1995).

The water-soluble fraction (WSF) of a petroleum product is a function of both the molecular class (alkane, alkene, or aromatic) and the molecular weight (number of carbon atoms) of its constituents. For the purpose of this paper, we will focus only on the petroleum constituents and not on the performance additives. Within a certain molecular class, lower molecular weight constituents usually tend to be more soluble. Published chemical research (Mackay and Shiu, 1992, Yaws et al., 1990) demonstrates that alkanes and alkenes have very low water solubilities (<10 mg/L) at molecular weights exceeding six carbon atoms (C_6) either as discrete molecules or as a component of complex petroleum products or crude oil. The monoaromatics (BTEX and substituted benzenes) have moderate water solubilities (>100 mg/L), with the lowest molecular weight (C_6) benzene having the highest relative solubility (1750 mg/L). The lower molecular weight PNAs have low to very low water solubilities (30 to <1 mg/L), with the higher molecular weight PNAs (e.g., benzo(a)pyrene[C_{20}]) being virtually insoluble in water (<0.01 mg/L). Based on this understanding of their chemical properties, we would expect that the WSF would be limited to a few petroleum constituents out of the thousands that make up the petroleum product or crude oil. Furthermore, the measurable portion of the WSF of a given product is a function not only of the solubility of each constituent, but also the mole-fraction of the constituent within the product and the partitioning coefficient of the constituent between water and the other organics in the product.

As discussed in Zemo and Synowiec (1995), the WSF of fresh petroleum products and crude oil has been investigated at laboratory conditions using gas chromatography (GC) in a qualitative manner by Bruya and Friedman (1992), and in a quantitative manner using EPA-approved GC and GC/MS methods by Thomas and Delfino (1991) and Chen et al. (1994). Potter (1996) investigated the WSF of several petroleum products using a direct-injection GC procedure. Bruya and Friedman (1992) found that the WSF of a tested crude oil, gasoline, and diesel was limited primarily to the low molecular weight alkanes (C_6), BTEX and substituted benzenes, and the smaller PNAs (e.g., naphthalenes, phenanthrene, and anthracene). Most significantly, the chromatograms of the WSF in each case was composed of discrete peaks that did not resemble the parent product (Figure 1).

The Thomas and Delfino (1991), Chen et al. (1994), and Potter (1996) findings are consistent with the qualitative findings of Bruya and Friedman (1992) and indicate that the WSF of the products tested (fresh gasoline, kerosene, and diesel [Thomas and Delfino]; fresh motor oil [Chen et al.]; and fresh gasoline, jet fuels, and diesel [Potter]) is limited primarily to the C_6 to C_{11} monoaromatics and the C_{10} to C_{14} PNAs, including naphthalene, methylnaphthalenes, acenaphthene, fluorene, phenanthrene, and anthracenes. Thomas and Delfino (1991) and Potter (1996) also reported low concentrations of phenol and methylated phenols in the

Figure 1. GC-FID chromatograms for various fresh petroleum products and their water-soluble fractions (WSF). Revised from Zemo and Synowiec (1995), after Bruya and Friedman (1992).

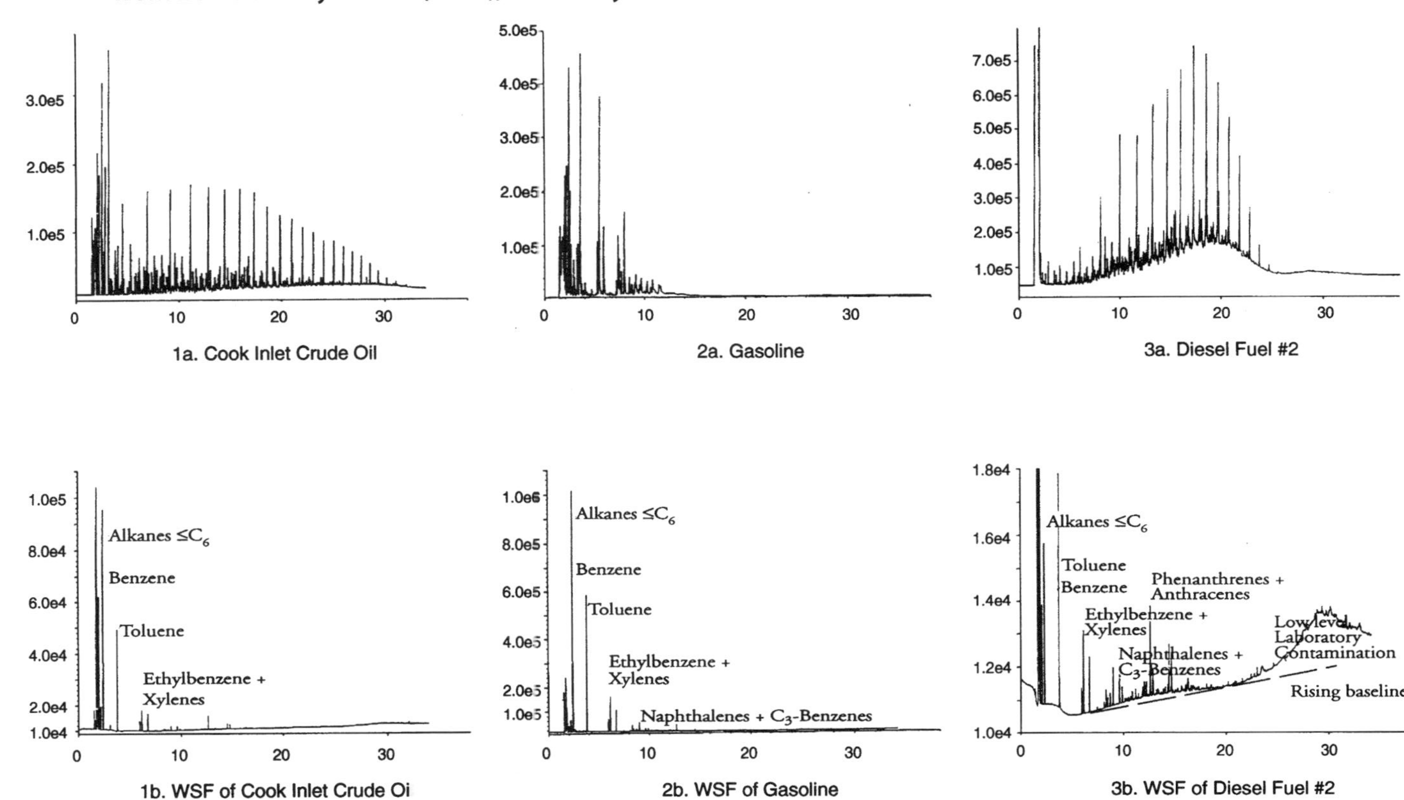

WSF of the products tested. The low-boiling alkanes were not included as target analytes in these studies. Chromatograms provided in Thomas and Delfino (1991) look similar to those of Bruya and Friedman (1992); the WSF of each product is composed of discrete peaks and does not resemble the parent product.

These studies provide clear evidence that the measurable WSF of fresh crude oil and fresh products is limited primarily to the low molecular weight alkanes ($\leq C_6$), the monoaromatics, and the PNAs having 14 carbons or fewer. This identification of the WSF of various products has great significance for interpretation of TPH analytical results from groundwater samples.

TPH (METHOD 8015M) ANALYTICAL METHOD AND POTENTIAL FOR POSITIVE INTERFERENCE

Method 8015M (TPH) is a GC analysis that quantifies an amount of volatile or semivolatile hydrocarbon compounds that elute within a selected boiling range or range of molecular weights. It is generally separated into purgeable and extractable fractions. It is fundamentally an aggregate rather than a constituent-specific analysis and has a number of interpretational difficulties. The most important interpretational difficulty is that a TPH analysis transmits no direct information about which petroleum constituents are present in the sample (alkanes, alkenes, aromatics). Moreover, TPH analyses are subject to a number of positive interferences, most commonly the incorporation of nonpetroleum hydrocarbons such as oxidized biodegradation by-products (such as alcohols) or biogenic materials (such as lipids) into the analytical result (Zemo et al., 1995).

Because of its nonspecificity and susceptibility to positive interferences, the use of Method 8015M has been demonstrated to be particularly unreliable for measurement of dissolved petroleum constituents in groundwater samples. Zemo and Synowiec (1995) showed that the TPH concentration of a groundwater sample resulting from constituents other than the $\leq C_6$ alkanes or $\leq C_{14}$ aromatics was from either or both of the following sources: (1) the sample contained nondissolved petroleum constituents, or (2) the sample contained soluble nonpetroleum hydrocarbons (such as polar biogenic materials or biodegradation products). Samples affected by either or both of these sources of interference do not provide an accurate assessment of dissolved-phase concentrations of petroleum in groundwater.

Nondissolved petroleum constituents are incorporated into water samples by passing a bailer or other sampling device through a sheen on top of the water column during sampling or by entraining petroleum that is sorbed onto sediment (turbidity) suspended in the water column inside the well. Nondissolved petroleum included in the sample will be extracted along with the water at the laboratory when using standard analytical procedures. Consequently, the "TPH" result for the groundwater sample will include these nondissolved constituents.

Soluble nonpetroleum hydrocarbons such as polar biogenic materials and biodegradation products are incorporated into water samples when wells are screened within or downgradient from petroleum-affected soil that is undergoing intrinsic biodegradation. Barcelona et al. (1995) identified by GC/MS numerous aliphatic and aromatic organic acids that are degradation metabolic intermediates of gasoline in groundwater samples. The Barcelona et al. study also pointed out that there are numerous other potential oxygenated metabolites, including phenols, alcohols, aldehydes, and hydroxy-aliphatic acids. Collection of such polar materials within the water sample is unavoidable because of their relatively high solubility. Method 8015M normally does not include a silica gel or other extract cleanup step to remove polar materials, consequently the "TPH" result for the groundwater sample will include these nonpetroleum constituents.

SUMMARY OF CASE STUDIES

To demonstrate the impact of these sources of positive interferences, filtering (to remove nondissolved petroleum sorbed to particulates) and/or silica gel cleanup (to remove polar biogenic materials) steps were added to the extractable TPH laboratory preparation of groundwater samples collected at numerous sites. These sites were similar in that they had elevated concentrations of extractable TPH in groundwater with no or very low detections of BTEX or PNAs. In addition, the chromatograms for samples collected at these sites had a characteristic "hump" pattern and not the discrete peak pattern found by Bruya and Friedman (1992) and Thomas and Delfino (1991). It should be noted that we expect similar interferences to the purgeable TPH analysis; however, we have not modified the conventional sample preparation because of the potential for loss of volatiles.

Filtering was accomplished by the laboratory using a glass fiber (0.7-micron) filter; several other filter materials were found to be unacceptable because of potential for sorption (see Foote et al., 1997). The samples were filtered prior to extraction. To remove the polar materials from the extract, the laboratory performed a silica gel cleanup based on EPA Method 3630B. Our experience demonstrated that a column cleanup is required; a cleanup based on EPA Method 418.1 (adding 3 grams of silica gel to the extract and shaking the mixture) was not acceptable because it frequently did not result in complete removal of polar biogenic material.

Comparisons of chromatograms and quantitative results between cleaned-up samples and duplicate samples that were conventionally prepared show clearly that the conventional TPH measurements are greatly affected by nondissolved and biogenic interferences and do not represent dissolved petroleum in the groundwater. Table 1 summarizes these results from 12 different sites affected by degraded crude oil, diesel, or Bunker C. Figures 2 and 3 show example chromatograms of samples before and after cleanup. As shown in the table, we have observed decreases in TPH concentration (after cleanup) of two to three orders of magnitude; in many cases TPH was not detected after sample cleanup was performed.

Table 1.   TPH (Method 8015M) Analytical Results for Groundwater Samples Before and After Cleanup

Site	Before Cleanup (Conventional) "TPHd" (μg/L)[a]	Filter and Silica Gel "TPHd" (μg/L)[b,c]	Duplicate Filter or Silica Gel Only[d] "TPHd" (μg/L)	Comments[e]
1	20,000	<1000	F only = 15,000 SG only = <1000	biogenic
2	110,000	1200	SG only = 55,000	grab samples with particulates, some pass filter
3	100	<50	SG only = <50	biogenic
4	200	<50	SG only = <50	biogenic
5	390	<50	SG only = <50	biogenic
6	6600	140	SG only = 140	incomplete cleanup
7	630	<50	--	biogenic
8	120,000	690	--	grab samples with particulates, incomplete cleanup
9	4500	750	--	biogenic, incomplete cleanup
10	1100	<50	F only = 390 SG only = <50	mostly biogenic
11	790	<50	SG only = <50	biogenic
12	1500	<50	SG only = <50	biogenic

[a] Analysis by USEPA Method 8015M, TPH as diesel; quantitation range varied among laboratories, typically C_8 to C_{30}.

[b] Filtered by laboratory using glass fiber filter (0.7-micron).

[c] Silica gel cleanup by either 418.1 or 3630B equivalents (see text).

[d] Blind duplicates run to discriminate between sources of interference.

[e] Major component of interference based on review of chromatograms.

Several important trends have become clear based on the number of sites we have studied so far: (1) where samples are collected from monitoring wells with very low turbidity, filtering makes virtually no difference to the analytical result and, therefore, may not be necessary (e.g., Zemo and Synowiec, 1995); (2) where samples are collected in the vicinity of petroleum-affected soil and are turbid, filtering is critically important although some nondissolved petroleum may still pass the 0.7-micron filter and influence the analytical result (e.g., Foote et al., 1997); (3) because biogenic interference has predominated at the vast majority of sites, extracts should be routinely cleaned up with silica gel; (4) silica gel cleanup may be incomplete at times (even with a column cleanup) due to either the polarity or the amount of biogenic material in the extract; (5) laboratory QA/QC must demonstrate that there is not negative bias due to sample cleanup (e.g., acceptable surrogate recoveries, no loss of petroleum constituents); and (6) only a review of the chromatograms can determine whether a cleanup was complete.

Figure 2. **GC-FID chromatograms for blind duplicate groundwater samples demonstrating removal of polar (biogenic) compounds by silica gel cleanup of extract.**

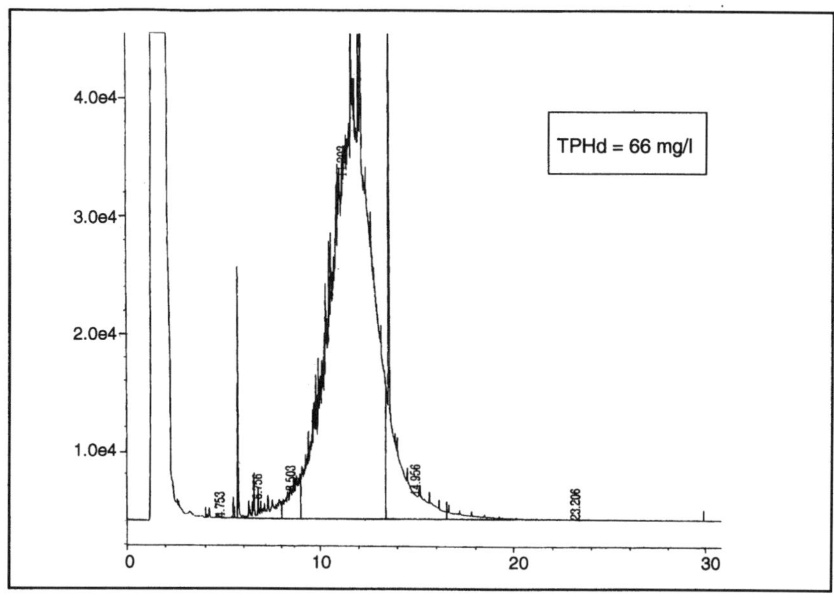

2a. Sample Before Cleanup

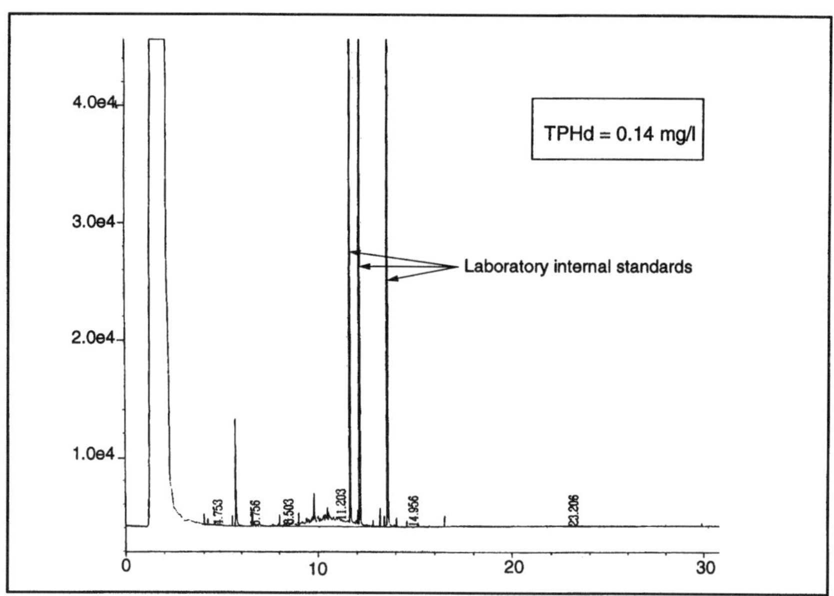

2b. Silica Gel Cleanup Only

Figure 3. **GC-FID chromatograms for blind duplicate groundwater samples demonstrating removal of nondissolved petroleum sorbed to particulates. Revised from Foote, et al. (1997).**

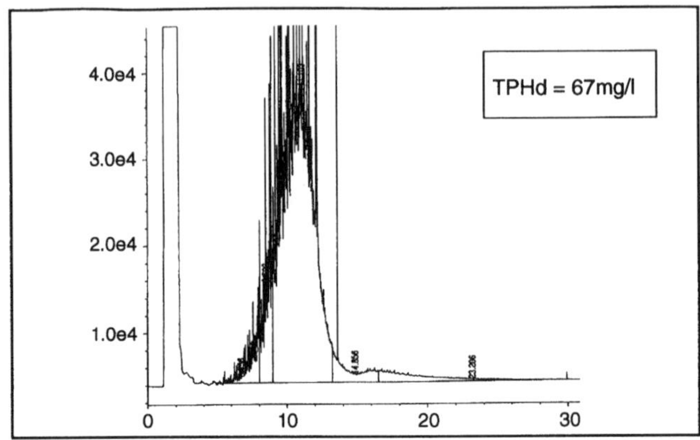

3a. Sample Before Cleanup

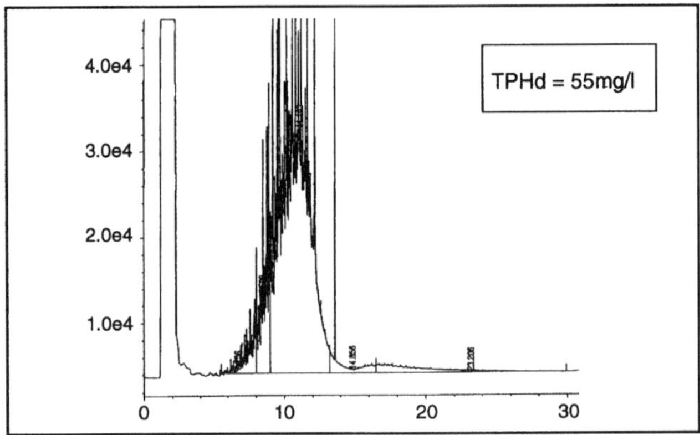

3b. Silica Gel Cleanup Only

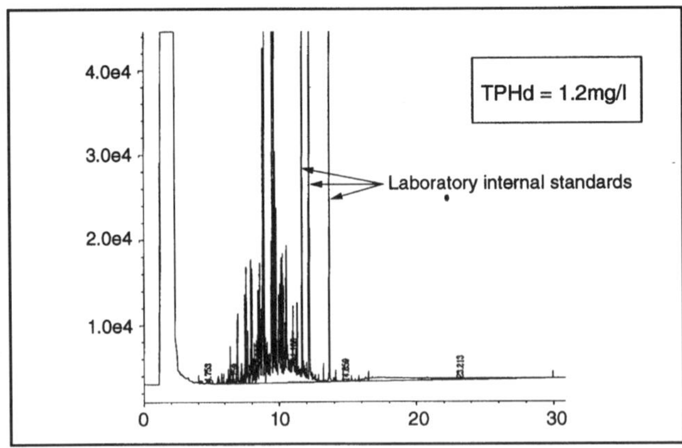

3c. Filtration and Silica Gel Cleanup

CONCLUSIONS

Our research has shown that TPH measurement of groundwater samples by Method 8015M can be subject to significant positive interferences and may not represent dissolved petroleum in groundwater. In many cases, the entire TPH measurement may be an artifact of these interferences. We conclude that conventional TPH measurements should not be used to assess dissolved petroleum in groundwater. If TPH measurements must be used due to regulatory requirements, samples should be cleaned up with silica gel and lab-filtered if turbidity is present.

REFERENCES

Barcelona, M.J., Lu, J., and Tomczak, D.M. 1995. Organic Acid Derivatization Techniques Applied to Petroleum Hydrocarbon Transformations in Subsurface Environments. *Ground Water Monitoring and Remediation,* Spring, 114-124.

Bruya, J.E. 1993. Petroleum Hydrocarbons: What Are They? How Much Is Present? Where Do They Go? Workshop notebook prepared for HAZMACON '93.

Bruya, J.E. and Friedman, A.J. 1992. Don't Make Waves: Analysis of Water Samples for Total Petroleum Hydrocarbons. *Soils,* January-February; 6-7, 46-47.

Chen, C.S-H., Delfino, J.J., and Rao, P.S.C. 1994. Partitioning of Organic and Inorganic Components from Motor Oil into Water. *Chemosphere,* 28(7), 1385-1400.

Foote, G.R., Zemo, D.A., Gallardo, S.M., Grant, M.J., Benson, B.T., and Bruya, J.E. 1996. Case Study: Interferences with TPH Analyses of Grab Groundwater Samples. Principles and Practices for Diesel Contaminated Soils, Volume 6, pp. 27-39. (Barkan, C.P.L, Kostecki, P.T. and Calabrese, E.J. , Eds.) Amherst, MA, Amherst Scientific Publishers.

Mackay, D. and Shiu, W.Y. 1992. Estimating the Multimedia Partitioning of Hydrocarbons: The Effective Solubility Approach, In: *Hydrocarbon Contaminated Soils and Groundwater,* Volume 2, pp. 137-154. (Calabrese, E.J. and Kostecki, P.T., Eds.). Chelsea, MI, Lewis Publishers, Inc.

Potter, T.L. 1996. Analysis of Petroleum-Contaminated Water by GC/FID with Direct Aqueous Injection. *Ground Water Monitoring and Remediation,* Summer, 157-162.

Thomas, D.H. and Delfino, J.J. 1991. A Gas Chromatographic/Chemical Indicator Approach to Assessing Groundwater Contamination by Petroleum Products. *Ground Water Monitoring Review,* Fall, 90-100.

Yaws, C.L., Yang, H-C., Hopper, J.R., and Hansen, K.C. 1990. 232 Hydrocarbons: Water Solubility Data. *Chemical Eng.,* April, 177-182.

Zemo, D.A., Bruya, J.E., and Graf, T.E. 1995. The Application of Petroleum Hydrocarbon Fingerprint Characterization in Site Investigation and Remediation. *Ground Water Monitoring and Remediation,* Spring, 147-156.

Zemo, D.A. and Synowiec, K.A. 1995. TPH Detections in Groundwater: Identification and Elimination of Positive Interferences. *Proceedings of the 1995 Conference on Petroleum Hydrocarbons and Organic Chemicals in Ground Water: Prevention, Detection, and Remediation*, pp. 257-271. NGWA/API, Houston, TX.

CHAPTER 5

Evaluation of a GC PID/FID Procedure for the Analysis of EPH

James F. Occhialini, David L. Gottshall, and **Robert A. Burke,** Camp Dresser & McKee Inc., Cambridge, Massachusetts

INTRODUCTION

There have been considerable efforts expended regarding the risk assessment and related analytical issues associated with petroleum-product contaminated environmental media. The evaluation of petroleum contaminated sites has been impeded by the lack of scientifically-based health standards and the limitations of commonly used analytical techniques. Analytical data concerning the distribution of aliphatic and aromatic hydrocarbons is required to facilitate a toxicological approach to evaluating the potential health effects associated with human exposure to petroleum hydrocarbons.

In August of 1995, the Massachusetts Department of Environmental Protection (MADEP) issued a draft procedure for the analysis of extractable petroleum hydrocarbons (EPH). The procedure was developed to measure the collective concentrations of extractable aliphatic and aromatic hydrocarbons within specified carbon ranges, as well as concentrations of selected polynuclear aromatic hydrocarbons (PAH) compounds. In the MADEP EPH procedure, the sample to be analyzed is extracted with methylene chloride, solvent exchanged into hexane and concentrated. The concentrated extract is then separated into aliphatic and aromatic fractions by silica gel column chromatography. Each individual fraction is then concentrated and analyzed by GC-FID. The aliphatic fraction chromatogram is collectively quantitated within two ranges: C_9 through C_{18}, and C_{19} through C_{36}. The aromatic fraction chromatogram is collectively quantitated within the C_{10} through C_{22} range. Individual concentrations of targeted PAH compounds are also determined from the aromatic fraction.

In this paper, the authors evaluate the suitability of a different procedure for EPH analysis. This procedure utilizes a new design, high temperature detector block GC, configured with a tandem photo ionization/flame ionization detector (PID/FID) to differentiate EPH contaminated samples into their aliphatic and aromatic components. The use of PID detection systems for petroleum analysis by gas chromatography has previously been limited to the lower boiling, gasoline range compounds because of the low operational temperature limitation of the PID design. Commonly used gas chromatographs equipped with a PID have maximum detector block operating temperatures in the range of 300°C. This relatively low detector temperature causes the higher boiling components of petroleum products to condense in the detector block and foul the photo ionization source. The Finnigan 9001 gas chromatograph used for this study was equipped with a newly-designed high temperature tandem PID/FID detector block with a maximum operational temperature of 350°C. This high temperature design allows the detector to be used for petroleum products analysis throughout the entire fuel and lubricating oil ranges.

OBJECTIVE AND EXPERIMENTAL DESIGN

The objective of this study is to determine whether the proposed procedure can meet the performance standards specified in the MADEP EPH procedure. The authors note that the method development work is still ongoing in the laboratory, and that any conclusions reached in this paper should be considered preliminary.

The analytical procedure described in this paper utilizes the difference in PID/FID relative response rather than silica gel column chromatography to differentiate between aromatic and aliphatic compounds. The PID detector responds to compounds that have ionization potentials less than the rated output of the detector's ionization source of UV light. In general, aliphatic compounds typically exhibit poor PID response, relative to the PID response of aromatic compounds. The higher the degree of ionization for a given compound, the greater the response. In the PID/FID procedure, the sample to be analyzed is extracted with methylene chloride and concentrated. The concentrated extract is then analyzed by GC-PID/FID. The quantitation of the C_9 through C_{18} and C_{19} through C_{36} aliphatic hydrocarbon ranges are determined by subtracting the total PID concentration from the total FID concentration for each hydrocarbon range. The aromatic C_{10} through C_{22} hydrocarbon range and targeted PAH compounds are quantitated based on calculated PID concentration only.

There are a number of potential advantages associated with this analytical procedure for EPH analysis. The silica gel fractionization procedure is not required for this method, which eliminates any potential sources of error associated with the recoveries or selectivity of the labor-intensive silica gel preparation step. Only one sample solvent extract is prepared using the PID/FID procedure which requires a single GC analysis rather than the two GC analyses currently used. The use of the PID for the quantitation of the PAH target compounds should increase the sensitivity of the method, which should approach the detection limits specified in the GW-1 category of the Massachusetts Contingency Plan 310 CMR 40.0000 April 5, 1996. Furthermore, the use of the dual PID/FID detector system will provide enhanced qualitative information for selected components due to the differences in relative response between the two detectors. The potential disadvantages associated with the PID/FID procedure were also investigated. The validity of the aromatic/aliphatic determination via PID/FID relative response was studied along with the need for increased chromatographic resolution and run times required to separate the EPH components in a single GC run.

The PID/FID method development effort was focused on the following experimental design considerations:

- Adhere to the letter and spirit of the MADEP EPH procedure to the extent possible
- Evaluate the selectivity versus sensitivity characteristics of PID ionization sources of different energies
- Evaluate chromatographic resolution versus sample capacity considerations
- Evaluate method performance using MADEP EPH performance criteria

The MADEP EPH procedure was designed to provide specific types of information regarding EPH composition and concentration. The primary experimental design parameter of the PID/FID EPH procedure was to provide the same basic information. Two PID ionization sources were evaluated for use with this method - a krypton 10.0 eV lamp and a xenon 9.6 eV lamp. A combined EPH standard mix composed of all of the EPH method calibration compounds was evaluated using each PID lamp. The 9.6 eV lamp was demonstrated to be the more appropriate ionization source for this application. The 9.6 eV lamp provided good sensitivity for all aromatic compounds but demonstrated virtually no response for the alkanes. The 10.0 eV lamp did provide somewhat greater sensitivity but it was not as selective as the 9.6 eV lamp in that it exhibited a response to some aliphatic compounds in addition to the aromatics.

The PID/FID procedure requires a greater degree of chromatographic resolution than the MADEP procedure due to the need to separate all EPH calibration compounds in one chromatographic run. The first column evaluated was a narrow bore J&W 30M, 0.25 mmID DB-5MS fused silica column with a 0.25 μm film thickness. This column provided excellent resolution for all calibration compounds under the initial temperature and electronic pressure control programming, with a total run time of less than 40 minutes. All but one of the PAH isomer pairs were resolved to the baseline, but column overloading was evident at calibration standard concentrations greater than 25 μg/mL. This column overloading condition so decreased the linear dynamic range of the procedure that this column proved to be unworkable for this application. The column was changed to a J&W 30M, 0.32 mmID DB-5MS fused silica column with a 0.5 μm film thickness in an attempt to improve column capacity without compromising resolution. This column did provide both the column capacity and chromatographic performance required for this procedure, but the increase in performance came at the expense of an increased running time that now totaled 65 minutes. Preliminary method performance data is discussed in the method performance section of this paper.

METHOD SUMMARY AND EXPERIMENTAL CONDITIONS

Aqueous samples prepared for method development purposes were spiked with the chloro-octadecane and ortho-terphenyl surrogate solution, serially extracted with methylene chloride and concentrated. The sample extracts were not solvent exchanged with hexane. A one liter sample volume was extracted with a final extract concentration of one milliliter. Soil sample matrices were not evaluated at this time, pending further evaluation of the instrumental procedure. As stated previously, a Finnigan 9001 gas chromatograph equipped with a high temperature, tandem PID/FID detector block was used for the analysis. The PID 9.6 eV lamp and the J&W 30M, 0.32 mmID DB-5MS (0.5 μm film thickness) fused silica column were used for all subsequent analyses after initial method development efforts proved these options to be the most suitable for the application. A split/splitless injection port was operated in the splitless mode at 285°C. Septum purge was employed at 0.75 minutes after injection. An extract injection volume of 1.0 uL was utilized. The GC oven temperature program used was 40°C hold for 1 minute, 5°C/minute to 250°C with no hold, 15°C/minute to 290°C with no hold, 20°C/minute to 320°C for a total program time of 65 minutes. The inlet pressure was maintained to provide a constant linear velocity of 50 cm/second. The inlet pressure is surged to 75 psi prior to and for 0.50 minutes after injection. The 25 μg/mL combined standard chromatogram is displayed in Figure 1. The individual aliphatic and aromatic standard chromatograms are displayed in Figures 2 and 3, respectively, as an indication of how each detector responds to each hydrocarbon classification. Figure 4 displays the PID/FID chromatogram obtained from the analysis of a Fuel Oil #2 reference standard.

Calibration was performed using the external standardization method. Five concentration levels were analyzed; 5.0, 10, 25, 38, and 50 ug/mL for target analytes. Surrogate analytes were analyzed at a level four times that of the targets. Aliphatic and aromatic hydrocarbons were analyzed together. The average response factor of each analyte from the FID was determined. The average response factor of each aromatic from the PID was determined. Aliphatics provided no response on the PID over the calibration range. The relative standard deviation for all average response factors were less than 25% for both detectors. The average of the average response factors for each target analyte within the three method-specified hydrocarbon ranges were determined on the PID. The average of the average response factors for each target analyte within the two method-specified aliphatic ranges were determined on the FID.

Quantitation is based upon two premises. Aliphatic and aromatic hydrocarbons, within their respective ranges, must respond roughly the same on the FID. Only aromatic hydrocarbons may respond to the PID. Concentrations for each

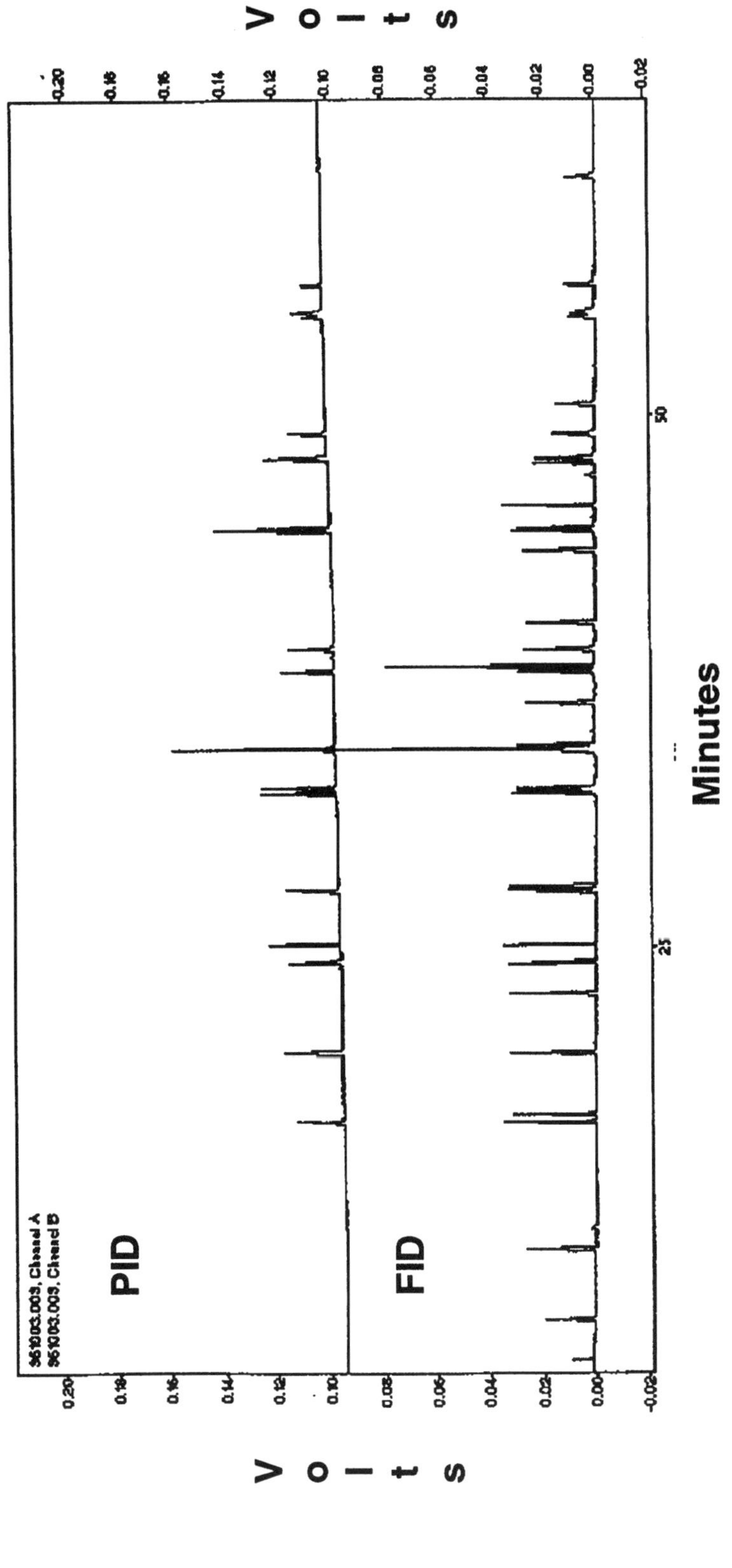

Figure 1. 25 µg/mL Combined Standard.

Figure 2.   50 µg/mL Aliphatic Standard.

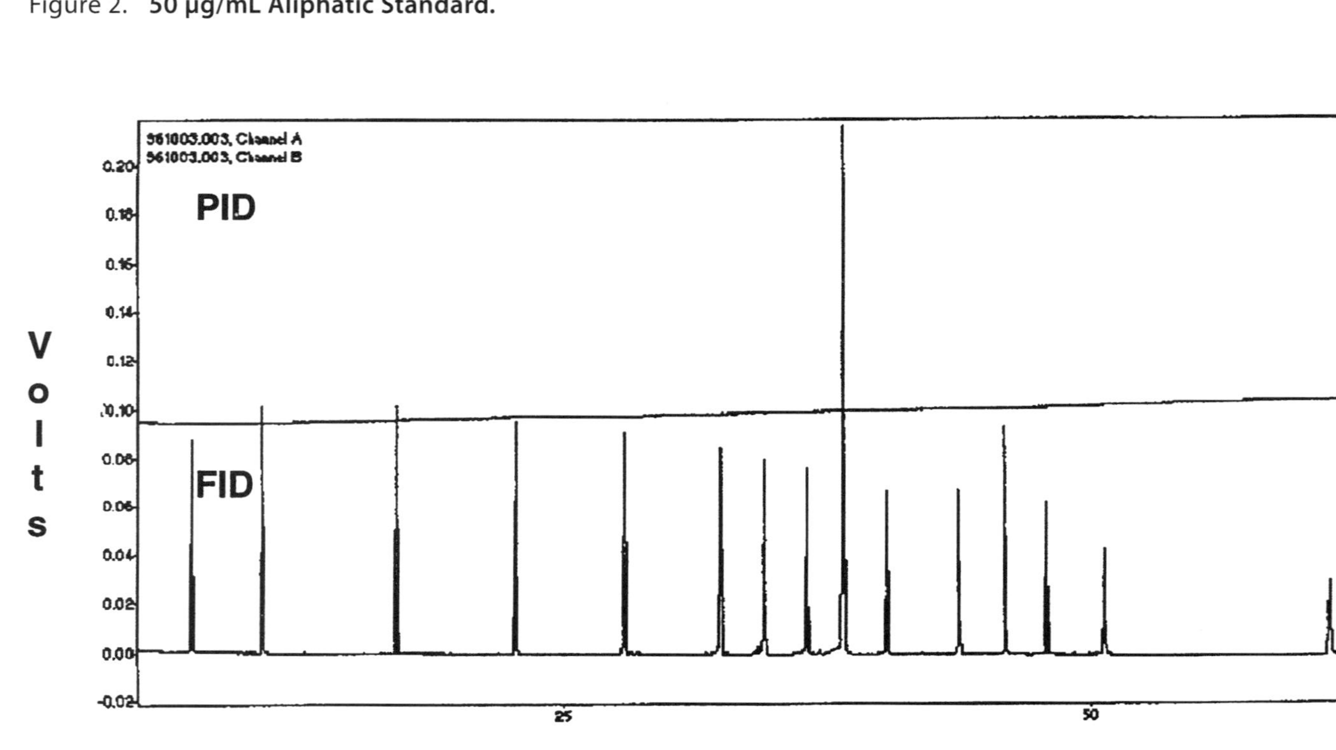

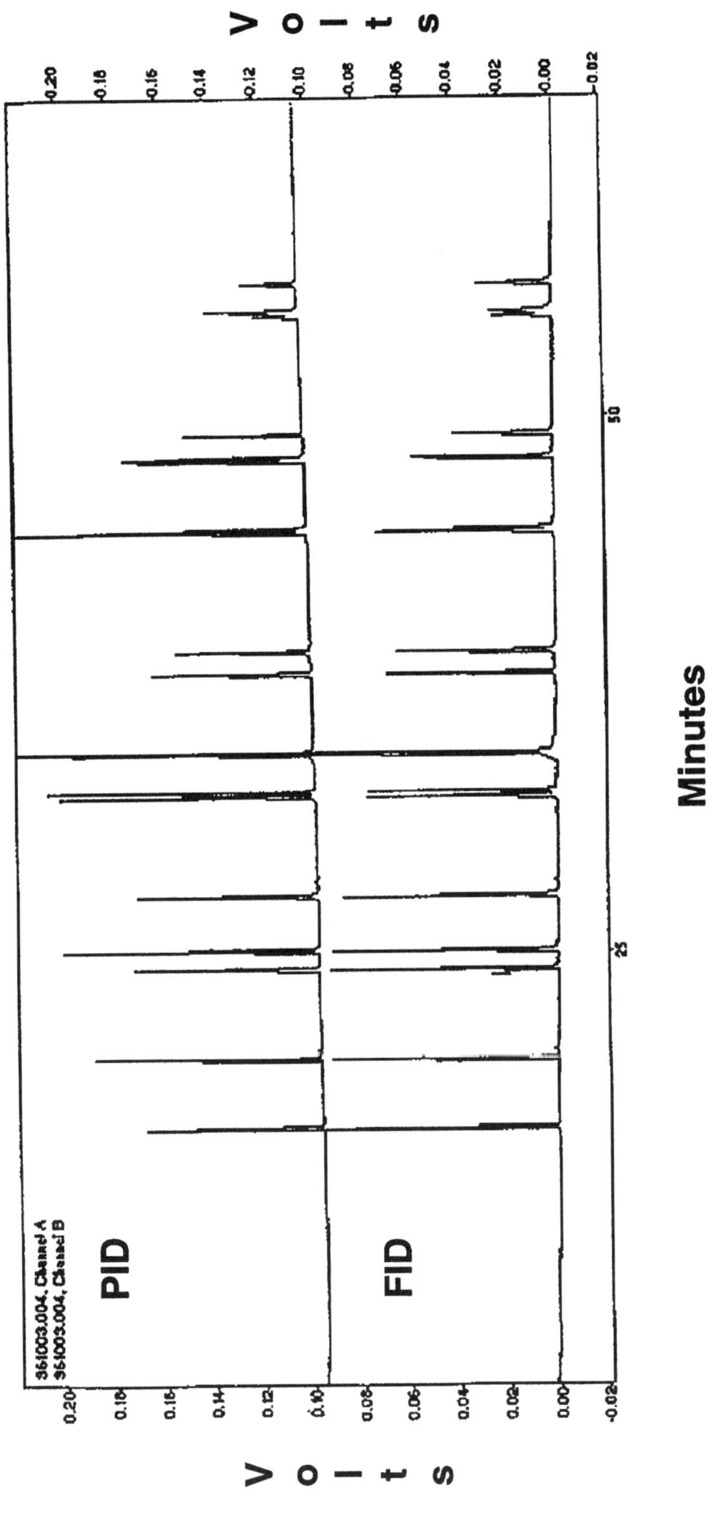

Figure 3. 50 µg/mL Aromatic Standard.

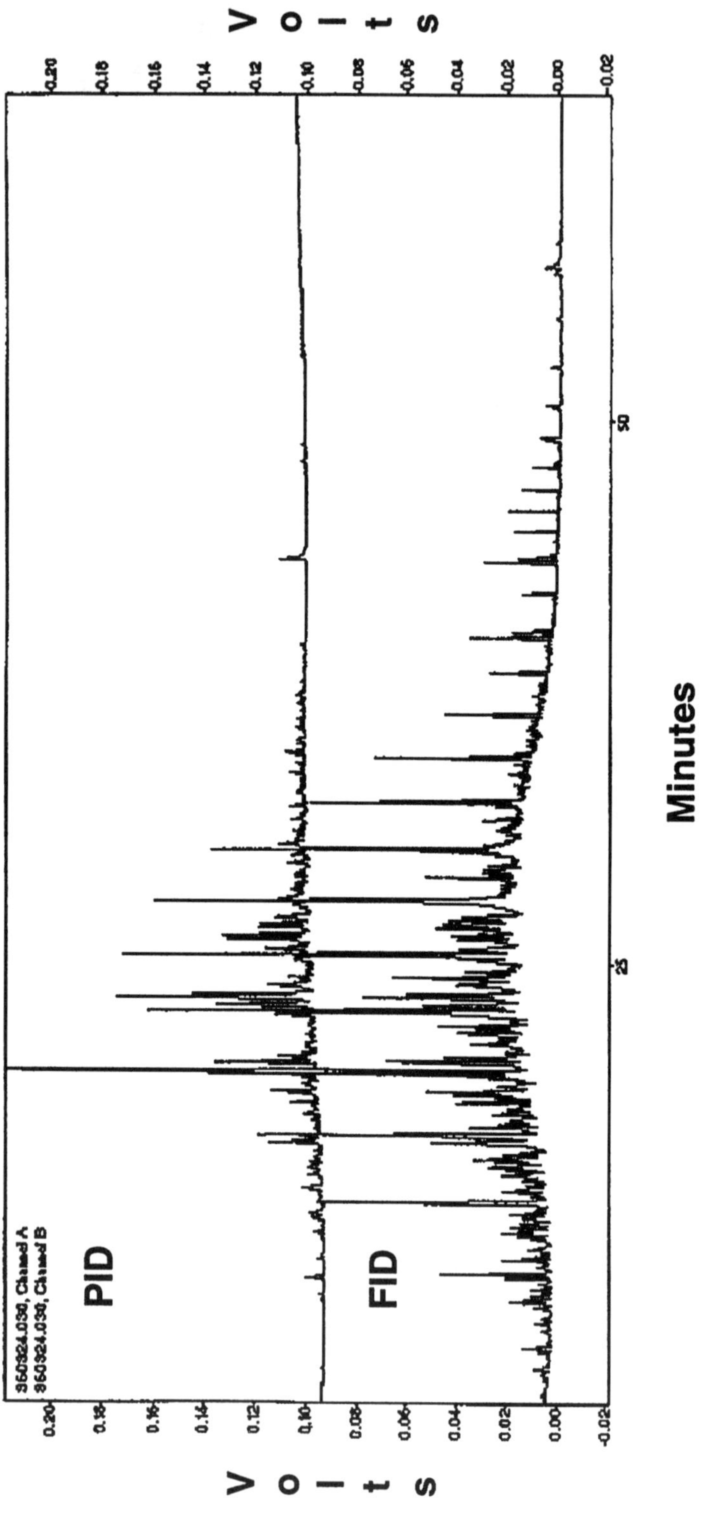

Figure 4. 2,000 µg/mL Fuel Oil #2.

range, on each detector, are determined by dividing the corrected area of the range (listed below) with the average of the average response factors for each target within that range.

FID Range   Corrected Area

C_9-C_{18} Collective area from 0.1 minutes before nonane to just before nonadecane less the area of ortho-terphenyl.

C_{19}-C_{36} Collective area from nonadecane to 0.1 minutes after hexatriacontane less the area of chloro-octadecane.

PID Range   Corrected Area

C_9-C_{18} Collective area from 0.1 minutes before nonane to just before nonadecane less the area of ortho-terphenyl.

C_{19}-C_{36} Collective area from nonadecane to 0.1 minutes after hexatriacontane.

C_{10}-C_{22} Collective area from 0.1 minutes before naphthalene to 0.1 minutes after benzo(g,h,i)perylene less the area of ortho-terphenyl and all identified individual PAHs.

From these calculations and the determination of individual PAHs from the PID, all of the required EPH reporting values can be obtained. Calculation summaries have been provided for 25 ug/mL mixed hydrocarbon standard, 50 ug/mL aliphatic hydrocarbon standard, and 25 ug/mL aromatic hydrocarbon standard (Tables 1, 2, and 3). Range calculations are presented below.

Range Calculations:

Aliphatic Hydrocarbons

C_9-C_{18} FID Concentration (C_9-C_{18}) minus PID Concentration (C_9-C_{18})

C_{19}-C_{36} FID Concentration (C_{19} C_{36}) minus PID Concentration (C_{19}-C_{36})

Aromatic Hydrocarbons

C_{10}-C_{22} PID Concentration (C_{10}-C_{22})

PAHs   PID Concentration of each individual PAH

Surrogates

Ortho-terphenyl   PID or FID Concentration

Chloro-octadecane   FID Concentration

When using average response factor quantitation, some positive bias is noted in the PID response at the upper end of the calibration range. While within the 25% RSD window, this positive bias can lead to both overestimation of aromatics and underestimation of aliphatic. Optimum accuracy is obtained by analyzing samples at a concentration that results in values near the midpoint of the calibration range. Regression analysis or the use of appropriate internal standards may be used to overcome this bias.

Table 1.   25 µg/mL Combined standard.

	AROMATICS C10 - C22 PAHs		ALIPHATICS C9-C18 Range		C19-C36 Range	
	Expected Value	Actual Value	Expected Value	Actual Value	Expected Value	Actual Value
FID Concentration	NA	NA	325	251	450	380
PID Concentration	425	352 (83%)	175	140	250	212
FID-PID Calculated Concentration	NA	NA	150	111 (74%)	200	168 (84%)

all values reported in ug/ml

Table 2.   50 µg/mL Aliphatic Standard Recovery Data.

	AROMATICS C10 - C22 PAHs		ALIPHATICS C9-C18 Range		C19-C36 Range	
	Expected Value	Actual Value	Expected Value	Actual Value	Expected Value	Actual Value
FID Concentration	NA	NA	300	311	400	414
PID Concentration	NA	NA	0	0	0	0
FID-PID Calculated Concentration	NA	NA	300	311 (104%)	400	414 (104%)

all values reported in ug/ml

Table 3.   25 µg/mL Aromatic Standard Recovery Data.

	AROMATICS C10 - C22 PAHs		ALIPHATICS C9-C18 Range		C19-C36 Range	
	Expected Value	Actual Value	Expected Value	Actual Value	Expected Value	Actual Value
FID Concentration	NA	NA	175	181	250	234
PID Concentration	425	427 (100%)	175	192	250	255
FID-PID Calculated Concentration	NA	NA	0	-11	0	-21

all values reported in ug/ml

Method Performance

Method performance was evaluated based on Section 10.2 of the MADEP EPH procedure, Minimum Instrument QC. In accordance with Section 10.2, the n-nonane peak was completely resolved from the solvent front, and both surrogate spike compound peaks were resolved to the baseline. All peaks of interest from the combined aliphatic/aromatic hydrocarbon standard are resolved to virtually baseline conditions. The benzo(b)fluoranthene/benzo(k)fluoranthene PAH isomeric pair exhibited the poorest chromatographic performance; however, the two compounds, as displayed in Figure 5 are still separated to a degree greater than 75% of baseline. The PID/FID method retention time and calibration data were compared with the Section 10.2 criteria, and determined to be compliant. The response ratio of C_{28} to C_{20} was monitored to demonstrate a control on mass discrimination. Calculated response ratios were consistently greater than the 0.85 minimum criteria specified in Section 10.2. The MADEP EPH procedure includes the analysis of the specified fractionization check solution to demonstrate the capability of properly fractionating aliphatic and aromatic components. The PID/FID procedure data, presented in Table 1, indicates that the method is functioning within the criteria specified in Section 10.3.

A series of four replicate reagent-water fortified blanks was prepared as an initial demonstration of capability (IDC) study at a concentration of 25 µg/L. Precision data, expressed as the average % RSD, was calculated to be 9.3% for the 17 aromatics compounds determined using the PID, and 10.7% for all 31 compounds using the FID. Accuracy data, expressed as percent recovery, was observed to be less than 60% for several of the lower boiling point aliphatic and aromatic compounds. Upon further review of the data, it was determined that the low percent recoveries obtained were due to an extract concentration problem that occurred during the sample preparation procedure. The extract concentration technique will be modified to include a less aggressive sample concentration technique. An additional IDC study will need to be performed to demonstrate acceptable recoveries of the lower boiling point compounds. A method detection limit (MDL) study was performed for the 17 PAH target compounds by preparing a series of 7 replicate reagent-water fortified blanks at a concentration of 1.0 µg/L. An increase in the level of sensitivity observed for the target PAH compounds, due to the use of the PID, was one of the goals of using this procedure for EPH analysis. The calculated MDLs are presented in Table 4 alongside the respective detection limits specified in the GW-1 category of the Massachusetts Contingency Plan 310 CMR 40.0000 April 5, 1996. In general, PAH target compound sensitivity using the PID appeared to be approximately the same order of magnitude as that observed using the FID. Due to lower signal-to-noise considerations, detectability on the PID appears to be approximately two to three times more sensitive than the FID. Further work needs to be performed regarding the optimization of PAH MDLs and the generation of practical quantitation limits suitable for use in MCP GW-1 evaluations.

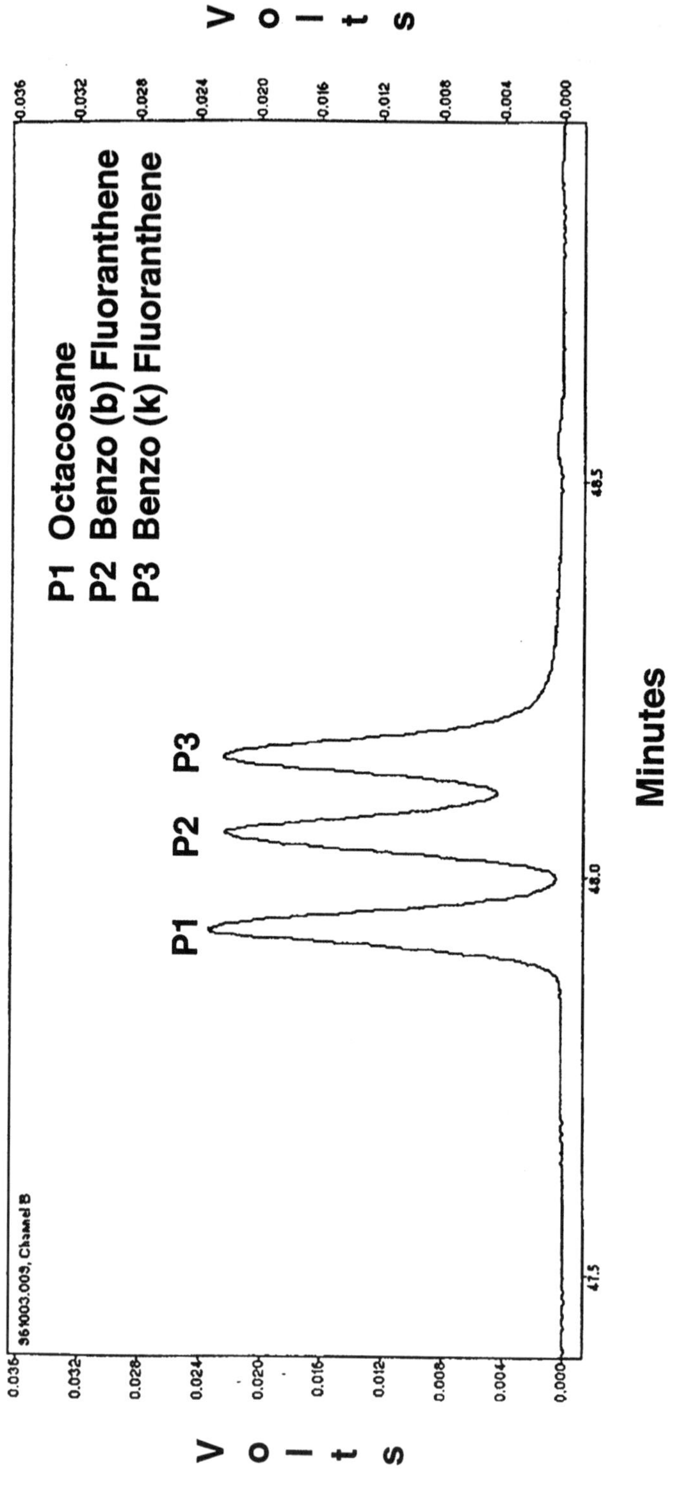

Figure 5. Resolution of critical pairs.

In summary, the preliminary work done to date of this writing indicates that the proposed GC PID/FID procedure may be a viable option for the determination of extractable petroleum hydrocarbons analysis. Pending the outcome of additional method development, the proposed procedure and associated documentation will be submitted to the MADEP for their review.

Table 4.   PAH Method Detection Limits.

Target Compound	PID MDL (ug/L)	MCP GW-1 Limits (ug/L)
naphthalene	0.12	20.
2-methylnaphthalene	0.25	10.
acenaphthylene	0.27	300.
acenaphthene	0.20	20.
fluorene	0.35	300.
phenanthrene	0.30	300.
anthracene	0.32	600.
fluoranthene	0.33	100.
pyrene	0.16	80.
benzo(a)anthracene	0.14	1.
chrysene	0.20	2.
benzo(b)fluoranthene	0.50	1.
benzo(k)fluoranthene	0.21	1.
benzo(a)pyrene	0.33	0.2
indeno(1,2,3-cd)pyrene	0.32	0.5
dibenzo(ah)anthracene	0.27	0.5
benzo(ghi)perylene	0.34	0.5

REFERENCES

Method for the Determination of Extractable Hydrocarbons (EPH), (Public Comment Draft 1.0), Massachusetts Department of Environmental Protection, Division of Environmental Analysis, Office of Research and Standards, Bureau of Waste Site Cleanup, August 1995.

Massachusetts Contingency Plan 310 CMR 40.0000 April 5, 1996.

CHAPTER 6

Application of Alkylcyclohexane Distribution Patterns for Hydrocarbon Fuel Identification in Environmental Samples

Isaac R. Kaplan and **Yakov Galperin**, Global Geochemistry Corporation, Canoga Park, California

INTRODUCTION

Petroleum hydrocarbons are probably the most frequently encountered contaminants in the environment. Among analytical methods currently used to identify fugitive crude and refined petroleum are those which focus on hydrocarbon group-type analysis, such as alkanes, volatile aromatic compounds (BTEX), polynuclear aromatic hydrocarbons (PAH), and polycyclic alkanes, often referred to as biomarkers. Alkane distribution patterns are generally found to be the most easily-measured and universal for fingerprinting purposes (Whittmore, 1979). Branched-chain hydrocarbons (isoprenoids), PAH, and biomarkers are also commonly used in forensic geochemistry for crude oil and heavy refined product characterization (Seifert and Moldowan, 1986; Douglas et al., 1992; Kaplan et al., 1996). However, their application is of limited significance for light and nondiesel containing middle distillates and in cases of mixed fuels contamination.

The serious limitation for the application of normal alkanes and BTEX arises from the fact that upon release into an environment, refined petroleum products are subject to various changes. In weathered environmental samples, most n-alkanes and light aromatic compounds could be lost, whereas distribution of more recalcitrant isoprenoid hydrocarbons, when present, often do not provide conclusive information on the source of fugitive fuel (Kaplan and Galperin, 1996). Furthermore, due to overlap of alkylbenzene and alkylnaphthalene patterns in fuels with similar hydrocarbon distributions, recognition of fuel type(s) present often becomes a challenging and formidable task.

It will be demonstrated in this paper that the cyclohexane homologous series of hydrocarbons also exhibits characteristic distribution patterns that allow for fingerprinting of fugitive fuels. The main advantage in utilizing alkylcyclohexane patterns for hydrocarbon fuel recognition is that cyclic compounds are more resistant to environmental alteration and could be detected in a sample even when most of the n-alkanes are degraded. For the above reason, the combination of two homologous series have been chosen as indicators of various refined fuels in the environmental samples. These are alkanes (straight-chain and branched) and alkylcyclohexanes. We describe how the combination of these homologous series, extending for alkanes from heptane to doeicosane and for cyclohexanes from methylcyclohexane to pentadecanylcyclohexane, is used for pattern recognition in complex environmental samples with multiple sources of refined fuel contamination.

ANALYTICAL METHOD

A Varian 3400 gas chromatograph was interfaced to a Finnigan Incos 50 quadrupole mass spectrometer equipped with electron ionization ion source. The ionization voltage was 70 eV; the ion source temperature 170°C. The full scan mode was utilized, scanning 50-350 amu at a rate of 1 scan/sec. A fused silica capillary column (DB-5, 30 m x 0.25 mm, J&W Scientific) was used for this study. The GC oven was programmed from the initial temperature of 30°C (5 min) and ramped at the rate of 8°C/min to the final temperature of 300°C (15 min). Injector and transfer line temperatures were set at 275°C. Reference fuels, obtained from commercial sources, and field samples were diluted 1:100 in methylene chloride and 1 μL aliquots were analyzed by direct injection on a splitless mode. Reconstructed ion chromatograms were acquired and mass chromatograms of m/z 85 (characteristic for alkanes) and m/z 83 (characteristic for alkylcyclohexanes) were extracted.

RESULTS AND DISCUSSION

Crude oil contains a wide range of hydrocarbons from light gases to heavy residue. At the refinery, crude oil is separated by distillation into three main products: naphtha, middle distillate, and bottoms fraction. Naphtha is mainly used for motor gasoline and processed further for octane improvement. The middle distillate can actually be separated into two categories consisting of kerosene range products (light-end) and diesel range products (heavy-end). The light-end products are used for specialty solvents (mineral spirits, Stoddard solvent, etc), certain jet fuels, and light diesel fuel (diesel #1). The heavy-end middle distillates are used for diesel fuel (diesel #2), some jet fuels, and heating oils.

Based on the systematic GC/MS study of different commercial and military fuels, conducted in our laboratory, we have found that in addition to a well-known alkane pattern for m/z 85 ion, cyclohexane homologous series compounds also exhibit a characteristic distribution pattern on the m/z 83 mass chromatogram. The rationale for selecting ion m/z 83 is evident from the mass spectrum of methylcyclohexane (Figure 1). The prominent peak m/z 83 is due to a loss of an alkyl substituent - a typical fragmentation pathway for alkylcyclohexanes. Because

Figure 1. **Mass spectrum of methylcyclohexane.**

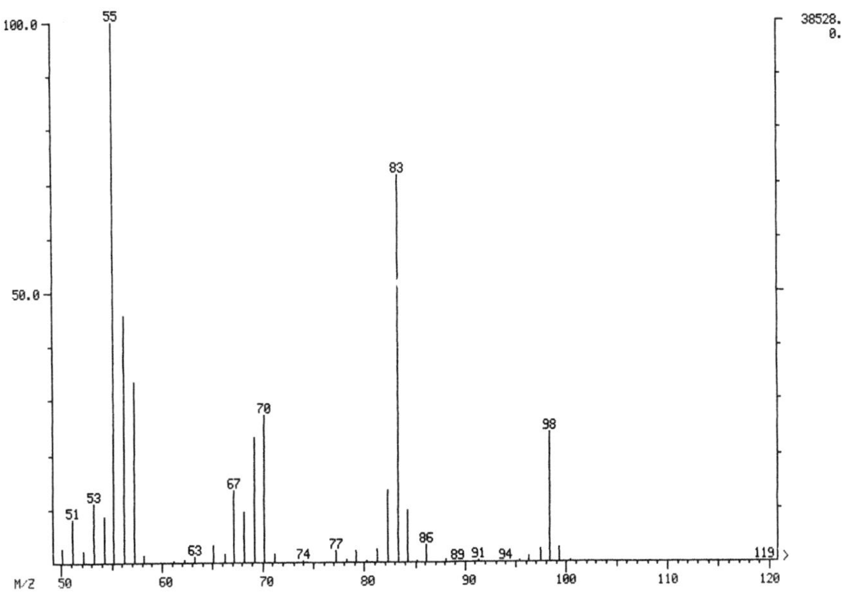

the alkane distribution patterns in different fuels are well-known (Senn and Johnson, 1987; Bruce and Schmidt, 1994), a short description of only alkylcyclohexane patterns for selected reference products is provided below.

Refined Reference Products

Gasoline m/z 83 mass chromatogram, shown in Figure 2, exhibits an asymmetric distribution pattern in the CH-1 (methylcyclohexanes) to CH-7 (heptylcyclohexane) range. Peak CH-1 is the most abundant, with peaks CH-2 to CH-7 rapidly decreasing in intensity.

Mineral spirits m/z 83 mass chromatogram, presented in Figure 3, shows a rather symmetrical distribution pattern with peaks CH-3 and CH-4 being dominant, while the CH-2 and CH-5 peak heights are much smaller.

Stoddard solvent alkylcyclohexanes pattern (Figure 4) in the CH-2 to CH-9 range demonstrates distribution maximizing at CH-5.

Naphtha produces alkylcyclohexanes distribution (Figure 5) in the range CH-1 to CH-6, and maximizes at CH-3.

Kerosene shows a characteristic distribution (Figure 6) in the range CH-1 to CH-9, and maximizes at CH-6.

Diesel Fuel #1 exhibits an alkylcyclohexanes pattern (Figure 7) in the range CH-1 to CH-14, and maximizes at CH-5.

Diesel Fuel #2 m/z 83 mass chromatogram (Figure 8) demonstrates a wide range of alkylcyclohexanes from CH-1 to CH-14, maximizing around CH-9 and CH-10 peaks.

Military Jet Fuel JP-5 (Figure 9) and jet fuel Jet-A (Figure 10) demonstrate a distribution pattern in the kerosene range (CH-1 to CH-9), with a noticeable difference in the maximum peak of distribution: CH-5 for JP-5 and CH-4 for Jet-A.

Military Jet Fuel JP-8 shows an asymmetric distribution pattern (Figure 11) in the diesel range (CH-1 to CH-14), but maximizes at CH-3.

Military Jet Fuel JP-4 m/z 83 mass chromatogram (Figure 12) is a kerosene-gasoline range fuel mixture, with a higher proportion of gasoline. Peak CH-1 is the most abundant in the alkylcyclohexanes distribution.

As can be seen from the description of m/z 83 mass chromatograms, alkylcyclohexane distribution patterns are as product-specific as are those for alkanes. Hence, they also can be used for fuel type identification in environmental samples in combination with and complementary to alkane distribution patterns.

Figure 2. **Gasoline m/z 83 mass chromatogram.**

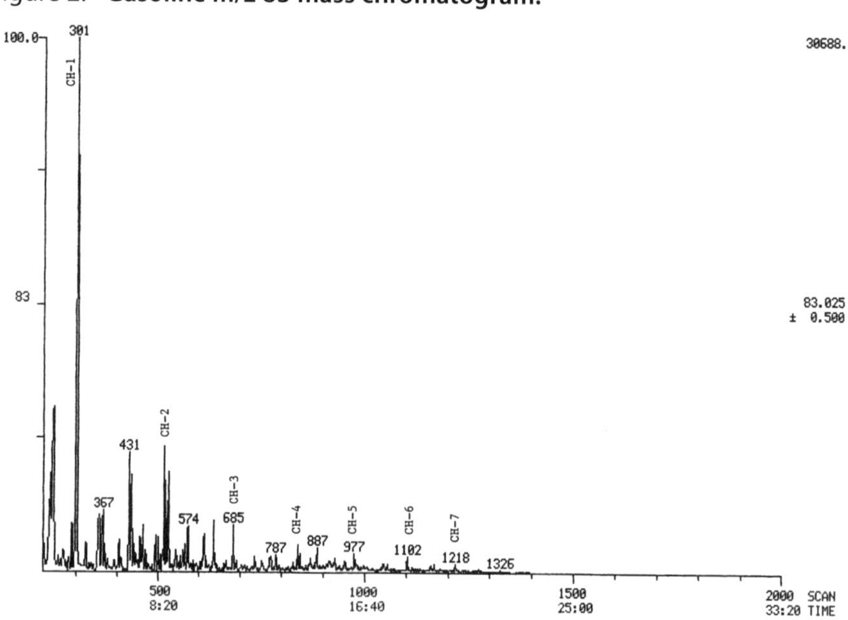

Figure 3. Mineral spirits m/z 83 mass chromatogram.

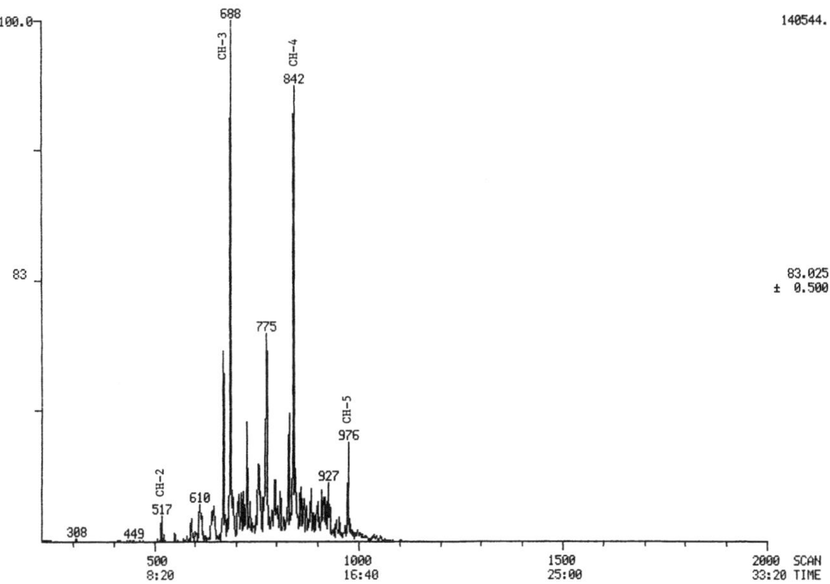

Figure 4. Stoddard solvent m/z 83 mass chromatogram.

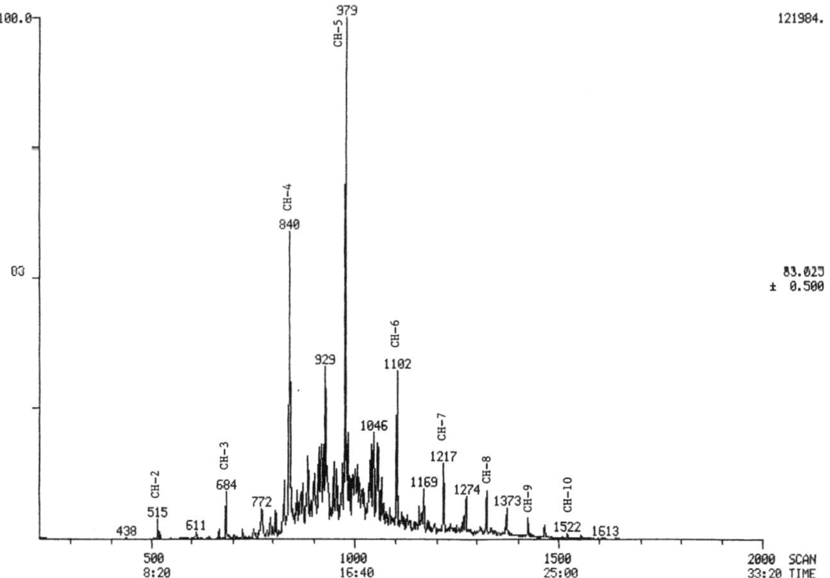

Figure 5. **Naphtha m/z 83 mass chromatogram.**

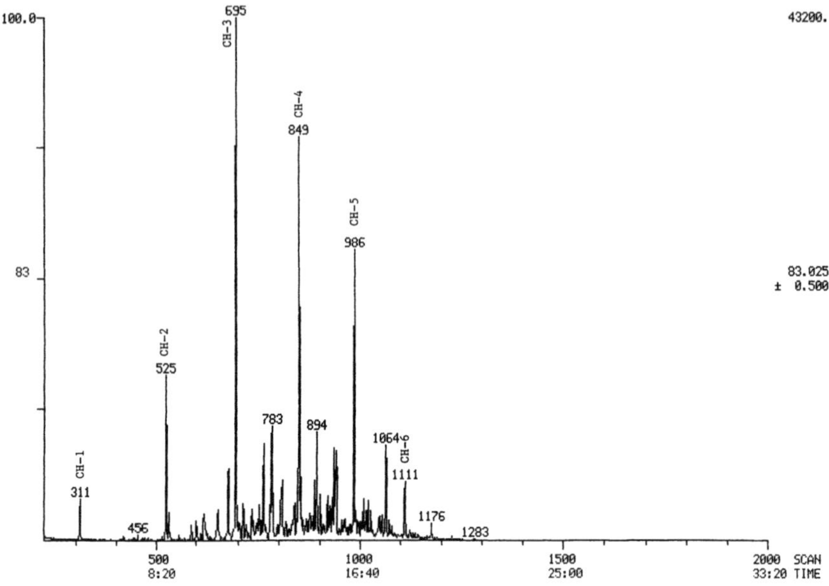

Figure 6. **Kerosene m/z 83 mass chromatogram.**

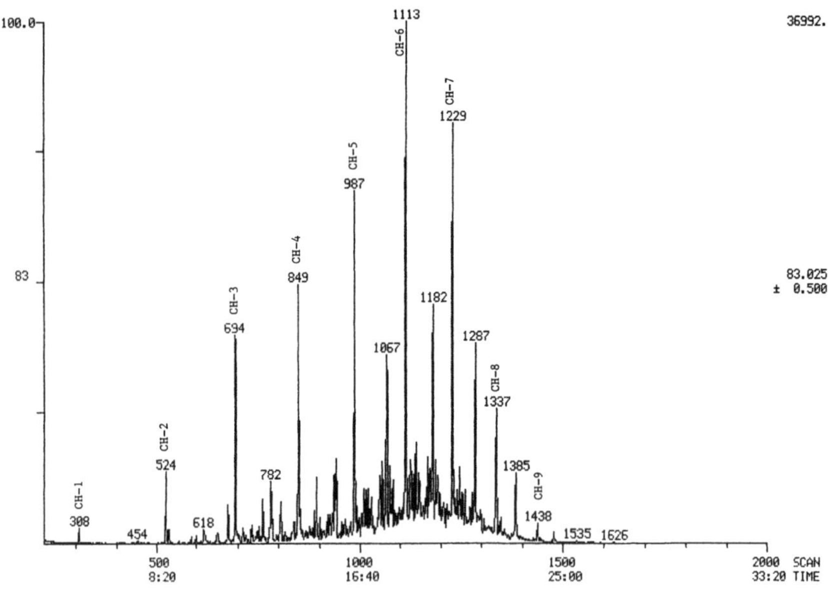

Figure 7. **Diesel fuel #1 m/z 83 mass chromatogram.**

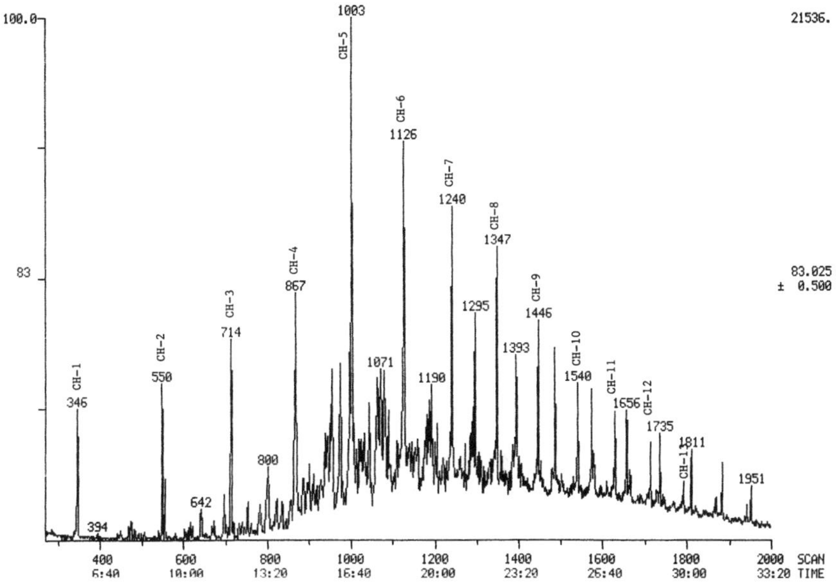

Figure 8. **Diesel fuel #2 m/z 83 mass chromatogram.**

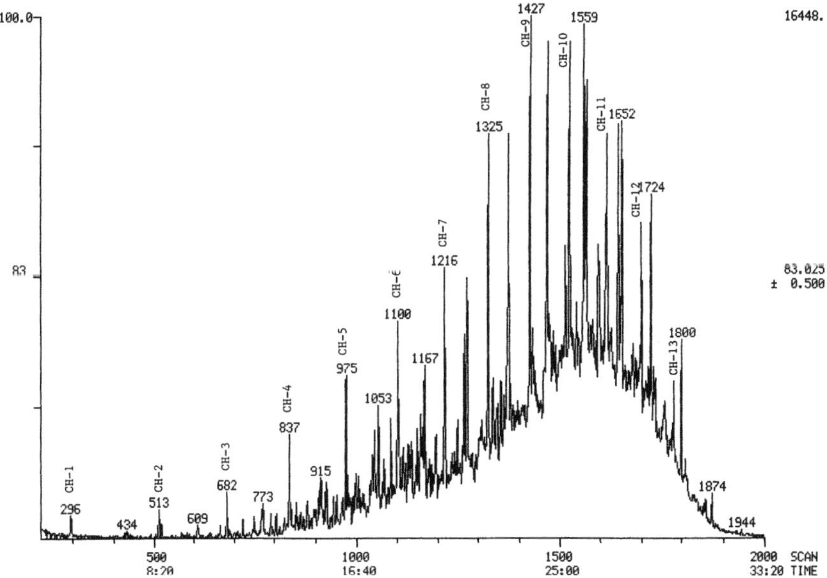

Figure 9.   JP-5 m/z 83 mass chromatogram.

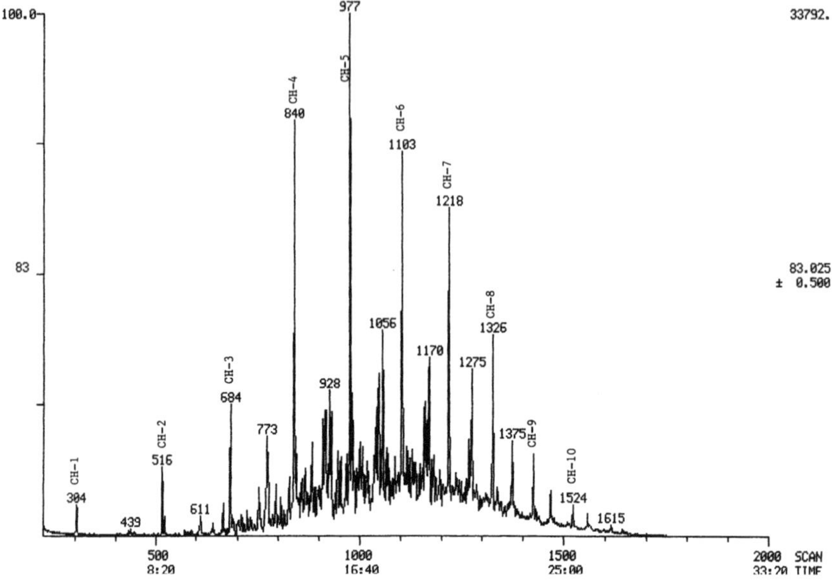

Figure 10. Jet-A m/z 83 mass chromatogram.

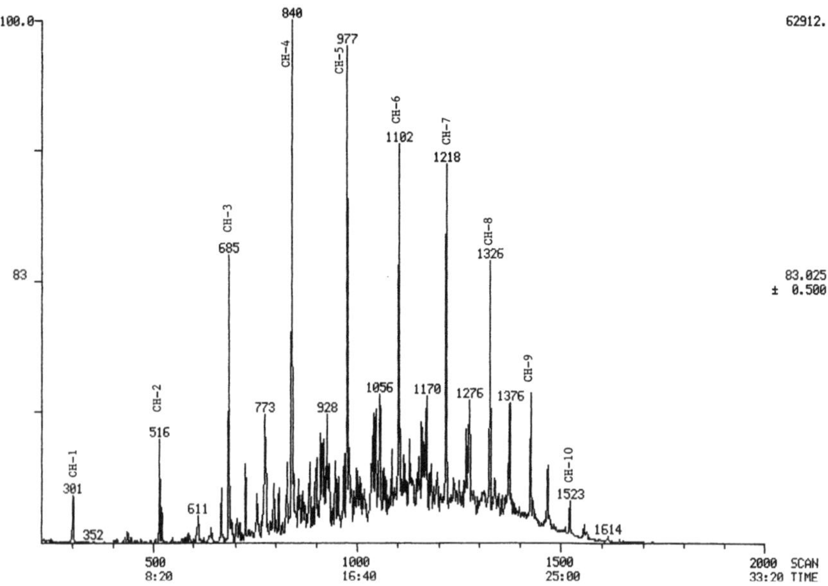

Figure 11. **JP-8 m/z 83 mass chromatogram.**

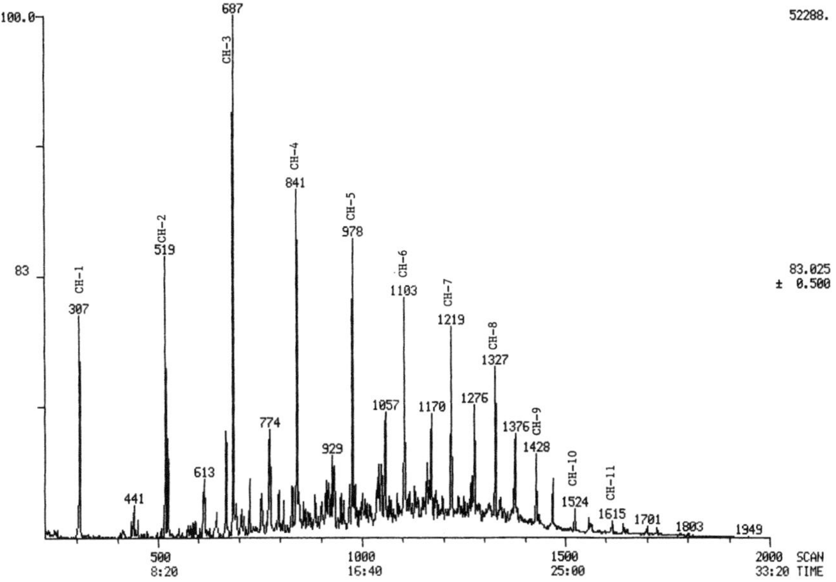

Figure 12. **JP-4 m/z 83 mass chromatogram.**

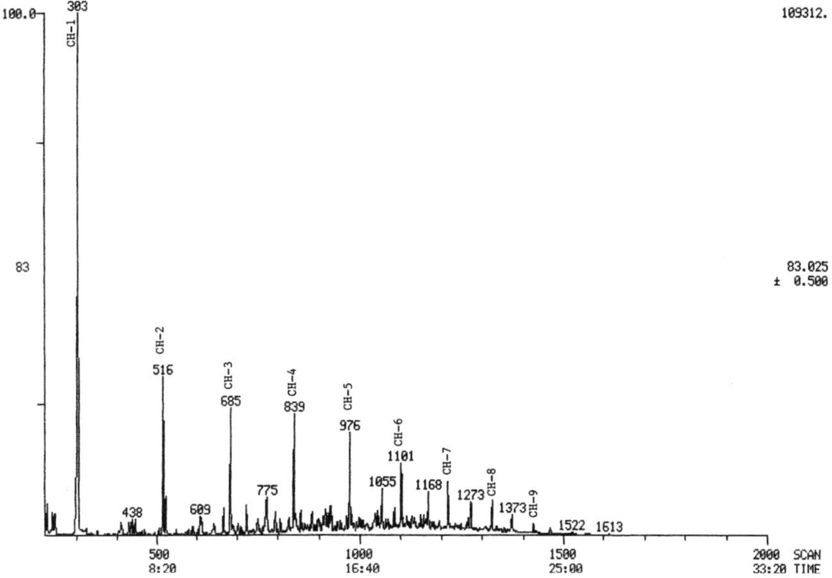

Fuel Types Characterization in Environmentally Weathered Samples

Any petroleum-related product, released into an environment, is being continually changed in its composition by numerous physical, chemical, and biological processes. These changes, usually referred to as "weathering," can cause alteration of molecular composition of the product, so that its original source is often hardly recognizable. Three case histories illustrating the application of alkylcyclohexane recognition patterns for weathered environmental samples are described below.

Case 1

Shown in Figure 13a, is a m/z 85 alkane mass chromatogram of a free-floating product. Due to the absence of a detectable amount of n-alkanes in this sample, the mass chromatogram is of limited value. However, when m/z 83 mass chromatogram is extracted from the same analytical run (Figure 13b), a clear distribution pattern of alkylcyclohexanes emerges. Comparison of this pattern with data for reference fuels allows one to conclude that the floating product consists of a weathered diesel fuel.

Case 2

The alkane mass chromatogram (m/z 85) of a free-floating product is presented in Figure 14a. Absence of n-alkanes, together with high abundance and distribution of isoalkanes suggests that this environmentally altered floating product represents a mixture of a middle distillate fuel (probably diesel) with some lighter petroleum-related product. The alkylcyclohexane distribution pattern (Figure 14b) demonstrates that the major component of the floating product is mineral spirits (pronounced CH-3 and CH-4 peaks) with a reduced amount of diesel fuel.

Case 3

Mass chromatograms of a more complex sample are shown on Figure 15 for another free-floating product. The alkane mass chromatogram (Figure 15a) demonstrates that the sample contains heavily weathered product(s) at the high-boiling end, because n-alkanes are depleted and only isoprenoid peaks in the C_{10}-C_{20} range are present. However, at the light end of the m/z 85 mass chromatogram, n-alkane peaks of more volatile hydrocarbons are detected. More detailed information can be derived from the m/z 83 mass chromatogram (Figure 15b). It shows a strong CH-1 peak, characteristic to gasoline, with peaks in the CH-2 to CH-9 range corresponding to a kerosene type fuel and peaks up to CH-15 indicative of diesel fuel.

Figure 13. Free-floating product mass chromatograms: (a) m/z 85 - alkanes and (b) m/z 83 - alkylcyclohexanes.

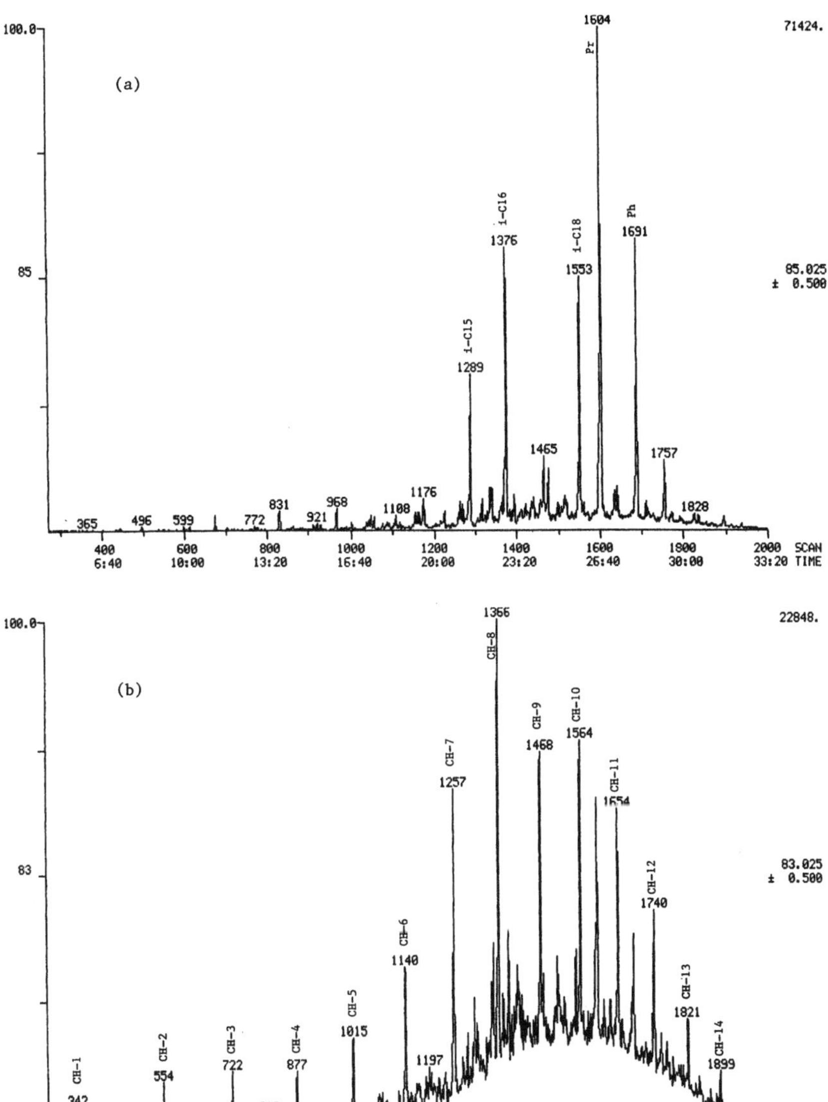

Figure 14. Free-floating product mass chromatograms: (a) m/z 85 - alkanes and (b) m/z 83 - alkylcyclohexanes.

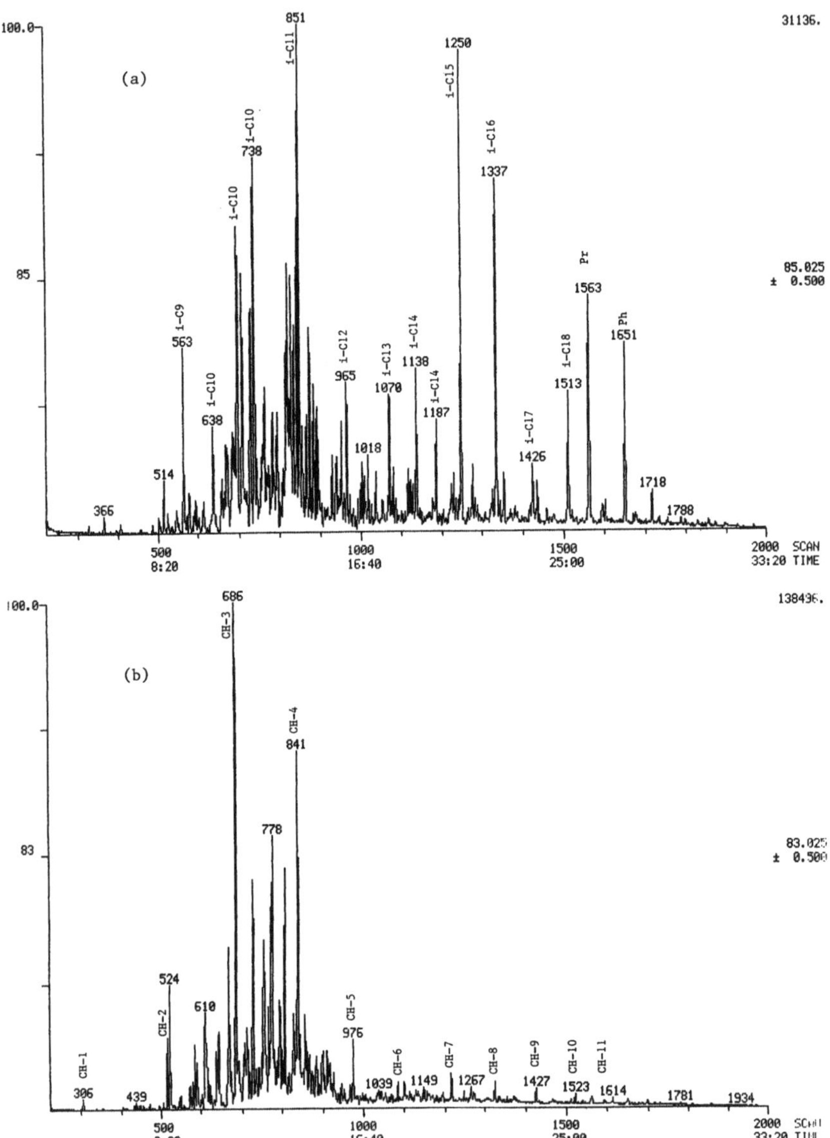

Figure 15. **Free-floating product mass chromatograms: (a) m/z 85 - alkanes and (b) m/z 83 - alkylcyclohexanes.**

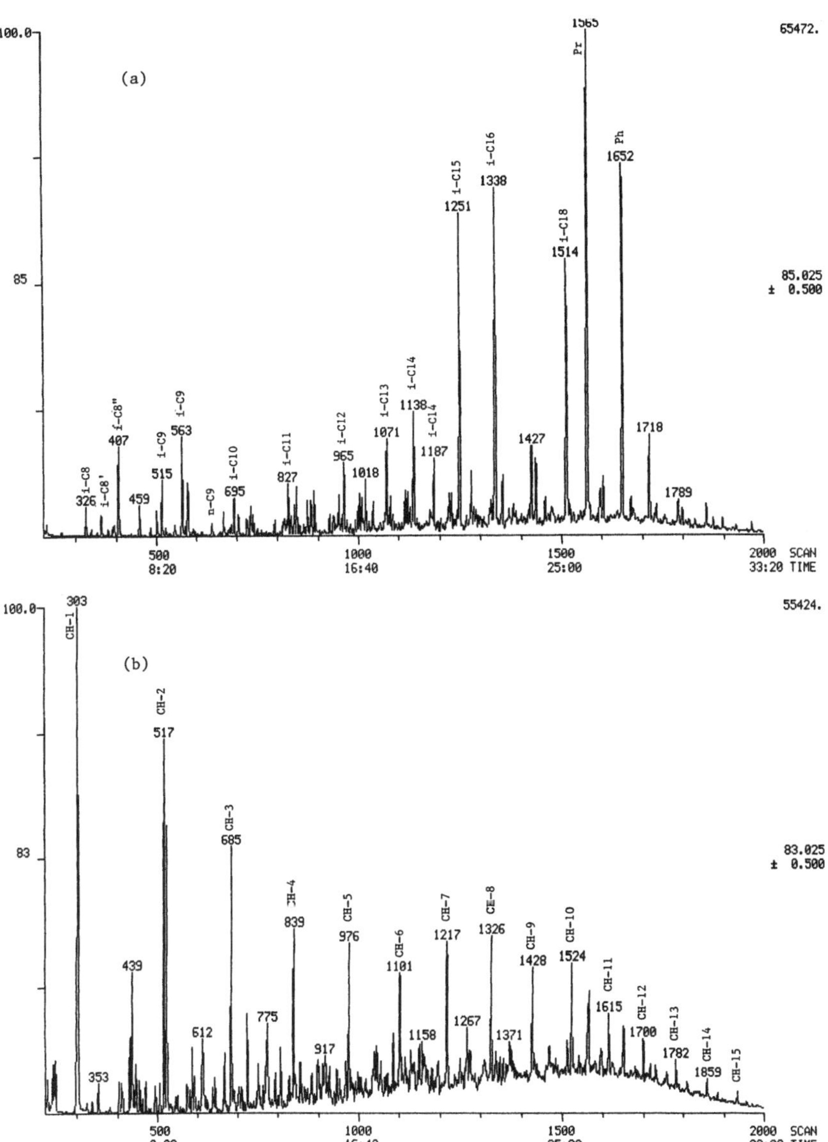

CONCLUSIONS

The cyclohexane homologous series of compounds exhibit product-specific distribution patterns in the m/z 83 mass chromatogram, which can be used for fuel type identification. It is demonstrated here that even for moderately altered environmental samples, when most of n-alkanes are removed, the alkylcyclohexane distribution pattern can provide valuable fingerprinting information. A combination of alkane and alkylcyclohexane patterns allows for fuel type identification in mixed environmental samples with multiple sources of contamination.

REFERENCES

Bruce, L.G. and Schmidt, G.W. 1994. Hydrocarbon Fingerprinting for Application in Forensic Geology: Review with Case Studies. *AAPG Bulletin.* 78, 1692-1710.

Douglas, G.S., McCarthy, K.J., Dahlen, D.T., Seavey, J.A., Steinhauer, W.G., Prince, R.C., and Elmendorf, D.L. 1992. The Use of Hydrocarbon Analyses for Environmental Assessment and Remediation. In: *Contaminated Soils: Diesel Fuel Contamination,* pp. 1-21 (Calabrese, E.J. and Kostecki, P.T., Eds.). Lewis Publishers, Chelsea, MI.

Kaplan, I.R. and Galperin, Y. 1996. How to Recognize a Hydrocarbon Fuel in the Environment and Estimate Its Age of Release. In: *Ground Water and Soil Contamination,* pp. 145-199. (Bois, T.J. and Luther, B.J., Eds.). John Wiley & Sons, New York.

Kaplan, I.R., Galperin, Y., Alimi, H., Lee, R.P. and Lu, S.T. 1996. Patterns of Chemical Changes During Environmental Alteration of Hydrocarbon Fuels. *Ground Water Monitoring and Remediation.* 16 (4), 113-125.

Seifert, W.K. and Moldowan, J.M. 1986. Use of Biological Markers in Petroleum Exploration. In: *Biological Markers in the Sedimentary Rocks* (Johns, R.B., Ed.). *Geochem. Geophys.* 24, 261-290.

Senn, R.B. and Johnson, M.S. 1987. Interpretation of Gas Chromatographic Data in Subsurface Hydrocarbon Investigation. *Groundwater Monitoring Review,* 7, 56-83.

Whittmore, I.M. 1979. High Resolution Gas Chromatography of the Gasolines and Naphthas. In: *Chromatography in Petroleum Analysis,* pp. 41-74 (Altgelt, K.H. and Gauw, T.H., Eds.). Marcel Dekker, New York.

CHAPTER 7

Regulatory Approval of the Use of a Hand-Held Field GC for Measurement of BTEX in Air Emissions from Remediation Systems

J. Patrick Byrnes, EnviroTrac Ltd., Deer Park, New York

Stephen L. Kane, Photovac Monitoring Instruments, Deer Park, New York

INTRODUCTION

A soil vapor extraction (SVE) system removes vapor-phase contaminants from the subsurface by applying a vacuum to induce air flow through the unsaturated (vadose) zone. Volatile compounds move from residual nonaqueous phase liquids (NAPL) and soil pore space water into the gaseous phase (Luhrs et al., 1994). These vapor phase contaminants are then either treated (captured by adsorption onto activated charcoal, for example) or emitted into the ambient air below a rate or amount permitted by the regulatory authority. The operator of the SVE (Responsible Party and its contractors) must obtain the appropriate emissions permit or enter into a Petroleum Spill Stipulation Agreement during system installation.

In the state of New York, the Department of Environmental Conservation, Divisions of Spills Management and Air Resources are jointly responsible for regulating and permitting all remedial activities at petroleum spill sites which involve SVE systems, air strippers, and cold-mix asphalt units (NYSDEC, 1993). During the operational startup period, which can be as long as 30 days, the operator of the SVE system must demonstrate to the state spill engineer by weekly monitoring that emission rates of benzene are less than 10 times the mandated limits. See Table 1. If emissions are greater than 10 times these limits, emissions treatment must be initiated.

During this startup period, the operator must adjust the SVE equipment for optimum performance, and bring the system into full compliance with these emission limits before full operational status is granted.

METHOD

EnviroTrac Ltd., Deer Park, NY, installed and/or managed 16 SVE systems for Shell Oil Products Company at various sites in the New York metropolitan area (NYSDEC Regions 1, 2, and 3) from December 1994 to May 1995. As part of the

Table 1. New York State Soil Vapor Extraction System Benzene Emission Limits

Stack Height (feet)	Air Flow (cu.ft./min.)	Emissions (PPM-V)	Emissions (mg/m3)	Emissions (lb./hr.)
15	50	8.00	26.360	0.00494
	100	4.00	13.180	0.00494
	150	2.66	8.787	0.00494
	200	2.00	6.590	0.00494
	250	1.60	5.272	0.00494
20	50	14.88	49.069	0.00919
	100	7.44	24.535	0.00919
	150	4.96	16.356	0.00919
	200	3.72	12.267	0.00919
	250	2.98	9.814	0.00919
25	50	24.10	79.458	0.01488
	100	12.05	39.729	0.01488
	150	8.03	26.486	0.01488
	200	6.02	19.864	0.01488
	250	4.82	15.892	0.01488
30	50	35.72	117.806	0.02206
	100	17.86	58.903	0.02206
	150	11.91	39.269	0.02206
	200	8.93	29.452	0.02206
	250	7.14	23.561	0.02206

startup operations, for the purpose of compliance with state guidelines, a series of whole air emission samples from each system was collected in 1-liter Tedlar gas sampling bags (SKC, Inc., Eighty-Four, PA) for subsequent analysis by an analytical laboratory using either EPA Method TO-3 (GC-FID/ECD with cryogenic preconcentration) or TO-14 (GC/MS or GC with multiple detectors). The total number of samples split was 78.

Samples were collected by connecting a gas sampling bag to the SVE system effluent sample port, located on the exhaust side of the blower. The sampling bag was connected to the inlet of the calibrated field gas chromatograph (GC), and a volume of the sample was aspirated and analyzed for BTEX. Maximum sample volume required for the field GC was only 35 milliliters, leaving nearly the entire bagged sample intact for use by the laboratory. Analysis time was 10 minutes per sample. Each sample was then packaged for transport to the laboratory and analyzed within 48 hours.

Field Instrument

The portable instrument used for the field measurements (SnapShot, PE Photovac, Markham, Ontario, Canada) is a handheld GC using a photoionization detector (PID) of 10.6 electron volts (eV). The unit weighs 9.7 pounds, and is powered for up to 6 hours by a removable lead/acid battery pack. An interchangeable Applications Module which contains a column, hardware and software is factory-programmed to measure a specific set of compounds over a predetermined range of concentrations. Sample concentrations found to be above or below the operating range of the instrument are reported (and datalogged) as (e.g., for benzene) ">50 PPM" or "<0.1 PPM." In this case, benzene, toluene, ethylbenzene, and the xylene isomers (ethylbenzene and xylenes are reported as total C_8 aromatics) can be simultaneously analyzed in 10 minutes.

The capillary column used for this application is a 10 meter, 0.53 mm I.D., 0.5 um film CPWax 52CB, a high polarity stationary phase selected for its ability to separate the target aromatics from the complex mixture of hydrocarbons normally present in petroleum fuels.

Carrier gas is supplied to the column from high pressure disposable cylinders of carbon dioxide, which thread into a receptacle on the underside of the instrument, and which will last for up to 24 hours. The flow rate of CO_2 is controlled by a series of pressure regulators and an on/off valve.

After an initial instrument warm-up period of approximately 10 minutes, a calibration was performed each day that testing was to occur. A portable cylinder of a certified mixture (+/- 2%) of the target compounds was connected to the inlet of the GC, and the internal sampling pump allowed to draw in a volume of this calibration gas (maximum 35 milliliters), flushing an internal sample loop. The cal gas sample was injected into the column through a series of valves, which also allowed the instrument to automatically perform a precolumn/backflush function. Once the instrument was calibrated (10 minutes), the collected gas samples were connected to the inlet and sequentially analyzed (Tables 2 and 3).

RESULTS AND DISCUSSION

Table 4 shows the collated results of all split sample testing (n = 78) on emissions from the SVE systems at all sites. Figure 1 shows a plot of the line of best fit through all data points. Correlation coefficient for the entire data set (total BTEX) was 0.9829. Figure 2 shows the same plot for samples in which reported values were ≤ 25 PPMv (n = 70). Correlation coefficient for these 70 samples was 0.8165. For both the inclusive data set and the low concentration data set for total BTEX emissions, greater than 90% of all data points fell between +/-25% of the "predicted" SnapShot result (line of best fit).

In 94% of samples in which either laboratory analysis or the SnapShot GC reported values above the Minimum Detection Limits (MDL), total BTEX was reported at a higher concentration by the SnapShot than by the laboratory. For

Table 2.   **Photovac SnapShot Monitoring Ranges Application Module #20**

Compound	Monitoring Range (PPMv)
Benzene	0.1 - 50
Toluene	1.0 - 200
Total C8 Aromatics	1.0 - 200

Table 3.   **Photovac SnapShot GC Calibration Gas Mixture**

Compound	PPMv
Benzene	4.51
Toluene	19.7
Ethylbenzene	21.5
m-Xylene	20.7
o-Xylene	22.8

the purpose of compliance with state SVE system emission guidelines, a false positive sample result is better than a false negative result, erring on the conservative side of the regulated limit and assuring compliance. In the remaining 6% of samples, in which reported laboratory results for Total BTEX were greater than SnapShot results, sample concentrations were at or below the MDL of the SnapShot. Laboratory MDLs were 0.1 PPMv for all compounds for the majority of samples split (two laboratories were used and had slightly different MDLs for these parameters).

It is probable that the Tedlar bag sampling method used to collect and transport air emission samples was responsible for apparent sample losses between the relatively short-term SnapShot analysis and laboratory analysis (after a longer transport and holding period). Posner and Woodfin (1986) reported VOC sample losses from Tedlar and other sample bag materials after an elapsed time of only 6 hours. Figure 3 (from Posner and Woodfin) shows up to a 10% loss of benzene (107 PPMv) after a 6-hour holding period in Tedlar.

Figures 4-9 show additional plots of individual compound sample results and their lines of best fit for each data set. A tabulation of all plotted results is shown in Table 5.

Analysis of these data indicates a positive correlation for individual compound parameters as well as for totaled sample results. A better data correlation was achieved for the heavier aromatics than for benzene, a probable result of the greater mobility of the lighter compound, leading to higher sampling container (Tedlar bag) losses than the C_7 and C_8 compounds.

Table 4. Results of All Split Sample Testing on Emissions at All Sites
(all values PPMv   ND = Not Detected)

Site	Date	Laboratory GC					SnapShot GC				
		Benzene	Toluene	Ethylbenzene	Total Xylenes	Total BTEX	Benzene	Toluene	Ethylbenzene	Total Xylenes	Total BTEX
#1	1/20/99	0.16	0.80	0.14	0.78	1.88	0.10	1.58	1.00	2.94	4.52
	2/15/99	0.31	5.32	0.83	3.23	9.69	0.34	6.89	1.38	5.58	14.19
	3/17/99	0.25	4.79	1.38	5.76	12.18	0.33	6.54	1.98	8.66	17.51
	4/14/99	0.40	7.50	1.10	4.10	13.10	0.41	10.50	1.80	7.28	19.99
	5/3/99	0.40	7.30	1.20	4.70	13.60	0.41	10.30	1.78	7.29	19.78
	6/23/99	0.60	8.10	1.20	5.90	15.80	0.46	8.36	1.54	7.32	17.68
	7/18/99	ND	4.30	1.10	3.60	9.00	0.21	5.30	1.25	5.95	12.71
	8/15/99	ND	3.20	ND	ND	3.20	0.19	2.75	2.02	2.20	7.16
#2	1/13/99	0.44	0.96	0.14	1.75	3.29	ND	1.70	ND	4.23	5.93
	2/14/99	ND	ND	ND	ND	ND	ND	ND	ND	ND	ND
	3/23/99	ND	ND	ND	ND	ND	ND	ND	ND	1.64	1.64
	4/4/99	ND	0.08	ND	0.09	0.17	ND	ND	ND	2.76	2.76
	6/8/99	ND	ND	ND	ND	ND	ND	ND	ND	2.60	2.60
	7/25/99	ND	ND	ND	ND	ND	ND	ND	ND	ND	ND
#3	12/29/98	0.13	0.13	ND	1.94	2.19	ND	ND	ND	4.29	4.29
	3/14/99	ND	ND	ND	1.06	1.06	0.20	ND	ND	3.29	3.29
#4	12/8/98	0.75	3.19	0.69	4.61	9.20	0.20	3.00	1.00	5.00	9.00
	1/28/99	2.38	8.78	1.38	7.37	19.90	1.60	11.00	2.00	9.00	24.00
	2/3/99	1.91	5.59	0.78	3.69	12.00	1.10	10.00	2.00	8.00	21.00
	3/15/99	8.15	26.60	2.53	15.21	52.50	ND	33.00	3.00	19.00	54.00
	4/12/99	9.40	31.00	3.90	21.00	65.30	ND	41.00	5.00	27.00	73.00
	5/11/99	11.00	62.00	6.60	42.00	121.60	ND	75.00	8.00	50.00	133.00
	6/6/99	11.00	36.00	3.80	24.00	74.80	0.20	48.00	4.00	30.00	82.00
	7/14/99	14.00	54.00	7.10	46.00	121.10	4.90	55.00	7.00	54.00	115.00
#5	3/17/99	ND	0.10	ND	0.60	0.70	ND	ND	ND	3.15	3.15
#6	1/26/99	ND	ND	ND	0.28	0.28	ND	ND	ND	ND	ND
	3/15/99	ND	ND	ND	ND	ND	ND	ND	ND	ND	ND
	6/1/99	ND	ND	ND	ND	ND	0.19	ND	ND	2.58	2.77
	7/25/99	ND	ND	ND	ND	ND	ND	ND	ND	2.69	2.69
#7	1/19/99	ND	ND	ND	0.45	0.45	0.27	1.07	ND	18.90	20.24
	2/16/99	ND	ND	ND	0.24	0.24	0.10	ND	ND	5.80	5.90
	3/2/99	ND	ND	ND	1.10	1.10	ND	ND	ND	4.04	4.04
	4/19/99	ND	ND	ND	0.20	0.20	0.13	ND	ND	1.68	1.81
	6/1/99	ND	ND	ND	ND	ND	ND	ND	ND	ND	ND
	6/23/99	ND	ND	ND	ND	ND	ND	ND	ND	ND	ND
#8	3/29/99	1.40	12.00	1.30	31.00	45.70	0.11	15.90	ND	40.60	56.61
#9	2/8/99	ND	0.27	0.37	1.24	1.88	1.47	5.38	ND	7.48	14.33
	3/7/99	0.13	0.13	ND	0.14	0.40	ND	1.84	ND	3.81	5.65
	5/2/99	0.20	0.20	0.20	3.80	4.40	0.30	3.08	ND	6.01	9.39
	6/23/99	0.10	1.00	0.20	2.40	3.70	ND	1.63	ND	4.33	5.96
	7/27/99	ND	ND	ND	ND	ND	ND	ND	ND	2.86	2.86
	8/3/99	ND	ND	ND	ND	ND	ND	1.00	ND	2.97	3.97
#10	5/16/99	0.30	0.70	0.10	2.00	3.10	ND	1.32	ND	5.16	6.48
#11	12/29/98	0.85	0.96	0.30	10.14	12.24	ND	1.70	ND	12.79	14.49
	1/17/99	0.47	0.35	0.09	6.22	7.13	0.11	ND	ND	7.76	7.87
	2/22/99	ND	0.51	ND	6.45	6.96	0.89	1.05	ND	7.62	9.56
	3/15/99	ND	0.13	ND	3.23	3.36	ND	ND	ND	5.83	5.83
	4/21/99	ND	ND	ND	2.30	2.30	ND	ND	ND	4.75	4.75
	5/11/99	ND	ND	ND	0.80	0.80	ND	ND	ND	3.10	3.10
	6/16/99	ND	ND	ND	0.80	0.80	ND	ND	ND	3.02	3.02
	7/12/99	ND	ND	1.00	3.90	4.90	0.21	ND	ND	8.93	9.14
	8/4/99	ND	ND	ND	8.70	8.70	0.67	ND	4.86	4.14	9.67
#12	2/24/99	ND	ND	1.15	ND	1.15	ND	1.06	ND	3.02	4.08
	7/21/99	ND	ND	ND	ND	ND	0.36	1.05	ND	2.40	3.81
#13	1/13/99	0.82	1.25	ND	8.53	10.60	ND	1.70	ND	11.00	12.70
	2/14/99	ND	0.11	ND	1.24	1.35	ND	ND	ND	2.59	2.59
	3/3/99	0.47	ND	ND	ND	0.47	ND	ND	ND	2.62	2.62
	4/18/99	ND	ND	ND	ND	ND	ND	ND	ND	ND	ND
	5/12/99	ND	ND	ND	ND	ND	ND	ND	ND	ND	ND
	6/8/99	ND	ND	ND	ND	ND	ND	ND	ND	ND	ND
	7/25/99	ND	1.10	ND	ND	1.10	ND	ND	ND	1.66	1.66
#14	4/19/99	11.00	40.00	5.60	46.00	102.60	11.90	60.20	8.59	63.90	144.59
	5/23/99	6.20	30.00	3.70	35.00	74.90	1.68	34.70	6.52	48.10	91.00
#15	1/17/99	0.88	2.26	0.35	4.61	8.09	0.20	2.85	ND	6.48	9.53
	3/9/99	ND	ND	ND	ND	ND	0.10	2.15	ND	4.34	6.49
	5/19/99	0.20	0.20	0.60	3.60	4.60	ND	1.43	ND	6.46	7.89
	6/9/99	0.30	0.20	0.80	4.20	5.50	ND	1.59	ND	6.92	8.51
	7/12/99	ND	ND	ND	4.10	4.10	ND	1.71	ND	8.08	9.79
	8/3/99	ND	2.20	ND	7.60	9.80	0.50	1.47	ND	8.43	10.40
#16	12/2/98	1.50	0.61	ND	4.15	6.26	ND	1.12	ND	5.44	6.56
	1/5/99	ND	0.11	ND	0.99	1.10	ND	ND	ND	3.22	3.22
	2/22/99	0.13	ND	ND	0.28	0.41	ND	ND	ND	2.61	2.61
	3/16/99	1.19	3.19	0.25	1.80	6.43	ND	4.26	ND	4.20	8.46
	4/21/99	ND	ND	ND	0.10	0.10	ND	3.75	ND	4.02	7.77
	5/13/99	ND	ND	ND	0.30	0.30	ND	ND	ND	2.69	2.69
	6/13/99	ND	ND	ND	1.40	1.40	ND	ND	ND	3.79	3.79
	7/12/99	2.50	ND	ND	6.50	9.00	ND	ND	ND	ND	ND
	8/4/99	ND	ND	ND	6.00	6.00	ND	ND	3.33	2.87	6.20

Figure 1.   **SVE system emissions, PPMv total BTEX.**

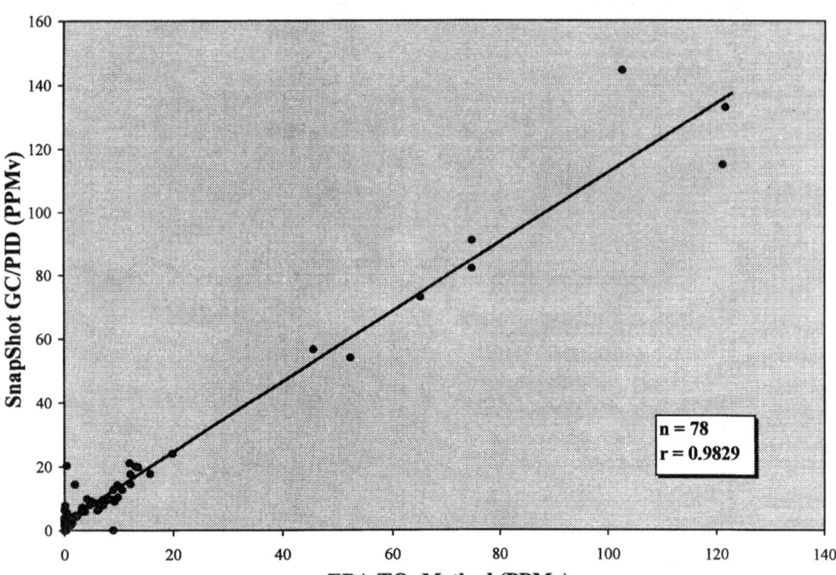

Figure 2.   **SVE system emissions, PPMv total BTEX (samples <25 ppm).**

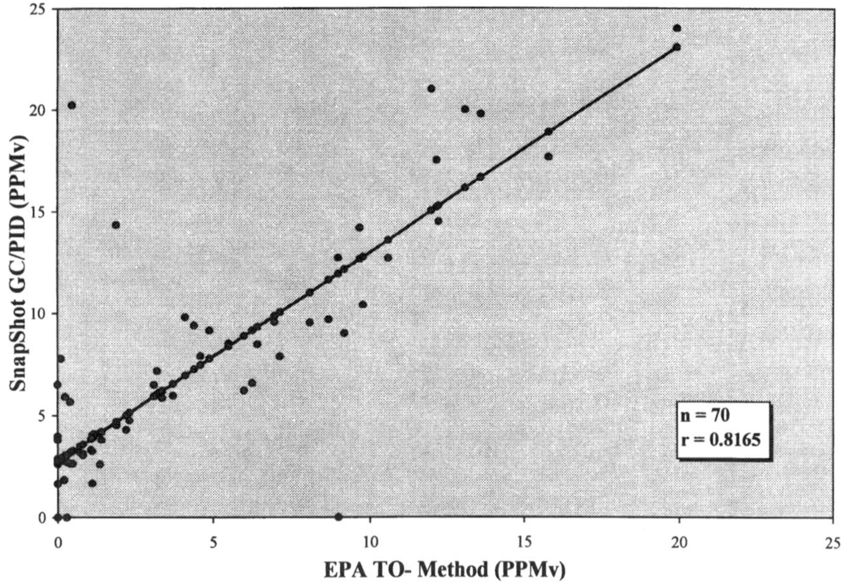

Figure 3. **Sample losses with time (benzene - 107 ppm).**

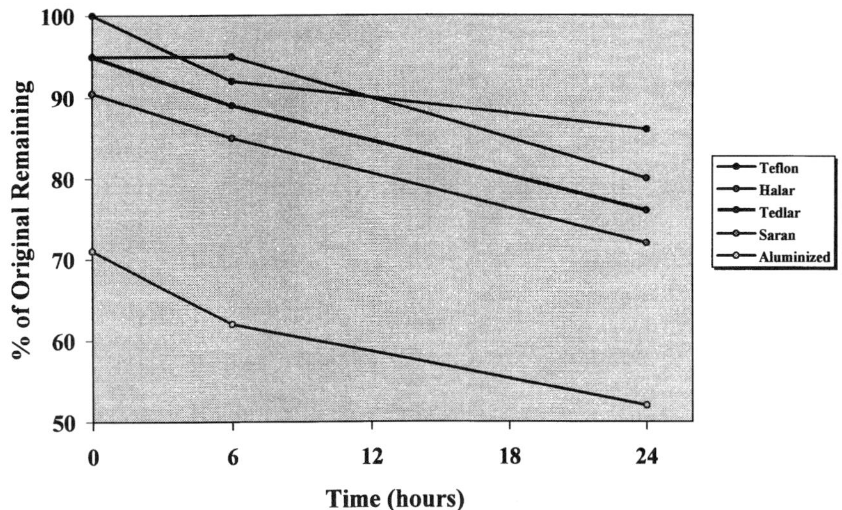

If it is assumed (for example) that each bagged sample experienced a 20% loss between SnapShot GC analysis and lab analysis, and the data are corrected for that loss, correlation of lab to field data for benzene improves to 0.6771 from 0.5532, while correlation of total BTEX experiences only a minimal change from 0.9829 (all data) to 0.9824 (corrected for hypothetical 20% loss). Benzene constituted only a small percentage of the total BTEX sample results; therefore, it did not significantly affect overall correlation for corrected data.

These sample comparison results were sufficient for the New York State DEC to accept the use of the SnapShot handheld GC in place of more expensive and time-consuming laboratory air analysis for routine BTEX emission measurements from soil vapor extraction systems. The SnapShot GC can be utilized for this application with no response or correction factor for field GC results, and the use of the unit is not site-specific.

EnviroTrac was able to bid and win remediation contracts in large part based on the vastly reduced monitoring costs of using the SnapShot GC for routine emission sampling. Although the cost of commercial laboratory air analysis has slowly been reduced due to competitive pressures, an approximate cost of $100 per sample would be reasonably correct. Using the SnapShot GC for measuring the identical parameters with near immediate sample results costs approximately $16.60 per sample, based on a sampling frequency of 5 analyses per day, and accounting for consumables such as carrier and calibration gases, and sampling bags. In addition, the rapid sampling turnaround time of the SnapShot GC allows remediation system adjustments to be made without the delays associated with "normal" laboratory sample turnaround.

Figure 4.   **SVE system emissions, PPMv benzene.**

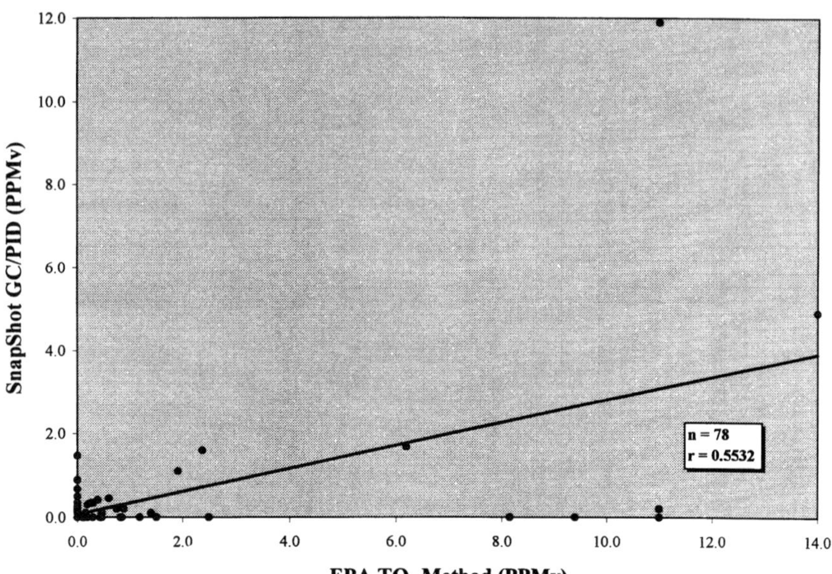

Figure 5.   **SVE system emissions, PPMv benzene (SnapShot samples >0.0 and <5.0).**

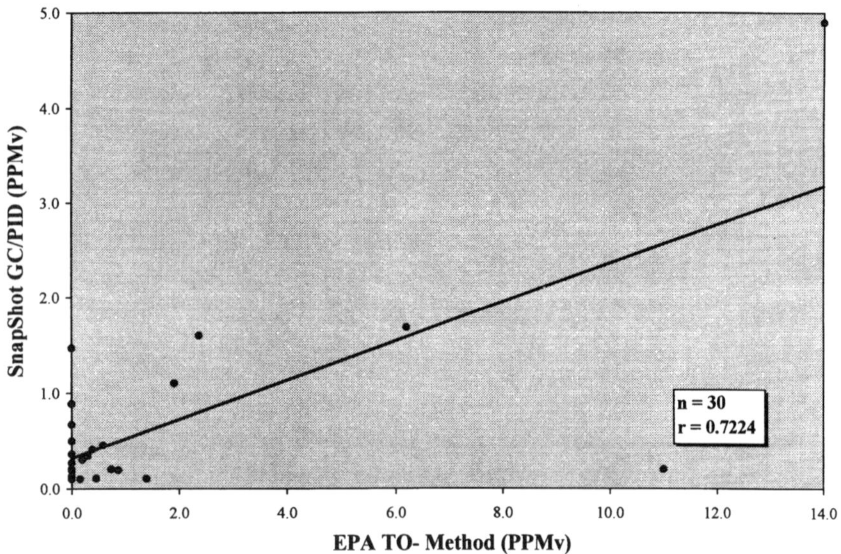

Figure 6. **SVE system emissions, PPMv toluene.**

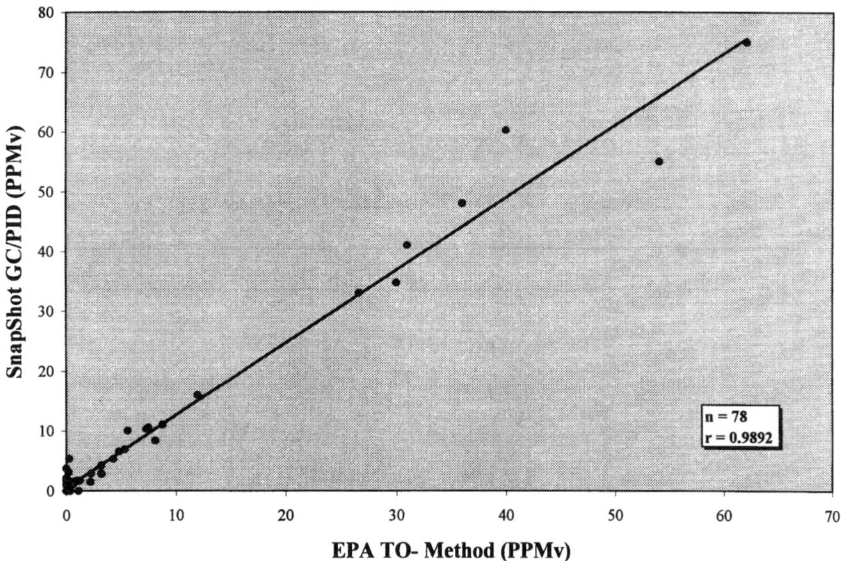

Figure 7. **SVE system emissions, PPMv ethylbenzene.**

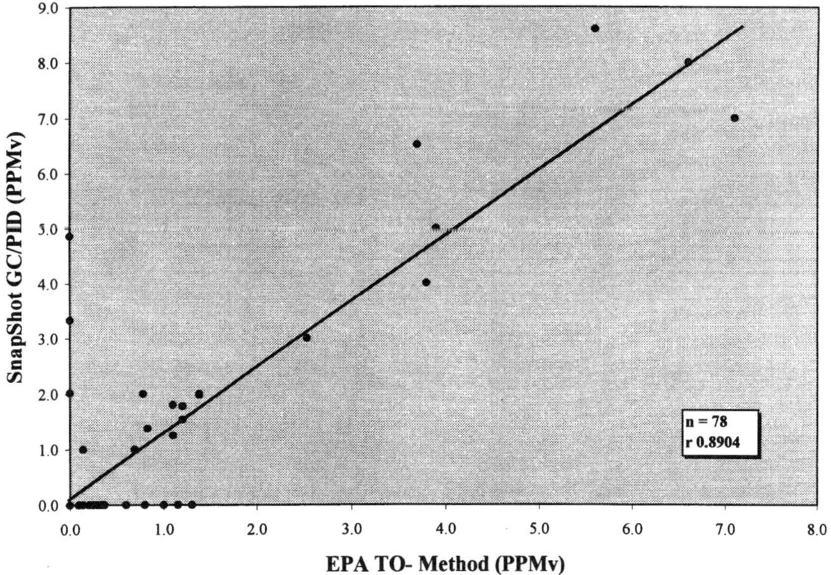

Figure 8.   SVE system emissions, PPMv total xylenes.

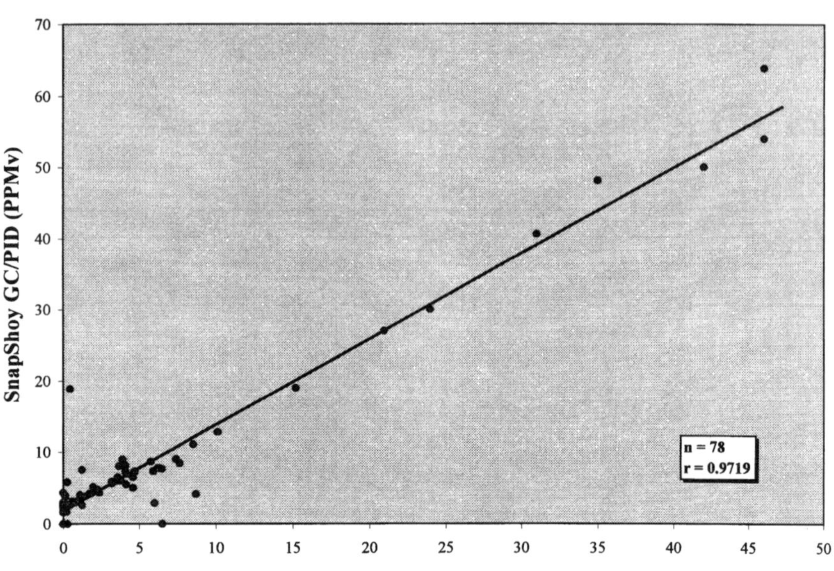

Figure 9.   SVE system emissions, PPMv total C$_8$ aromatics.

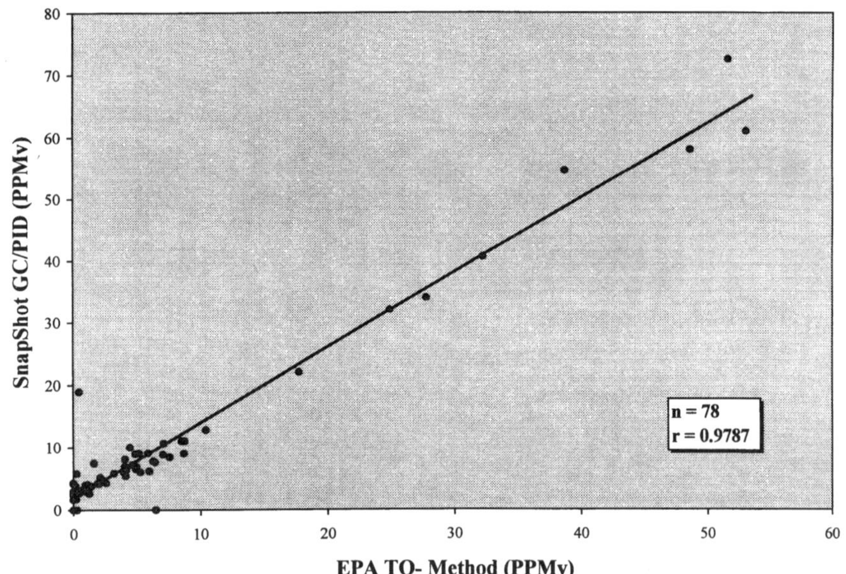

Table 5.   **Correlation Coefficients for Plotted Data**

Parameter	# of Samples (n)	Correlation (r)
Total BTEX	78	0.9829
Total BTEX (< 25 PPM)	70	0.8165
Benzene	78	0.5532
Benzene (> 0.0 & < 5.0 PPM)	30	0.7224
Toluene	78	0.9892
Ethylbenzene	78	0.8904
Total xylenes	78	0.9719
Total C8 aromatics	78	0.9787

In the interim between completion and publication of this study, Envirotrac has installed and is managing a soil vapor extraction system at a former dry cleaning site under the Superfund program where trichloroethylene (TCE) and tetrachloroethylene (PCE) are being removed from the subsurface. They have received similar regulatory approval from New York state for use of the SnapShot in place of laboratory air analysis for these compounds at this site as well.

REFERENCES

Luhrs, R.C., Dresel, P.E., and Pyott, C.J. 1994. Geochemical Interpretation for Remedial Success, workshop presented at the Eighth National Outdoor Action Conference and Exposition, Minneapolis, MN, May 24, 1994.

NYSDEC. 1993. NYS Department of Environmental Conservation, Division of Spills Management and Division of Air Resources. Memorandum of Understanding, Technical Guidance for Regulating and Permitting Air Emissions from Air Strippers, Soil Vapor Extraction Systems and Cold-Mix Asphalt Units, April 13, 1993.

Posner, J.C., and Woodfin, W.J.. Sampling with Gas Bags I: Losses of Analyte with Time. *Appl. Indus. Hyg.* 4, 163-168.

CHAPTER 8

Passive Volatilization of Gasoline from Unsaturated Soil

Tej Gidda, Abigail Salb, Lubna Hussain, Warren H. Stiver, and Richard G. Zytner, School of Engineering, University of Guelph, Ontario, Canada

INTRODUCTION

Gasoline contamination in soil can be the result of surface spills or leakage from underground storage tanks (USTs). This is a significant environmental problem, as 10 to 25% of all gasoline underground storage tanks (USTs) at commercial facilities, farms, and government agencies are leaking (USEPA, 1991; Environment Canada, 1987). A consequence of the released gasoline is contaminated soil, and potentially contaminated air and groundwater.

Atmospheric contamination presents both an explosion and a health risk. The health risk is related to some of the hazardous constituents that comprise the 200-component mixture of gasoline (Donaldson et al., 1992). The volatilization rate of gasoline from soil controls the atmospheric concentrations and thus the risk. In addition, several remediation strategies that are effective in the field, rely on volatilization as the primary removal mechanism. The calculation of risks and remediation timing, effectiveness, and cost are limited because, in general, the volatilization rate of gasoline from soil is not well quantified.

To address the deficiency in data, a laboratory research program was undertaken to collect experimental data on the passive evaporation rate of gasoline and its components in two soils: silt loam and a loamy sand. Experiments were conducted under subzero (Celsius) conditions to simulate the Canadian climate, while water was varied from dry to slightly wet. A 10-component synthetic gasoline was formulated to approximate the chemical composition of gasoline and allow monitoring of gasoline on a total and individual component basis as a function of both depth and time. This paper will present and discuss the findings of the research.

BACKGROUND

Considerable evidence exists that effective remediation of gasoline contaminated sites can be accomplished using one of a variety of volatilization based techniques (Newton, 1990). The most common technique is soil vapor extraction (SVE) but bioventing and passive volatilization are also applied. SVE is a popular remediation technique for soil and groundwater remediation, as the induced vacuum effectively removes the more volatile compounds including benzene, toluene, ethylbenzene, and xylenes. Passive volatilization is a low-cost alternative, as it relies on natural mechanisms, and can be used as a first response tool for surface and subsurface spills of multicomponent petrochemical spills. Although there is ample field evidence of the applicability of a volatilization based remediation technique, there is limited quantification of the factors that control the overall rate. As a result, calculating overall effectiveness, timing, and cost for a particular site is difficult.

Existing experimental data on the movement of gasoline in unsaturated soil is generally limited to dry simple soils. Dominguez-Laseca et al. (1990) observed that 64% of gasoline evaporates from beach sand after 25 min. Jin and O'Connor (1990) demonstrated that volatilization governs the fate of toluene in soil and that the entire mass is lost in 10 days. Baehr (1989) has shown that long-term groundwater contamination can occur from the gasoline vapors emanating from a release. Galin et al (1990) reported that 240 h were required for kerosene to attain 40% concentration at a depth of 75 mm in loamy sand. Donaldson et al. (1992) reported that 32 days were required for petrochemical hydrocarbons to attain 61% of the initial concentration at a depth of 20 cm in a sandy loam soil.

Information on the effects of changing water content, soil type, and temperature are limited. Batterman et al. (1992) showed that soil gas humidities of less than 30% resulted in significant retardation of hydrocarbon vapor transport. Acher et al. (1989) showed that at moisture contents at or approaching the field retention capacity of the soil, vapor transport is completely inhibited. Galin et al. (1990) found that kerosene volatilization was significantly greater from a dune or loamy sand than from a silty loam. Johnson and Perrott (1991) conducted experiments on gasoline vapors at gasoline stations, and determined that temperature was the most significant influence on volatilization.

Arthurs et al. (1995) have demonstrated the importance of a wicking behavior in the overall volatilization from dry soils. Additional studies by Smith et al. (1994) showed that increasing moisture content reduces both the rate of volatilization and wicking. Smith's study was limited to a silt loam soil at room temperature.

Due to the paucity of experimental information, models have been developed to describe the behavior of gasoline in soil, the majority of which describe SVE (Gierke et al, 1990; Ostendorf, 1990; Hunt et al., 1988; Jury et al., 1990; Lingineni and Dhir, 1992; and Benson et al., 1993). These models have shown that the soil properties affect the rate of migration and that volatilization through diffusion is a significant transport mechanism. The models have also demonstrated that the presence of nonequilibrium transport results in the tailing of the contaminant response curve.

METHODS AND MATERIALS

The research was completed using "real" soils under various environmental conditions. This section describes these soils, the conditions studied, the procedures followed, and details the composition of the synthetic gasoline used.

Synthetic Gasoline

Synthetic gasoline was used to spike the soils to simplify the analysis and observation of trends. Through consultation with Imperial Oil (1994), the composition given in Table 1 was selected based on chemical cost, handling safety, ease of chemical analysis (GC identification), boiling point range and total mixture volatility. The n-hexadecane included in the mixture is not a normal component of gasoline, but was added as a nonvolatile tracer to monitor the wicking movement of the gasoline.

Table 1.   Synthetic Gasoline Composition

Component	Volume %	Mass %
Isopentane	28	23.9
Methyl-t-butyl Ether (MTBE)	2	2.0
Hexane	20	18.1
Methylcyclopentane	5	5.2
Isooctane	13	12.4
Toluene	15	17.9
m-Xylene	10	11.9
1,3,5-Trimethylbenzene	5	5.9
Naphthalene	1	1.6
Hexadecane	1	1.1

The synthetic gasoline mixture has a volatility of 34.3 kPa at 25°C and an estimated vapor pressure of 46 kPa at 37.8°C. Typical gasoline has a volatility of approximately 30 kPa at 20°C in the summer months, with a Reid vapor pressure of approximately 65 kPa (Kirk-Othmer, 1982).

Soil and Soil Columns

The two soils used in the experiments were Elora Silt Loam (ESL) and Delhi Loamy Sand (DLS). These were obtained from two University of Guelph research farms. Their characteristics are given in Table 2.

The soil experiments were conducted using 250 mm long acrylic cylinders (73 mm OD, 64 MM ID) with top and bottom lids as shown in Figure 1. The bottom lid has a pinhole opening to maintain atmospheric pressure at the bottom of the

Table 2.   **Soil Characteristics**

Characteristic	Delhi Loamy Sand	Elora Silt Loam
Sand % by wt.	86.5	34.0
Silt % by wt.	9.0	50.1
Clay % by wt.	4.5	15.9
Organic matter % by wt.	1.2	2.5
CEC (cmol+/kg)	8.3	10.2
Bulk density g/cm3	1.5	1.3
Field capacity % vol.	40	48

column, while the top lid has an opening of 49 mm to allow volatilization to occur. Ten 20 mm long segments and one 50 mm long segment with internal diameters of 49 mm were then inserted into the outer sleeve. These segments allowed easy sampling at consistent depths relative to the surface of the column.

Each column was packed in four equal lifts of contaminated soil and gradually compacted with a plunger to achieve the desired bulk density given in Table 2. All mixing and packing was performed in such a manner that gasoline handling losses would be minimized.

Experimental Procedures

Prior to packing the soils into the columns, 50 g of soil were spiked with varied mixtures of water and synthetic gasoline. The ratios of water and synthetic gaso-

Figure 1.   **Soil column.**

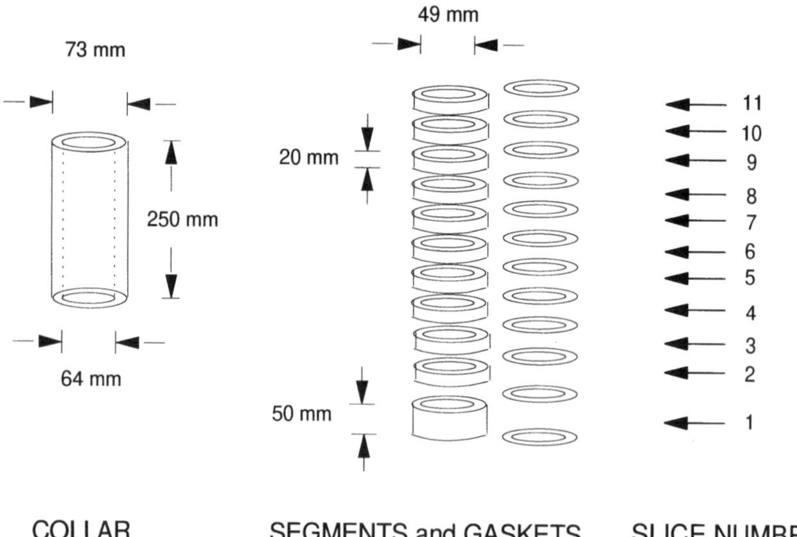

line were varied such that the retention capacity of the soil was not exceeded (to minimize liquid redistribution at time zero) and to maintain a constant soil-air volume fraction. A total of nine combinations of water and synthetic gasoline were used, as given in Table 3.

Table 3.   **Mass Fractions of Gas and Water for Cold Weather Experiments**

Experiment #	Soil Used	Mass of Gasoline %	Mass of Water %
1	ESL	17.5	3.5
2	ESL	13.0	8
3	DLS	14.5	0.5
4	DLS	10.0	5.0

After spiking the soil and packing it in the soil columns as mentioned in the previous section, the columns were placed on the roof of the Engineering Building in a well-ventilated enclosure and allowed to volatilize. The enclosure was necessary to avoid the effects of precipitation on the columns. The temperature in the soil columns and ambient air were monitored using thermocouples and a data logger. The average temperature of the ambient air for the entire course of the experiments was -3.5°C.

Starting at 0 d and continuing for up to 14 d, the soil columns were destructively sampled at various time intervals (0, 1, 2, 6, 24, 48, 72, 120, 240, and 335 h). From each segment in the column, approximately 3-4 g of soil was taken for analysis. Thus, for each of the four experiments, there were ten independent soil columns started.

Sampling Analysis

Following the specified duration of volatilization, the columns were destructively sampled by taking approximately 3-4 g of soil from the top of each segment and placing it into 10 mL of dichloromethane solvent in a 15 mL borosilicate vial with a Teflon-lined silicone septum.

The vials were then placed on a multi-wrist action shaker (Lab-line #3589) for 5 minutes at maximum agitation to extract the samples. The samples were then allowed to settle by gravity for approximately one hour, after which time a fraction of the separated solvent was transferred to a 2-mL sample vial. The vials were then placed immediately into the autosampler queue for analysis. However if the number of vials exceeded 25, the excess vials were placed into the refrigerator for storage at 4°C until analysis could proceed the next day. Overall, there were 10 compounds measured per sample, 11 samples per column and 10 columns per experiment for 4 experiments in total.

The chemical concentrations in the soil were determined using a Hewlett Packard 5890 Series II Gas Chromatograph equipped with a Flame Ionization Detector (GC-FID). The temperature program used was 4 minutes at 40°C, a ramp of 15°C/min to 140°C, then 30°C/min to 260°C, with a final hold time of 2 minutes. Calibration was by means of external standards. Detection limits ranged from 3 to 24 µg of chemical per g of soil.

RESULTS AND DISCUSSION

Results of the experiments demonstrated differences in gasoline volatilization behavior as a function of water content and soil type, particular to a cold-weather environment. These differences are described in the following sections, with emphasis on overall volatilization behavior, individual component behavior, and total gasoline concentration with depth.

Overall Gasoline Volatilization Behavior

Figure 2 demonstrates the cumulative surface gasoline flux and the mass fraction of gasoline lost as a function of time for Elora Silt Loam under dry conditions. The cumulative surface gasoline flux is based on the change in total amount of gasoline in the entire column over the time period between the two sampling points. Cumulative total gasoline volatilization is based on the concentration difference between the time zero column and the column analyzed at a particular point in time.

Figure 2. Elora Silt Loam dry gasoline flux and mass fraction gas lost as a function of time.

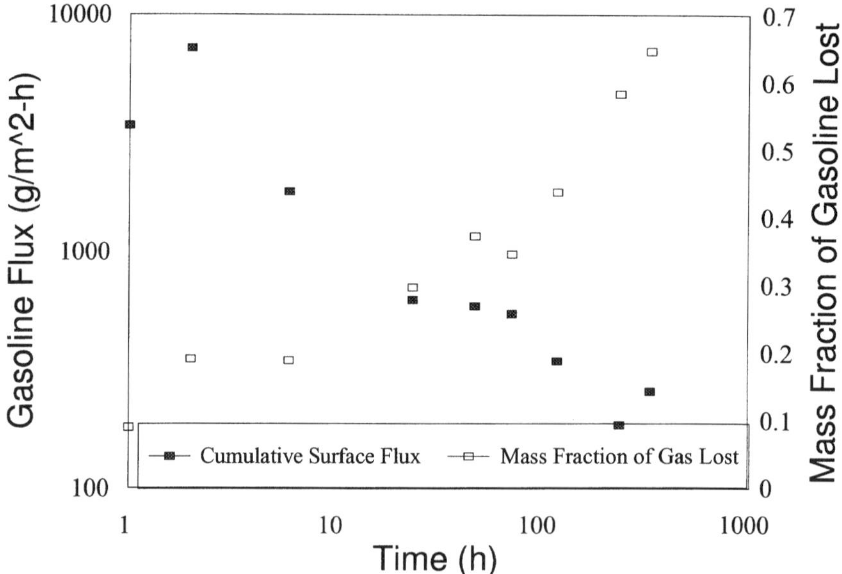

The time zero columns were utilized to assess the handling losses that are inevitable with soil columns when using highly volatile constituents. The losses occur both in column preparation and in column dismantling. The tests completed in the laboratory showed that the average handling losses measured was 32%. This magnitude of loss is high and demonstrates the importance of measuring the initial condition rather than assuming the handling losses were zero and calculating the initial conditions. Provided the handling losses are quantified and are consistent, they will not create errors in the analysis. Poor quantification of the handling losses will lead to errors in the volatilization fluxes measured for the first sampling column (1 hour point in time). This seems evident in Figure 2, as the flux at the 2 hour point exceeds the flux for the first hour. Inconsistencies in the handling losses will also lead to noise in the measured data as a function of time. In Figure 2 and in the other figures displaying the results, it is evident that some noise exists but that it is not excessive.

In the setup of the columns, the soil temperature started near 20°C as the columns were assembled in a laboratory fumehood. Upon placement on the roof, the columns cooled to subzero temperatures in approximately 1 hour. Thus, this cooling period would add an additional uncertainty to the 1 hour sample. In the future, the columns should be precooled prior to enabling an evaporative process.

As expected, gasoline flux decreases with time. The higher fluxes initially are the result of the fresh, highly volatile gasoline being present right at the surface of the soil. As time proceeds, the volatility of the gasoline diminishes and the total gasoline content at the surface diminishes. Both processes lead to a slowing of the total evaporative process and a reduction in the observed flux. The curve of the cumulative total gasoline fraction lost is simply an integration of the flux data.

Figure 3 illustrates flux versus time for all four of the experiments. The behavior witnessed in the dry Elora Silt Loam experiments is consistent for all the experiments (decreasing flux with time). In general, soil under dry conditions experienced greater fluxes than soil under wet conditions, with the Delhi Loamy Sand showing the highest fluxes. This is consistent with the lower initial gasoline content used in the wet experiments, and with the greater porosity of the Delhi Loamy Sand.

Mass fraction of gasoline lost versus time is displayed in Figure 4. With time, there is a gradual convergence of the gas fraction lost for all the experiments. After two hours the loamy sand was losing gasoline at twice the rate of the silt loam. However, following two weeks of evaporation, the two soils and two water contents had similar losses ranging from 63 to 74%. This convergence results from one system that is initially evaporating faster, losing its more volatile components and having the total gasoline content diminish at the surface. Then over time, the soil type and water content influences take hold, slowing down the evaporation and essentially allowing the slower system time to catch up.

Figure 3. Total surface gasoline flux versus time.

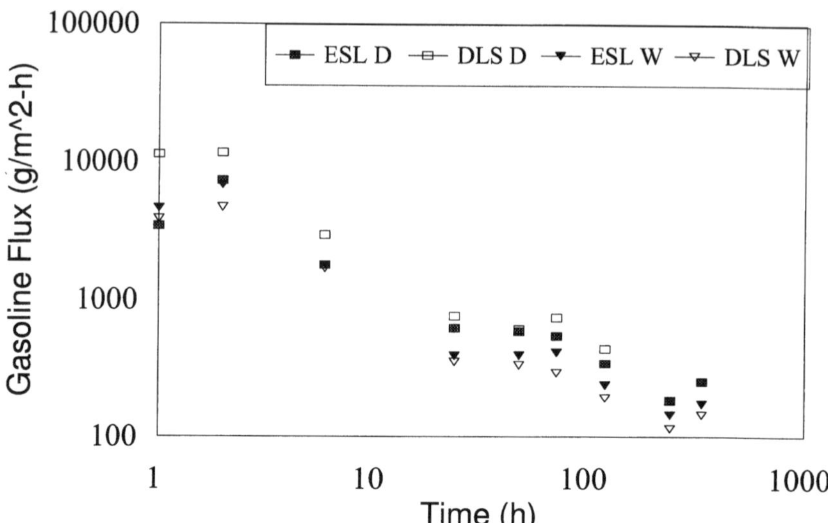

Figure 4. Mass fraction of total gasoline lost versus time.

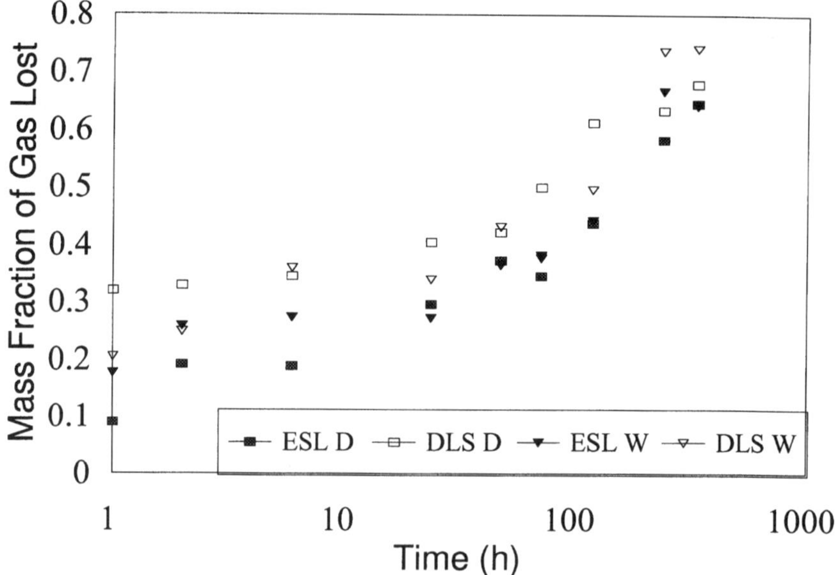

Figure 5 presents total cumulative gasoline flux as a function of the mass fraction of gas lost. Again, the dry Delhi Loamy Sand exhibited the highest fluxes. Initially, fluxes varied significantly with the gas fraction lost; however, there is clear convergence after approximately 40% of the initial gasoline has been lost. This convergence shows that while the initial evaporation rates are dependent on soil and water content, the gasoline flux becomes independent of soil type or wetting conditions.

Individual Component Behavior with Depth

Figure 6 shows the concentration of isopentane as a function of depth in the column for the dry Elora Silt Loam at various times. Isopentane is the most volatile component of the synthetic gasoline used, and behaves in a simple manner. Concentration decreases with time, and losses are greater nearer the surface of the soil.

A similar plot of concentration versus depth is given for hexadecane in Figure 7. Hexadecane is not a normal component of gasoline; however, it was added as a nonvolatile tracer. Figure 7 shows that hexadecane is accumulating at the top of the column, in particular, in the topmost slice. As a nonvolatile tracer, hexadecane will be an indicator of bulk movement of liquid gasoline in the soil. Accumulation at the top of the column implies that there is a wicking action transporting gasoline in the immiscible phase toward the top of the column. This wicking action increases the volatilization of all gasoline components by providing an additional pathway for all gasoline components to reach the surface of the soil.

Figure 5.   **Total surface gasoline flux versus mass fraction of gasoline lost.**

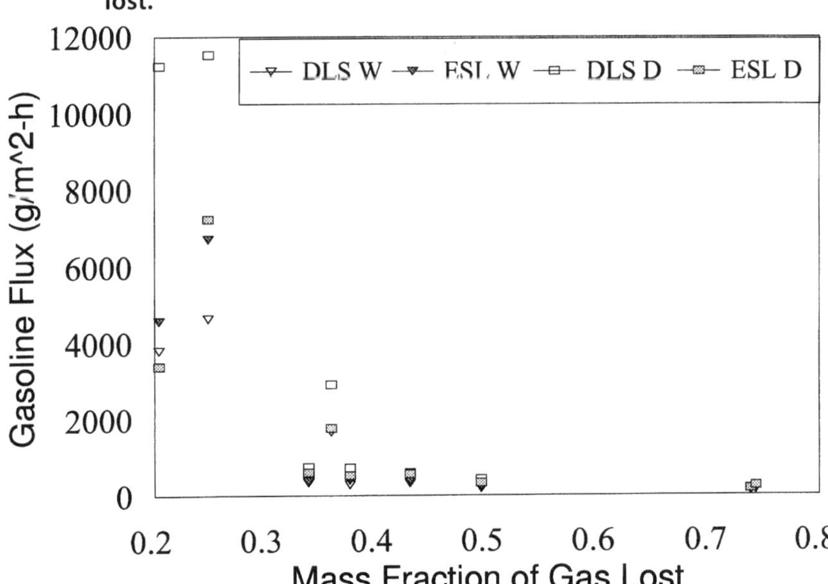

Figure 6.   **Elora Silt Loam dry isopentane concentration versus depth.**

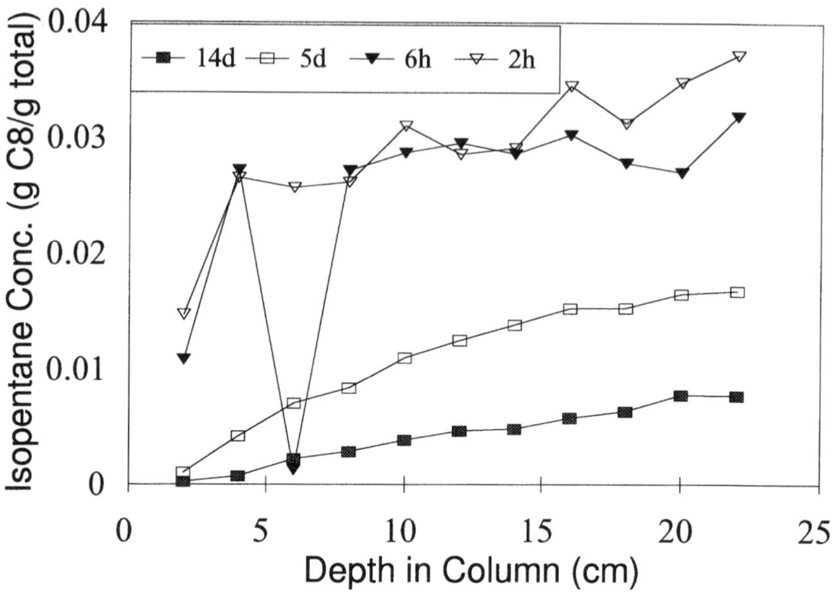

Figure 7.   **Elora Silt Loam dry hexadecane concentration versus depth.**

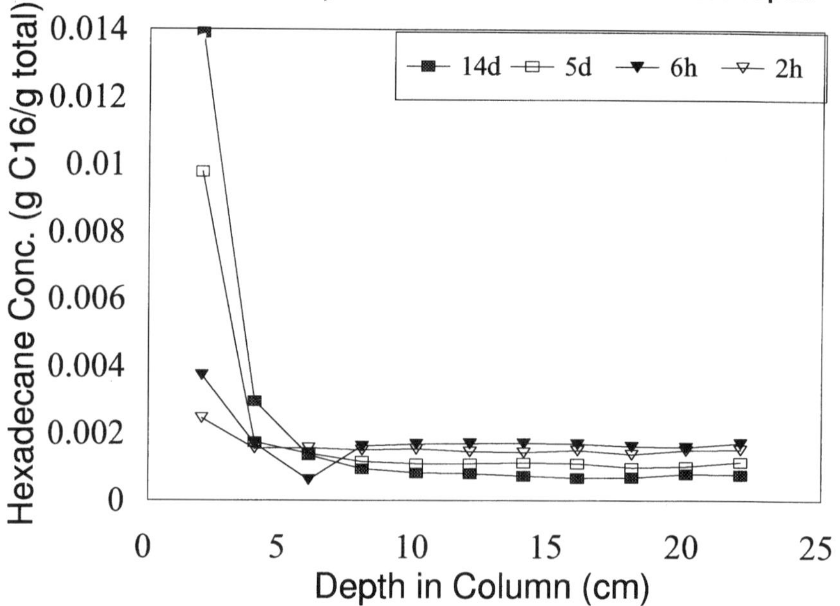

To confirm the hexadecane behavior, the flux trend for naphthalene was isolated in Figure 8. Although naphthalene will not act as a fully conservative tracer, its behavior is similar and is clearly accumulating at the surface, which can only be explained by a wicking process.

Total Gasoline Behavior with Depth

Figure 9 illustrates the total gasoline behavior with depth for Elora Silt Loam under dry conditions. Clearly, there is a gain in total gasoline concentration at the top of the column for all the times plotted. Similar behavior at the top of the column was observed for both soils under both wet and dry conditions.

It is clear from Figure 9 that the gasoline continues to accumulate at the surface as time proceeds. This does not make sense since accumulation of gasoline in the topmost portions of the column is contradictory to the observation of wicking of gasoline to the surface of the soil. That is, for wicking to occur, the concentration of gasoline at the bottom of the column should be higher than the concentration of gasoline at the top of the column. However, in the case detailed by Figure 9, there is more gasoline at the surface as compared to the bottom of the column, indicating that the driving force for bulk movement is down and not up.

A possible explanation for this contradiction is related to the cold temperature. As temperature decreases, the solubility of the higher molecular weight components of the gasoline mixture is likely to diminish. Thus, it is possible that at the colder temperatures some of the components solidify. This factor becomes

Figure 8.   **Elora Silt Loam dry naphthalene concentration versus depth.**

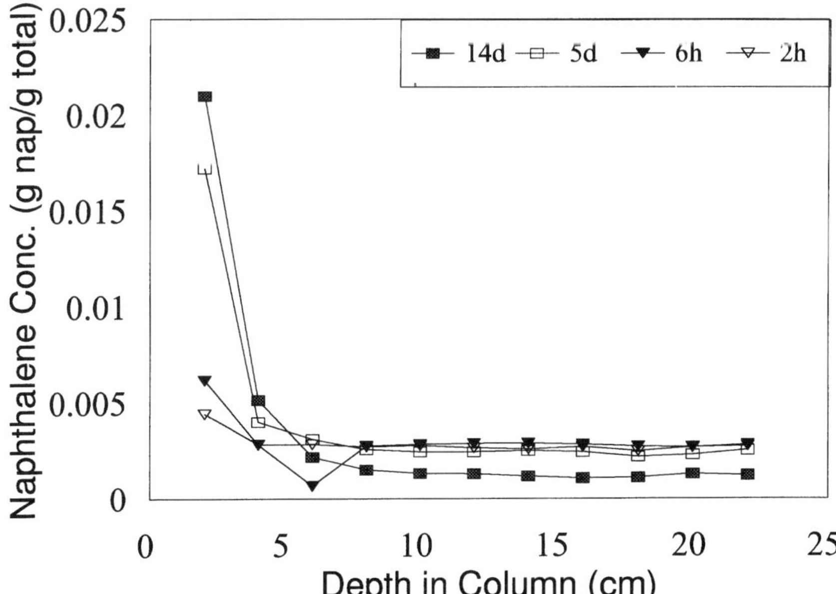

Figure 9. **Elora Silt Loam dry gasoline concentration versus depth.**

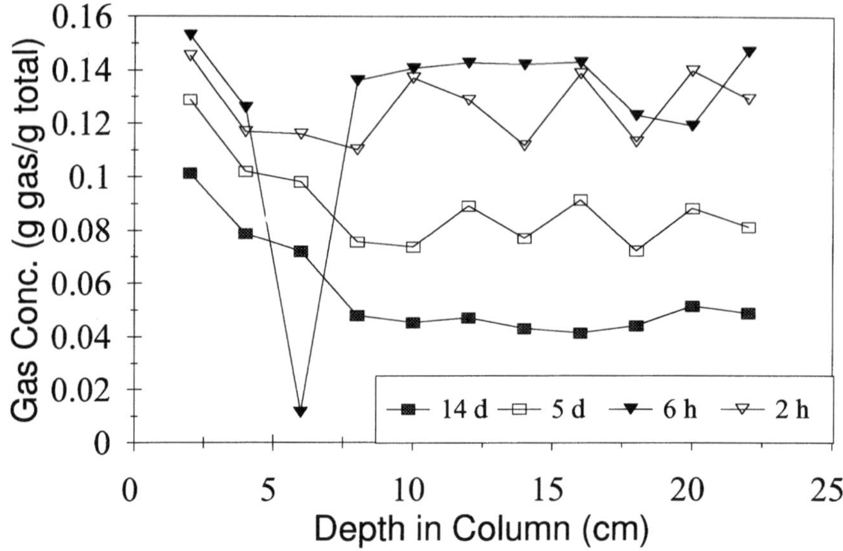

more important in terms of the chemical mixture that exists at the surface of the soil during the evaporation process. The composition of the surface contamination after 14 days includes 18% hexadecane and 20% naphthalene. At these high concentrations it is even more likely for these two compounds to partially solidify. If a substantial fraction of the contamination existing at the surface solidifies, then the liquid contamination at the surface may be less than the liquid contamination at greater depths in the soil. Thus, a liquid contamination gradient remains and is favorable to a wicking upward process. The liquid gradient is masked by the total contamination analyzed for each soil sample, which includes both solids and liquids at each depth.

Potential Implications to Gasoline

The accumulation through freezing phenomenon observed at the top of the column could be significant for gasoline contamination of soil in cold weather climates. Although gasoline does not contain hexadecane, it does include high melting point components such as naphthalene.

To test the potential for freezing of the less volatile components in these mixtures, freezing tests were conducted on an unweathered gasoline fuel. Samples were placed in open beakers, in a -2°C ice bath, in a standard laboratory fumehood. As the gasoline volatilized, the mixture was observed for evidence of any solidification.

By the time that the initially unweathered gasoline had lost 57 to 66% by evaporation, a significant portion of the residual gasoline had solidified. Frozen masses were approximately 15% of the initial gasoline content or almost half of the residual gasoline. These results indicate that the freezing behavior seen in the synthetic gasoline evaporation experiments will also occur with real gasoline. As a result of the freezing, the wicking process will be enhanced, which will ultimately enhance the overall volatilization rate.

CONCLUSIONS

Observations have shown the importance of soil type, water content, and temperature on passive volatilization rates. Dry soils initially exhibit greater volatilization fluxes than do wet soils, and Delhi Loamy Sand initially demonstrates a greater volatilization flux than does Elora Silt Loam. However, gasoline flux ceases to be a factor of soil type or water content after 40% of the original gasoline is lost. Tracking nonvolatile components and total gasoline demonstrated immiscible phase movement, through a wicking process, also significantly contributes to volatilization by moving the gasoline to the soil surface. At the surface, freezing of the nonvolatile components leads to an increase in the overall gasoline concentration while maintaining the liquid gradient required for continued wicking.

These results have implication on risk assessment following a gasoline release, where the movement of the immiscible phase via wicking must also be taken into consideration when doing a site investigation and evaluating remedial options.

ACKNOWLEDGMENTS

The authors would like to thank Dr. Larry Lawlor of Esso Petroleum Canada for his contributions on this project. Funding for this research was provided by Imperial Oil Ltd. and Natural Science and Engineering Council of Canada.

REFERENCES

Acher, A.J., Boderie, P., and Yaron, B. 1989. Soil Pollution by Petroleum Products. I. Multiphase Migration of Kerosene Components in Soil Columns, *J. Contaminant Hydrol.* 4, 333-345.

Arthurs, P., Stiver, W.H., and Zytner, R.G. 1995. Passive Volatilization of Gasoline from Soil, *J. Soil Contamination* 4, 123-135.

Baehr, A.C. 1989. Selective Transport of Hydrocarbons in the Unsaturated Zone Due to Aqueous and Vapour Phase Partitioning, *Water Resour. Res.* 23, 1926-1938.

Batterman, S., Kulshrestha, A., and Chang, H. 1992. Hydrocarbon Vapor Transport in Low Moisture Soils, *Environ. Sci. Technol.* 29, 171-180.

Benson, D.A., Huntley, D., and Johnson, P.C. 1993. Modelling Vapour Extraction and General Transport in the Presence of NAPL Mixtures and Non-Ideal Conditions, *Groundwater* 31, 437-445.

Dominguez-Laseca, F., Bergueiro Lopez, J.R., and Riveria Julia, A. 1990. Evaluation of the Effects of Hydrocarbon Spills. IV. Evaporation on Beach Sand, *Ing. Quim.* (Madrid), 22, 133-138.

Donaldson, S.G., Miller, G.C., and Miller, W.W. 1992. Remediation of Gasoline Contaminated Soil by Passive Volatilization, *J. Environmental Quality* 2, 94-102.

Environment Canada. 1987. *Leaking Underground Storage Tank Newsletter*, Vol. 1(4).

Galin, T.S., McDowell, C., and Yaron, B. 1990. The Effect of Volatilization of the Mass Flow of a Non-Aqueous Pollutant Liquid Mixture in an Inert Porous Medium: Experiments with Kerosene, *J. Soil. Sci.* 41, 631-641.

Gierke, J.S., Hutzler, N.J., and Crittenden, J.C. 1990. Vapour Transport in Unsaturated Soil Columns: Implications for Vapour Extraction, *Water Resour. Res.* 26, 1529-1547.

Hunt, J.R., Geller, J.T., Sitar, N., and Udell, K.S. 1988. Subsurface Transport Processes for Gasoline Components, *1988 Joint ASCE-CSCE Environ. Engineering Conf.*, pp. 536-543, Vancouver, BC.

Imperial Oil. 1994. Personal communication with Dr. Larry Lawlor, Products and Chemical Division, May 12.

Jin, Y. and O'Connor, G.A. 1990. Behaviour of Toluene in Sludge Amended Soil, *J. Environmental Quality*, 19, 573-579.

Johnson, R. and Perrott, M. 1991. Gasoline Vapor Transport Through a High-Water-Content soil, *J. Contaminant Hydrol.* 8, 317-334.

Jury, W.A., Russo, D., Streile, G., and El Abd, H. 1990. Evaluation of Volatilization by Organic Chemicals Residing Below the Soil Surface, *Water Resour. Res.* 26, 13-20.

Kirk-Othmer. *Encyclopedia of Chemical Technology*, Vol. 11, 3rd ed., Toronto, Wiley Interscience.

Lingineni, S. and Dhir, V.K. 1992. Modelling of Soil Venting Processes to Remediate Unsaturated Soils, *ASCE J. Environmental Engineering* 118, 135-152.

Newton, J. 1990. Remediation of Petroleum Contaminated Soils, *Pollution Engineering*, Dec, 46-52.

Ostendorf, D. 1990. Long Term Fate and Transport of Immiscible Aviation Gasoline in the Subsurface Environment, *Water Sci. Technol.* 22, 37-44.

Smith, M., Stiver, W.H., and Zytner, R.G. 1994. The Effect of Varying Water Content on Passive Volatilization of Gasoline from Soil, *49th Annual Industrial Waste Conference, Purdue University, May 9-11*, pp. 111-116.

USEPA. 1991. OUST Corrective Action 45-Day Study Team Report, Office of Underground Storage Tanks, U.S. Environmental Protection Agency, Washington, DC.

CHAPTER 9

Using Dielectric Track Detectors to Evaluate Alpha Activity of Contaminated Soils Samples

Yefim I. Knizhnik, Consultant, Philadelphia, Pennsylvania

Victor S. Prokopenko and **Vladimir V. Tokarevsky**, Interdisciplinary Scientific and Technical Center "Shelter," Chernobyl City, Ukraine

Mikhail V. Artsimovich, Institute for Nuclear Research, Kiev, Ukraine

INTRODUCTION

The choice of dielectric track detectors (DTDs) for registration of alpha particles in the biosphere objects is determined by some essential advantages of the method as compared to other techniques. The most important of them are as follows: insensitivity to gamma and beta radiation; a high spatial resolution and detection efficiency; a low intrinsic background; possibility for making the DTD of an arbitrary form, for its location in places difficult for approach, and for long-term information storage; simple and reliable data processing; a low cost (Knizhnik et al., 1991).

The main methodological difficulty of widespread use of DTDs in field investigations of α-active soils is converting from the number of recorded tracks to the specific activity of the soil or the density of surface contamination, for example, in Ci/km^2, widely used for constructing maps of radioactive contamination of a territory. The problem essentially reduces to substantiating the representativeness of the measurements, since the surface, analyzed by the detector, has an area of ~1 cm^2 and a depth of up to several tens of microns. Extension of the measurement results to the required area and depth must be metrologically sound (Knizhnik et al., 1994).

We undertook this study with two main objectives:

1. to deduce an equation for estimation of a specific alpha activity of a soil by use of the dielectric track detector technique;
2. to carry out an experimental verification of the equation by use of certified samples of soil with known alpha activity.

THEORY

The density ρ of α particles tracks etched on the surface of the DTD is determined by equation (Durrani and Bull, 1987):

$$\rho = 0.25nRcos^2\,\theta\,, \tag{1}$$

where n is the number of α decays per unit volume of the soil sample (i.e., alpha source); R is the range of an alpha particle in the source material; and θ is the critical etch angle of the DTD.

The sandy soil is essentially a dispersed mixture. Considering a model of the soil in the form of a small close-packed spheres of silicon dioxide separated by air gaps, it is easy to show that the range of an α particle R_{SD} in such a sphere is approximately equal to the range R_{AG} in the air gap. Therefore, the real range of an a particle in such a dispersed mixture is:

$$R \approx 2R_{SD}. \tag{2}$$

Dividing both sides of Equation 1 by the product of the exposure time t of the detectors by the source density d and using Equation 2 and the fact that the evaluated specific alpha activity of the source is $A_e = n/(td)$, we obtain the expression:

$$A_e = \frac{2\rho}{R_{SD} \cdot cos^2\theta \cdot t}\,. \tag{3}$$

(The dimensions of the appearing quantities in Equation 3 are as follows: ρ - track/cm^2, A_e - kBq/kg, R_{SD} - g/cm^2, and t - sec).

EXPERIMENTAL PROCEDURE

To check the procedure for measuring absolute α activity of a soil and to verify the theory by experiment, we employed samples of fine-grained sandy soil with known α activity: reference Sample No. 1 (prepared at the V.G. Khlopin Radium Institute, certified specific activity $A_c = 4.54$ kBq/kg), uniformly saturated with [241]Am, and reference Sample No. 2 (prepared at the Institute of Nuclear Power, Academy of Sciences of Byelarus, $A_c = 3.73$ kBq/kg), uniformly saturated with [239]Pu (Table 1).

The unique set of characteristics - the wide energy range (0.1-20 MeV) of alpha particles forming identifiable tracks in CR-39 plastic and the high detection efficiency (0.95) with which they are recorded, low intrinsic background and critical etch angle, high isotropy and optical transparency - of the CR-39 plastic as a material for DTDs (Durrani and Bull, 1987; Marenny, 1987; Palfalvi et al., 1988; Knizhnik et al., 1991; Knizhnik et al., 1994) makes it applicable for the purpose to be achieved.

Four CR-39 plastic track detectors with a total area of 2.95 cm^2 were chemically etched (6 M NaOH, 65°C, 6-8 h) after being in contact for 25 h with the reference Sample No. 1. Visual examination and calculation of the number of tracks under an optical microscope with X400-500 magnification revealed 4619 tracks (with 450 background tracks). Similar experiments with four detectors, having a total area of 3.65 cm^2, and the reference Sample No. 2 revealed 3259 tracks (66 background tracks).

RESULTS AND DISCUSSION

The results of measurements and calculations (Table 1) show that the simplified model of soil is almost ideally suited for a fine-grained sandy soil, uniformly saturated with ^{239}Pu. The larger difference of the evaluated value A_e from the certified value A_c in the case of Sample No. 1 is probably due to the relatively less uniform (in comparison with Sample No. 2) distribution of alpha emitters over the volume, apparently due to the higher mobility and solubility of ^{241}Am (than ^{239}Pu) in soil components (Hanson, 1980).

It should be noted that under real conditions the distribution of α-emitting radionuclides is not uniform, either over the soil surface or over the depth in soil. For this reason, field measurements must include a check of their depth profile at several characteristic locations of the area being analyzed. The contribution of radon and thoron to the number of tracks recorded by the detector must also be estimated.

Table 1. **Measured and Evaluated Characteristics of Reference Samples of Fine-Grained Sandy Soil, Uniformly Saturated with Alpha Radiating Actinides, and Dielectric Track Detectors (DTDs) after a Contact with the Samples of Soil**

Characteristic	Soil Reference Sample No.	
	1	2
Mass of sample, g	57.77	59.0
Soil-saturating isotope	^{241}Am	^{239}Pu
Certified specific alpha activity A_c, kBq/kg	4.54	3.73
Energy of primary outgoing alpha particle, MeV	5.49	5.16
Range of alpha particle in silicon dioxide R_{SD}, mg/cm^2	5.881	5.349
Duration of contact between DTDs and sample t, h	25	25
Critical each angle θ of DTD of CR-39 plastic, °	10	10
Surface density of tracks ρ, track/cm^2	1413	875
Evaluated specific alpha activity A_e, kBq/kg	5.50	3.75
$(A_e - A_c)/A_c$, %	21.1	0.5

CONCLUSIONS

1. An equation has been obtained to evaluate the specific alpha activity A_e (kBq/kg) of a soil by use of dielectric track detector (DTD) technique.

2. The experimental verification showed that the equation is almost ideally suited for a fine-grained sandy soil, uniformly saturated with ^{239}Pu. The larger difference of the estimated value of A_e from the certified value of A_c in the case of the soil sample, uniformly saturated with ^{241}Am, is probably due to the relatively less uniform (in comparison with the sample saturated with ^{239}Pu) distribution of alpha emitters over the volume, apparently due to the higher mobility and solubility of ^{241}Am (than ^{239}Pu) in soil components.

REFERENCES

Durrani, S.A. and Bull, R.K. 1987. Solid State Nuclear Track Detection Principles, Methods, and Applications. AERE, Harwell, United Kingdom, Pergamon Books Ltd.

Hanson, W.C., Ed. 1980. Transuranic Elements in the Environment. U.S. Department of Energy.

Knizhnik, Y.I., Stetsenko, S.G., and Tokarevsky, V.V. 1991. Estimation of Soil Surface Alpha Activity and Alpha Contamination by Polymer Track Detectors. *Nucl. Tracks Radiat. Meas.* 19, 765-768.

Knizhnik, Y.I., Prokopenko, V.S., Stolyarov, S.V., and Tokarevsky, V.V. 1994. CR-39 Plastic Track Detector for Field Investigations of α-Active Soils. *Atomic Energy* 76(2), 105-111.

Marenny, A.M. 1987. Dielectric Track Detectors in Radiation-Physical and Radiobiological Experiments. Moscow, Energoatomizdat (in Russian).

Palfalvi, J., Lancsarics, G., and Sagi, L. 1988. Alpha Spectrum of "Hot Particles" Determined by CR-39 SSNTD. *Nucl. Tracks Radiat. Meas.* 15, 779-782.

PART III
RISK ASSESSMENT

CHAPTER 10

Human Health Risk Assessment: A Comparison Between the USEPA Approach and the Québec Approach

Daniel Morin and **Richard Desbiens**, D'Aragon, Desbiens, Halde associés ltée, Montréal, Québec, Canada

Serge Barbeau, City of Montréal, Environmental Engineering Division, Montréal, Québec, Canada

INTRODUCTION

Over the past few decades, outmoded industrial complexes have been decommissioned throughout North America. A number of these industrial complexes, through their day-to-day operations, contaminated surrounding soil and groundwater. The City of Montréal wants to reuse some of these former industrial lots for various purposes, mainly industrial and residential redevelopment.

The policy currently in force at the Québec Ministry of the Environment and Wildlife (MEF) for managing contaminated sites is based on specified levels of contamination or generic criteria inspired by the Dutch "ABC" criteria (MEF, 1988). This kind of approach is usually conservative, resulting in high remediation costs. As a consequence, industrial and residential developers may be moving outside Montréal to more pristine settings or to areas where legislation is less stringent. This trend contributes to undesirable urban sprawl and to potential contamination of virgin lands.

The City of Montréal, in collaboration with MEF, has designed and implemented a two-year pilot project to look at an alternative site-specific approach that would take into account the effects of contaminated sites on human health and the environment. This approach must protect human health and the environment and result in more cost-effective cleanups than through the use of generic criteria.

Although other aspects were covered during the pilot project, the focus of this paper is human health risk assessment. At six sites, D'Aragon, Desbiens Halde associés ltée (DDH) compared the approach developed by the U.S. Environmental Protection Agency (USEPA) for Superfund sites (USEPA, 1989) with a new, much more complex approach proposed by MEF in 1996. MEF guidelines are provided in two separate documents totaling over 850 pages (MEF, 1996a; MEF, 1996b).

This paper discusses the main differences between the USEPA and the Québec approaches, and their implications for the management of contaminated sites.

QUÉBEC APPROACH

Types of Risk Assessment

Two types of human health risk assessments (HHRA) are proposed by MEF (MEF, 1996a; MEF, 1996b): a baseline HHRA and a detailed HHRA.

The main objective of a baseline HHRA is to verify the presence or the absence of significant potential risks. When the findings of this first conservative screening level shows a significant potential risk, two options are available. In the first option, the risk assessor can perform a detailed HHRA to characterize the risk with more "precision" and less uncertainty. The detailed HHRA is conducted only for the chemicals or the exposure pathways identified as posing a potential problem during the baseline HHRA. It can involve the use of probabilistic methods, uncertainty analysis, or the thorough review of toxicological data. The second option is to propose mitigation measures. This option can be used if the costs associated with risk reduction are not important or if, according to the professional judgment of the risk assessor, the detailed HHRA would probably not change significantly the findings of the baseline HHRA.

To help perform a baseline HHRA, MEF has produced a software containing a database on almost 100 chemicals with their physicochemical, environmental, and toxicological properties, predefined exposure scenarios with predefined values, and fate and transport screening models. The use of this software should, according to MEF, help allocate time and resources efficiently by pointing out the chemicals and exposure pathways that must be covered in a detailed HHRA.

The MEF have no technical guidelines for a detailed HHRA because they consider a detailed HHRA to be based mostly upon professional judgment rather than on a rigid approach.

Steps for the Risk Assessment

MEF methodology is divided into the four typical HHRA steps:
- toxicological characterization;
- exposure assessment;
- risk assessment;
- risk evaluation.

Those four steps will be described briefly for the baseline and the detailed phase of the HHRA.

Toxicological Characterization

During a baseline HHRA, the toxicological characterization consists of gathering quantitative (Reference Doses [RfDs] and Slope Factors) and qualitative information on the toxicological properties for a given chemical.

At first, if the data are derived from animal studies, MEF converts the animal dose into a human equivalent dose. The basis of the interspecies scaling used is the physiological time which takes into account the metabolic rate and the body weight. This method of interspecies scaling is, according to MEF, more conservative than the one based only on the body weight, but less conservative than the one using the skin surface area.

For noncarcinogens, RfDs are developed by applying uncertainty and modifying factors to the toxicological and epidemiological data. MEF has developed in-house uncertainty factors which depend on the number of factors applied and the correlations among those factors. According to this approach, the uncertainty factors used to develop RfDs do not automatically have a value of 10, as is often seen in other governmental agencies.

For carcinogens, a modified version of the Model-Free Approach to Low-Dose Extrapolation (MFX) by Krewski et al.(1986) is used to convert toxicological animal data into Slope Factors when there are no epidemiological data available.

For a baseline HHRA, the RfDs and Slope Factors are provided to the risk assessor by MEF. A comparison by DDH has been made between the RfDs and Slope Factors found in IRIS (1995) and HEAST (1995) and those developed by MEF. In general, less than 1 order of magnitude differentiate the US numbers from those of MEF. Differences arise mainly for effects associated with Polycyclic Aromatic Hydrocarbons (PAHs), where MEF considers that all PAH compounds are carcinogens. Also, for each chemical, RfDs and Slope Factors are available for inhalation and ingestion with specific MEF values. The inhalation pathway is not covered in such detail in IRIS (1995) and HEAST (1995).

For a detailed HHRA, toxicokinetic modeling can be performed to convert dose from animal to human. Also, the toxicodynamics of chemicals can be considered to see if the effects observed in animals can be transposed to human. The amount of effort here is considerable.

Exposure Assessment

For a baseline HHRA, exposure modeling can accomplished through screening models with a deterministic approach. The modeling is based on predefined scenarios according to the primary land use of the study area. The predefined exposure scenarios ease the HHRA modeling exercise since one does not need to establish, for each computation, the value of the variables. The exposure patterns vary significantly according to the land use variables. This approach may be, on the other hand, criticized for its lack of flexibility.

Four pre-defined scenarios have been established by MEF:

- residential;
- agricultural;
- commercial;
- industrial.

Each scenario is characterized by a large set of variables. Each variable can be classified into two categories: those variables that are dependent on the exposure scenarios, or those that are dependent on both age and sex. The first category includes the following variables:

- age classes of exposed populations;
- exposure pathways;
- exposure duration in the study zone;
- time spent indoors and outdoors.

For exposed populations, 17 classes ranging from 0 to 80 years are considered for residential and agricultural scenarios. For commercial and industrial scenarios, 4 classes are taken into account, ranging from 20 to 60 years. For exposure pathways, a maximum of 21 pathways can be considered for an agricultural scenario, 18 for residential, 11 for industrial, and 8 for a commercial scenario. Ingestion of breast milk is one of the exposure pathways that must be considered for residential and agricultural settings.

The second category, variables dependent on the age and sex of the human receptor, covers also a large spectrum of variables like weight, skin surface area, ingestion and inhalation rate, and so on.

The data are available from MEF for men, women, and both sexes. According to MEF, the predefined values that are proposed should not overestimate the exposure, as is done in the worst-case scenario and in reasonable maximum exposure scenario (MEF, 1996b). An overestimation is always possible, but would not happen systematically.

Depending on the effects that must be characterized (cancer or noncancer effects), the average daily dose to be taken into account differs. For cancer effects, the basis is an incremental probability of developing cancer. In this case, the average daily dose is based upon the exposure from chemicals identified at the site. But for noncancer effects, a comparison is made between the average daily dose and the RfD. In this case, the background dose resulting from the day-to-day exposure to chemicals, before being exposed to a contaminated site, is added to the average daily dose from the site. This is done because the definition of a RfD refers to total exposure and not to the incremental exposure due to the contaminated site itself.

In a detailed HHRA, a stochastic approach can be considered to take into account the exposure variability among the population at risk, using the Monte Carlo simulations for example.

Risk Assessment

In a baseline HHRA, risk indexes are computed for both noncancer and cancer effects. The acceptability of risk is based upon the RfD for noncancer effects and the incremental risk for cancer effects.

For carcinogens, one index is calculated by adding the dose accumulated for each age class during lifetime. As for USEPA, an index (= average daily dose x Slope Factor) is calculated for:

- each chemical;
- each exposure pathway;
- aggregated exposure pathways.

For noncarcinogens, an index is calculated for each age class identified. A gross risk index (= [average daily dose + background dose]/RfD) is also calculated. In this index, the background is taken into account. For many chemicals, the background dose is higher than the RfD. In such a case, the gross risk index is always higher than 1, independent of the nature or contribution from the contaminated site. In this case, MEF proposes an adjusted risk index (= average daily dose/[background dose x 1%]). With this index, MEF allows a 1% excess of the background dose when it is higher than the RfD. In the USEPA approach, the background dose is not taken into account in the same way as the MEF approach.

For detailed HHRA, the risk assessment can be performed through an exhaustive and critical look at the data compiled during the toxicological characterization step.

Risk Evaluation

For a baseline HHRA, this step is incorporated during the risk assessment step by preestablishing levels of acceptable risk.

According to MEF guidelines, an acceptable risk is:

- for noncancer effects: no effects (below 1);
- for cancer effects: an incremental risk of 10E-04.

In a detailed HHRA, the acceptability of risk will take into account the benefits in terms of public health and costs associated with the reduction of risk.

COMPARISON BETWEEN THE USEPA APPROACH AND THE QUÉBEC APPROACH

Six industrial sites were investigated for testing these new methods. According to generic criteria, widely used in Québec, the sites considered involved over 9 million $CDN in remediation costs. On all the sites, a baseline HHRA and an ecological assessment were performed by DDH.

For human health, the USEPA and MEF approaches were compared. The costs dropped from 9 to 2 million $CDN when the USEPA risk assessment approach was used. The costs remained the same (9 million $CDN) when the MEF baseline HHRA approach was used.

Many differences separate the two approaches, but the main one is in the risk characterization of noncancer effects. By adding the background dose, the risk is not acceptable in all cases, even when an adjusted index is calculated.

When the background dose is not taken into account, the results are approximately the same for both approaches, but the MEF approach generates more information on the variability of the exposure because of the number of age classes taken into account.

Following this test at six Montréal sites, MEF has decided to examine the issue when the background dose is higher than the RfDs. This examination is still going on, and will most likely emerge with a new way to take into account this problem from a technically and economically sound basis.

CONCLUSION

Besides the problem of how background dose should be integrated into the baseline HHRA for noncancer chemicals, the Québec baseline HHRA approach showed some interesting departures from the normal HHRA; namely:

- the modifications for the development of RfDs (uncertainty factor which depends on the number of factors applied and the correlations among those factors);
- the number of exposure pathways considered;
- the age classes to be taken into account, which in some way gives information on the variability of the exposure.

For risk assessors working in Québec, this approach, with minor modifications, will be an improvement for the management of contaminated sites, over generic criteria, a conservative and costly approach in terms of remediation goals, as we have seen over a number of years of practice experience.

ACKNOWLEDGMENT

The authors would like to thank the personnel at DDH who have worked on this project: C. Diemer, M.-J. Duquette, C. Gaudreault, R. Lemoine, C. Marcotte, F. Ouellet, M.-J. Poulin, and R. St-Germain, and the following City of Montréal staff: C. Guay, P. Legendre, and L. Vézina. The authors would like to thank also Mr. David Malcolm, from Malroz Engineering in Kingston (Ontario), for reviewing the manuscript.

REFERENCES

HEAST. 1995. Health Effects Assessment, Summary Tables. EPA 540/R-95-036 PB95-921199. United States Environmental Protection Agency, Office of Research and Development and Office ™of Emergency and Remedial Response, Washington, DC, May 1995.

IRIS. 1995. Integrated Risk Information System (IRIS). National Technical Information Service, Springfield, 1995.

Krewski, D., D. Murdoch, and A. Dewanji. 1986. Statistical Modeling and Extrapolation of Carcinogenesis Data. In: *Modern Statistical Methods in Chronic Disease Epidemiology*, pp. 259-282. (S.H. Moolgavkar and R.L. Prentice, Eds.). New York, John Wiley and Sons.

MEF. 1996a. Lignes directrices pour la réalisation des analyses de risques toxicologiques. Version préliminaire pour consultation. Gouvernement du Québec, Ministère de l'Environnement et de la Faune, Direction des laboratoires, Groupe d'analyse de risque, Sainte-Foy, Juin 1996.

MEF. 1996b. Guide technique pour la réalisation des analyses de risques toxicologiques. Version préliminaire pour consultation. Gouvernement du Québec, Ministère de l'Environnement et de la Faune, Direction des laboratoires, Groupe d'analyse de risque, Sainte-Foy, Juin 1996.

MEF. 1988. Contaminated Sites Rehabilitation Policy. Envirodoc 880100. Gouvernement du Québec, Ministère de l'Environnement, Direction des substances dangereuses, Sainte-Foy, February 1988.

USEPA. 1989. Risk Assessment Guidance for Superfund. Volume I: Human Health Evaluation Manual. Interim Final. OSWER Directive 9285.7-01a. United States Environmental Protection Agency, Office of Emergency and Remedial Response, Toxics Integration Branch, Washington, DC, September 1989.

CHAPTER 11

Dermal Bioavailability of Soil-Aged Nickel in Male Pig Skin in Vitro

Mohamed S. Abdel-Rahman, Pharmacology and Physiology Department, New Jersey Medical School, University of Medicine and Dentistry of New Jersey, Newark, New Jersey

Gloria A. Skowronski, Pharmacology and Physiology Department, New Jersey Medical School, University of Medicine and Dentistry of New Jersey, Newark, New Jersey

Rita M. Turkall, Pharmacology and Physiology Department, New Jersey Medical School, and Clinical Laboratory Sciences Department, School of Health Related Professions, University of Medicine and Dentistry of New Jersey, Newark, New Jersey

INTRODUCTION

Chemical bioavailability is a critical factor in toxicity. However, bioavailability studies of chemicals in soil are usually conducted on soil that has been treated with chemical immediately before exposure of the animal model. In actuality, chemicals diffuse over a period of time, from the external surface of soil particles to internal and more remote sites within the soil matrix so that chemicals become increasingly desorption-resistant, a process that has been termed "aging." The fraction of a chemical that is not aged is likely to disappear by microbial degradation, leaching, or volatilization (Alexander, 1995; Alexander et al., 1995).

In our laboratory, oral (in vivo) or dermal (in vivo and in vitro) studies showed that brief contact with a sandy or a clay soil significantly reduced the bioavailability of benzene (Abdel-Rahman and Turkall, 1988), m-xylene (Skowronski et al., 1990), trichloroethylene (Kadry et al., 1991), phenol (Skowronski et al., 1994), and arsenic (Abdel-Rahman et al., 1995). While this data is useful for estimating the risk from exposure to newly contaminated soil, the chemicals in hazardous waste sites have been in the soil for many years. Therefore, there is a definite need for assessing the effects of chemical aging on bioavailability. Nevertheless, there is only one known study in the literature that correlates bioavailability with the time that a chemical resided in soil. Investigators reported that less 2,3,7,8-tetrachlorodibenzo-p-dioxin (TCDD) was absorbed orally or dermally by rats from soil that had been in contact with the chemical for 8 d than for 10-15 h (Poiger and Schlatter, 1980). Other bioavailability studies have been conducted on TCDD in soil. However, it is not known from these studies if there was a decrease in bioavailability prior to the time the aged chemical in soil was tested (Shu et al., 1988; Umbreit et al., 1988; Lucier et al., 1986; and Bonaccorsi et al., 1984).

The purpose of this study was to compare the dermal bioavailability of nickel aged in soil for 6 mo versus nickel added immediately to soil and pure nickel (in the absence of soil). The key factors (e.g., soil composition, and temperature) that affect the bioavailability of the chemical were also examined. This information will be more accurate and can be used to derive environmentally acceptable endpoints (EAEs) for soils. An EAE is the concentration of a chemical in soil that does not adversely affect human health (Alexander et al., 1995). If chemical aging in soil reduces bioavailability, the risk of toxicity is decreased and less cleanup of a con-taminated site is necessary.

Nickel was studied because of its prevalence in the environment and because of its frequency of detection at hazardous waste sites. Although nickel is present naturally in the earth's crust and releases occur from windblown dust and volcanic eruptions, it is estimated that 42.5 million kg nickel/year are released into the envi-ronment from man-made sources, with most releases being to soil. People who live near or work at facilities that produce stainless steel and other nickel-containing alloys, oil- and coal-fired power plants, and incinerators, as well as persons who live in the vicinity or work at the sites that receive the wastes from these facilities, may be exposed to high levels of nickel in soil. Soil generally contains between 4 and 80 ppm of nickel. However, up to 9,000 ppm can be found near industries where nickel is extracted from ore (ATSDR, 1995a). The metal has been identified in at least one-half of the 1,400 hazardous waste sites that have been proposed for inclusion on the National Priorities List (NPL) of highest priority sites for possible remedial action (ATSDR, 1995b). It is also possible that nickel from wastes sites can leach and contaminate groundwater.

Several million workers worldwide are exposed to nickel. The most serious effects of nickel, cancer of the lung and nasal sinus, have occurred in people who have breathed nickel dust while working in nickel refineries or nickel processing plants. However, allergic contact dermatitis is the most common health effect of nickel in humans and once an individual is sensitized, even minimal contact with nickel by any route of exposure may elicit a reaction. Some sensitive individuals develop asthma following exposure to nickel (ATSDR, 1995a).

METHODS

Chemicals, Animals, and Soil

[63]Nickel chloride, 12.63 mCi/mg specific activity, 99.9% radiochemical purity, was purchased from DuPont-New England Nuclear (NEN) Products, Boston, MA.

Male York-Hampshire pigs (40-60 lb) were purchased from Cook College Farm, Rutgers University, New Brunswick, NJ.

Two soils were examined. The Atsion sandy soil containing 90% sand, 2% clay, and 4.4% organic matter was collected from an aquifer drill site of the Cohansey sand formation near Chatsworth in south central New Jersey. The Keyport clay soil containing 50% sand, 22% clay, and 1.6% organic matter was collected from the Woodbury formation near Moorestown in southwestern New Jersey. Both soils are representative of soil types widely distributed in the eastern, southern, and southwestern United States (USDA, 1977; USDA, 1972).

Chemical Aging in Soil

Radiolabeled nickel was added to each soil which had been oven-dried at 80°C overnight. The concentration was 2.4 ppm of nickel chloride in sandy or clay soil with ethanol as the vehicle. Chemical and soil were mixed thoroughly to ensure uniform distribution of nickel in soil and to allow the vehicle to evaporate. Spiked soil was then added to Teflon-sealed glass vials and stored in the dark at room temperature (21°-25°C) or refrigerated (2°-4°C) to determine the effect of temperature to account for seasonal and geographical variation on chemical aging. The test compound was aged for 6 mo.

Penetration Studies

Whole pig skin was obtained from the costo-abdominal areas of sacrificed pigs. After the skin was washed with water and lightly shaved, it was placed in ice-cold HEPES-buffered (25 mM) Hanks' balanced salt solution (HHBSS), pH 7.4, containing 1% dextrose and gentamicin sulfate (50 mg/L) and was transported on ice to the laboratory where it was immediately prepared for diffusion cells according to Bronaugh and Stewart (Bronaugh and Stewart, 1985). The skin was used 2-3 h after the sacrifice time.

Excised skin was cut to a thickness of 200 μm with a dermatome (Padgett Electro-Dermatome Model B, Padgett Instruments Inc., Kansas City, MO). Circular pieces of dermatomed skin were then mounted into Teflon flow-through diffusion cells (Crown Bio Scientific, Inc., Somerville, NJ). The exposed skin surface area was 0.64 cm². Diffusion cells were maintained at 35°C (yielding a skin surface temperature of 32°C) in an aluminum block holder, which was heated by a circulating water bath. The dermal side of each skin sample was bathed with receptor fluid at a flow rate of 5 mL/h by a multichannel peristaltic cassette pump (Manostat, New York, NY). The HHBSS receptor fluid, containing gentamicin sulfate (50 mg/L) and 10% fetal bovine serum (Sigma Chemical Co., St. Louis, MO) was aerated continuously with oxygen (Collier et al., 1989). After skin samples were allowed to equilibrate for 1 h, [63]nickel chloride was applied to the stratum corneum surface of the skin either alone in 10 μL of ethanol vehicle, immediately after the addition of sandy or clay soil, or aged in each of the two soils for 6 mo. The chemical dose was 113. 8

ng/cm^2 containing 0.92 µCi of radioisotope and the weight of the soil samples was 30 mg. Perfusate was collected in scintillation vials at 15 min intervals up to 1 h, at 1.5 and 2 h, then at 2 h intervals up to 16 h post-dosing, using an automated fraction collector.

At the conclusion of the 16 h study, unabsorbed chemical was washed off from the skin surface once with 1 mL of a 1% aqueous soap solution (Ivory Liquid, Procter & Gamble, Cincinnati, OH) and twice with 1 mL distilled water. Cell tops were rinsed in the same manner. Skin samples were completely solubilized in Solvable (Packard Instruments Co., Inc., Meriden, CT) to determine the binding capacity of the skin. Chemical that remained bound to soil was extracted by heating soil residues with 1 mL of 0.5 N hydrochloric acid in a boiling water bath for 0.5 h followed by shaking at room temperature for 1 h. Radioactivity in receptor fluids, skin digests, skin and cell washes, and soil residue extracts was counted in Formula 989 liquid scintillation cocktail (Packard) by liquid scintillation spectrometry (LS 7500, Beckman Instruments Inc., Fullerton, CA). Sample quench was corrected by using the H-ratio method.

Data Analysis

All data were reported as the mean $\pm$ SEM. Statistical differences between treatment groups were determined by one-way analysis of variance (ANOVA) with Scheffe's test except for soil residues which were determined by Student's independent t-test. A p value <0.05 was considered significant.

RESULTS

After nickel was aged in soil for 6 mo, the amount of soil-sorbed nickel that penetrated skin was significantly decreased by either soil compared to nickel in brief contact with soil or to chemical without soil. This was supported by the significantly lower amount of nickel penetrating into receptor fluid as percent of the initial dose of nickel which permeated skin within a designated time interval (Figures 1 and 2). Sixteen hours after skin was treated, the total percent of radioactivity in the receptor fluid for the chemical aged in sandy soil at 21°-25°C was 90% lower than for pure nickel or nickel added immediately to sandy soil (Figure 1). With clay soil, the percentage of radioactivity in the receptor fluid was also significantly reduced by 85% when nickel was aged versus pure nickel or soil treated immediately with nickel (Figure 2). However, the quantity of receptor fluid radioactivity was the same for pure nickel and nickel in contact with soil for a short time (Figures 1 and 2).

A summary of the amount of nickel-derived radioactivity that penetrated pig skin into receptor fluid as well as the binding capacity of nickel to skin is shown in Tables 1 and 2. The amount of radioisotope covalently bound to skin was four-fold lower after chemical aging at 21°-25°C versus nickel adsorbed to sandy soil for a short time but 20-fold lower than nickel alone (Table 1). The total amount

Figure 1.   **The effect of aging in sandy soil (21°-25°C) on the penetration of [63]nickel through male pig skin into receptor fluid. Skin was treated in vitro with 0.92 µCi [63]nickel alone, immediately after the addition of 30 mg soil, or aged in the same amount of soil for 6 mo. The chemical dose was 113.8 ng/cm^2. Values (means $\pm$ S.E.M.) represent the percent of the initial dose collected in the receptor fluid from 0 time until the indicated time from 10-17 replicates per treatment from 3-4 pigs.**

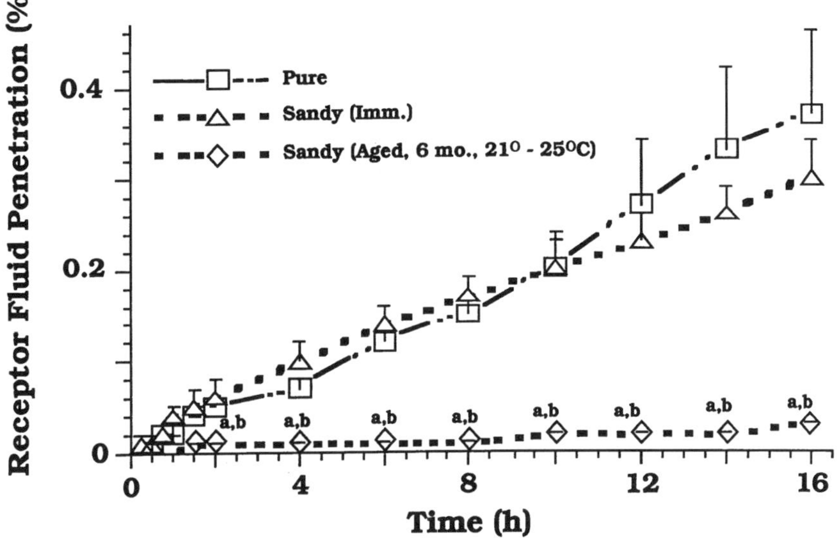

^{a} **Significantly different from pure nickel (p < 0.05)**
^{b} **Significantly different from nickel added immediately to sandy soil**

of chemical that is available for distribution to the body (total penetration) is the sum of the percent initial dose in the receptor fluid and bound to skin (Bronaugh et al., 1989). Total penetration was significantly decreased by 75% when nickel was aged in sandy soil than when nickel was added immediately to sandy soil. However, compared to pure nickel, total penetration was 95% lower after nickel aging (Table 1). At the same time, a significant decrease of nickel was observed in the skin wash at the end of the experiment. A major part of the nickel dose remains bound to soil. This fraction of the dose (42% for immediate addition to soil and 65% for 6 months aging) is dependent on the time the metal resides in the soil and indicates that the organic matter in sandy soil binds nickel more strongly with time. Similar results were recorded for clay soil. The skin binding capacity of nickel aged in clay soil was 4% of the initial dose relative to 12% when nickel was freshly added to clay soil and 58% for pure nickel (Table 2). Aging significantly decreased total penetration of nickel by 70% versus nickel added immediately to clay soil,

Figure 2.   The effect of aging in clay soil (21°-25°C) on the penetration of [63]nickel through male pig skin into receptor fluid. Skin was treated in vitro with 0.92 µCi [63]nickel alone, immediately after the addition of 30 mg soil, or aged in the same amount of soil for 6 mo. The chemical dose was 113.8 ng/cm^2. Values (means $\pm$ S.E.M.) represent the percent of the initial dose collected in the receptor fluid from 0 time until the indicated time from 11-17 replicates per treatment from 3-4 pigs.

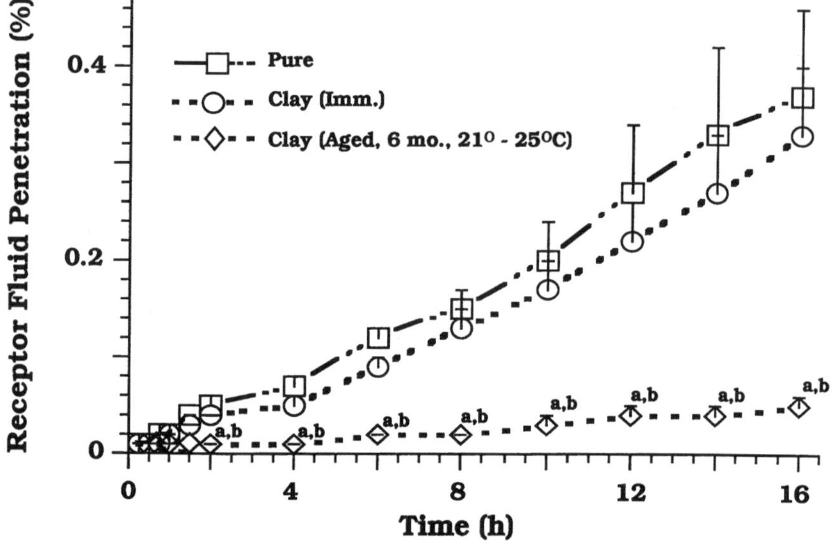

^{a} Significantly different from pure nickel ($p < 0.05$)
^{b} Significantly different from nickel added immediately to clay soil

and 95% for nickel alone. Since the bulk of the pure chemical penetrated skin, the amount of the radioisotope in skin wash was less (34%) than when chemical was aged (43%) or in contact with clay soil for a short time (73%). Conversely, more radioactivity was recovered from the clay soil residue after aging (49%) than after short-term treatment (17%). Less than 5% of the administered dose for all treatments was recovered in the cell washes (Tables 1 and 2). When a comparison was made to assess the total penetration of nickel aged in either soil at 21°-25°C and 2°-4°C, it was found that the dermal bioavailability of nickel was similar at both temperatures (Figure 3). Extending the aging time of nickel to 1 year did not further decrease the dermal bioavailability of the metal (data not shown).

DISCUSSION

Organic matter and the type and amount of clay minerals in soil are two of the key factors which are most likely to influence the retention of chemicals in soil. Although the bioavailability of nickel is similar for the two soils after aging, the

Table 1. ^{63}Nickel Penetration after Aging in Sandy Soil

	Pure Nickel	+Soil	
		Immediate	Aged (6 mo.)
Receptor fluid	0.4 ± 0.1[a]	0.3 ± 0.1	0.03 ± 0.0[b,c]
Bound to skin	57.6 ± 2.2	11.2 ± 1.0[b]	3.1 ± 0.7[b,c]
Total penetration	57.9 ± 2.2	11.5 ± 1.0[b]	3.1 ± 0.7[b,c]
Skin wash	34.3 ± 2.0	40.3 ± 3.5	27.5 ± 0.8[b]
Bound to soil	-	42.4 ± 3.5	64.7 ± 1.7
Cell wash	2.7 ± 0.6	3.5 ± 0.9	4.3 ± 0.6

[a] Values (mean ± S.E.M.) represent the percent of the initial dose recovered at the end of the 16 h study (n= 10-17 replicates per treatment from 3-4 pigs). Chemical was aged in sandy soil at 21°-25°C.

[b] Significantly different from pure nickel (p < 0.05)

[c] Significantly different from nickel added immediately to sandy soil

Table 2. ^{63}Nickel Penetration after Aging in Clay Soil

	Pure Nickel	+Soil	
		Immediate	Aged (6 mo.)
Receptor fluid	0.4 ± 0.1[a]	0.3 ± 0.1	0.05 ± 0.0[b,c]
Bound to skin	57.6 ± 2.2	12.1 ± 2.0[b]	3.7 ± 0.5[b,c]
Total penetration	57.9 ± 2.2	12.4 ± 2.0[b]	3.8 ± 0.5[b,c]
Skin wash	34.3 ± 2.0	73.4 ± 2.8[b]	43.2 ± 2.4[b,c]
Bound to soil	-	17.0 ± 2.8	49.4 ± 2.6[c]
Cell wash	2.7 ± 0.6	0.6 ± 0.2	2.2 ± 0.7

[a] Values (mean ± S.E.M.) represent the percent of the initial dose recovered at the end of the 16 h study (n= 10-17 replicates per treatment from 3-4 pigs). Chemical was aged in clay soil at 21° -25°C.

[b] Significantly different from pure nickel (p < 0.05)

[c] Significantly different from nickel added immediately to clay soil

higher amount of nickel remaining bound to sandy soil than to clay soil suggests that sandy soil, having a greater organic content, was a more effective adsorbent of nickel than clay soil. However, pH is also an important parameter in the retention of nickel. When Harter (1983) studied the adsorption and desorption of nickel from pH adjusted soils, he found that the quantity of nickel retained was dependent upon the pH of the soil sample. The retention of nickel by soil increased when

Figure 3.   Temperature effects on the total penetration of [63]nickel aged in soil. Skin was treated in vitro with 0.92 µCi [63]nickel alone, immediately after the addition of 30 mg soil, or aged in the same amount of soil either at 21°-25°C or 2°-4°C for 6 mo. The chemical dose was 113.8 ng/cm^2. Values (means ± S.E.M.) represent the sum of the initial dose in the receptor fluid and bound to skin 16 h after dosing from 10-17 replicates per treatment from 3-4 pigs.

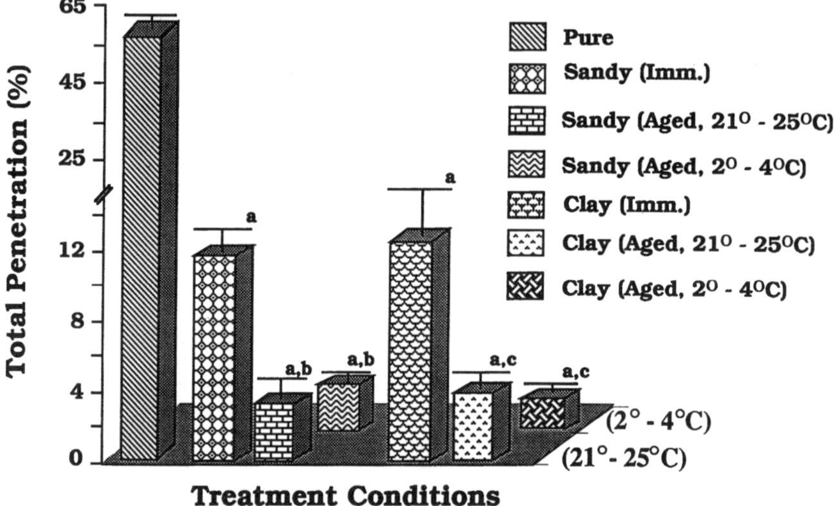

a Significantly different from pure nickel (p < 0.05)
b Significantly different from nickel added immediately to sandy soil
c Significantly different from nickel added immediately to clay soil

the pH was at or greater than 7. Similarly, after King examined the factors that affect metal retention of 13 soil series representing a range of soils in the southeastern United States, he concluded that soil pH was the most important factor affecting adsorption of nickel (King, 1988).

It is evident from the results of this study that the reduction in nickel bioavailability after aging will have a direct impact on regulatory decisions regarding the risk from exposure and the need for soil remediation. The risk from dermal exposure to pure nickel is high because more than 50% of the nickel dose forms a reservoir in skin from which it can be absorbed into the systemic circulation. Recent contamination of soil with nickel lowers the risk, while aging significantly decreases the risk even more. There is an inverse relationship between bioavailability and the EAE. The lower the bioavailability of chemical, the higher is the EAE. For example, for the purpose of discussion if an appropriate EAE for pure nickel might be 5 ppm, the EAE would increase to 20 ppm and 100 ppm, respectively, for nickel added immediately to soil and aged in soil based on the approximately 4-fold and 20-fold decrease in bioavailability for the respective treatments (Tables 1 and 2). Such differences are important because they will affect the health hazard and the cleanup of contaminated sites.

ACKNOWLEDGMENT

This research was supported in part through funding from the Hazardous Substance Management Research Center and the New Jersey Commission on Science and Technology.

REFERENCES

Abdel-Rahman, M.S., and Turkall, R.M. 1988. Determination of Exposure of Oral and Dermal Benzene from Contaminated Soils. In: *Petroleum Contaminated Soils,* Volume 1, pp. 301-311. (Kostecki, P.T. and Calabrese, E.J., Eds.). Chelsea, MI, Lewis Publishers.

Abdel-Rahman, M.S., Skowronski, G.A., Kadry, A.M., and Turkall, R.M. 1996. Soil Decreases the Dermal Bioavailability of Arsenic in a Chemical Mixture in Pig. In: *Contaminated Soils,* Volume 1, pp. 461-472. (Calabrese, E.J., Kostecki, P.T., and Bonazountas, M., Eds.). Amherst, MA, Amherst Scientific Publishers.

Alexander, M. 1995. How Toxic Are Toxic Chemicals in Soil? *Environ. Sci. Technol.* 29, 2713-2717.

Alexander, M., Goldstein, L., Pauwels, S., Edwards, D., Zaborsky, O., Menzie, C., Heiger-Bernays, W., Montgomery, C., Nakles, D., Loehr, R., and Webster, M. 1995. *Draft Report for Gas Research Institute Workshop on Environmentally Acceptable Endpoints in Soil: Risk - Based Approach to Contaminated Site Management Based on Availability of Chemicals in Soil.* Washington, DC.

(ATSDR) Agency for Toxic Substances and Disease Registry. 1995a. HazDat Database, U.S. Department of Health and Human Services, Public Health Service, Atlanta, GA.

(ATSDR) Agency for Toxic Substances and Disease Registry. 1995b. Toxicological Profile (Draft) for Nickel, pp. 1-9,141-178. U.S. Department of Health and Human Services, Public Health Service, Atlanta, GA.

Bonaccorsi, A., di Domenico, A., Fanelli, R., Merli, F., Motta, R., Vanzate, R., and Zapponi, G.A. 1984. The Influence of Soil Particle Adsorption on 2,3,7,8,-Tetrachlorodibenzo-p-Dioxin Biological Uptake in the Rabbit. *Arch. Toxicol.* Suppl. 7, 431-434.

Bronaugh, R.L., and Stewart, R.F., 1985. Methods for *In Vitro* Percutaneous Absorption Studies. IV. The Flow-Through Diffusion Cell. *J. Pharm. Sci.* 74, 64-67.

Bronaugh, R.L., Stewart, R.F., and Storm, J.E. 1989. Extent of Cutaneous Metabolism During Percutaneous Absorption of Xenobiotics. *Toxicol. Appl. Pharmacol.* 99, 534-543.

Collier, S.W., Sheikh, N.M., Sakr, A., Lichtin, J.L., Stewart, R.F., and Bronaugh, R.L. 1989. Maintenance of Skin Viability During *In Vitro* Percutaneous Absorption Metabolism Studies. *Toxicol. Appl. Pharmacol.* 99, 522-533.

Harter, R.D. 1983. Effect of Soil pH on Adsorption of Lead, Copper, Zinc, and Nickel. *Soil Sci. Soc. Am. J.* 47, 47-51.

Kadry, A.M., Skowronski, G.A., Turkall, R.M., and Abdel-Rahman, M.S. 1991. Kinetics and Bioavailability of Soil-Adsorbed Trichloroethylene in Male Rats Exposed Orally. *Biol. Monit.* 1, 75-86.

King, L.D., 1988. Retention of Metals by Several Soils of the Southeastern United States. *J. Environ. Qual.* 17, 239-246.

Lucier, G.W., Rumbaugh, R.C., McCoy, Z., Hass, J., Harvan, R.D., and Halbro, P. 1986. Ingestion of Soil Contaminated with 2,3,7,8,-Tetrachlorodibenzo-p-Dioxin (TCDD). *Fund. Appl. Toxicol.* 6, 364-371.

Poiger, H. and Schlatter, C. 1980. Influence of Solvents and Adsorbents on Dermal and Intestinal Absorption of TCDD. *Food Cosmet. Toxicol.* 18, 477-481.

Shu, H.D., Paustenbach, J., Murray, J., Marple, L., Brunk, B., Dei Rossi, D., Webb, A.S., and Teitelbaum, T. 1988. Bioavailability of Soil-Bound TCDD: Oral Bioavailability in the Rat. *Fund. Appl. Toxicol.* 10, 648-654.

Skowronski, G.A., Turkall, R.M., Kadry, A.M., and Abdel-Rahman, M.S. 1990. Effects of Soil on the Dermal Bioavailability of m-Xylene in Male Rats. *Environ. Res.* 51, 182-193.

Skowronski, G.A., Kadry, A.M., Turkall, R.M., Botrous, M.F., and Abdel-Rahman, M.S. (1994). Soil Decreases the Dermal Penetration of Phenol in Male Pig *In Vitro. J. Toxicol. Environ. Health.* 41, 467-479.

Umbreit, T.H., Hesse, E.J., and Gallo, M.A. (1988). Bioavailability and Cytochrome P-450 Induction from 2,3,7,8-Tetrachlorodibenzo-p-Dioxin Contaminated Soils from Times Beach, Missouri and Newark, New Jersey. *Drug Chem. Toxicol.* 11, 405-418.

(USDA) U.S. Department of Agriculture. 1972. National Cooperative Soil Survey: Official Series Description, Keyport Series, Soil Conservation Service, Washington, DC.

(USDA) U.S. Department of Agriculture. 1977. National Cooperative Soil Survey: Official Series Description. Atsion Series, Soil Conservation Service, Washington, DC.

CHAPTER 12

Cost-Effective Remediation Through the Establishment of Background

John K. Ghirardi, Ian M. Phillips, and **Ileen S. Gladstone**, GEI Consultants, Inc., Winchester, Massachusetts

INTRODUCTION

This paper will focus on the establishment of background as a means of achieving regulatory closure. The authors present a case study illustrating that when regulated by risk-based cleanup standards, the establishment and attainment of background contaminant concentrations can be a cost-effective means of achieving regulatory closure. Attainment of background is particularly effective in urban settings where contamination may be attributed to conditions other than from a contaminant release such as atmospheric deposition or historical fill materials.

Background concentrations of contaminants are those concentrations that are omnipresent in the local environment and are not attributable to a particular release of hazardous materials. The Massachusetts Department of Environmental Protection further explains that elevated levels of contaminants may be a result of "geologic or ecologic conditions, atmospheric deposition of industrial process or engine emissions, fill materials containing wood or coal ash, and/or petroleum residues that are incidental to the normal operation of motor vehicles" (310 CMR 40.0006).

Establishing that a site contains contaminant concentrations that are consistent with the background contaminant concentrations means that the site does not pose any more risk than the surrounding area, given similar uses. If the site does not pose additional risk, the regulatory end point for remediation can be established or achieved.

There are four steps involved in establishing and attaining background as a cleanup goal: (1) establish background concentrations and distributions; (2) conduct remedial actions as necessary; (3) construct a compliance data set; and (4) test the compliance distribution against the background distribution. Figure 1 is a flow chart of this process. This paper will describe these steps in detail and conclude with a case study to illustrate the application of the proposed method and source of potential cost savings.

Figure 1. **Establishing and attaining background as a cleanup goal.**

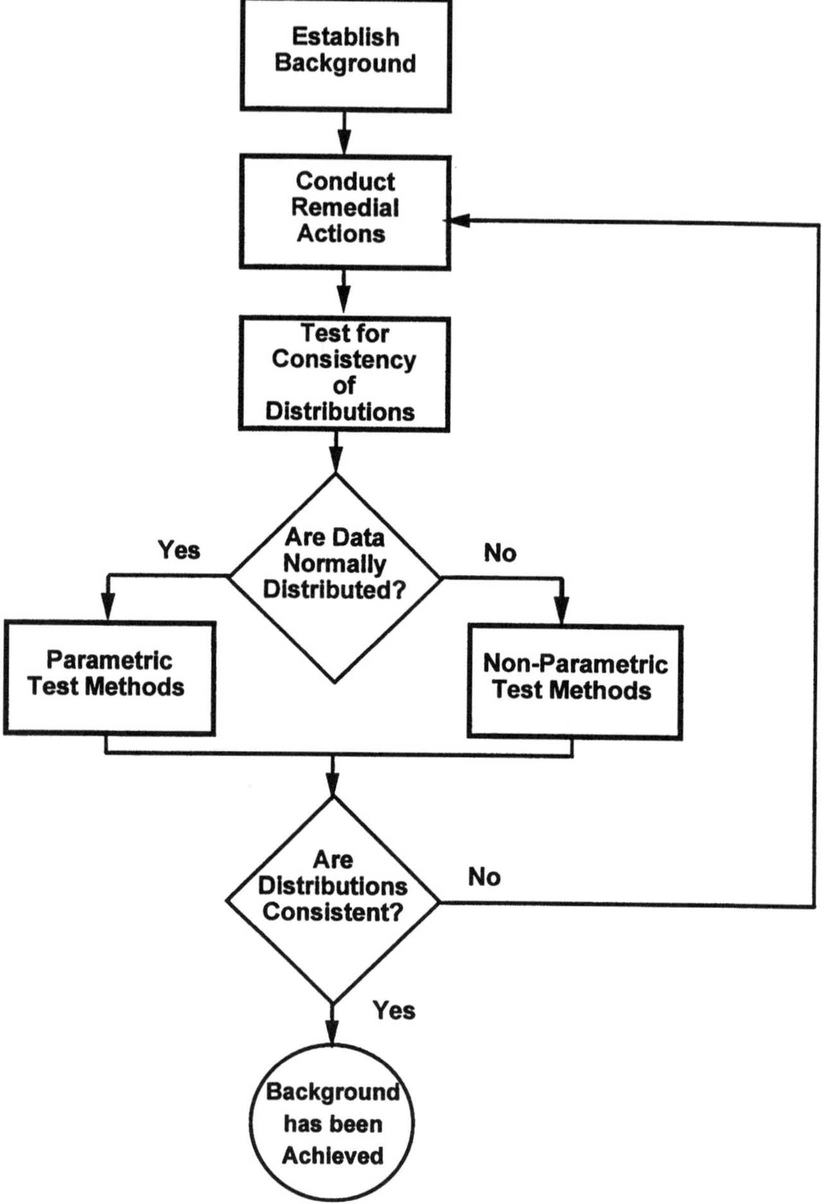

ESTABLISHING BACKGROUND

The establishment of background may be further broken down into two subtasks (Figure 2): (1) Data Sources and Comparability; and (2) Statistical Analysis.

Figure 2.   **Establishment of background.**

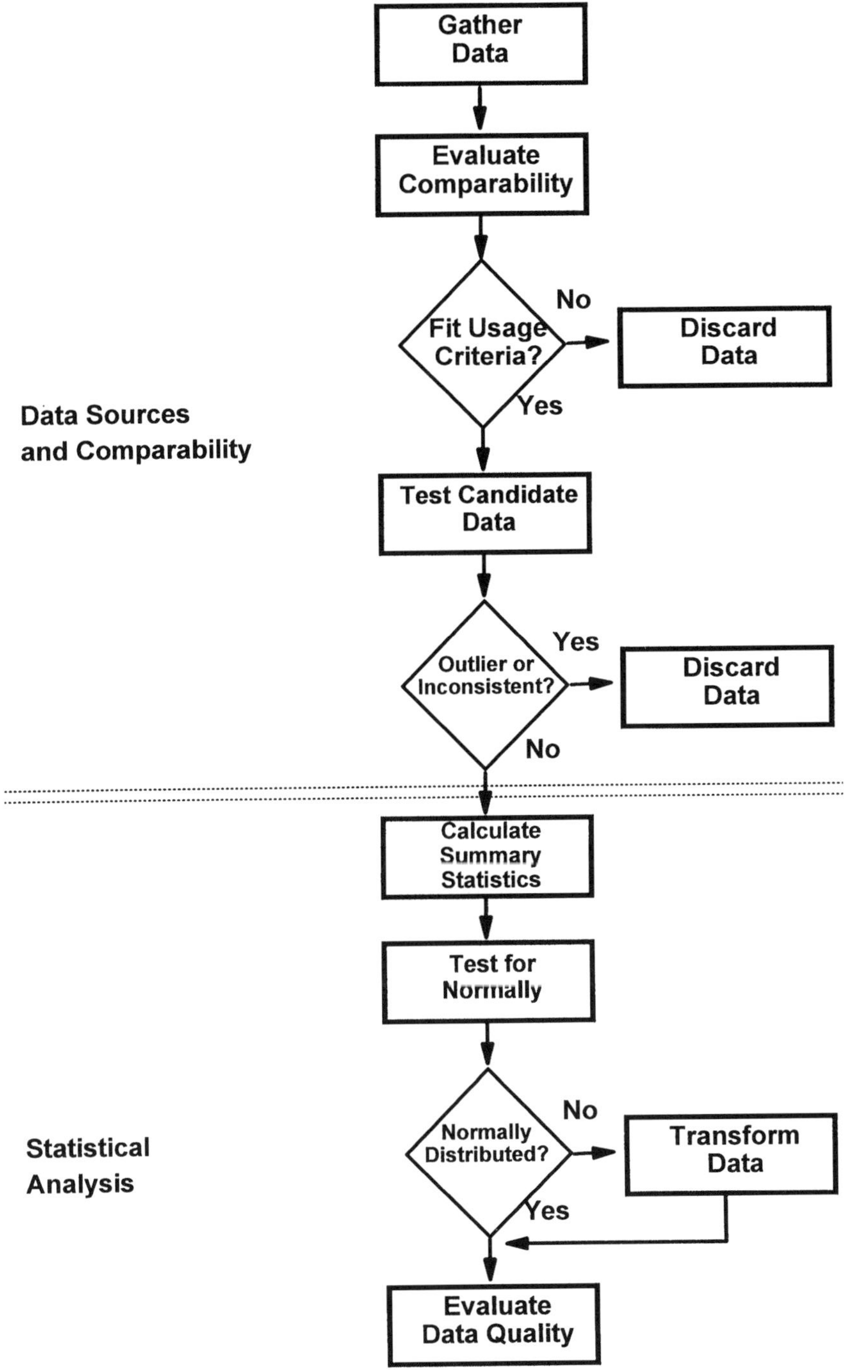

Data Sources and Comparability

In general, background areas are upgradient of the area of interest, have no historical releases of the contaminant of concern, and have comparable geologic settings to the compliance area. Sources of data may include nearby due diligence real estate investigations, hazardous waste site investigations, published values, and field investigations conducted for the sole purpose of establishing background.

The data set gathered from areas considered to be background must be reviewed using preestablished criteria to judge comparability. Differences in the sampling and analysis procedures, heterogeneity in the media, and contaminant releases can introduce variability in the measured concentration and must be minimized. All data that do not comply with the preestablished criteria are eliminated from further consideration. Data comparability criteria used to establish background should be based on physical characteristics of the environment and on the methods of data collection and analysis. For example, when evaluating the comparability of environmental aspects, the factors to be considered include: medium, stratigraphy, depositional environment, depth, and physical properties. Comparability is also influenced by how the data are collected. To be used, the data should be gathered using the same or similar collection procedures and followed by similar analysis methods. Comparability of collection times may also be an important consideration.

Once the data are gathered and screened for comparability, the background data set as a whole must be reviewed for outliers and other inconsistencies. Outliers are those data that seem to lie at concentrations much greater or much less than the remainder of the data set. Outliers may be extreme values of the background distribution or may not be part of the background distribution at all. It is more protective of human health to remove outlying values of high concentration from the background data set, since such values will increase the mean and variability. Plotting the data can suggest which data are outliers, and numerical tests may be performed to confirm these suspicions.

Statistical Analysis

The summary statistics can be calculated from the background data set after the removal of outliers. Most spreadsheet programs will calculate summary statistics including: mean, minimum, maximum, variance, standard deviation, skewness, and coefficient of variation. The summary statistics are used to evaluate the normality and the significance of the variability of the data, and to test the compliance distribution against the background distribution.

It is also important to test whether the data are normally distributed because the use of many common hypothesis test methods is predicated on the assumption of normality of the data. For normal distributions, 84% and 98% of the population lies within one and two standard deviations of the mean on a cumulative basis, respectively. A rule-of-thumb proposed by the authors to judge the significance of the magnitude of the variability for environmental data is based on the coefficient of variation. If the coefficient of variation of the data set is greater than 1, there is a large amount of variability in the data. In such a case, the comparability of the background data set must be reviewed again with a more critical eye to evaluate the source(s) of variability. It may be true that the background distribution has a large variability due to the nature of the contaminant (i.e., immobile, stable compounds like PCBs), the depositional mechanism, or some noncomparable data that is adding artificial variability to the data set.

The final statistical parameters that should be considered are confidence, power, and minimum detectable relative difference (MDRD). In this case, the confidence and power are measures of the probability of correctly concluding that the site concentrations are consistent with background and the probability of correctly concluding that the site concentrations are greater than background, respectively. The minimum detectable relative difference is the smallest difference that can be detected between the site and background contaminant levels. The EPA has published guidance on the minimum acceptable values for confidence, power, and MDRD in the Guidance for Data Useability in Risk Assessment (EPA, 1992). If the number of background data does not provide the desired confidence, power, and MDRD, more data must be collected.

CONSTRUCTING A COMPLIANCE DATA SET

Once the remediation has been conducted, the compliance site must be tested for consistency with background. In order to conduct such a test, the compliance data set must be scrutinized in the same manner as the background data set (i.e., ensuring data are comparable, testing data for outliers and other inconsistencies, and calculating and evaluating the statistical parameters of the compliance data set). It is important to use sample collection and analysis procedures that will ensure comparability and minimize the potential for artificial variability.

TESTING THE COMPLIANCE DISTRIBUTION AGAINST THE BACKGROUND DISTRIBUTION

There are a number of parametric and nonparametric statistical methods that can be used to test whether the compliance data set is consistent with background. Commonly-used parametric methods include the Student t-test and z-test and can be found in any basic statistics text. Parametric techniques should only be used with normally-distributed populations or a "large" sample size. For distributions other than normal, nonparametric methods should be used. Commonly-used nonparametric methods include the sign test, the Wilcoxon (1945) signed-rank test, and the Fisher (1935) permutation test.

CASE STUDY

Site Information

The site is located in an urban industrial setting in the Boston area. A program of soil borings and test pits was conducted for the foundation design of a new manufacturing building. From the ground surface downward, the soils at the site consisted of the following strata: miscellaneous urban fill, organic deposits, silt, clay, glacial till, and bedrock. The urban fill layer ranged from about 7 to 21 feet thick and consisted mainly of sand and gravel with ash, cinders, wood chips, and other debris. Laboratory analysis of the soil samples collected in the urban fill layer indicated lead was present in a discrete portion of the site at concentrations as high as 51,600 mg/kg.

Remedial actions were conducted to remove the area of elevated lead concentrations from the site. The goals were to achieve a condition of no significant risk by reducing the concentrations of lead in the environment to background levels.

Data Sources and Comparability

Lead concentration data were gathered from five nearby areas to evaluate the lead concentration distribution in background soils. The data sources were:

- the geotechnical investigation for the building foundation design;
- soil characterization for the dry well removal from an adjacent building;
- a portion of the right-of-way characterization for the Central Artery; and
- two investigations pursuant to real estate transactions for adjacent properties.

To create one background data set from the several investigations, the following comparability criteria for data use were established:

- include all samples that were collected from the fill layer;
- include all samples that were collected using split spoon sampling methods from an interval of 2 feet or less or composited by weight from a larger interval;
- include all samples that were analyzed by EPA-approved test methods at DEP-certified laboratories; and
- use duplicate samples as separate data points since each value is equally likely.

Statistical Analysis

After applying the comparability criteria to the data gathered from the five nearby areas, a background data set was constructed for the fill layer consisting of 74 values (see Table 1). The background data set was then reviewed for outliers and other inconsistencies. Based on the history of the site and surroundings, none of the locations were known to be sources or to contain releases of lead. However, two of

Table 1. Background Data on Lead in Fill

Point #	Concentration (mg/kg)	Depth (ft)	Soil Description
1	< 7.00	5-7	Sandy Silt
2	< 7.00	5-7	Silty Sand
3	< 8.00	14-16	Silty Sand
4	7.87	4-6	Sand, Ash & Brick
5	16.0	0-5.4	Fill, Clay, Gravel & Rubble
6	25.0	5-7	Silty Clay
7	36.0	0.5-2.5	Silty Sand
8	45.0	0.5-2.5	Sand
9	46.0	7-8	Silty Sand
10	57.0	5-7	Sand, Ash & Brick
11	66.0	2-4	Sand, Ash, Coal & Gravel
12	76.0	11-13	Silty Gravel
13	76.3	5-7	Sand, Cinders & Gravel
14	80.0	0.5-2.5	Silty Sand
15	82.0	7-8	Sand, Ash & Brick
16	85.0	1-3	Silty Sand
17	86.6	10-12	Sand, Brick & Shells
18	103	5-6	Sand
19	110	4-8	Sand & Gravel
20	110	5-7	Sand, Cinders & Gravel
21	114	1.5-3.5	Sand & Gravel
22	122	0.5-9	Fill Composite
23	133	0.5-2.5	Sand, Ash & Gravel
24	141	0.3-2.3	Silty Sand
25	150	4-6	Silty Sand, Ash & Brick
26	154	0.5-2.5	Silty Sand
27	168	4.5-6.5	Sand, Clay & Gravel
28	170	0.5-2.5	Silty Sand
29	185	5-7	Sand, Ash & Brick
30	185	4-6	Sand, Ash & Brick
31	188	2-4	Sand, Gravel & Silt
32	211	7-9	Cinders & Sand
33	245	0.5-9	Fill Composite
34	245	2-4	Sand, Gravel & Silt
35	250	9-11	Silty Sand, Gravel & Shells
36	256	4.5-6.5	Sand, Clay & Gravel
37	260	1-3	Sand
38	282	1-3	Sand & Brick
39	284	1-3	Sand & Brick
40	298	0.5-2.5	Sand, Ash & Brick
41	317	0.5-2.5	Sand & Ash
42	320	0-5.4	Fill, Clay, Org Silt & Scrap Metal
43	340	14-16	Silty Sand, Gravel & Wood
44	373	5-7	Silty Sand
45	381	1-3	Sand
46	390	2-4	Sand, Ash, Coal & Gravel
47	403	4.5-6.5	Sand, Brick, Coal & Wood
48	408	4-6	Sand, Ash, Coal & Shells
49	435	2-4	Sand, Ash, Coal & Gravel
50	534	10-12	Sand & Leather
51	553	4.5-6.5	Sand, Brick, Coal & Wood
52	562	5-7	Sand, Cinders & Gravel
53	605	5-7	Silty Sand
54	634	8-9	Sand, Ash & Brick
55	686	24-26	Sand
56	722	5-7	Sand, Cinders & Wood
57	843	7-9	Cinders & Sand
58	858	5-7	Silty Sand & Shells
59	865	2-4	Sand, Ash, Coal & Gravel
60	895	5-7	Sand, Cinders & Wood
61	916	9-11	Silty Sand, Clay & Peat
62	920	0-9	Fill, Silty Sand & Scrap Metal
63	1,010	8-9	Sand, Ash & Brick
64	1,110	4-6	Sand, Ash, Coal & Shells
65	1,340	5-7	Silty Sand
66	1,350	7-8	Sand, Ash & Brick
67	1,500	4-6	Sand & Ash
68	1,730	5-7	Sand, Cinders & Wood
69	1,760	9-11	Silt, Gravel & Wood
70	2,700 (a)	10-12	Silty Sand, Brick & Shells
71	2,910 (a)	1-7	Fill Composite
72	4,410 (a)	5-7	Sand, Cinders & Wood
73	11,000 (a)	5-7	Sand, Brick & Gravel
74	12,100 (a)	1-7	Fill Composite

(a) These data tested as outliers.

the data points (11,000 and 12,100 mg/kg) appeared to be outliers since neither the results of duplicate analysis nor the remainder of the data points support these two data points. The results of outlier testing using EPA's SCOUT software marked the five points of greatest concentration as outliers, suggesting that these points were biasing the background concentration. These five outlying data points were removed from the background data set.

Results of normality testing using SCOUT indicated that the background data were not normally distributed. To obtain normally-distributed data, the background data set was transformed using the square root transformation:

$$X' = \sqrt{X - \min(X) + 1x10^{-5}}$$

where: X' is the transformed concentration,
          X is the measured concentration, and
          min(X) is the minimum concentration value in the data set.

The transformed background data set passed the Kolmogorov-Smirnov test of normality ($\alpha = 0.05$). Figure 3 is a probability plot of these data.

The summary statistics of the transformed background data set are presented in Table 2. The statistical parameters of power, confidence, and minimum detectable relative difference (MDRD) were also examined. After setting the desired power to 95% ($\beta = 0.05$), the confidence level and the MDRD of the transformed background data set were 85% and 19.5%, respectively. The power, confidence,

Figure 3. **Probability plot of transformed background data.**

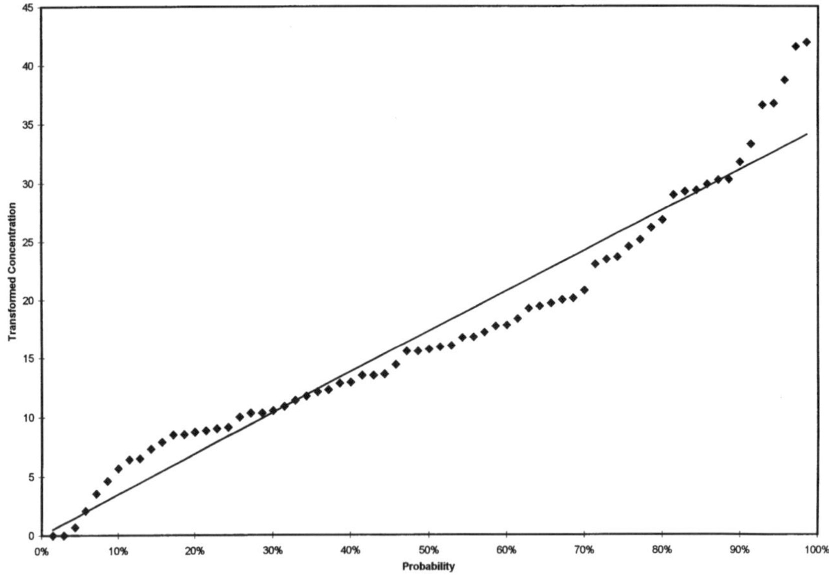

and MDRD are all within acceptable limits (EPA, 1992). Therefore, detected lead concentrations at the site had to be greater by 19.5% in order to be distinguished from background.

Table 2. Comparison of Summary Statistics [a,b]

Statistic	Background Data Set			Compliance Data Set	
	All Data	Without Outliers	Transformed	All Data	Transformed
Average	824.86	404.63	17.26	712.40	23.86
Variance	3,781,765.07	183,893.00	104.68	206,867.60	154.34
Standard Deviation	1,944.68	428.83	10.23	454.83	12.42
Coefficient of Variation	2.36	1.06	0.59	0.64	0.52
Skewness	4.85	1.54	0.57	-0.94	-1.17
Kurtosis	25.04	1.88	-0.26	-1.26	-0.35
Minimum	3.50	3.50	0.00	4.00	0.00
Maximum	12,100.00	1,760.00	41.91	1,090.00	32.95
Range	12,096.50	1,756.50	41.91	1,086.00	32.95
Count	74	69	69	10	10

[a] Nondetects are assigned a value of ½ of the detection limit

[b] Transformation = SQRT(Xi-Xmin+1e-5)

Remedial Actions

Employing the background data set and the associated statistics, soils from the site were excavated, stockpiled, and analyzed until the distribution of lead concentrations in the remaining soils was consistent with the background distribution. To meet this standard, approximately 290 cubic yards of soil were disposed of off-site.

Construction of the Compliance Data Set

The data set for samples collected from the side walls and bottom of the excavation at the completion of the remedial actions included 10 values and represents the distribution of lead concentration remaining at the site. These data are referred to as the compliance data set (see Table 3). Lead concentrations in the soils of the excavation limits ranged from less than the detection limit of 8 mg/kg to 1,090 mg/kg.

Table 2 contains the summary statistics of the compliance data set. Although the data did not pass the normality tests (due to too few data points), use of the normal distribution was assumed to be valid since the background data set showed that the lead concentration in this vicinity can be modeled by the normal distribution. The data were then transformed for comparability to the background distribution. The summary statistics of the transformed compliance data set were used to calculate a representative normal distribution.

Testing the Compliance Distribution Against the Background Distribution

The compliance distribution was compared with the background distribution to evaluate whether or not the concentrations of lead remaining at the site were consistent with background using the t-test for differences in the means of two

Table 3.   **Compliance Data Set**[a]

Point #	Concentration (mg/kg)	Depth (ft)
1	<8	7.5
2	87	10
3	104	7.5
4	781	10
5	978	11
6	1,000	10
7	1,000	5
8	1,020	11
9	1,060	7
10	1,090	5

[a] All data were collected from the miscellaneous urban fill

populations with unknown and unequal variances (Haan, 1977). Test results indicate that the lead distribution in the site soils is statistically consistent with the background distribution even though the mean of the compliance data set is higher than that of the background data set. Figure 4 contains a graphical comparison of the background and compliance distributions.

Remediated Soils

A total of 290 cubic yards of soil were removed from the site for recycling because the lead concentrations were not consistent with background. Concentrations of lead in these soils ranged above 1,200 mg/kg as high as 51,600 mg/kg. Approximately 750 cubic yards of soil were reused at the site because the lead concentrations were consistent with background. Lead concentrations in the soils reused at the site ranged from 708 to 1,200 mg/kg.

CONCLUSIONS

Based solely on the volume of soil that was reused on-site from the excavation with concentrations of lead between 300 mg/kg (the regulatory limit) and 1,200 mg/kg, the authors estimate a savings of $50,000 in off-site disposal costs. However, additional benefits of establishing and attaining background were also important, but were more difficult to quantify. The use of background allowed us to cease excavation before we reached published Massachusetts cleanup standards of 300 or 600 mg/kg. The established distribution indicated that most of the background values were higher than 300 mg/kg, and a significant portion were higher than 600 mg/kg for the urban fill material in the vicinity of the site. An additional benefit was that the background distribution is applicable for any future encounters with soils having lead concentrations that exceeded 300 to 600 mg/kg at the site.

Figure 4. **Background and compliance distributions.**

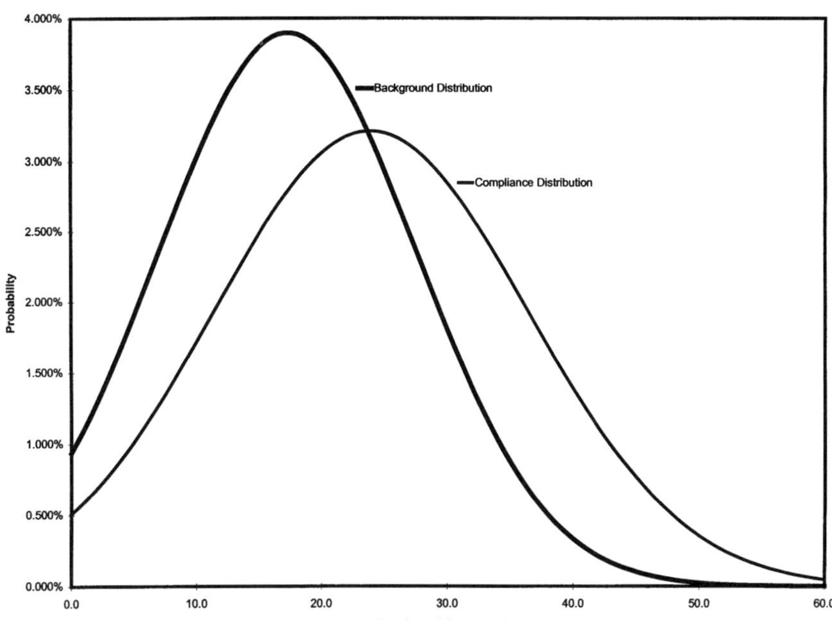

REFERENCES

Haan, C.T. 1977. Chapter 8, Hypothesis Testing. Statistical Methods in Hydrology, pp. 168-178. Ames, IA, Iowa State Press.

Massachusetts Department of Environmental Protection. 1996. Massachusetts Contingency Plan, 310 CMR 40.0000. Boston, MA, Bureau of Waste Site Cleanup.

U.S. Environmental Protection Agency. SCOUT software program. Las Vegas: Environmental Monitoring Systems Laboratory.

U.S. Environmental Protection Agency. 1992. Guidance for Data Useability in Risk Assessment. Washington, D.C. Office of Emergency and Remedial Response.

CHAPTER 13

Ecological Risk Assessment at an Alaska Airport

Jeffrey A. Peterson, **Mooncalf Reilly** and **Thomas L. Foster**, EMCON, Portland, Oregon

Priscilla Anderson Zieber, EMCON, Bothell, Washington

INTRODUCTION

Potential ecological effects of historic releases of a variety of chemicals (e.g., petroleum products, DDT, PCBs) to soil, sediment, and surface water bodies near an Alaskan airport were evaluated using assessment procedures consistent with recent U.S. Environmental Protection Agency (USEPA) guidance for ecological risk assessment (USEPA, 1992). This common-sense approach involves the following interrelated steps: (1) create specific assessment objectives based on a preliminary review of available data, (2) identify ecological receptors and model exposure to chemicals in nature, (3) review toxicity data and estimate the ecological effects of chemical exposure, and (4) integrate the exposure and effects assessments to predict the likelihood of adverse effects.

Potential chemical effects on four suites of ecological receptors were assessed using slightly different assessment methods. Risks to wildlife (birds and mammals) were assessed by modeling potential chemical effects on correlates of fitness (mortality, growth, reproduction) of individual members of select indicator species. The quotient method (QM), which evaluates the ratio of the ecological receptor intake to a "safe" dose, was used to characterize risks chemical exposures may pose to wildlife. Ratios above 1.0 suggest a potential for adverse effects. Risks to fish and other aquatic organisms were assessed by comparing waterborne chemical concentrations with either USEPA ambient water quality criteria for the protection of freshwater biota (WQC), or water-risk based concentrations (WRBCs) developed for chemicals with no established criteria using information from standardized aquatic toxicity tests. Potential risks to benthic invertebrates and other sediment-dwelling organisms were evaluated by comparing chemical concentrations in sediment with a variety of sediment quality criteria (SQC) or guidelines developed by a number of regulatory agencies (WDE, 1995). Similarly, risks to vascular plants were estimated by comparing chemical concentrations in soil with soil risk-based concentrations (SRBCs) estimated using results of plant toxicity tests. Very little published information was available describing chemical effects on plants, so plants underwent a limited evaluation.

The assessment focused on estimating the direct effects of chemicals to individual members of wildlife species, but did not attempt rigorous estimates of potential population-level effects. Defensible estimates of chemical impacts on vertebrate populations require information about population size, mortality rates, and birth rates, which are usually available for only a few managed fish and game species. Also, the highest quality and most plentiful data available are from studies of the effects of chemical exposures on the health and reproduction of individuals. However, information on typical home range sizes and densities of indicator species were compiled to aid in qualitative assessments of the number of wildlife individuals that may be exposed to contaminated environmental media.

Below we give a description of the general process used to evaluate risks that exposure to contaminated media may pose for several types of ecological receptors using both terrestrial and aquatic habitats. Predicting the effects of perturbations to any complex ecological system is wrought with uncertainty. Neither the theoretical constructs, nor the empirical relationships, are available to accurately predict the effects of diffuse, historical chemical releases to ecosystems. While the ecological assessment synthesizes much of the available exposure and ecological effects data, all estimates of potential effects should be viewed with healthy skepticism.

COPC Selection

Some metals were screened out as chemicals of potential concern (COPCs) in soil, surface water, or sediment based on comparisons with background concentrations. Chemicals that did not exceed WQC or SQC were excluded as COPCs in surface water and sediment, respectively.

Fractionation Approach for Petroleum-Impacted Media

Mixtures of chemicals such as gasoline range organics (GRO) and diesel range organics (DRO) were not analyzed for individual constituents because of technical and economic constraints. Instead, surrogate chemicals were selected to represent the chemical, physical, and toxicological properties of fractions in each mixture. Petroleum products were separated into fractions by chemical structure and size (i.e., aliphatic or aromatic compounds, and a variety of carbon chain lengths). Surrogate chemicals were chosen based on the availability of published human health toxicity data and chemical and physical properties. When ecological toxicity data were not available for some surrogate chemicals, toxicity data for structurally similar chemicals were used instead. It was conservatively assumed that the composition of weathered petroleum products in soil would be the same as the composition of the fresh, unweathered petroleum products.

Fate and Transport Modeling

Some chemically-impacted aquifers were hydraulically connected to surface waters, and groundwater from these aquifers was modeled from each source to the nearest surface water body. Runoff of contaminated surface soil to surface water bodies and partitioning between surface water and sediment were also modeled.

Ecological Receptors

Both aquatic and terrestrial species inhabit the contaminated sites. As a result, two food chains were developed, one including terrestrial receptors and the other freshwater aquatic receptors. Lists of potential ecological receptors, including major plant species and virtually all of the vertebrates (fish, birds, and mammals) that are permanent or seasonal residents of the area were compiled. Using information on food habits of wildlife species collected from several sources (Whitaker, 1980; Ehrlich et al., 1988; USEPA, 1993), species were assigned to a single trophic level on the basis of the most common food type consumed during the breeding season. Species were categorized by food habits (trophic level) and habitat-use patterns, thus allowing for qualitative evaluation of the number and types of receptors exposed to particular media and chemicals. Although several species of special concern were known to frequent the region, no state or federal threatened or endangered species, or species of special regulatory concern, were expected to have significant exposure to chemicals at the various sites.

Selection of Indicator Species

Indicator species were selected that had (1) diets, rates of food and water ingestion, and foraging behaviors likely to give high exposure to chemicals in impacted media; (2) adequate data on determinants of exposure (e.g., ingestion rates, diet); and/or (3) high social or economic value. At least one bird and one mammal from each trophic level in both the terrestrial and aquatic food chains were used to model exposure for other species with similar trophic positions. Trophic levels included primary producers (plants), primary consumers (herbivores), secondary consumers (insectivores and carnivores), and tertiary consumers (certain fish or bird predators). All of the species in the USEPA's Wildlife Exposure Factors Handbook (USEPA, 1993) that live in the region were included as indicator species because the data available for estimating exposure and ecological effects were particularly good. The moose (*Alces alces*) was selected because of its high social and economic value in the region.

EXPOSURE ASSESSMENT

Ingestion and inhalation are the usual pathways by which wildlife are exposed to chemicals in contaminated media. Ingestion routes include uptake of chemicals in impacted food or water, and incidental or intentional ingestion of contaminated soil or sediment during foraging, burrowing, or grooming. Insufficient data were available to evaluate dermal or inhalation uptake.

Estimates of the daily dose (adjusted for body mass) of chemicals that wildlife are expected to receive if they feed and drink exclusively at an impacted site for several months were made by estimating food, water, and soil or sediment ingestion rates. Depending on data availability, either empirical estimates of actual food and water ingestion rates (USEPA, 1993), or allometric relationships (Nagy, 1987; USEPA, 1993) were used to estimate daily food and water consumption of indicator species. Recently compiled data on soil and sediment ingestion rates as a function of food intake rates were used to model this exposure pathway for wildlife (Beyer et al., 1994). Because incidental soil or sediment ingestion rates were available for only a few birds and mammals, data from closely related species with similar feeding habits were used to model soil or sediment ingestion of indicator species lacking data.

Bioaccumulation

Several commercial databases were searched for literature describing how COPCs move through food chains. Unfortunately, few data or studies appropriate for calculating bioaccumulation factors (BAFs) in natural settings were identified. Aquatic BAFs for insectivorous and piscivorous species were determined for each COPC in aquatic systems using data or procedures outlined in the Great Lakes Water Quality Initiative (58 FR 72:2104). BAFs for COPCs in terrestrial habitats were estimated using simple relationships for soil-to-plant and plant-to-animal chemical transfer (Travis and Arms, 1988) based on physical properties of COPCs (e.g., log K_{ow}), and measurements of chemical transfer from soil to crop plants and plants to cattle tissue. However, the regression equations presented by Travis and Arms (1988) were modified to calculate the feed-to-meat transfer in lean wildlife species. It was assumed that aquatic plants uptake chemicals from sediment in the same manner as terrestrial plants uptake chemicals from soil.

Estimating Wildlife Exposure

Models estimating wildlife exposure assumed that all food, water, soil, or sediment consumed by wildlife was from contaminated sites, and that animals do not selectively forage on uncontaminated foods. It was assumed that animals are chronically and continuously exposed to impacted media and foods.

The general equation for estimating chronic daily chemical intake of indicator species from ingestion of food was as follows (Peterson and Nebeker, 1992):

$$CDI_f = C_f \times IR_f \tag{1}$$

where:

CDI_f = chronic daily intake of COPC from the ingestion of food (mg/kg/d)

C_f = modeled concentration (mg/kg) of COPC in food (i.e., plant, herbivore, or insectivore/carnivore tissue)

IR_f = average daily amount of food consumed (kg/kg/d)

Chronic daily chemical intake from ingestion of water was estimated as follows:

$$CDI_w = C_{sw} \times IR_w \qquad (2)$$

where:

CDI_w = chronic daily intake of COPC from the ingestion of surface water (mg/kg/d)

C_{sw} = measured or modeled waterborne concentration of COPC (mg/L)

IR_w = average daily volume of water consumed (L/kg/d)

Also, chronic daily chemical intake from sediment or soil ingestion was estimated as follows:

$$CDI_s = C_s \times IR_s \qquad (3)$$

where:

CDI = chronic daily intake of COPC from the ingestion of sediment or soil (mg/kg/d)

C_s = concentration of COPC in sediment or soil (mg/kg)

IR_s = average daily amount of sediment or soil ingested (kg/kg/d)

Finally, the general equation for estimating total daily chemical intake was as follows:

$$CDI_t = CDI_f + CDI_w + CDI_s \qquad (4)$$

TOXICITY ASSESSMENT

A thorough search of the literature on the toxicity of COPCs to plants, fish, and wildlife yielded few appropriate, high-quality data. Uncertainty factors (UFs) were used, along with the highest quality toxicological data available, to estimate threshold doses or concentrations protective of plants, fish, and wildlife.

Derivation of Toxicity Values

The toxicity assessment is intended to derive the highest chronic daily COPC intake rate or highest concentration that would have no adverse effects on correlates of fitness (e.g., growth, mortality, reproduction). WRBCs and SRBCs were developed using data from standardized aquatic and plant toxicity tests, respectively. The wildlife approach was analogous to human health risk assessment. Toxicity data were normalized by body mass and expressed as a daily dose. The resulting dose is called the toxicity reference value (TRV), and it is similar to the reference dose (RfD) used in human health assessments.

UFs in TRV and WRBC Extrapolation

Some of the uncertainties associated with extrapolation of toxicity data include:

- Interspecies: Absorption, metabolism, distribution, excretion, and toxicity might differ between two species. A UF of 1 to 8 for terrestrial species was applied to test data depending on the relatedness of test and indicator species, and a UF of 10 was used for aquatic species if toxicity data was available for fewer than three species.
- Endpoint: The best available endpoint may not be as protective as the no observed adverse effect level (NOAEL). A UF of 1 to 50 was applied to toxicity data depending on endpoint and exposure duration (Calabrese and Baldwin, 1993; USEPA, 1994; 40 CFR 132, Appendix D, Sections IIIG and IIIH).
- Exposure duration: The study might have been conducted over an acute or a subchronic duration, rather than the preferred chronic duration (see endpoint above).
- Database: The toxicological database might not be sufficient to evaluate all of the chemical's potential health effects. Uncertainty associated with data quality was addressed qualitatively in terms of high, medium, or low confidence.

Little guidance on the use of UFs to assess risks to fish, wildlife, and other ecological receptors has been proposed (Calabrese and Baldwin, 1993). Furthermore, our understanding of the genetic, physiological, and temporal determinants of responses to contaminants are rudimentary for many taxa. As a result, it is not yet possible to develop rigorous UFs that can be used to reliably evaluate toxicity thresholds for most organisms, and many UFs were somewhat subjective. Uncertainty associated with interspecies extrapolations was reduced as much as possible by conducting a thorough literature review and selecting the most sensitive species. It was assumed that extrapolating toxicity data between species that are closely related is more reliable than extrapolating between more distant species (Calabrese and Baldwin, 1993). For many chemicals, mammalian toxicity data were found, but avian toxicity data were not. Although the metabolic profiles of birds and mammals are expected to be substantially different, excluding a chemical due to lack of toxicity data amounts to assigning it zero toxicity. We felt it was appropriate to perform interclass extrapolations because several studies have shown close correlations in chemical concentrations producing adverse effects in distantly related species (Doherty, 1983; LeBlanc, 1984; Calabrese and Baldwin, 1993). If responses to chemical exposures between taxa are correlated, then implicitly assuming chemicals have no effect (by not estimating risks of exposure) fails to incorporate potentially useful and meaningful information.

Estimates of chronic NOAELs were extrapolated from less-protective endpoints using UFs based on a combination of several approaches in the literature and the regulations. When multiple NOAEL values were available for different species in the same family, the NOAEL for the most sensitive species was chosen to represent that family. When multiple NOAEL values were available for different endpoints in the same species (e.g., growth and reproduction), the NOAEL for the most sensitive endpoint was chosen.

Toxicity data for the aquatic environment were not extrapolated for specific indicator species. Rather, the WRBC for the most sensitive species was chosen to represent a safe concentration for all aquatic species. That involved extrapolating to a variety of classes of both invertebrates and vertebrates. Based on the approaches of the USEPA Office of Water and National Pollutant Discharge Elimination System, a UF of 10 was applied to the WRBC if the toxicological database included fewer than three species (two vertebrate and one invertebrate).

RISK CHARACTERIZATION

Risks to Wildlife

TRVs were used in the quotient method to assess the potential for adverse impacts to indicator species. The quotient used to estimate the risk chemical exposure poses to wildlife was calculated as follows:

$$HQ = CDI \,/\, TRV \tag{5}$$

where:

HQ = hazard quotient (unitless)

CDI = chronic daily chemical intake (mg/kg/day)

TRV = toxicity reference value, representing the "safe" dose of the chemical for the specific environmental receptor (mg/kg/day)

HQs were derived for exposures to food, surface water, and soil or sediment, and they were summed to estimate a total HQ for the chemical. A hazard index (HI) representing the sum of the individual HQs for each chemical was calculated for each indicator species. In the absence of information regarding the synergistic (more than additive) or antagonistic (less than additive) effects of COPCs, it was assumed that the potential toxicities of multiple chemicals were additive (Watkin and Stelljes, 1993).

An HQ or HI greater than one indicates a potential for adverse health effects to species with food habits similar to the indicator species. The greater the number by which an HQ or HI exceeds one, and the more chemicals with HQs above one, the more likely is an adverse effect. However, the risk estimates involve considerable uncertainty, and an HI above one does not indicate that adverse effects will necessarily occur.

Home Range Adjustments

Most wildlife in the region tended to range over areas larger than the contaminated sites to meet daily food and water requirements. Therefore, more realistic HI estimates were made by considering the home range sizes of the indicator species. It was assumed that wildlife feed and drink equally over all areas in their home range. If the area of an impacted site represented only a fraction of an animal's home range, it was assumed that only a portion of ingested food, water, soil, and sediment was impacted. For example, if the area of impacted soil was 20% of the home range size of an indicator species, the total HI for the species at the site would be reduced by multiplying by 0.2.

Aquatic Organisms and Risk to Plants

Risks for fish and other aquatic organisms were evaluated by comparing the exposure point concentration (EPC) for each COPC in surface water with established WQC or the estimated WRBC. It was inferred that fish and other aquatic organisms would be at risk if waterborne COPC concentrations were above these criteria. Similarly, risks to plants were evaluated by comparing EPCs for each COPC in surface and root zone soil with SRBCs.

Population Density

The literature was reviewed for estimates of average population densities of each indicator species. By comparing the spatial extent of contamination with estimates of wildlife population density, a crude estimate of the number of individuals that may be exposed in a given year can be made. Although inadequate for rigorous evaluation of potential population-level effects, the process may assist risk managers in attempts to place estimated risks to individuals into perspective.

SUMMARY OF RESULTS

Small Water Bodies

At several localities, small-bodied, insectivorous and piscivorous wildlife were potentially at risk from ingestion of food containing petroleum constituents. HIs for aquatic wildlife foraging from these sites were often below 1, but at some sites HIs were as high as 4,000. Similarly, waterborne concentrations of some petroleum-related chemicals were above WRBCs, suggesting that aquatic organisms are potentially at risk from chemical exposures.

In general, small-bodied birds and mammals were estimated to be at greater risk than larger species because small species tended to have higher metabolic rates and thus, higher food and water ingestion rates. Although detoxification and depuration rates may also be higher in small birds and mammals, these processes were not evaluated. Also, smaller species tended to have small home ranges and were estimated to obtain a higher proportion of food and water resources from contaminated areas. Estimated risks were higher for aquatic predators because chemical constituents of petroleum were expected to bioaccumulate based on their physical properties. Importantly, the data available for estimating the fate and toxicity of petroleum constituents were probably more limited than for any other class of COPCs. As a result, uncertainty associated with estimated risks resulting from exposure to petroleum constituents was particularly high.

Large Water Bodies

Insectivorous and piscivorous birds and mammals from the aquatic food chain were potentially at risk from ingestion of food and sediment containing lead, petroleum components, and mercury. Because detected concentrations of total mercury were assumed to be methylmercury, a substance that bioaccumulates in aquatic food chains and is particularly harmful to wildlife, estimated risks to some wildlife were high. Fish and other aquatic organisms were potentially at risk from exposure to mercury and petroleum-related compounds in surface water because waterborne concentrations exceeded their respective WQC and WRBCs.

Soil

At one site, small, herbivorous, ground-feeding mammals were potentially at risk from ingestion of plants that had accumulated PCB-1260 from soil, with HIs ranging from less than 1 to 7. At another locality, small-bodied, ground-feeding birds and mammals were potentially at risk from ingestion of food and soil contaminated with dioxins and petroleum hydrocarbons, with a high HI of 14. Because large carnivores tended to forage over areas much larger than the contaminated sites, bioaccumulation of PCBs and dioxins were not expected to pose important risks for higher trophic level species. Plants were potentially at risk from exposure to petroleum-related chemicals in soil, on the basis of exceedances of estimated SRBCs. However, visual inspection of plants growing in these localities revealed no obvious signs of plant stress or important changes in plant species composition.

UNCERTAINTY

The ecological risk assessment was much more complicated than the human health risk assessment for these sites, and correspondingly, has much higher uncertainty. In our opinion, the following are some of the most significant sources of uncertainty in the risk assessment:

- fractionation approach for petroleum
- fate and transport modeling
- inability to evaluate inhalation and dermal exposures (especially for burrowing species)
- bioaccumulation modeling
- primary use of rodent toxicity data to evaluate COPC toxicity to other wildlife
- lack of toxicity data for petroleum constituents

Portions of the risk assessment required extensive modeling. The following are site-specific information that would have improved confidence in the risk estimates:

- analysis of the composition of petroleum in soil, surface water, and sediment
- chemical analyses of tissue samples from the terrestrial and aquatic food chains
- more extensive characterization of chemical types and concentrations in surface water, soil, and sediment

Despite the uncertainties, the risk assessment provided important information to assist risk managers and other parties in governing the site.

ACKNOWLEDGMENTS

Several EMCON staff assisted in various activities associated with the assessment, but the help of K. Ginn, K. Graham, M. Jones, B. Pugh, and K. Yost was especially appreciated.

REFERENCES

Beyer, W.N., Conner, E., and Gerould, S. 1994. Estimates of Soil Ingestion by Wildlife. *J. Wildl. Manage.* 58, 375-382.

Calabrese, E.J. and Baldwin, L.A. 1993. *Performing Ecological Risk Assessments.* Boca Raton, FL, Lewis Publishers.

Doherty, F.G. 1983. Interspecies Correlations of Acute Aquatic Median Lethal Concentration for Four Standard Testing Species. *Environ. Sci. Technol.* 17, 661-685.

Ehrlich, P.R., Dobkin, D.S., and Wheye, D. 1988. *The Birder's Handbook: A Field Guide to the Natural History of North American Birds.* New York, Simon & Schuster/Fireside Books.

LeBlanc, G.A. 1984. Interspecies Relationships in Acute Toxicity of Chemicals to Aquatic Organisms. *Environ. Toxicol. Chem.* 3, 47-60.

Nagy, K.A. 1987. Field Metabolic Rate and Food Requirement Scaling in Mammals and Birds. *Ecol. Monogr.* 57, 111-128.

Peterson, J.A., and Nebeker, A.V. 1992. Estimation of Waterborne Selenium Concentrations That are Toxicity Thresholds for Wildlife. *Arch. Environ. Contam. Toxicol.* 23, 154-162.

Travis, C.C. and Arms, A.D. 1988. Bioconcentration of Organics in Beef, Milk, and Vegetation. *Environ. Sci. Technol.* 22, 271-274.

USEPA. 1992. Framework for Ecological Risk Assessment. U.S. Environmental Protection Agency, Risk Assessment Forum, Washington, DC. EPA/630/R-92/001.

USEPA. 1993. Wildlife Exposure Factors Handbook. U.S. Environmental Protection Agency, Office of Research and Development, Washington, DC. EPA/600/R-93/187b.

USEPA. 1994. Region 10 Supplemental Guidance for Ecological Risk Assessments: Amendments to the August 1991 EPA Region 10 Supplemental Risk Assessment Guidance for Superfund. Prepared for USEPA Region 10, Seattle, WA, by Christine Chew and Dean Boening, ICF Kaiser Engineers, Inc., Seattle, WA. Draft. October 14.

Watkin, E.G. and Stelljes, M.E. 1993. A Proposed Approach to Quantitatively Assess Potential Ecological Impacts to Terrestrial Receptors from Chemical Exposure. In: *Environmental Toxicology and Risk Assessment*, Volume 2. (Gorsuch, J.W., Dwyer, F.J., Ingersoll, C.G., and La Point, T.W., Eds.). American Society for Testing and Materials, Philadelphia. ASTM STP 1216.

WDE. 1995. Summary of Guidelines for Contaminated Freshwater Sediments. Washington State Department of Ecology. March.

Whitaker, J.O., Jr. 1980. *The Audubon Society Field Guide to North American Mammals*. New York, Alfred A. Knopf.

PART IV—REMEDIATION

CHAPTER 14

Surfactant-Enhanced Mass Transfer Rates of Residual Hydrocarbons in Porous Media

Rajesh T. Kukreja, Dalibor Hodko, and Oliver J. Murphy, Lynntech, Inc., College Station, Texas

INTRODUCTION

The contamination of soil with organic solvents, fuels, oils, and other hydrocarbons has prompted intensive study in analyzing the transport of nonaqueous phase liquids (NAPL) in subsurface environment. The hydrocarbons enter the soil by waste disposal and accidental spills and remain trapped owing to capillary forces (Wilson et al., 1990). Since the solubility of NAPLs in water is very low, the removal of trapped NAPLs offers a formidable task. The residual hydrocarbons can be retained for very long periods in the subsurface environments, and their presence poses a significant long-term threat to aquifer quality. An understanding of mass transfer rates for solubilizing residual hydrocarbons is not only critical in the prediction of dispersion in ambient groundwater, but also for successfully predicting the removal of residual organics by enhanced pump-and-treat.

Surfactants have shown promise of greatly increasing the efficiency of removing residual NAPLs. Surfactants are classified according to the overall electrostatic charge as anionic, cationic, nonionic, and amphoteric. The surfactants are known to increase solubility of organics in the aqueous phase as well as increase the mobility of entrapped NAPLs (Sabatini et al., 1995). At a certain concentration called critical micellar concentration (CMC), the surfactant forms micelles, enclosing the hydrophobic compounds. This increases the aqueous solubility of NAPLs drastically. In addition, the interfacial tension between the NAPL and the aqueous phase is found to be the lowest when the surfactant concentration reaches CMC. The addition of surfactants can lead to formation of a middle phase emulsion (Shiau et al., 1994) that will reduce the interfacial tension between the water and the NAPL phase, and increase the mobility of NAPL blobs.

Micromodels can be employed to examine the mass transfer interaction between the trapped NAPL blobs and the flowing surfactant solution. Micromodels are physical models of pore network and have been effectively used as a visualization tool to gain an understanding of transport mechanisms in soil (Conrad et al., 1992). A predetermined pore network is chemically etched onto two glass pieces that are mirror images of each other. The two glass pieces are fused together to form a micromodel. The use of micromodels enables the direct observation and quantification of the solubilization or mobilization of trapped blobs within the network.

The main objectives of the current study were (a) to observe the solubilization and mobilization of residual hydrocarbons in micromodels, and (b) to quantify the removal rates of NAPL blobs attained with flushing of different aqueous surfactants under various operating conditions. The temporal changes occurring in the shapes and sizes of residual NAPL ganglia can be visualized by using a charge-coupled device (CCD) camera. The colored digital images can be captured and analyzed to evaluate changes in surface areas. In addition to quantifying the blob removal rates, the visualization technique can also be used to identify effective surfactants for several representative light NAPLs and dense NAPLs, and to study the effects of varying surfactant concentration, flow rate, and porous medium properties on the removal of residual NAPL contaminants.

EXPERIMENTAL APPROACH

A homogeneous micromodel with well-structured diamond-shaped pore network, as shown in Figure 1, was selected for the visualization study. The porosity and hydraulic conductivity measured for this micromodel are listed in Table 1. Toluene was chosen as a representative NAPL. To select effective surfactants for the visualization study, several surfactants were screened by performing bench-scale solubility experiments. The surfactants that yielded reasonably higher solubility of toluene in aqueous solutions of surfactants than that in water, were used to study the removal of residual toluene in the micromodel. They are listed as follows,

(i) Anionic - Sodium dodecyl sulfate (SDS)
(ii) Non-ionic - Polyoxyethylene sorbitan monooleate (Tween 80), polyoxyalkylated fatty acid ester (Adsee 799) and fatty acid alcohol (Witconol 2648).

Experimental Setup

Figure 2 shows the experimental test setup for visualization of residual NAPL in the micromodel. It essentially consisted of a syringe pump, a micromodel, and a CCD camera to capture the images of NAPL blobs in the micromodel. The image-capturing system was comprised of a Pulnix ½″ CCD RGB camera, a Coreco Occulus-TCX frame grabber, and a Pentium computer with necessary peripherals. The true color images were captured by the frame grabber periodically. The digital images were analyzed with Image-Pro Plus software to determine changes in surface area and size of the NAPL blobs. The micromodel was placed on a polypropylene plate, below which was a fluorescent lamp employed to illuminate the viewing area.

Figure 1.   Micromodel employed for the visualization experiments.

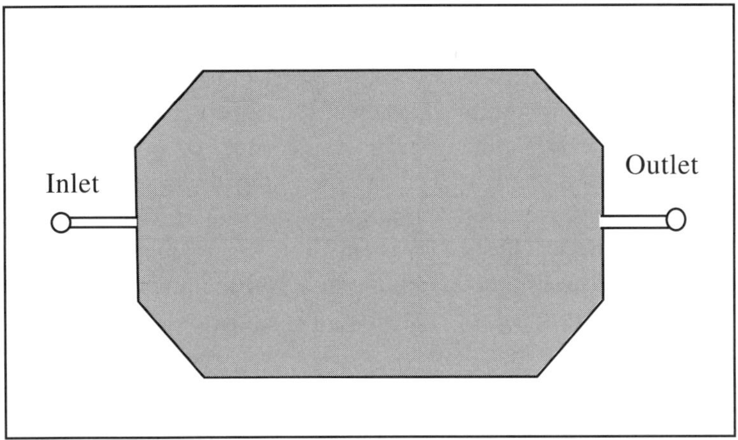

Table 1. Properties of Micromodel

Parameter	Micromodel
Pore Volume (cm3)	0.13
Porosity	0.28
Hydraulic conductivity (mm/s)	0.20

Figure 2.   Experimental test apparatus.

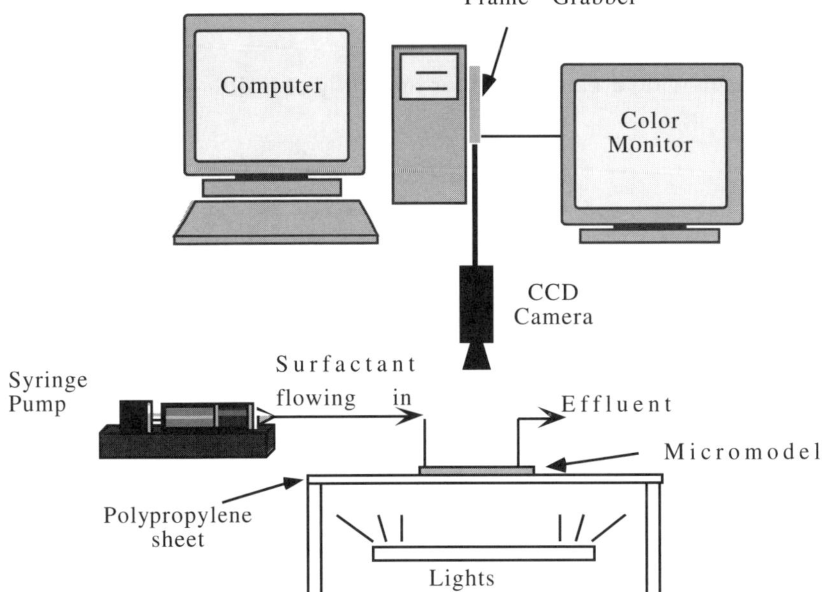

Experimental Procedure

To study the temporal changes in residual saturations of NAPLs in the micromodel, it was necessary to select a viewing area that was unaffected from the boundary effects imposed in the micromodel. A viewing area of 9 mm x 7 mm was selected near the center of the micromodel, as depicted in Figure 3.

The micromodel was initially degassed and saturated with deionized water. The dye Oil Blue N was mixed at 0.35 gm/liter of the organic contaminant (toluene). The dyed NAPL was passed through the micromodel at a controlled flow rate. The micromodel was flushed with water until the residual saturation of toluene was attained. The immobile organic liquid in the micromodel represents trapped NAPL in a saturated zone. An image depicting the initial residual saturation was captured and stored for analysis. The aqueous solution of the surfactant was passed at desired flow rates using a syringe pump. The imaging software was activated to capture images at preselected time intervals. The experiment was continued until most of the residual toluene was removed from the viewing area of the camera.

Data Analysis

The digital images captured during the course of the experiment were analyzed for residual saturation of blue-dyed organic contaminant. To determine the saturation of toluene, it was necessary to calculate the total pore volume and the volume occupied by the NAPL blobs. The pore volume of the NAPL blobs was determined by summing up the area of blue region using the Image-Pro Plus software. The total pore volume was determined separately after the end of the experiment by removing the residual organic contaminant and by filling the micromodel with deionized water dyed by blue food processing color. The same

Figure 3.   **Viewing area selected for the study of residual NAPL saturation.**

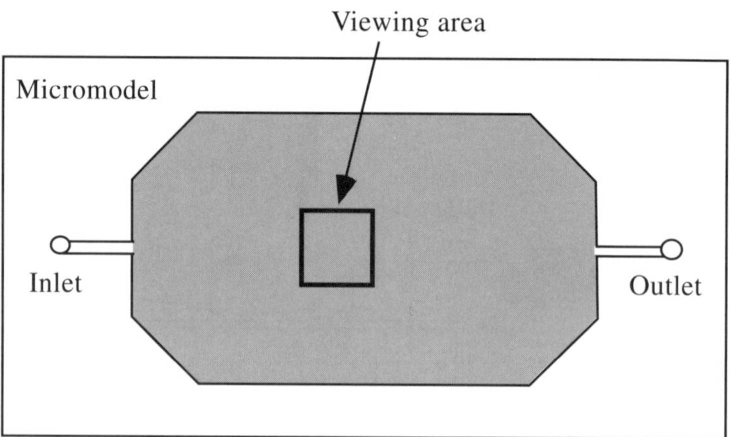

viewing area as before was selected and the total pore volume was determined by summing the areas of blue colored regions. The initial residual saturation was determined from the ratio of area of initial blue NAPLs to the total area of the pores as indicated by the blue food processing color.

The 24-bit (3 colored channels of 8-bit) RGB images captured periodically during the course of the experiment were analyzed for the residual saturation. This was accomplished by extracting either the red channel from the RGB images or by converting the RGB model images into the HSI (hue-saturation-intensity) model and then extracting the hue channel from the HSI images. The resulting single channel (red or hue) image was an 8-bit monochrome image and was analyzed for residual saturation.

RESULTS

Initial Residual Saturation

Experiments were conducted to examine the size and shape of the blob of residual NAPL blobs when water was passed at different flow rates through the micromodel filled with organic hydrocarbons. Figure 4 shows the captured images of the micromodel when water was passed at 1.2, 6, 60, and 150 mL/hr. To calculate the initial residual saturation at different flushing velocities, toluene was dyed with Red Oil and water with blue food color. As seen, flushing with lower flow rates of water yields higher residual saturation of NAPL blobs with an interconnected, long chain of blobs. Increasing the flow rates overcomes the capillary forces of trapped NAPL ganglia, decreasing the residual saturation and yielding sparsely-distributed individual circular blobs.

Removal Rates with Water Flushing

Figure 5 shows the rate of toluene removal in the micromodel by flushing with water. This case was used as a reference case to evaluate the efficacy of surfactants in removing NAPL. The NAPL saturation, C, at any given instant is expressed as a percentage of initial residual saturation, C_o (attained prior to flushing with surfactant), on the Y-axis of the graph. Water was flushed with 0.6 mL/hr through each micromodel. An approximate calculation corresponding to this flow rate yielded an average velocity of 1.8 m/day. This average velocity is comparable to a realistic average velocity of 1 to 2 m/day of natural groundwater flow. The removal of NAPL in the micromodel resulting from water flushing was found to be slow and incomplete due to low aqueous solubility of toluene. The visualization experiment revealed that the removal of residual toluene blobs occurred entirely due to slow solubilization, which was evident by the shrinkage of NAPL blobs. No mobilization effect was observed with water flushing at the rate of 0.6 mL/hr. Since the blobs trapped in large pores have greater surface area exposed to water flow, these blobs shrank (by dissolving) faster than those trapped in narrow channels or smaller pores.

Figure 4. **Variation in residual saturation of toluene in the micromodel attained by flushing water at different flow rate. The dark black color represents the trapped residual NAPL and the light gray color represents the area filled with water.**

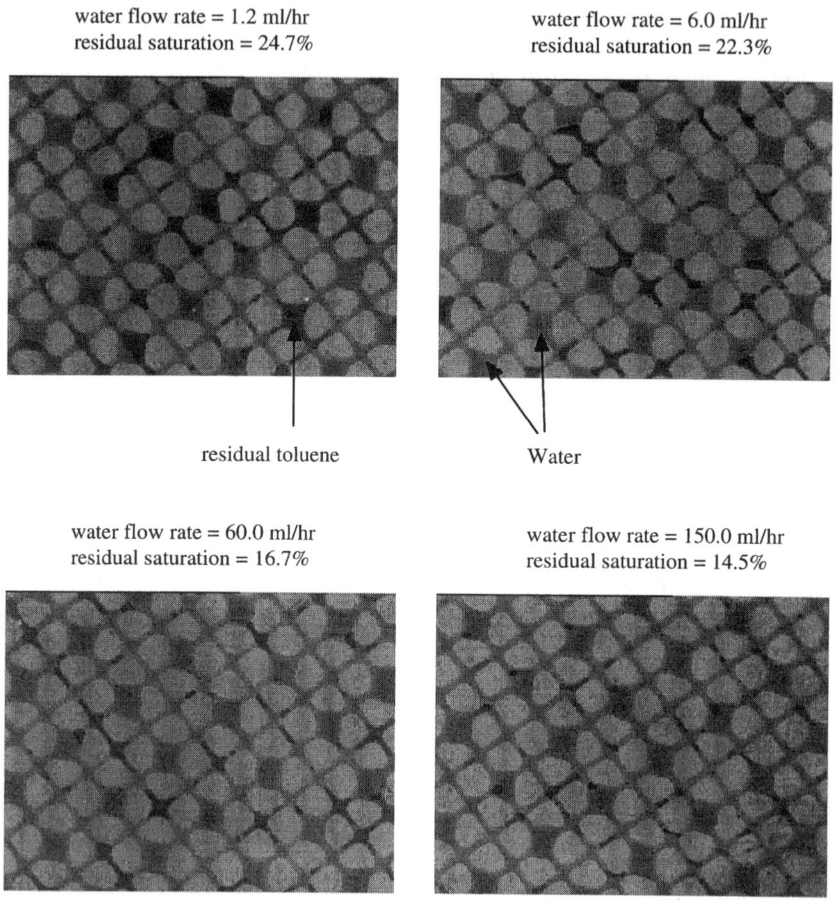

Removal Rates with Surfactant Flushing

Figure 5 also shows the removal rate of residual toluene in the micromodel attained by flushing with different surfactants; namely, SDS, Tween 80, Adsee 799, and Witconol 2648. As seen, Witconol 2648 outperformed all other surfactants in flushing out residual toluene. It removed residual blobs in about 6 hours as compared to 24 hours with aqueous SDS. Figure 5 also reveals that the removal of residual organic contaminant is increased with increasing solubility of toluene in aqueous surfactant solution. As the solubility of toluene increased in the order of

SDS < Tween 80 < Adsee 799 < Witconol 2648,

the removal rate of toluene was found to increase in the same order.

The visualization experiments using micromodels clearly elucidate different mechanisms responsible for the enhanced removal of the organic contaminant through the flushing of surfactants. With the introduction of aqueous SDS solution into the micromodel, the trapped elongated blobs in narrow channels were immediately released (or mobilized) into larger pores forming spherical-like blobs (see Figure 6a). This occurred because of the reduction in surface tension of elongated blobs with the aqueous SDS flushing, and thereby overcoming of the capillary pressure induced in narrow channels.

Snap-off mechanism was also observed between the blobs that were interconnected through narrow channels (see Figure 6b). Snap-off occurs when the ratio of pore to throat (or channel) cross-sectional area is high. The large spherical or ellipsoidal-like toluene blobs remained trapped in pores and slowly shrank in size owing to their enhanced solubilization by aqueous SDS. Once the size of the blob was sufficiently reduced, typically lower than the pore channel size, the blobs were mobilized and carried away along with mainstream aqueous SDS flow.

Figures 7 and 8 show the images depicting the entire viewing area at different times in the micromodel with the flushing of aqueous 1 wt% SDS and Witconol 2648, respectively. The removal of toluene with SDS was due to the combined effect of mobilization and solubilization, while the removal of toluene with Witconol 2648 was observed to be occurring entirely due to the solubilization of trapped

Figure 5. **Removal rates of toluene blobs with different 1 wt% aqueous solutions of surfactant flushed through the micromodel at the rate of 0.6 ml/hr**

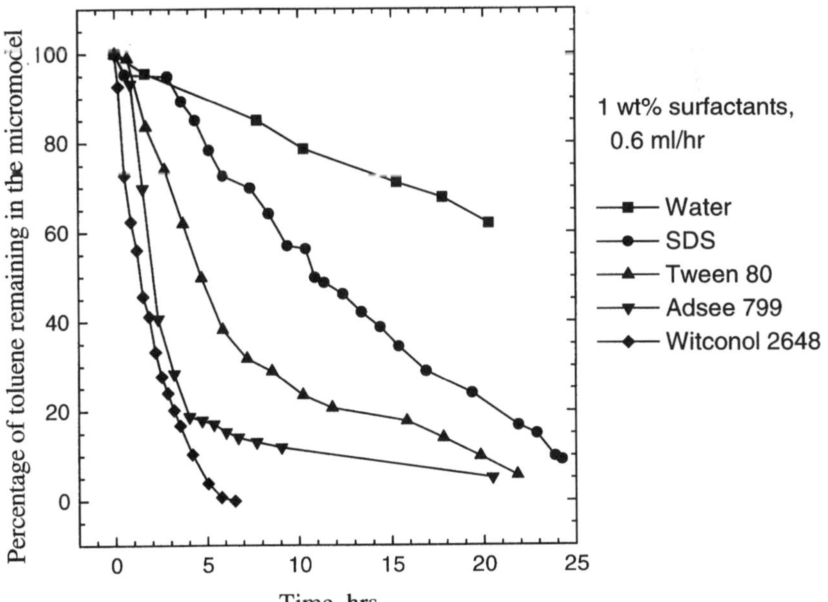

NAPL blobs. While the blobs were observed to move along with the mainstream flow in the case of SDS flushing, there was no mobilization or snap-off with Witconol flushing. The solubilization of toluene was clearly evident with the formation of grayish residue immediately upon the introduction of aqueous Witconol. In spite of increased mobilization caused by SDS, Witconol performed better owing to the extremely high solubility of toluene in Witconol 2648 (~ 80 times higher than that in water) than that in SDS solution (~ 8 times higher than that in water).

Figure 6. **Images depicting different mechanism responsible for mobilization and solubilization resulting in removal of toluene with aqueous SDS flushing.**

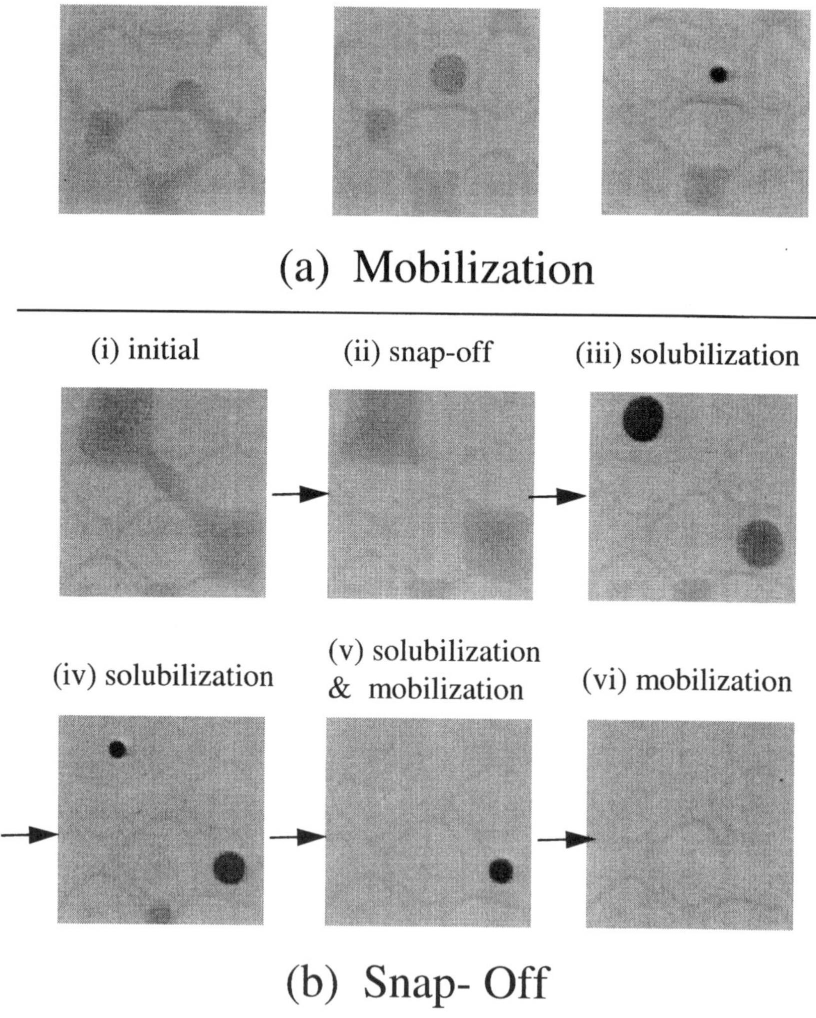

Figure 7. **Images displaying residual toluene in micromodel at different times with flushing of 1wt% aqueous Sodium Dodecyl Sulfate (SDS) at the flow rate of 0.6 ml/hr.**

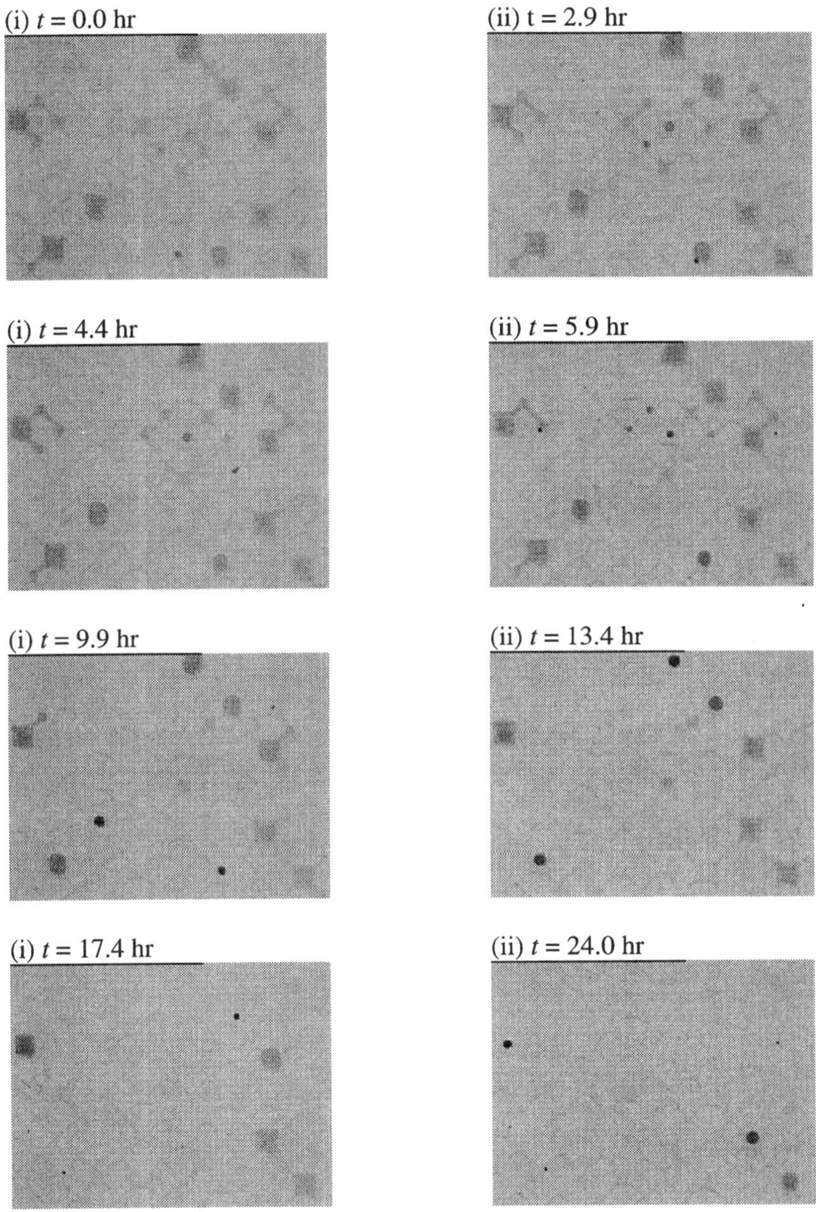

Figure 8. **Images displaying residual toluene in micromodel at different times with flushing of 1 wt% aqueous Witconol 2648 at the flow rate of 0.6 ml/hr.**

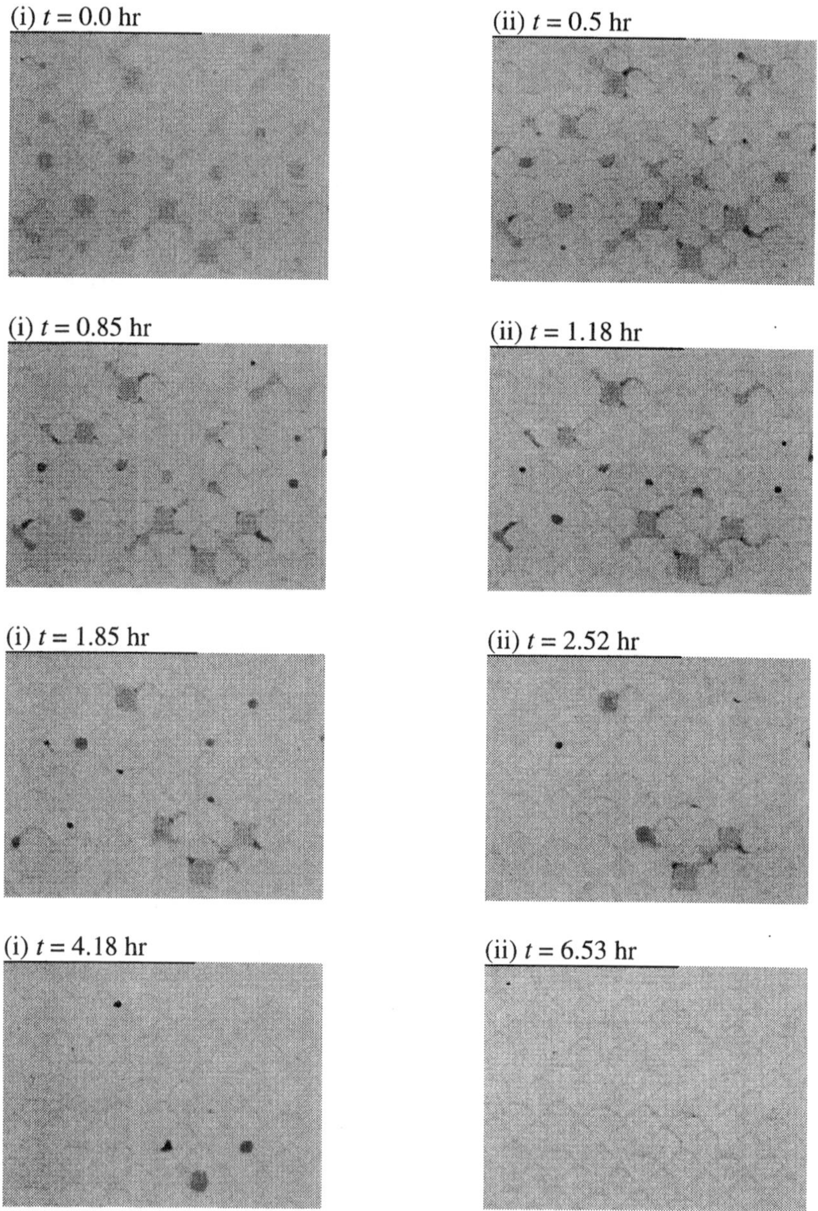

Effect of Surfactant Concentration

Figure 9 shows the effect of varying surfactant concentration from 0 to 2% in aqueous Witconol 2648 solution on the removal rate of residual toluene. In each case, the aqueous solution was passed at the flow rate of 0.6 mL/hr in micromodel. As seen, the removal of toluene was significantly enhanced by employing dilute surfactant solution with concentration as low as 0.1% in comparison to pure water. With 0.1 wt% Witconol 2648, the residual toluene was removed in about 23 hours. Increasing the concentration to 1 wt% of Witconol, the toluene removal was significantly faster and was completed in about 6 hours. However, increasing the concentration further to 2% did not cause any significant improvement in the removal rate of residual toluene.

Effect of Flow Rate

Figure 10 shows the effect of increasing the flow rate of aqueous 0.1 wt% Witconol 2648 in the micromodel on the removal of residual toluene. Increasing the flow rate ten-fold from 0.6 mL/hr to 6 mL/hr significantly increased the solubilization of toluene. The residual toluene was completely removed in about 7 hours with 6 mL/hr, as against 24 hours with 0.6 mL/hr. This was expected, as the increased pore flow velocity resulting from the increased flow rate enhanced the mass transfer rate between the surfactant and the residual toluene blobs, resulting in an enhanced removal of residual organic contaminant.

Increasing Surfactant Concentration vs. Increasing Flow Rate

In order to compare the advantages of increasing surfactant concentration against increasing the total flow rate of the surfactant, the total surfactant consumption in each case was calculated and compared. Increasing the surfactant concentration from 0.1 to 1 wt% (at a constant flow rate of 0.6 mL/hr) in Figure 9, or increasing the flow rate from 0.6 to 6 mL/hr (at constant 0.1 wt%) in Figure 10 increases the total Witconol consumption. The total surfactant consumption to remove the residual toluene saturation is as shown in Figure 11. Although the toluene removal is achieved at a faster rate (in about 6 hours) by increasing velocity or concentration, the surfactant consumption is increased more than 3 times than by flushing with lower surfactant concentration (0.1 wt%) at lower velocity (0.6 mL/hr).

This result indicates that there exists an optimum value of the flow rate and the surfactant concentration that results in the most favorable condition for the equilibrium between surfactant micelles and the contaminant blobs, so as to achieve the faster solubilization of NAPL blobs. The above results also have direct implication upon the economics of site cleanup. As shown, the residual contaminant cleanup could be achieved with surfactant concentration as low as 0.1 wt% at low surfactant velocity effectively and economically.

Figure 9. **Removal rates of residual toluene in micromodel B by flushing with different concentrations of Witconol 2648 at the rate of 0.6 ml/hr**

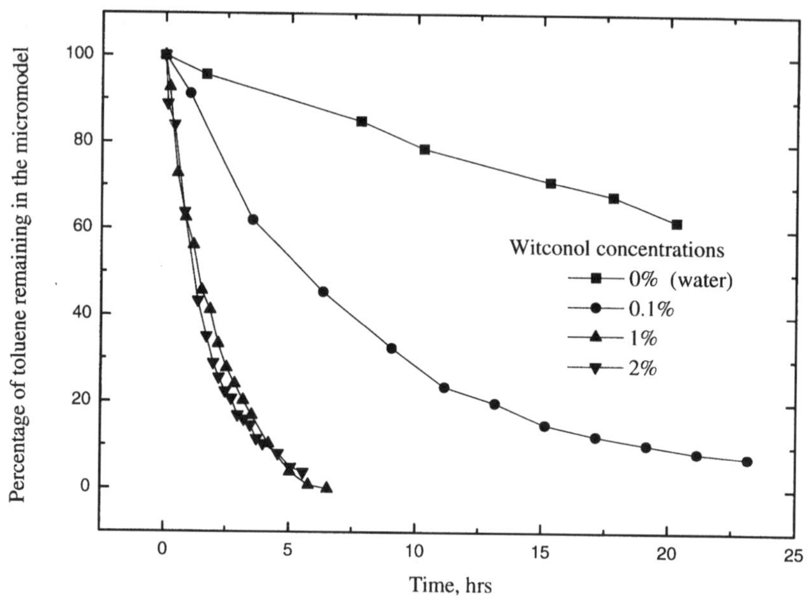

Figure 10. **Removal rates of residual toluene in micromodel B by flushing with different aqueous 0.1 wt% Witconol 2648 at different flow rates.**

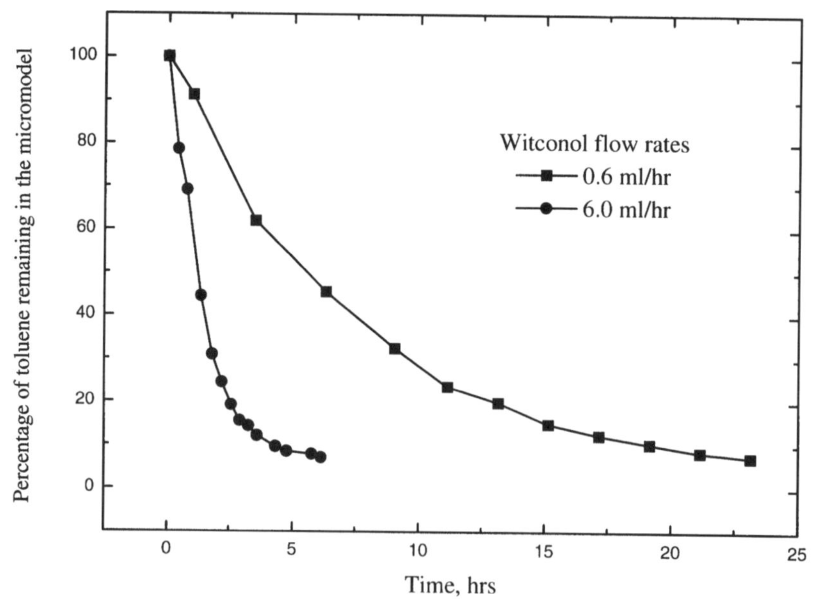

Effect of Pulsed Flow

Another experiment was carried out in order to examine the effectiveness of pulsed flow (or interrupted flow) on the removal of residual toluene. It was believed that an optimum residence time (during which the flow is interrupted) could be found that will be sufficient to result in equilibrium between micelles and toluene blobs. This equilibrium attained between micelles and the contaminant may enhance solubilization and/or mobilization with lower consumption of surfactant.

As seen in Figure 12, flushing the aqueous 1 wt% aqueous Witconol, with 3 minutes on and 10 minutes off pumping periods, significantly improved the removal of organic contaminant in terms of the total amount of the surfactant consumption. Pulsating the aqueous surfactant flow increases the residence time of the surfactant in contact with the residual NAPL blobs. This, in effect, results in efficient solubilization of the residual NAPL blobs with minimum surfactant consumption.

CONCLUSIONS

The micromodels can be successfully employed to visualize and quantify the removal rates of trapped residual NAPL blobs. The visualization technique can also be used to identify the most effective surfactant and to investigate the mechanisms responsible for the NAPL removal. The effects of several operating parameters, pertinent to the contaminated site and the aqueous surfactant, can be examined.

The experimental study conducted revealed that the nature of initial residual NAPL saturation strongly depended upon the flushing velocity of water. Low flushing velocity resulted in high saturation with long interconnected blobs, whereas high flushing velocity resulted in sparsely-distributed small spherical blobs and low residual saturation.

The comparison of *solubilizing* surfactants reveals that a surfactant with higher solubility for organic hydrocarbons causes faster removal of residual organics. The aqueous solution of Witconol 2648 displayed a significantly higher solubility for toluene and resulted in faster removal of residual toluene blobs. Sodium dodecyl sulfate (SDS) was found to mobilize surfactants through snap-off and release of trapped blobs from narrow channels into large pores by drastically reducing interfacial tension and by overcoming capillary pressures.

Increasing the velocity or surfactant concentration of the aqueous flow increased the rate of trapped NAPL removal at the expense of additional surfactant consumption. Significant reduction in surfactant use was accomplished with reduced flushing velocity and reduced surfactant concentration in the aqueous phase. Pulsed or interrupted flow of aqueous surfactant solution resulted in lower surfactant consumption than with continuous flow.

Figure 11. The effectof increasing concentration of velocity of aqueous Witconol on the removal of residual toluene.

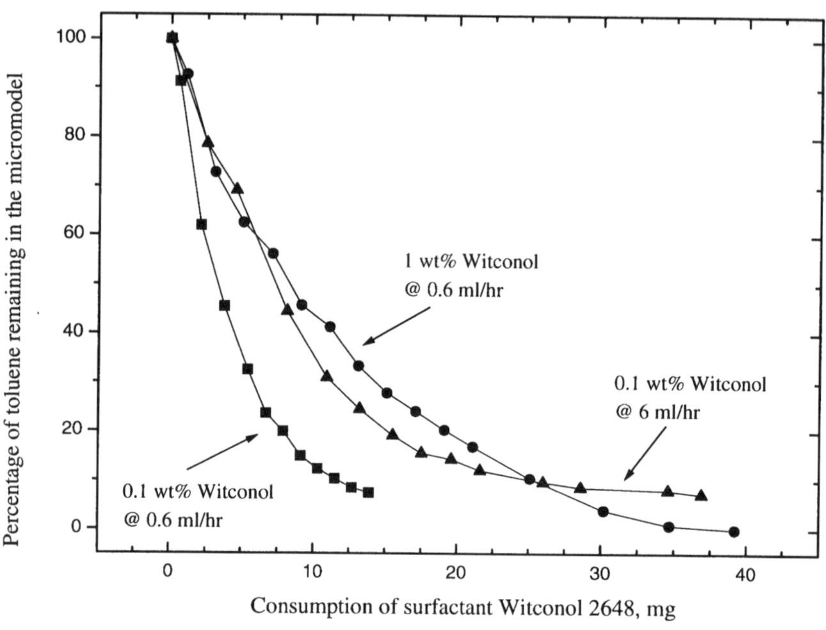

Figure 12. The effect of pulsing (or interrupting) aqueous Witconol flow on the removal of residual toluene.

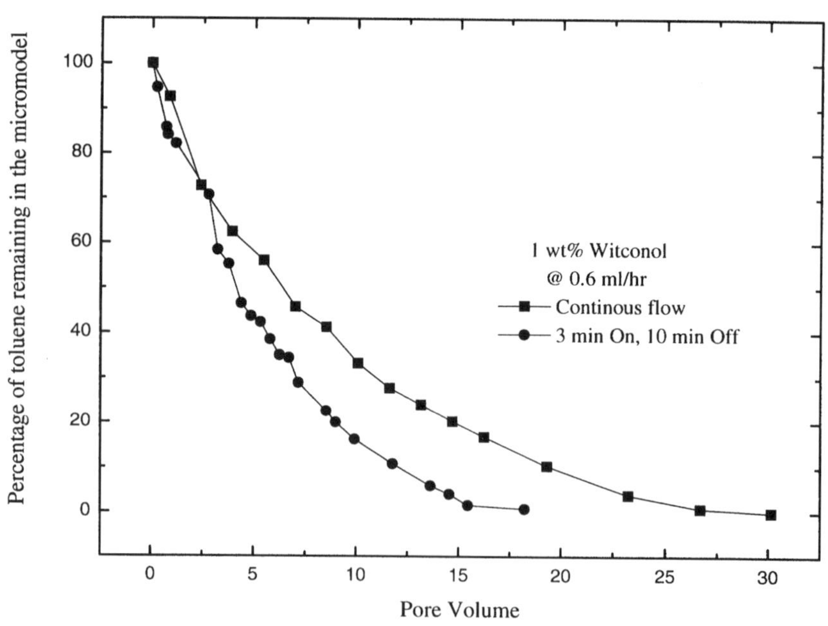

ACKNOWLEDGMENTS

The current study was supported through SBIR contract (F08637-95-C-6019) by Tyndall Air Force Base. The authors wish to express gratitude to Dr. M. Y. Corapcioglu for providing micromodels and for useful technical discussions pertaining to the project. The authors also acknowledge Ms. S. Chowdhury, a graduate student at Texas A&M University, for her help in conducting partial experiments at Lynntech, Inc.

REFERENCES

Conrad, S.H., Wislon, J.L., Mason, W.R., and Peplinski, W.J. 1992. Visualization of Residual Organic Liquid Trapped in Aquifers. *Water Resour. Res.* 28, 467-478.

Sabatini, D.A., Knox, R.C., and Harwell, J.H. 1995. Emerging Technologies in Surfactant-Enhanced Subsurface Remediation. In: *Surfactant-Enhanced Subsurface Remediation*, pp. 1-8. (Sabatini, D.A., Knox, R.C., and Harwell, J.H., Eds.). Washington, D.C., American Chemical Society.

Shiau, B-J., Sabatini, D.A., and Harwell, J.H. 1994. *Ground Water*, 32, 561-569.

Wilson, J.L., Conrad, S.H., Mason, W.R., Peplinski, W., and Hagan, E. 1990. Laboratory Investigation of Residual Liquid Organics from Spills, Leaks, and the Disposal of Hazardous Wastes in Groundwater, EPA/600/6-90/004.

CHAPTER 15

Unique Cleanup Strategy Devised for Malden Mills Fire Site

Evan T. Johnson P.E., LSP, Tighe & Bond, Inc., Westfield, Massachusetts
John E. Main and Connie E. Morton, Malden Mills Industries, Inc., Lawrence, Massachusetts

INTRODUCTION

On December 11, 1995, fire ravaged three major buildings at the Malden Mills complex in the town of Methuen, Massachusetts. The fire injured 23 people and caused the complete destruction of the company's Main Mill, and Process and Monomac Buildings. The fire began in the early evening of December 11 and, due to the sheer magnitude of the conflagration, was still smoldering over a week after it started. In addition to fire personnel who responded to the blaze, representatives from the Massachusetts Department of Environmental Protection (DEP) were also on site to evaluate whether any releases of oil or hazardous materials (OHM) had occurred as a result of the fire. Two initial spills of a caustic solution and an oil, (heat transfer fluid) from the boiler room, were immediately observed at the time of the fire. As a result of these releases, Malden Mills was directed to hire a Massachusetts Licensed Site Professional (LSP) to oversee the remediation of the two known spills and to determine whether any other releases may have occurred at the site. An Environmental Engineer from Tighe & Bond Consulting Engineers of Westfield, Massachusetts was hired by Malden Mills to perform the LSP services for the company.

The regulations governing "releases" of OHM are the Massachusetts Contingency Plan (MCP). The MCP distinguishes between emergency and historical releases, and establishes reporting time frames for notification to the DEP. Notification for historical releases of OHM are typically required within 120 days of obtaining "knowledge" of the release. The department must be notified of emergency responses within 2 or 72 hours, depending upon the type of release. Due to the identification of releases of OHMs (the caustic solution and heat transfer fluid), a 2-hour notification was required for this release under the requirements of the MCP; these releases are treated and remediated as Immediate Response Actions (IRA).

Initial cleanup responses were undertaken on the following day (December 12, 1995), and included vacuuming the caustic liquid for ultimate disposal off-site. The second release, the heat transfer fluid, was contained with absorbent material, and left in place pending investigations of the fire scene by the Massachusetts State Fire Marshal's Office. With these two initial response actions complete, it was the task of Malden Mills and their LSP to identify whether additional releases may have occurred in or under the rubble. Once identified, a detailed approach for identifying and removing these materials in a manner consistent with public health and safety requirements was required. Finally, it was the task of all involved to complete these activities in a timely and cost-effective manner.

PROJECT APPROACH

In order to determine whether other contaminants were even present at the site, Malden Mills and the LSP developed an inventory of chemicals, products, equipment, and other materials which might contain oil and hazardous substances and which were present on-site prior to the fire. In order to accomplish this task, the Malden Mills Environmental Health and Safety Group developed individual site plans of each of the three buildings destroyed in the fire. Working with managers of the various departments in each of the buildings, inventories of stored materials and equipment were collected and located on each of the site plans. Among the items that were included on the site plans were the following:

- Mercury chart recorders (meters): Malden Mills maintained these instruments for measuring steam flow. Each one of these instruments had a switch that contained a cylinder of liquid mercury.
- Radiation sources: Malden Mills also maintained instruments for measuring fabric thickness which contained cesium or krypton radiation sources.
- Metal-complexed organic dyes: Fabric at the mill was colored with a variety of dyes which contained a variety of metals.
- Transformer oils: Several large and small transformers containing various quantities of PCB-contaminated oil and non-PCB-contaminated mineral oil dielectric fluid (MODF) were located throughout the three-building complex.
- Machine oils: Machines located throughout the three buildings contained a variety of oils including cutting oil, machine oil, lubricating oils, gear oil, and other lubricants which may have been released to the floors of the three buildings.
- Heat Transfer Fluid (HTF): Three boilers located in a boiler room adjacent to the west side of the Monomac Building heated the HTF which was used for drying and curing in the flock process.
- Batteries: A series of lead-based batteries were located throughout the three mills. The batteries were stored in charging areas for powering the forklifts that were used throughout the plants.

- Acids and bases: As evidenced by the release of caustic solution during the initial notification, a series of vats and other containers containing acidic and basic solution were located throughout the three buildings.
- Latex: Although non-hazardous, latex vats, tanks, and pits were scattered throughout the mill complex and contained latex material used to provide an adhesive base for flock fabric and as backing for woven fabrics.

As noted above, each of the areas in the plant where these wastes were stored was located on a working site plan. Throughout the project, the site plan was updated as new data became available or as additional materials were identified. In addition, the site plans served as a field "checklist" to identify those wastes that were located and removed.

Concurrent with the development of the site plans and shortly after the fire was extinguished, representatives from Malden Mills determined that the plant would be rebuilt on a portion of the same site. Consequently, Malden Mills and the LSP, working with a remediation and demolition contractor, began the task of preparing site demolition, site remediation, and site health and safety plans for the cleanup. A series of meetings were held in the first weeks after the fire with representatives of the local wastewater treatment plant, the state Department of Public Health, and the DEP to come up with a reasonable approach that was cost-effective yet sensitive to the environmental issues associated with the fire.

In addition to the potential contaminants known to have been in the building prior to the fire, cleanup contractors and the LSP toured the site with state representatives to establish an approach for handling the mixture of ash, concrete, steel, intact raw materials, and partially incinerated substances which were left on the site. Specifically, a working definition for what constituted "soil" versus what was ash and other fire "debris" had to be established. Similarly, given the difficult task of identifying wastes in a partially burned state, a further distinction was made between what were OHMs and what was merely fire debris. Only with these definitions established could a plan for the proper disposal of all of those items be finalized.

Initial response actions, as noted above, included the emergency response conducted at the caustic and heat transfer fluid spill sites. The caustic spill was identified in a loading dock at the southern end of the Main Mill Building. The fire had destroyed a portion of a tank, causing the liquid to discharge to a depressed area near a combined sewer drain at that site. The storm drain was sealed and a remediation crew was brought in to pump the caustic solution to tanker trucks for ultimate disposal. Concurrently, representatives of Malden Mills notified the Greater Lawrence Sanitary District (GLSD) of the release and informed them that a portion of this spill may have entered the sewer system. Meetings were held with treatment plant operators to discuss this release in greater detail and to incorporate GLSD officials in the remediation plans in the event that additional liquids were encountered which might end up in the sewer system. A notification process was established with the GLSD to account for these types of potential releases as demolition progressed.

The second release, as discussed previously, included a release of heat transfer fluid from three boilers located at the western side of the Monomac Building. Temporary measures were put in place to isolate this spill and contain it from further migration. Once the release was satisfactorily isolated and contained, the state Fire Marshal's office determined that no further remediation was to be conducted at this location until the spring of 1996. Periodic inspections were conducted in the area throughout the remediation period to ensure that the release did not spread beyond the area where it had been contained.

Once the initial spills were addressed, the LSP and Malden Mills Environmental Health and Safety staff began designing the remediation process. Due to the magnitude of the cleanup and in order to maintain a safe working environment, a series of preconstruction meetings were held with each of the on-site contractors to establish protocols and approaches for the remediation of the three-building area. The cleanup was overseen by a representative from Tighe & Bond who worked directly with Kidder Building and Wrecking and Draghi Environmental Services.

The teams assembled by the environmental remediation and demolition firms were instructed that the cleanup would occur on a "inch-by-inch" basis with specific attention paid to those areas which had been identified by Malden Mills on the site plans as potentially containing OHMs. In order to maintain control of areas where potential OHMs were found and to provide an area for subsequent confirmation testing, a grid was established over the entire property. The grids were established based on the known locations of potential OHMs that were identified by Malden Mills. With inspection by the LSP and staff from Malden Mills, the demolition contractor began the task of removing rubble material from the site. Concurrently, steel and other metals were segregated for recycling. Given the type of steel present, Malden Mills was able to defray demolition costs with the money paid for the steel and other metals sent to the recycling facility. Each piece of steel was visually inspected prior to loading for recycling, to be certain that hazardous materials were not present.

The fire at Malden Mills occurred in mid-December and demolition and remediation work began almost immediately after the fire was extinguished. Thus, the work was begun in and continued through the winter season. Under normal circumstances, this is a difficult time to engage in any outdoor activities. The winter of 1995-96 in the northeast was anything but normal; with below average temperatures, record snowfall, and relentless wind, the conditions posed their own set of challenges to the cleanup.

The first task that was accomplished was the clearing of driveways between the foundations of the Main Mill and Process Building. Removal of wastes and debris in this area allowed us to provide a clear path for the ingress and egress of equipment, personnel, the removal of waste, and to gain access to an area where transformers were known to have been located and which were believed to have released transformer fluids during the fire.

After a portion of the driveway had been cleared, the transformers were encountered and oil was found on the pavement between the two former building foundations.

A total of nine transformers were identified, three of which were found to be leaking. The leaking oil was contained on the pavement and absorbed by the remediation crew. The damaged transformers were then loaded onto a containerized flatbed for removal from the site. Visibly stained debris was also collected and placed in a lined roll-off for subsequent transport for disposal. Testing of the residual ash and debris in that area was performed in order to determine if PCBs were present in the residual materials on the site.

With the remediation of this last known release, the emergency response activities were complete and the LSP and Malden Mills began development of the final detailed sampling approach for submittal to the DEP. Work crews continued the demolition under the direction of the environmental staff, but a more detailed "plan" was required to satisfy the DEP that an environmentally sound approach was in place. Using a review of the materials that had been established as being located at the site, and through a literature search of fire waste characteristics, the following approach was devised and submitted to the DEP as an IRA Plan.

The LSP petitioned the DEP to exclude the analysis of polycyclic aromatic hydrocarbons (PAH). Historical data indicated that PAHs may be present simply from the combustion of organic materials unrelated to OHMs that were present on the site. The prospect of "false positives" for PAHs would lead to a slowdown of work to confirm the source, expensive testing, and unnecessary disposal of noncontaminated fire debris.

The sampling protocol would include the RCRA eight metals (arsenic, barium, cadmium, chromium, lead, mercury, selenium, and silver) and three additional metals (cobalt, antimony, and copper). These latter metals were known to have been present on the site in significant quantities as a result of historical research conducted by representatives from Malden Mills. In addition to testing for these metals in the "total" form, if the result was less than 20 times the Toxicity Characteristic Leaching Procedure (TCLP) concentration for that metal, the sample was considered nonhazardous. If the 20-times rule was exceeded, an actual TCLP test was performed on the debris sample; the purpose of this approach was, again, cost saving for analytical work and waste disposal.

In addition to metals analysis, the selected testing procedure also included analysis for the hazardous constituents: corrosivity (pH), ignitability (°F), and reactivity (both cyanide and sulfide). A failure of any of these analyses or on the TCLP metals analysis would require a segregation of that sample area and retesting to determine what portion of the area included hazardous waste.

The LSP developed a definition of rubble/demolition debris to distinguish it from soil on the site. By reviewing data and pre-fire site plans, it was determined that no "soil" would be present at the site. The three buildings were surrounded by either paved or cobble roadways and, with the exception of a portion of the Monomac and Process Buildings basements, included concrete slab floors in all of the buildings. Therefore, all materials at the site were characterized as rubble/demolition debris, and that determination was conveyed to the DEP. This distinction is key, because soil is specifically defined in the MCP as a medium for which cleanup standards are developed. By avoiding a set cleanup standard, the debris could be handled by testing it as either hazardous or nonhazardous. Other site specific approaches included the following:

1. A distinction was also made with the DEP between fire contamination and historical contamination that existed on the site prior to the fire. A site assessment by another LSP had been underway at the property prior to the fire. This assessment included releases of selected PAHs and oils to the soil and groundwater beneath at least a portion of the three-building site. For the purpose of developing a distinction between the ongoing assessment activities and the IRA, a physical distinction between fire contamination and historical contamination was developed. Specifically, IRA activities included all debris from the ground surface and continuing downward to the basement of each of the three buildings. All releases below the floors of the buildings were considered part of the prior site assessment. This latter distinction was critical because three LSPs were working on the cleanup simultaneously. The first LSP, from Tighe & Bond, was involved in the fire-related emergency response activities. A second LSP was on-site representing Gheraghty & Miller, the firm overseeing the historical site assessment activities in the soil and groundwater beneath the buildings. Finally, a third LSP, from Stone & Webster, was overseeing the construction activities of the new building on the property.

2. The approach presented to the DEP included a proposal to conduct site-specific sampling and analysis at those areas where known or potential OHMs were expected to be encountered. For example, in the vicinity of the former transformers in the driveway between the Main Mill and Process Building, the final cleanup sampling included analysis for PCBs which may have been released from the transformers that were located at that portion of the site.

3. Finally, the IRA plan included a proposal for confirmation sampling to close out the IRA. In each area of the aforementioned grids, the residual debris was collected in a nine-part composite sample and analyzed for TCLP metals and hazardous waste characteristics. In selected grids where other known contaminants were either suspected or encountered during cleanup, those individual parameters were also tested to confirm adequate remediation.

The cleanup and sampling strategy defined above was presented to the DEP in the form of an IRA plan. The Monomac Building was not included in this initial plan because it was the subject of further investigation by the state Fire Marshal's office. The remediation of the Monomac Building is being conducted, however, on the same model under the same plan. The DEP approved the conceptual approach of conducting site-specific investigations of areas where contaminants were known to have been stored, collecting site-specific analytical data in areas where releases were observed and cleanup was required, and providing a confirmation cleanup sample in each grid of the site as demolition activities were completed. In addition, the DEP requested that the LSP provide ongoing reports of data collected from the site and updates on the progress of the remediation.

As demolition proceeded, the known wastes identified at the site were stockpiled or contained and removed from the site for appropriate disposal. However, large quantities of brick, ash, rubble, and other debris from the fire, which were tested and did not contain concentrations of contaminants that exceeded the hazardous characteristics, TCLP, or Reportable Concentrations (under the MCP), were stockpiled on the site. This material is typically characterized from a regulatory standpoint as either demolition waste or asphalt, brick, and concrete (ABC) material. Typical disposal alternatives include landfilling at a demolition waste site. However, due to the sheer volume of the material present, it was infeasible and extremely costly to consider removing all of this material from the site for disposal. Consequently, Malden Mills and the LSP prepared a Beneficial Use Determination (BUD) submittal to the Solid Waste Division of the DEP. The BUD called for the reuse of the nonhazardous rubble and demolition debris in the footprint of the former buildings prior to construction of the new building. Tighe & Bond and Malden Mills worked closely with the Solid Waste Division of the DEP, who provided rapid approval of the BUD for the reuse of this material.

Additional waste that was encountered in the building during the demolition process included dozens of 55-gallon drums, some of which were partially destroyed and others that were intact despite being in the buildings during the fire. These barrels were either overpacked, placed into lined roll-offs, or removed intact to a hazardous waste disposal facility. The detection and disposal of the radioactive devices (cesium and radium meters) were overseen by a Nuclear Regulatory Commission licensed representative from Affrex of Pittsburgh, PA and the Massachusetts Department of Public Health. These units, when they were encountered, were placed in brick-lined barrels pending evaluation prior to proper disposal off-site at a nuclear waste disposal facility. The remainder of the materials, which included mercury chart recorders, dyes, oils, and batteries, were properly contained and taken to an OHM staging area for proper storage prior to disposal at an appropriate facility.

Although a large volume of the rubble and other debris was segregated from the site for reuse under the BUD, a portion of rubble, waste, and ash was found to be contaminated through direct contact with spilled oils and other wastes. This material was typically segregated and tested prior to being stored in lined roll-off containers for proper disposal off-site. Liquid wastes were also encountered in several locations where they had ponded or pooled on solid surfaces, in vaults in the basement, or on paved areas of the site. These wastes were either absorbed or vacuumed to containers and manifested off-site. Finally, some of the liquid wastes that were detected on the site were discharged directly to the sewer for treatment, with prior approval, at the wastewater treatment plant.

Throughout the course of remediation, the LSP and on-site health and safety personnel evaluated the risks associated with the releases that occurred at this site. Three potential media were impacted by the release and, in each case, the risks were evaluated. The first impacted medium was the air. Initial accounts from the night of the fire indicated breathing problems and an acrid odor that was emanating from the fire site. Air sampling was conducted within days of the fire, and no traces of organic chemicals or other possible irritants were detected. Background research indicated that the possible source of throat irritation reported during the night of the fire may have been by-products (such as pyrolle) generated by the combustion of nylon which was located throughout the buildings. An air testing report was submitted to the Massachusetts Department of Public Health and local Boards of Health, and no further air impacts were detected during the cleanup.

The second potentially impacted medium included surface and groundwater. Because the fire was contained on pavement or in building foundations, any liquids that mixed with waste and ash, including water used to fight the fire, was discharged to the wastewater treatment plant slowly over the first few weeks of the cleanup. A "spike" of low pH water was detected at the treatment plant on the morning following the night of the fire; however, the cause of this problem (acid from the initial spill leaching into the sanitary sewer) was discussed with GLSD and no further impacts of the release were detected throughout the cleanup of the site.

Finally, the last potential medium was soil. Again, because the wastes were contained in the building and on the pavement, no soil was impacted as a result of the fire.

The remediation of the majority of debris and materials encountered at the fire site was completed on August 23, 1996. A single area, which has been contained by a chain-link fence, is currently being held as evidence and cannot be remediated. Once this area has been released, the demolition and remediation staff will return to the site and finish that cleanup and issue a Response Action Outcome (RAO) to close the IRA emergency response activities of the MCP for the Malden Mills property.

Table 1 identifies wastes which were removed from the site.

Table 1. **Waste Removal Summary**

CATEGORY	QUANTITY
SOLIDS/SLUDGE	
Special wastes: including latex solids, oily solids, dyes, pigments, ash and bricks	10,400 cubic yards
Mercury solids	Four (4) 55 - gallon drums
Paint sludge	275 gallons
Batteries	8,210 pounds
Non-PCB Transformers	6,110 pounds
Fluorescent light bulbs	90 cubic yards
RCRA empty drums	135 cubic yards
Corrosive solids	fifteen 55 - gallon drums
Other solids	30 cubic yards
LIQUIDS	
Wastewater	In excess of 20,000 gallons
Oily water	15,000 gallons
Corrosive liquids	2,000 gallons
Waste oil and Non-PCB transformer oil	1,900 gallons
Ethlylene Glycol	165 gallons
PCB WASTES	
Transformers	36,459 pounds
Oil-soaked absorbent/ash/debris	Twelve 55 - gallon drums
Fluorescent light ballasts	19,000 pounds
PCB oil	1,815 gallons

SUMMARY

The devastating Malden Mills fire impacted the lives of hundreds of employees of the mill, injured 23 people, and destroyed three major mill buildings. Within eight months, a major remediation had been substantially completed and construction was underway to rebuild on the site of the fire. A clock tower, which had been damaged but not destroyed during the fire, was saved as a reminder of the grandeur of the former mills and is being currently incorporated into the construction of the new mill, which was completed in the spring of 1997.

In retrospect, the following key lessons can be observed from this experience. The first is the concept of soil versus fire debris. Although thousands of tons of waste were generated, the activities occurred on or near pavement or within the foundations of the building, and no "soil" was impacted as a result of this release. This distinction was key, because soils are specifically governed by the Massachusetts Contingency Plan. By distinguishing between soil and fire debris, the cleanup and disposal of the waste was accomplished through the use of a BUD and by normal recycling of metals and the disposal of hazardous materials which were encountered. Further, by identifying potential sources of PAHs as "fire generated" and not from OHMs, the analytical and disposal costs were greatly reduced.

The second lesson to be taken from this experience involved the plan for sampling and identifying debris inside the building. The copious records maintained by Malden Mills are a tribute to the conscientious management of their plant. This high-quality documentation allowed the LSP to present to the DEP a reasonable and cost-effective approach for sampling the residual debris without unnecessary analytical parameters and with minimal samples. Identifying those known contaminants limited the primary parameter list to hazardous waste characteristics and the RCRA eight, plus three additional metals that were identified on the site. The remainder of the analytical work was done on a site-specific and cost-effective basis.

The third lesson to be taken from the experience involved the oversight of the remediation activities. By establishing a detailed site and sampling plan in the early phases of cleanup, the protocols for proceeding through the remediation activities were firmly established and consistently followed. This allowed for a good working relationship between the remediation contractor, the demolition contractor, and the LSP overseeing the activities, and between these groups and the Department of Environmental Protection. Although the "inch-by-inch" approach appeared laborious in the early stages, the efficiency of identifying wastes and the quality site plan developed by Malden Mills allowed this process to proceed in a relatively efficient manner despite extremely tough working conditions and poor weather through the first four months of site cleanup.

With respect to risks associated with the site, by identifying early in the process that media governed by the MCP were not impacted (groundwater and soils), all risks were fully covered in the health and safety plan that was developed in the initial stages of activities. Each contractor was aware of their responsibilities associated with the health and safety. Further, health and safety representatives were present on the site full-time to oversee the activities and isolate wastes as they were excavated from the site. In short, we were able to change the approach from the remediation of a hazardous waste site to the more straightforward cleanup of a building which contained waste materials. Again, the attention to detail with respect to health and safety in the early stages of the project allowed the process to proceed smoothly.

Finally, through detailed research in the initial stages of the types of contaminants present, and through the implementation of a BUD for the disposal of the waste, the most cost-effective disposal options were developed. As described above, the remedial activities are essentially complete, and the new building was completed in the spring of 1997, just over one year after the devastating fire occurred. The fire was a tragic event in many ways; however, through diligent startup activities, the site problems were addressed in a reasonable amount of time, employing cost-saving measures.

CHAPTER 16

Enhanced Removal of Gasoline Constituents from the Capillary Fringe Utilizing Radio Frequency Heating

Raymond S. Kasevich and **Stephen L. Price**, KAI Technologies, Inc., Portsmouth, New Hampshire
Dan Wiberg and **Mark Johnson**, DAHL & Associates, Inc., St. Paul, Minnesota

INTRODUCTION

In March 1996, KAI Technologies, Inc. (KAI) and DAHL & Associates, Inc. (DAHL) conducted a joint remediation project at a former retail petroleum marketing facility in Blaine, Minnesota. Petroleum hydrocarbon-impacted soil and groundwater are present at this site due to release(s) from an underground storage and dispensing system. The site has undergone investigation and remediation since the early 1980s, under several remedial strategies and consultants. Remedial activities through March 1996 have resulted in the removal of significant volumes of product, but residual levels of impact remained at concentrations exceeding the Minnesota Pollution Control Agency's Health Risk Limits (HRLs). The technologies utilized during this study were soil vapor extraction (SVE), groundwater ventilation (air sparging), and radio frequency heating (RFH).

The site is located in an intermingled commercial/residential area of the city. It rests atop an expansive Pleistocene glacial outwash plain known as the Anoka Sandplain. The Anoka Sandplain consists primarily of brown, fine to medium grained sand and covers a large area including many of the northern suburbs of the Twin Cities Metropolitan Area. The Sandplain is typically present from ground surface to depths of greater than 80 feet.

Groundwater is typically found under unconfined conditions within 20 feet of the ground surface, and flow characteristics are affected significantly by local surface water bodies and groundwater recharge zones. Groundwater beneath the subject site ranges from approximately 10 to 15 feet below grade and flows west/northwest, toward a small lake located roughly one-quarter mile in that direction.

BACKGROUND

Following closure of underground storage tanks (USTs) in 1988, a remedial investigation (RI) was conducted at the site which defined an area of impacted soil and groundwater covering much of the northern half of the property. New underground tanks were installed within the same tank basin immediately following removal of the old USTs. The results of the RI, in light of regulatory remediation guidelines, indicated that active remediation was warranted at the site. In 1991, a groundwater pump-and-treat and soil vapor extraction remediation system was installed and remediation activities were initiated.

In the fall of 1992, the newer tanks and associated piping were removed and the property was placed into a commercially dormant state. An additional release associated with product piping in the dispenser island area was identified during the 1991 removal activities. The area impacted from this release was outside the area of influence of the initial remediation system. Additional vapor extraction structures were installed to address this area. Figure 1 illustrates the layout of the remediation system and the site. Due in part to the presence of significant volumes of impacted soil below the water table, and to substantial increases in groundwater elevation, a groundwater ventilation (air sparging) structure was installed in 1993. Immediately following startup of this portion of the system, elevated soil vent system effluent concentrations were observed. Concentrations quickly diminished; however, elevated dissolved concentrations persisted in the groundwater.

A limited subsurface investigation was conducted near the former dispenser islands to evaluate system performance in that area. Data collected during the investigation indicated impact in the capillary fringe and phreatic zones, while limited remaining impact was identified in the vadose zone. Figure 2 is a cross-sectional diagram which represents petroleum hydrocarbon impact in the vicinity of the former dispenser islands prior to the study. This location was selected for the RFH test project due to the continuing presence of petroleum constituents in that area which exceeded regulatory HRLs.

Background Sampling and System Configuration

Prior to initialization of site activities, all underground utilities beneath the site were located and mapped. The existing remediation system at the site was shut down throughout the duration of this study. The target area consisted primarily of the structures shown in Figure 1. All borings, vents, and the RFH applicator well were completed per applicable well codes.

Prior to initializing the study, Geoprobe borings were advanced to determine pretest baseline conditions in the study area. Both soil and groundwater samples from these probes were laboratory-analyzed for benzene, ethyl benzene, toluene, and xylenes via EPA Method 8020 and for Gasoline Range Organics (GRO). The results are represented graphically in Figure 2 and are tabulated in Tables 1 and 2. Groundwater samples were also collected from a monitoring well (MW-11) and five vent system structures prior to the test.

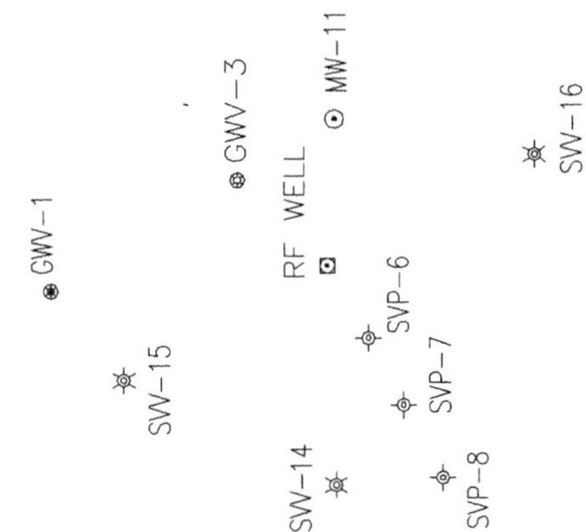

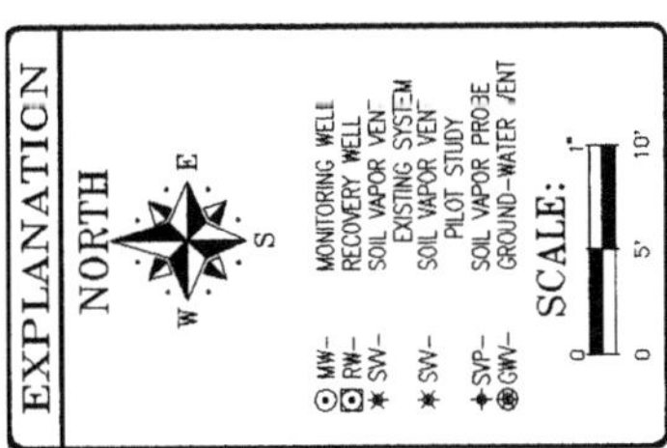

Figure 1. Layout of remediation system and site.

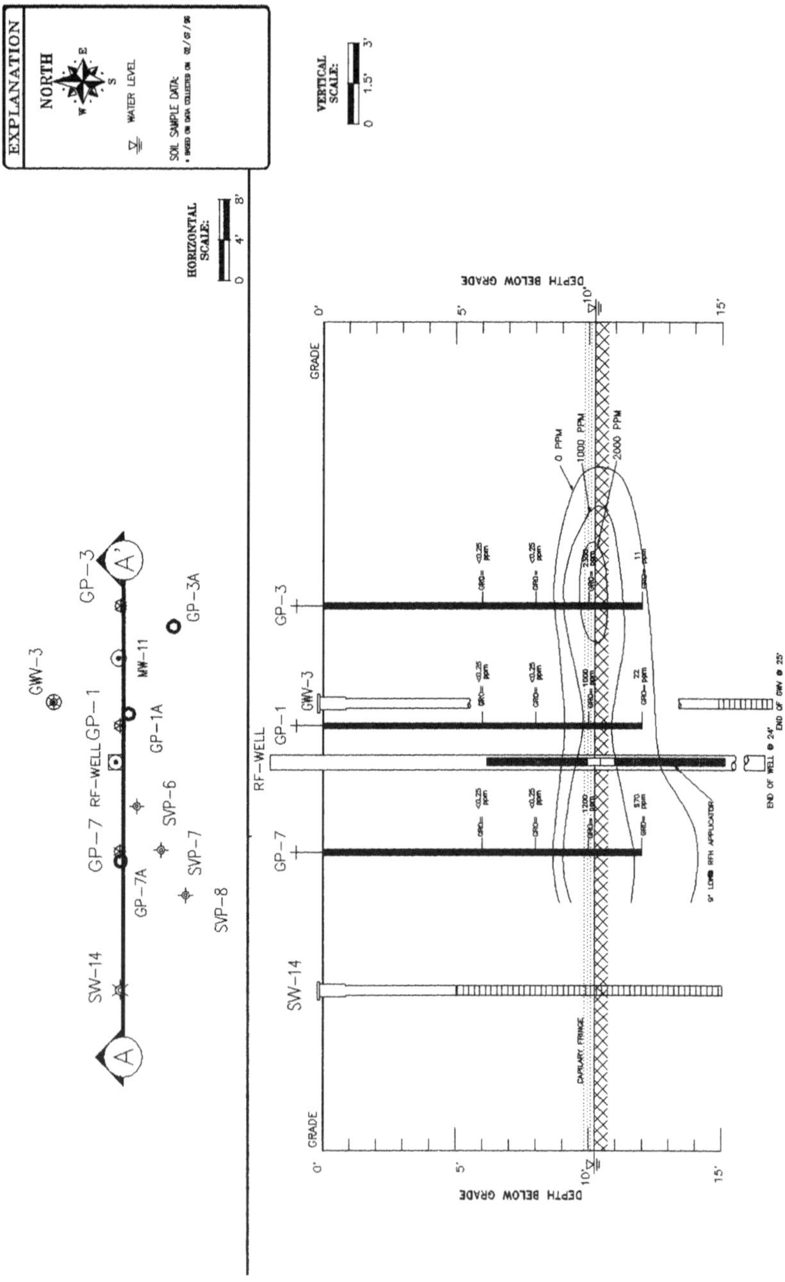

Figure 2.   Cross section of remediation system structure.

The greatest soil concentrations observed were within an approximate three-foot layer situated between 9 and 12 feet below ground surface (bgs). The area of greatest groundwater impact roughly corresponded to this same area. The water table was encountered at 10 feet bgs. The impacted areas, the capillary fringe and uppermost phreatic zone, are often the most challenging to remediate with conventional technologies and were thus selected as the target for the study.

The pilot study was positioned within the zone of greatest impact and consisted of one RFH well, two soil vapor vents (SVV), three soil vapor probes (SVP), and one groundwater vent (GWV). All pilot structures are indicated in Figure 1. The RFH well was constructed of 7-inch diameter, 30-foot long, fiberglass casing which housed the RFH applicator. The 3.5-inch diameter, 9-foot long RFH applicator was located from 6 feet bgs to 15 feet bgs, so that it would straddle both the vadose and phreatic zones and direct most of the RF energy into the capillary fringe. The two SVVs were completed with 10-foot screens positioned to intersect the water table to provide not only vapor extraction but also groundwater sampling points. The single groundwater vent was constructed with a 3-foot screen positioned roughly 15 feet below the water table. All structures were positioned to assure (1) optimum capture of volatilized hydrocarbon vapors, (2) optimum monitoring configuration during the test, and (3) flexibility of function during the test.

The equipment used to operate the SVE and GWV system was housed in a mobile remediation unit. The trailer-contained unit included all blowers, compressors, piping, and gauges necessary for the operation of the SVE and GWV systems. The mobile unit was also equipped with a programmable logic controller (PLC) which allowed remote access and control at the site via telemetrics.

The equipment used to supply the power for the RFH was also housed in a mobile unit. The RFH trailer contained computers for power control, data collection, instrumentation, and RF generator remote operation. Three-phase AC power for running the RFH system was provided by a diesel generator.

COMPUTER MODELING

To gain a better understanding of the RFH process, computer models based on the ideal gas law and finite difference time domain (FDTD) electromagnetic analysis were employed to predict the effect of RFH on contaminant removal and to determine the radiation pattern of the applicator. The ideal gas law model which was based on previous work (Johnson, et al) provides a way of predicting recovery rates by SVE of weathered gasoline constituents through knowledge of their vapor pressures. Numerical results indicated a significant increase in the vapor pressure of the contaminant results from an increase in soil temperature (approximately an order of magnitude for every 50°C increase) (Cox, 1923) using RFH. Figure 3 compares the removal rates of weathered gasoline by SVE from soil at ambient temperature (50°F) and after RFH has increased the soil's temperature to 150°F. The initial amount of weathered gasoline, 400 gallons, was determined

Table 1.   Blaine, MN Site Groundwater Analytical Data

WELL #	DATE	B	E	T	X	GRO
MW-11	03/27/96	88.9	31.1	19	77.7	990
	04/04/96	88.6	11.2	6	75.9	880
	04/08/96	36.4	7.1	86.5	142	590
SVV-15	03/27/96	4820	599	5940	4590	22600
	04/04/96	97.2	31.8	291	449	2690
	04/08/96	327	83.8	1540	2171	7020
SVV-16	03/27/96	2200	288	1904	1282	13500
	04/04/96	709	144	1280	974	4390
	04/08/96	113	19.5	161	211.5	1390
SVP-6	03/27/96	2160	720	4640	5830	36300
	04/04/96	125	17.4	27.8	194.3	940
	04/08/96	62.2	7.7	130	96.5	1510
SVP-7	03/27/96	2380	296	1060	2325	12000
	04/04/96	129	35.6	74.4	208.6	1240
	04/08/96	21.7	3.5	15.5	74.5	570
SVP-8	03/27/96	893	188	147	1064	5920
	04/04/96	171	3.4	161	194.5	1290
	04/08/96	15.3	5.2	16.8	102	310
GP-1	02/07/96	2900	520	270	800	7200
GP-1A	04/12/96	110	650	3400	7600	37000
GP-2	02/07/96	38	34	130	130	620
GP-2A	04/12/96	5	4	25	28	140
GP-3	02/07/96	210	<50	230	3300	29000
GP-3A	04/12/96	<1	<1	2	1	<100
GP-4	02/07/96	590	160	65	110	1600
GP-4A	04/12/96	<1	<1	3	2	<100
GP-5	02/07/96	2.2	<1	1.6	5.3	110
GP-5A	04/12/96	<1	<1	<1	<1	<100
GP-6	02/07/96	4.5	<1	1.6	1.3	<100
GP-6A	04/12/96	1	<1	<1	<1	<100
GP-7	02/07/96	1700	220	700	680	4600
GP-7A	04/12/96	2	6	11	16	140

Explanation:
All results are expresed in micrograms per liter (ug/L) which is equivalent to parts per billion (ppb)

B = benzene	MW = monitoring well
E = ethyl benzene	SVV = soil vapor vent
T = toluene	SVP = soil vapor probe
X = xylene	GP = Geoprobe test boring
GRO = gasoline range organics	

Table 2. Blaine, MN Site Site Soil Sample Analytical Results

Sample Location	DATE	Sample Depth	B	E	T	X	TPH as GRO	TPH as FO
GP-1	02/07/96	6'	<0.005	<0.005	<0.005	<0.005	<0.25	<0.25
		8'	<0.005	<0.005	<0.005	<0.005	<0.25	<0.25
		10'	<1.0	5.9	11	72	1000	<.250
		12'	<0.25	<0.25	0.61	2.5	22	<12.
GP-1A	04/12/96	6'	NA	NA	NA	NA	NA	NA
		8'	NA	NA	NA	NA	NA	NA
		10'	0.84	0.8	0.84	0.8	89	<125
		12'	0.28	<0.25	<0.25	0.54	5.8	<13.
GP-2	02/07/96	6'	<0.005	<0.005	<0.005	<0.005	<0.25	<0.25
		8'	<0.005	<0.005	<0.005	<0.005	<0.25	<0.25
		10'	<25.	<25.	100	260	3200	<1200
		12'	5.7	23	60	120	1400	<620
GP-2A	04/12/96	6'	NA	NA	NA	NA	NA	NA
		8'	NA	NA	NA	NA	NA	NA
		10'	0.46	2.1	2.2	8.4	71	<0.25
		12'	0.19	1.7	0.89	2.5	59	<12
GP-3	02/07/96	6'	<0.005	<0.005	<0.005	<0.005	<0.25	<0.25
		8'	<0.005	<0.005	<0.005	<0.005	<0.25	<0.25
		10'	<25.	48	230	260	2300	<1200
		12'	0.84	<25.	1.1	1.7	11	<12.
GP-3A	04/12/96	6'	NA	NA	NA	NA	NA	NA
		8'	NA	NA	NA	NA	NA	NA
		10'	0.8	5.7	49	210	1000	<13000
		12'	8.3	13	40	110	540	<250
GP-4	02/07/96	6'	<0.005	<0.005	<0.005	<0.005	<0.25	<0.25
		8'	<0.005	<0.005	<0.005	<0.005	<0.25	<0.25
		10'	0.26	<25.	1.2	1.8	7.2	<12.
		12'	1.6	0.33	0.35	0.57	<5.0	<12.
GP-4A	04/12/96	6'	NA	NA	NA	NA	NA	NA
		8'	NA	NA	NA	NA	NA	NA
		10'	4	<1	23	87	320	<50
		12'	5.6	5.6	32	25	120	<120
GP-5	02/07/96	6'	<0.005	<0.005	<0.005	<0.005	<0.25	<0.25
		8'	<0.005	<0.005	<0.005	<0.005	<0.25	<0.25
		8'	<0.005	<0.005	<0.005	<0.005	<0.25	<0.25
		10'	<0.005	<0.005	<0.005	<0.005	<0.25	<0.25
GP-5A	04/12/96	6'	NA	NA	NA	NA	NA	NA
		8'	NA	NA	NA	NA	NA	NA
		10'	<0.005	<0.005	<0.005	<0.005	<0.25	<0.25
		12'	<0.005	<0.005	<0.005	<0.005	<0.25	<0.25
GP-6	02/07/96	6'	<0.005	<0.005	<0.005	<0.005	<0.25	<0.25
		8'	<0.005	<0.005	<0.005	<0.005	<0.25	<0.25
		10'	<0.005	<0.005	<0.005	<0.005	<0.25	<0.25
		12'	<0.005	<0.005	<0.005	<0.005	<0.25	<0.25
GP-6A	04/12/96	6'	NA	NA	NA	NA	NA	NA
		8'	NA	NA	NA	NA	NA	NA
		10'	<0.005	<0.005	<0.005	<0.005	<0.25	<0.25
		12'	<0.005	<0.005	<0.005	<0.005	<0.25	<0.25
GP-7	02/07/96	6'	<0.005	<0.005	<0.005	<0.005	<0.25	<0.25
		8'	<0.005	<0.005	<0.005	<0.005	<0.25	<0.25
		10'	<5.0	12	26	110	1200	<250
		12'	5.5	19	89	143	970	<250
GP-7A	04/12/96	6'	NA	NA	NA	NA	NA	NA
		8'	NA	NA	NA	NA	NA	NA
		10'	<5.	26	10	34	240	<250.
		12'	<5.	18	57	110	430	<250.

Explanation:

All results are reported in milligrams per kilogram (mg/kg) which is equivalent to parts per million (ppm).

B = benzene E = ethyl benzene

T = toluene X = xylene

GP = Geoprobe test boring TPH as FO = total petroleum hydrocarbons as fuel oil using the 1,000 ft³ soil volume (Hoag and Marley, 1986). The result shows that under the same conditions the SVE system will complete recovery of this gasoline approximately four times faster when the soil is heated by RFH to 150°F. The FDTD model (REMCON, 1994) running on a Silicon Graphics 132 MHz workstation was used to calculate the electromagnetic field produced by the RFH applicator. The FDTD method is particularly useful for modeling the electric field and SAR (specific absorption rate in watts per kilogram of soil) patterns produced by an RFH system as it allows the electromagnetic test environment to be completely specified. During this test the RFH applicator straddled three different materials (i.e., the vadose zone, capillary fringe, phreatic zone) and was also surrounded by an air-filled borehole.

Two FDTD runs were conducted, one in which the conductivity of the phreatic zone was set at 9.63 mS/m, and another where it was increased by a factor of 5 to 50 mS/m. A comparison between Figures 4 and 5 demonstrates how the actual SAR pattern depends on the electrical conductivity. A greater focus of energy on the capillary fringe occurs when the phreatic zone electrical conductivity is significantly larger than that of the vadose zone. This information was helpful in positioning the applicator for maximum heating effect in the capillary fringe. (Note: OdB on the SAR Pattern Plots equals 10.2 watts/kg absorbed in the soil.)

Figure 3. **Comparison of removal rates at ambient temperature (50°F) and after RFH (150°F).**

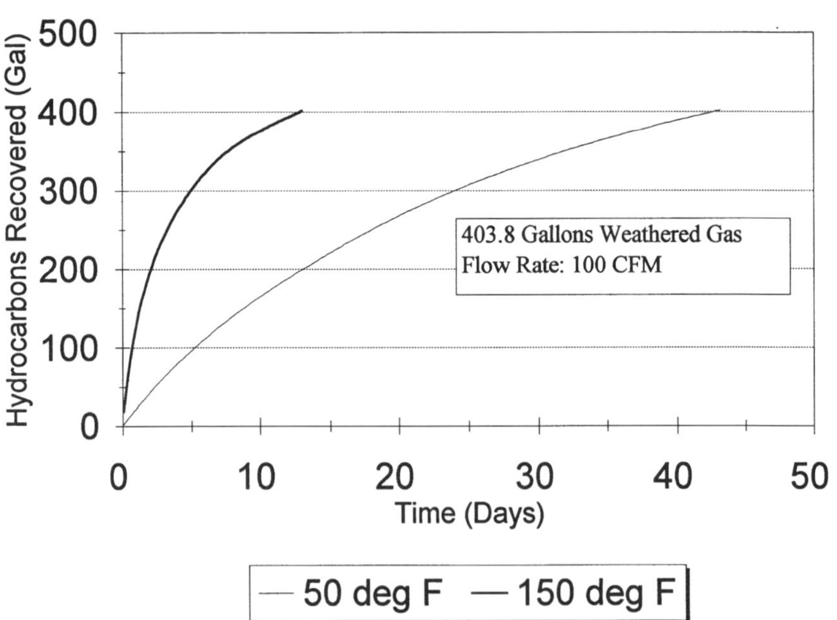

Figure 4. **RFH SAR pattern calculated by FDTD method (phreatic zone = 9.63 mS/m).**

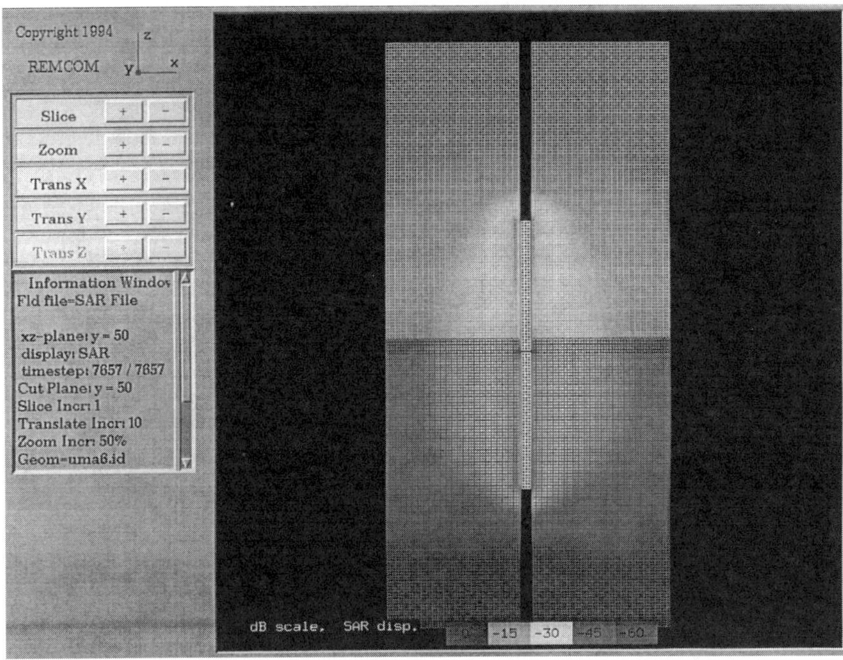

Figure 5. **RFH SAR pattern calculated by FDTD method (phreatic zone = 50.0 mS/m).**

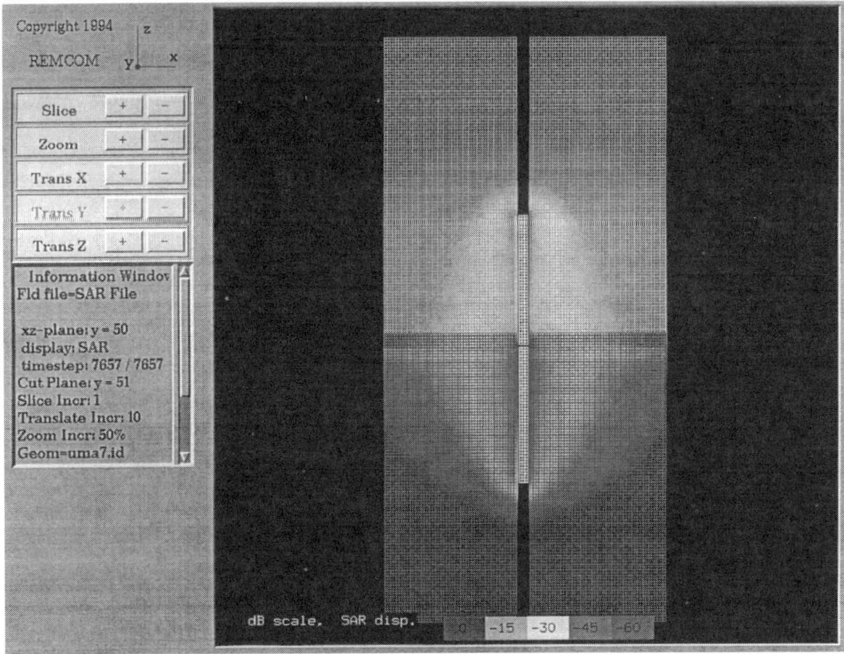

DESCRIPTION OF PILOT TEST

On March 5, 1996, the SVE portion of the system was started with an anticipated three-week heating period. The two SVVs were set at the same flow rate to maintain a uniform vacuum at the RFH well. A flame ionization detector (FID), calibrated to a methane standard, was used to periodically measure the SVE vapor stream effluent concentrations. After FID measurements peaked and declined to baseline, RFH at 27.12 MHz was started at 5 KW on March 14.

Temperature at various monitoring points was periodically measured as a function of depth and time. Numerous fluctuations in effluent vapor stream concentrations were observed during the ensuing three weeks of the test; these are graphically represented in Figure 6. Figure 7 indicates temperature fluctuations with respect to time both at the RFH applicator center section and at a 5-foot radius from the applicator.

Soil temperatures were raised from an ambient of roughly 8°C to greater than 100°C in the immediate vicinity of the RFH applicator and to 40°C five feet from the applicator, at a depth of 8.5 feet bgs. Generally, as soil temperatures increased, effluent vapor concentrations increased.

Figure 6.   **RFH/SVE remediation system performance parameters.**

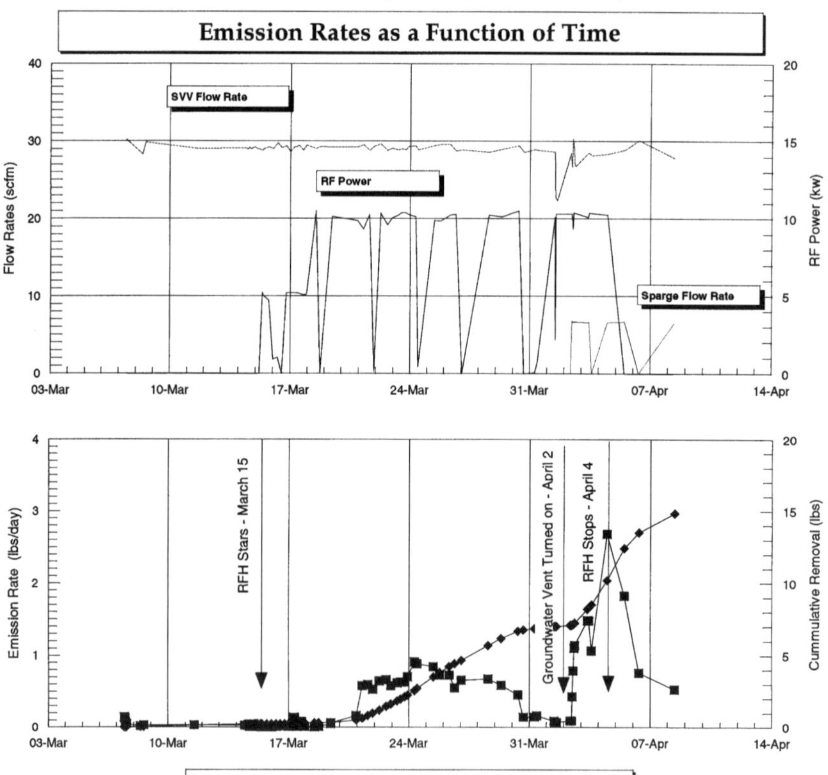

As the vapor concentrations began to plateau, the groundwater vent was turned on-line on April 2; vapor concentrations immediately began to increase. Following this action, soil temperatures quickly dropped, an evident effect of additional moisture in the vadose zone; however, within 24 hours temperatures began to increase again. The test was completed and power to the RFH applicator was shut down on April 4, 1996. Data collected after the power was shut down

Figure 7. RFH/SVE remediation system performance parameters.

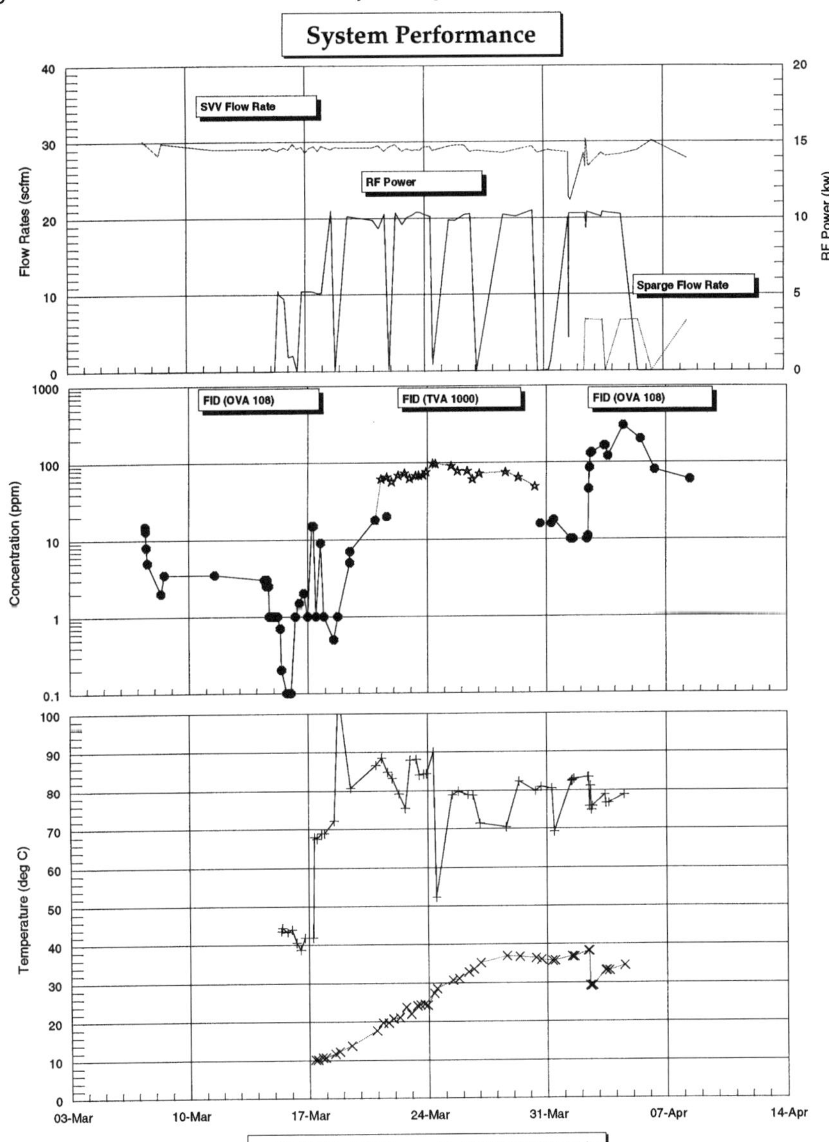

indicated that residual heat resulted in elevated SVE effluent readings for approximately one week following shutdown before returning to baseline levels. A total of 2,300 KW hours was delivered to the soil at 27.12 MHz at power levels which ranged between 5 and 10 KW.

Following shutdown on April 4, a second round of groundwater samples was collected from the structures referenced earlier. A third round of samples was collected from these structures four days later, on April 8, 1996. The results of all of these sampling events are included in Table 1.

On April 12, a second round of Geoprobe borings, immediately adjacent to the first round of borings and designated with an "A" following the probe number (e.g., GP-1A was advanced immediately adjacent to GP-1), were advanced to determine post-test soil and groundwater conditions. The results are included as Table 2.

Results

The three-week study resulted in significant decreases in both soil and groundwater impact. These reductions are attributed to accelerated volatilization of hydrocarbons upon introduction of RFH.

Groundwater concentrations were reduced by one or two orders of magnitude in most sampling locations. The most dramatic decrease was evident in the area of GP-3/3A, where Gasoline Range Organics (GRO) levels were reduced from 29 parts per million (ppm) to less than 0.1 ppm. Significant decreases were also achieved in SVV-15, SVV-16, SVP-6, and SVP-7 (refer to Table 2 for full results).

Soil concentrations revealed similar reductions in most cases (refer to Figure 2 and Table 1), the most dramatic being in the area of GP-2/2A, where GRO concentrations were reduced from 3.2 ppm to 0.71 ppm at a depth of 10 feet bgs. Similar reductions were also observed at GP-1/1A and GP-7/7A.

Interestingly, the data also indicates increases in GRO soil impact at GP-4/4A at 10 and 12 feet bgs, and at 12 feet bgs in GP-3/3A. These increases are attributed to hydrocarbon redistribution induced during the study or to heterogeneities in contaminant distribution inherent at such sites. Similarly, groundwater samples from GP-1/1A indicated increases in GRO as well as ethyl benzene, toluene, and xylenes. This increase, the only increase in groundwater impact apparent post-test, is attributed to volatilization and dissolution of contaminants previously trapped beneath the water table in the form of a nonaqueous phase liquid (NAPL). Given GP-1/1A's proximity to the RFH applicator well, this result in not surprising, and would likely have been buffered had the study been extended or the technology applied over a longer period of time, allowing for full volatilization of the NAPL present. The three-week study resulted in significant decreases in both soil and groundwater impact in the area of the study.

An estimated total mass of approximately 15 pounds of hydrocarbons was removed from the subsurface during the course of the test.

Discussion

The introduction of RF energy within and adjacent to the capillary fringe at this site resulted in mobilization and volatilization of petroleum hydrocarbons occurring in the form of NAPL held in soil retention or beneath the water table. The study verified what thermodynamic principles dictate, that liquids will preferentially undergo a phase transfer to a gaseous state at elevated temperatures.

At this site, elevating soil and groundwater temperatures from a relatively low ambient of 8°C to 100°C near the applicator had marked effects on hydrocarbon removal rates. Despite the limited duration of the test, significant reductions in soil and groundwater impact were achieved. Given a longer period and/or more expansive area of RFH application, even greater results could be anticipated.

The effect of groundwater ventilation on the RFH/SVE system deserves further analysis. Though initially resulting in a cooling effect, the GWV provided dramatically increased SVE effluent levels. The cooling effect was overcome by the RFH energy in a relatively short time frame, indicating no obviously detrimental effects when combining the three technologies.

A necessity in utilizing these three technologies is adequate vapor capture and careful system control. Indiscriminate application of RFH could result in an uncontrolled vapor accumulation in underground structures or utilities. Considerations for air discharge monitoring and control are also important. The technologies result in accelerated removal rates which will likely require some means of post-vapor treatment in most cases.

The technology heats most soil types uniformly, regardless of grain size (discounting the relatively negligible effect of varying soil electrical conductivities which affects the soil's ability to absorb RF energy). Since vacuum inducement and vapor capture are very dependent on soil permeability, uniform recovery at sites where soils are more heterogeneous than the subject site should be considered. In soils where adequate capture vacuum cannot be achieved, contaminant volatilization and recondensation following RFH application may occur in the vadose zone. In addition, accelerated dissolution of NAPL in the phreatic zone, if not addressed via GWV, could result in greater levels of groundwater impact in the area of application.

Conclusions

The combination of RFH, SVE, and GWV provides an effective means of accelerating in-situ volatilization of petroleum hydrocarbons from the vadose and phreatic zones. At the subject site, significant hydrocarbon concentration reductions were observed in both soil and groundwater in a very short time period.

REFERENCES

Cox, E.R. 1923, *Industrial Eng. Chem.*, 15, 592.

Hoag, G.E. and Marley, M.C. 1986. Gasoline Residual Saturation in Unsaturated Uniform Aquifer Materials, *J. Environ. Eng.*, 112(3). West Hollow Research Center, Houston, TX 77251-1380.

Hough, B.K. 1969. *Basic Soils Engineering*, 2nd ed. Rolland Press Co.

JAYCOR, San Diego, CA, January 12, 1981.

Johnson, P.C., Kemblowski M.W., Colhart, J.D., Byers, D.L., Stanley, C.C. A Practical Approach to the Design, Operation, and Monitoring of In-Situ Soil Venting Systems. Shell Development/Shell Oil Company, WestHollow Research Center, Houston, TX 77251-1380.

Kunz, K. and Luebbers, R. 1993. *The Finite Difference Time Domain Method for Electromagnetics*, Boca Raton, FL, CRC Press.

Mallon C., et al. Low Field Electrical Characteristics of Soil, Theoretical Notes, Note 315.

Pearce, J. and Zuluaga, A. 1996. Electrical Properties of Soils at Radio Frequencies, University of Texas and University of Illinois, presented at the 31st Annual IMPI Symposium, Boston, MA, July 1996.

REMCOM, Inc. 1994. User's Manual for XFDTD for Electromagnetic Calculations, State College, PA.

CHAPTER 17

Remediation of Dioxin Contaminated Soil Enhanced by Chemical Oxidation Pretreatment

Jimmy Kao, Geophex, Ltd., Raleigh, North Carolina

Andrew Dasinger, Sykorsky Aircraft, Connecticut

Sharon C. Long, Civil and Environmental Engineering, University of Massachusetts, Amherst

Christopher C. Lutes, Acurex Environmental Corporation, Research Triangle Park, North Carolina

Steve J. Gagnon, U.S. Army Corps of Engineers, Waltham, Massachusetts

INTRODUCTION

Many hazardous waste sites are contaminated with organochlorine compounds, including polychlorinated dibenzo-*p*-dioxins (dioxins). Dioxins, especially 2,3,7,8-tetrachlorodibenzo-*p*-dioxin (TCDD), are considered to be extremely toxic environmental pollutants. Dioxins can enter the environment from numerous sources including municipal solid waste incinerator emissions, pulp and paper mill waste discharges, and through the manufacture and use of organochlorine pesticides. Once dioxins are in the environment, they tend to persist, as they have been shown to be resistant to biodegradation and tend to become strongly bound to soil organic matter.

Incineration has been the technology relied upon to treat dioxin-contaminated soils. However, incineration is becoming more politically unpopular and is both energy intensive and relatively costly. Chemical oxidation is becoming a viable alternative for treatment of low-biodegradable organic wastes. Hydrogen peroxide (H_2O_2) oxidation in the presence of ferrous iron (Fe^{2+}) (known as the Fenton Reagent) has been highly effective in oxidatively destroying highly-chlorinated, environmentally persistent compounds. In fact, Fenton's Reagent has been proposed as the most effective oxidizing agent for organochlorine compounds (Carberry, 1994). The chain reactions induced by Fenton's Reagent first produce the highly reactive hydroxyl radical, $\cdot OH$:

$$H_2O_2 + Fe^{2+} \rightarrow Fe^{3+} + \cdot OH + OH^-$$

Hydroxyl radical generation is enhanced at low pH. Hydroxyl radicals are a highly reactive species that tend to be nonspecific reactants capable of reacting with dioxins, when they are present in the system. Several reaction pathways have been proposed by researchers (Rugge et al., 1993). Figure 1 shows two of the possible cleavage patterns during the dioxin degradation. The resulting products include chlorophenols (CPs), chlorobenzenes (CBs), and lower chlorinated dioxins. These products are more water soluble and more biodegradable, and therefore can be more easily removed from soil systems through biodegradation or carbon adsorption compared with incineration or soil washing methods.

The goal of this study was to investigate whether a two-stage, slurry-phase treatment strategy could be employed to decontaminate soil containing TCDD. The first stage would involve treatment with Fenton's Reagent to chemically dechlorinate and partially oxidize TCDD, forming CPs, CBs, and other less-chlorinated compounds more amenable to biodegradation. Thus, the first step in the treatment under investigation is in essence a chemical "softening" step, and is not intended to achieve complete chemical destruction. The chemically treated soils

Figure 1. Schematic diagram of symmetrical and asymmetrical OH cleavage of TCDD.

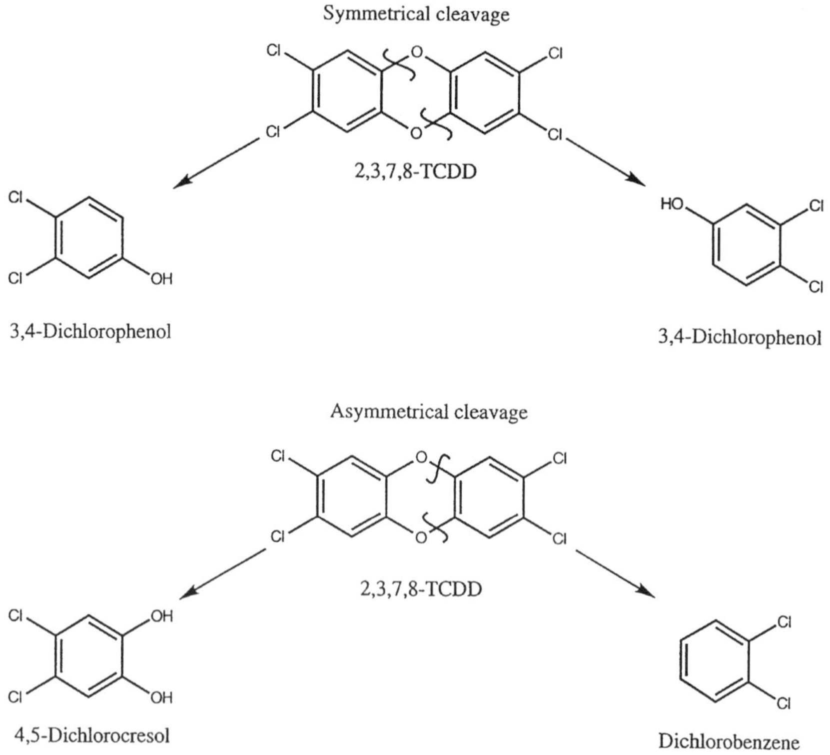

would then be transferred to a second reactor, seeded with acclimated microbial consortia, to biodegrade the residual oxidation by-products. Any residual H_2O_2 would serve as an electron acceptor source for microbial activity.

OXIDATION EXPERIMENTS

Soil Spiking Procedures

To produce the dioxin contaminated soils used in this study, a sandy soil was spiked with TCDD. To prepare the spiked soil, 2 kg of uncontaminated soil was homogenized in a stainless steel container. The soil was spiked with 25 mg of TCDD dispersed in 320 mL of a 45/55% (v/v) acetone/hexane solution. The soil was then further homogenized. The solvents were allowed to evaporate from the soil by placing the container of spiked soil in a fume hood, thus leaving behind the dioxin in the soil at a theoretical concentration of 12,500 pg/g.

The effectiveness of the spiking procedure was verified through extraction and analysis of 10 g subsamples immediately after preparation of the spiked soil (replicates 1 through 4) and again two months after preparation (replicate 5). The procedures for soil extraction and TCDD analysis are described in the following sections. Results indicate that spiking of the soil was homogeneous, and recovery was very reproducible (Table 1). The average recovery was 54% with a coefficient of variation of 7.3. Less than 100% recovery of a spiked compound is an expected result due to irreversible binding at certain soil surface sites.

Table 1. **Extraction Efficiency of TCDD Spiked Soil**

Replicate	TCDD Recovered (pg/g)	% TCDD Recovered
1	5,998	48
2	6,284	50
3	6,875	55
4	6,252	50
5	7,134	57
Average	6,509	54

A high concentration of TCDD was added to another batch of soil to confirm the types of products produced. Seventy-five mL of TCDD stock solution (with a concentration of 10 mg TCDD/mL hexane) in an acetone/hexane mixture was equally dispensed into 300 g of previously spiked soil. The acetone/hexane was allowed to evaporate. The final TCDD concentration in this soil was 2,512,500 pg/g.

The effectiveness of the second spiking procedure was verified through extraction and analysis of two 10-g subsamples immediately after preparation of the spiked soil. The results are presented in Table 2. The data indicate that the TCDD was again spiked homogeneously throughout the soil. The recovery efficiency was low for this experiment (14% on average). This might be due to solubility limitations of the TCDD in the spiking solution. However, these high-concentration spiked soils were used for oxidation by-product analysis, and a high recovery efficiency was not necessary.

Table 2. Extraction Efficiency of High-Concentration TCDD Spiked Soil

Replicate	TCDD Recovered (pg/g)	% TCDD Recovered
1	370,340	15
2	318,801	13
Average	344,571	14

Soil and Water Extraction Procedures

A slurry-phase reaction was used for oxidation experiments. The soil extraction procedure included internal standardization, overnight Soxhlet extraction with toluene (16 to 24 hours), Snyder column concentration, and nitrogen gas vaporization to a final volume of 50 mL. For the liquid phase samples, liquid-liquid extraction techniques were applied to extract TCDD. Liquid was mixed with 10 mL toluene, shaken for approximately one minute, then the upper toluene phase was drawn. This procedure was repeated five times to achieve complete TCDD extraction from the liquid phase. After the extraction, collected toluene was passed through a sodium sulfate column to remove any water from the toluene solution, then concentrated down to 50 mL for gas chromatography (GC) analysis, using Snyder column concentration and nitrogen gas evaporation. A known quantity (100 mL) of isotopically-labeled internal standards (^{13}C-TCDD and ^{13}C-1,2,3,7,8,9-hexachlorodibenzo-p-dioxin) was also added before the extraction to measure the recovery efficiency.

In later experiments, the goal was to validate that the hypothesized oxidation by-products were actually being produced. For these analyses, liquid-liquid extraction was used to extract CPs and CBs from both the soil and liquid phase after the oxidation process (EPA Method 8040). For the soil-phase extraction, 10 g of sodium sulfate and 100 mL of internal standards were added to the soil, followed by acetone/methylene chloride (50/50) extraction. After five passes of liquid-liquid extraction, the collected solvent was concentrated down to one mL for GC analysis. For the liquid-phase extraction, sulfuric acid was used to adjust the sample to a pH of 1 to 2, then methylene chloride was used to perform the liquid-liquid extraction using a separatory funnel. After five passes of extraction, the volume was concentrated down to one mL for GC analysis, using a K-D concentrator and nitrogen gas evaporation.

Analytical Methodology

A high resolution GC/electron capture detection methodology with internal standardization was utilized, similar to methods employed previously by Nestrick et al. (1980). Calibration curves were developed for TCDD, 2-monochlorodibenzo-p-dioxin (MCDD), 2,7-dichlorodibenzo-p-dioxin (DCDD), and 2,3,7-trichlorodibenzo-p-dioxin (TriCDD), using varying concentrations from 50 to 2,000 pg/mL, and constant amounts of internal standards (250 pg/mL). Calibration was later repeated at a higher range (500 to 50,000 pg/mL) to allow analysis of high-concentration samples. All native and isotopically labeled PCDDs were obtained from Cambridge Isotope Laboratory, Andover, Massachusetts.

A 30 m ¥ 0.25 mm DB-5 capillary column with a 0.25 mm film was used to separate compounds. The injector temperature was 290°C and the detector temperature was 300°C. The oven temperature was programmed to increase from 150°C (5 min) to 290°C (35 min) at 3°C/min, with a carrier gas (He) flow of 1 mL/min (15 psig) and make-up gas (N_2) flow of 30 mL/min.

CP and CB concentrations were analyzed on a high resolution GC with a flame ionization detector (FID) (EPA Method 8040). A 30 m x 0.25 mm DB-5 capillary column with a 0.25 mm film was used to separate compounds. The injector temperature was 250°C and the detector temperature was 250°C. The oven temperature was programmed to increase from 80°C (0 min) to 150°C (12 min) at 4°C/min.

Five CP and CB standards were prepared to develop the calibration curves for compounds including 2-CP, 3+4-CP, 3,4-CP, CB, and 1,2-dichlorobenzene (1,2-DCB). Concentrations in these standards varied from 50 to 8,000 pg/mL. A phenol compound mixture (Absolute Standard, Inc.) was also obtained, which included 13 different compounds [4-chloro-3-methylphenol (4-C-3-MP), 2-CP, 2,4-dichlorophenol, 2,6-dichlorophenol, 2,4-dimethylphenol (2,4-DMP), 2,4-dinitrophenol, 2-methyl-4,6-dinitrophenol, 2-nitrophenol, 4-nitrophenol, pentachlorophenol, phenol, 2,4,6-trichlorophenol, and 2,3,4,6-tetrachlorophenol]. This mixture was prepared using a 1:100 dilution in 2-propanol, and analyzed by GC/FID to identify the by-product peaks after oxidation.

H_2O_2 concentrations were measured using a Hach titration kit. Ferrous ion and total iron were measured using a Hach test kit and a spectrophotometer. Dissolved oxygen (DO) and pH values were monitored using a Corning DO meter and a Fisher pH meter, respectively. These analyses were used to study the oxidation conditions.

Oxidation Experiment Procedures

Each oxidation reactor was prepared in a 40-mL volatile organic analysis (VOA) vial, and contained 10 g of TCDD-spiked soil and Fe^{2+} solution. Oxidation conditions were varied by using different concentrations of ferrous sulfate ($FeSO_4 \cdot 7H_2O$) and H_2O_2. In this study, three different H_2O_2 concentrations (2%,

0.2%, and 0.02%, which corresponds to 1, 0.1, and 0.01 mL of a 30% H_2O_2 solution in vials) were used to evaluate the effects on the oxidation. The pH was adjusted to 3.5 by adding H_2SO_4, and the final liquid volume was 16 mL. A total of six oxidation experiments were conducted in this study. Table 3 summarizes the conditions imposed for each oxidation experiment.

Table 3. **Summary of Conditions Used for Oxidation Experiments**

Expt. No.	Recovered TCDD (pg/g)	Fe2+ (mg/L)	H2O2 (mg/L)	Reaction Time (hr)	Rotator
1	6,352	277	2,000	0, 3, 6, 12, 24	not used
2A	6,352	277	20,000	0	used
2B	6,352	277	20,000	1	used
2C	6,352	1,385	20,000	1	used
2D	6,352	277	20,000	6	used
3A	7,134	2,770	20,000	6	used
3B	7,134	277	2,000	6	used
4[a]	7,134	2,770	20,000	2	used
5A	334,571	2,770	20,000	6	used
5B[b]	334,571	0	0	6	used
6	334,571	1,108	8,000	5	used

[a]Sonicator used, [b]Control and no-oxidation.

After a predetermined reaction time (which varied from 0 to 24 hours), the soil and supernatant in each sample were separated and analyzed for TCDD and inorganic indicators (Fe^{2+}, total iron, H_2O_2, and pH). The 0-hour reaction time represents conditions immediately (i.e., less than one minute) after the addition of Fenton's Reagent. TCDD in supernatant was analyzed in oxidation Experiments 3 to 5. By-products in soil and supernatant were analyzed in Experiments 5 and 6. Oxidation reactions for all experiments were quenched by adding concentrated H_2SO_4 to each vial to lower the pH to below 2. To enhance the TCDD removal rate, a rotator was used to agitate the oxidation vials for Experiments 2 through 6.

In this first batch experiment, oxidation effectiveness was evaluated for five reaction time periods (0, 3, 6, 12, and 24 hours). The second oxidation experiment was conducted to assess the effect of H_2O_2 and Fe^{2+} doses on the TCDD removal efficiency. The third experiment was designed to investigate whether an even higher dose of Fe^{2+} catalyst, in combination with a longer reaction time (more than 3 hours) would increase TCDD removal. The fourth experiment was designed to assess the effect of sonication treatment on enhancing removal of TCDD from soils during oxidation. The fifth experiment was conducted to identify TCDD by-products after oxidation. The high-concentration TCDD-spiked soil was used in an attempt to amplify by-product GC peaks so that they could be detected. In this experiment, four vials containing 10 g of respiked soil and 16 mL of water were prepared as control vials. After six hours of equilibration, we analyzed the TCDD and by-product concentrations in both the soil and water phases. These control vials were used to delineate the percentage of TCDD and by-product distributed in the soil and water phases in the absence of Fenton's Reagent.

The sixth experiment was conducted by starting with 0.1 mL of 30% H_2O_2 and 16 mL of Fe^{2+} (5 mM) with a one-hour oxidation period. After one hour of reaction time, the same amounts of H_2O_2 and Fe^{2+} were added in each vial to activate the second oxidation process. This process was repeated three times after the first oxidation. This procedure allowed for greater TCDD oxidation and allowed for the appearance of by-products to be confirmed and identified in the soil and water phases after oxidation.

OXIDATION EXPERIMENT RESULTS

Table 4 presents the results for Experiment 1. When the oxidized samples are compared with the blank samples (Table 1), the oxidation results indicate that approximately 50 to 60% of the extractable TCDD was removed from the treated soil. Analytical results show that H_2O_2 and Fe^{2+} concentrations dropped significantly within the first 6 hours of reaction, then stopped for the remaining 18 hours. Furthermore, approximately 61% of the extractable TCDD was removed within one minute (0-hour vial), with one-third of the H_2O_2 and two-thirds of the Fe^{2+} depleted. This indicates that the oxidation reaction occurs rapidly, with no lag period.

Table 4. **Results of Oxidation Experiment 1**

Reaction Time (hour)	TCDD in Soil (pg/g) [a]	Fe2+ before (mg/L) [b]	Fe2+ after (mg/L) [c]	Total Fe (mg/L) [d]	H2O2 before (mg/L) [b]	H2O2 after (mg/L) [c]	TCDD Removal (%) [e]
0	2,503	277	90	290	2,000	1350	60.6
3	1,685	277	17	230	2,000	320	73.5
6	4,255	277	9	210	2,000	140	33.0
12	2,776	277	7.5	235	2,000	140	56.3
24	2,920	277	7	220	2,000	140	54.0

[a]TCDD in Soil (pg/g) = the amount of TCDD remaining in the spiked soil after oxidation. [b]Fe^{2+} and H_2O_2 before (mg/L) = the theoretical concentration of Fe^{2+} and H_2O_2 in each vial before oxidation. [c]Fe^{2+} and H_2O_2 after (mg/L) = the concentration of Fe^{2+} and H_2O_2 in each vial after oxidation. [d]Total Fe (mg/L) = the amount of Fe^{2+} and Fe^{3+} in each vial after oxidation. [e]TCDD Removal (%) = the percentage of TCDD removal from soil = {1 - [the amount of TCDD in soil after oxidation divided by the amount of recovered TCDD from spiked soil (6352 pg/g)]} x 100.

No significant TCDD removal occurred after the initial rapid oxidation [time 0 (less than one minute)]. Therefore, it can be concluded that part of the H_2O_2 depletion after time 0 was caused by the reaction with Fe^{2+}, and most of the depletion was caused by the dissociation into water and oxygen. The pH dropped from 3.5 to 2.5 as a result of the oxidation reaction. This indicates that chloride ions were removed from TCDD, generating HCl in the solution. Because different reaction vials were sacrificed at each time, variability between subsamples may cause the 6-hour vial to appear to have less TCDD removal than the 0-hour vial.

The results for Experiment 2, which varied Fenton's Reagent concentrations, are presented in Table 5. Upwards of 80% of the extractable TCDD was removed in the 25 mM (1,385 mg/L) Fe^{2+} treatment after one hour of reaction, and 5 mM (277 mg/L) Fe^{2+} treatment after 6 hours of reaction. Approximately 70% of the extractable TCDD was removed in the 5 mM Fe^{2+} treatment after one hour of reaction. These data indicate that high concentrations of Fenton's Reagent (20,000 mg/L H_2O_2 and 1,385 mg/L Fe^{2+}) and longer reaction times (longer than one hour) are more effective than the oxidation conditions tested in oxidation Experiment 1 (2,000 mg/L H_2O_2 and 277 mg/L Fe^{2+}).

Table 5. **Results of Oxidation Experiment 2**

Reaction Time (hour)	TCDD in Soil (pg/g) [a]	Fe^{2+} before (mg/L) [b]	Fe^{2+} after (mg/L) [c]	Total Fe (mg/L) [d]	H_2O_2 before (mg/L) [b]	H_2O_2 after (mg/L) [c]	TCDD Removal (%) [e]
0	1,639	277	20	210	20,000	5,000	74.2
1A	1,890	277	25	150	20,000	1,800	70.2
1B	1,028	1,385	150	1,000	20,000	400	83.8
6	622	277	23	160	20,000	3,800	90.2

[a], [b], [c], [d], and [e] (please refer to Table 1).

The results for Experiment 3, which utilized a higher dose of catalyst (20,000 mg/L H_2O_2 and 2,770 mg/L Fe^{2+}) in addition to a long reaction time, are presented in Table 6. An average of 80% TCDD removal was observed in higher-dosed vials, and approximately 85% removal was observed in the lower-dosed (2,000 mg/L H_2O_2 and 277 mg/L Fe^{2+}) vials. These results indicate that there is a threshold of Fenton's Reagent concentration, above which oxidation efficiency is not significantly enhanced. An oxidation time of up to 6 hours provides adequate removal. Both the soil and water phases were analyzed in this experiment. Using the amount of recovered TCDD from the TCDD-spiked soil (7,134 pg/g), approximately 25% of the TCDD was desorbed from the soil phase to the aqueous phase, 20% of the TCDD remained bound to the soil particles, and 55% of the TCDD was oxidized for the high H_2O_2 and Fe^{2+} addition reactors. Approximately 24% of the TCDD was desorbed from the soil phase to the aqueous phase, 15% of the TCDD remained bound to the soil particles, and 61% of the TCDD was oxidized for the low H_2O_2 and Fe^{2+} addition reactors. Therefore, the removal of TCDD from the soil particles was demonstrated to be through either oxidation or desorption.

Table 6. Results of Oxidation Experiment 3

Reaction Time (hour)	TCDD in Soil (pg/g) [a]	$Fe^{2}+$ before (mg/L) [b]	$Fe^{2}+$ after (mg/L) [c]	Total Fe (mg/L) [d]	H_2O_2 before (mg/L) [b]	H_2O_2 after (mg/L) [c]	TCDD Removal (%) [e]
0	1,639	277	20	210	20,000	5,000	74.2
1A	1,890	277	25	150	20,000	1,800	70.2
1B	1,028	1,385	150	1,000	20,000	400	83.8
6	622	277	23	160	20,000	3,800	90.2

[a, b, c,] and [d]: please refer to Table 1 (concentration in the water phase (pg/mL) was converted to pg/g of soil). [e]TCDD Oxidized (pg/g) = the amount of oxidized TCDD. [f]: % = the percentage of TCDD in soil (or water) or the amount of oxidized TCDD = the amount of TCDD in soil (or water) after oxidation or the amount of oxidized TCDD divided by the amount of recovered TCDD from spiked soil (7,134 pg/g).

Experiment 4 investigated the effect of sonication on oxidation efficiency. Theoretically, the high-frequency sound waves created by ultrasonic instruments can generate millions of small bubbles that violently implode. The energy released by these imploding bubbles may act to physically separate TCDD from the soil surface and increase the availability of TCDD to Fenton's Reagent, thus enhancing oxidation efficiency. A similar percentage of TCDD removal was achieved, approximately 86%, compared with results from oxidation Experiment 3. Fe^{2+} was reduced from 2,770 mg/L to 75 mg/L after oxidation. H_2O_2 concentrations were reduced from 20,000 mg/L to about 600 mg/L. This consumption by oxidation reagents was similar to those measured in Experiment 3. Approximately 16% of the TCDD was desorbed from the soil phase to the aqueous phase after oxidation, 14% of the TCDD remained bound to the soil particles, and 71% of the TCDD was oxidized. Thus, these data indicate that ultrasonic treatment does not significantly enhance TCDD removal.

Table 7 presents the results from Experiment 5, which was designed to confirm the oxidation of TCDD to lesser chlorinated by-products. Fe^{2+} and H_2O_2 were reduced to 32 mg/L and 25 mg/L, respectively. These changes in oxidation reagents were higher than those measured in oxidation Experiment 3, as expected, since the TCDD concentration oxidized was three orders of magnitude higher than those used in all the other oxidation experiments. Using the amount of recovered TCDD from the high-TCDD spiked soil, approximately 54% of the TCDD was desorbed from the soil phase to the aqueous phase, 33% of the TCDD remained bound to the soil particles, and 14% of the TCDD was oxidized. The analyses for the control reactions are also presented in Table 7. A total of 338,575 pg/g of TCDD remained in the soil phase, and only 5,821 pg/g of the TCDD was released to the water phase. Compared with the recovered TCDD (344,571 pg/g), approximately 2% of TCDD dissolved in the water phase, and 98% of TCDD remained in the soil phase. Results of the control experiments indicate that without Fenton's Reagent, TCDD did not desorb from the soil. However, with the addition of Fenton's Reagent, TCDD desorbed from the soil phase, and was released to the liquid phase.

Table 7. Results of Oxidation Experiment 5

Reaction Time (hour)	TCDD in Soil (pg/g)[a]	TCDD in Water (pg/g)[a]	Fe^{2+} before (mg/L)[b]	Fe^{2+} after (mg/L)[c]	Total Fe (mg/L)[d]	H$_2$O$_2$ before (mg/L)[b]	H$_2$O$_2$ after (mg/L)[c]	TCDD Oxidized (pg/g)[e]
6A	96,553	207,830	2,770	39	2,410	20,000	30	40,188
6B	129,591	160,803	2,770	25	2,280	20,000	20	54,177
average	113,072	184,317	2,770	32	2,345	20,000	25	47,182
%[f]	32.8	53.5						13.7
6A-control	338,575	3,570	0	0	0	0	0	2,426
6B-control	na[g]	8,072	0	0	0	0	0	na
average	338,575	5,821	0	0	0	0	0	na
%[f]	98.3	1.7						0

[a, b, c, d,] and [e]: please refer to Table 4. [f]: % = the percentage of TCDD in soil (or water) or the amount of oxidized TCDD = the amount of TCDD in soil (or water) after oxidation or the amount of oxidized TCDD divided by the amount of recovered TCDD from spiked soil (334,571 pg/g). [g]: na = not available because of experimental errors.

Some of the oxidation by-products (CPs and CBs) were analyzed in both the soil and water phases. Both soil and water chromatographs did not show clear CP and CB peaks after oxidation. However, peaks corresponding to MCDD, DCDD, and TriCDD were larger. Since the initial TCDD concentration was high (2.5 mg/g), only 14% of TCDD was oxidized after the oxidation process. It is unlikely that significant by-product formation would be detected due to the lesser sensitivity of GC/FID analytical techniques. It can be hypothesized that the formation of the dechlorination products MCDD, DCDD, and TriCDD, rather than the cleavage products CPs and CBs, were caused by an insufficient quantity of Fenton's Reagent. To confirm the CP and CB peaks, a sequential oxidation experiment was conducted.

After sequential oxidation was applied in Experiment 6, chromatographs between the oxidation and control samples were compared. 1,2-DCB and 3+4-CP peaks were identified. This indicates that 1,2-DCB and 3+4-CP are two of the TCDD oxidation by-products. The peaks for CB, 2-CP, and 3,4-DCP were still not clearly resolved. However, CB and 3+4-CP are the expected by-products from the subsequent dechlorination of 1,2-DCB and 3,4-DCP, respectively. Therefore, it can be concluded that CB and 3,4-DCP are also oxidation by-products. The chromatograph of the phenol mixture (Absolute Standards, Inc.) indicates that two additional compounds, 2,4-DMP and 4-C-3-MP, may also be oxidation by-products. Since cresol isomers (o-cresol, m-cresol, and p-cresol) are the expected by-products from the subsequent oxidation of 2,4-DMP and 4-C-3-MP, they may also be oxidation by-products. Furthermore, several other significant peaks were observed on the chromatographs. They are believed to be other benzene, phenol, cresol, and catechol-related chloroaromatic compounds, or some simple straight-chain by-products.

Since MCDD, DCDD, and TriCDD are the by-products of TCDD oxidation at the very early stages, they will be subsequently degraded under the sequential oxidation process, and will not accumulate in the reactor.

ENRICHMENT AND BIODEGRADATION STUDY

Culture Enrichment and Acclimation

TCDD-contaminated soils were used as microbial inocula, and enriched by acclimation to CPs. A mineral buffer medium provided essential inorganic nutrients for microbial growth and was the matrix for microbial pollutant degrader enrichments. The mineral medium consisted of (milligram per liter of H_2O) $(NH_4)_2SO_4$, 100; K_2HPO_4, 174; KH_2PO_4, 136; $CaCl_2$, 0.4; $FeCl_3 \cdot 6H2O$, 0.3; H_3BO4, 0.03; concentrated HCl, 0.25, $MgSO_4.7H_2O$, 3; $MnSO_4 \cdot H_2O$, 0.09; and $NaMoO_4.2H_2O$, 0.01. The inocula of 100 g was placed into a 1-liter Erlenmeyer flask with 500 mL of sterile mineral buffer media containing 2-CP, 3-CP, 4-CP, and 3,4-CP. The final concentration for each CP was 0.05 mM (0.2 mM total).

Plate Counts for CP-Specific Degraders

To estimate the number of CP degraders present in the enrichments, plate counts were conducted for CP-specific degraders. In this enumeration experiment, 10-fold serial dilutions (10^{-3} to 10^{-9}) of the enrichment cultures were plated onto ultrapure agar (1.5% w/v) plates (Noble agar, Difco) using the spread plate method (APHA, 1992) and exposed to a mixture of CP vapors as the sole carbon source. These plates were placed inverted into sealed glass or stainless steel dessicators. Each dessicator contained a beaker with CP mixture, and was incubated at 20°C in the dark.

Microcosm Experiment

To assess the potential capabilities of the enriched cultures to degrade dioxin oxidation by-products in a seeded reactor, a microcosm experiment (CP biodegradation experiment) was conducted. Two 125-mL aliquots of the enrichment were transferred to sterile 250 mL centrifuge bottles. The aliquots were centrifuged at a speed of 1700xg for 15 minutes at 4°C. The supernatant was discarded to minimize interference of CPs and metabolites from the enrichment flasks, which would otherwise be carried into the biodegradation microcosms. The pellets containing the microorganisms were resuspended in sterile mineral buffer media and the two pellets from the same enrichment cultures were combined in a sterile container.

Microcosms were set up in ultraclean and sterile 40-mL VOA vials. The microcosms were prepared using the same mineral buffer media used for enrichment (9.5 mL), and injected with the washed enrichment pellets (20 mL), and amended to a final concentration of 0.10 mM CP mix (0.5 mL). The total volume of the VOA vial is approximately 44 mL. Thus, a 14 mL headspace remained in each biodegradation microcosm to provide an oxygen supply. Triplicate samples were prepared. All microcosms were inverted, to form a water seal, and placed on a platform shaker at 100 rpm. The microcosms were incubated in the dark at 25°C before analysis. Abiotic controls were prepared using 0.5% (w/v) final concentration of sodium azide to inhibit metabolism. At the appropriate times, the microcosm contents were solvent-extracted and CP concentrations were analyzed.

Plate Counts and Biodegradation Results

The numbers of CP-specific degraders were counted between 10^3 to 10^4 colony-forming units (cfu)/mL after 6 weeks of adaptation, and went up to 1.7×10^7 after 12 weeks of adaptation. Results indicate that the microbial communities became acclimated, and sizable microbial consortia were observed after three months of incubation. The results of the biodegradation experiment demonstrated that the enrichment culture contained microbial consortia capable of degrading the CP mixture under aerobic conditions. Complete degradation was observed after 17 days of incubation (Table 8).

Table 8. CP Degradation Results in Microcosm Containing Acclimated TCDD-Contaminated Soil Culture

Day	2-chlorophenol (mg/L)	3+4-chlorophenol (mg/L)	3,4-dichlorophenol (mg/L)
0	4.7a	9.4	4.6
17 (Control)	5.4	10.3	4.6
17 (Live)	< 0.05	< 0.05	< 0.05

[a]All data are average of triplicate.

CONCLUSIONS

In the first part of this study, Fenton's Reagent has been used to oxidize TCDD to less-chlorinated by-products. After a series of oxidation experiments, it was found that Fenton's Reagent can be used as both an oxidant and a solubilizing agent. It can oxidize TCDD to less-chlorinated by-products, and it appears to lower the soil/water distribution coefficient, shifting the sorption equilibrium, and releasing adsorbed TCDD for dissolution in the aqueous phase. This shift will increase the aqueous concentration of TCDD, thereby increasing the treatment efficiency. From an engineering point of view, soluble TCDD is much easier to remove by either adsorption, oxidation, or other means. In the second part of this study, active microbial consortia using CPs as the carbon source have been acclimated. Those adapted bacteria can mineralize CPs with a high degradation rate. The microbial consortia can be further acclimated by the identified oxidation by-products to increase the degradation efficiency. Results of this study indicate that the two-stage (oxidation and biodegradation) treatment process can be developed into a promising on-site remediation technology to clean up soils contaminated with TCDD and other environmentally persistent compounds.

REFERENCES

American Public Health Association (APHA). 1992. Standard Methods for the Examination of Water and Wastewater, 18th ed.

Carberry, J.B. 1994. Enhancement of Bioremediation by Partial Preoxidation. In: *Remediation of Hazardous Waste Contaminated Soils*, pp. 543-597. (Wise, D.L. and Trantolo, D.J., Eds.), New York, Marcel Dekker, Inc.

Nestrick, T.J., Lamparski, L.L., and Townsend, D.I. 1980. Identification of Tetrachlorodibenzo-p-dioxin Isomers at the 1-ng Level by Photolytic Degradation and Pattern Recognition Techniques. *Anal. Chem.* 52, 1865-1874.

Rugge, C.D., Ahlert, R.C., and O'Connor, O.A. 1993. Development of Bacterial Cultures which can Metabolize Structural Analogs of Dioxin. *Environ. Prog.* 12, 114-122.

U.S. Environmental Protection Agency. 1986. EPA Method 8040, SW-846 in Test Methods for Evaluating Solid Wastes, Vol. IB, Field Manual Physical/Chemical Methods, EPA, November 1986.

CHAPTER 18

Integrating Soil and Groundwater Remediation for Rapid Site Closure

John E. Hunt and **John T. Lee**, Barr Engineering Company, Minneapolis, Minnesota

Robert A. Paschke, 3M Environmental Technology & Services, St. Paul, Minnesota

INTRODUCTION

In 1981, a state pollution control agency (PCA) received complaints regarding the presence of suspicious 55-gallon drums in a rural setting near an automobile scrapyard (see Figure 1). Follow-up to the complaints revealed the presence of drums containing waste solvents in two general locations on the property. The PCA determined from the property owner that the drums had been transported to the site by a now defunct hazardous materials disposal and recycling company in the late 1970s. The PCA subsequently conducted an emergency removal action and investigation at the site and began the process of examining records of the defunct company to determine the name(s) of the generator(s) of the waste.

The Isanti Sites Trust (Trust) is comprised of seven companies that had contracted with the defunct disposal firm for hazardous waste disposal and recycling services. Recognizing the potential for lengthy litigation and seeking to address site contamination problems in the most cost-effective manner possible, Trust members negotiated a unique settlement for the site with the PCA. The Trust offered to complete the remedial investigation and implement approved remedial actions at the site. In exchange, the PCA agreed to take over operation of the remedial systems if site cleanup goals were not met within 24 months of the effective start date of treatment. The settlement provided the PCA and Trust members with a more timely and less litigious response for investigation and remediation of the site.

SITE CONDITIONS

The area identified by the investigation as most impacted was less than 1 acre in size. Prior to remediation, the area was littered with automobile parts and other

Figure 1. **Site location map, Isanti Rumpel Site.**

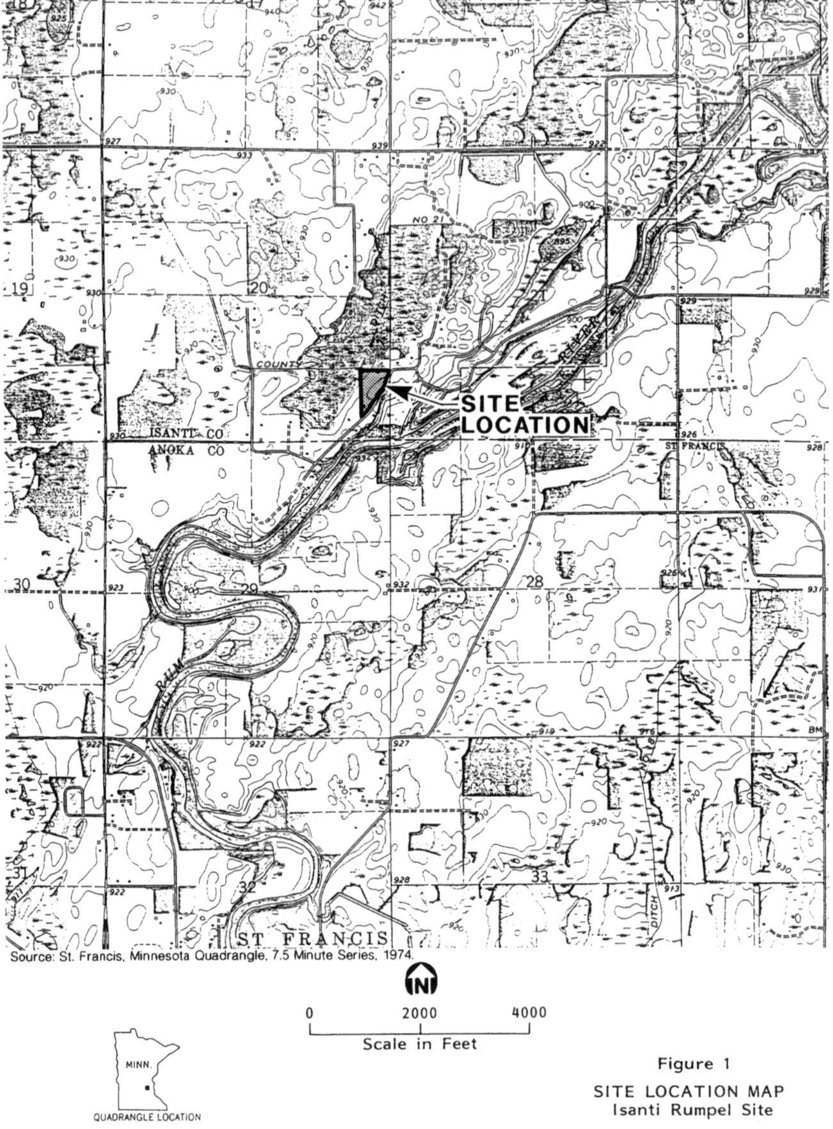

Source: St. Francis, Minnesota Quadrangle, 7.5 Minute Series, 1974.

miscellaneous scrap. The water table under the site was within 1 meter of the
ground surface, and site soils consisted of approximately 5 to 7 meters of silty sand
alluvium which lay over a sandy clay stratum. The impacted surficial aquifer discharged to a wetland which eventually discharged to the Rum River, a state-desig-
nated wild and scenic river. The proximity of groundwater to the ground surface
and the need to remove contaminants from the entire layer of surficial soils presented a unique challenge.

REMEDIAL DESIGN AND IMPLEMENTATION

Recognizing that the PCA would not approve a remedial action that could not reasonably bring the site to closure within 24 months, the Trust directed Barr Engineering Company to design a remedial system that addressed residual contamination in both the soils and groundwater. The waste solvents were a mixture of volatile organic compounds (VOCs) with a variety of physical and chemical properties, suggesting a multiphase approach toward the remediation. The most feasible alternative not only had to remove and/or degrade the residual contaminants at the site, but needed to minimize the potential for volatilization and/or odors in the surrounding neighborhood. After ruling out excavation of impacted soils for off-site disposal and traditional groundwater pump-and-treat, a remedy comprised of direct contaminant removal via groundwater pumping, vadose zone venting, and surficial aquifer sparging was selected. Dewatering the primary treatment area would facilitate more rapid contaminant removal via soil venting, while oxygenation of the surficial aquifer via sparging would stimulate natural biodegradation of amenable compounds.

A groundwater extraction system (GES) consisting of horizontal wells was placed around the primary treatment area to prevent contaminant migration and to dewater the area to facilitate soil venting in the newly created vadose zone. Groundwater removed by the GES passed through granular activated carbon (GAC) prior to discharge to the county ditch adjacent to the site. A vapor extraction system (VES) consisting of three horizontal vents located approximately 2 meters below ground surface, placed through the primary treatment area, was installed. A centrifugal blower extracted approximately 680 m^3/hr (~400 cfm) of soil gas at approximately 0.07 atm (~30" water) vacuum for treatment by GAC prior to atmospheric discharge. The air injection system (AIS) consisted of six sparge points designed to introduce a total of 85 to 170 m^3/hr (~50 to 100 cfm) of air at the base of the surficial aquifer, stripping additional volatile compounds, and providing oxygen for aerobic degradation of the more water-soluble solvent constituents. This venting/sparging combination represented one of the first licensed applications of the U.S. Department of Energy's patent for horizontal well technologies.

After receiving a state groundwater appropriation permit, an NPDES permit, and the U.S. Department of Energy patent license for horizontal well technology for the site, the remedial systems were constructed in the summer of 1993 (see Figure 2). Standard construction methods were used to install subsurface piping and manholes. Groundwater extraction began in August 1993; soil venting and air injection in September 1993.

Figure 2. Site plan, Isanti Rumpel Site.

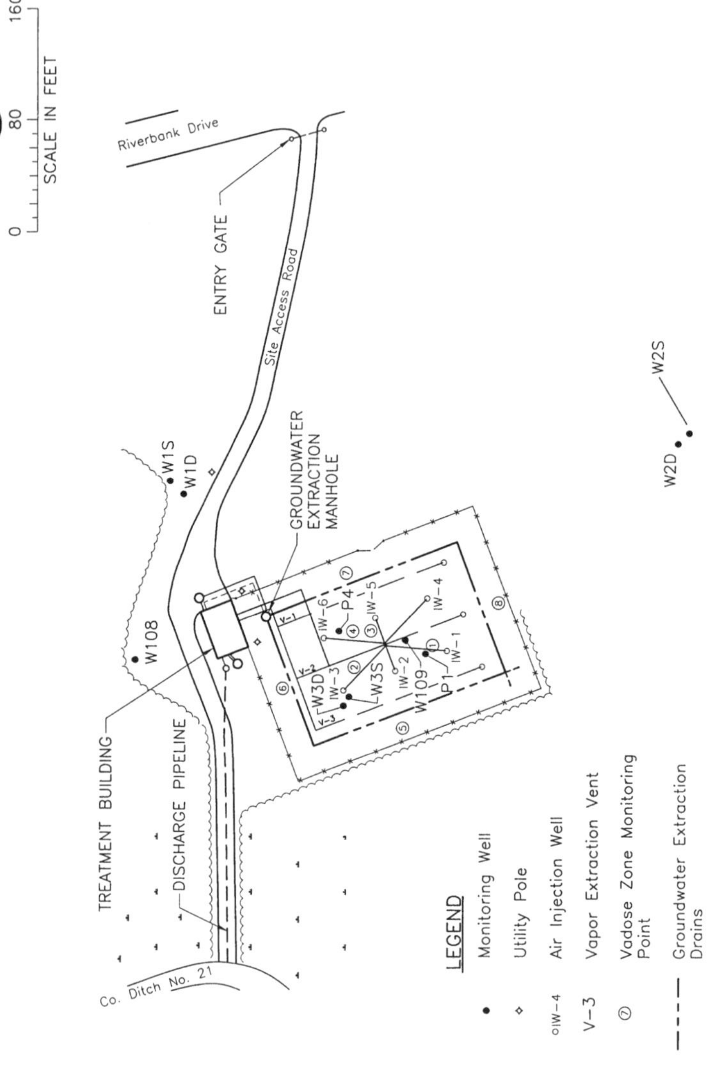

OPERATION AND MONITORING

The negotiated settlement for the site called for a period of startup monitoring intended to demonstrate that the remedial systems were operating as designed, followed by periodic site monitoring during the 24-month operating period. Data collected during startup monitoring included water level measurements in monitoring wells, soil gas vacuum and flow rate data, and air injection rates. Final verification of cleanup would be confirmed through groundwater samples collected from seven monitoring wells located both inside and outside of the primary treatment area, and soil samples collected from selected locations within the primary treatment area. Site soils could contain no more than 1 ppmv total VOCs, and site groundwater was required to meet applicable Recommended Allowable Limits (RALs) established by the state for private drinking water supplies.

As shown on Figure 3, the concentration of total volatiles in groundwater removed by the GES dropped dramatically during the startup monitoring period. Concentrations of total volatiles measured in the soil gas removed by the VES showed a similar trend. The PCA reviewed the startup data and approved commencement of the 24-month operating period as of February 1994. Collection of system monitoring data continued on a monthly basis, with groundwater samples collected quarterly. Using this multiple-media approach, groundwater quality in the primary treatment area improved quickly. Table 1 summarizes analytical data for groundwater samples collected from monitoring wells within the primary treatment area just prior to system startup (7/93), four months after startup (12/93), one year after startup (8/94), and just prior to site closure (2/96).

Figure 3. **Total VOCs in groundwater vs. time.**

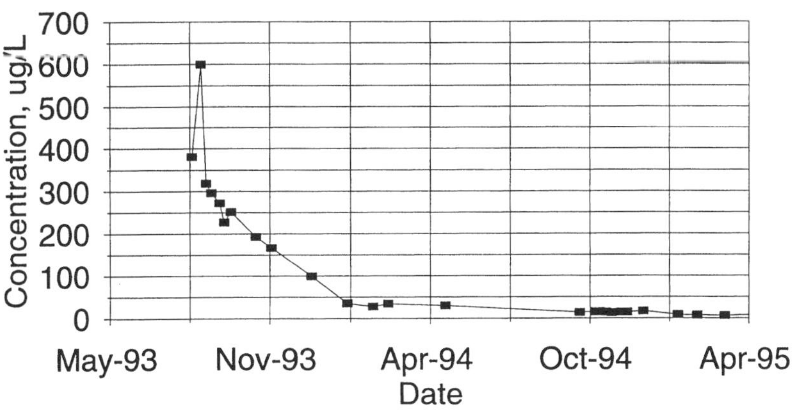

Table 1. **Groundwater Monitoring Isanti Rumpel Site
(all concentrations in ug/L)**

	07/30/93	12/15/93	08/29/95	02/26/96
W109				
Acetone	65 b	—	<50	<50
Allylchloride	<1.0	—	<1.0	<1.0
Benzene	5.2	—	<1.0	<1.0
Bromobenzene	<1.0	—	<1.0	<1.0
Bromochloromethane	<1.0	—	<1.0	<1.0
Bromo dichloromethane	<0.3	—	<0.3	<0.3
Bromoform	<0.5	—	<0.5	<0.5
Bromomethane	<5.0	—	<5.0	<5.0
Butylbenzene	10	—	<1.0	<1.0
Sec-butylbenzene	<1.0	—	<1.0	<1.0
Tert-Butylbenzene	<1.0	—	<1.0	<1.0
Carbon tetrachloride	<1.7	—	<1.7	<1.7
Chlorobenzene	1.2	—	<1.0	<1.0
Chlorodibromomethane	<2.5	—	<2.5	<2.5
Chloroethane	<1.0	—	<1.0	<1.0
Chloroform	<1.5	—	<1.5	<1.5
Chloromethane	<5.0	—	<5.0	<5.0
o-Chlorotoluene	<1.0	—	<1.0	<1.0
p-Chlorotoluene	<1.0	—	<1.0	<1.0
1,2-Dibromo-3-chloropropane	<10	—	<10	<10
1,2-Dibromoethane	<0.2	—	<0.2	<0.2
Dibromomethane (methylene bromide)	<5.0	—	<5.0	<5.0
1,2-Dichlorobenzene	<0.2	—	<0.2	<0.2
1,3-Dichlorobenzene	<1.5	—	<1.5	<1.5
1,4-Dichlorobenzene	0.3	—	<0.2	<0.2
1,1-Dichloroethane	87	—	<1.0	<1.0
1,2-Dichloroethane	14	—	<0.3	<0.3
1,1-Dichloroethylene	11	—	<1.0	<1.0
1,2-Dichloroethylene, cis	290	—	<0.2	<0.2
1,2-Dichloroethylene, trans	1.6	—	<0.2	<0.2
Dichlorodifluoromethane	<5.0	—	<5.0	<5.0
Dichlorofluoromethane	<5.0	—	<5.0	<5.0
1,2-Dichloropropane	8.6	—	<1.0	<1.0
1,3-Dichloropropane	<1.0	—	<1.0	<1.0
2,2-Dichloropropane	<1.0	—	<1.0	<1.0
1,1-Dichloro-1-propene	<0.5	—	<0.5	<0.5
cis-1,3-Dichloro-1-propene	<0.5	—	<0.5	<0.5
trans-1,3-Dichloro-1-propene	<0.5	—	<0.5	<0.5

Ethyl benzene	460	—	<1.0	<1.0
Ethyl ether	<1.0	—	<1.0	<1.0
Hexachlorobutadiene	<1.0	—	<1.0	<1.0
Cumene	26	—	<1.0	<1.0
P-Cymene	150	—	<1.0	<1.0
Methyl Ethyl Ketone	<5.0	—	<5.0	<5.0
Methyl Isobutyl Ketone	620	—	<5.0	<5.0
tert-Butyl Methyl Ether	<1.0	—	<1.0	<1.0
Methylene Chloride	<5.0	—	<5.0	<5.0
Naphthalene	22	—	<1.0	<1.0
Propylbenzene	22	—	<1.0	<1.0
Styrene	<1.0	—	<1.0	<1.0
1,1,1,2-Tetrachloroethane	<0.5	—	<0.5	<0.5
1,1,2,2-Tetrachloroethane	<1.2	—	<1.2	<1.2
Tetrachloroethylene	16	—	<1.0	<1.0
Tetrahydrofuran	<5.0	—	<5.0	<5.0
Toluene	3900	—	<1.0	<1.0
1,2,3-Trichlorobenzene	<1.0	—	<1.0	<1.0
1,2,4-Trichlorobenzene	<1.0	—	<1.0	<1.0
1,1,1-Trichloroethane	630	—	<2.0	<2.0
1,1,2-Trichloroethane	16	—	<1.2	<1.2
Trichloroethylene	16	—	3.3	5.7
Trichlorofluoromethane	<1.0	—	<1.0	<1.0
1,2,3-Trichloropropane	<1.0	—	<1.0	<1.0
Trichlorotrifluoroethane	<5.0	—	<5.0	<5.0
1,2,4-Trimethylbenzene	42	—	<1.0	<1.0
1,3,5-Trimethylbenzene	54	—	<1.0	<1.0
Vinyl chloride	<1.0	—	<1.0	<1.0
m and p Xylene	2200	—	<1.0	<1.0
o-Xylene	850	—	<1.0	<1.0
Sum volatile organics	9500	—	3.3	5.7

b=Potential false positive value based on blank data validation procedure.

Table 1 (cont.)

	07/30/93	12/15/93	08/29/95	02/26/96
W3S				
Acetone	<50	<50	<50	<50
Allylchloride	<1.0	<1.0	<1.0	<1.0
Benzene	3.1	<1.0	<1.0	<1.0
Bromobenzene	<1.0	<1.0	<1.0	<1.0
Bromochloromethane	<1.0	<1.0	<1.0	<1.0
Bromo dichloromethane	<0.3	<0.3	<0.3	<0.3
Bromoform	<0.5	<0.5	<0.5	<0.5
Bromomethane	<5.0	<5.0	<5.0	<5.0
Butylbenzene	8.5	<1.0	<1.0	<1.0
sec-Butylbenzene	<1.0	<1.0	<1.0	<1.0
tert-Butylbenzene	<1.0	<1.0	<1.0	<1.0
Carbon Tetrachloride	<1.7	<1.7	<1.7	<1.7
Chlorobenzene	<1.0	<1.0	<1.0	<1.0
Chlorodibromomethane	<2.5	<2.5	<2.5	<2.5
Chloroethane	<1.0	<1.0	<1.0	<1.0
Chloroform	<1.5	<1.5	<1.5	<1.5
Chloromethane	<5.0	<5.0	<5.0	<5.0
O-Chlorotoluene	<1.0	<1.0	<1.0	<1.0
P-Chlorotoluene	<1.0	<1.0	<1.0	<1.0
1,2-Dibromo-3-chloropropane	<10	<10	<10	<10
1,2-Dibromoethane	<0.2	<0.2	<0.2	<0.2
Dibromomethane (methylene bromide)	<5.0	<5.0	<5.0	<5.0
1,2-Dichlorobenzene	<0.2	<0.2	<0.2	<0.2
1,3-Dichlorobenzene	<1.5	<1.5	<1.5	<1.5
1,4-Dichlorobenzene	<0.2	<0.2	<0.2	<0.2
1,1-Dichloroethane	110	1.8	2	<1.0
1,2-Dichloroethane	2.2	<0.3	<0.3	<0.3
1,1-Dichloroethylene	6.5	<1.0	<1.0	<1.0
1,2-Dichloroethylene, cis	220	1.1	1.1	0.7
1,2-Dichloroethylene, trans	1.6	<0.2	<0.2	<0.2
Dichlorodifluoromethane	<5.0	<5.0	<5.0	<5.0
Dichlorofluoromethane	<5.0	<5.0	<5.0	<5.0
1,2-Dichloropropane	2.4	<1.0	<1.0	<1.0
1,3-Dichloropropane	<1.0	<1.0	<1.0	<1.0
2,2-Dichloropropane	<1.0	<1.0	<1.0	<1.0
1,1-Dichloro-1-propene	<0.5	<0.5	<0.5	<0.5
cis-1,3-Dichloro-1-propene	<0.5	<0.5	<0.5	<0.5
trans-1,3-Dichloro-1-propene	<0.5	<0.5	<0.5	<0.5
Ethyl benzene	180	<1.0	<1.0	<1.0

Ethyl ether	<1.0	<1.0	<1.0	<1.0
Hexachlorobutadiene	<1.0	<1.0	<1.0	<1.0
Cumene	7.3	<1.0	<1.0	<1.0
P-Cymene	65	<1.0	<1.0	<1.0
Methyl ethyl ketone	16 b	<5.0	<5.0	<5.0
Methyl isobutyl ketone	300	<5.0	<5.0	<5.0
tert-Butyl methyl ether	<1.0	<1.0	<1.0	<1.0
Methylene chloride	<5.0	<5.0	<5.0	<5.0
Naphthalene	10	<1.0	<1.0	<1.0
Propylbenzene	9.1	<1.0	<1.0	<1.0
Styrene	<1.0	<1.0	<1.0	<1.0
1,1,1,2-Tetrachloroethane	<0.5	<0.5	<0.5	<0.5
1,1,2,2-Tetrachloroethane	<1.2	<1.2	<1.2	<1.2
Tetrachloroethylene	<1.0	<1.0	<1.0	<1.0
Tetrahydrofuran	<5.0	<5.0	<5.0	<5.0
Toluene	1600	<1.0	<1.0	<1.0
1,2,3-Trichlorobenzene	<1.0	<1.0	<1.0	<1.0
1,2,4-Trichlorobenzene	<1.0	<1.0	<1.0	<1.0
1,1,1-Trichloroethane	150	<2.0	<2.0	<2.0
1,1,2-Trichloroethane	2.6	<1.2	<1.2	<1.2
Trichloroethylene	1.3	1	2.1	0.5
Trichlorofluoromethane	<1.0	<1.0	<1.0	<1.0
1,2,3-Trichloropropane	<1.0	<1.0	<1.0	<1.0
Trichlorotrifluoroethane	<5.0	<5.0	<5.0	<5.0
1,2,4-Trimethylbenzene	14	<1.0	<1.0	<1.0
1,3,5-Trimethylbenzene	22	<1.0	<1.0	<1.0
Vinyl chloride	<1.0	<1.0	<1.0	<1.0
m and p Xylene	900	<1.0	<1.0	<1.0
o-Xylene	330	<1.0	<1.0	<1.0
Sum volatile organics	4000	3.9	5.2	1.2

b=Potential false positive value based on blank data validation procedure.

Table 1 (cont.)

	07/30/93	12/15/93	08/29/95	02/26/96
P1				
Acetone	<50	<50	<50	<50
Allylchloride	<1.0	<1.0	<1.0	<1.0
Benzene	<1.0	<1.0	<1.0	<1.0
Bromobenzene	<1.0	<1.0	<1.0	<1.0
Bromochloromethane	<1.0	<1.0	<1.0	<1.0
Bromo dichloromethane	<0.3	<0.3	<0.3	<0.3
Bromoform	<0.5	<0.5	<0.5	<0.5
Bromomethane	<5.0	<5.0	<5.0	<5.0
Butylbenzene	<1.0	<1.0	<1.0	<1.0
sec-Butylbenzene	<1.0	<1.0	<1.0	<1.0
tert-Butylbenzene	<1.0	<1.0	<1.0	<1.0
Carbon tetrachloride	<1.7	<1.7	<1.7	<1.7
Chlorobenzene	<1.0	<1.0	<1.0	<1.0
Chlorodibromomethane	<2.5	<2.5	<2.5	<2.5
Chloroethane	<1.0	<1.0	<1.0	<1.0
Chloroform	<1.5	<1.5	<1.5	<1.5
Chloromethane	<5.0	<5.0	<5.0	<5.0
o-Chlorotoluene	<1.0	<1.0	<1.0	<1.0
p-Chlorotoluene	<1.0	<1.0	<1.0	<1.0
1,2-Dibromo-3-chloropropane	<10	<10	<10	<10
1,2-Dibromoethane	<0.2	<0.2	<0.2	<0.2
Dibromomethane (methylene bromide)	<5.0	<5.0	<5.0	<5.0
1,2-Dichlorobenzene	<0.2	<0.2	<0.2	<0.2
1,3-Dichlorobenzene	<1.5	<1.5	<1.5	<1.5
1,4-Dichlorobenzene	<0.2	<0.2	<0.2	<0.2
1,1-Dichloroethane	<1.0	23	1.3	<1.0
1,2-Dichloroethane	<0.3	<0.3	<0.3	<0.3
1,1-Dichloroethylene	<1.0	1.3	<1.0	<1.0
1,2-Dichloroethylene, cis	0.8	1.5	<0.2	<0.2
1,2-Dichloroethylene, trans	<0.2	<0.2	<0.2	<0.2
Dichlorodifluoromethane	<5.0	<5.0	<5.0	<5.0
Dichlorofluoromethane	<5.0	<5.0	<5.0	<5.0
1,2-Dichloropropane	<1.0	<1.0	<1.0	<1.0
1,3-Dichloropropane	<1.0	<1.0	<1.0	<1.0
2,2-Dichloropropane	<1.0	<1.0	<1.0	<1.0
1,1-Dichloro-1-propene	<0.5	<0.5	<0.5	<0.5
cis-1,3-Dichloro-1-propene	<0.5	<0.5	<0.5	<0.5
trans-1,3-Dichloro-1-propene	<0.5	<0.5	<0.5	<0.5
Ethyl benzene	2.2	2.2	<1.0	<1.0

Ethyl ether	<1.0	<1.0	<1.0	<1.0
Hexachlorobutadiene	<1.0	<1.0	<1.0	<1.0
Cumene	<1.0	<1.0	<1.0	<1.0
p-Cymene	<1.0	<1.0	<1.0	<1.0
Methyl ethyl ketone	<5.0	<5.0	<5.0	<5.0
Methyl isobutyl ketone	<5.0	<5.0	<5.0	<5.0
tert-Butyl methyl ether	<1.0	<1.0	<1.0	<1.0
Methylene chloride	<5.0	<5.0	<5.0	<5.0
Naphthalene	<1.0	<1.0	<1.0	<1.0
Propylbenzene	<1.0	<1.0	<1.0	<1.0
Styrene	<1.0	<1.0	<1.0	<1.0
1,1,1,2-Tetrachloroethane	<0.5	<0.5	<0.5	<0.5
1,1,2,2-Tetrachloroethane	<1.2	<1.2	<1.2	<1.2
Tetrachloroethylene	<1.0	<1.0	<1.0	<1.0
Tetrahydrofuran	<5.0	<5.0	<5.0	<5.0
Toluene	<1.0	<1.0	<1.0	<1.0
1,2,3-Trichlorobenzene	<1.0	<1.0	<1.0	<1.0
1,2,4-Trichlorobenzene	<1.0	<1.0	<1.0	<1.0
1,1,1-Trichloroethane	<2.0	<2.0	<2.0	<2.0
1,1,2-Trichloroethane	<1.2	<1.2	<1.2	<1.2
Trichloroethylene	<0.5	340	10	6.7
Trichlorofluoromethane	<1.0	<1.0	<1.0	<1.0
1,2,3-Trichloropropane	<1.0	<1.0	<1.0	<1.0
Trichlorotrifluoroethane	<5.0	<5.0	<5.0	<5.0
1,2,4-Trimethylbenzene	<1.0	<1.0	<1.0	<1.0
1,3,5-Trimethylbenzene	<1.0	<1.0	<1.0	<1.0
Vinyl chloride	<1.0	<1.0	<1.0	<1.0
m and p Xylene	3.1	3.4	<1.0	<1.0
o-Xylene	1.2	<1.0	<1.0	<1.0
Sum Volatile Organics	7.3	370	11	6.7

b=Potential false positive value based on blank data validation procedure.

Table 1(cont.)

	07/30/93	12/15/93	08/29/95	02/26/96
P4				
Acetone	<50	<50	<50	<50
Allylchloride	<1.0	<1.0	<1.0	<1.0
Benzene	2.2	1.8	<1.0	<1.0
Bromobenzene	<1.0	<1.0	<1.0	<1.0
Bromochloromethane	<1.0	<1.0	<1.0	<1.0
Bromo dichloromethane	<0.3	<0.3	<0.3	<0.3
Bromoform	<0.5	<0.5	<0.5	<0.5
Bromomethane	<5.0	<5.0	<5.0	<5.0
Butylbenzene	4.1	20	<1.0	<1.0
sec-Butylbenzene	<1.0	7	<1.0	<1.0
tert-Butylbenzene	<1.0	<1.0	<1.0	<1.0
Carbon Tetrachloride	<1.7	<1.7	<1.7	<1.7
Chlorobenzene	<1.0	2.1	<1.0	<1.0
Chlorodibromomethane	<2.5	<2.5	<2.5	<2.5
Chloroethane	2.7	<1.0	3	<1.0
Chloroform	<1.5	<1.5	<1.5	<1.5
Chloromethane	<5.0	<5.0	<5.0	<5.0
O-Chlorotoluene	<1.0	<1.0	<1.0	<1.0
P-Chlorotoluene	<1.0	<1.0	<1.0	<1.0
1,2-Dibromo-3-chloropropane	11	11	<10	<10
1,2-Dibromoethane	<0.2	<0.2	<0.2	<0.2
Dibromomethane (methylene bromide)	<5.0	<5.0	<5.0	<5.0
1,2-Dichlorobenzene	<0.2	<0.2	<0.2	<0.2
1,3-Dichlorobenzene	<1.5	<1.5	<1.5	<1.5
1,4-Dichlorobenzene	<0.2	0.2	<0.2	<0.2
1,1-Dichloroethane	110	58	18	1.1
1,2-Dichloroethane	2.7	<0.3	<0.3	<0.3
1,1-Dichloroethylene	9.9	17	<1.0	<1.0
1,2-Dichloroethylene, cis	250	270	5.7	1.8
1,2-Dichloroethylene, trans	0.8	0.8	<0.2	<0.2
Dichlorodifluoromethane	<5.0	<5.0	<5.0	<5.0
Dichlorofluoromethane	<5.0	<5.0	<5.0	<5.0
1,2-Dichloropropane	3.6	4.2	1.3	<1.0
1,3-Dichloropropane	<1.0	<1.0	<1.0	<1.0
2,2-Dichloropropane	<1.0	<1.0	<1.0	<1.0
1,1-Dichloro-1-propene	<0.5	<0.5	<0.5	<0.5
cis-1,3-Dichloro-1-propene	<0.5	<0.5	<0.5	<0.5
trans-1,3-Dichloro-1-propene	<0.5	<0.5	<0.5	<0.5
Ethyl benzene	150	380	3	<1.0

Ethyl ether	<1.0	<1.0	<1.0	<1.0
Hexachlorobutadiene	<1.0	<1.0	<1.0	<1.0
Cumene	7.7	20	<1.0	<1.0
P-Cymene	59	100	<1.0	<1.0
Methyl ethyl ketone	5.3	10	<5.0	<5.0
Methyl isobutyl ketone	160	110	<5.0	<5.0
tert-Butyl methyl ether	<1.0	<1.0	<1.0	<1.0
Methylene chloride	<5.0	<5.0	<5.0	<5.0
Naphthalene	9.7	12	<1.0	<1.0
Propylbenzene	8.8	25	<1.0	<1.0
Styrene	<1.0	<1.0	<1.0	<1.0
1,1,1,2-Tetrachloroethane	<0.5	<0.5	<0.5	<0.5
1,1,2,2-Tetrachloroethane	<1.2	<1.2	<1.2	<1.2
Tetrachloroethylene	22	21	<1.0	<1.0
Tetrahydrofuran	<5.0	<5.0	18	<5.0
Toluene	540	1100	<1.0	<1.0
1,2,3-Trichlorobenzene	<1.0	<1.0	<1.0	<1.0
1,2,4-Trichlorobenzene	<1.0	<1.0	<1.0	<1.0
1,1,1-Trichloroethane	280	930	<2.0	<2.0
1,1,2-Trichloroethane	2.2	8.1	<1.2	<1.2
Trichloroethylene	19	19	4.7	3.5
Trichlorofluoromethane	<1.0	<1.0	<1.0	<1.0
1,2,3-Trichloropropane	<1.0	<1.0	<1.0	<1.0
Trichlorotrifluoroethane	<5.0	<5.0	<5.0	<5.0
1,2,4-Trimethylbenzene	8.8	15	<1.0	<1.0
1,3,5-Trimethylbenzene	22	44	<1.0	<1.0
Vinyl chloride	<1.0	<1.0	<1.0	<1.0
m and p Xylene	770	1400	1	<1.0
o-Xylene	300	700	1.1	<1.0
Sum Volatile Organics	2800	5300	56	6.4

b=Potential false positive value based on blank data validation procedure.

SITE CLOSURE

Based on VES operational data and soil gas analytical data, the Trust proposed to collect confirmatory soil and soil gas samples from within the primary treatment area in March 1995. As expected, the results indicated that vadose zone soils were in compliance with site cleanup goals. By August 1995, groundwater analytical data indicated that contaminant concentrations in site groundwater had decreased to below cleanup goals for the site. The PCA approved the Trust's proposal to shut down the remedial systems contingent on the Trust conducting three additional quarters of followup groundwater monitoring. Contaminant concentrations remained below cleanup goals for the three additional quarters, and the Trust proposed the site for closure in April 1996. The PCA approved site closure and removed the site from the state's Permanent List of Priorities in June 1996. The remedial systems at the site were removed or abandoned in place by the end of September 1996.

SUMMARY

The remediation of a waste solvent release site was hastened by the integration of multiple remedial technologies aimed at both site soils and groundwater. This integration allowed the Isanti Sites Trust to fulfill an aggressive time table established through negotiations with the state pollution control agency and avoid protracted litigation over the site, while protecting human health and the environment in this rural residential area.

CHAPTER 19

The Selective Recovery of Nonaqueous Phase Liquids (NAPLs) from Groundwater Using Oleophilic Suction Lysimetry

Christopher C. Barker and **Clifford J. Bruell**, Department of Civil and Environmental Engineering, University of Massachusetts, Lowell

Ralph S. Baker, ENSR Consulting and Engineering, Acton, Massachusetts

INTRODUCTION

An innovative method of recovering nonaqueous phase liquids (NAPLs) from contaminated vadose zone soil was proposed by Dr. Ralph Baker in 1993. This technology is called oleophilic suction lysimetry (OSL) and utilizes an oil recovery lysimeter which is covered by an oleophilic, hydrophobic membrane. The lysimeter is placed into the ground to intercept a region of soil contaminated by NAPL. Vacuum pressures are applied to the lysimeter to draw NAPL through the membrane. The recovered NAPL is then conveyed from the lysimeter to the ground surface. OSL is unique as a remediation technology because it focuses on the selective removal of pure NAPL contaminant and excludes the removal of air and water.

BACKGROUND

NAPL which migrates through the vadose zone can become trapped within the soil pores, held by capillary forces. Capillary forces are a result of interfacial tension between liquid phases, and the wettability of the soil particles. Generally, forces holding NAPL within soil pores increase as the soil pore size decreases (Wilson et al., 1990). OSL is based on the development of vacuum pressures which can overcome capillary forces. To develop a pressure gradient within NAPL in a soil, contact between the oleophilic membrane and interconnected soil pores containing NAPL is required. These interconnected pores can be visualized as ganglia of NAPL extending out in the soil from the location of the membrane (Baker, 1993). Figure 1 is a diagram of a membrane in contact with NAPL held within a soil matrix. When a vacuum pressure is applied at the membrane, pressure gradients develop in the NAPL ganglia and NAPL is thus pulled toward and through the membrane.

Figure 1.　**Diagram showing NAPL located within soil pores in contact with an oleophilic membrane.**

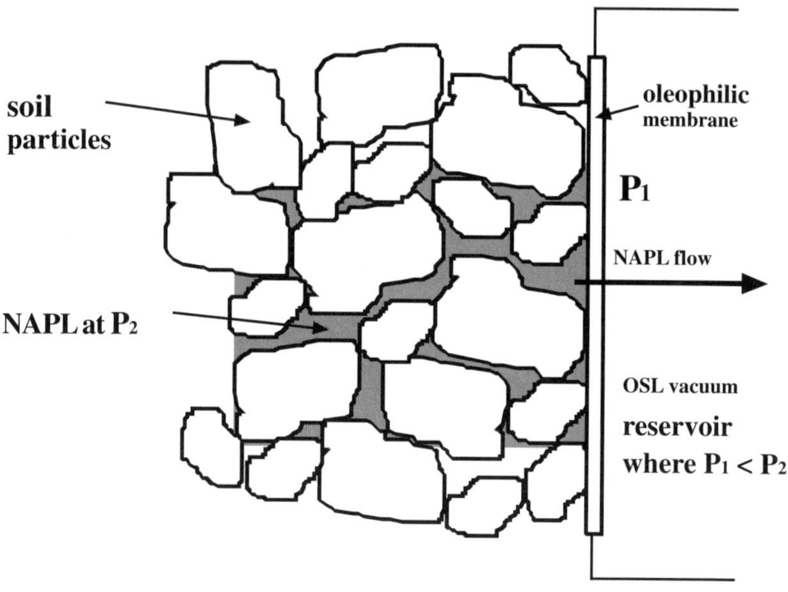

Since pore sizes vary in soil surrounding the membrane, forces holding the NAPL within the pores also vary. Applying a vacuum pressure and then increasing it in a stepwise manner would allow NAPL to be removed from larger pores first, and then from increasingly smaller pores as long as there is continuity within interconnected pores that remain NAPL-filled.

Exceeding the air entry pressure of the membrane during OSL would cause a flow of air through the soil and into the membrane, thus greatly reducing the effectiveness of OSL. Therefore, it was theorized that an oleophilic membrane with a high air-entry pressure would be most desirable for OSL.

MATERIALS AND METHODS

Twenty porous, oleophilic hydrophobic membranes were tested for their air-entry pressure. A Teflon® membrane, Tetratech model 6525, (Tetratec Corp., Feasterville, PA) had an air-entry pressure of 17.2 psi and was observed to be physically suitable for use in soil. Therefore, it was used in all subsequent experiments.

General dimensions of a bench-scale, two-dimensional (2-D) chamber, utilized to examine the recovery of NAPL from the vadose zone, are given in Figure 2. The 2-D chamber had three rows of five tensiometer ports each. The numbers assigned to ports in the middle row are subsequently explained. The bottom section of the U-shaped metal spacer had holes which allowed water to be added to the 2-D chamber through an opening in the side of the spacer. The OSL device (hereafter referred to as the extraction device) displayed in Figure 3, was positioned inside the 2-D chamber to contact the NAPL contaminated soil.

Figure 2. **Diagram of 2-D chamber. Tensiometer ports in middle row are numbered.**

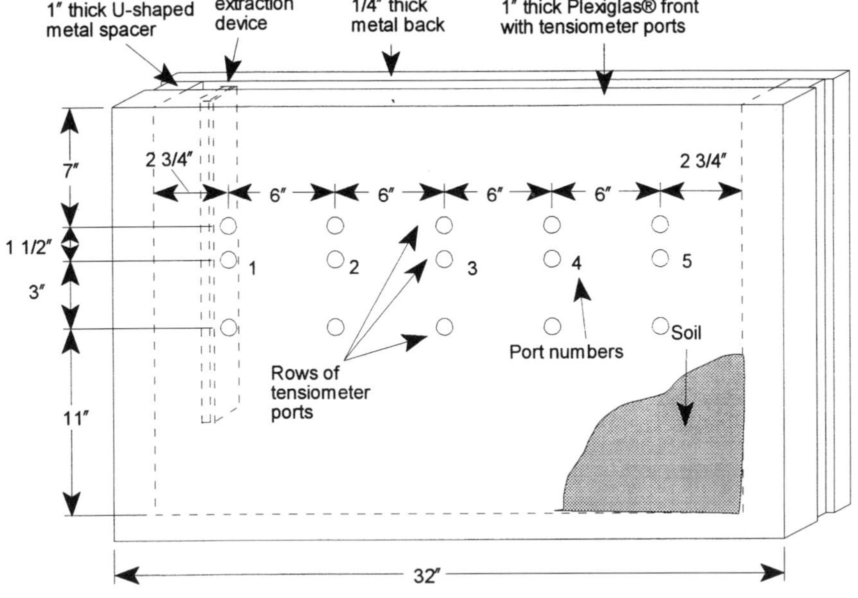

Figure 3. **Diagram of oleophilic extraction device showing (a) plan, (b) section, and (c) end views.**

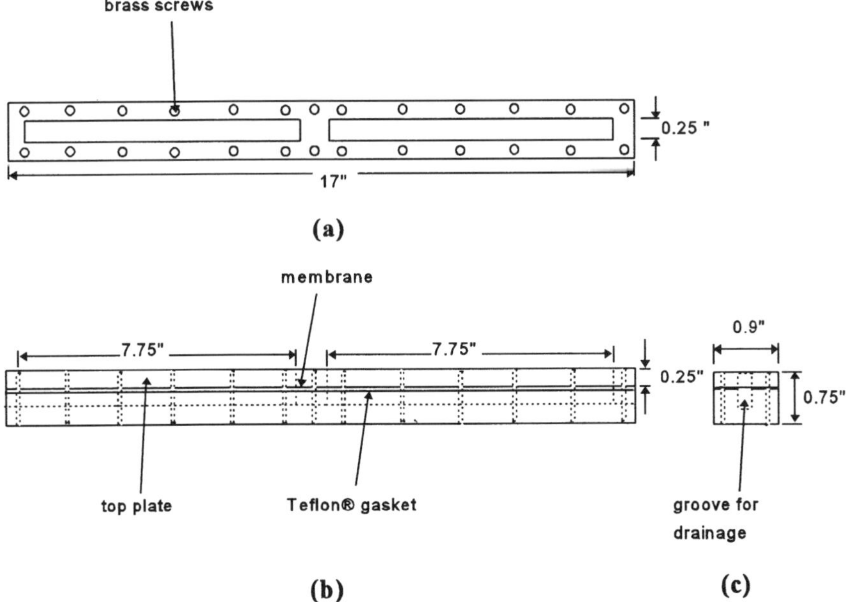

The soils used in the experiments were a coarse sand, loam, and diatomaceous earth. A description of the soils is presented in Table 1. The 2-D chamber was placed on a vibrating table and the soil of interest was packed in approximately two-inch layers with a tremie tube. After adding each layer, the 2-D chamber was vibrated for 30 seconds while pressure was applied to the top of the soil layer with a 1-inch thick metal piston.

Table 1. Properties of Soils Used in Experiments

Soil	Median Particle Diameter (mm)	Bulk Density (g/cm³)	Particle Density[a] (g/cm³)	Porosity (cm³/cm³)
coarse sand	0.520	1.61	2.64	0.39
loam	0.160	1.51	2.59	0.41
diatomaceous earth	0.015	0.32	2.32	0.86

[a] Data was obtained from soil analyses provided by Daniel B. Stephens & Associates, Inc., 1994.

The contaminant used was reagent grade dodecane (Aldrich Chemical Company, Milwaukee, WI). In experiments with sand and loam, a water table was established after the 2-D chamber was packed with soil. The water table elevation was adjusted to keep dodecane between the top and bottom rows of ports.

Capillary rise in diatomaceous earth was determined to be in excess of three feet; therefore, introducing a water table in diatomaceous earth would cause high water saturation of the soil in the region between the top and bottom rows of ports. Dodecane injected into this region might become trapped and/or forced up by water to the top region of soil where tensiometers would not be present to monitor changes in dodecane tensions. Therefore, rather than introducing a water table, diatomaceous earth was mixed with water to a moisture content of 50% by mass (15.7% volumetric moisture content) and then packed into the 2-D chamber. To prevent dodecane from distributing to the bottom of the 2-D chamber, the bottom of the chamber was elevated to two inches below the bottom row of tensiometer ports.

In the experiment with coarse sand, a screened recovery well (1/4" O.D., 3/8" I.D.) was positioned vertically in the 2-D chamber to allow gravity drainage of dodecane prior to vacuum extraction. This was done to simulate conditions in which OSL is used to recover residual NAPL only after free product has been removed by conventional recovery wells. In subsequent experiments (i.e., loam and diatomaceous earth), a screeneW recovery well was not used. Gravity drainage through the OSL device was used instead, as an initial NAPL removal process.

Oil tensiometers were inserted into soil within the 2-D chamber to measure the tensions of the oil held within the soil pores. By monitoring dodecane tensions in the 2-D chamber, it might be possible to monitor relative dodecane distribution within the soil (Lenhard and Parker, 1987). Theoretically, regions of soil from which dodecane is removed would display higher dodecane tensions since remaining dodecane would be held more tightly.

Oil tensiometers were constructed by attaching oleophilic membrane disks to the ends of glass tubes. Each glass tube had a 9 mm O.D., a 6.5 mm I.D., and a length of 10 cm. Each tensiometer was attached to a manometer and then both were filled with dodecane. The tensiometer tip was held at the elevation of the 2-D chamber port into which it would be inserted and allowed to equilibrate to atmospheric pressure. After this process of being "zeroed," the tensiometers were inserted into the 2-D chamber ports and adjusted to ensure contact between soil and membrane tips. Dodecane flowed into or out of the tensiometers depending on whether the pressure of the dodecane held within the soil was higher or lower than atmospheric pressure.

To ensure that dodecane resided in the soil region between the top and bottom rows of tensiometer ports, dodecane was added through a perforated insertion tube (1/8″ O.D., 1/25″ I.D.) positioned horizontally in the soil, midway between the middle and bottom row of ports. Following charging of the chamber, dodecane was allowed to distribute, and tensiometer readings were monitored. When dodecane tensions were relatively steady, gravity drainage of dodecane through the screened well (in coarse sand) was initiated. When this drainage ceased, gravity drainage through the extraction device was allowed.

Vacuum extraction of dodecane then began. Vacuum pressure was applied stepwise in an attempt to slowly remove dodecane from increasingly smaller pores. During the experiments with loam and diatomaceous earth, the vacuum pump was turned off periodically to prevent overheating of the motor.

When the experiment with diatomaceous earth ended, soil samples were removed from the chamber and analyzed for dodecane and water concentrations. Dodecane was extracted from soil samples using methylene chloride. Dodecane concentrations were determined using a Hewlett Packard 5970A gas chromatograph (GC) with flame ionization detector (FID). Total liquid content was gravimetrically determined by comparison of wetted and dried samples. Water concentrations were calculated from total liquid and dodecane concentrations.

RESULTS

Coarse Sand

The extraction device was positioned vertically so that the surface of the membrane was approximately 1.9 inches to the left of the centerline of the nearest tensiometer ports (port number 1 in Figure 2). The water table was established 4.5 inches below the middle row of ports.

A total of 500 mL of dodecane was added to the soil. Gravity drainage yielded 327.0 mL of dodecane through the screened well and 31.6 mL through the extraction device. Of the remaining (i.e., residual) 141.4 mL of dodecane in the coarse sand, 17.3 % was recovered by vacuum extraction. Figure 4 shows the percent dodecane recovered and corresponding vacuum pressure head applied versus the cumulative time that vacuum pressure was applied. The rate of recovery eventually decreased and became essentially zero even as vacuum pressure was increased.

It is theorized that the number of connected, dodecane-filled soil pores in contact with the membrane decreased rapidly as dodecane was removed from the larger soil pores and snap-off occurred. Consequently, the percentage of dodecane that can be recovered by OSL in a coarse grain soil appears to be limited by the soil pore size distribution and a further stepwise increase in the applied vacuum pressures does not have any beneficial effect.

Figure 5 is representative of dodecane tension behavior in each type of soil. This figure specifically shows dodecane tensions in coarse sand at the middle row of ports when vacuum pressure was applied to the membrane continuously throughout the experiment. Port numbers shown in Figure 2 indicate tensiometer locations (no tensiometer was located in port 3). Results from tensiometer data were inconclusive, i.e., changes in soil conductivity were not apparent. Dodecane tensions generally increased gradually during experiments at each tensiometer location in experiments with all three soils. Dodecane tensions were not noticeably higher near the membrane than in regions of soil located further from the membrane.

Figure 4.   **Percent recovery of dodecane from coarse sand and applied vacuum pressure head versus cumulative time of applied vacuum.**

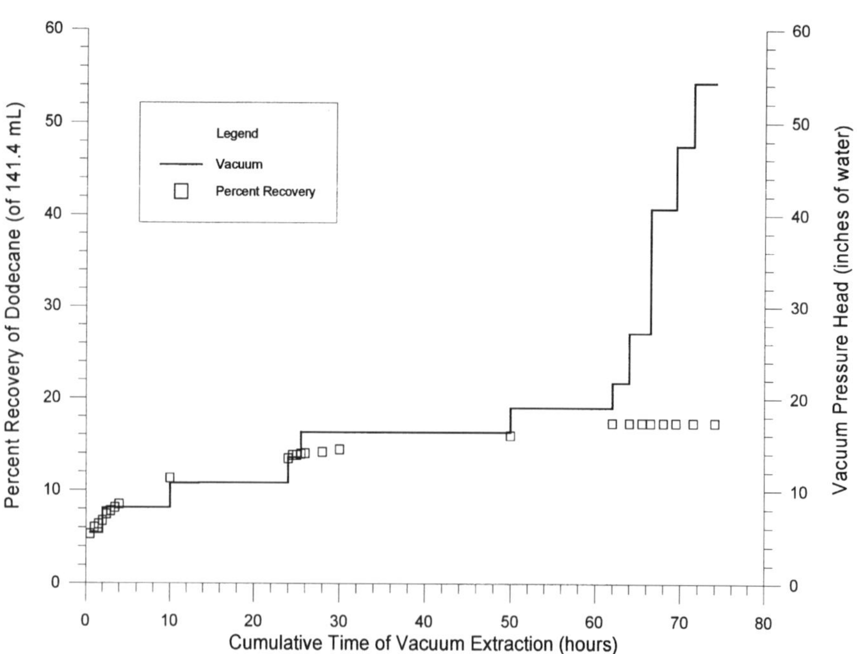

Figure 5. **Dodecane tensions in the middle row of tensiometer ports and applied vacuum pressure head versus cumulative time of experiment with coarse sand.**

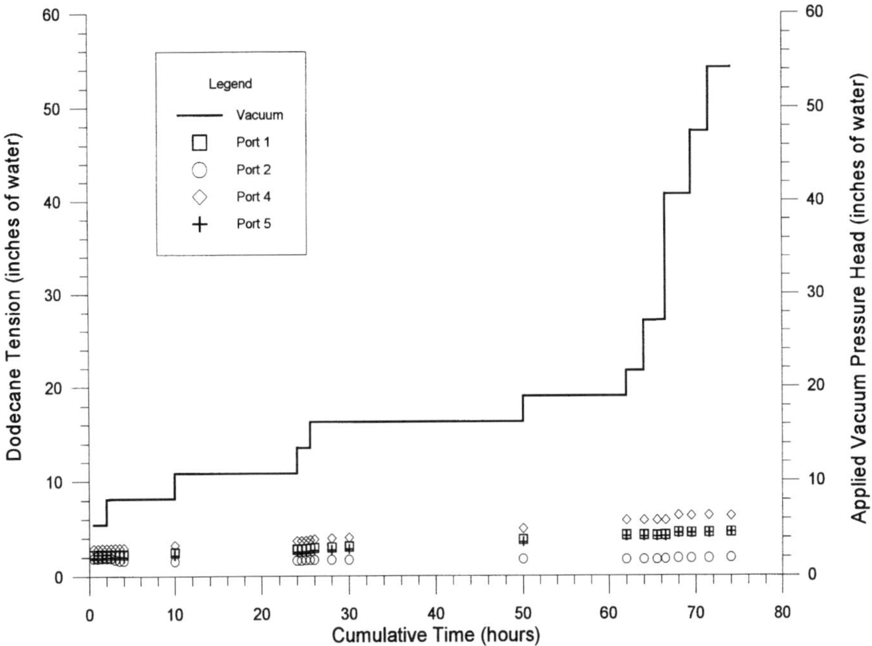

Loam

A water table was established 10 inches below the middle row of tensiometer ports. Of 1,000 mL of dodecane added to the soil, none drained by gravity through the extraction device. It was theorized that increasing the vacuum pressure before the rate of dodecane recovery had become asymptotic, could cause dodecane snap-off and result in decreases in soil conductivity near the membrane. Higher dodecane tensions near the membrane might be indicative of lower soil conductivity. Therefore, in loam and subsequent diatomaceous earth experiments the extraction device was positioned approximately 0.5 inches from the midpoint of the nearest tensiometer ports so that soil tensions near the membrane could be monitored.

Figure 6 shows a series of stepwise increases in vacuum pressure head with corresponding rates of dodecane recovery decreasing and becoming asymptotic. Generally, when vacuum pressure was increased, recovery of dodecane began again. Vacuum pressure was applied for approximately 900 hours and the length of the entire experiment was approximately 1,350 hours (due to the pump being turned off periodically). It is likely that low rates of dodecane recovery from the loam (generally less than 1 mL per hour of applied vacuum) were due to a small percentage of interconnected, dodecane-filled pores. A total of 31.7% of the residual 1,000 mL of dodecane was recovered.

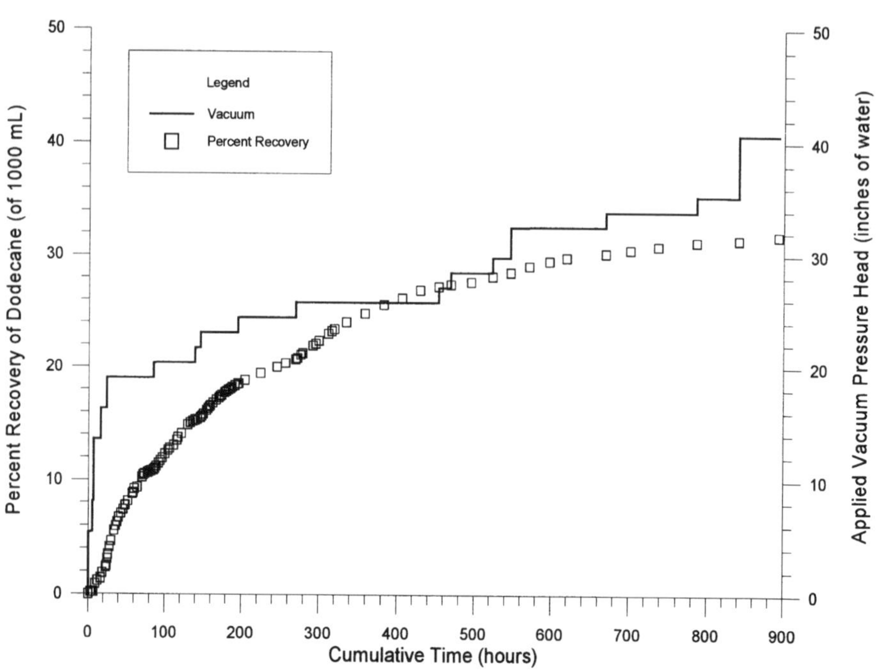

Figure 6.   **Percent recovery of dodecane from loam and applied vacuum pressure head versus cumulative time of applied vacuum.**

Diatomaceous Earth

Of 1,500 mL of dodecane added to the soil, none drained by gravity through the extraction device. Figure 7 shows percent dodecane removed from diatomaceous earth versus the time of applied vacuum pressure. An initial vacuum pressure was applied and increased stepwise. After 46.25 hours, only 8.1 mL of dodecane had been recovered. The experiment was interrupted and an additional 975 mL of dodecane was added through the perforated insertion tube. Percent recovery is therefore based on a total of 2,475 mL of dodecane added.

Figure 7 shows that a vacuum pressure was applied until the rate of dodecane recovery from the soil had slowed. The experiment was terminated before the rate of recovery had approached zero during the period when a vacuum pressure of 27.8 inches of water was applied. The percentage of dodecane recovered was 13.8% after 1,300 hours of applied vacuum pressures. Although the rate of recovery at the end of the experiment was higher than recovery rates achieved at lower applied vacuum pressures used in this experiment, it was thought that ultimately the rate of dodecane recovery would become zero, and continuing the experiment would not have resulted in an overall recovery percentage of dodecane significantly greater than 13.8%.

Figure 7.   **Percent recovery of dodecane from diatomaceous earth and applied vacuum pressure head versus cumulative time of applied vacuum.**

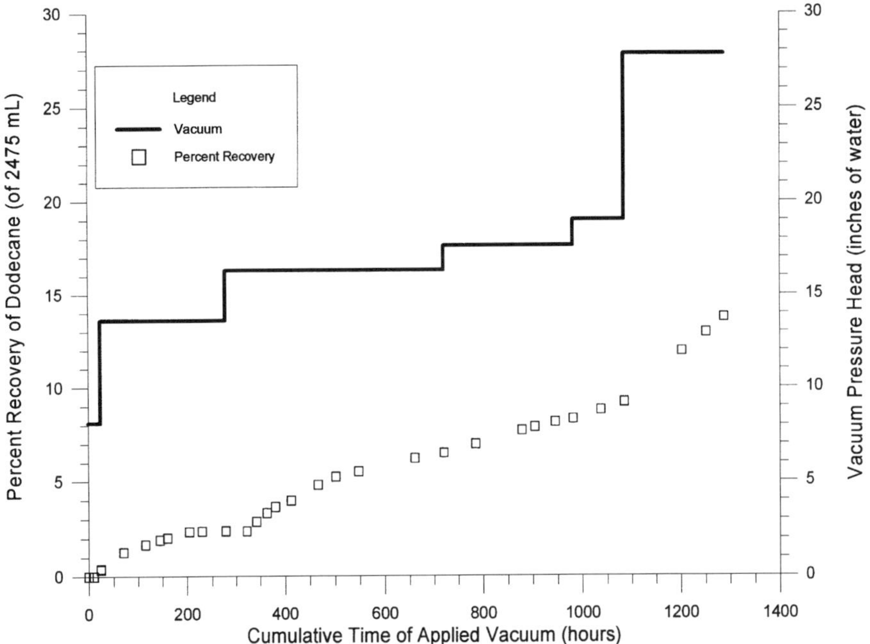

Figure 8 shows contours of dodecane saturation (determined as the volume of dodecane divided by the volume of voids) in the diatomaceous earth at the end of the experiment. Locations from which the samples were removed are represented by solid dots. Dodecane saturation contours running within the region of soil up to 7 inches above the chamber's bottom and out to 9 inches to the right of the membrane are essentially vertical. Based on the orientation of these contours it appears that dodecane moved horizontally toward the membrane. Other dodecane saturation contours run essentially horizontal and show that dodecane saturations generally increase when moving down from the soil surface to the bottom of the chamber. This is likely a result of gravity drainage.

The vertical dodecane saturation contours near the membrane decrease in value from 0.50 to 0.40 dodecane saturation approximately 2.5 inches to the right of the membrane. Dodecane saturations in the vicinity of the membrane are higher than dodecane saturations in most other locations; therefore, it appears that, at the end of the experiment, the conductivity of soil in this region had not decreased to where it was limiting the recovery of dodecane. If the experiment had continued, dodecane saturation near the membrane would likely have further decreased. Although dodecane had distributed toward the membrane, the rate of recovery had been too low to warrant continuation of the experiment.

Figure 8.   **Plot of dodecane saturations inside the 2-D chamber at the end of the experiment with diatomaceous earth. The solid dots indicate the locations from which samples were removed for analysis.**

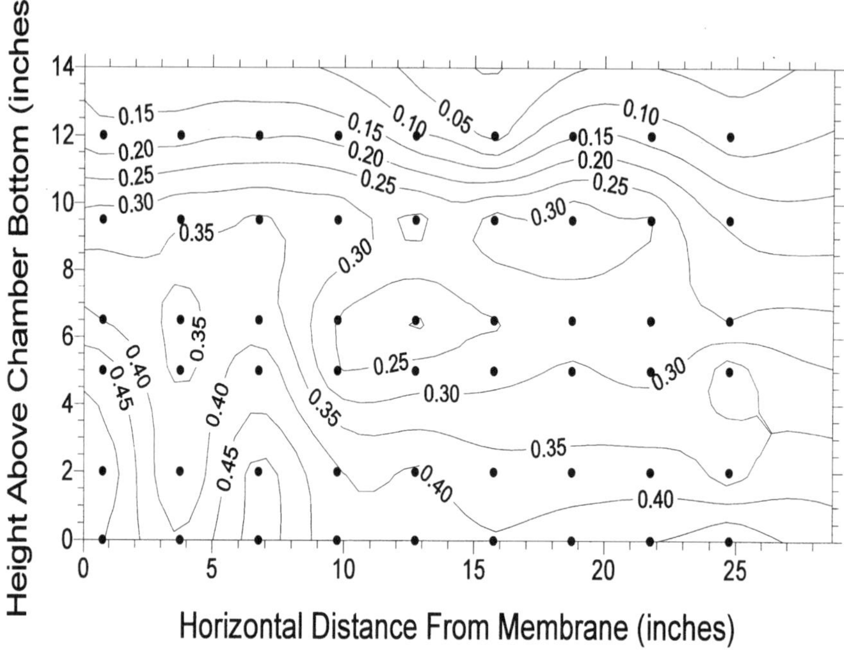

Figure 9 shows water saturations in the chamber at the end of the experiment. Water saturation contours also run vertically near the membrane. Water saturations generally decrease where dodecane saturations increase, and increase in regions where dodecane saturations decrease. This may be a result of complementary redistribution of water as dodecane was recovered.

CONCLUSIONS

When this investigation began it was expected that high vacuum pressures would be applied to the membrane to remove NAPL held at high tension in small soil pores. However, after performing dodecane recovery experiments it became apparent that much lower vacuum pressures are used in practice to avoid snap-off and decreases in soil conductivity.

Recovery of dodecane from coarse sand was low; i.e., 17.3%, which was expected given the soil's relatively large pore sizes. Loam, which had a greater percentage of small pore sizes, provided slightly higher recovery of dodecane; i.e., 31.7%. The highest recovery of dodecane from diatomaceous earth was 13.8%, which was disappointing since it had the smallest pore sizes. On the other hand, it may not be realistic to expect to be able to recover residual NAPL from the vadose zone as a liquid.

Figure 9.   **Plot of water saturations inside the 2-D chamber at the end of the experiment with diatomaceous earth. The solid dots indicate the locations from which samples were removed for analysis.**

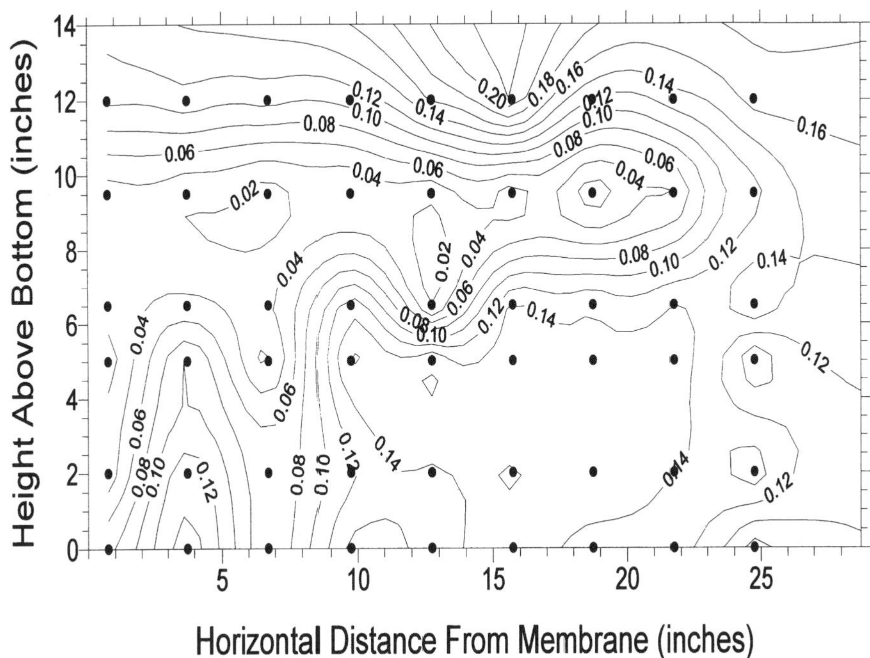

Analysis of soil samples removed from the 2-D chamber at the end of an experiment with diatomaceous earth indicated that dodecane distributed toward the membrane during OSL. It appears that water within the soil pores redistributed rather than blocking the movement of dodecane toward the membrane.

It was demonstrated that oleophilic suction lysimetry (OSL) could be used in a soil environment to selectively remove NAPL such as dodecane from wet soils. Evaluations of OSL in a 2-D test chamber illustrated the limitations of the technology. Dodecane recoveries appear to be too low to make OSL attractive as an in situ soil remediation technology. It is theorized that OSL can only remove limited quantities of NAPL before snap-off occurs within soil pores, limiting further NAPL mobilization. Snap-off limits the radius of influence and quantity of NAPL that can be recovered using OSL technology.

ACKNOWLEDGMENTS

This research was supported by a grant from The Research and Development Program of ENSR Consulting and Engineering, Acton, MA. This research was conducted at the University of Massachusetts Lowell Center for Environmental Engineering and Science Technologies (CEEST).

The authors would like to thank David Rondo and Frank Modica of the University of Massachusetts Lowell for their technical assistance, and graduate student Krishna Mummareddi for his help in this investigation.

REFERENCES

Baker, R.S. 1993. Request for Proposal: Suction-Enhanced Recovery of Non-Aqueous Phase Liquids, ENSR Consulting and Engineering, Acton, MA.

Laboratory Analysis of Soil Hydraulic Properties of Bruell Loam, Coarse Flint Shot, and Diatomaceous Earth Samples. 1994. Daniel B. Stephens & Associates, Inc., Albuquerque, NM.

Lenhard, R.J. and Parker, J.C. 1987. Measurement and Prediction of Saturation-Pressure Relationships in Three-Phase Porous Media Systems, J. Contaminant Hydrology, 1: 407-424.

Wilson, J.L., Conrad, S.H., Mason, W.R., Peplinski, W., and Hagen, E. 1990. Laboratory Investigation of Residual Liquid Organics from Spills, Leaks, and the Disposal of Hazardous Wastes in Groundwater, U.S. Environmental Protection Agency, EPA Publication 600/6-90/004.

CHAPTER 20

Removal Mechanisms Associated with the Remediation of Jet Fuel Contaminated Soil and Groundwater: A Case Study

Robert J. Roth, Terra Vac, Windsor, New Jersey

Nicholas C. Pressly, Pressly Associates, Inc., Brookhaven, New York

Marvin Kirshner, Port Authority of New York and New Jersey, New York City

INTRODUCTION

Contamination of soil and groundwater by jet fuel exists at most airports, including metropolitan and military facilities. Jet fuel is typically stored in aboveground storage tanks and to a lesser extent, underground storage tanks, which have historically leaked jet fuel into the subsurface. At most commercial airports, the jet fuel is pumped via high-pressure underground piping to the terminals. At the terminals, the piping is manifolded to distribute the jet fuel to hydrant valves at each gate where the aircraft are fueled from a hydrant pit. Leaks in the fuel distribution system and overflow at the hydrant pits have been the entry points for jet fuel into the subsurface. At airports where soil and groundwater monitoring have been performed, the analyses show contamination of the subsurface by jet fuel at the fuel storage areas and at the hydrant pits where jet fuel is floating on the groundwater surface. The thickness of the layer can range from a sheen to several feet.

The remediation technologies that are currently employed to address the problem include organic phase extraction by pumping with subsequent aboveground treatment, aqueous phase extraction with subsequent aboveground treatment, or total fluids pumping with subsequent aboveground treatment. These pump-and-treat technologies address the gross contamination in the capillary and saturated zones. Additionally, air sparging (AS) is used to air strip volatile organic compounds (VOCs) from the groundwater and to provide oxygen to stimulate aerobic biodegradation of the adsorbed and dissolved phase VOCs and semivolatile volatile organic compounds (SVOCs). To extract the VOCs generated during AS, soil vapor extraction (SVE) is used which conveys the extracted vapors to the surface for subsequent treatment. Where the contamination is only in the soil, SVE can be used to remove the VOCs by volatilization and to stimulate aerobic biodegradation by moving oxygen through contaminated vadose zone soils. The later is referred to as bioventing and in most instances requires lower vapor extraction rates.

In the design of a remediation system that addresses jet-fuel contaminated soil and groundwater, the design engineer should be aware of the removal mechanisms that will occur during the remediation and the role that each mechanism plays. Additionally, the design engineer should be aware of the timing of each mechanism to optimize the system performance. When the above-mentioned technologies are employed for the remediation of jet-fuel contaminated soil and groundwater, the predominant removal mechanisms are organic phase removal by liquid extraction, aqueous phase removal by liquid extraction, vapor phase extraction, aqueous phase/vapor phase transfer, and biodegradation in the saturated and unsaturated zone. This paper documents the remediation of a jet fuel contaminated site where four separate areas were remediated over a two-year period with an evaluation of the removal mechanisms monitored during the remediation of each area.

SITE BACKGROUND

The site is the decommissioned Terminal One at John F. Kennedy International Airport in Queens, New York. Terminal One is located in the Central Terminal Area on the southern portion of the airport. The building was constructed in 1959 and operated as a passenger terminal until February of 1990. The soil and groundwater on the airside of Terminal One are contaminated with jet fuel which spilled and/or leaked from the hydrant fueling system. A small fraction of the contamination is also from underground storage tanks which leaked motor oil, heating oil, or ethylene glycol.

The Port Authority of New York and New Jersey (Port Authority) completed several investigations which implemented the installation of soil borings, monitoring wells, and hydropunches for the collection of soil and groundwater samples. The results show that the soils are contaminated with VOCs, predominately ethylbenzene and xylenes, and with SVOCs, predominately base neutral compounds. Ethylbenzene and xylene concentrations ranged from ND to 30,000 mg/kg and ND to 34,000 mg/kg, respectively. SVOC concentrations ranged from 0 to 33 mg/kg and were predominately naphthalene, 2-methylnaphthalene, and phenanthrene. The analyses also show low concentrations of ethylene glycol. Soil samples were also tested for total heterotrophs and petroleum hydrocarbon degrading bacteria. Populations averaged 10^6 CFU/g soil and 10^5 CFU/g soil, respectively. The results of the groundwater analyses show contamination with benzene, toluene, ethylbenzene, and xylenes (BTEX) and SVOCs. The BTEX concentrations ranged from ND to 40 mg/L. SVOC concentrations ranged from ND to 800 mg/L and were predominately naphthalene, 2-methylnaphthalene, and acenaphthalene. Existing levels of nitrogen and phosphorus were determined to be sufficient for sustaining growth.

In general, the site is covered by 1 to 1.5 feet of reinforced concrete pavement. The soils beneath the pavement are fine to medium sands with trace quantities of silt to depths ranging from 11 to 14 feet below grade. This stratum is underlain by a thin low-permeability clayey-peat layer. The depth to groundwater ranges from 6 to 8 feet below grade.

The demolition of the existing terminal began in the first quarter of 1995 and took approximately six months. The construction of the new terminal is scheduled for approximately two years after completion of the demolition. The Port Authority conducted pilot tests, and provided system layout and design parameters. The Port Authority retained Terra Vac to provide the final design, install and operate the system before demolition and during construction of the new terminal. The remediation was divided into two phases. Phase 1 focused on Areas 1 and 2 (Figure 1) of the site and was scheduled for six months of operation. Phase 2 focused on Areas 3 and 4 and was scheduled for approximately one year of operation.

Figure 1. **Site plan.**

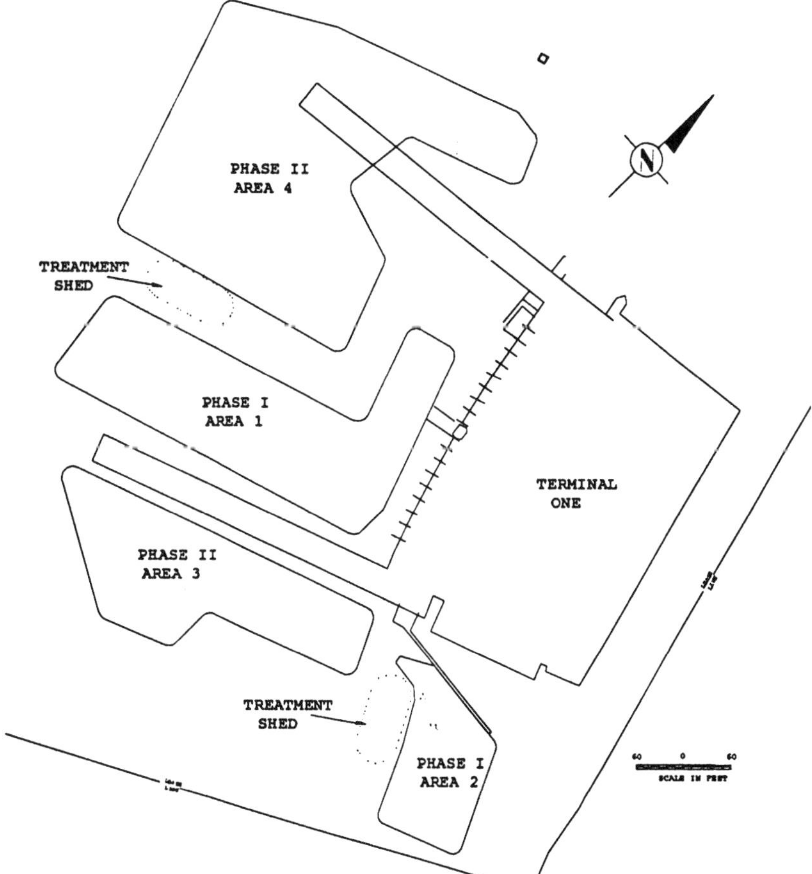

Based on the results of the pilot study, the Port Authority required that the proposed remediation system be capable of providing a vacuum of 14 inches of mercury (Hg) with a vapor flow rate of 20 cfm from each dual phase extraction (DPE) well, and a yield of 3 gpm. The Port Authority also required that the air sparging system be capable of providing at each AS well an average air flow rate of 7 cfm at 3 psi. The water treatment system discharge limits as set by the New York State Department of Conservation (NYSDEC) were not to exceed 250 ppb for VOCs, 50 ppb for PAHs, 15 ppm for oil and grease, and 45 ppm for suspended solids. Additionally, the vapor phase benzene discharge was not to exceed 0.02 lb per hour.

SYSTEM DESIGN AND INSTALLATION

Based on the contaminants detected in the groundwater and the discharge limits, the following unit processes were selected by Terra Vac: groundwater extraction, oil/water separation, iron oxidation, clarification, air stripping, and LPGAC. The treatment system equipment was sized for 500 gpm at Areas 1 and 4, and 250 gpm at Areas 2 and 3.

To prevent iron fouling of the stripper packing media, the effluent from the oil/water separator was treated by an iron oxidation tank which used diffused aeration, deaeration, and coagulation/flocculation.

An inclined plate clarifier was installed to remove suspended solids and the flocs formed during the iron oxidation and coagulation/flocculation process. The discharge from the clarifier flowed to the centralized groundwater recovery system tank, where a portion was recirculated to the DPE wells, while a portion was conveyed to the air stripper. The effluent from the air stripper was pumped through filters before a LPGAC adsorption system for polishing. This was necessary to remove the SVOCs from the water that were not removed by the air stripper, and to a lesser extent, remove residual VOCs that were not removed by the air stripping process. The unit processes that were used for the extracted vapor treatment systems include air/water separation, soil vapor extraction, catalytic oxidation (Area 1 only), and VPGAC adsorption. Due to low VOC extraction rates experienced during operations, the catalytic oxidation unit at Area 1 was removed after 60 days and replaced with VPGAC adsorbers.

The number of wells installed in each area for the remediation was the following: 36 DPE and 21 AS wells in Area 1; 15 DPE and 8 AS wells in Area 2; 21 DPE and 12 AS wells in Area 3; and 54 DPE and 33 AS wells in Area 4.

MONITORING THE SYSTEM PERFORMANCE

Samples of the extracted vapors were taken intermittently at each area, and analyzed using an on-site gas chromatograph. Vapor samples were analyzed for VOCs, and specifically for BTEX. Samples of the extracted vapors were taken from the main header. Analyses were done by a gas chromatograph (GC) equipped with a Flame Ionization Detector (FID). Several grab samples were analyzed by GC/mass spectrophotometer (GC/MS). Air flow rates were measured at the time of vapor sampling with a self-averaging annubar.

Samples of the extracted vapors from the main headers were taken with a vacuum pump, collected in Tedlar® bags, and analyzed for CO_2 and O_2 to monitor the biological oxidation of the hydrocarbons. Vapor samples are analyzed for CO_2 and O_2 with a Gastech CO_2/O_2 analyzer.

To confirm the zone of vacuum influence of the DPE wells in each area, subsurface measurements were taken during the use of one DPE well, and utilizing surrounding DPE wells as monitoring points. Subsurface vacuum readings were taken with a water manometer. Samples of the air stripper off-gases and groundwater treatment system effluent were analyzed monthly.

RESULTS OF MONITORING AT AREA 1

Vapor-Phase Hydrocarbon Removal

The vapor-phase monitoring results for Area 1 are presented in Table 1. The GC analyses showed numerous peaks, but only quantified BTEX as a small fraction. This is supported by the GC/MS analyses which showed the predominant compounds to be branched alkanes, branched alkenes, cyclohexanes, and methyl cyclohexanes. The VOC concentrations for the extracted vapors increased after startup, reached a maximum and decreased thereafter. The vapor-phase VOC and BTEX extraction rates are presented in Figure 2. The results show that the BTEX extraction rate remained relatively constant for the first 23 days of operations, while reducing to nondetectable levels thereafter. After 96 days of operations, approximately 1,067 lb of vapor phase VOCs were extracted.

Aqueous-Phase Hydrocarbon Removal

The volume of groundwater extracted from startup to Day 132 (shutdown) of Area 1 operations was 1,563,500 gallons. The groundwater extraction rate during this period averaged 11,850 gpd for 31 wells (an average well yield of approximately 0.3 gpm per well).

Table 1.   **Area 1 Performance.**

Date	Run Time of DPE (days)	System Vapor Extract. Rate (scfm)	System Vacuum (in Hg)	VOC Conc. (ppm)	VOC Extract. Rate (lbs/day)	Cum. Mass of VOCs Removed (lbs.)
10-13-94	0					0-Start
11-7-94	23	485	11	88	15	347
11-8-94	24	485	11	104	18	358
12-15-94	39	217	4	202	16	607
1-12-95	65	238	1.5	46	4	862
3-30-95	96	220	2	118	9	1067

Figure 2. Vapor-phase VOC/BTEX extraction rates, Area 1.

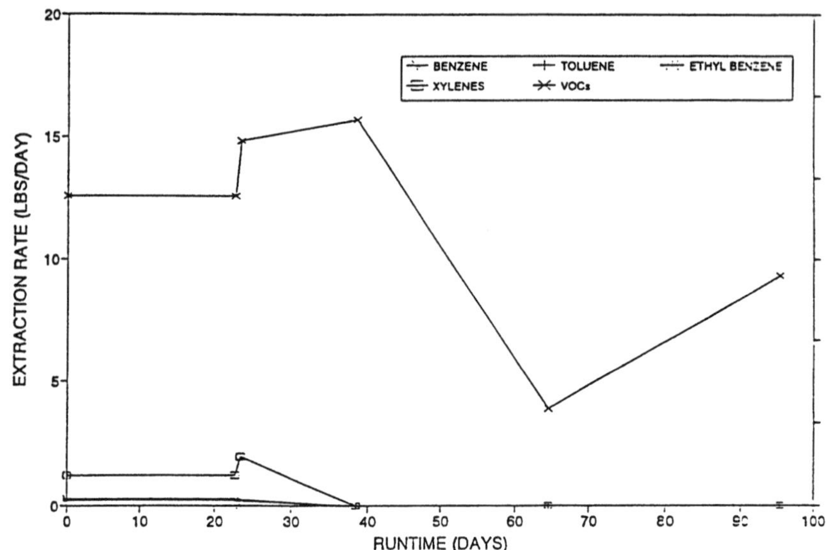

After 132 days of operations, 13,375 gallons of recovered jet fuel and 10,605 gallons of sludge were collected and transported off-site. The mass of the non-aqueous phase and aqueous phase hauled off-site was 80,560 lb and 90,000 lb, respectively. The total organic carbon (TOC) concentration of the sludge ranged from 12,000 to 88,240 mg/L. The mass of TOC (as jet fuel) associated with the aqueous phase was calculated to be 4,075 lb. Therefore, 84,630 lb of hydrocarbons was removed from the extracted groundwater.

The removal rate of hydrocarbons associated with the removal of jet fuel and sludge increased to a maximum of 910 lb/day after startup and decreased to 404 lb/day after 122 days of operation.

The mass of aqueous-phase hydrocarbons removed by LPGAC was estimated using the mass of LPGAC used and applying an adsorptive capacity of 15%. The mass of aqueous-phase hydrocarbons removed by LPGAC is estimated to have been 3,690 lb.

Biological Oxidation of Hydrocarbons

The CO_2 and O_2 concentrations in Area 1 were measured periodically to monitor the biological oxidation of the hydrocarbons. The results of the monitoring data are depicted in Figure 3. The CO_2 concentration in vapor samples from the main header was used to estimate the mass of hydrocarbons biologically oxidized. The CO_2 data were normalized assuming a background CO_2 concentration of 1.4% (the lowest concentration measured at the wellheads). The mass of CO_2 extracted was calculated using the soil vapor extraction rate and the CO_2 concentrations, and integrated for each sampling event. From startup to Day 96 of op-

Figure 3. CO_2/O_2 monitoring results, Area 1.

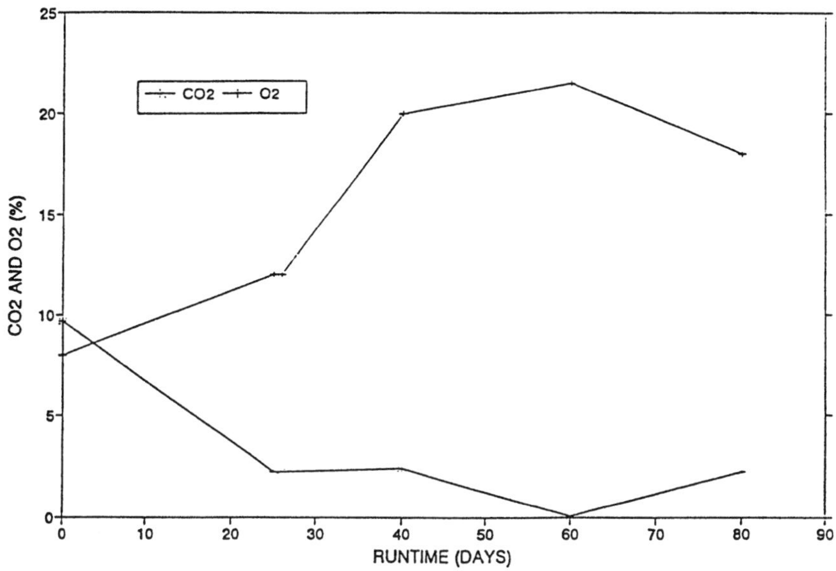

erations, the mass of CO_2 generated is calculated to have been 10,870 lb. The mass of hydrocarbons biologically oxidized is based on the following oxidation of hexane: $C_6H_{14} + 9.5\,O_2 = 6\,CO_2 + 7\,H_2O$. Using this relationship, approximately 1 lb of hydrocarbons (expressed as hexane) are oxidized for every 3 lb of CO_2 generated. A 50% yield was used to estimate the mass fraction of elemental carbon utilized for building cellular mass. Therefore, it is estimated that approximately 7,170 lb of hydrocarbons were biologically oxidized during SVE operations.

Total Hydrocarbon Removal

Cumulative vapor-phase, aqueous phase, biological, and total hydrocarbon removal versus operating time is depicted in Figure 4. After 132 days of operations, approximately 96,600 lb of hydrocarbons have been removed by SVE, groundwater extraction, and biological oxidation.

RESULTS OF MONITORING AT AREA 2

Vapor-Phase Hydrocarbon Removal

The vapor-phase monitoring results for Area 2 are summarized in Table 2. In general, the VOC extraction rates decreased during system operations, while the VOC concentrations for the extracted vapors increased after startup, attained a maximum, and generally decreased thereafter. The VOC and BTEX extraction rates are depicted in Figure 5. As in Area 1, the results show that the BTEX extraction rate remained relatively constant for the first 114 days of operation, while concentrations of BTEX after this point were nondetectable. In general, the VOC extraction

Figure 4. **Total hydrocarbon removal, Area 1.**

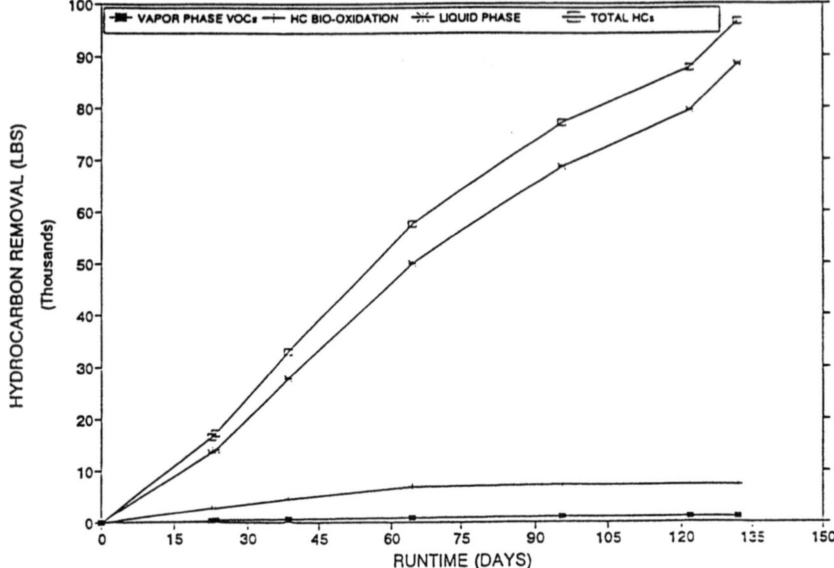

Table 2. **Area 2 Performance.**

Date	Run Time of DPE (days)	System Vapor Extract. Rate (scfm)	System Vacuum (in Hg)	VOC Conc. (ppm)	VOC Extract. Rate (lbs/day)	Cum. Mass of VOCs Removed (lbs.)
7-12-94	0					0 -Start
7-15-94	3	384	7	292	40	101
9-22-94	54	225	8	592	48	2,495
9-23-94	55	227	8	560	46	2,541
11-7-94	98	284	7.5	147	15	3,844
11-8-94	99	257	10	196	18	3,860
12-15-95	114	327	7	152	18	4,045
1-12-95	140	230	6	137	11	4,425

rate decreased during Area 2 operations. After 140 days of operations, approximately 4,425 lb of VOCs were extracted from the contaminated soils at Area 2.

Aqueous-Phase Hydrocarbon Removal

The volume of groundwater extracted from startup to day 174 (shutdown) of operations was approximately 1,470,560 gallons. The groundwater extraction rate during this period ranged from 5,000 to 27,000 gpd, with an average of 10,500 gpd for 15 wells (an average well yield of approximately 0.5 gpm per well).

After 174 days of operation, 2,078 gallons of recovered jet fuel and 7,860 gallons of sludge were collected and transported off-site. The mass of the nonaqueous phase hauled off-site was 16,150 lb. The mass of TOC associated with the aque-

Figure 5.   Vapor-phase VOC/BTEX extraction rates, Area 2.

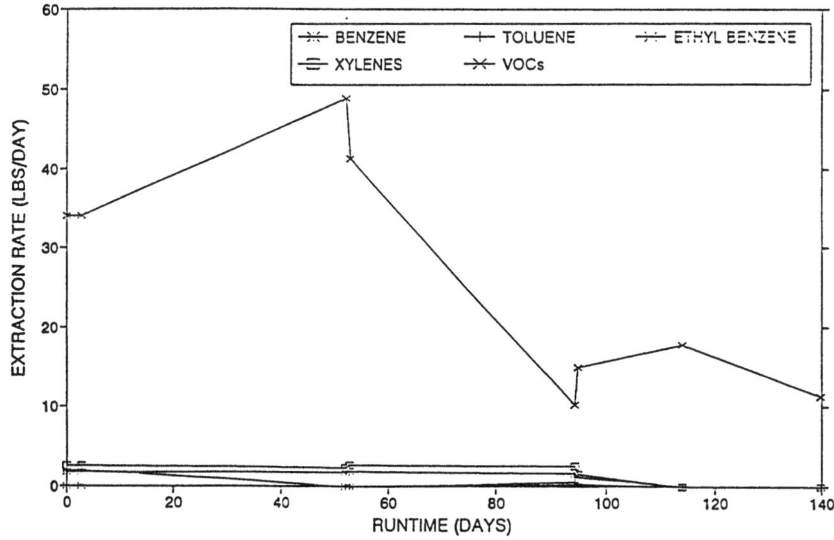

ous phase was calculated to be 2,714 lb. Assuming the TOC is from the recovered jet fuel, a total of 18,860 lb of hydrocarbons were removed from the extracted groundwater by physical separation.

The removal rate of hydrocarbons associated with the removal of jet fuel was 74 lb/day at Day 33, and 31 lb/day at Day 97, a decrease in approximately 2/3 lb/day of recovered jet fuel during the operating period.

The removal rate of hydrocarbons associated with the sludge was as high as 30 lb/day in the initial stages of treatment, but was 3 lb/day after 3 months of operations. This was most likely due to the decreased TOC in the subsurface. The mass of aqueous-phase hydrocarbons removed by LPGAC was estimated to have been 2,460 lb.

Biological Oxidation of Hydrocarbons

The CO_2 and O_2 monitoring results are depicted in Figure 6. The CO_2 in vapor samples from the main header was measured periodically and used, as discussed above, to estimate the mass of hydrocarbons biologically oxidized. It is estimated that approximately 53,380 lb of hydrocarbons were biologically oxidized during SVE operations.

Total Hydrocarbon Removal

Cumulative vapor-phase, aqueous-phase, biological, and total hydrocarbon removal versus operating time is depicted in Figure 7. After 174 days of operations, approximately 79,680 lb of hydrocarbons have been removed by SVE, groundwater extraction, and biological oxidation.

Figure 6. CO_2/O_2 monitoring results, Area 2.

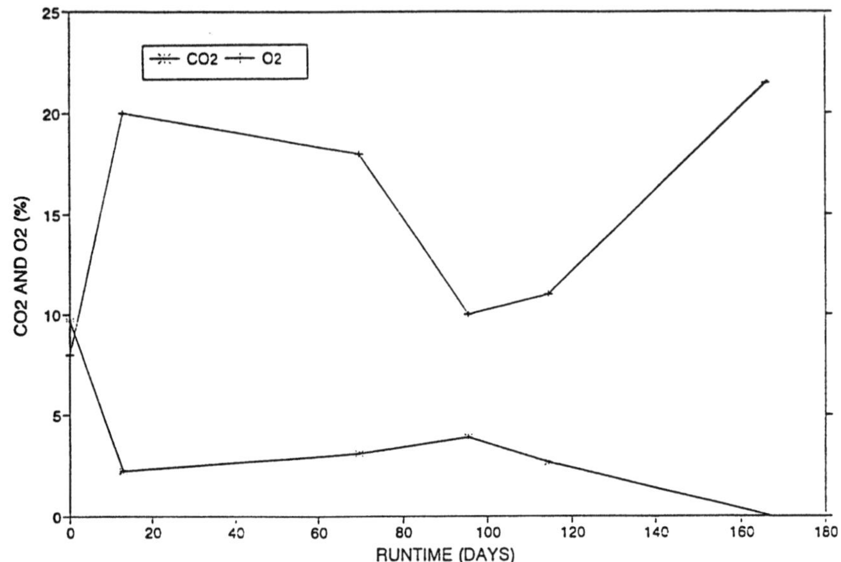

Confirmatory Groundwater Sampling/Closure

Confirmatory groundwater samples were collected and analyzed for selected analytes to verify the effectiveness of the remediation. The data show that BTEX concentrations ranged from ND to 140 mg/L and SVOC concentrations ranged from ND to 74 mg/L. The data show that the TPH concentration was reduced by four orders of magnitude. Prior to the abandonment of the DPE and AS wells in each individual area, the wells were used to recharge the aquifer with clean water to inhibit the migration of contaminants from adjacent areas not yet remediated.

A summary of the mass of petroleum hydrocarbons removed by the different mechanisms at all four areas is presented in Table 3. The results show that in Areas 1 and 4, organic phase extraction was the major removal mechanism, followed by biological oxidation. In contrast, at Areas 2 and 3, the results show that biological oxidation was the major removal mechanism followed by organic phase extraction. This is consistent with the remedial investigation results which showed a greater thickness of floating product at Areas 1 and 4 compared to 2 and 3.

DISCUSSION

Vapor-phase BTEX/VOC Extraction

The results of the extracted vapor analyses from Area 2 show that for BTEX, xylenes had the highest concentration. This is consistent with the remedial investigation data which showed xylenes having the highest concentration in both the soil and the groundwater samples.

Figure 7. **Total hydrocarbon removal, Area 2.**

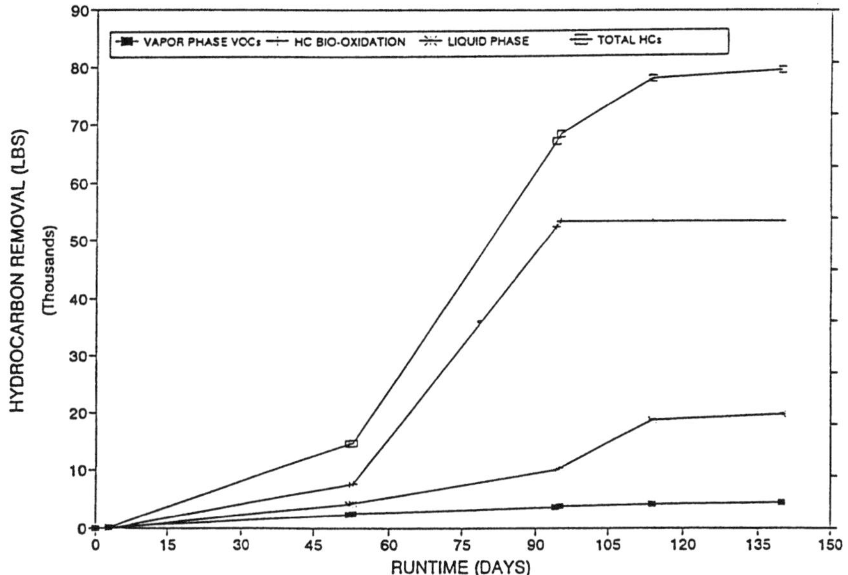

Since the concentrations of BTEX in the extracted vapors remained relatively constant, the decrease in the VOC extraction rate was, therefore, primarily due to the decrease in the extraction of other VOCs. Other VOCs could include C5 through C10 compounds (as detected during the pilot study); branched alkanes, branched alkenes, cyclohexanes, and methyl cyclohexanes (from GC/MS analyses), other VOCs associated with jet fuel, and/or intermediate volatile products from the biodegradation of the jet fuel.

The constant vapor-phase BTEX extraction rates suggested a significant mass of BTEX in the unsaturated zone. The partitioning of adsorbed phase BTEX from the soil particles likely attained a steady state with the extraction of vapor-phase BTEX. Since the data showed an increase in vapor-phase BTEX extraction rates during the initial phases of the remediation, vapor transport was most likely predominately advective. However, the steady-state vapor-phase extraction of BTEX suggests a diffusion limitation.

Biological Oxidation

The reduction in CO_2 levels in the extracted vapors during SVE operations could indicate the following: the hydrocarbons being utilized by the subsurface microorganisms as a substate might be limiting; the moisture content in the soil might have been reduced by groundwater extraction/soil vapor extraction to levels that are less than optimum or inhibitory to the subsurface microorganisms; or, nutrients could have been utilized to levels that were limiting.

Analyses for sulfate in the influent to the groundwater treatment showed concentrations up to 16 mg/L. The average concentration of sulfate from groundwater

Table 3.   Summary of Performance of Each Area After Closure.

Removal Mechanism	Area 1 lbs. (% of Total)	Area 2 lbs.(% of Total)	Area 3 lbs.(% of Total)	Area 4 lbs.(% of Total)
Soil Vapor Extraction (a)	1,070 (1)	4,430 (6)	9,370 (6)	7,800 (7)
Biological Oxidation (b)	7,280 (7)	53,380 (67)	100,000 (68)	24,890 (22)
Organic-Phase Extraction (c)	89,880 (83)	16,150 (20)	33,460 (23)	72,980 (65)
Aqueous-Phase Extraction (d) From Sludge	3,880 (4)	2,890 (4)	2,890 (2)	6,040 (5)
Aqueous-Phase Extraction from adsorption onto LPGAC (e)	6,150 (6)	2,460 (3)	630 (0.4)	1,260 (1)
Aqueous-Phase Extraction from Air Stripper Treatment (f)	50 (0.05)	50 (0.06)	50 (0.03)	40 (0.03)
Total Mass Extracted	108,310	79,360	146,400	113,010
Volume of Groundwater Extracted (gal.)	1,563,500	1,470,600	2,013,800	3,206,800

Note:
   a. Based on total VOCs.
   b. Based on CO2 concentration, hexane oxidation and biological yield of 50%.
   c. Based on actual weight of material disposed.
   d. Based on TOC of subnatant.
   e. Based on LPGAC used and 15% adsorptive capacity.
   f. Estimated using concentration of VOCs in air stripper exhaust.

samples from Area 2 ranged from 150 to 1100 mg/L. The average concentration of sulfate prior to the remediation was less than 1 mg/L. This shows the oxidation of sulfides and/or other reduced sulfur compounds created by anaerobic activity within the organic peat layer as a result of air sparging. The oxidation was likely due to autotrophic and/or heterotrophic sulfur-oxidizing bacteria of the genus Thiobacillus in an aerobic environment. The presence of Thiobacilli was verified by microscopic investigations which identified gram-negative rod-shaped bacteria. The oxidation of reduced sulfur compounds results in the formation of hydrogen ions, which in turn results in a reduction of the pH. This is supported by the fact that groundwater analyses from Area 2 showed a pH as low as 2 near the termination of remediation. This reaction most likely inhibited the petroleum hydrocarbon oxidizing microorganisms. In addition, the oxidation of FeS could also have contributed to the presence of sulfate.

The numbers of petroleum hydrocarbon degrading bacteria (PHDB) in the unsaturated and saturated zones prior to and during SVE/AS in Area 3 were used to estimate the mass of petroleum hydrocarbons biologically oxidized during the remediation. The results of the bacterial enumeration show that the number of PHDB at the onset of SVE/AS averaged 2.56 x 10^5 CFU/g-soil, while after 106 days of SVE/AS, the number of PHDB averaged 10^{10} CFU/g-soil in the unsaturated zone and 10^8 CFU/g-soil in the saturated zone.

To estimate the change in biomass (Δx) from $t = 0$ to $t = 106$ days, the zone of influence for SVE and AS was multiplied by the Δx and converted to a mass. Assuming a yield coefficient of 0.5, the mass of petroleum hydrocarbons biologically oxidized in the unsaturated and saturated zones is estimated to be 91,400 lb and 5,000 lb, respectively for a total of 96,400 lb.

The mass of petroleum hydrocarbons biologically oxidized based on CO_2 data and hexane oxidation, after 106 days was estimated to be 63,000 lb. This correlation was considered to be consistent with the biomass approach, since both results are within the same order of magnitude. Since the contamination significantly varied between locations in the unsaturated zone and only a limited number of soil samples were used for bacterial enumeration, the authors rely more upon the CO_2 approach in this case. However, the challenge still exists to account for background CO_2 which also varies with time.

SYSTEM MODIFICATIONS

During Phase 1, there were accumulations of floating scum and foam on the coagulation/flocculation tanks, inclined plate clarifier, CGRS tank, and air stripper. Microscopic analyses of the scum revealed oil globules with diameters ranging from 10 to 100 mm and bacteria. The bacteria were predominately bacilli with grazing protozoa and metazoa. It is suspected that the floating scum was partly due to the entrainment of gas bubbles from the respiration of bacteria to the oil globules. It is also suspected that the oil was emulsified from the groundwater extraction process, and the agglomeration of small globules to form larger globules was partly inhibited by the bacterial cells and/or biological surfactants.

The presence of emulsified oil caused high organic loading to the LPGAC adsorbers, which required excessive use of LPGAC. The Phase 2 systems were modified by the addition of a 20,000-gallon holding tank between the CGRS tank and the air stripper. This afforded a 1 to 2 day holding time, in lieu of the 1 to 2 hour design holding time, to maximize oil/water separation. The resultant subnatant from the holding tank had significantly lower turbidity, suspended solids, and oil content which, in turn, reduced the organic loading to the LPGAC adsorbers.

CONCLUSIONS

The monitoring results from the operations of the soil and groundwater remediation system at each area showed the removal of hydrocarbons by three mechanisms: soil vapor extraction, groundwater/free product extraction, and biodegradation. The vapor-phase removal rate decreased with time due to the initial extraction of VOCs with concentrations at or near saturation, extraction of VOCs with higher vapor pressures, and a transition in vapor transport from primarily advective to one that is diffusion-limited.

The aqueous-phase removal rate for the recovered jet fuel decreased with time. The higher removal rates of the recovered jet fuel during the initial phases of operations were most likely due to the extraction of free product which was floating on the groundwater surface. The subsequent decrease in the recovery of jet fuel from the aqueous phase may be due to the slower subsurface velocity of jet fuel relative to the velocity of the groundwater, which is in turn due to the higher viscosity of jet fuel and retardation of migration as a result of adsorption/desorption on the soil particles.

The data suggest that the mechanism which had the greatest effect on remediating the soil and groundwater in Areas 2 and 3 was biodegradation. The mass of hydrocarbons removed by biodegradation was relatively low in the initial stages of remediation, which is typical for most facultative microorganisms that require time to acclimate to an aerobic environment. The hydrocarbon removal rate associated with biodegradation increased after approximately two months of operation and is the major mechanism for the removal of hydrocarbons from the subsurface. Based upon CO_2/O_2 monitoring, the log-growth phase of the subsurface microorganisms was approximately two months before entering the log-growth phase where a substantial increase in hydrocarbon removal was achieved. At Areas 1 and 4, groundwater extraction appeared to be the primary mechanism for hydrocarbon removal. This is partly explained by the fact that there was more free product in these areas compared to Areas 2 and 3. Therefore, at sites where there is a considerable volume of free product, groundwater extraction will, at least initially, be the primary mechanism, whereas at sites where there is only a thin lens of free product, biological oxidation will be the primary mechanism. It should be noted that demolition and construction adjacent to Areas 3 and 4 removed the pavement. This likely resulted in greater air infiltration and subsequent dilution of CO_2 that was measured at the inlet of the blowers.

In the design of systems for similar sites, the relative importance of each of these mechanisms should be considered for determining the size of the treatment equipment, frequency of disposal of residuals, and potential problems that could arise from changing the ecology of the unsaturated zone, capillary zone, and saturated zone. Additionally, monitoring data should be used during operations to maximize the performance of the remediation system as the subsurface ecology changes.

ACKNOWLEDGMENTS

The authors wish to acknowledge Bruce Frumer of the Port Authority, who managed the construction and operations of the four systems.

CHAPTER 21

A Treatability Study of the Physical Stabilization of Oil Field Drilling Wastes and Tank Bottoms

John C. McMillan, Linda Kelley, and **Zeynep G. Ungun**, ICF Kaiser Engineers, Inc., Oakland, California

INTRODUCTION

A treatability study was conducted to obtain sufficient data to develop an optimum design mix for stabilization of the sump materials at the TCL State Superfund Site (site) in Wilmington, California. Stabilization of the sump materials was prescribed in the Remedial Action Plan (RAP) for the TCL Consent Order Study Area (ICF Kaiser, 1996a) approved by the Department of Toxic Substances Control (DTSC) in April 1996. The stabilization treatability study (ICF KE, 1996b) was approved by the DTSC in May 1996.

The site is located within a crude oil production facility that was acquired by the Port of Long Beach (Port) in 1994. From 1951 to 1971 the TCL Corporation operated a series of shallow impoundments or "sumps" throughout the site for the disposal of oil field production wastes. Wastes inconsistent with the operating agreement were known to be placed in the sumps and the site was included on the California Superfund List in 1983. Remediation, which includes physical stabilization of the wastes, is being conducted under a Consent Order with DTSC oversight. Following remediation, the site is planned to be redeveloped and used as a container terminal.

PROJECT DESCRIPTION

Approximately 25 sumps, covering 35 acres and containing 530,000 cubic yards (cy) of sump materials are scheduled to be excavated and stabilized as part of the remediation and redevelopment of the site. The balance of the 162-acre site is comprised of roadways, oil production wells and facilities, and piping corridors. The majority of the site is below sea level and protected by levees and drainage facilities. Most of the sump surfaces are at elevations consistent with the road grades, although materials were placed up to 10 feet above road grade in some of the sumps. The average thickness of materials placed in the sumps is 9.5 feet.

Groundwater occurs 3 to 4 feet below road grades; thus much of the sump material is currently below groundwater.

The sump materials are dark gray to black fine-grained soils with moderate to low plasticity. Some of the sump soils possess very low strength relative to other sump soils. The sump soils are often visibly oily and contain viscous tarry materials. On average, the materials placed in the sumps include 3.5 feet of fill material overlying the sump soils, 4 feet of oil sump (OS) soils, and 2 feet of very low strength (VLS) oil sump soils. The VLS material generally has a higher water content and lower dry density than the OS material.

The remedy, consistent with the RAP, includes excavation of the sump soils, placement of an embankment base layer over the road grades of up to 4 feet thick, followed by stabilization and placement of the excavated sump soils atop the embankment base layer. The stabilized sump soils will then be above current groundwater levels. The stabilized materials will be covered with a minimum of 5 feet of imported fill and a 2-foot thick pavement section. The stabilized soils will be left in place and serve as the embankment fill upon which the container terminal will be constructed.

The strength goal for the stabilized sump material is 15 pounds per square inch (psi) as determined by the Unconfined Compressive Strength (UCS) test (ASTM D2216) after 7 days of curing. This UCS corresponds to a shear strength of approximately 7.5 psi or 1,000 pounds per square foot (psf). Additional criteria are that the stabilized sump soils not have undesirable consolidation behavior, and that these materials not be subject to excess fluid release under anticipated loading conditions. The design mix must also be cost-effective and use additives that are locally and commercially available in sufficient quantities.

OBJECTIVES AND SCOPE

The primary objective of the treatability study was to identify a target design mix for the OS and VLS materials at the site. Another objective of the treatability study was to establish a correlation between pocket penetrometer resistance, which can be measured 1 day following stabilization, and 7-day UCS. Because sump soils are scheduled to be stabilized daily and in areas overlying the previous day's work, this correlation will allow an early indication of whether the stabilization criteria will be met in the absence of the 7-day UCS test results. To assist in prescribing the design mix for materials during remediation, correlations between percent cement required to achieve the desired strength and wet density and moisture content were also formulated.

The treatability study included collection of representative sump soil samples from the site, followed by laboratory testing of the samples collected. To ensure that the samples collected were representative and unbiased, a stratified sampling approach was used in selecting the sample locations.

The laboratory testing was completed in two phases: (1) the screening phase and (2) the optimization phase. The screening phase evaluated the relative strength performance of various types and percentages of additives. Additives including cement, lime, flyash, and kiln by-products were evaluated singly or in combination at percentages ranging from 2.5 to 10% for each additive constituent. A composite sump soil sample was used for the screening study. The main objective of the screening phase was to identify the more promising additive type(s) for evaluation in the optimization phase. It was anticipated that use of a composite sample would yield this information in a cost-effective manner.

The optimization phase was conducted to evaluate the effectiveness of the most promising additive, which was identified to be cement, on a variety of discrete OS and VLS sump material samples collected from the site. Due to the inherent variability in strength and handling characteristics of the sump soils, the OS and VLS samples were randomly selected from those collected during the field program. A sample of the lowest strength VLS material encountered during field sampling was also included in the optimization phase. In this way, both the statistical approach of random sampling and the goal of developing a successful design even for the worst conditions were achieved. The UCS of the OS samples that appeared to have the potential to achieve the desired strength through direct recompaction (i.e., without prior stabilization) were also evaluated in the optimization phase of testing. In addition to the UCS tests, liquid release tests and consolidation tests were also performed on a subset of the samples evaluated for UCS.

SCREENING PHASE

The objective of the screening phase was to assess the relative performance of various design mixes and to identify the most cost-effective designs for evaluation in the optimization phase. A composite sump material sample was used for preparing all of the design mixes so the UCS results for the various design mixes could be compared. Table 1 contains a matrix of the additives and additive percentages evaluated in the screening study.

The cement used was Type II Portland and the flyash used was Type F. Cement kiln dust (CKD) and a coke calciner plant kiln by-product (KBP) were also

Table 1. Screening Phase Design Mixes.

Design Mix	2.5 percent (each additive)	5 percent (each additive)	7.5 percent (each additive)	10 percent (each additive)
Cement		X	X	X
Cement and KBP	X	X	X	
Cement and CKD	X	X	X	
Cement and Flyash	X	X	X	
Lime and Flyash	X	X	X	
Lime		X	X	X

used as additives in the study. Type II cement and KBP had been successfully used in the physical stabilization of 80,000 cy of sump soils during the development of the first 30-acre portion of the TCL site.

UCS Testing

As indicated in Table 1, single additive mixes were formulated by adding 5, 7.5, and 10% dry additive based on wet weight of the sump material. Dual additive mixes were formulated by adding 2.5, 5, and 7.5% of each additive, thereby providing total additive percentages of 5, 10, and 15%. The mix designs were selected based on previous stabilization experience with sump soils at an adjacent 30-acre parcel of the TCL site. In addition to the design mixes containing additives, an untreated, remolded composite sample was evaluated for UCS and was compared with the treated results.

Replicates of each design mix were fabricated and tested in the screening phase to assess the reproducibility of the laboratory in fabricating samples. Three samples of each design mix were fabricated and tested. Given the above, a total of 18 design mixes and one untreated sample were evaluated. Thus, 57 cylindrical samples were fabricated and subjected to UCS testing as part of the screening phase. The UCS test results for each design mix following 7 days of curing are presented in Figure 1.

Screening Phase Results

The UCS results from the screening phase were used to identify a promising design mix and prepare the scope for the optimization phase testing. Of the design mixes tested in the screening phase, cement alone as an additive yielded the greatest strength improvement per unit cost of additive and the optimization phase was thus focused on evaluating the effectiveness of cement as the sole stabilization

Figure 1.   **7-Day UCS versus design mix.**

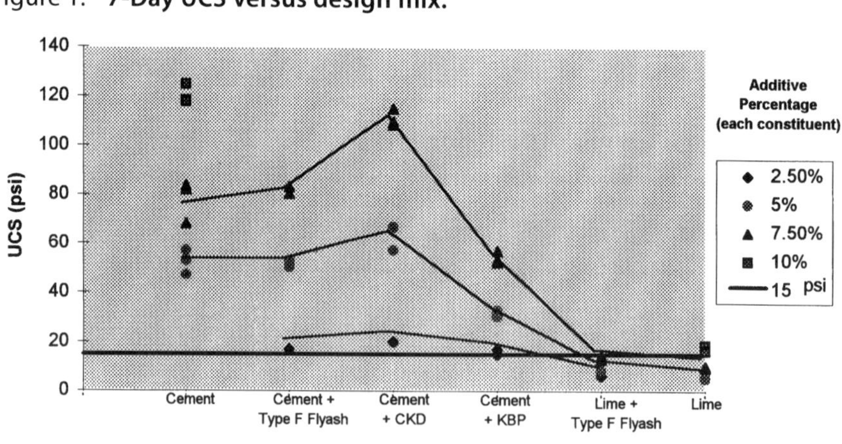

additive. Interpretation of the screening study and rationale for the cement-only additive selection for the optimization phase follows.

The UCS test results for the 5 and 7.5% cement mixes are similar to those for the 5 and 7.5% cement plus the same percentage Type F flyash. This indicates that the flyash does not contribute significantly to the stabilization of the sump soils. The addition of CKD does produce higher strengths than with cement alone; however, the CKD did not provide a cost-effective substitute for cement on a strength gain versus additive cost basis for this project application.

The combination of KBP and cement produced a lower strength after 7 days than cement alone at the same per component additive rate. Thus, the KBP appears to in some way inhibit or retard the strength gain of the cement. KBP was consequently not considered further for this project application. The UCS results for the lime and lime with Type F flyash mixes were marginally acceptable at relatively high additive percentages. Based on the higher cost of quick lime relative to cement at this project location and the potential safety hazards associated with handling quick lime, lime was not considered further as a stabilization additive.

There are several additional benefits to selecting a design mix consisting of cement only. First, the necessity for preblending additives is eliminated. Second, there are not likely to be supply limitations on cement. Last, CKD and flyash may contain heavy metals, and the potential for introducing these additional metals to the stabilized soils is eliminated if CKD and flyash are not used.

OPTIMIZATION PHASE

The chief objective of the optimization phase was to formulate cement additive percentages for the OS and VLS sump soils considering the inherent variability in the sump materials (e.g., water content, oil content, material strength and density). The optimization phase scope was designed to:

- Identify target cement addition percentages for stabilization of the OS and VLS sump soils. These targets will be used at the beginning of remediation and the percentage of cement added may be modified as remediation progresses and performance data become available.
- Identify possible correlations between the percentage of cement required to achieve the UCS criterion and sump material properties that can be measured in the field. These correlations will be used to modify the cement addition rate during remediation as sump materials with varying properties are encountered.
- Establish possible correlations between 1-day pocket penetrometer resistance and 7- day UCS for use in the field during remediation. This information will be used to assess the performance of the stabilization mix design and mixing and compaction methods shortly after placement.

- Assess whether the target cement percentages identified for the OS and VLS materials will yield undesirable liquid release or consolidation behavior, and thus require design mix modification.

The following tests were performed to achieve the optimization phase testing objectives:

- UCS and pocket penetrometer tests on discrete sump material samples stabilized with varying percentages of cement; and
- Liquid release tests (LRTs) and consolidation tests on stabilized sump material samples.

To assess the variation in UCS test results obtained, discrete samples of the VLS and OS sump materials were combined with varying percentages of cement. The composite sample was also used in the optimization phase as a baseline sample. The majority of the samples fabricated were tested for UCS following 7 days of curing. However, replicate cylinders of three samples (one OS, one VLS, and the composite sample) were also tested after 14 and 21 days of curing. Pocket penetrometer readings after 1, 3, and 7 days of curing were obtained for each of the samples.

Additional components of the optimization study included performing LRTs and consolidation tests to assess the nature and quantity of fluids that could be generated by post-placement consolidation of the stabilized sump soils. The prevention of such migration is consistent with the RAP objective of reducing potential migration of chemicals of concern. The results of these tests were evaluated in conjunction with the results of the UCS tests to assess whether the cement addition rates that yield a UCS value above the 15 psi goal would also produce materials with acceptable liquid release or consolidation behavior. This information was used to assess whether the prevention of fluid loss might become the controlling criteria in distinguishing small differences in additive percentages. The consolidation data were collected for use in this study as well as in the geotechnical design of the container terminal surface facilities.

Design Mix UCS Testing

UCS testing was conducted on discrete samples of OS and VLS materials stabilized with varying percentages of cement. Pocket penetrometer readings also were collected for each of the discrete samples after 1, 3, and 7 days of curing. For the samples cured for 14 and 21 days, pocket penetrometer data also were obtained at 14 and 21 days. The UCS results for the OS and VLS samples for varying cement percentages added are presented in Figures 2 and 3, respectively. For the samples tested, strengths generally increase somewhat linearly with increasing percent cement added. However, the VLS sample plots do not indicate correlations as linear as the OS samples.

For the samples tested, the strength criterion for discrete OS samples was met with cement additive percentages ranging from less than 1% to 5% (see Figure 2). With the exception of the sample collected from Sump T-22 (S10-VLS), the discrete VLS samples meet the strength criterion with the addition of 3 to 7% cement (see Figure 3). Based on the site investigation data collected from 1992 to present, Sump T-22 contains the lowest strength materials of the sumps at the site. The UCS versus cement additive percentage results for samples collected from Sump T-22 indicate large percentages (estimated to exceed 20%) would be required to stabilize these materials. It was noted that these samples, and the sample S20-VLS, did not contain sufficient soil to effect cementing. It is anticipated that materials from Sump T-22 and materials with little soil content will be blended with OS material prior to stabilization to cost-effectively stabilize these soils and to aid in achieving the strength goal. Although the data are not presented in this paper, UCS generally increases 30 to 50% when the curing duration is extended to 21 days. The exception is for VLS sump soil with a low cement additive percentage (eg. 4%) where strength gain after 7 days is minimal.

Liquid Release Testing

The objective of the liquid release tests was to evaluate the propensity of the treated samples to release fluid under worst-case anticipated loading conditions. The data have been used to assess the need to modify the intended design mix in order to reduce the potential for fluid generation in the embankment fill. The samples used for these tests were cured for 21 days prior to testing and the LRTs were performed using a modified EPA Method 9096 procedure. The samples were mounted in a flexible-walled permeameter cell (see ASTM D5084), and 20 psi confining pressure was applied pneumatically to the sides of the samples. No deviator pressure was directly applied to the top or bottom of the samples, as tubing from the sample ends was used to allow liquid release.

To estimate the percentage of fluid released from the samples, the samples were carefully weighed and measured before and after testing. The percent liquid release was calculated to be the weight loss of the samples during the test divided by the initial wet weight of the samples.

Plotting the UCS against the percent fluid released for the samples tested indicates that liquid release decreases with increasing strength. Approximately 1.75% fluid release corresponds to a 7-day UCS of 15 psi. The percentage fluid release decreases by 50 to 75% with a 5 to 10 psi increase in UCS. It is anticipated that the percent cement added to the sump materials during remediation will produce stabilized materials with strength that exceed 15 psi, thus diminishing the fluid release estimates.

Figure 2. 7-Day UCS versus percent cement - OS samples.

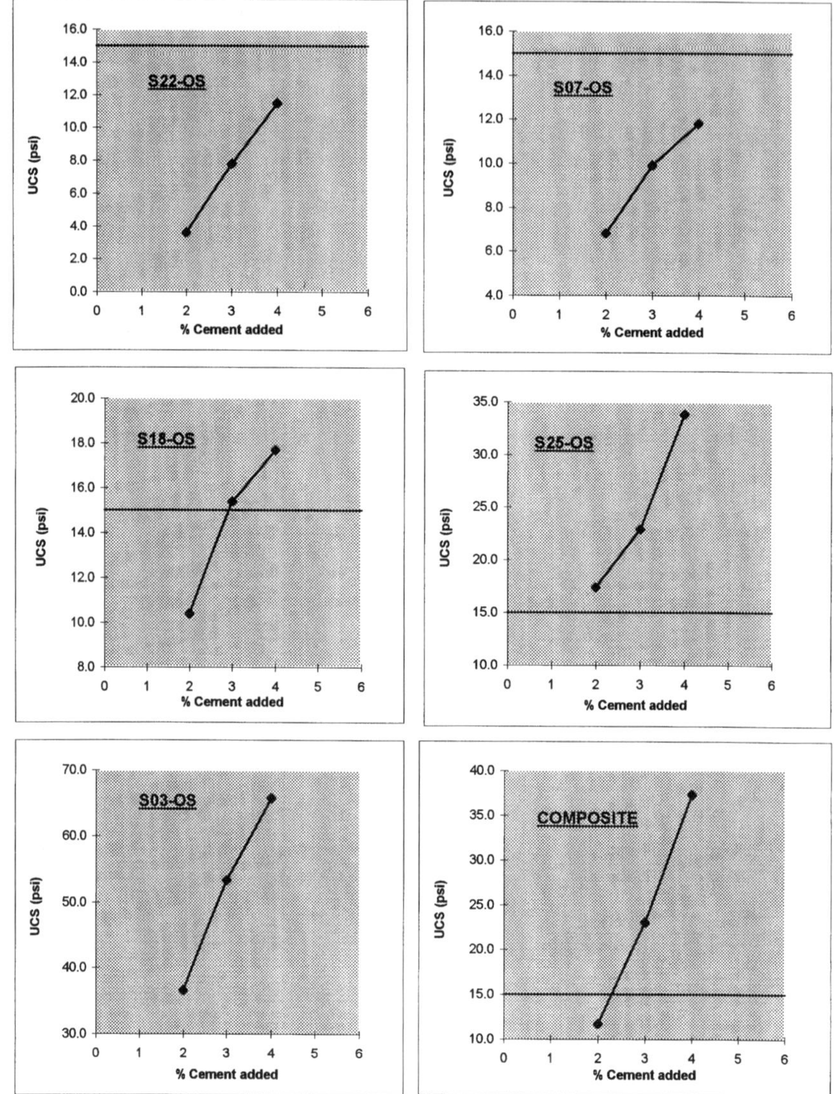

Consolidation Testing

Consolidation tests on stabilized sump soil samples following 21 days of curing were completed per ASTM D2435. To replicate field placement conditions, the samples were not saturated prior to testing; however, the degree of saturation of the stabilized materials is estimated to exceed 80% for all of the samples tested.[1]

---

[1]The percent saturation for each sample was based on a specific gravity of 2.65, which may be high for the sump soils containing oil.

Figure 3. **7-Day UCS versus percent cement - VLS samples.**

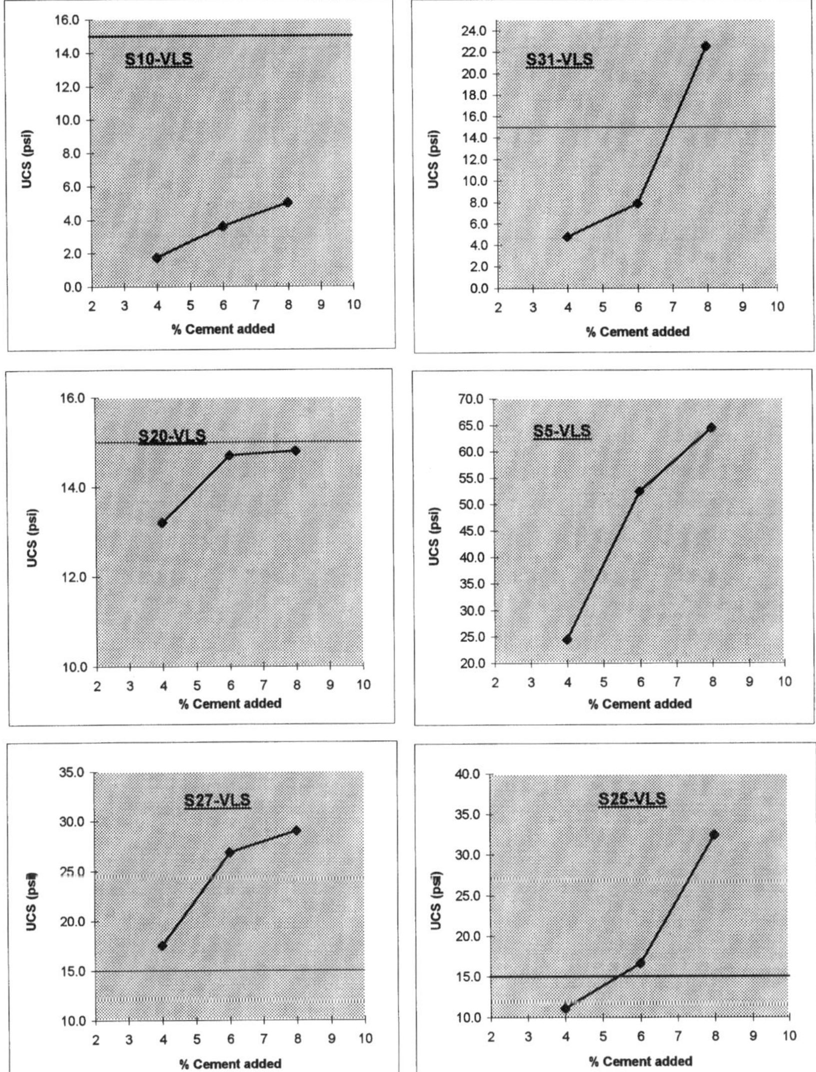

Maximum total settlements due to consolidation of the sump material have been estimated for a thickest embankment section considering the fill loads and dead loads of 4 to 5 empty containers stacked on site. The estimated settlements are approximately 0.5 inches for these load conditions sand are estimated to be less for the thinner embankment sections toward the north end of the planned container terminal development. Additional settlement of the embankment fill due to consolidation of the underlying native soils are anticipated at this site, but were not been estimated as part of the study.

DESIGN MIX AND UCS CORRELATIONS

One of the objectives of this study was to formulate correlations that can be used in the field to ensure that the stabilized sump materials meet the desired strength goal. Two types of correlations have been formulated:

- Design mix correlations that can be used to specify the percent cement required based on physical characteristics; and
- A UCS correlation that will provide an early indication that the stabilized materials will meet the desired 7-day UCS criterion.

UCS Versus Wet Density and Moisture Content

Based on the UCS design mix testing results, several physical material characteristics appear to affect the percentage of cement required to achieve the strength criterion for the stabilized soils. These characteristics include: material classification and gradation, in situ and remolded consistency, density and moisture content. Based on material consistency, sump materials have been segregated into OS and VLS materials. The VLS materials, which generally have Unified Soils Classification System symbols of ML, CL, or SM, are very oily and/or tarry, and are very soft. These VLS materials require greater cement additive percentages than the OS materials. The OS materials have similar general classifications, as the VLS materials, but the OS materials are soft to medium stiff in consistency in place.

Linear regression analyses on the percentage of cement required to meet the strength criterion relative to remolded wet density and moisture content of the OS and VLS material samples were performed (see Figures 4 and 5).

A strong linear relationship exists for the OS samples for both remolded wet density and moisture content. A relatively weak linear relationship can be approximated for the VLS samples.

To establish the correlations, the percentage of cement required to meet the strength goal was estimated for each sample based on the data shown in Figures 2 and 3. Using a linear regression analysis on the strength and wet density data presented on Figure 4, it appears that remolded sump materials with wet densities below approximately 90 pounds per cubic feet (pcf) can only be stabilized with cement additive rates greater than 8%. For these materials, it appears that the soils solid content of the sump soils is too low to enable the material to be effectively stabilized, regardless of the percentage of cement applied. Soils with low wet densities may have to be preblended with higher density OS material prior to stabilization. Table 2 presents a summary of the estimated cement percent required for various untreated remolded wet densities of sump materials.

A moisture content correlation with percent cement required to meet the strength goal also has been formulated from the data collected. The correlation using oven-dried moisture contents is presented in Figure 5. As shown in Figure 5, the percent cement required to meet the stabilization strength criteria increases with increasing moisture content. The correlation data presented in Table 3 can be used as a guide to design mix selection in the field.

Figure 4. **Percent cement versus remolded wet density.**

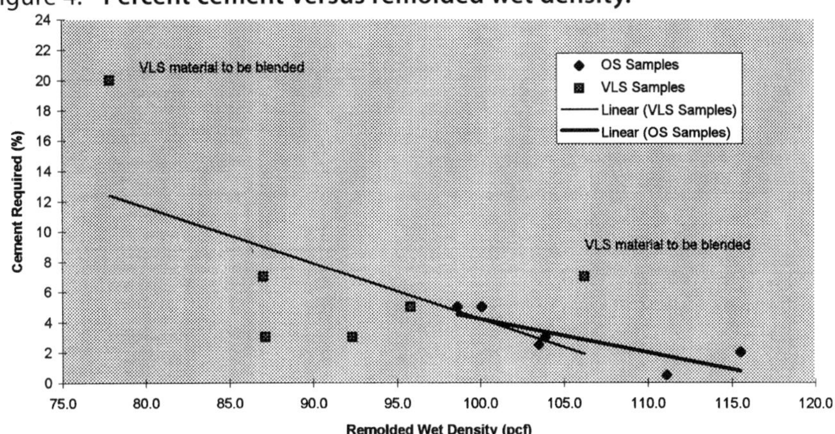

Figure 5. **Percent cement versus oven-dried moisture content.**

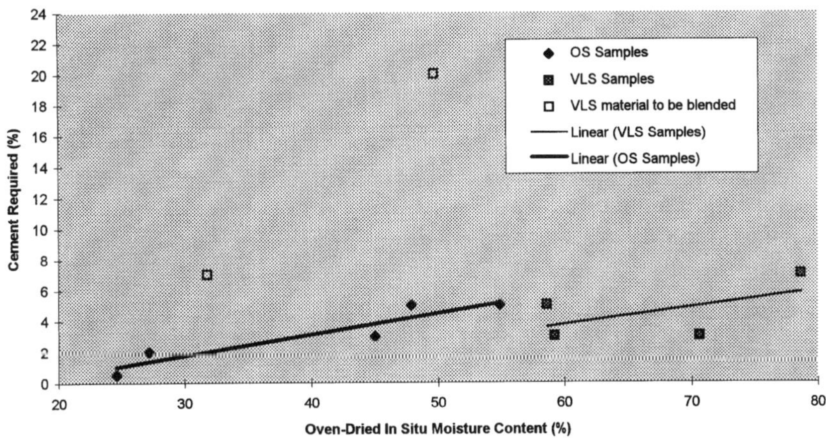

Table 2. **Wet Density Correlation .**

Remolded Wet Density (pcf)	Percent Cement Required
90 to 95	6 to 8
95 to 100	4 to 6
100 to 110	2 to 4
110 or greater	2

Table 3.   **Moisture Content Correlation.**

Oven-dried Moisture Content (%)	Percent Cement Required
35 or less	2
35 to 45	2 to 4
45 to 60 for OS material	4 to 6
45 to 75 for VLS material	4 to 6
>75	6 to 8

UCS Versus Pocket Penetrometer Resistance

In addition to the design mix correlations to wet density and moisture content, a UCS correlation for the stabilized sump materials has been formulated with pocket penetrometer resistance. Pocket penetrometer readings taken in the field after the stabilized materials have cured for 1 day may be used to assess whether the materials will meet the 7-day UCS criterion. This information will serve as an early indicator of the success of the design mix selection and of the mixing and compaction methods employed.

As shown in Figure 6, 1-day pocket penetrometer resistance increases with increasing 7-day UCS for both the OS and VLS materials. To achieve a 7-day UCS of 15 psi, the 1-day pocket penetration resistance of the stabilized materials should be approximately 1 ton per square foot (tsf) and 3 tsf for the VLS and OS materials, respectively. The higher percentages of cement in the stabilized VLS soils may explain the lower allowable 1-day pocket penetration resistance requirement for the VLS materials.

TARGET DESIGN MIX SELECTION

Target design mixes for stabilization of the OS and VLS materials were formulated. These design mixes represent the percent cement required to meet the 7-day UCS criterion. To meet the engineering requirements for the site, it was stated in the treatability report that the 7-day UCS criterion of 15 psi be met 90% of the time

Figure 6.   **7-Day UCS versus pocket penetrometer reading.**

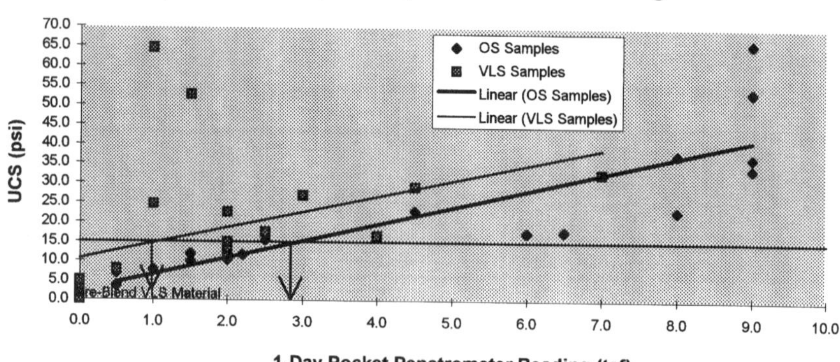

in the upper fill, and that the criterion be met 70% of the time in the lower fill. The upper fill is defined as the top 7 feet of material below final surface grades. The lower fill is any material placed below that horizon.

Based on the available data, the percent cement required to meet the strength criterion for the upper and lower fill for both the OS and VLS materials has been formulated. These target percentages were estimated by calculating the percentage of the total number of samples exceeding the UCS goal. As only 6 discrete OS and 6 discrete VLS samples were considered in the optimization phase of this study, the target design mixes will be modified during remediation as performance data become available. For the lower embankment fill, the target design mixes for OS and VLS materials are 4 and 7% cement, respectively. An additional 2% cement is estimated to be sufficient to meet the requirements for stabilized material placed in the upper fill with the desired reproducibility. It is currently planned that stabilized materials be placed only in the lower fill section.

FIELD ENGINEERING

The percentage of cement added to the sump materials excavated from the site during remediation will be selected in the field based on the data collected during this treatability study. A field engineer will be on site during the entire remediation project and will specify the percentage of cement to be added to the sump materials immediately following excavation. The field engineer will use the target additive percentages presented in the previous section for prescribing the cement addition rates. Slight modifications in the design mixes may also be made by the field engineer using the design mix correlations. The field engineer will also use the UCS versus pocket penetrometer resistance correlation to assess the strength performance of the stabilization process in real time. It is anticipated that the correlations developed in this treatability study will be continually updated as data become available during remediation.

The field engineer will be equipped to test the remolded wet density and pocket penetration resistance in the field. Oven-dried moisture contents may be tested at an off-site laboratory. Design mixes and/or mixing and compaction methods may be evaluated and adopted by the engineering team based on apparent stabilization performance. The remediation contractor will collect and test samples of the stabilized sump materials for 7-day UCS. These data will be provided to the field engineer, and modifications in the design mixes may also be made based on these results.

ACKNOWLEDGMENTS

Mr. Andrew Austin of Diversified Minerals, Inc. of Ventura, California supplied additive samples, and additive content and availability information. His assistance is gratefully acknowledged. Mr. Davis Scharff of Union Pacific Resources Company provided valuable input on the formulation of the site remedy. Drs. Geraldine

Knatz and Robert Kanter, and Mr. Kendall Robinson of the Planning Department of the Port of Long Beach guided site redevelopment and remedial planning and are gratefully acknowledged. Mr. David Burditt of the Engineering Department of the Port provided much essential engineering input in the remedial planning process, for which we are sincerely grateful.

REFERENCES

ICF KE (ICF Kaiser Engineers, Inc.). 1996a. Remedial Action Plan, TCL Consent Order Study Area, Wilmington, CA, April 4.

ICF KE (ICE Kaiser Engineers, Inc.). 1996b. Stabilization Treatability Study Report (Draft), TCL Consent Order Study Area, Pier A Remediation, Wilmington, CA, April 1.

CHAPTER 22

A New Approach to Unexploded Ordnance Removal

Christine J. Shields and **Gregory M. Gibbons**, Parsons Engineering Science, Inc. Oak Brook, Illinois

INTRODUCTION

In these times of federal budget cutbacks, the cost of ordnance contamination cleanup on former military installations has become a growing concern for the Department of Defense. The traditional ordnance cleanup action involved 100% subsurface removal. A typical cost for a subsurface removal action to a 2-foot depth is approximately $6,000 per acre for the land survey and unexploded ordnance (UXO) removal action. The U.S. Army Corps of Engineers estimates that tens of thousands of acres will require remediation at a cost of billions of dollars.

The current objective of ordnance removal actions is to remove ordnance to a level that will reduce the risk to the public, not necessarily to remove all ordnance from the site. One alternative is the completion of an investigation targeted at the collection of a statistically significant data set that can be used to estimate the amount of ordnance remaining on site. The methodology behind the statistical approach is to accept a small amount of uncertainty in characterizing the individual grids in exchange for a much greater understanding of the contamination of the overall site. Once the site information has been collected, an Engineering Evaluation and Cost Analysis (EE/CA) is prepared, evaluating the need to perform remediation and the remedial options available. Through this process, the expensive removal efforts are limited to the necessary sites and areas.

THE CAMP GRANT RIFLE RANGE UXO REMOVAL PROJECT

Parsons Engineering Science, Inc. (Parsons ES) is conducting an EE/CA for the former Camp Grant Rifle Range located near Rockford, Illinois. The work is being conducted under contract with the Huntsville Corps as part of the Defense Environmental Restoration Program for Formerly Used Defense Sites. The Huntsville Corps has taken the lead on the cleanup of UXO at former military installations already turned over to the public.

The activities at the former rifle range included a Time-Critical Removal Action (TCRA) encompassing a surface removal of ordnance in 170 of the 312 acres. A subsurface investigation and partial removal action was performed in selected 100-by-200-foot grids during the EE/CA investigation. The data generated through the two actions was used to determine remedial alternatives for the EE/CA.

The EE/CA investigation included the use of a statistical sampling approach and computer risk analysis. The sampling was performed using GridStats and SiteStats, an innovative statistical package designed for UXO investigations projects, developed by QuantiTech of Huntsville, Alabama. The risk analysis was conducted with the Ordnance and Explosives Cost-Effectiveness Risk Tool (OE*Cert*).

SITE DESCRIPTION

The former Camp Grant is located near Rockford, Illinois. Camp Grant was established in July of 1917 and designated as an infantry replacement and training camp in the spring of 1918. Camp Grant was used as a National Guard training ground between World War I and World War II. With the onset of World War II, Camp Grant was reactivated as a troop induction center, training camp, and a German Prisoner of War internment camp. The camp was deactivated in approximately 1954. In 1955, Mr. Seth B. Atwood purchased the rifle range portion of Camp Grant and donated it to the Rockford Park District (USACE, 1992).

The portion of Camp Grant used as a rifle range currently is owned by the Rockford Park District, and is known as Atwood Park. The Nature Awareness Center at the park operates day- and week-long camping programs for outdoor environmental and ecology awareness classes for young people. Approximately 4,000 young people participate in these programs annually. The park is open to the general public for camping, picnicking, biking, walking, hiking, canoeing, and cross-county skiing. The park district performs some recreational construction on the property (i.e., installation of recreational equipment, utilities and road improvements, trails, fencing, other park buildings, etc.). The majority of the park programs are operated on the north side of the park property, with only a few programs conducted on the south side.

Periodically, throughout the time the park has been in operation, UXO has been found by park officials and the public. Ordnance found on the property included small arms ammunition (.30 and .45 caliber), 37 millimeter (mm) projectiles, 3-inch Stokes trench mortars, pineapple-type hand grenades, and rifle grenade casings. UXO found on the property has been detonated on site or removed by the local bomb squad or the Fort McCoy Explosive Ordnance Demolition (EOD) Unit.

SITE LAYOUT

The former rifle range is shown in Figure 1. The former rifle range is divided into northern and southern sectors by the Kishwaukee River. During the rifle range operation, the area north of the river was the impact zone and the area south of the river was the firing line. An overshoot area north and east of the rifle range was designated as the range danger zone. Currently, the majority of the former range danger zone is residential housing. The property line between the former rifle range and the residential housing is wooded and undeveloped. A portion of the eastern former range danger zone currently is used as the Blackhawk Valley Campground. The areas included in the EE/CA are:

1. north of the river (northern sector);
2. south of the river (southern sector);
3. Kishwaukee River (river sector);
4. Former range danger zone sector; and
5. Blackhawk Valley Campground sector.

SITE INVESTIGATION FIELD ACTIVITIES

Traditionally, the remedial option for former ordnance sites was 100% removal. Under the new approach to UXO removal, the sites are first characterized by determining the amount of individual and public risk. The first stage of determining the risk at a site is to adequately characterize the UXO density.

For this project, the site characterization was accomplished by analyzing the data generated by the TCRA and EE/CA investigation activities. The TCRA consisted of a surface removal action in the northern sector and in the potential short-shot area in the southern sector. The EE/CA investigation consisted of an intrusive investigation of a statistically determined set of anomalies found in selected areas of the entire property to a depth of 2 feet below ground surface. Magnetometers were used to detect changes in the earth's ambient magnetic field (anomalies) caused by ferrous materials. Located UXO, UXO-related scrap, and household debris was removed from the investigated anomalies. Although no investigation was performed in the former range danger zone or the campground, the TCRA and EE/CA investigation data were used to complete the risk evaluation and evaluate the need to perform remedial activities in these areas.

Time-Critical Removal Action Activities

The main focus of the TCRA was to conduct a surface removal in the northern sector and in the area in the southern sector across from the major impact area on the northern bank. The TCRA included a surface UXO location, removal, and disposal in 150 acres in the northern sector and 20 acres in the southern sector.

Northern Sector

In the northern sector, ten 3-inch Stokes mortar rounds and one 37-mm projectile were recovered and disposed of by detonation in place. Numerous other pieces of UXO-related scrap, approximately 1,260 pounds, were recovered. The UXO-related scrap included 140 sand-filled expended practice rounds. The sand-

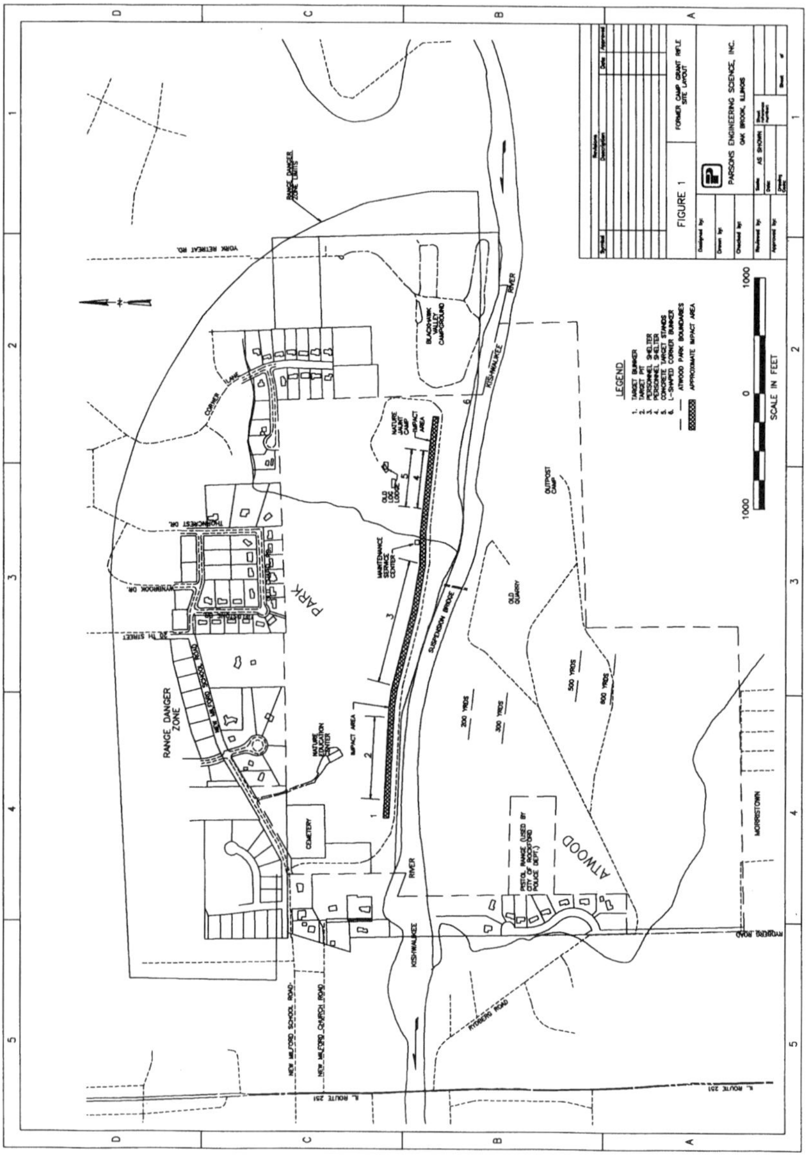

Figure 1. Former Camp Grant rifle range site layout.

filled rounds were moved to an on-site quarry and detonated with a shape charge to ensure they did not contain any high-explosives. UXO and UXO-related scrap was recovered up to the northern and eastern property lines. The findings of the TCRA in the northern sector were used to establish the lateral density of UXO and UXO-related scrap in the northern sector.

Southern Sector

In the southern sector, no UXO or UXO-related scrap was located on the surface of the TCRA grids.

Engineering Evaluation and Cost Analysis Field Investigation Activities

The EE/CA investigation was conducted in the southern and northern sectors, and the river. The northern and southern sector investigation included a subsurface UXO location, removal, and disposal in selected grids. The river investigation included only a geophysical investigation; no intrusive activities were conducted. The main purpose of the EE/CA investigation was to determine UXO density in the southern sector, typical UXO penetration depth in the northern sector, and anomaly density in the Kishwaukee River.

The EE/CA investigation utilized sequential sampling tests to characterize a 100-by-200-foot grid as homogeneous. The sequential sampling tests are part of the GridStats sampling tool developed by QuantiTech, under contract with the Huntsville Corps. The need for statistical sampling tools was determined during early investigations of UXO sites. A 100-by-200-foot grid at a UXO-contaminated site potentially contains thousands of subsurface anomalies that are not UXO items. The investigation of every anomaly in such a grid could take weeks and generate a considerable amount of information regarding the particular grid; however, such a grid investigation would generate little information regarding the rest of the site. The methodology behind the investigation is based on the statistical sampling and characterization of an individual grid. The purpose of the statistical anomaly investigation was to establish the sample size for each grid to estimate the UXO density within statistical error bounds. The grid characterization then facilitates a much greater understanding of the entire site while reducing the number of anomalies that require investigation. The methodology behind the statistical grid and sector sampling is to accept a small amount of uncertainty in characterizing the individual grids in exchange for a much greater understanding of the contamination of the overall site. Sequential sampling techniques are used under this approach to minimize the overall cost (QuantiTech, 1994).

Northern Sector

Nine 100-by-200-foot grids were investigated in the northern sector. Each grid was geophysically surveyed and the total subsurface anomaly count tallied. A statistically significant number of anomalies were intrusively investigated.

UXO, UXO-related scrap, and household debris were removed from the investigated anomalies. Two 3-inch Stokes mortar rounds and one 37-mm projectile were recovered and disposed of by detonation in place. Numerous pieces (approximately 99 pounds) of UXO-related scrap were recovered.

Southern Sector

Twenty-two 100-by-200-foot grids were investigated in the southern sector. The grids investigation was completed utilizing GridStats. Additionally, since the purpose of the EE/CA investigation in the southern sector was to determine UXO density in the sector, a sector characterization tool was utilized. In this way, the sector was characterized based on the investigation of a smaller sampling set. The statistical tool, SiteStats, directed the UXO team to continue the grid investigation until the sector was characterized to an acceptable confidence level.

No UXO or UXO-related scrap was recovered from the subsurface in the southern sector. One piece of UXO-related scrap was recovered from the surface of one grid.

River Sector

In the river, five grids were geophysically surveyed. Grids near either property border, by the major impact area in the northern sector, and in the center of the river were investigated. The geophysical investigation was completed by securing ropes on either river bank and sweeping the river bottom from the northern to the southern river bank. Additionally, the river bank was examined for UXO items that may have surfaced due to erosion.

Fifty anomalies were detected in the river grids. Some of the anomalies were too large to be UXO, based on the UXO recovered in the northern sector.

RISK EVALUATION

Risk Evaluation Model

A risk evaluation was completed to estimate the potential for risk to the public due to exposure to UXO remaining in the sectors. OE*Cert* was developed by QuantiTech for Huntsville Corps for use as a quantitative risk evaluation tool of ordnance contamination for UXO sites across the country. In the OE*Cert*, risk is determined by the potential for public exposure to both surface and subsurface UXO. The risk estimate is based on a Poisson process (QuantiTech, 1994).

Unexploded ordnance exposure is defined as an individual being in the proximity of surface or subsurface UXO (depending on the activity type), regardless of the individual's awareness of the UXO presence. Although an exposure does not correlate to an accident or injury, it is used to assess the potential for public risk from UXO.

Understanding the UXO concept of risk evaluation and the difference to the risk assessment performed in conjunction with hazardous waste projects is a fundamental issue in UXO remediation projects. The U.S. Army Corps has adopted similar technology in their programs - risk, risk evaluation, and exposure - used in performing an EE/CA. This causes confusion with individuals familiar with the RCRA (Resource Conservation and Recovery Act) program, which uses similar terminology. A crucial element of an UXO project is to, early-on, develop an understanding of the terminology, definitions, and concepts in the individuals conducting the work and regulators reviewing the findings.

The OE*Cert* was developed to provide a means to compare remediation alternatives for an ordnance-contaminated site. The OE*Cert* results can be used to prioritize ordnance remediation projects for funding by allowing a comparison of potential risk. The prioritization is based on the maximum risk reduction for the total cost of the remedial effort. The public exposures are approximated based on individuals performing specific activities within UXO-contaminated sites. The expected number of remaining UXO items is based on the UXO density determined in the TCRA and the EE/CA intrusive sampling data. The expected number of exposures is determined based on the ordnance density, activities on the site, and population performing the activities.

Northern Sector Risk Evaluation

The maximum depth of UXO recovered during the two investigations was approximately 12 inches, while other investigated anomalies were found to a maximum of 20 inches. Over 97% of all anomalies investigated were less than or equal to the 12-inch depth.

Analysis of the available data (both the TCRA and the EE/CA intrusive samples) would lead to characterizing this northern sector as one or more clusters (subareas) differentiated only by levels of UXO subsurface contamination. This would lead to the division of the northern sector into east and west partitions (approximately 100 acres in the eastern partition and 50 acres in the western partition).

According to the historical information regarding property use, the northern sector was used as the impact or target zone. Based on the removal action, investigation findings, and historical information, there is likelihood that subsurface UXO items remain on the northern sector of the property with the eastern half of the northern sector being the most likely area to contain subsurface UXO items. Based on the findings of the TCRA and EE/CA efforts, an estimated 0.17 UXO items per acre may remain in the western partition, and 1.7 UXO items per acre may remain in the eastern partition (QuantiTech, 1996).

The public usage of this park is significant with its ongoing educational nature awareness programs and its close proximity to the city of Rockford's population and nearby residential areas. The northern sector of the park currently is used for hiking, cross-country skiing, picnicking, camping, biking, children playing, and archaeology investigations (historical artifacts search). Future park-related construction activities involving ground intrusion activities may include installation of recreational equipment, utilities and road improvements, trails, fencing, and other park buildings. In summary, activities may occur that may pose both surface and subsurface UXO exposures to the public. The risk resulting from exposure to UXO during these activities was analyzed in the OE*Cert*.

Based on the estimated remaining quantities of UXO from the EE/CA and TCRA results and on the type and quantity of activities expected to be present in this sector, some risk of public exposure to UXO is to be expected in the northern sector. Approximately 26 exposures (22 exposures in the eastern partition and 4 exposures in the western partition) may be expected annually based on the OE*Cert* model calculations (QuantiTech, 1996). As noted, an exposure is defined as the public being within the proximity of an UXO item and does not equate to risk of injury. Compared to other former military institutions, the former Camp Grant Rifle Range is among the lowest expected exposure risk grouping.

Southern Sector Risk Evaluation

The analysis of the field investigations indicate there is extremely low probability that UXO, from the previous firing range activities, is expected in this sector. Although some potential for small localized burial pits may exist, the field investigation results and the available historical records do not provide any information regarding the possible location.

According to the historical information regarding the property use, the south side of the property was the firing line, while the impact zone was on the north side. As a result, the potential for short-shots is the only risk in the southern sector. Both the EE/CA intrusive sampling and the surface TCRA in the northeast portion of the southern sector found no indications of UXO or related scrap or fragments to verify the existence of this short-shot risk.

Based on the investigation findings and the historical property use, it is unlikely that any UXO items are present; therefore, no UXO items are expected by the OE*Cert* model and no annual exposures expected (QuantiTech, 1996).

Kishwaukee River Segments Risk Evaluation

During the EE/CA investigation, 50 subsurface anomalies were identified in the five river investigation segments. These five segments covered approximately 20% of the river sector area. None of the anomalies were intrusively investigated. Land areas on the northern river bank were found to be contaminated with surface and subsurface UXO. It can be expected that a portion of these 50 anomalies may likely be both surface and subsurface UXO. According to the historical information regarding the property use, the river appeared to be the division between the southern sector firing line and the northern sector impact zone. The potential for short-shots would appear the likely rationale for potential contamination in the river sector. Some of the anomalies were single-point contacts, while others appeared to span up to 25 feet. These longer contacts are not likely to be UXO, based on the type of UXO recovered in the northern sector. The results of the OE*Cert* report indicate that 0.56 UXO items may be expected per acre in the river.

Public activities in the river sector include swimming (and tubing), canoeing, and wading. Although these water activities are easily accessible by the park visitors, they do not have as high public usage as the other park activities identified in the northern or southern sectors of the park. These water activities are likely to involve much less surface contact (bottom of the river) by the public, and it also is unlikely that the activities will intrude below the river bottom surface. The results of the analysis of the investigation findings and this OE*Cert* indicate that the potential for public exposure to UXO in the river bottom is minimal, due to the low probable density and minimal activities (QuantiTech, 1996).

Former Range Danger Zone Risk Evaluation

After review and assessment of the TCRA and EE/CA investigations along with the historical records, it is possible to make some assumptions about the potential for UXO in portions of the former range danger zone next to the northern park boundary. For the most part, the range danger zone is improved property. Areas have been cleared and houses constructed. There is a section of undeveloped, wooded property along the northern border of the park. For this evaluation, the area of concern is only the undeveloped area along the northern park border. This area, like the northern sector, will be discussed in eastern and western partitions.

With some surface UXO found during the TCRA activities along the northern sector boundary, it is likely that some additional items may be on the other side of the park boundary. It is also likely that previous property improvements, e.g., grading and seeding or excavating for house foundations by the adjacent property landowners may have already eliminated the UXO items.

According to the historical information regarding the property use, the former range danger zone appeared to be the safety area outside the primary impact area. The potential for long-shots would appear the likely rationale for potential contamination in this sector. The terrain also appears to be a factor, as these longer shots seem to follow the area topographical features through the northeastern park area.

Public usage in this area on the park northern boundary area includes individuals taking shortcuts to the park and residential backyard activities. Although the number of potential participants is limited, these activities are likely to be both surface and ground intrusive.

The OE*Cert* indicated approximately 0.19 UXO items per acre may be expected in the western partition and 1.8 UXO items per acre may be expected in the eastern partition of the former range danger zone. No exposures are expected in either partition due to the very low probable ordnance density, minimal activities, and few public participants (QuantiTech, 1996).

Campground Risk Evaluation

After review and assessment of the TCRA and EE/CA investigations, along with a review of the historical records, it is possible to make some assumptions about the potential for UXO in portions of the campground.

Surface UXO was recovered along the eastern park boundary next to the campground area. Although the historical target line stops at the eastern boundary, it does not appear from the TCRA surface activities that the UXO and associated scrap and fragment also stops. Additionally, there is no terrain feature that abruptly prevents the items from crossing the current park boundary. Given these assessments, it is likely there is a potential for some UXO items to exist in the campground. Improvements in the campground may have eliminated some of these UXO items during initial clearing phases.

According to the historical information regarding the property use, this area was in the former range danger zone outside the primary impact area. The potential for long-shots and off-azimuth shots would appear the likely rationale for potential contamination in this sector. The OE*Cert* estimates that 1.8 UXO items per acre may exist in the campground.

Public usage in the campground includes the same activities as identified in the northern sector of the park, with camping as the primary activity. Although the number of potential participants is limited, these activities are likely to be both surface and ground intrusive. If no further action is performed, OE*Cert* indicates that approximately 15 exposures to UXO may occur annually (QuantiTech, 1996).

ENGINEERING EVALUATION AND COST ANALYSIS

The EE/CA is the USEPA-specified remedial alternative analysis mechanism for non-time-critical removal actions to be remediated under Superfund. This USEPA model was developed as an expedited approach to compliance with the cleanup requirements of the Comprehensive Environmental Response, Compensation and Liability Act (CERCLA). The non-time-critical removal action represents a primary tool for accomplishing early actions and can be applied to a broad array of response actions. Although the former Camp Grant Rifle Range is not a National Priority List site, the removal action is being conducted in conformance with CERCLA.

Approach

The purpose of the EE/CA was to evaluate remedial alternatives for the five sectors on or surrounding the former rifle range (Parsons ES, 1996). The five sectors included in the evaluation are:

1. Northern sector (both the entire 150 acres and the eastern 100 acres);
2. Southern sector;
3. Kishwaukee River;

4. Former range danger zone; and
5. Blackhawk Valley Campground.

For the evaluation, four basic alternatives were identified for the properties. The possible alternatives include:

1. No further remedial action;
2. Institutional controls in the form of public awareness;
3. Surface removal; and
4. Subsurface removal.

Alternatives were selected for each sector based on the investigation findings and the risk evaluation for the sector. A general description of the alternatives follows.

No additional activities would be conducted on the property with the implementing of the no further remedial action alternative. The property would remain in the current condition and access would be unlimited. This alternative was evaluated for all sectors as a baseline.

Institutional controls include public awareness or education of the potential for exposure to UXO. Institutional controls in the form of public awareness were evaluated as a separate alternative where appropriate, and in conjunction with other removal action alternatives.

The surface clearance alternative includes ordnance location, removal, and disposal on the surface of the sector. The surface clearance would be conducted until removal or property boundaries were met or until the UXO finds diminish to zero. The surface clearance alternative was evaluated for those sectors, as appropriate, in which a surface removal action had not already been conducted. Even after the removal action, there is still a potential for subsurface UXO to remain on the properties and for exposure to UXO. Institutional controls in the form of public awareness should be implemented as part of this alternative. Precautions should be exercised in the event of any future construction.

The subsurface clearance alternative includes ordnance location, removal, and disposal to a depth of 2 feet in the sector. The subsurface removal would be conducted until the removal or property boundaries were met or until the UXO finds diminish to zero. Even after the removal action, there is still a potential for UXO to remain on the properties and for exposure to UXO. Institutional controls in the form of public awareness should be implemented as part of this alternative. Precautions should be exercised in the event of any future construction.

Alternative Analysis

Alternatives were identified for each sector and evaluated against three criteria: effectiveness, implementability, and cost. Each alternative was evaluated based on its own merit and against the other alternatives identified for the sector.

For two sectors, the southern sector and the river, only the no further action alternative was identified. For these two sectors, an alternative analysis was not completed. In the northern sector, former range danger zone and campground, the analysis indicated that the no further action alternative was not protective of human safety and health, and the environment. Additionally, in the former range danger zone and the campground, the institutional controls alternative was determined to be not protective of human safety and health, and the environment.

Three alternatives remained for the northern sector after the application of the effectiveness alternative. Since a surface removal was conducted during the TCRA, the surface clearance alternative was not evaluated for the northern sector. These remaining alternatives are:

1. Institutional controls/public awareness;
2. Subsurface clearance in selected areas (the eastern 100 acres); and
3. Subsurface clearance in the entire sector (150 acres).

The remaining three alternatives were evaluated against the implementability and cost criteria. Preliminary recommendations were based on the criteria analysis. Additional input to the implementability criteria will be incorporated after the public comment period.

Two alternatives remained for both the former range danger zone and the campground after the application of the effectiveness alternative. For both sectors, these alternatives are:

1. Surface clearance; and
2. Subsurface clearance.

The two remaining alternatives were evaluated against the implementability and cost criteria. Preliminary recommendations were based on the criteria analysis. Additional input to the implementability criteria will be incorporated after the public comment period.

Recommendations

The preliminary alternative recommendation was based on the application of the effectiveness, implementability, and cost criteria. After the preliminary recommendations have been agreed upon by the U.S. Army Corps and the Illinois Environmental Protection Agency, the EE/CA will be opened for public review and comment. Final recommendations for each sector will be completed after the public comment has been received.

The most cost-effective alternative that is protective of public safety for the northern sector is institutional controls. This alternative is implementable and will effectively manage the risk on the property. There is no cost associated with the implementation of this alternative since the park already has a public awareness program in place. This alternative includes educating the park patrons as to the possibility of exposure to UXO. Precautions should be taken in the event that future construction is performed.

For both the southern sector and the Kishwaukee River, the preliminary recommendation is no further action. There is no cost or action associated with this alternative.

The most cost-effective alternative that is protective of public safety for the former range danger zone and the campground is a surface clearance. This alternative includes a surface clearance in the undeveloped areas of the former range danger zone, along the park and property border, and in the campground along the property border. Even after the implementation of this alternative, the property owner and public should be aware that some risk of exposure to subsurface UXO still exists, and precautions should be exercised in the event of future construction. The total cost to implement this alternative in the former range danger zone is $38,800. The total cost to implement this alternative in the campground is $32,400. This alternative can be implemented with minimal disturbance to the property owner and based on the rifle range investigation findings, will effectively manage the risk in these areas.

CONCLUSIONS

Former Camp Grant Rifle Range

The purpose of the EE/CA was to determine the remedial action which best manages the risk remaining for the former rifle range and the surrounding properties. Based on the analysis, there is minimal risk remaining in the area resulting from ordnance. Under the traditional approach, the 100% clearance effort would have cost, at a minimum, $2 million for the entire former rifle range, former range danger zone, and campground. The total cost surface removal in the northern sector and short-shot area of the southern sector and the EE/CA was $900,000. The results of the EE/CA indicate that approximately another $70,000 will be spent on additional remedial actions, for a total cost of less than $1 million for the rifle range and the surrounding properties.

Additionally, the OE*Cert* results can be used to rank the former rifle range relative to other UXO cleanup projects for funding prioritization purposes. To date, the Huntsville Corps has conducted the UXO cleanup projects similar to the former Camp Grant Rifle Range at 16 former military institutions.

Unexploded Ordnance Remediation Projects

The use of statistical sampling techniques and a risk reduction benefit/cost analysis will significantly reduce the cost of UXO remediation efforts while protecting the public safety. Considering the number of potential UXO sites, and the cost of investigation and cleanup, a rational, limited action approach is essential.

The Huntsville Corps has been charged with developing and implementing the risk-based decision-making approach. The use of the developed statistical sampling techniques (GridStats and SiteStats) and an UXO risk evaluation model (OE*Cert*) is in the early stages of application. Approximately 16 sites have been or are under evaluation and/or cleanup using this approach. The results to date have been encouraging. This approach allows for rational decision-making and direct comparison of risk reduction from various alternatives. It also allows comparison of the risk posed by one site relative to other sites, allowing the U.S. Army Corps to prioritize funding.

As a new technology, it is imperative to educate those using these tools and reviewing findings in the concept and application of the risk-based approach. The unique use of terminology common to the UXO and RCRA programs has caused confusion. This further emphasizes the need to educate the interested community.

This approach successfully addresses risk posed by UXO-contaminated sites. This risk-based approach will provide the U.S. Army Corps with the tools needed to address the many UXO-contaminated sites in a logical and cost-effective manner.

REFERENCES

QuantiTech, 1994. OEW Risk Mitigation Prioritization, Huntsville, AL, June QuantiTech, 1996. Former Camp Grant Rifle Range OECert Analysis Final Report, Huntsville, AL, August.

USACE (U.S. Army Corps of Engineers), Rock Island District, 1992. Archive Search Report and Conclusions and Recommendations for the Former Camp Grant Rifle Range, Rock Island, IL, August.

Parsons ES (Parsons Engineering Science, Inc.). 1996. Engineering Evaluation and Cost Analysis Draft Final Report, Oak Brook, IL, September.

CHAPTER 23

Extremely Rapid PCE Removal from Groundwater with a Dual-Gas Microporous Treatment System

William B. Kerfoot, K-V Associates, Inc., Falmouth, Massachusetts

INTRODUCTION

A system using ozone/air injected periodically in conjunction with a pulsing pump has been demonstrated to reduce PCE-contaminated groundwater from 1,000 ppb to less than 1 ppb in three week's time. The rate of decay was found to be a ten-fold reduction per week in monitoring wells located 5 to 15 feet from the specially-designed sparging well.

This Chapter discusses the treatment of a 60-foot wide, 15-foot deep, and 150-foot long plume of tetrachloroethene (PCE) extending from an urban dry cleaner. Three special treatment wells were installed (Figure 1). Monitoring from seven wells was performed weekly during and after treatment.

The unique process appears capable of operating without vapor control. The balancing of ozone concentration to rate of extraction by the microbubbles resulted in hardly detectable levels of PCE in the vapor control system. In addition, by the use of microbubbles, the rate of ascent of the bubbles was designed to allow oxidation reactions to be completed before the bubbles reached the unsaturated (vadose) zone.

PROCESS DESCRIPTION: IN-SITU AIR STRIPPING WITH MICRO-ENCAPSULATED OZONE

The K-V Associates, Inc. (KVA) process of C-Sparging,™ (in-situ air stripping with micro-encapsulated ozone[1]) combines two unit operations. First, fine bubbles with a high surface to volume ratio are injected into the saturated zone to extract dissolved chlorinated solvents out of contaminated groundwater. Second, ozone contained within the bubbles reacts extremely rapidly in the gaseous phase to decompose the solvents into end products consisting of carbon dioxide (CO_2), very dilute hydrochloric acid (HCL) and water. The reaction detoxifies groundwater

---

[1] Patent Pending.

Figure 1.   **The C-Sparger™ system showing in-well pulse pump and spargepoint combined with a lower two-inch spargepoint.**

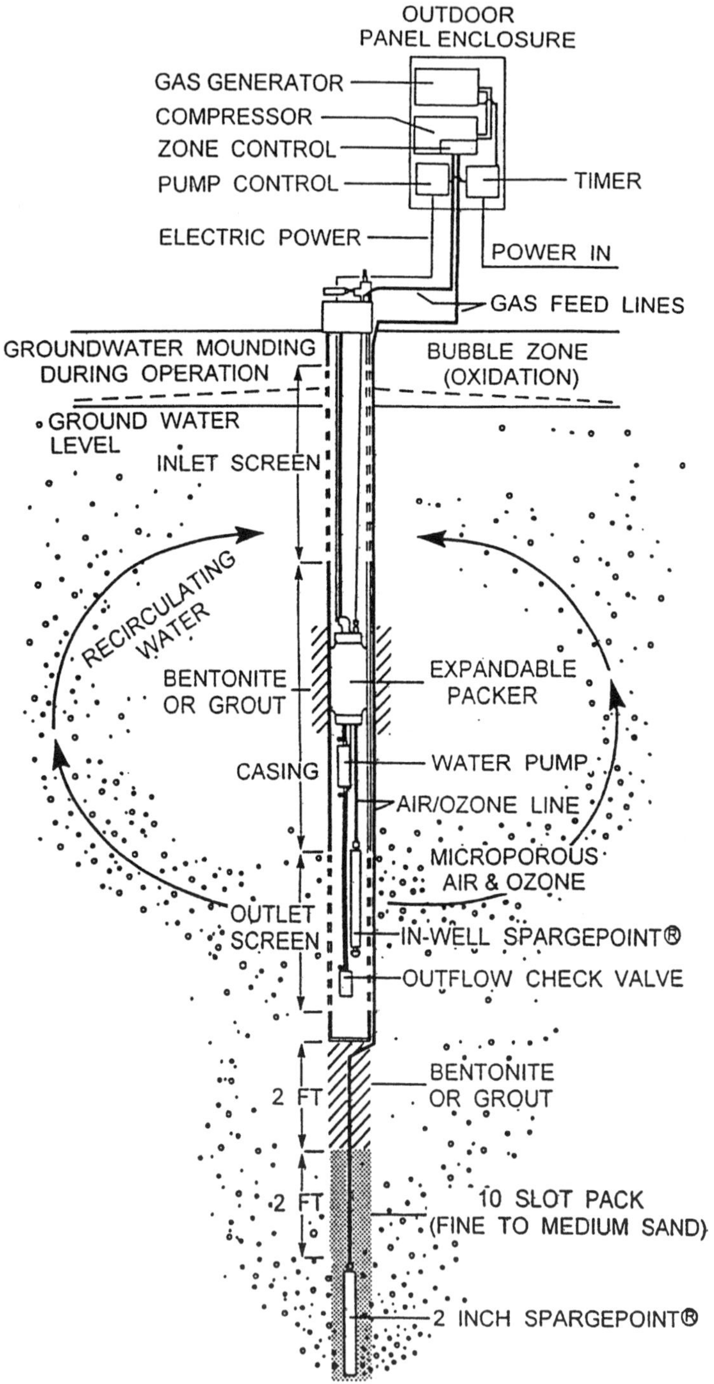

containing VOCs like PCE, TCE, and 1,1-DCE rapidly to below drinking water standards without producing harmful by-products. It is important to note that the reaction is accomplished with very low ozone concentrations (molar ratios) compared to VOC concentrations in the groundwater.

The treatment system injects fine bubbles below and into the main VOC plume (Figure 1). Pulsed injection of air/ozone through a micro-channeled diffuser (Spargepoint®) introduces fine bubbles into the bottom of the plume region, which move upward through the contaminated water. Within the central core area of the plume, a second Spargepoint®, combined with the intermittent operation of a submersible pump, displaces the vertically-moving bubbles sideways to maximize dispersion and contact. By periodically pulsing this process, groundwater is circulated from top to bottom of the sparge well, allowing effective capture and treatment of the VOC-impacted saturated zone.

SITE DESCRIPTION

The site is located in a downtown shopping center on Main Street, in Falmouth, Massachusetts. The region was formed on a glacial outwash plain and contains a thick (250-foot plus) sand and gravel sole-source aquifer (LeBlanc et al., 1986). Surface area mainly consists of a large asphalt-paved parking lot behind the Main Street businesses. Depth to groundwater ranges 12 to 13 feet, depending upon rainfall, which averages 42 inches per year (Table 1).

Table 1.   Well and Groundwater Elevations, April 1, 1991

Well	Well Elevation (ft)	Depth to Groundwater	Groundwater Elevation (ft)
MW-1	16.81	12.54	4.27
MW 2	16.03	11.88	4.15
MW-3	16.30	12.15	4.15
MW-4	16.46	12.35	4.11
MANHOLE	16.61		

Soil Composition

The subsurface soils consist of predominantly fine to medium sand, occasionally laced with cobblestone and coarse sand layers. A typical grain size distribution is shown in Figure 2. Coarse sand and gravel account for about 31% of weight, medium sand for 31%, and fine sand for about 35% of total weight. Fine soil fractions of silt and clay account for less than 3% by weight.

Groundwater Elevation

Groundwater elevation was measured by surveying the well elevation relative to the sewer system manhole cover and then measuring depth to groundwater in each

Figure 2.   An example (MW-1) of particle size distribution of the fine to medium sand found across the site area.

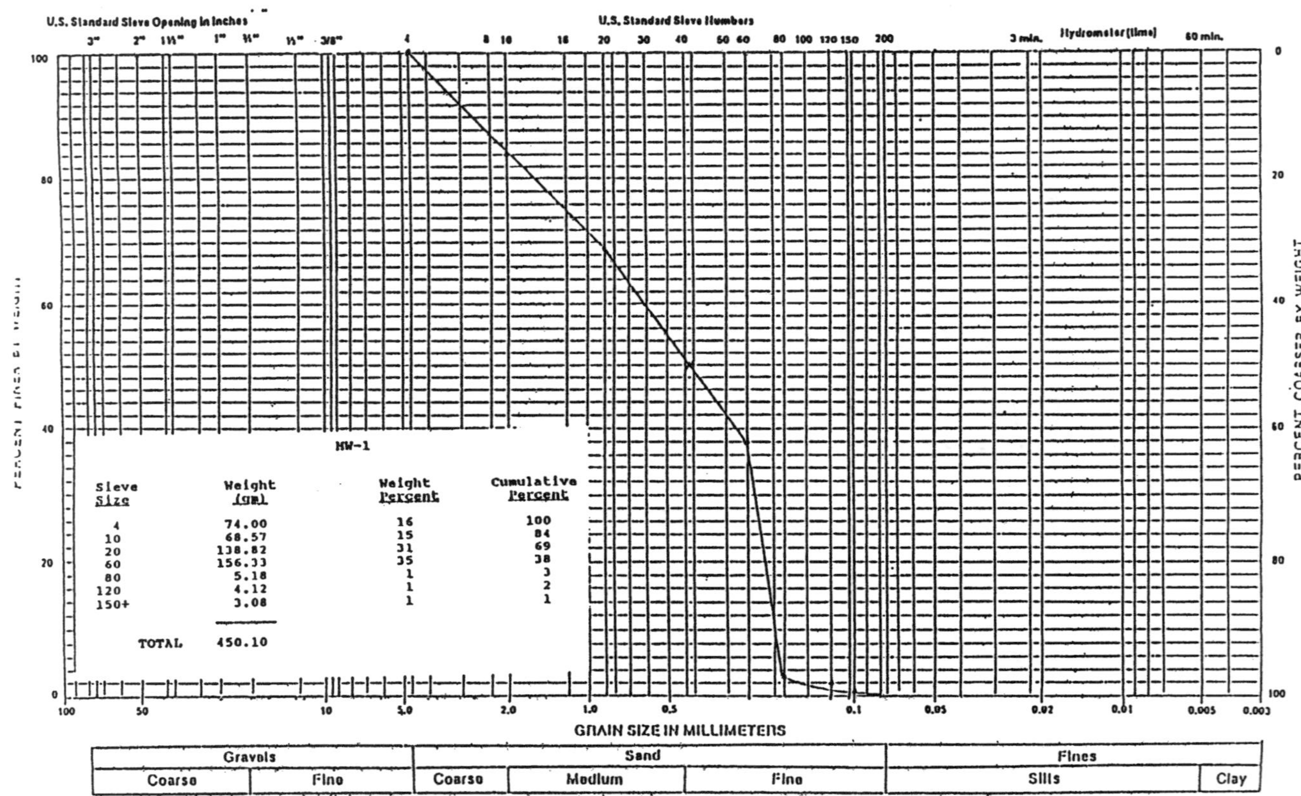

well. Groundwater elevations measured on March 19, 1991, are shown in Table 2. The water table is contoured and shown in Figure 3. Direction of groundwater flow is southeast. The gradient at the site varies from .005 to .0015.

Groundwater Flow Measurements

Groundwater flow direction and rate were measured in well MW-3 using the KVA Model 40 GeoFlo groundwater flowmeter. The flowmeter works on a heat-pulse methodology, and is accurate to 0.1 ft/day horizontal flow, +5° direction (Kerfoot, 1988). Recommended protocol is to take three measurements per well, top, middle, and bottom. Magnetic declination for the area is 15°. Data for MW-3 is presented in Table 2.

Table 2. **Flow Direction and Rate**

Well	Depth	Direction	Flow rate	
			Machine Units"	Ft/Day
MW-3	16.5 ft	242°	28	1.3
	17.5	250°	25	1.1
	18.5	317°	10	.4

TREATMENT SYSTEM

A single C-Sparger™ master unit with three in-well units was installed along the historical plume region (spargewells A, B and C) (Figure 4). The first installation was a double Spargepoint® installation with a dual screen, as shown in Figure 1. The second installation consisted of a dual-screened four-inch well with casing between, but no lower spargepoint. The third installation was a double spargepoint/ dual screen, but no in-well unit was used and only air/ozone supplied to the lower spargepoint.

The general construction consisted of four-inch casing leading to a five-foot screen 1.5 feet above the water table, a blank casing which was bentonite-sealed in the annular space to prevent short-circuiting, and a lower five-foot screen (10 slot). Below this was placed a support section, five feet of two-inch PVC, holding a two-inch spargepoint (18 inches long), with a compression fitting and one-half inch polypropylene tubing leading up to the wellhead region.

TREATMENT RESULTS

Initial results of the treatment process were monitored by three procedures: (1) groundwater samples taken at weekly intervals from purged monitoring wells and analyzed on a gas chromatograph using the 25% headspace procedure developed by Spittler, et al. (1985), (2) groundwater samples obtained by purging monitor- ing wells and submitting samples to certified analytical laboratories by EPA Method

8010 for volatile organics (Table 3), and (3) groundwater samples obtained from purging the monitoring wells and analyzed for selected cations and anions [alkalinity, total dissolved solids (TDS), chloride, sulfate, and nitrate] (Table 4).

A rapid decrease in tetrachloroethene (PCE), trichloroethene (TCE), and cis-dichloroethene (1,2 DCE) was observed in monitoring wells (Figure 5). The rate was related to proximity of the monitoring wells to the sparging wells. The rate of decrease was also more rapid with higher molecular weight: PCE> TCE> DCE. The headspace analyses showed a quicker response than did the laboratory samples, but both were closely correlated (Table 3). The rapid decrease in HVOCs

Figure 3. Groundwater contours across the site and flow direction/rate obtained from MW-3 on the site.

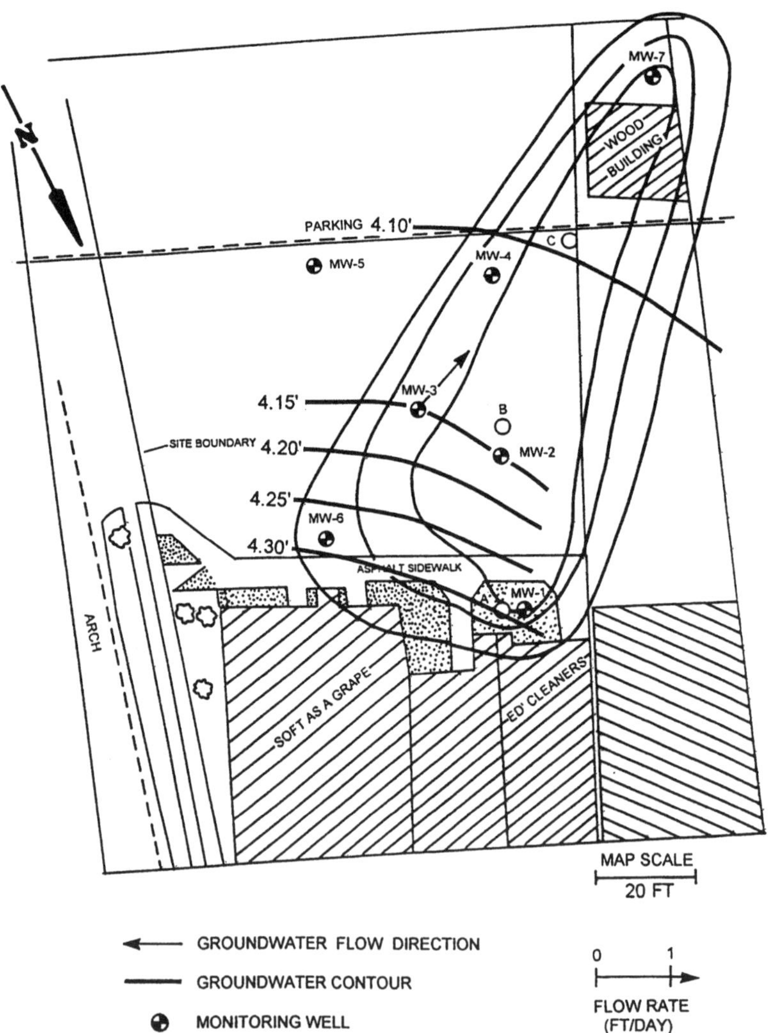

was very similar to that observed at a treatment site in Carson City, Nevada (Kerfoot, McCulloch, and Connors, 1996).

Table 4 shows a relationship between rates of PCE removal and changes in anion/cation balance. The proximity of monitoring wells to sparging wells allowed them to be segregated into heavy, moderate, and light C-Sparging treatments. Sparge well C and monitoring wells MW-1 and MW-4 were closest to C-Sparging sources. A significant increase in alkalinity, TDS, chloride, and sulfate was observed.

Generally heavier C-Sparging shows elevated hardness (carbonate production), chloride, sulfate, and TDS. The total amount of increase of chlorides in the

Figure 4.   **Plume region showing isopeth contours of equal concentration of total HVOCs. The spargewell locations (A, B, C) are indicated with their expected zones of influence.**

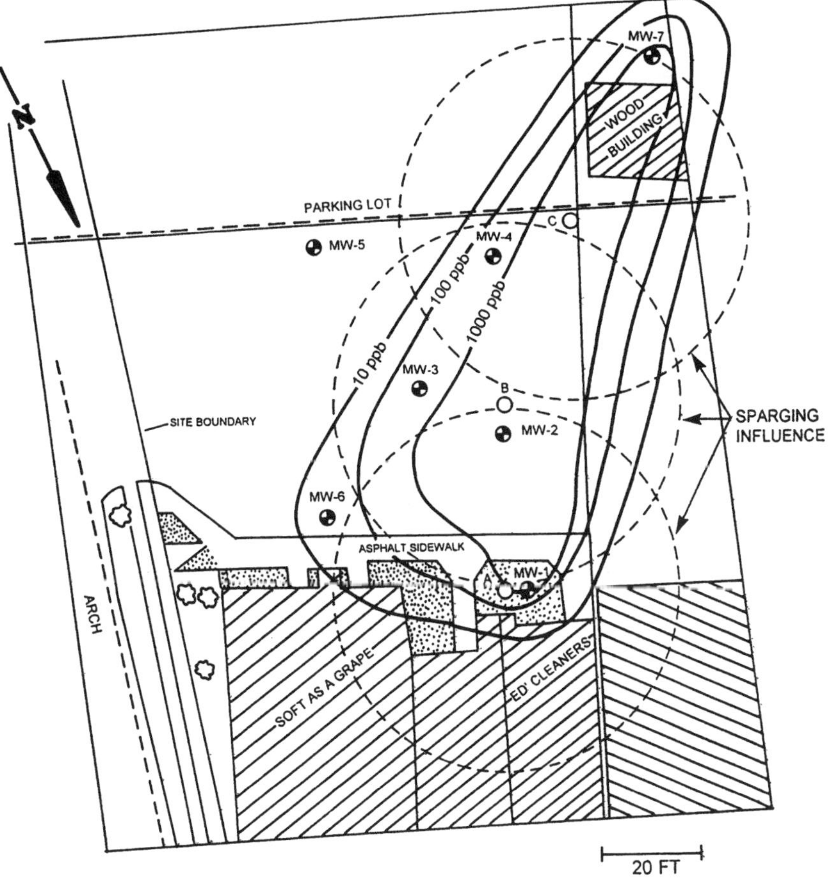

Table 3. Groundwater Samples Analyzed by EPA Method 8010

Concentrations (ppb) in Monitoring Well MW-1	Before Treatment		After Continuing Treatment	Rebound Test Turned off 1.5 months
Compound	12-29-95	02-02-96	04-01-96	07-01-96
Tetrachloroethene (PCE)	430	11	BRL	3.3
Trichloroethene (TCE)	50	2	BRL	BRL
CIS-1,2 Dichloroethene (DCE)	14	2	BRL	BRL
	12-29-95	02-02-96		07-01-96
Tetrachloroethene (PCE)	200	88		90
Trichloroethene (TCE) 1,2	29	18		11
CIS-1,2 Dichloroethene (DCE)	45	43		3.2
trans-1,2 Dichloroethene (1,2 DCE)	1	BRL		BRL
		MW-3		
	12-13-95	02-02-96	04-01-96	07-01-96
Tetrachloroethene (PCE)	7	12	4	17
Trichloroethene (TCE)	4	8	BRL	13
CIS-1,2 Dichloroethene (1,2 DCE)	6	5	BRL	BRL
	12-13-95	MW-4	04-01-96	07-01-96
Tetrachloroethene (PCE)	50	77	38	910
Trichloroethene (TCE)	19	25	6	1500
CIS-1,2 Dichloroethene (1,2 DCE)	79	45	22	29
trans-1,2 Dichloroethene	1	BRL	BRL	BRL
Vinyl Chloride	BRL	BRL	BRL	28
		MW-7		
	05-95	02-02-96	04-01-96	07-01-96
Tetrachloroethene (PCE)	4400	180	38	2200
Trichloroethene (TCE)	1400	65	7	990
CIS-1,2 Dichloroethene (1,2 DCE)	28	190	28	25

Table 4. Relationship of Degree of Treatment to Observed Decrease in PCE and Selected Anion/Cation Changes.

	VOC (PCE) (in ppb)		Alkalinity	TDS	Hardness	Chloride mg/l	Sulfate	Nitrate
	Start	End						
Heavy C-Sparge								
KVC	2000	BDL	6.0	186	59	67	33	0.6
MW1	400	BDL	4.0	163	44	50.1	29.5	1.2
MW4	3000	38	10.0	185	68	52	21	4.3
Moderate C-Sparge								
MW3	600	BDL	8.0	178	53	39.1	18.5	3.6
MW6	40	BDL	4.0	193	54	56.4	20	4.6
MW7	4000	38	9.0	167	53	45	24	3.8
Light C-Sparge								
MW2	300	100	1.0	79	14	31	6	1.0
(shielded from bubbles - old septic leaching pit)								
Upgradient (or outside of influence)								
MW5 (2)			4.0	85	19	22	8.9	1.8

dissolved fraction exceeded the rise expected (up to 40 mg/L-ppm) from decomposition of PCE (1,000 ppb to nondetect). This suggests some adsorbed liquid was reacted with. Decomposition products of C-Sparging are CO_2, water (HOH), and dilute HCl. Hardness measures equivalent $CaCO_3$ concentrations (as mg/L). Alkalinity is the measurement of capacity of water to accept protons, usually imparted by bicarbonate compounds of the natural water.

CONCLUSIONS

Ed's Cleaners in Falmouth, Massachusetts exhibited a plume region containing over 1,000 ppb PCE, roughly 70 feet wide and 200 feet long, which persisted for over four years. The region was treated with a KVA C-Sparger utilizing three in-well units. The observed concentrations dropped rapidly, reducing the entire region to

Figure 5.   **Observed decrease in concentration of PCE, TCE, and DCE observed in groundwater samples analyzed by GC headspace procedure taken from purged monitoring wells at two distances (5 feet and 35 feet) from spargewells.**

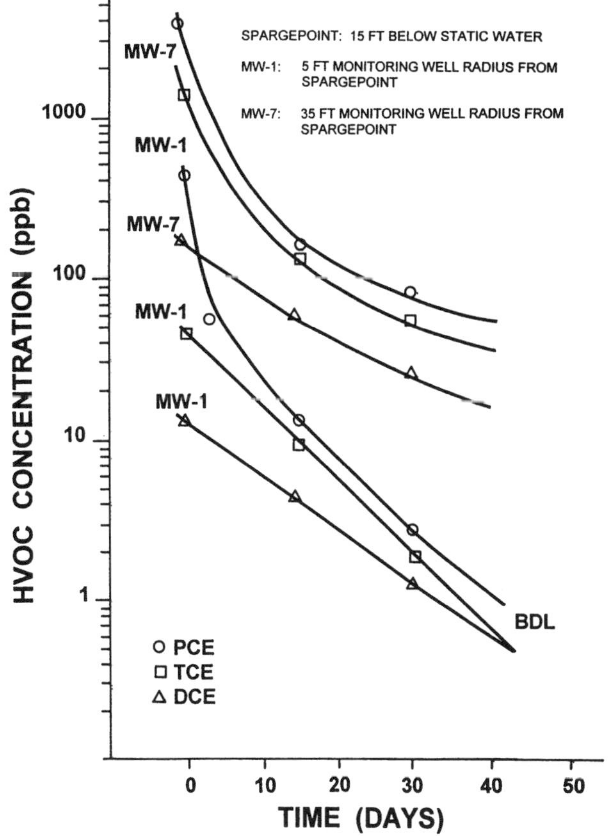

near drinking water standards (5 ppb) after two months of treatment. A concentrated region, located in the center of an abandoned cesspool, resisted reduction. After two months of maintaining a concentration near 250 ppb PCE, two injected Spargepoints® were placed in, and slightly below, the septic unit. Within two weeks of treatment, water samples showed less than 5 ppb PCE.

Three dissolved chlorinated solvents were observed on the site: (1) PCE - tetrachloroethene, (2) TCE - trichloroethene, and (3) DCE - dichloroethene. The TCE and DCE were breakdown products of anaerobic bacterial activity generated near the septic system. All three dissolved chlorinated solvent products rapidly decreased with treatment without any transfer from one compound to another. Cessation of treatment for 1.5 months did result in significant rebound in concentration VOCs within a septic source (abandoned cesspool) and downgradient from a catchbasin which had received liquid solvent at some point in time.

The technique of cessation of treatment to check for rebound appears to be a powerful means of locating secondary sources on a site, as well as a means of rapid removal of dissolved HVOCs.

Water sampling was performed across the site to determine if any unwanted degradation products occurred. A slight rise in pH occurred, from pH 6.0 to 5.5. Breakdown products are very dilute HCL and CO_2. No evidence of epoxide formations with dichloro acetate or format were observed.

REFERENCES

Kerfoot, W.B. 1988. Monitoring Well Construction and Recommended Procedures for Direct Groundwater Flow Measurements Using a Heat-Pulsing Flowmeter. In: *Groundwater Contamination: Field Methods, ASTM STP 963, A6.* (Collins, A.G. and Johnson, A.I., Eds.).

Kerfoot, W.B., McCulloch, W., and Connors, J. 1996. The Use of Micropores for Fine Bubble Creation in the Removal of PCE in Groundwater. Proceedings of the Sixth West Coast Conference on Contaminated Soils and Groundwater, Hyatt Newporter, California, The Association for the Environmental Health of Soils (AEHS), School of Public Health, University of Massachusetts, Amherst, MA 01003.

LeBlanc, D. et al. 1986. Groundwater Resources of Cape Cod, Massachusetts, U.S. Geological Survey Atlas, HA-62.

Spittler, T., Clifford, W.S., and Fitch, L.G. 1985. A New Method for Detection of Organic Vapors in the Vadose Zone. Proceedings of the Characterization and Monitoring of the Vadose Zone, National Water Well Association, Dublin, Ohio.

CHAPTER 24
Remediation System Technologies Costs Evaluation as Compared to Air Sparging and Pump-and-Treat Technologies

Michael Wright, Tyree Organization Ltd., Westborough, Massachusetts 01581

INTRODUCTION

The main objective is the comparison of remediation technologies for the adsorbed and residual phase gasoline contaminants in the saturated and smear zones using air sparging technology versus pump-and-treat methods. Costs associated with the installation and monthly maintenance of both systems are compared. Air sparging is the far more aggressive of either system for remediating dissolved phase hydrocarbons (HC) in the groundwater. Air sparging allows for a quicker remediation of the soils and groundwater, and therefore a cheaper cost over time.

The groundwater pump-and-treat system which was originally installed on the site consisted of three groundwater recovery wells, a product skimmer, a stripping tower, and a post-treatment aeration stripper. Later, a Soil Vapor Extraction System (SVES) consisted of three horizontal legs connected to five vapor extraction wells. The pump-and-treat system and vapor extraction system effectively removed all free-phase HC from the groundwater. However, influent-dissolved HC levels remained elevated throughout its operation.

Evaluation of the contaminant mass removed compared to the continuing costs of the pump-and-treat system indicated that the system was not cost-effective after the first year of operation. Rather than investing further expenses to continue operation of the pump-and-treat system, pilot testing of an air sparging system was investigated.

An air sparge pilot test was conducted on site. Evaluation of groundwater elevation changes, dissolved oxygen concentration changes, soil vapor pressure changes, and volatile organic vapor concentration changes were monitored throughout the duration of the air sparge pilot test. Results of the pilot testing were then used to prepare a Remedial Action Plan.

The Air Sparge/Vapor Extraction system was initialized in August 1995. During the first six months of operation, the groundwater dissolved contaminant levels decreased significantly compared to previous levels reported in the operation of the pump-and-treat system.

Remediation associated with properly designed air sparging systems can eliminate extensive long-term costs associated with pump-and-treat methods.

BACKGROUND

In response to contamination detected during underground storage tanks removals in 1989, Tyree installed a groundwater pump-and-treat system and soil vapor extraction system (SVES) at the project site. In September 1989, Tyree began operation of the groundwater pump-and-treat and in April 1990 Tyree began the soil vapor extraction system to remove petroleum hydrocarbons from soil and groundwater underlying the project site. The project site currently contains a total of 13 groundwater monitoring wells (five of which are connected to the vapor extraction system) and three groundwater recovery wells. The soil vapor extraction system is valued into three separate loops, Loop A, Loop B and Loop C, which are manifolded together into a 2 horsepower (H.P.) regenerative blower. The monitoring wells and recovery wells are shown on the Site Plan, Figure 1.

Operation of the existing groundwater pump-and-treat system and SVES has effectively cleaned up contamination in the unsaturated soils (vadose zone) but has not significantly affected dissolved groundwater contamination. Because groundwater pump-and-treat and SVES technology are not the most aggressive remediation technology for addressing contaminants present in saturated soils, Tyree recommended that a more aggressive remediation technology (i.e., air sparging) be evaluated for the project site.

The geology underlying the project site is composed primarily of sands and gravel, and no impermeable barriers (i.e., clay/silt layers) are known to exist. The permeability of the underlying sands was thought to be relatively high, which is favorable for air sparging technology. Depth to the groundwater table ranges seasonally from approximately 13 to 15 feet below grade, which is great enough to allow capture of air sparging vapors in the vadose zone using a standard soil vapor extraction system. Based upon these favorable conditions, air sparging/vapor extraction appeared to be an appropriate remedial technology for the project site.

Consequently, air sparging and soil vapor extraction pilot testing were recommended for the project site. One nested air sparge pilot test point (SP-1), one groundwater monitoring well (MW-14), and four vapor monitoring probes (VP-1 through VP-4) were installed on the project site in March 1995. The installations were performed by using a truck-mounted drill rig equipped with hollow stem augers. The drilling and installations were supervised by a Tyree hydrogeologist.

To provide appropriate subsurface data and to minimize the number of additional groundwater monitoring wells required to perform the pilot test, the

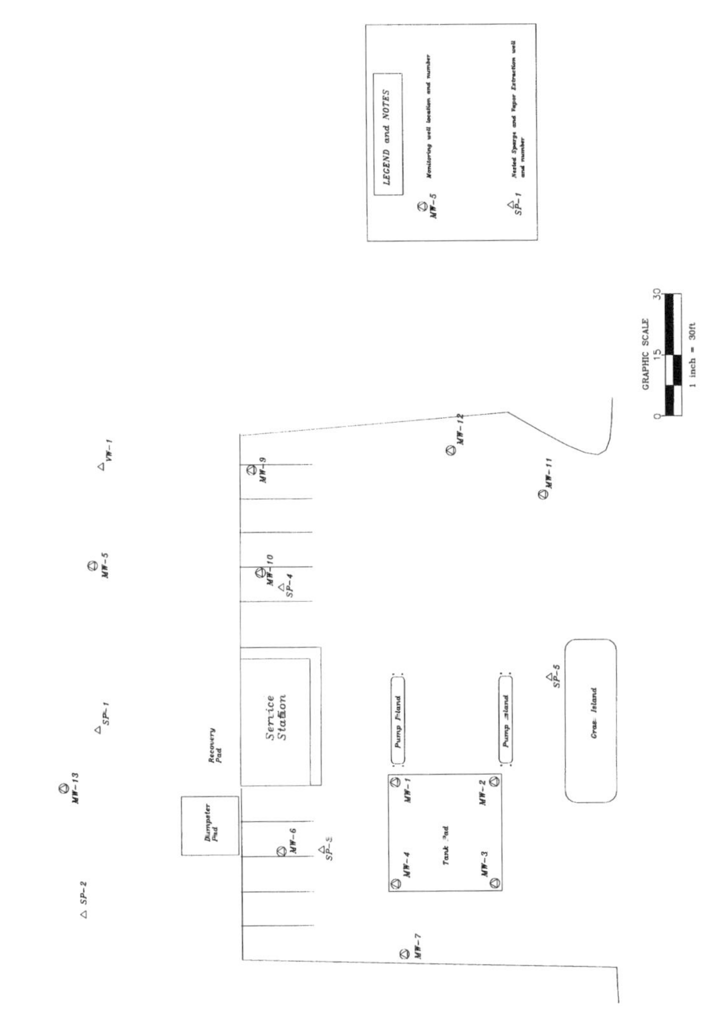

Figure 1. Monitoring wells and recovery wells.

nested air sparge pilot test point was installed within the existing groundwater contaminant plume and in the vicinity of existing groundwater monitoring wells.

After calculating radii of influence of the air sparge test well, five additional nested air sparge pilot test points (SP-2 through SP-5) were intended to be installed on the site as shown on Figure 1. Due to the presence of large boulders on the eastern half of the property, the air sparge well designed to be installed by MW-5 could not be installed. Instead only a vapor .extraction well was installed

A remediation trailer consisting of a 15 H.P. vapor extraction blower and a 25 H.P. air sparge injection blower was installed at the site and connected to each of the nested air sparge/vapor extraction wells. The system was set up on a timer so that the system would cycle on and off in 12-hour intervals.

Historic BTEX Concentrations

The BTEX concentrations for the site are presented in Figure 2, Historic BTEX Concentrations. Information throughout the pump-and-treat systems operation can only be based upon the influent levels obtained from the system sampling. No groundwater monitoring well data was ever collected (other than gauging data) from the wells. At the end of 1994, in an effort to evaluate the performance of the pump-and-treat system, all of the site monitoring wells started to be sampled on a quarterly basis. At this time the pump-and-treat system was shut down, and the background groundwater contamination levels were evaluated.

COST EVALUATION OF THE REMEDIAL SYSTEMS

Pump and Treat

In 1989, there were not many remedial technologies that were more widely accepted than pump-and-treat technology. So the immediate response to the finding of floating product on the groundwater was to install eight groundwater monitoring wells 5 to 10 feet into the groundwater table and three recovery wells 45 to 50 feet into the groundwater table. Due to the geology of the site containing numerous boulders, only one of the recovery wells was installed to the required depth to allow for proper flow.

The remediation system consisted of a 35 ft X 1 ft fiberglass stripping tower with two carbon effluent polishing drums. The system was designed to pump at 1 to 10 gallons per minute, yielding a 15 to 25 foot radius of influence.

Monthly costs for the monitoring of the system, as required by the Connecticut Department of Environmental Protection (CTDEP), determined the monthly costs. The costs listed in Table 1 for the monthly monitoring include weekly site visits, to gauge monitoring wells, and inspection of the system plus a monthly sampling of the treatment system influent and effluent for EPA Method 602 with MTBE.

Figure 2. **Historic BTEX concentrations.**

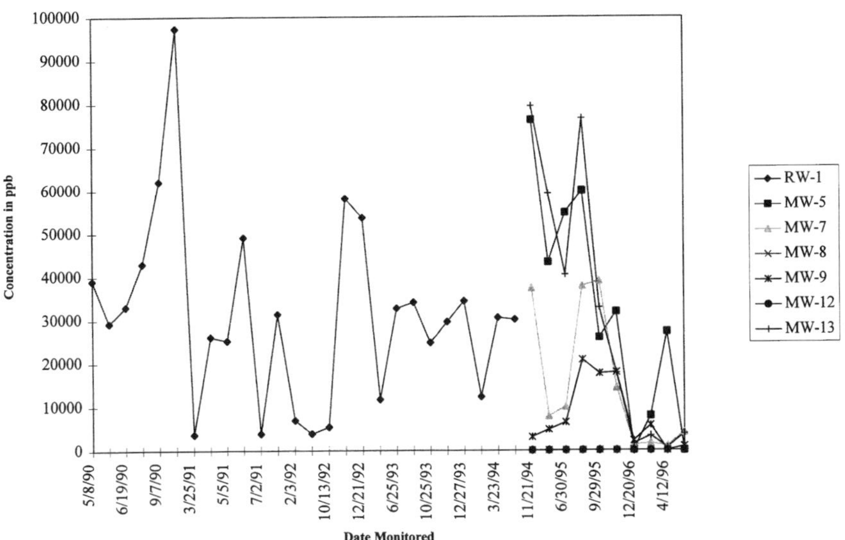

A monthly cost for the cleaning of the air stripper due to iron fouling had been averaged over the period of pump-and-treat operation. Generally, the cleaning of the system was performed every four months.

Disposal costs for used carbon was averaged for a monthly period by adding all costs for disposal and trucking and dividing through the period of operation of the pump and treat.

Table 1. **Installation Costs for Pump-and-Treat System**

Item	Cost	
Remediation equipment	$ 24,885.00	
Preliminary pilot testing	$ 3,890.00	
Permitting	$ 18,150.00	
Installation	$ 51,938.00	
Drilling	$ 23,900.00	
Monthly monitoring	$ 2,848.00	
Iron fouling cleaning	$ 750.00	
Carbon removal	$ 250.00	
First year costs		$ 168,189.00
Every year thereafter		$ 46,176.00

Soil Venting System

The SVES system was installed during the tank installation and consisted of three 4-inch vapor extraction legs. The system was not connected to a blower until March of 1991. The system costs, therefore, were primarily incurred by the tank removal/installation process and could not be broken out. So the costs in Table 2 only reflect the regenerative blower costs and the labor hours to install and house the motor.

Table 2. **Installation Costs for Soil Venting System**

Item	Cost
Remediation equipment	$ 6,845.00
Installation	$ 2,674.00
Electrical	$ 1,605.00

Modification to Pump and Treat

In 1993 the system was redesigned utilizing two low-profile treatment systems to replace the remediation tower and carbon polish units. The primary reason for the rebuilding of the system was to increase GPM so that the radius of influence could be increased to speed up the remediation. Primarily, the costs from Table 3 were kept to a minimum by utilizing most of the hardware from the former pump-and-treat system.

Table 3. **Installation Costs for Pump-and-Treat Modification**

Item	Cost
Remediation equipment	$ 19,344.00
Electrical	$ 6,543.00
Shed	$ 2,545.00

Sparge System

By 1995, air sparging technology had gained significant acceptance by the CTDEP and was well known as an aggressive form of remediation for BTEX contamination. The site was ideal for air sparging, with no sensitive receptors on the site and no free-phase floating product present.

The costs presented in Table 4 for the monthly operation of the air sparge primarily consisted of weekly gauging for depth to water and dissolved oxygen concentrations, plus monthly sampling of the monitoring wells. A quarterly air sample of the soil vapor extraction system's effluent stack by NIOSH 1501 was divided into the monthly costs.

Table 4. **Installation Cost for Air Sparge**

Item	Cost	
Remediation equipment		
Electrical installation	$ 6,543.00	
Pilot testing	$ 3,599.00	
Well installation	$ 12,579.00	
Startup	$ 4,200.00	
Trenching and paving	$ 18,968.00	
Monthly monitoring	$ 2,446.00	
First year costs		$ 94,585.00
Every year thereafter		$ 29,352.00

After five months of gauging and inspection of the system, the frequency had been changed from weekly to monthly. The cost savings were not taken into consideration for ease of comparison to the pump-and-treat monthly schedule.

Cost Comparison

Pump-and-treat technology as compared to the air sparging technology is primarily more expensive. The costs of the systems and their individual installations are relatively equal. But down-the-road costs of monthly monitoring, carbon disposal, and system operation and maintenance for the pump-and-treat system are not as practical as the air sparge costs. The aggressive nature of the air sparge generally promotes a faster site cleanup time (generally two to three years), compared to the more timely pump-and-treat (approximately five years). As shown in Table 5, the long-term costs associated with the remediation cleanup for the site were extensive with the pump-and-treat, while air sparging was much cheaper (and more effective).

Mass Removals Analysis

Two graphs comparing the mass removal rates of HC in pounds (lb) per day are presented for the pump-and-treat system and the air sparge system. The groundwater pump-and-treat system mass removal graph (Figure 3) compares the removal of hydrocarbons treated by the system to the influent and effluent levels in ppb. The levels were then taken and converted into pounds per day, taking into account time, flow, and temperature. The air sparge mass removal graph (Figure 4) compares the effluent emissions of the soil vapor extraction system connected to the nest air sparge/vapor extraction wells. The readings were taken by a photoionization detector (PID) in ppm and were then compared on a quarterly basis to the effluent emissions sampling of NIOSH 1501 for BTEX.

Figure 3.   **Mass removal graph, pump-and-treat.**

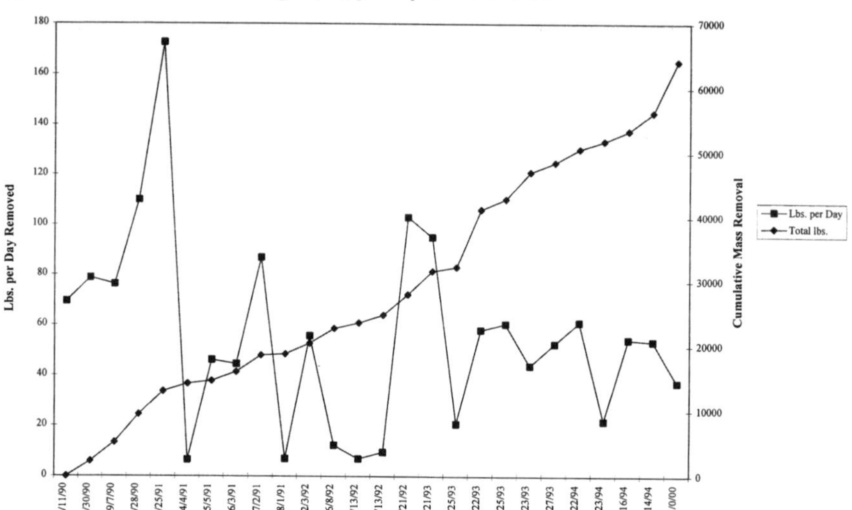

Figure 4.   **Mass removal graph, air sparging.**

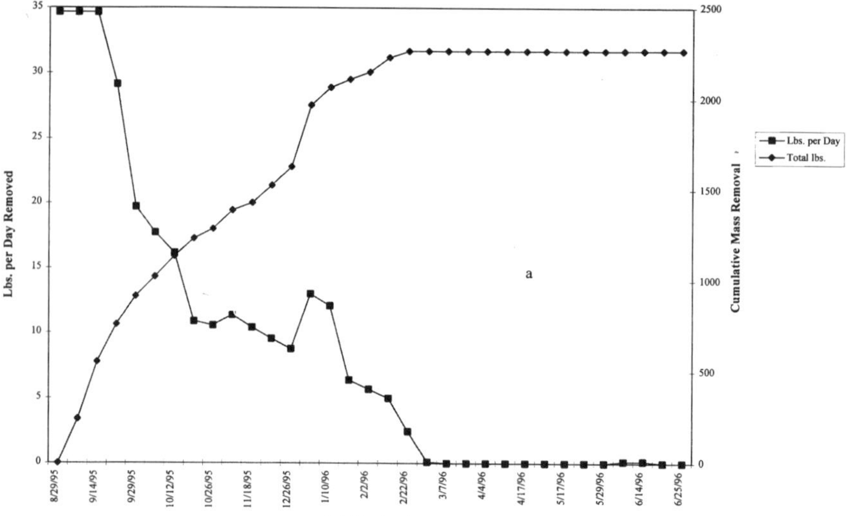

Table 5. **Cost Analysis Total Operation**

Technology	Time of Operation	Projected Cost
Pump-and-treat	5 years	$ 392,449.00
Air sparge	2 years	$ 123,937.00

The groundwater pump-and-treat system mass removal averages a total of 45 pounds per day removal of hydrocarbons throughout its five-year operation. At the termination of the pump-and-treat system's operation it can be seen that the total pounds removed remains fairly constant, and was showing little to no signs of reaching the end of its remediation.

The air sparge mass removal graph averages only 7.5 pounds per day removal cumulative for the entire site (all five vapor extraction wells), with a total mass removed of only 1680 pounds until site cleanup was achieved. Although the removal rates differ, the cleanup of the HC was most likely enhanced by the biodegradation occurring over the site due to the increased levels of dissolved oxygen, as shown in Figure 5.

Figure 5.   **Dissolved oxygen trend graph.**

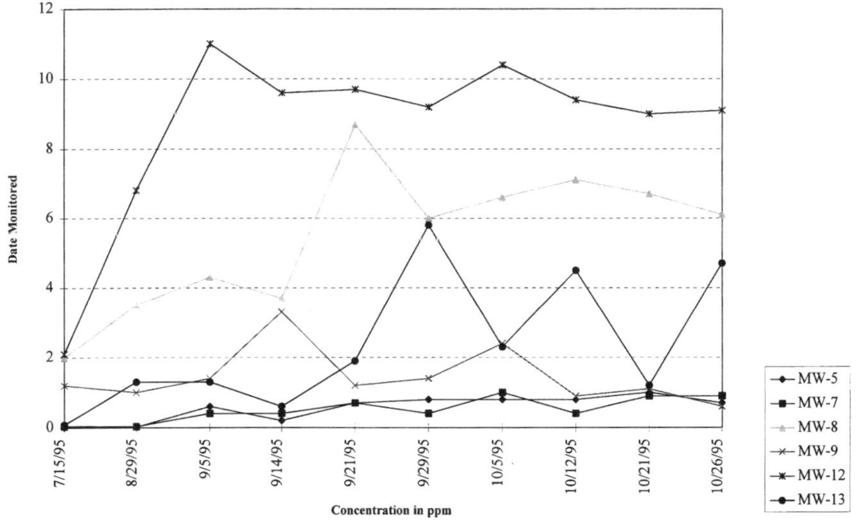

CHAPTER 25

High Energy Radiation Destruction of Polycyclic Aromatic Hydrocarbons (PAH) in Soil Wash Water Containing Surfactants

Vivek Galav and **Thomas D. Waite**, Department of Civil, Architectural, and Environmental Engineering, University of Miami, Coral Gables, Florida

Charles N. Kurucz, Department of Management Science & Industrial Engineering, University of Miami, Coral Gables, Florida

William J. Cooper, Drinking Water Research Center, Florida International University, Miami, Florida

INTRODUCTION

Soil contamination is one of the major environmental problems facing the United States. The United States Environment Protection Agency (USEPA) has, in its National Priority List (NPL), identified more than 1,200 hazardous waste sites as the most serious in the country, and it is estimated that the NPL will continue to grow by approximately 50 to 100 sites per year (USEPA, 1991). These contaminated sites have been found to contain a wide range of chemicals such as polycyclic aromatic hydrocarbons (PAHs), pesticides, polychlorinated biphenyls (PCBs), and heavy metals. The contamination is a result of decades of uncontrolled disposal of waste and chemicals into the ground and surface waters. The contamination is not just limited to surface and subsurface soils but also to marine sediments, which account for more than one-third of the polluted sites in the U.S. (Speight and Gill, 1993).

The current technologies available for treating contaminated sites vary depending on the type of soil, contaminants present, and other factors. For example, in the case of surface soil contamination, the soil can be excavated and incinerated or sent to a designated landfill for final disposal. For sites containing large volumes of contaminated soil, i.e., too large to be excavated economically, in-situ washing of soil by water is becoming a popular technology. However, washing contaminated sites in situ with water has its limitations; i.e., organic pollutants such as PAHs and PCBs are strongly hydrophobic in nature and hence cannot be washed with water only. A survey conducted by the USEPA identified only hydrophilic organic compounds in soils and groundwater mainly because these compounds tend to be washed from soil by infiltrating rain (USEPA, 1985). Therefore, in-situ washing with water is not usually an effective method for washing hydrophobic pollutants from soils; therefore, surfactants are needed to enhance washing efficiency.

Adding surfactants to improve the in-situ soil washing poses some potential problems. For example, the surfactants used may not be environmentally safe, i.e., they may be toxic or hazardous themselves and may not be removed from the subsurface water by natural or anthropogenic processes. There are other concerns associated with in-situ washing of contaminated soils; e.g., surfactants disperse in soil-clay particles because of their surface active properties which can lead to the clogging of the soil pore space and to the diversion of the surfactant solution from the contaminated zone (Abdul and Gibson, 1991).

Excavating contaminated soil and washing it at the site with a surfactant solution, and placing the soil back, is gaining popularity. A number of laboratory studies have been conducted to study the washing of contaminants from soil by surfactants (Abdul and Gibson, 1991; USEPA, 1985; USEPA, 1993). According to the results of these studies, aqueous surfactant solution is a potentially useful method for the washing of contaminated soils. However, for the cost-effective application of this technology, on-site destruction of the contaminants and reuse of surfactants is imperative.

Past studies have shown that high-energy electron beam (E-beam) irradiation is effective in removing PAHs, PCBs, and other hazardous organic compounds from aqueous solutions. Kurucz et al. (1991) noted that 10 organic compounds were successfully removed from wastewater using E-Beam. Other studies (Nickelsen et al., 1994; Cooper et al., 1994; Nickelsen et al., 1992) have shown that high-energy E-beam irradiation is in fact effective in removing many organic pollutants from aqueous solutions, even in the presence of more than 5% solids. Based on the results of the aforementioned studies, it is expected that organic compounds will be removed from wash water solution by the electron beam. If the aqueous surfactant solutions are not affected by electron-beam irradiation, it is possible to reuse the aqueous surfactant solution, while destroying the toxic organics.

RADIATION CHEMISTRY OF ELECTRON BEAM IRRADIATION OF AQUEOUS SYSTEMS

When an energetic particle, for example, an electron from an accelerator or a photon from ^{60}Co, interacts with water, it ionizes the water molecule. Generally, the electron liberated in this ionization event has sufficient energy to ionize further molecules. This leads to the formation of clusters of ions, called spurs, along the path of the ionizing particle. When this radiation is absorbed by water, all the molecules may not be ionized; i.e., some are excited to higher states from which they dissociate, autoionize, or fall back to their ground state.

The electrochemical processes described above are generally completed in about 10^{-12} seconds after the ionization event. The species, produced as a result of ionization, are e^-_{aq}, HO^-, $H_3O^+_{aq}$, H , and H_2O_2, and molecular fragments which are in thermal equilibrium with the bulk medium. Next, the radiolysis products begin to diffuse and a fraction of them interact to form molecular or secondary radical products, while the remaining get distributed into the solution. This so-called "spur expansion" is complete by about 10^{-7} seconds after the radiation.

Based on the above it can be seen that high-energy E-beam irradiation is effective for removing organic pollutants from aqueous solutions. If a soil wash surfactant solution, containing organic pollutants, is irradiated it is expected that the toxic organic compounds will be removed when a sufficient dose is applied. If the radiation does not affect aqueous surfactant solution, then it is possible to remove organic pollutants from wash surfactant solution by E-beam irradiation and re-use the aqueous surfactant solution. In this laboratory study, the feasibility of high-energy E-beam irradiation as an effective and innovative technique for removing PAHs from surfactants, used for washing PAHs contaminated soils, and the possibility of recycling them has been evaluated.

MATERIALS AND METHODS

Particle Size Distribution of Ottawa Sand and Silt

Ottawa sand was obtained commercially and the silt was obtained from a south Florida location. The particle size distribution of the sand and the silt (Figure 1) was determined by sieve analysis (Das, 1989).

Soil Contamination Procedure

Figure 1. **Particle size distribution curve of soils used in study.**

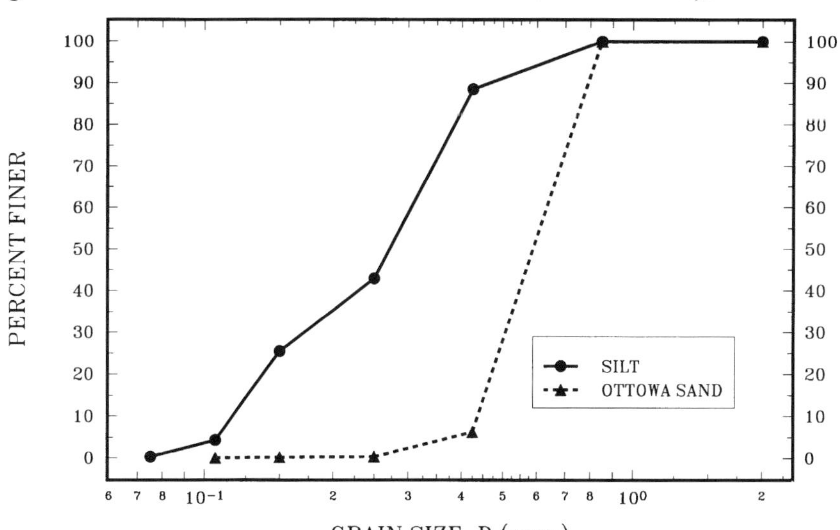

A 60-gram sample of soil (sand or silt) was contaminated by mixing a desired amount of Phenanthrene in pesticide grade methylene chloride obtained from Fisher Chemical Company, Atlanta, GA. Phenanthrene (F.W. 178.23) was obtained from Aldrich Chemical Company, Inc., Milwaukee, WI. Methylene chloride solvent was allowed to evaporate at room temperature. During the evaporation of methylene chloride, the soils were mixed at regular intervals to ensure that the contaminants were distributed uniformly.

QUANTIFICATION OF PHENANTHRENE

High Pressure Liquid Chromatography

A Waters 510 High Pressure Liquid Chromatograph (HPLC) equipped with a Model 486 tunable absorbance detector and a Millennium 2010 data system, from Millipore Corporation, was used for measuring the concentration of Phenanthrene. Chromatographic quantification of Phenanthrene was carried out by filtering 10 μL of irradiated surfactant solutions through a 0.45 μm filter and injecting the filtrate onto a C-18 reverse phase column. The column was 150 mm long and 3.9 mm i.d. Elution was carried out by out by pumping acetonitrile and water (70:30 v/v) isocractically at a flow rate of 1.0 mL/min.

Selection of Surfactants

In preliminary studies, six surfactants were tested to study their ability to resist destruction by radiation. After analyzing the spectra of all the six irradiated surfactants, Triton X-100, an ethoxylated alcohol from Union Carbide Corporation, was selected for the study.

Quantification of Surfactants

The UV spectrum of Triton X-100 was obtained by taking 10 mL of surfactant sample (the pH of the surfactant was adjusted to ~7) in a 40-mL screw cap vial and adding 3 mL of ammonium cobalt thiocyanate reagent and 3.5 gm of sodium chloride. Ammonium cobalt thiocyanate reagent was obtained by dissolving 62 gm of reagent grade ammonium thiocyanate (NH_4CN) and 28 gm of reagent grade cobalt nitrate hexahydrate [$Co(NO_3)_2.6H_2O$] in 100 mL of distilled water. The contents of the 40 ml vial were shaken to dissolve the sodium chloride. The solution was then allowed to stand for 15 minutes and 15 mL of methylene chloride was added to the 40 mL vial. The vial was capped and shaken for another 5 minutes at 300 rpm on a gyrorotating shaker. The vial was allowed to stand until the organic layer separated. The organic layer was collected and transferred to a 1-cm cell and its absorbance spectrum determined on a Hewlett-Packard 8452 Diode Array Spectrophotometer. The spectrum of the surfactant was obtained relative to the baseline of distilled water. Figures 2 and 3 show the UV spectra of 1% and 2% solutions of Triton X-100 solution, respectively.

LABORATORY PROCEDURE

Recycling of Triton X-100

Samples (60-gm) of soil (sand or silt) were placed in several 500-mL Erlenmeyer flasks. The soils were then contaminated with Phenanthrene concentration of 270 mg of Phenanthrene per kg of the soil, as described above.

Figure 2.   **Spectrum of 1% solution of Triton X-100.**

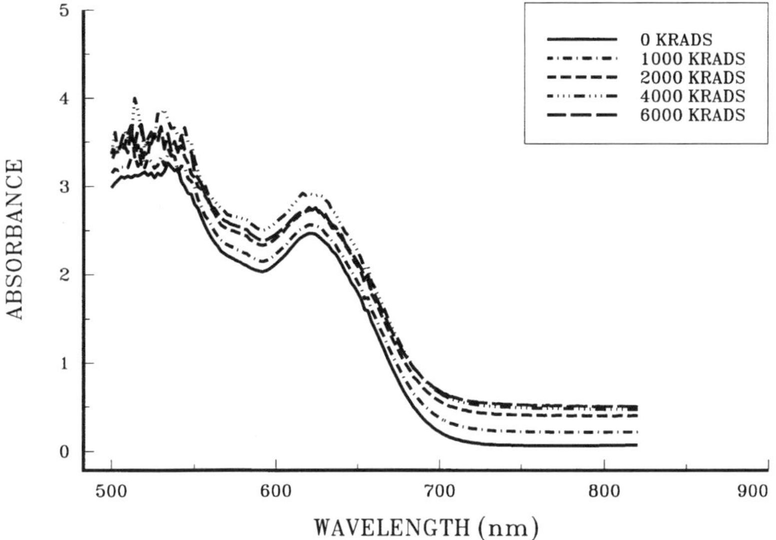

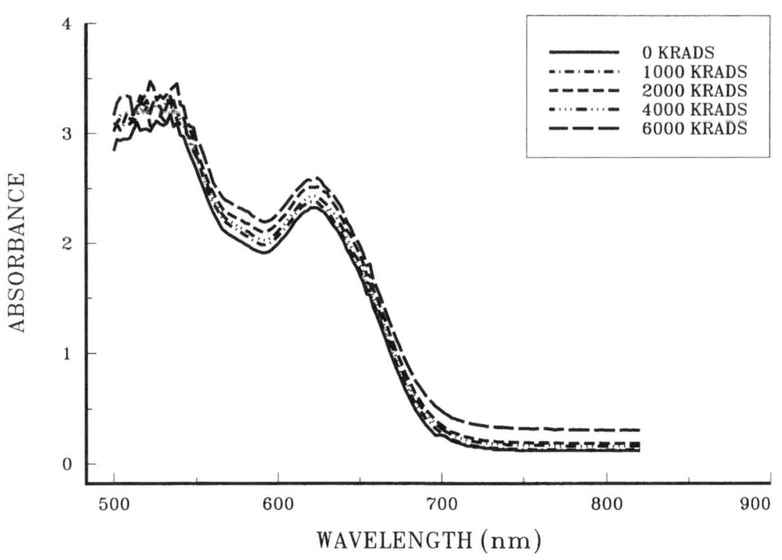

In five flasks containing contaminated soil, 180 mL of Triton X-100 solution (1% or 2%) was added and in two, 60 mL of methylene chloride was added. All seven flasks were shaken for 30 minutes at 200 rpm. The wash Triton X-100 solution was then transferred to a 120-mL vial and stirred on a magnetic stirrer for three minutes to ensure that the surfactant was mixed thoroughly. Ten milliliters of the solution were transferred to a 40-mL vial to analyze the washing efficiency by measuring residual contaminant with HPLC, and the remaining was irradiated at 2 Mrads; i.e., the predetermined optimum irradiation dose.

Ten milliliters of the irradiated Triton X-100 solution were used for HPLC analysis to determine the percent removal of Phenanthrene due to irradiation, and the remaining was used to wash soil contaminated with approximately the same concentration of Phenanthrene as before. The above procedure was repeated, and in all the same surfactant solution was irradiated and used three times to wash soil contaminated with approximately the same concentration of Phenanthrene as in the first washing. The surfactant volume to soil weight ratio was maintained constant at 3:1.

RESULTS AND DISCUSSION

Optimum Dose of Irradiation

In order to determine optimum dose of radiation required to destroy the Phenanthrene, Ottawa sand was spiked with three different concentrations of Phenanthrene, and was washed with aqueous solutions of Triton X-100. The wash solutions of Triton X-100 were irradiated at six different doses: 0, 0.5, 1, 2, 3, and 4 Mrads. The optimum dose of radiation refers to the dose at which more than 85% of Phenanthrene is removed and any further increase in dose does not result in significant increase in removal. Figure 4 shows the removal of three different concentrations of Phenanthrene from surfactant solution at the above-mentioned five radiation doses. Table 1 summarizes the optimum dose of irradiation for three different concentrations of Phenanthrene in sand. It is seen from Figure 4 that as the concentration of the contaminant in the soil increases, so does the required dose of radiation for the removal contaminant in the wash water solution.

Table 1.   **Optimum Dose of Irradiation**

Concentration of Phenanthrene (mg/kg)	Optimum Dose of Irradiation (Krads) [a]
270	2000
550	3000
1110	4000

[a] Dose required to destroy 85% of compound.

Figure 4. **Optimum dose of radiation: 2% solution of Triton X-100 used for washing sand contaminated with three concentrations of phenanthrene.**

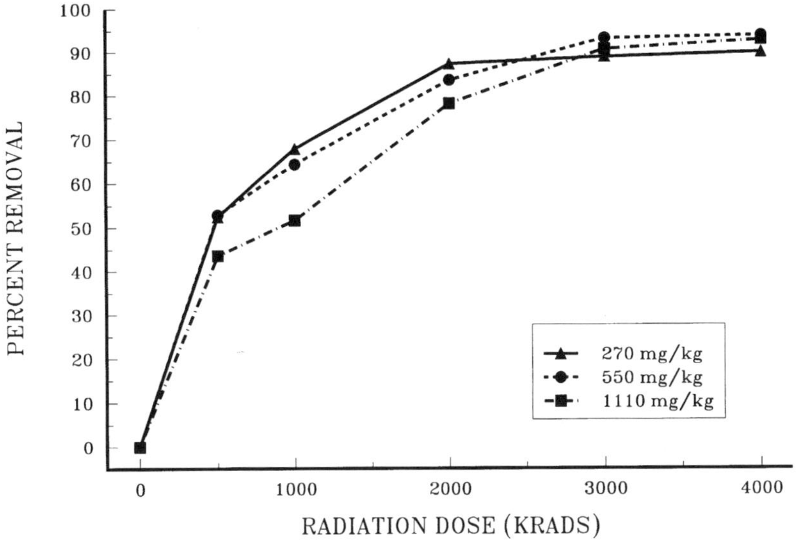

Table 2 shows the results obtained when Ottawa sand contaminated with Phenanthrene was washed with 1% and 2% solutions of Triton X-100. It is observed that the 2% solution of Triton X-100 was more effective in washing/removing Phenanthrene from the sand as compared to the 1% solution. It is also observed from Table 2 that the successive washing efficiencies of the same Triton X-100 solution remain approximately the same for all the three washings. The small change in the washing efficiency is within experimental error.

Table 2. **Successive Washing Efficiencies of Triton X-100 Solutions [a]**

Washing No.	1 % Triton X-100		2 % Triton X-100	
	Sand (%)	Silt (%)	Sand (%)	Silt (%)
1	66.44	69.95	72.69	72.11
2	65.70	62.70	77.05	65.78
3	67.90	47.25	72.41	59.95

[a] Concentration of Phenanthrene in sand or silt approximately 270 mg/kg.

Table 3 shows the percent removal of Phenanthrene from 1 and 2% wash surfactant solutions after irradiation at 2 Mrads. It is seen that the percent removal of Phenanthrene from successive wash solutions due to irradiation follows a decreasing pattern. This can be explained by the fact that after the wash surfactant solution was irradiated at the optimum dose, some Phenanthrene remained in the wash solution; therefore, as the amount of Phenanthrene in the wash solution increased, the same amount of radiation removed less Phenanthrene.

Table 3.   Removal of Phenanthrene From Triton X-100 Wash Solutions Used For Successive Washings of Sand or Silt At A Dose of 2 Mrads [a]

Washing No.	1 % Triton X-100		2 % Triton X-100	
	Sand (%)	Silt (%)	Sand (%)	Silt (%)
1	93.41	67.31	97.76	76.69
2	72.74	46.17	85.22	46.37
3	59.97	16.94	72.24	15.78

[a] Concentration of Phenanthrene in sand or silt approximately 270 mg/kg.

Table 2 shows the results of washing silt with 1% and 2% of Triton X-100 solutions. It is observed that the first washing efficiency of Phenanthrene from silt is about 72% and 70% for 2% and 1% solutions of Triton X-100, respectively. Both these washing efficiencies are less than first washing efficiencies of Phenanthrene from Ottawa sand. This can be explained by the fact that silt particles are smaller in size, resulting in better adsorption of Phenanthrene, which in turn results in less removal of Phenanthrene by the surfactant solution at the same concentration, as compared to sand.

It is also observed that the successive washing efficiencies of the same Triton X-100 solution in silt, unlike that of sand, decrease. It should be noted that neither the silt nor the sand wash solutions of Triton X-100 were filtered prior to irradiation. This was done in order to represent a typical field situation. Unlike sand, silt formed a slurry with the Triton X-100 solution where silt particles would remain suspended and some silt was transferred after washing and remained in suspension. After irradiation, the silt was observed to settle. When contents of the vial were shaken after irradiation, some Phenanthrene adsorbed to the silt particles probably desorbed into the surfactant solution, indicating a lesser removal of Phenanthrene due to irradiation, which in turn affected the results of the second washing (see Table 3).

Surfactant Spectra

Figures 2 and 3 show, respectively, the spectra of 1 and 2% solutions of Triton X-100. The concentration of a surfactant is proportional to the absorbance, so if the concentration of surfactant changes due to irradiation, it can be determined from the spectra of the surfactant obtained after irradiation. The absorbance of the Triton X-100 solution is also directly proportional to the number of ethyloxide units in solution (Union Carbide Corporation, 1993). It is observed from Figures 2 and 3 that as the radiation dose is increased, the absorbance of Triton X-100 solutions increases. Therefore, the number of ethyloxide units are somehow increasing, indicating that some radiation by-products are being formed. It is not certain if the surfactant solution is actually being broken down or if its concentration is changing. However, it can be concluded that the change in activity, if any, is not significant up to 6 Mrads of dose because the spectra of aqueous solutions of Triton X-100 does not deviate appreciably from the zero dose (nonirradiated) spectra.

CONCLUSIONS

Based on the results of this study, the activity of a surfactant solution does not change appreciably due to irradiation, as the washing efficiency remains approximately the same. Therefore, 1 and 2% solutions of Triton X-100 have been successfully recycled for washing a PAH from two soil types, while most of the PAH was destroyed by irradiation.

It is seen from the results that the successive washing efficiencies of 1 and 2% solutions of Triton X-100 used for sand remains approximately 67% and 72%, respectively; the little variation in the washing efficiency is within experimental errors. However, the successive washing efficiencies of 1 and 2% solutions of Triton X-100 used for washing Phenanthrene from silt decreased considerably; from approximately 70% to 47% and 72% to 60% for 1 and 2% solutions of Triton X-100, respectively. This decrease in washing efficiency was the result of silt forming a slurry with the wash solution which showed lesser removal of Phenanthrene by radiation.

The washing efficiency of Phenanthrene from both the soils was more with 2% solution than with 1% solution of Triton X-100. Average washing efficiency of Phenanthrene from sand was approximately 74% by 2% as compared to 67% by 1% solution of Triton X-100. For silt, the washing efficiency of Phenanthrene was approximately 66% by 2% as compared to 60% by 1% solution of Triton X-100. Therefore, the washing efficiency of a contaminant from a soil depends on the concentration of the surfactant solution.

It is seen from these results that as the amount of residual Phenanthrene increases in the successive wash solutions, the same amount of radiation removes less Phenanthrene. This implies that the amount of radiation dose required to remove Phenanthrene from Triton X-100 is proportional to the amount of Phenanthrene in the surfactant solution.

REFERENCES

Abdul, A.S. and Gibson, T.L. 1991. Laboratories Studies of Surfactant-Enhanced Washing of Polychlorinated Biphenyl from Sandy Material. *Environ. Sci. Technol.* 25, 665-671.

Cooper, W.J., Meacham, D.E., Nickelsen, M.G., Lin, K., Ford, D.B., Kurucz, C.N., and Waite, T.D. 1994. The Removal of Tri- (TCE) and Tetrachloroethylene (PCE) from Aqueous Solution Using High Energy Electrons. *J. Air Waste Manage. Assoc.* 43, 1358-1366.

Das, B.M. 1989. Soils and Rocks. In: *Soil Mechanics Laboratory Manual*, Second Edition, pp 19-27. San Jose, CA, Engineering Press, Inc.

Kurucz, C.N., Waite, T.D., Cooper, W.J., Nickelsen, M.G., Lin, K.J. 1991. Treatment of Hazardous Industrial Wastewater and Contaminated Groundwater Using Electron Beam Irradiation. 8th International Conference, Chemistry for the Protection of the Environment, Lubin, Poland.

Nickelsen, M.G., Cooper, W.J., Lin, K., Kurucz, C.N, and Waite, T.D. 1994. High Energy Electron Beam Generation of Oxidants for the Treatment of Benzene and Toluene in the Presence of Radical Scavengers. *Water Res.* 28(5), 1227-1237.

Nickelsen, M.G., Cooper, W.J., Kurucz, C.N, and Waite, T.D. 1992. Removal of Electron Irradiation. *Environ. Sci. Technol.* 26(1), 144-152.

Speight, A.E. and Gill, K.O. 1993. Charting the Course of Coastal Cleanups. *Water Environ. Technol.* 42-48.

Union Carbide Corporation. 1993. Industrial Chemicals, 39 Old Ridgebury Road, Danbury, CT 06818-0001.

(USEPA) U.S. Environment Protection Agency. 1985. Treatment of Contaminated Soils with Aqueous Surfactants. Report EPA 600-S2-85.

(USEPA) U.S. Environment Protection Agency. 1991. National Priorities List Sites: Florida. Solid Waste and Emergency Response. EPA/540/8-91/025.

(USEPA) U.S. Environment Protection Agency. 1993. BIOGENESIS™ Soil Washing Technology: Innovative Technology Evaluation Report. EPA/540/R-93/10.

CHAPTER 26

Managing Contaminated Soils on the Central Artery Tunnel Project

John F. Zipeto, Bechtel Environmental, Inc., Boston, Massachusetts

Stephan T. Roy, GZA GeoEnvironmental, Inc., Newton Upper Falls, Massachusetts

INTRODUCTION

The purpose of this presentation is to illustrate an approach for managing large quantities (14 million cubic yards) of contaminated soils which is necessary during the construction of the Central Artery/Tunnel Project (CA/T Project) in Boston, Massachusetts, one of the largest infrastructure projects ever undertaken in an urban environment. This approach allowed the regulatory community to establish working parameters such that the Project Designer could incorporate handling of contaminated soils into the construction documents and provide the Contractor with a mechanism for performing the work without costly delays due to these environmental issues. The implementation of the management plan for contaminated soils is also discussed, as are the results of the preliminary and subsequent analytical testing of the contaminated soils. Results to date indicate that the approach taken for the management of these contaminated soils has allowed the project to proceed in a timely manner without significant delays due to encountering adverse environmental conditions within the CA/T project footprint.

BACKGROUND

The Central Artery/Tunnel Project currently being constructed in Boston, Massachusetts, is the largest and most complex highway project ever undertaken in the core of an American city. The work is being undertaken by the Massachusetts Highway Department (MHD) with funding from the Federal Highway Administration (FHWA). This $7.9 billion project, building or reconstructing 7.5 miles of urban highway (approximately 50% in tunnels), will replace Boston's existing elevated central artery (I-93) with a modern underground expressway traversing downtown Boston. This project will also extend the Massachusetts Turnpike (I-90) to Logan International Airport through a new tunnel under Boston

Harbor, completing the last link in the U.S. interstate highway system. In addition, the CA/T Project will carry through traffic northward across the Charles River and make connections to regional and local roadways.

The underground central artery will be constructed while maintaining the existing elevated highway in service until the underground replacement is open to traffic. The new underground highway will be eight-to-ten lanes wide, compared to the six lanes available on the existing elevated roadway. The number of on-ramps and off-ramps will be reduced from 27 to 14, helping to separate the vehicles passing through Boston from those whose destination is downtown.

The removal of the existing elevated highway will create 27 acres of open space in the heart of the city, making way for broad, tree-lined boulevards, parks, and limited commercial and residential development. Boston is one of the most "walkable" cities in America, and the CA/T Project will enhance pedestrian access along the Freedom Trail, the HarborWalk, Quincy Market, and other popular historical and cultural attractions. Neighborhoods will again be joined and enjoy new public space, housing opportunities, and access to the waterfront.

The new third harbor tunnel which opened for commercial traffic on December 15, 1995, was built using twelve giant concrete filled steel tubes, each measuring 325 feet in length, which were submerged and connected end to end in a trench 50 feet wide and nearly three-quarters of a mile long. The tunnel is four lanes wide, two lanes eastbound and two lanes westbound, increasing the prior cross-harbor capacity from four to eight lanes. The tunnel will enable airport-bound drivers from the west and south, who comprise about 70% of airport traffic, to bypass downtown Boston entirely.

According to the 1991 Final Supplemental Environmental Impact Report (MDPW, 1990), the CA/T Project was committed to maximizing reuse and recycling while minimizing the disposal of excavated materials. Accomplishing the CA/T Project requires the management of nearly 14 million cubic yards (mcy) of this excavated material, including 11.75 mcy of land-based excavated soil and 2.25 mcy of dredged harbor sediments. The excavated materials are comprised of 8.15 mcy of historic fills and silt, 3.25 mcy of clay and glacial till, and 0.35 mcy of other materials. The majority of excavated materials, nearly 11.6 mcy, have been targeted for reuse. The primary beneficiaries would include the project with this material being used as backfill, and in-state landfills in need of closure where CA/T urban fill and clay/till would be used as closure and capping materials. These landfills are located in several Massachusetts communities as well as at Spectacle Island, an unlined landfill located in Boston Harbor which will be developed into a recreational facility. The CA/T excavation work began in 1991 and is expected to be completed in 2004.

REGULATORY CONSTRAINTS

Massachusetts General Law Chapter 21E, Section 4, authorized the Massachusetts Department of Environmental Protection (MDEP) to take or arrange for response actions to releases or threats of release of oil and hazardous materials (OHM) as it deems necessary. The MDEP has the sole authority and discretion to determine the appropriate extent and nature of the response actions consistent with the Massachusetts Contingency Plan (MCP), cited at 310 CMR Part 40.0000, as the agency's regulations for implementing Chapter 21E. The MDEP determined that their regulatory authority under MGL Chapter 21E extended to materials excavated during construction of the CA/T Project. The MDEP's decision was the key factor that catapulted the CA/T Project toward a massive testing and tracking program for the approximately 14 mcy of excavated material.

An important aspect of the MCP is that it allows for the implementation of short-term measures to address imminent hazards. The MDEP expanded its policy areas to address "Interim Measures" (MDEP, 1990); that is, responses to environmental conditions which do not constitute an imminent hazard and would not significantly benefit the site if lengthy and detailed analyses or remedial action were performed. This policy allowed for the removal of contaminated media to facilitate planned construction activities and further allowed the MHD to incorporate within the construction contract specifications identified "Clearance Areas" to be removed and disposed of during the course of construction activities. These removal activities were acceptable as response actions per Chapter 21E.

PROJECT STRATEGY

A major challenge to the Massachusetts Highway Department (MHD) was to craft an excavated materials program that would allow the CA/T Project to minimize cost impacts, maintain project schedule, and enable the MDEP to approve response actions (per MGL 21E requirements) for the excavation, transport, and placement of excavated material. Essentially, the CA/T Project was required to implement procedures for the MDEP to receive and approve test results, receive approval for transport of 21E-regulated materials, and satisfy the construction schedule to foresee project completion in the year 2004.

Anticipating the need for a comprehensive assessment of soil conditions within the project area to satisfy state regulatory requirements, the MHD completed a subsurface investigation/limited soil sampling and analysis program (Cortell, 1989) in 1989 in order to compile an environmental database for soils within the proposed CA/T project limits. Soil samples were recovered from test borings advanced during the study to assess the presence of 13 EPA priority pollutant metals, total petroleum hydrocarbons (TPH), sodium, and soluble chlorides. In addition, a selected number of samples were analyzed for volatile (VOCs) and semivolatile organics (SVOCs), and polychlorinated biphenyls (PCBs). The study concluded that there were no serious constraints in disposing excavated materials at upland landfill sites, including Spectacle Island, and at sites approved for ocean disposal. The report (Cortell, 1989) also recommended that additional studies would be required to classify soils for disposal, especially in areas where high concentrations of metals, petroleum hydrocarbons, and polycyclic aromatic hydrocarbons (PAHs) were found (MDPW, 1989).

DEVELOPING THE PROJECT-WIDE STANDARD OPERATING PROCEDURE

To develop a project-wide sampling and analysis standard operating procedure, CA/T evaluated the soil quality data compiled during the 1989 Cortell study and identified key pollutants for further assessment based upon the frequency of occurrence, toxicity, and persistence in the environment. With MDEP's acceptance of the limited sampling/testing protocol in hand, the CA/T Project was able to proceed with a comprehensive and streamlined characterization program to assess soil and groundwater quality conditions project- wide.

A more detailed and systematic assessment of the CA/T alignment was also needed to comply with several other requirements of the MCP. The MCP describes the requirements and procedures for discovery notification, assessment, and response to releases and threats of release of oil or hazardous materials. The process involves a phased approach to the assessment of site conditions and implementation of remedial measures. The standard of care for property assessments includes preparing site assessments, conducting subsurface investigations, developing remedial alternatives, preparing plans and specifications, and implementing the remedial action plan. A series of standardized procedures manuals were prepared to streamline the approval and acceptance of the soil characterization program. The procedures addressed assessment and characterization, sampling and analysis, quality assurance, and emergency response.

Several strategies for streamlining the excavation, sampling, testing, and placement of excavated materials were informally discussed and debated with the MDEP in an effort to achieve an acceptable balance for adherence to project schedule and compliance with the MCP requirements for assessing and tracking Chapter 21E regulated materials and segregating clearance and nonclearance materials for proper management and placement.

The April 1992 MDEP correspondence formalized several months of discussion for establishing de facto soil quality standards for placing 500,000 cubic yards of granular excavated materials at Governors Island located adjacent to active runways at Logan International Airport. The excavated material was intended to assist the Massachusetts Port Authority (MASSPORT) in developing airport operations facilities at this location. MASSPORT had a critical interest in the testing program since the Logan Airport area was deemed a 21E disposal site and any additional deterioration of site conditions was an unacceptable issue for liability of future response actions. MHD had an equal incentive to minimize contamination since future cleanup actions resulting from current CA/T soil placement could result in uncertain future liability.

CA/T correspondence in May 1992 to MDEP detailed the methods and procedures for tracking and analyzing excavated materials. A surveyed grid system was developed at Governors Island so that temporary stockpiles could be located if laboratory test results exceeded agency approved limits. The application of a "just-in-time" strategy for directing excavated materials to disposal locations was especially beneficial to enable construction activity to proceed on schedule. Biweekly progress reports, with soil quality summaries and brief statistics, were forwarded for MDEP review and acceptance.

The culmination of several weeks of intensive negotiation resulted in a Memorandum of Understanding (MOU) between the MHD and the MDEP. The agreement provided a regulatory framework for the CA/T Project which paralleled procedures set forth in the MCP. The agreement stipulated that any excavated materials containing levels of constituents exceeding concentrations above naturally occurring background levels is a release of hazardous materials under MGL Chapter 21E and that the excavation, handling, and disposal of such materials pose a threat of release which is subject to regulation by MDEP pursuant to MGL Chapter 21E.

The agreement also formalized MHD's position that areas within the CA/T alignment which contain releases of oil and hazardous materials pursuant to MGL 21E would be defined as Clearance Areas. In addition, MHD concurred on the adequacy of sampling and analysis procedures described to ensure that releases or threats of releases were adequately identified and remediated. The critical designation of clearance and nonclearance areas resolved the problem by categorizing excavated materials in several classifications with attendant differences in management strategies. Clearance material would be removed and disposed of off-site as part of the construction project. The agreement also required MHD to manage nonclearance materials on site "to ensure the protection of the health and safety of the public and the environment and in a manner approved by DEP" (MOU, 1993).

The basic methodologies for sampling and testing of all nonclearance excavated materials was formalized for all materials scheduled for reuse on the project, such as placement at Governors Island and Spectacle Island. The basic volume for stockpile sampling was set at 1,000 cubic yards (cy) with sufficient composite samples collected based upon end use: 3 samples per 1,000 cy for off-site landfill daily cover; 2 samples per 1,000 cy for placement at Governors Island and Spectacle Island; and 1 sample per 1,000 cy and for use as project backfill.

PROJECT DESIGN

Based on the agreements between MHD and the MDEP, contract specifications were developed to implement the goal established by MHD; namely, to reuse and recycle excavated materials to the maximum extent possible. Contractors were directed to segregate several types of excavated materials in order to expedite the transfer of specific materials and waste streams with a minimum of handling. The contractors were immediately responsible for managing and disposing of solid waste, demolished road surface materials, clay, organic wastes, demolition debris, large rock (>36" in diameter), glass, metal, and other miscellaneous wastes. The remaining excavated materials; granular and nongranular urban fill, interface clay, and rock (< 36" in diameter) could be immediately transported to a Materials Testing Site (MTS), such as Subaru Pier in South Boston, for stockpiling, sampling and testing.

The MTS Contractor was made responsible for implementing a comprehensive sampling and testing program. MHD provided specification language and required the contractor to provide a Chemical Quality Management and Sampling Plan (CQMSP). The CQMSP detailed sampling procedures, analytical methodologies, the QA/QC program objectives and design, the qualifications of all responsible analytical staff, data reporting procedures, and preparation of bills of lading for MDEP approval. This approach was crucial in encouraging the contractor to retain the services of qualified environmental engineering consultants and state certified laboratories. The contractor was also required to submit certification statements indicating that the laboratory test results being reported on soil stockpiles were represented in the chain-of-custody forms submitted as documentation for the analytical test results for specified stockpiles at the MTS.

PROJECT IMPLEMENTATION

The primary MTS to date on the CA/T Project has been at the Subaru Pier site in South Boston. Soil management at Subaru Pier began in September 1992 and continues today. Since the beginning of the work at this location, excavated materials have been brought to Subaru Pier from the various CA/T Project active construction locations. Through the end of 1995, approximately 1,500,000 cubic yards of material was managed in accordance with the project-wide procedures and CQMSP. This material has been reused as either project backfill or transported to Spectacle Island/Governors Island. A small portion of this material (less than 10%) was disposed of off-site.

As part of the project requirements, the contractor's environmental consultant prepared the CQMSP for review, comment, and approval by the MHD and the MDEP. The CQMSP included a description of the soil stockpile management plan including soil stockpile sampling, coordination of analytical testing, reporting requirements, and determination of final soil disposition. Soil stockpile samples are analyzed to determine whether they meet the project reuse criteria. Materials meeting these criteria are reused, while nonconforming materials are transported and disposed of at approved off-site facilities.

The total soil stockpiling area at Subaru Pier is approximately 350,000 square feet. Soils excavated from various project locations are brought to Subaru Pier, then placed into large stockpiles. The stockpiles are constructed on 20-mil polyethylene surrounded with haybales. For sampling purposes the large stockpiles are divided into "cells," each being approximately 1,000 cubic yards in size, by surveying the top of stockpile elevation then calculating an appropriate grid spacing to provide the required cell volume. Wooden stakes mark off the cell corners. Each cell has a unique identification number and is located on a Stockpile Location Plan which is updated on a regular basis. The material remains in the stockpile until completion of the analytical testing. Based on the analytical results, the material in each cell is categorized and disposed of in accordance with the CQMSP.

Soil samples are collected at a frequency of one to three per stockpile cell based on anticipated reuse options. Each sample is a composite of up to 10 grab samples from up to four test pits excavated from the top of the stockpile cell. Soils are collected with stainless steel sampling equipment and field composited. Samples for VOCs are not composited. Between sample collection, the sampling equipment is decontaminated following standard protocols identified in the CQMSP.

Based on the analytical results of the initial soil characterization investigation (Cortell, 1989), appropriate constituents were identified for further assessment during construction activities. In most cases, soils were analyzed for VOCs via EPA Method 8240; TPH-IR via Method 418.1; PCBs via Method 8080; total PAHs via Modified 8100; and selected metals (arsenic, cadmium, chromium, mercury, and lead) via appropriate SW-846 Methods. QA/QC samples were performed at a rate of 10% duplicates (full suite of analyses), 5% rinsates (full suite), and 5% trip blanks (VOCs only). Analytical testing was performed by certified Massachusetts laboratories identified in the CQMSP and approved by the MHD and MDEP. Soil samples which resulted in constituent concentrations that exceeded the "20 times" rule of thumb for total metals were additionally assessed for the toxicity characteristic via TCLP for that metal. Results for any of the constituents over the reuse criteria required off-site disposal and additional facility-specific testing.

Data reporting and documentation begins with the Stockpile Location Plan which is generated after the stockpile is surveyed and the 1,000 cy cells are laid out. The Stockpile Location Plan indicates the cell ID, their sampling status, and the results from any analytical testing. The Stockpile Location Plan is updated within 24 hours of receiving new information, and distributed to the contractor and other appropriate project personnel. As a follow-up, the analytical data are tabulated by date sampled and stockpile ID. For soils to be reused, the data are submitted on a weekly basis with a current Stockpile Location Plan and bill of lading for MHD signatures and MDEP approval prior to transport.

Soil to be disposed of off-site requires facility-specific analytical testing. This information is submitted directly to the facility for review and approval. Copies of the data are also sent to project personnel for review and acceptance. Approvals from the disposal facility and project personnel are normally achieved within 48 hours from their receipt of submittal.

REVIEW OF ANALYTICAL DATA

From September 1992 through December 1995, over 1,500,000 cy of excavated materials have been stockpiled at material testing sites, where they were sampled, tested, and reused on the CA/T Project. To date, approximately 2,800 analytical tests have been performed on project material from four major work areas. The majority of these samples have been from the South Boston and East Boston areas.

The most common method of reviewing large quantities of data is via descriptive statistics; that is, performing statistical analyses which result in single values which describe the whole data set. Selected descriptive statistics are shown in Table 1. These values include: the number of samples analyzed; the percentage of samples indicating detectable concentrations above the detection limit (DL); the minimum/maximum/mean values of the data set reported in mg/kg (ppm); and the confidence level at which the mean is established. These statistical data are provided for both the project-wide soil characterization study (MDPW, 1989) and the unpublished results through 1995 collected by the MTS contractors from samples collected and analyzed from the excavated materials stockpiled at CA/T Project material testing sites. A review of these data sets will provide information regarding the chemical nature of the urban and historical fills in this section of Boston, as well as providing a test for the validity of the overall soil management/characterization approach taken on the CA/T Project. For comparison to regulatory levels, appropriate soil constituent concentration standards referenced in the MCP (Method 1; S-1/GW-2 or GW-3) are also included in Table 1.

The inherent differences between the two data sets should be clearly understood; approximately three times as many analytical tests have been performed on the MTS stockpile soils versus tests performed during the initial soil characterization. Secondly, the initial soil characterization spans the entire CA/T Project footprint, while the soil stockpile results represent analytical results from soils primarily originating within the East Boston/South Boston areas. Lastly, since the initial soil characterization was a "true" soil investigation, the method detection limits were as normally experienced for these constituents tested. The soil stockpile testing, however, is essentially a large-scale "soil screening" procedure based on the observed results of the initial site characterization. Therefore, the DLs were generally not set as low as in the initial investigation. These limitations have all affected to some extent the data that is presented in Table 1.

Table 1. **Analytical Data Statistical Summary**

Statistics	TPH	Total As	Total Cd	Total Cr	Total Hg	Total Pb	Total VOCs	Total PCBs	Total PAHs
Project Wide Initial Limited Site Investigation Analytical Results									
Number of Samples Analyzed	991	754	756	756	785	850	NA	NA	873
Percent Detects	81%	98%	43%	93%	45%	87%	NA	NA	62%
Minimum (mg/kg)	10	0.25	0.10	1.0	0.045	0.050	NA	NA	0.080
Maximum (mg/kg)	49,000	99	25	530	23	11,000	NA	NA	3,000
Mean (mg/kg)	158	5.26	0.50	13.5	0.15	51	NA	NA	2.75
Confidence Level (95.0%)	20	0.3	0.1	0.5	50.9	6	NA	NA	0.45
Stockpile Soil Characterization Analytical Results									
Number of Samples Analyzed	2,753	2,813	2,813	2,813	2,813	2,813	2,008	2,009	2,813
Percent Detects	89%	36%	53%	99%	16%	97%	4%	6%	59%
Minimum or DL (mg/kg)	40	7.5	0.75	0.75	0.75	5.0	0.01	0.050	0.35
Maximum	5,410	132	315	4,107	41.9	3,264	0.76	7.9	2,036
Mean (mg/kg)	337	8.01	2.44	31.4	ND	65.3	ND	0.050	12.20
Confidence Level (95.0%)	16.06	0.30	0.26	4.09	NA	3.88	NA	NA	1.71
Applicable MCP Method 1 Soil Standards									
Soil S-1, GW-2 or GW-3	500	30	30	1,000	10	300	NA	2	NA
Upper Concentration Limit	10,000	300	800	10,000	600	6,000	NA	100	NA

With these issues understood, it is evident from this analytical data that the percent detects in both data sets were similar with the exception of As and Hg. The variation between data sets for these two constituents is likely due to the difference in the elevated DLs set for the MTS soil stockpile data. For some constituents, such as TPH, PAH, Cd, and Cr, the soil stockpile data have higher mean concentrations than that of the initial soil characterization data. This is likely due to the heavily industrial characteristics of the fills in the South Boston area, where historical fill dates back over 250 years. Other constituents, such as As and Pb, have mean concentrations of similar magnitude. This may be the result of the more widespread occurrence of these constituents, as well as the known sources of these constituents not unique to South Boston.

The data were also reviewed to determine if the presence of one constituent at an elevated concentration correlated to the presence of a second constituent at a similarly elevated concentration. A correlation value of 1.0 indicates an exact match of concentration change for constituent A in direct proportion to constituent B. A correlation value of -1.0 indicates an inverse (equal and opposite) relationship between constituents A and B. A value at or around 0.0 indicates no clear correlation. These values can be thought of as a percent correlation between constituents. A review of the data indicates there is no correlation between any constituents greater than 20%.

With regard to regulatory levels within these contaminated soils, the mean values from all constituents in both data sets are well within the MCP Method 1; S-1/GW-2 or GW-3 Soil Standards which are normally applied to the urban Boston area. This provides significant evidence that the analytical data from the initial soil characterization investigation and the MTS stockpile soils are statistically similar from a regulatory perspective. This also points to the fact that the conclusions drawn from the initial soil characterization investigations as they have been applied to developing the soil management approach were correct. The project approach has worked successfully to date and continues to work without adversely affecting the project schedule, costs, or impact on human health or the environment.

ACKNOWLEDGMENTS

The authors wish to thank the Massachusetts Highway Department for allowing the use of the soil analytical data for this publication.

REFERENCES

Bechtel/Parsons-Brinckerhoff, May 1996, FHWA Task Team Review.
Cortell, J.M., and Associates, Inc., May 1989, Soil Characterization Report.
(MDEP) Massachusetts Department of Environmental Protection, Massachusetts Contingency Plan, 310 CMR Part 40.000.

(MDEP) Massachusetts Department of Environmental Protection, August 1990, Interim Measures Policy, #WSC-131-90.

(MDEP) Massachusetts Department of Environmental Protection, April 23, 1992, Correspondence to Peter M. Zuk, Project Director, CA/T Project, from Steven G. Lipman, Boston Harbor Coordinator.

MDEP/MHD Memorandum of Understanding, February 5, 1993.

(MDPW) Massachusetts Department of Public Works, November 1989, Service Contract No. 90079-M025A, Right-of-Way Assessment and Remediation Services (ROWARS), Central Artery/Third Harbor Tunnel Project.

(MDPW) Massachusetts Department of Public Works, November 1990, Final Supplemental Environmental Impact Report.

(MHD) Massachusetts Highway Department, May 14, 1992, Correspondence to Steven G. Lipman, P. E., Boston Harbor Coordinator, Massachusetts Department of Environmental Protection, from Peter M. Zuk, Project Director, CA/T Project.

CHAPTER 27

Dual Phase ("Hi-Vac") Extraction: An Effective Remediation Alternative for Enigmatic Geologic Conditions (Case Studies)

Michael A. Lamarre, Mobil Business Resources Corporation, Inwood, New York

Timothy D. Foster and **David H. Lucas**, Land Tech Remedial, Inc., Monroe, Connecticut

INTRODUCTION

Through the history of time with service stations, underground storage tanks (USTs) have evolved from the original single-wall steel USTs, to the single-wall fiberglass USTs to the current double-wall steel and/or double-wall fiberglass USTs. When, by chance, tanks and/or lines have been found to be leaking, or when an overfill occurs without the proper overspill containment, petroleum-based hydrocarbons are introduced into the subsurface, with the potential need for installing a remedial system to remedy the situation. These hydrocarbons travel through the subsurface, some adsorbing in the overburden, some dissolving into the groundwater, and some adsorbing and becoming trapped in the saturated zone ("smear zone"). Starting from the late 1980s until the present, current remedial practices include the usage of Groundwater Recovery and Treatment ("Pump-and-Treat"), Soil-Vapor Extraction ("SVE"), Air Sparging and recently, the usage of Dual Phase or "Hi-Vac" Extraction. This paper will discuss the basics of the four commonly utilized remedial options noted above, further explore the advantages of the utilization of Hi-Vac Extraction and show some case studies and the results associated with the installation and activation of Hi-Vac systems.

REMEDIAL ALTERNATIVES

As previously noted, there are four remedial options which are primarily utilized on service station and/or bulk storage facilities: Pump-and-Treat, SVE, Air Sparging, and Hi-Vac. Let us discuss each of these options in further detail.

Groundwater Recovery and Treatment (Pump-and-Treat)

Pump-and-treat systems are primarily utilized for hydraulic control and are generally ineffective in remediating unsaturated zones, dissolved-phase plumes, and/

or trapped hydrocarbons in the saturated zone. Pump-and-treat utilizes a down-well pump to accumulate and remove water from a recovery well. From there, the water is directed to the surface via piping where it is either treated and/or directly discharged to a permitted discharge point. The recovery well(s) is usually located either in the source area to stop the plume from migrating or at the downgradient edge of the plume to contain the contaminants from migrating downgradient of a property boundary. In a tight formation, with a hydraulic conductivity of approximately 10^{-7} cm/s to 10^{-9} cm/s, the area of influence ("capture zone") is minimal with a large drawdown in the well(s), whereas in sandy formations, with a hydraulic conductivity of approximately 10^{-1} cm/s to 10^{-3} cm/s, the capture zone is quite large, with a minimal drawdown in the well and the surrounding area.

Soil-Vapor Extraction (SVE)

SVE is utilized primarily in the remediation of overburden or unsaturated soil, including small thickness of Separate-Phase Hydrocarbons (SPH). SVE systems have little to no influence on any dissolved phase plumes, with the exception of removing the continual source in the overburden, nor any influence on hydrocarbons trapped below the water table or smear zone. SVE involves the use of a vacuum which is placed on either vertical or horizontal wells to draw air through the subsurface to the wells. The resulting vacuum gradient causes the soil gas to migrate through the soil pores toward the extraction wells. Volatile organic compounds (VOCs) are volatilized and transported through the subsurface by the migrating soil vapors, where the SVE blower then moves the air from the subsurface to above-grade to where it is either treated and/or discharged into the atmosphere, based upon regulatory limits. As a rule of thumb, SVE systems are applicable where intrinsic soil permeabilities range from 10^{-5} cm^2 to 10^{-8} cm^2, from gravel to silty sand material, respectively. In tighter soil conditions, the availability of air movement with "normal" vacuum is minimal, therefore overworking the equipment, obtaining minimal run-time and a minimal area of influence.

Air Sparging and BioSparging

Air sparging is an effective means of adding oxygen into the subsurface while also volatilizing adsorbed and dissolved hydrocarbons from the saturated zone. Air sparging is usually combined with SVE, but in some instances the SVE is eliminated due to the introduction of low flow/low pressure air into the subsurface with little risk of stray vapors ("BioSparging"). BioSparging is primarily used to increase the oxygen content of the water, which will increase the natural biological activity in the area. Both of these techniques are generally used in soils with intrinsic permeabilities greater than 10^{-9} cm^2, and can obtain an average radius of influence of 20 feet. In lower permeable soils, the radius of influence is reduced and the effectiveness is minimized because in a lower permeable soil, greater pressures are required to introduce the air into the subsurface, and the soil conditions allow for little movement of air and oxygenated water through the subsurface.

Dual Phase or "Hi-Vac" Extraction

Dual Phase or "Hi-Vac" is a relatively new technology which utilizes a combination of pump-and-treat and SVE, primarily in lower intrinsic soil permeability conditions, 10^{-10} cm^2 to 10^{-13} cm^2, silt to a glacial till, respectively, including bedrock. Through the utilization of a high vacuum (9" to 25" Hg), there is an increase in "available air" to be recovered in the SVE system, as well as an increase in the water yield in the recovery wells. The increased water yield is attributed to the enhancement of a pressure gradient applied to the Hi-Vac wells. It is also believed that through the utilization of Hi-Vac in bedrock situations, SPH that is trapped in fractures can be volatilized and removed, therefore removing the continual source. In this remedial option, there are two different setups of equipment: (1) utilization of a liquid ring pump to remove both liquid and water via drop tubes from any given well, utilized on sites where groundwater is less than 20 feet below grade, (see Figure 1), or (2) utilization of groundwater recovery pumps and a high vacuum extraction blower from the same well (see Figure 2). A liquid ring pump utilizes water in a cavity that is oblong in shape, in which impellers inside the cavity are spun from the drive of the motor, forcing the water to the outside of the cavity to generate a vacuum, whereas the other extraction blowers are usually a form of regenerative blowers to generate the high vacuum.

This remedial option has been an effective alternative where other remedial systems alone have not. Some advantages of these systems include: (1) the ability to remove both liquid and vapor through the same effluent stream, minimizing equipment, controls, and time of remediation, (2) the applicability of utilizing these systems for emergency response, where a drop tube can be installed in wells and recover separate-phase hydrocarbons ("SPHs"), control the dissolved plume migration as well as control any vapor migration, (3) increase groundwater yields for larger capture zone, (4) expose and remediate the "smear zone," and (5) to remediate difficult geologic conditions that previously were considered long-term remediation projects in a shorter time frame.

CASE STUDIES - SITES WITH PREVIOUS REMEDIAL SYSTEMS

This section will show three separate case studies where Hi-Vac has been applied after other remedial technologies have been applied and failed.

Active Service Station in Orange County, New York

This service station is located adjacent to a reservoir, and is actually on a fill/marsh area that was once part of the reservoir. The geology of the site consists of highly compacted glacial till, with an intrinsic permeability of approximately 10^{-12} cm^2. Soil and groundwater contamination was found during a tank removal, with no exact known time of release. In the early 1990s, a pump-and-treat system was installed along the downgradient edge of the property, with the anticipation of capturing the groundwater at the property line. The system was designed to recover

Figure 1. Hi-Vac with liquid ring pump and drop tubes.

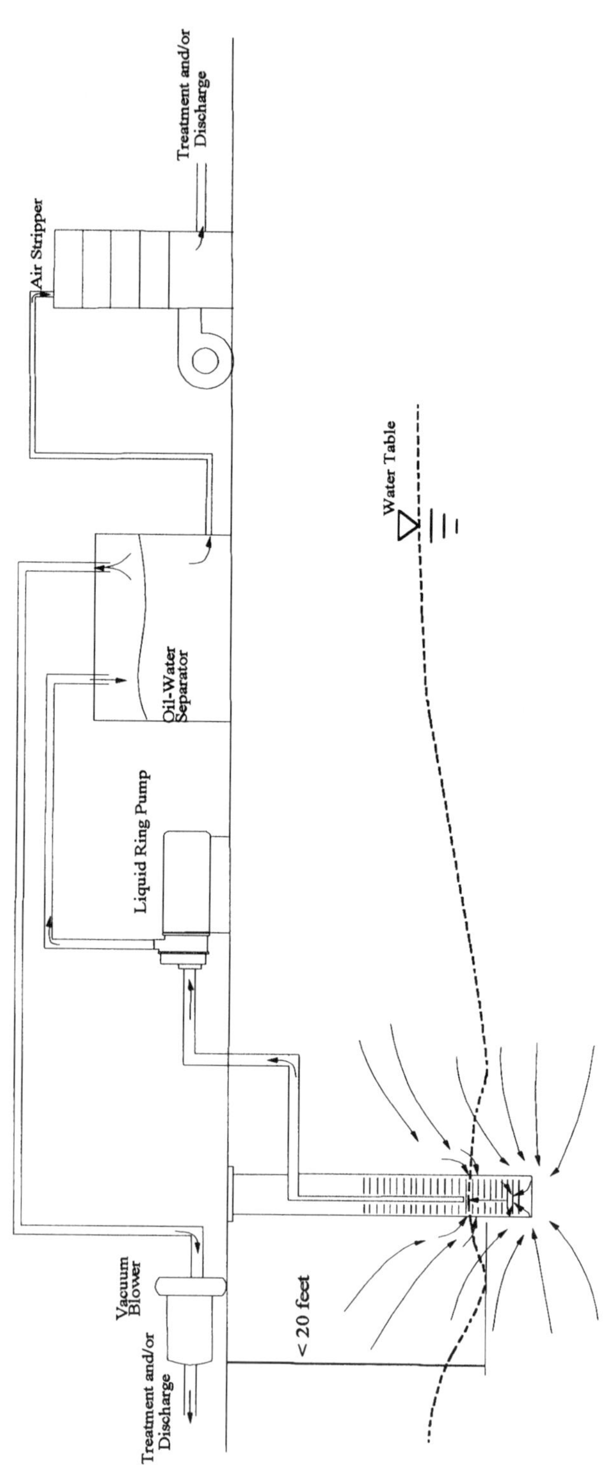

Figure 2. Hi-Vac with groundwater extraction pumps and high vacuum blower.

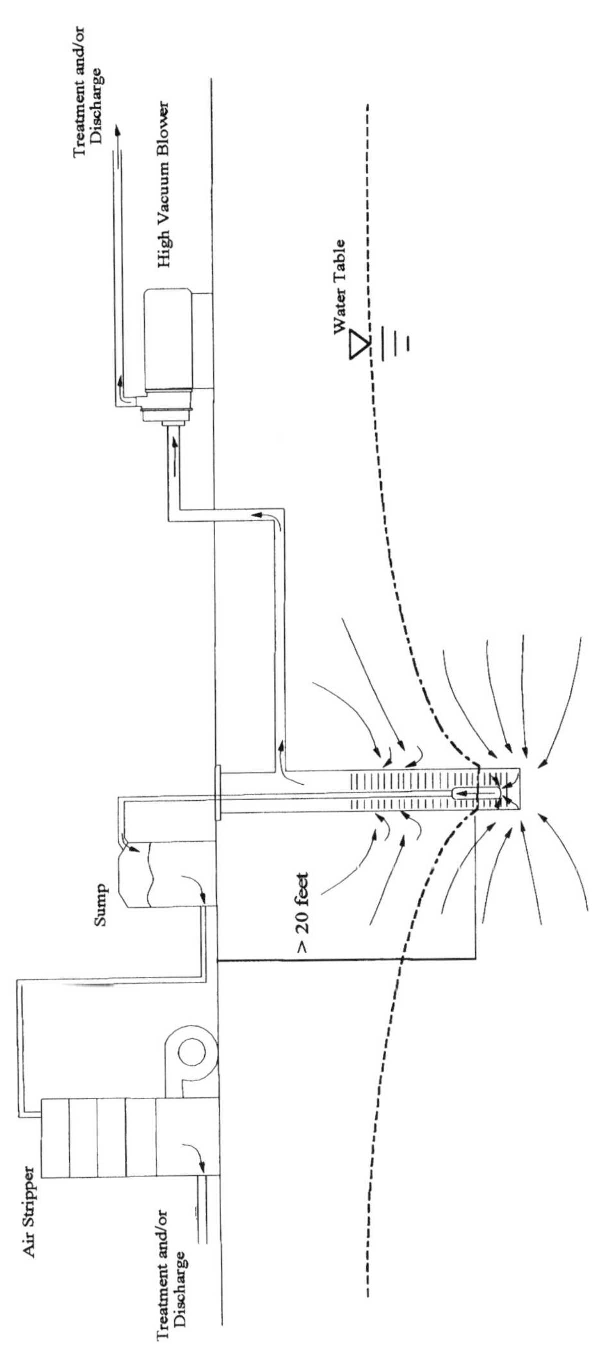

approximately five gpm with a capture zone which included the entire frontage of the site. The system operated for nearly three years with an average groundwater recovery rate of 0.19 gpm and a minimal capture zone. Through the two years it was operating, it recovered less than 10 pounds of hydrocarbons while processing 332,690 gallons of water.

After the pump-and-treat system was deemed ineffective, an extensive GeoProbe assessment was performed to further determine geologic conditions and the horizontal and vertical extent of the soil contamination. Based upon these discoveries, many different companies, who utilized various technologies, were solicited to find the appropriate remedial system. In 1995 it was determined that a Hi-Vac system, with some soil excavation and removal during a capital upgrade of the site, would be the most effective system for this site.

At the end of 1995, the Hi-Vac system, with nine sparge points for potential bioenhancement with the Hi-Vac system, was installed and started in February 1996. This system utilizes seven extraction wells, two-three hp water-sealed liquid ring pumps to remove both liquid and vapors, one-inch diameter drop tubes with pitless adapters in the wells, an oil-water separator, and carbon for treatment of the recovered water. Since the system has been operational (six months at the time of this paper), it has treated nearly 600,000 gallons of water and recovered more than 1,400 pounds of hydrocarbons, 140 times more than what was recovered by the previous system in quarter of the time. In addition, groundwater concentrations in the former hot zones have been reduced by 90% (see Figures 3 and 4, Table 1)

Active Service Station in Westchester County, New York

This service station is located in lower Westchester County on a site in which bedrock is found from two to six feet below grade. The bedrock consists of gneiss, which is fairly competent and fractured. During the July 4th weekend in 1993, multiple feet of SPH were discovered in an on-site monitoring well. Emergency response action plans were to remove SPH via a "vac" truck, to install various monitoring and recovery wells, and to install a pump-and-treat system to gain hydraulic control. During the operation of the pump-and-treat system, SPH was still detected in small quantities in various wells on the site, and over time it was evident that this system would not be able to recover and/or remediate these areas.

In March 1995, a Hi-Vac pilot test was performed utilizing a three-hp liquid ring pump on an existing monitoring well. The recovery well was sealed to minimize "short circuiting," and both liquid and vapor were pulled from the well utilizing a drop tube, and sent to an oil-water separator for required treatment and discharge. After two hours into the pilot test, influence was observed 30 feet from the test well, where the existing pump-and-treat system showed no capture zone greater than 15 feet (see Figure 5). In addition, a flame ionization detector (FID) was utilized to monitor vapor readings from the oil-water separator and was rendered useless due to continual "flame outs," where the concentration exceeded the

Figure 3. Site plan, Mobil service station, Orange County, New York.

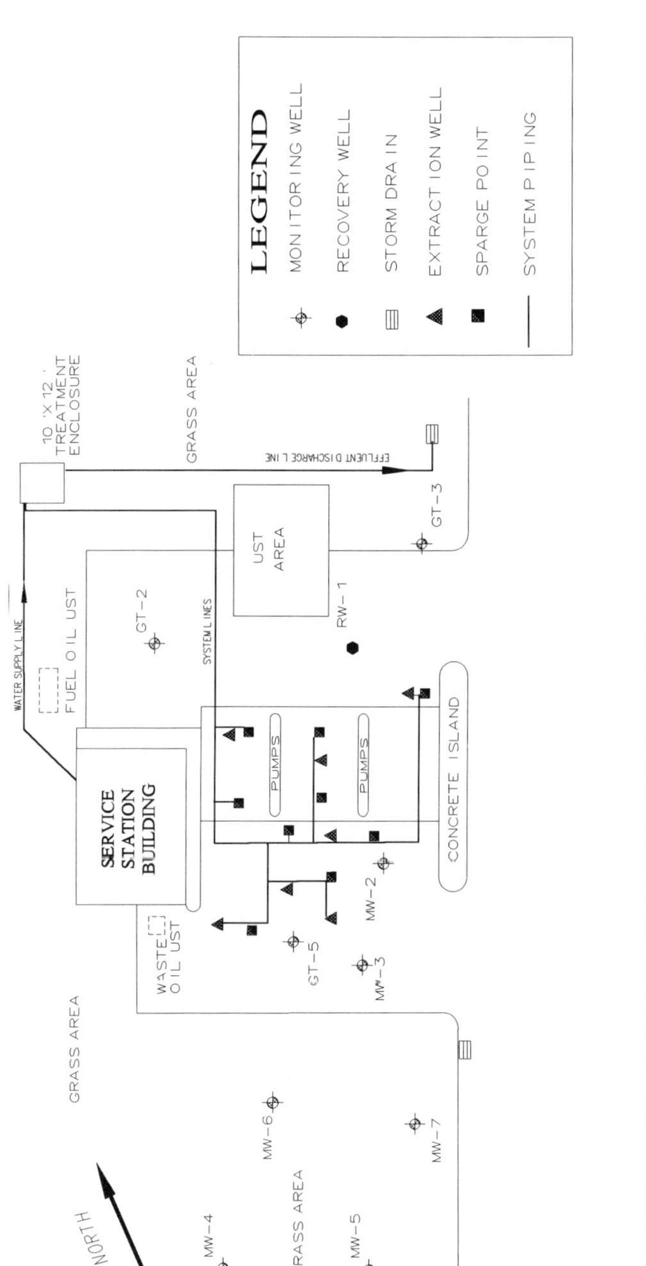

Figure 4. Piping and instrument diagram, Mobil service station, Orange County, New York.

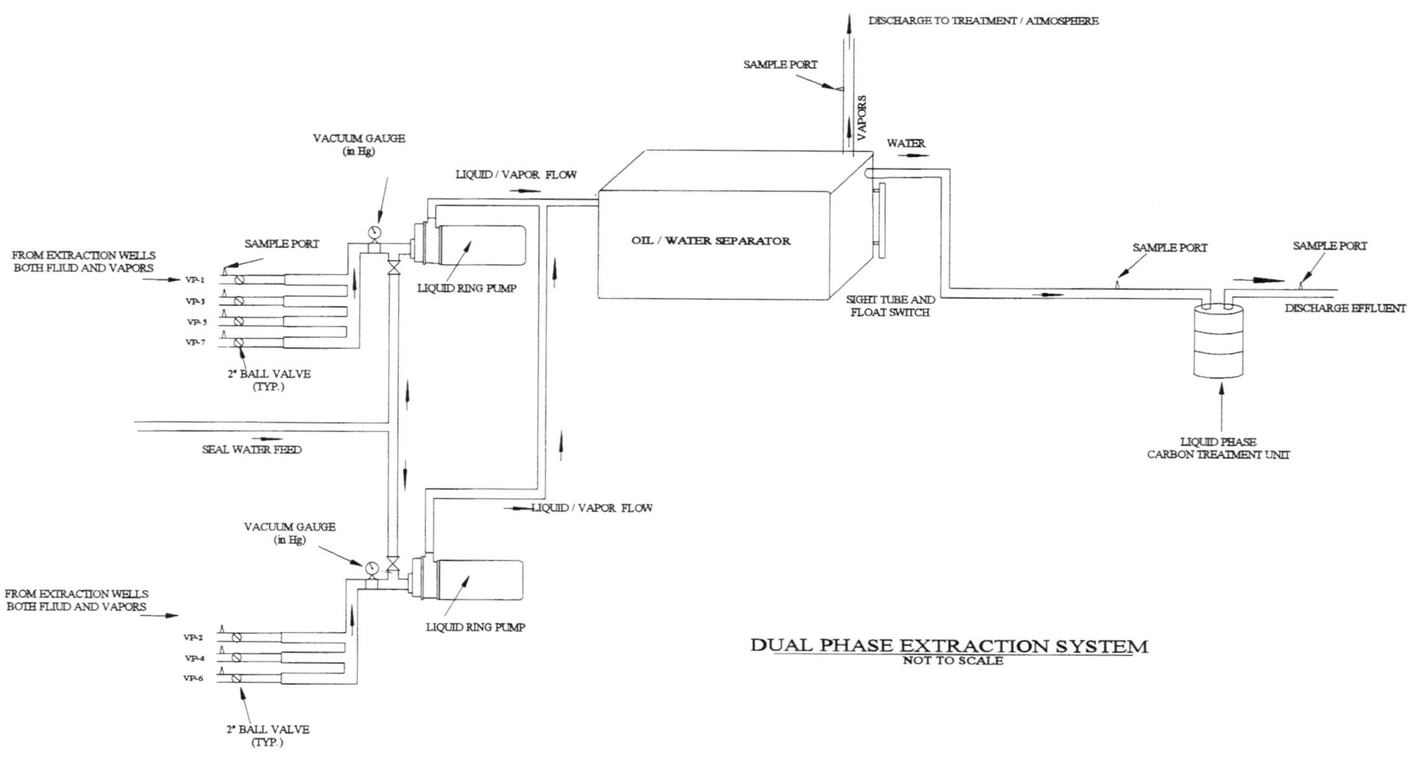

Table 1. Mobil Service Station, Orange County, New York [a]
(Pounds of Hydrocarbons Extracted)

Month	SVE Air Flow (cfm)	SVE Total BTEX (ppmv)	Groundwater Recovery Gallons Treated This Period	Groundwater Recovery Influent Total BTEX (ppb)	Mass of Hydrocarbons Removed (lb) Liquid	Mass of Hydrocarbons Removed (lb) Dissolved	Mass of Hydrocarbons Removed (lb) Vapor	Cumulative Hydrocarbons Removed(lb)
Before Hi-Vac								10.84
2/1/96	55	175.85	780	1,690.00	0.00	0.0733	7.33	18.24
2/2/96	55		1,432		0.00	0.1347	21.33	39.71
2/8/96	55		9,848		0.00	0.9253	126.00	166.63
2/21/96	55		9,720		0.00	0.9133	214.66	382.21
2/26/96	55		32,600		0.00	3.0633	105.33	490.60
3/29/96	55	62.40	199,515	905.00	0.00	10.0393	238.67	739.31
4/17/96	55	97.80	103,385	2,935.00	0.00	16.8706	222.67	978.85
4/23/96	55		15,160		0.00	2.4740	72.67	1,053.99
5/31/96	55	114.00	15,102	33.50	0.00	0.0280	150.07	1,204.09
6/30/96	55	42.30	128,238	29.60	0.00	0.2113	151.87	1,356.17
7/23/96	55	50.50	105,900	2.80	0.00	0.0067	181.27	1,537.45

1,000 ppm capacity of the FID. These results indicated that Hi-Vac would be a suitable option. During the installation phase, sparge points were also installed for biological enhancement later in the project.

In June 1995, the Hi-Vac system was installed and started. The system utilizes four extraction wells, a five-hp water-sealed liquid ring pump to remove both liquid and vapors, one-inch drop tubes on pitless adaptors, an oil-water separator with a product drum, a rotary lobe blower to recover and remove vapors from the oil water separator, a compressor for future sparging, and carbon for both liquid and vapor polishing. All of this equipment, with the exception of the rotary lobe blower and vapor phase carbon, is located in a 16 foot long by 8 foot wide trailer which has the ability to be moved from site to site for future use or emergency response. Since the Hi-Vac system has been operational (12 months), the system has recovered 24 times more hydrocarbons in half the time that the pump and treat system removed them, alone (1,011 pounds of hydrocarbons recovered to date). In addition, the Hi-Vac system has influenced and removed the remaining SPH in the wells (see Figures 5 and 6, Table 2).

Inactive Service Station in Bronx, New York

This site is located in the northern region of Bronx, New York, nearly 100 yards from the Harlem River. The geology of this site consists of bedrock varying from one to five feet below grade across the site. The overburden material is a mix of fill and debris, with groundwater at the site averaging 20 feet below grade (in the bedrock). The bedrock in this area primarily consists of gneiss and schist, which appears to be fairly competent with intermittent sand stringers. The tank hole at

Figure 5. **Site plan, Hi-Vac and pre-Hi-Vac, Mobil service station, Westchester County, New York.**

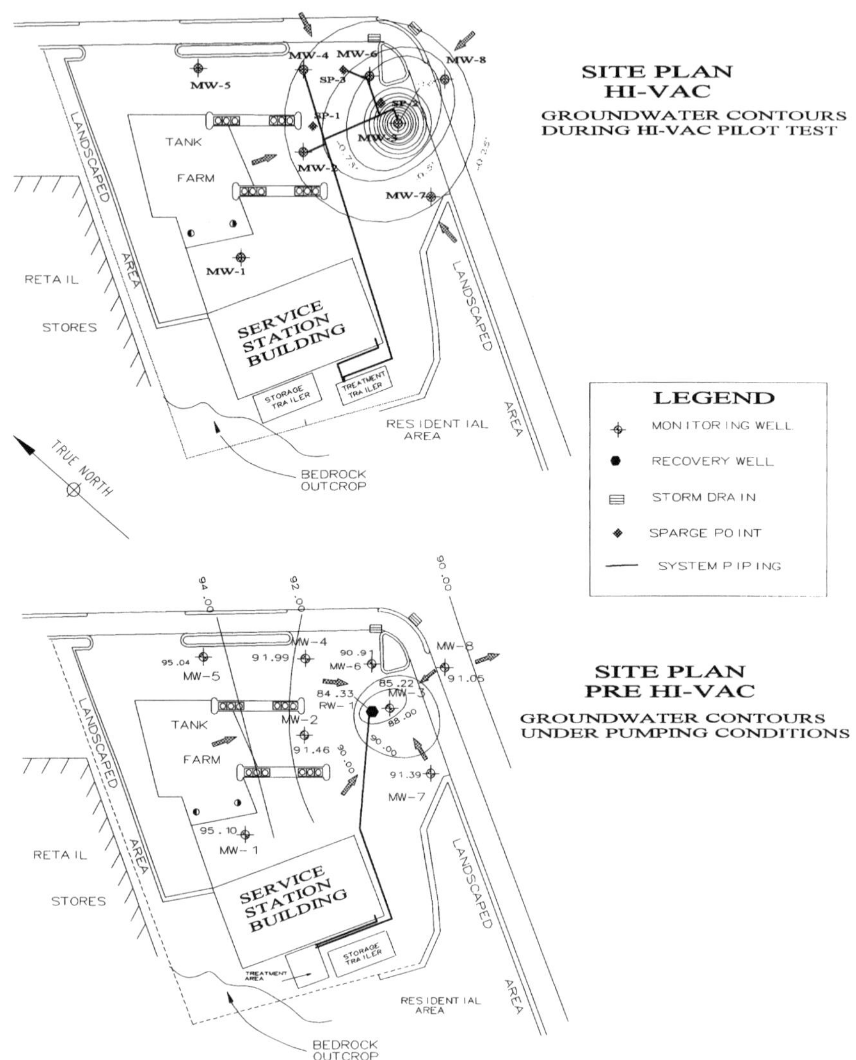

Figure 6. Piping and instrument diagram, Mobil service station, Westchester County, New York.

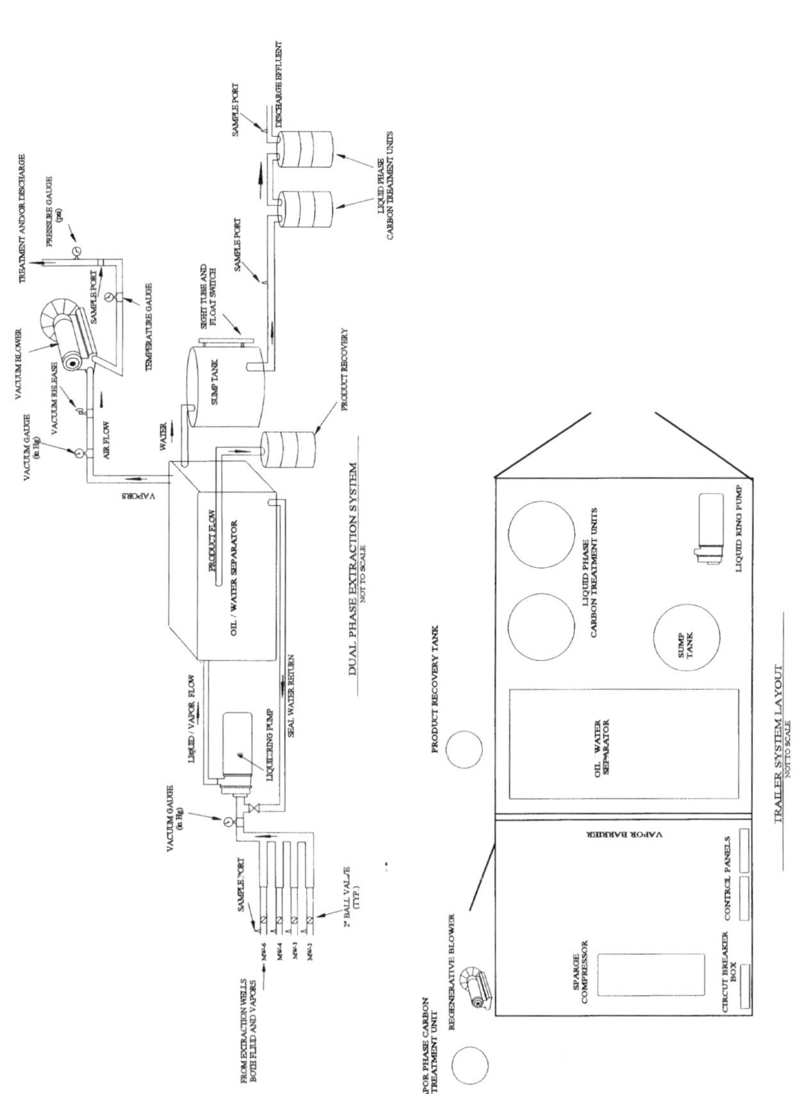

**Table 2.   Mobil Service Station, Westchester County, New York [a]
(Pounds of Hydrocarbons Extracted)**

Month	Air Flow (cfm)	Total BTEX (ppmv)	Gallons Treated This Period	Influent Total BTEX (ppb)	Liquid	Dissolved	Vapor	Cumulative Hydrocarbons Removed (lb)
	SVE		Groundwater Recovery		Mass of Hydrocarbons Removed (lb)			
7/13/93			11,232	20,660	1.0	1.9353	0.00	2.94
9/9/93			10,892	20,040	3.0	1.8204	0.00	11.87
11/2/93			6,230	12,760	3.0	0.6630	0.00	17.91
12/1/93			8,171	7,160	3.5	0.4879	0.00	21.89
3/14/94			14,270	0	3.0	0.0000	0.00	27.60
6/3/94			2,253	1,981	3.0	0.0372	0.00	32.27
9/16/94			11,800	1,970	1.5	0.1939	0.00	35.87
3/10/95			5,980	1,957	1.0	0.0976	0.00	38.77
5/1/95			6,470	506	1.0	0.0273	0.00	40.39
7/6/95	35	1,900	New Meter	1,573	0.0	0.0066	5.43	46.32
7/31/95	35	220	44,990		0.0	0.5902	62.82	109.73
9/29/95	50	200	68,430	3	0.0	0.0019	94.63	251.21
10/10/95	35	550	39,660	60	0.0	0.0198	69.10	320.32
1/4/96	55	500	129,640	6	0.0	0.0064	278.19	606.51
5/31/96	35	300	6,538	191	0.0	0.0104	74.24	739.84
6/18/96	50	200	252	92	0.0	0.0002	17.68	757.51
7/2/96	50	455	13,135		0.0	0.00	253.65	1,011.17

[a] This is only a portion of the original table.

this location was blasted into the bedrock, along with the product piping lines Contamination appears to have been in the former tank farm area, in which any releases entered the bedrock immediately.

Due to the complexity of remediating bedrock, numerous assessments were performed. As a result, there are 20 wells located on-site. In the early 1990s, a pump-and-treat system was installed on-site utilizing an eight-inch recovery well and a two-pump recovery system. SPH was present in at least 10 wells from 1991 until 1994, where the remedial system was recovering little SPH, with the majority of the SPH being recovered via hand bailing (144 gallons total recovered until March 1994). Utilizing the pump-and-treat system, there was an apparent communication observed between wells, where two recovery wells could obtain a one-foot drawdown 30 feet away.

In August 1993, a Hi-Vac pilot test was performed. A liquid ring pump with drop tubes was set in a former recovery well. During the testing period, a drawdown in the groundwater of two feet was observed in most wells, with as much as four and one-half feet observed in one well (see Figure 7). In addition, SPH, which was already present in some wells, was increasing in thickness. The effluent air concentrations were as much as 387 ppm during a 90-minute test, as well as achieving 100% LEL during total testing period. Based upon this information, a Hi-Vac system was designed and installed in late 1993.

The system was started in March 1994. The system utilized a 15-hp water-sealed liquid ring pump, eight extraction wells, a sealed oil-water separator, and carbon for liquid polishing. Vapors were accumulated in vapor-liquid separator and treated utilizing a 100-cfm electric-preheat CATOX unit. During its initial operation, effluent air reading ranged from 1,000 ppm to a high of 6,100 ppm utilizing a FID. In addition, most of the remaining SPH had been recovered through volatilization, including some SPH that was trapped in fractures opened by the increased drawdown and vacuum in the area. Problems with run time of the system in 1995 required a modification to the system (minimal water recovery was insufficient to keep the liquid ring pump cool and maintain the seal). Even with these problems, the system was able to remove more than 4,000 pounds of hydrocarbons in 15 months.

In October 1995, the liquid ring pump was removed from the treatment shed. In its place, two groundwater depression pumps and a seven and one-half hp rotary lobe blower, capable of 15 inches of Hg for vacuum, was installed. Once the SVE portion of the system was put back on-line, removal rates started to increase even without starting the pump-and-treat system (this area of the Bronx has shown a two- to three-foot depression in groundwater elevation over the past two years). Since the system has been restarted with the new blower, the system has recovered an additional 2,000 pounds of hydrocarbons in eight months, with no SPH remaining in the wells. In the upcoming months, the pump-and-treat portion of the system will be started to gradually lower the water table and remove any additional hydrocarbons trapped in fractures beneath the water table (see Figures 7 and 8, Table 3).

SITES WITHOUT PREVIOUS REMEDIAL DATA

In this section, we will discuss two sites, located across the street from each other, which had no previous remedial systems.

Active Service Station in Queens, New York

This station and adjacent parcel are located in Queens, New York, on an area that is approximately 50 feet above sea level, but less than one-half mile from Flushing Bay. The geology in the area consists primarily of sand, pebbles, and cobbles to approximately 30 feet below grade, with average density (blow counts between six and 17 blows per six inches). From 30 to 45 feet below grade, the geology changes to till with silt, clay, and cobbles, with a high density (blow counts between 30 and 40 blows per six inches). From 45 feet and below, the geology turns to a coarse sand with some pebbles and cobbles, but with the same density as the area above. Groundwater in the area ranges from 45 to 50 feet below grade. Contamination in the overburden on the service station property varies in depth across the site, most likely due to different source areas as well as greater horizontal than vertical soil permeability. In addition to the soil contamination, SPHs have been detected

Figure 7. Site plan, Mobil service station, Bronx, New York.

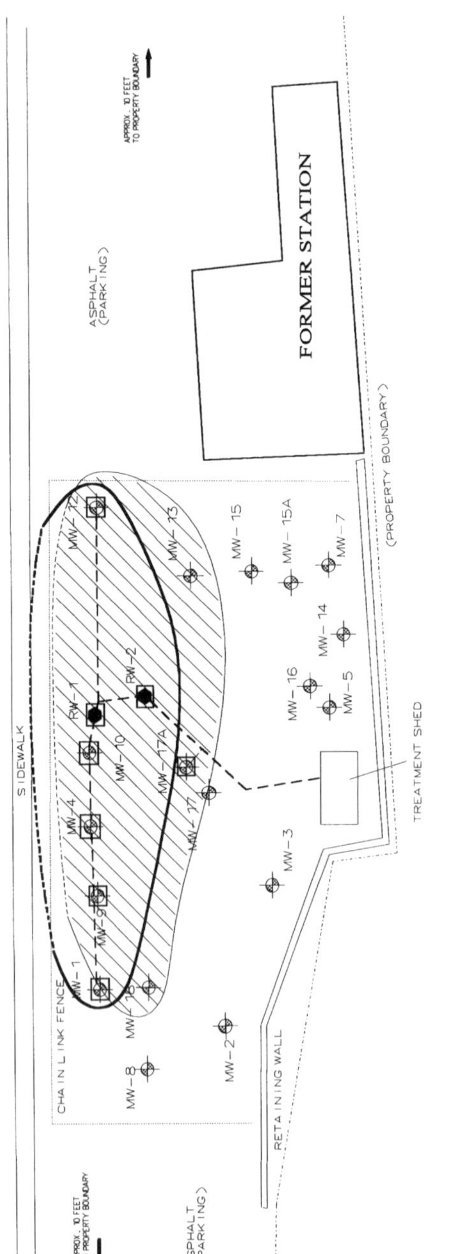

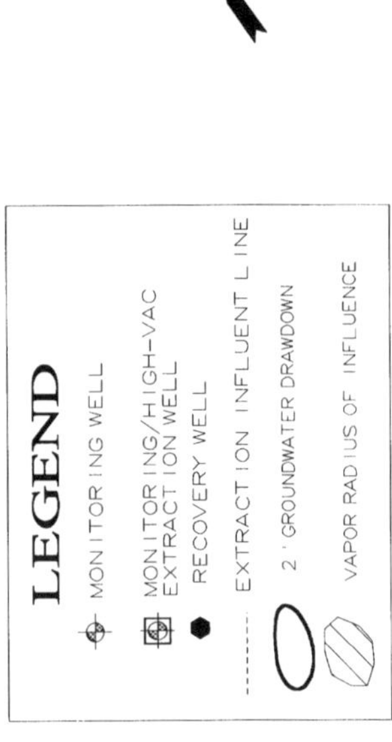

Figure 8. Piping and instrument diagram, Mobil service station, Bronx, New York.

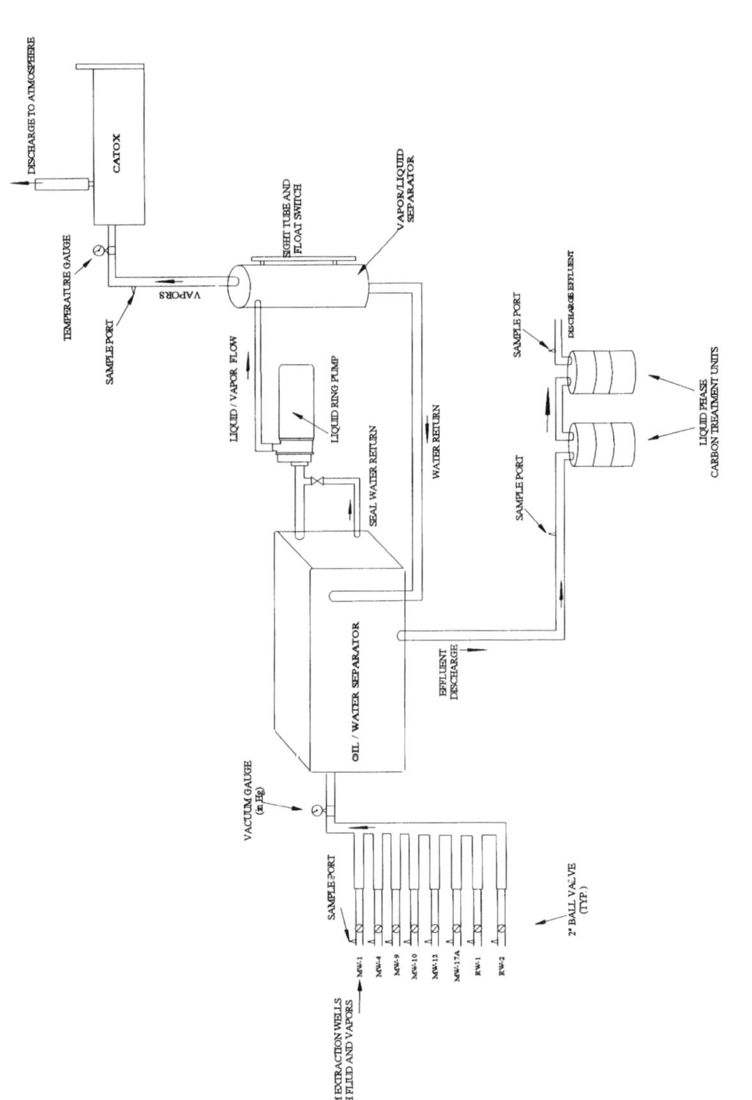

**Table 3. Mobil Service Station, Bronx, New York [a]
(Pounds of Hydrocarbons Extracted)**

Month	Air Flow (cfm)	Total BTEX (ppmv)	Gallons Treated This Period	Influent Total BTEX (ppb)	Liquid	Dissolved	Vapor	Cumulative Hydrocarbons Removed(lb)
	SVE		**Groundwater Recovery**		**Mass of Hydrocarbons Removed (lb)**			
Before Hi-Vac								864.00
3/31/94	68	5320	2,537		0.00	0.0000	24.59	888.78
4/1/94	121	3220	1,729	8660	0.00	0.1250	71.52	960.42
5/19/94	116	1500	3,230		0.00	0.0000	106.46	1,855.89
6/20/94	107	6100	178	13	0.00	0.0000	53.25	2,152.89
8/11/94	109	6000	6,197	822	0.00	0.0420	924.79	3,536.83
9/26/94	87	4000	1,380	1226	0.00	0.0140	126.33	3,686.28
10/14/94	120	2800	478	56	0.00	0.0000	50.25	4,052.38
12/16/94	120	950	936	298	0.00	0.0020	148.80	4,530.76
1/23/95	120	700	431		0.00	0.0000	15.99	4,619.56
2/6/95	115	800	66	16	0.00	0.0000	16.26	3,681.24
3/22/95	115	550	209	48	0.00	0.0000	5.16	4,793.45
5/23/95	100	520	0		0.00	0.0000	20.50	4,855.18
6/15/95	110	650	0		0.00	0.0000	28.19	4,911.57
Oct '95			System off for retro-fit					
11/1/95	155	380			0.00	0.0000	3.20	4,914.77
12/4/95	155	120			0.00	0.0000	21.75	5,165.55
1/19/96	155	560			0.00	0.0000	448.47	5,694.09
2/2/96	120	250			0.00	0.0000	97.90	5,838.97
3/5/96	120	360			0.00	0.0000	352.42	6,211.58
4/12/96	120	5			0.00	0.0000	2.74	6,283.66
5/3/96	100	220			0.00	0.0000	150.76	6,434.42
6/18/96	120	300			0.00	0.0000	540.38	6,974.81

[a] This is only a portion of the original table.

on-site at thickness greater than 10 feet, but this has been considered an anomaly based upon the locally tight soil conditions creating a pseudo-artesian effect.

Several pilot tests were performed at the site prior to any Hi-Vac tests, specifically SVE and slug tests. Results of the SVE test concluded that utilizing a ten-hp rotary lobe blower, a ten-foot radius of influence could be obtained (this would require 19 SVE points, in which this system would not address SPH at these thicknesses). The slug test showed minimal rebound over a two-hour time period, indicating the test was inconclusive. On July 22, 1993, a Hi-Vac pilot test was performed utilizing a groundwater depression pump to control water uplift/recovery and a liquid ring pump to determine vacuum influence. During the pilot test, groundwater depression was observed 42 feet from the test well, along with vacuum influence being observed in a well as far as 64 feet from the test well. In addition, the effluent air had readings at or above 500 ppm on a FID and a steady 2 gpm from the groundwater depression pump in the test well (see Figure 9).

In May 1994, a Hi-Vac system was installed. The system includes the ability to pump five recovery wells, three at a time, pull a vacuum on nine individual wells utilizing a 15-hp liquid ring pump and the off-gas being treated by a 200-cfm electric-preheat CATOX. The water is transferred from the wells to a 250-gallon oil-water separator, product being gravity fed to a 500 gallon aboveground tank, and a low-profile air stripper to treat the groundwater. During the first five months, effluent air concentrations were recorded as high as 6,500 ppm on a FID. From the system's inception until now, including hand-bailing SPH from the wells, there have been more than 15,800 pounds of hydrocarbons recovered in 26 months and nearly 1,000,000 gallons of water treated (see Figures 9 and 10, Table 4).

Adjacent Parcel in Queens, New York

This property is located hydraulically downgradient from the above-discussed service station, with similar geology and groundwater elevations. The primary difference between the two sites is that this parcel has little to no overburden contamination, only the residual SPH that adhered to the soils during the seasonal groundwater fluctuation. On this parcel, SPH has been detected in multiple wells with thicknesses greater than three feet (potential again for pseudo-artesian conditions), with a remedial action plan to address the SPH, dissolved phase plume and the remaining SPH, trapped below the water table in the smear zone.

Based upon the preliminary SVE pilot test and slug test performed on the service station property, it was decided to also perform a Hi-Vac pilot test on this property. On July 29, 1993, the pilot test was performed utilizing the same setup as found on the service station property: groundwater depression pump control groundwater uplift/recovery in the well and a liquid ring pump to determine vacuum influence. The results of this test, pumping from one well with the assistance of the high vacuum applied simultaneously, groundwater depression was observed over a majority of the site with recovery of 0.75 gpm and vacuum influence was also observed in all wells (see Figure 9).

As a result of this test, a Hi-Vac system was installed in June 1994 and was operational in July 1994. The system designed for this site was identical to the one installed on the service station, except that this location has the ability to pump from four wells and apply a vacuum at seven wells. Effluent air concentrations were as high as 2,000 ppm on a FID within the first month, with sustained concentration between 100 and 300 ppm. This system will be modified in the near future with the deletion of the liquid ring pump and the insertion of a seven and one-half hp regenerative blower, capable of delivering 100 inches of water vacuum. The reason for the change is that the liquid ring pump was having difficulty maintaining a consistent electric supply and difficulty keeping calcium scaling from the impellers (resulting from water hardness). After 24 months of operation, the system, with the addition of SPH from hand-bailing wells, has removed more than 5,000 pounds of hydrocarbons, and there is no more SPH in the wells (see Figures 9 and 10, Table 5).

Figure 9.   **Site plan, Mobil service station, Queens, New York.**

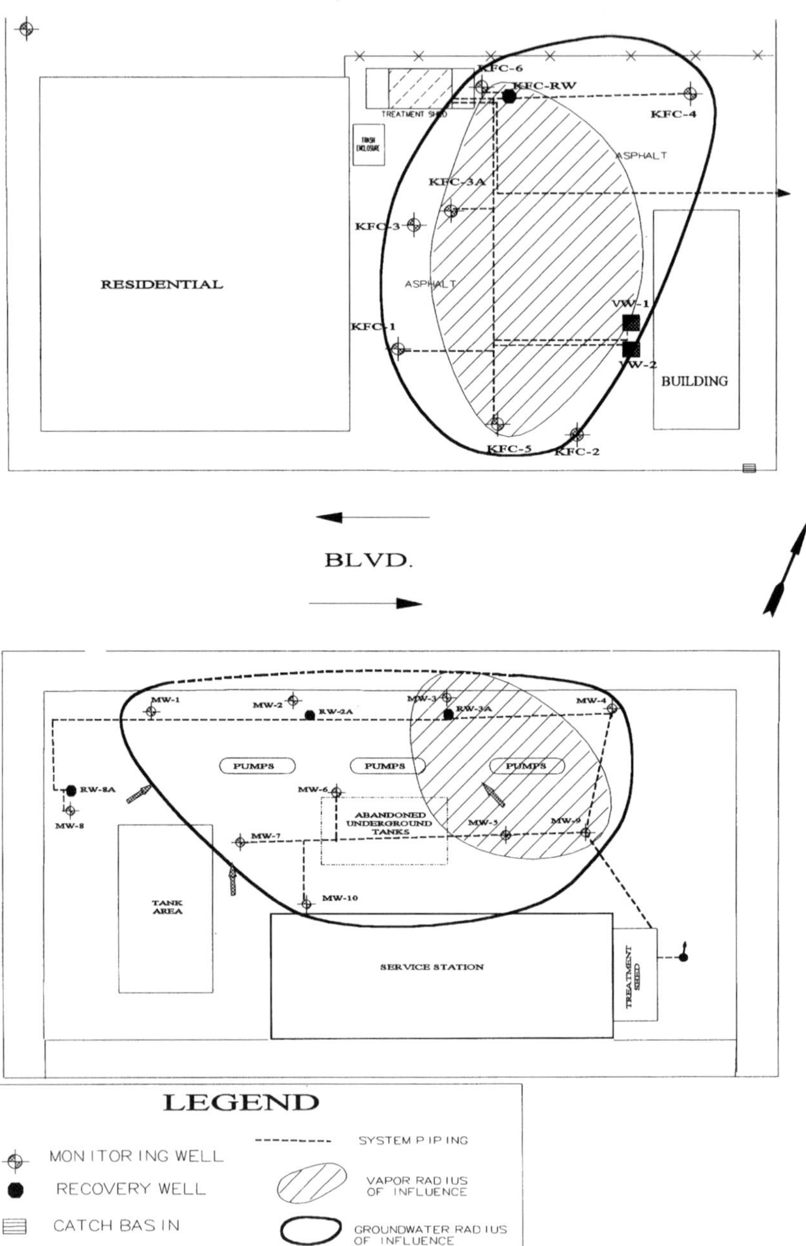

Figure 10. **Piping and instrument diagram, Mobil service station, Queens, New York.**

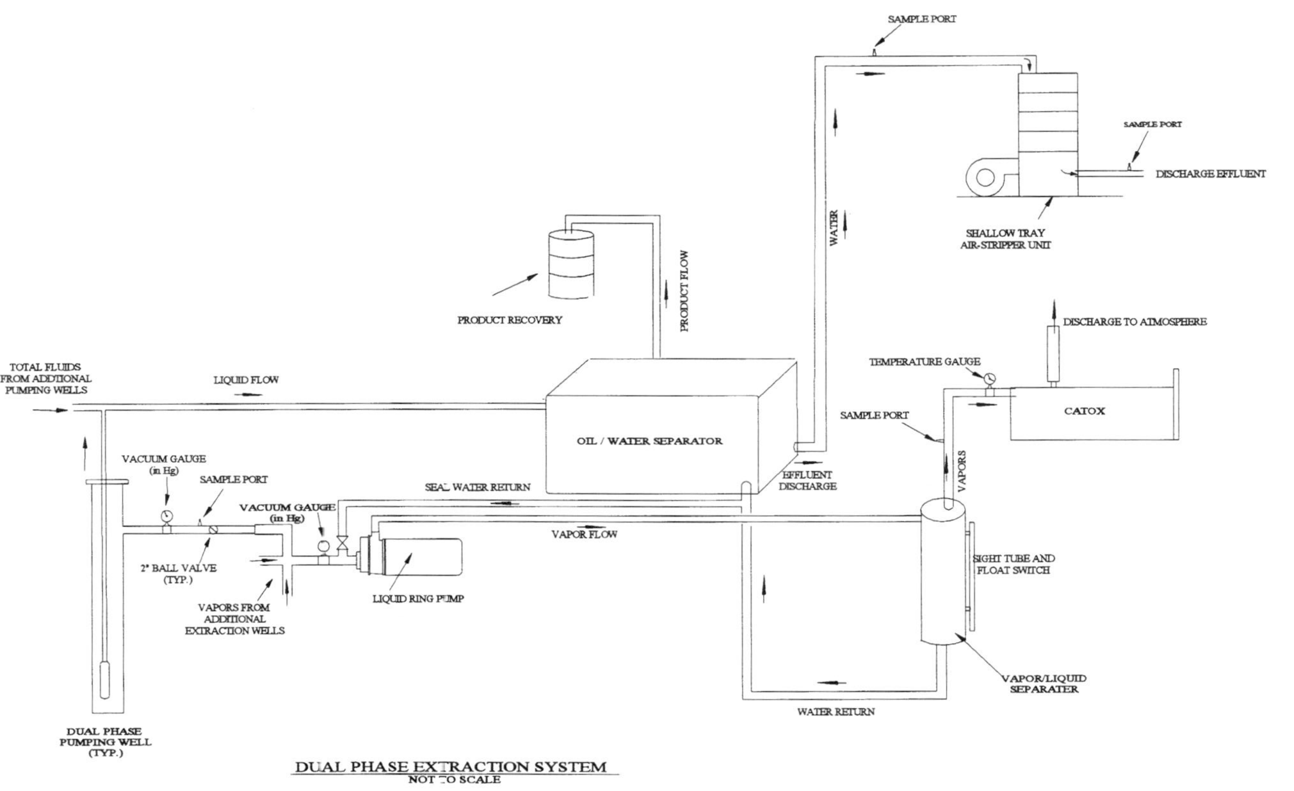

Table 4.   **Mobil Service Station, Queens, New York** [a]
            (Pounds of Hydrocarbons Extracted)

Month	SVE Air Flow (cfm)	SVE Total BTEX (ppmv)	Groundwater Recovery Gallons Treated This Period	Groundwater Recovery Influent Total BTEX (ppb)	Mass of Hydrocarbons Removed (lb) Liquid	Mass of Hydrocarbons Removed (lb) Dissolved	Mass of Hydrocarbons Removed (lb) Vapor	Cumulative Hydrocarbons Removed(lb)
5/10/94	68	1,600	3,206	10,090	570.0	0.27	4	575
6/6/94	57	1,800	10,055		0.0	0.59	278	1,023
7/11/94	109	1,800	5,326		41.0	0.31	115	1,961
8/23/94	130	1,500	6,362	7,080	0.0	0.38	318	2,639
9/21/94	130	4,500	13,872	920	384.0	0.11	2,863	6,321
10/17/94	126	6,500	12,899	11,090	0.0	1.19	1,114	8,273
11/10/94	55	2,100	7,617	23,100	0.0	1.47	42	9,027
12/5/94	126	840	13,652	666	92.0	0.08	235	9,618
1/12/95	130	1,100	7,889	1,983	0.0	0.13	420	10,578
2/27/95	110	1,800	281	1,854	24.0	0.00	13	11,668
3/22/95	126	1,500	138	6,280	0.0	0.01	18	12,164
4/12/95	100	1,000	23,240	56	0.0	0.01	159	12,641
5/2/95	100	1,000	3,533	100	0.0	0.00	159	13,115
6/1/95	110	400	10,202	2,149	0.0	0.18	70	13,439
7/7/95	120	600	11,377	307	0.0	0.03	115	13,699
8/16/95	122	750	8,388	436	3.0	0.03	146	14,000
9/5/95	120	550	9,375	1,020	0.0	0.08	105	14,105
10/10/95	110	780	188	6,484	0.0	0.01	136	14,310
11/29/95	110	400	10,602	11,501	3.0	1.02	70	14,454
12/29/95	100	400	27,608		3.5	0.00	64	14,600
1/24/96	100	400	62	8,807	0.0	0.00	97	14,740
2/19/96	100	220	9,979	6,791	0.0	0.57	72	14,893
3/16/96	100	230	1,173	8,855	0.0	0.09	45	14,960
4/17/96	100	568	7,559	12,943	0.0	0.82	93	15,228
5/24/96	100	450	266	27,670	0.0	0.06	20	15,558
6/30/96	96	200	54,997	2,733	0.0	1.25	69	15,645
7/29/96	120	200	77,988		0.0	0.00	227	15,874

CONCLUSIONS

It is apparent that there are many different remedial alternatives for petroleum impacted soil and groundwater; it's a matter of applying the appropriate remedial technologies for the specific geologic conditions. It is also rare to find homogeneous geologic conditions across an entire site, so professional judgment and appropriate pilot testing is required.

For tight soils with low intrinsic permeability, Dual Phase or Hi-Vac Extraction is an extremely effective remedial option. Even for bedrock, Hi-Vac has proven to assist in remediating this difficult condition. The ability to remove both liquid

**Table 5. Mobil Service Station, Queens, New York, Adjacent Parcel [a]
(Pounds of Hydrocarbons Extracted)**

Month	SVE Air Flow (cfm)	SVE Total BTEX (ppmv)	Groundwater Recovery Gallons Treated This Period	Groundwater Recovery Influent Total BTEX (ppb)	Mass of Hydrocarbons Removed (lb) Liquid	Mass of Hydrocarbons Removed (lb) Dissolved	Mass of Hydrocarbons Removed (lb) Vapor	Cumulative Hydrocarbons Removed (lb)
7/25/94	130	420	166	52,100	0	0.72	2	3
8/18/94	130	2,000	10,426	43,600	0	37.91	594	953
9/19/94	130	900	14,801	5,360	30	6.62	326	1,742
10/17/94	130	500	38,227	28,900	0	92.140	569	2,404
11/15/94	130	1,100	4,495	2,110	0	0.790	173	2,681
12/6/94	130	200	2,393	365	0	0.070	5	2,763
1/20/95	130	650	10,220	1,960	6	1.670	67	2,904
2/27/95	130	900	25,669	1,863	6	0.400	772	3,763
3/6/95	130	168	7,962	4,590	0	0.300	50	3,814
4/25/95	130	120	3,685	2,000	25	0.060	31	3,989
5/9/95	130	40	2,962	2,000	0	0.050	83	4,082
6/23/95	130	120	4,922	4,147	0	0.170	71	4,272
7/20/95	130	140	5,682	4,570	0	0.220	1	4,370
8/29/95	130	150	7,914	5,202	0	0.340	52	4,470
9/19/95	130	240	2,862	5,000	0	0.120	10	4,502
10/10/95	130	120	14,393	21,540	0	2.590	51	4,627
11/10/95	130	420	2,031	16,110	0	0.270	33	4,757
12/4/95	130	320	3,377	16,510	0	0.460	68	4,925
1/24/96	130	78	3,904	650	0	0.540	30	5,064
2/23/96			1,317	15,100	0	0.180	0	5,071
3/29/96			5,096	13,060	0	0.700	0	5,072
4/12/96			4,550	12,840	0	0.630	0	5,073
5/31/96			16,327	16,590	0	2.250	0	5,075
6/30/96			5,892	19,900	0	0.810	0	5,076

[a] This is only a portion of the original table.

and vapor from the same well, and sometimes in the same effluent stream, with the increased water and vapor yield due to the application of a high vacuum, is an attractive alternative. Utilizing this technology can reduce the amount of money spent on equipment and reduce the amount of money spent on operation and maintenance because this technology can expedite the remediation process. Unless the geology of the site consists primarily of sand where a high vacuum is not required to enhance vapor or groundwater recovery rates, Hi-Vac is a cost-effective technology that has been proven to work.

REFERENCES

Freeze, R.A. and Cherry, J.A. 1979. *Groundwater*, Edgewood Cliffs, NJ, Prentice-Hall, Inc.

Site Cleaning By Vacuum Extraction. 1993. *Indus. Lubr. Tribology*, 45(2), 9-10.

Trowbridge, B.E. and Ott, D.E. 1994. The Use of In-Situ Dual Vacuum Extraction™ for Remediation of Soil and Groundwater, *Petro-Safe '94*, pp. 126-141, Houston, TX, January.

Wellensiek, M.R. and McCreanor, K.T. 1993. Accelerated Implementation of a Dual-Phase Soil Vapor Extraction System, *Superfund XIV Conference and Exhibition*, pp. 981-989. Washington D.C., Nov/Dec '93.

PART V—BIOREMEDIATION

CHAPTER 28

Bioremediation of Petroleum Contaminated Soils at Loring Air Force Base, Maine

Thomas R. Wood, Bechtel Environmental, Inc., Oak Ridge, Tennessee

Peter Forbes, Air Force Base Conversion Agency, Loring, Maine

John Mueller, Air Force Center for Environmental Excellence, Loring, Maine

INTRODUCTION

Bechtel Environmental, Inc. (Bechtel), contracted (Prime Contract No. F41624-94-D-8072) by the Air Force Center for Environmental Excellence (AFCEE) as the removal/remedial action contractor for Loring Air Force Base (AFB), performed removal actions that included bioventing of petroleum contaminated subsurface soil at Loring AFB in Limestone, Maine. This chapter is a summary of the full report that describes the implementation of the bioventing removal action work in Operable Units (OUs) 5, 8, 9, 10, and 11 performed during the 1995 construction season by Bechtel and its subcontractors.

The intent of the bioventing removal actions at OUs 5, 8, 9, 10, and 11 is to remove the organic compounds in the soil that are above preliminary remediation goals (PRGs) by enhancing the natural degradation of organic compounds using microbiological organisms. Eventually, confirmatory samples will be collected from the biovented areas, with the analytical results compared to applicable PRGs to verify the success of the removal action. Periodic monitoring of the areas will indicate when these confirmation samples will be taken, which should be after two years of operation.

Scope and Purpose

- Documenting that the bioventing removal action remedy has been implemented at OUs 5, 8, 9, 10, and 11 and providing verification that design and performance standards were met at these OUs, where applicable.
- Notifying the U.S. Environmental Protection Agency (USEPA) and the Maine Department of Environmental Protection (MDEP) that the bioventing removal actions at OUs 5, 8, 9, 10, and 11 were implemented in accordance with:

- Final Design Analysis Report, OUs 5, 9, 10, and 11 (URS, 1995a)
- Final Engineering Evaluation/Cost Analysis Operable Units 5, 9, 10, and 11 (URS, 1995b)
- Design drawings and technical specifications for OUs 5, 9, 10, and 11 (URS, 1995c–f)
- Preliminary and draft design drawings for OU 8, Fire Training Area (URS, 1995g)
- Final Engineering Evaluation/Cost Analysis Operable Unit 8 (URS, 1995h)
- Removal Action Work Plan, Bioventing at OUs 5, 8, 9, 10, and 11 (Bechtel, 1995)
- Operation and Maintenance Plan for Bioventing at OUs 5, 8, 9, 10, and 11 (Bechtel, 1996a)

This report was developed in accordance with USEPA guidance documents Close Out Procedures for National Priorities List Sites (USEPA, 1995) and Superfund Removal Procedures Removal Response Reporting: POLREPs and OSC Reports (USEPA, 1994).

The work items addressed in this report can be summarized as bioventing of petroleum contaminated soil (OUs 5, 8, 9, 10, and 11), which includes the following work:

- drilling and installing air injection wells
- drilling and installing monitoring points
- testing in-situ air permeability
- installing power to each bioventing area
- installing insulated buildings at each location to house the equipment
- installing skid-mounted bioventing equipment, air piping, and associated control systems
- startup and operation of the bioventing systems installed

Other removal action activities (i.e., excavation and disposal of contaminated soil) are not within the scope of this report. These activities have been documented under separate removal/remedial action reports. (See the removal action report for excavations in OUs 5, 8, 9, 10 and 11; Bechtel, 1996b.)

Except as detailed in this report, Bechtel has not independently verified historical information or prior reports upon which this report is based. Information available as of the issue date was used. Further information and/or changes in the condition of the property over time, arising from natural conditions or activities on or around the property, may alter or invalidate the conditions described herein. This report has been prepared by Bechtel for the exclusive use of AFCEE and the Loring Air Force Base Conversion Agency (AFBCA). Any person who wishes to use any information in this report for any other purpose should independently verify that the information is current and accurate for their purpose.

Site Description and Background

Loring AFB is in Aroostook County in the northeastern corner of Maine, about 3 miles from the Canadian border. The base occupies about 9,000 acres, mostly in the town of Limestone, Maine. The base was constructed between 1946 and 1953, and improvements were made throughout its operational life. Most recently, the base was part of the Air Combat Command. On September 30, 1994, it was officially closed and is now the responsibility of AFBCA. Loring AFB is designated as Operating Location M by AFBCA. Figures 1 and 2 show the location of the base and the major OUs.

Specific information regarding the site characterization for OUs 5, 8, 9, 10, and 11 (e.g., site description, surface waters, ecological characteristics, geology, and hydrogeology) is contained in the design analysis report (DAR) (URS, 1995a), the removal action work plan (Bechtel, 1995), and various remedial investigation (RI) reports cited in these documents. Additional information is provided on the referenced drawings (URS, 1995d-g), and summary figures for each site are provided in Figures 3 through 6.

Summary Description of Operable Units 5, 8, 9, 10, and 11

The bioventing installations planned at each OU included in this removal action are briefly outlined below. (Note: Not all sites or removal action activities within OUs 5, 8, 9, 10, and 11 are included in this report. Additional bioventing installations are planned by the Air Force at this location.) The work included installation of injection wells, monitoring points, and equipment and connection of piping.

The specific sites where the bioventing systems were installed are shown in Table 1.

Table 1.   Locations of Bioventing Systems

OU	Site	Approx. Area (ft2)	Primary Contaminants
5	Former Jet Engine Test Cell (FJETC)	21,050	TPH, Benzene, Toluene, Xylenes
8	Fire Training Area (FTA)	20,000	TPH, Benzene, Toluene, Xylenes
9	Power Plant Drainage Pipe (PPDP)	24,250	TPH
10	Entomology Shop (ES)[a]	7,275	TPH
11	Vehicle Maintenance Building (VMB)	25,480	TPH, Benzene

[a] being installed in 1996.

Figure 1.   Loring Air Force Base location map.

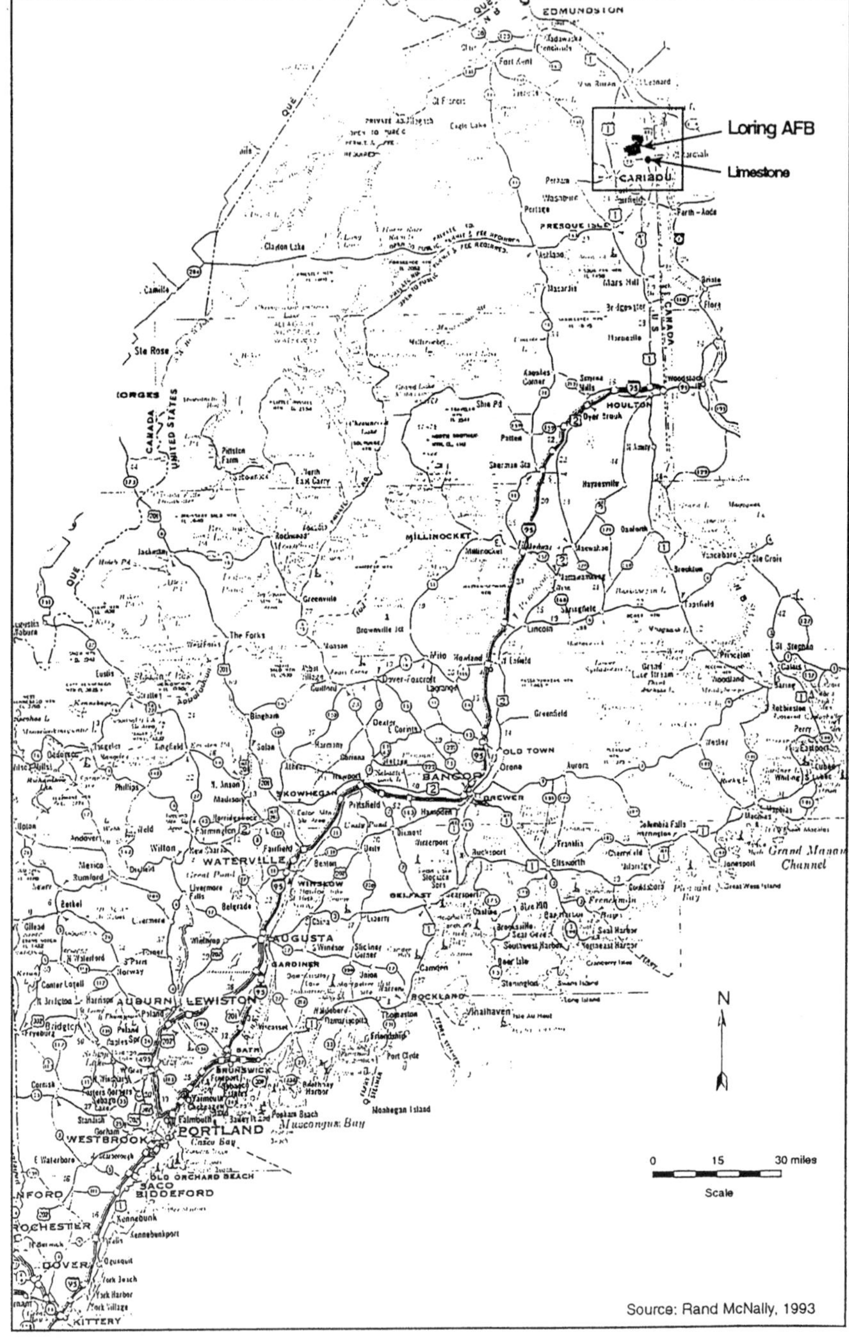

Regulatory Setting

The bioventing removal actions being performed at Loring AFB are being implemented under the Comprehensive Environmental Response, Compensation, and Liability Act (CERCLA), Section 104. Loring AFB was listed on the National Priorities List in 1991. An engineering evaluation/cost analysis report (EE/CA) has been prepared for the OUs involved with this removal action. The removal action design is based on a number of previous studies done at this site. A complete list of references is included in the bioventing work plan. These previous activities include:

- RI report for OU 5 (summarized by Camp, Dresser, and McKee Consultants, Inc. [CDM] in 1995) involving separate investigations by CDM 1994;

Figure 2. **Operable units at Loring AFB.**

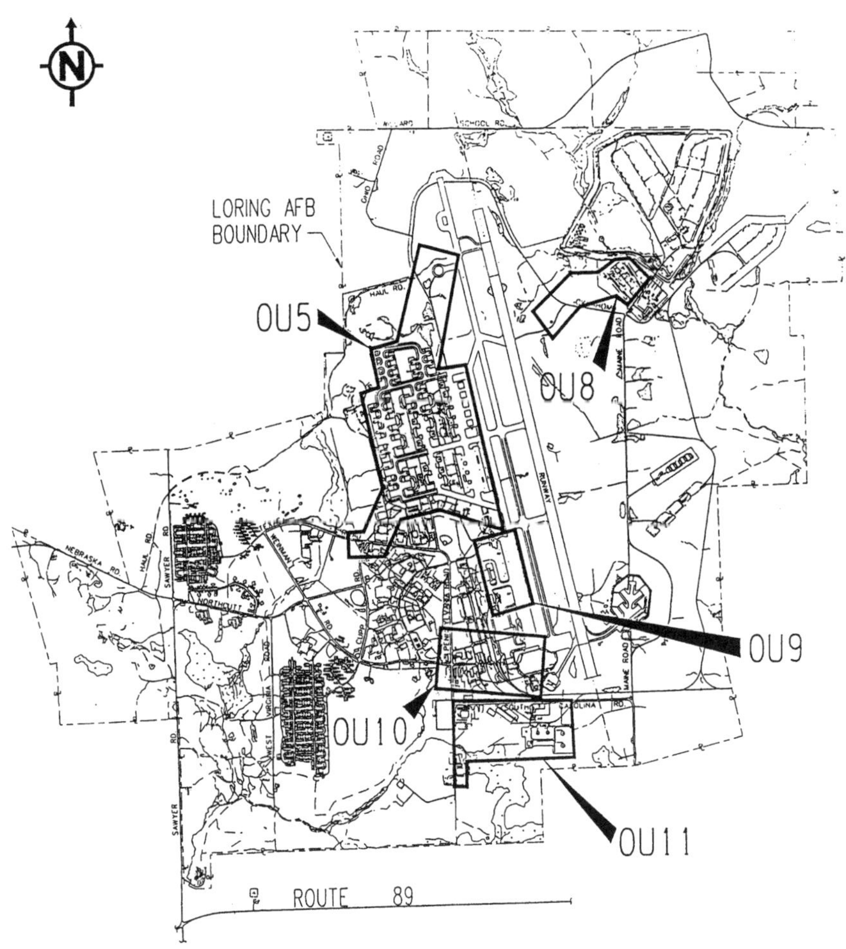

Figure 3.   **Biovent system layout** (former jet engine test cell).

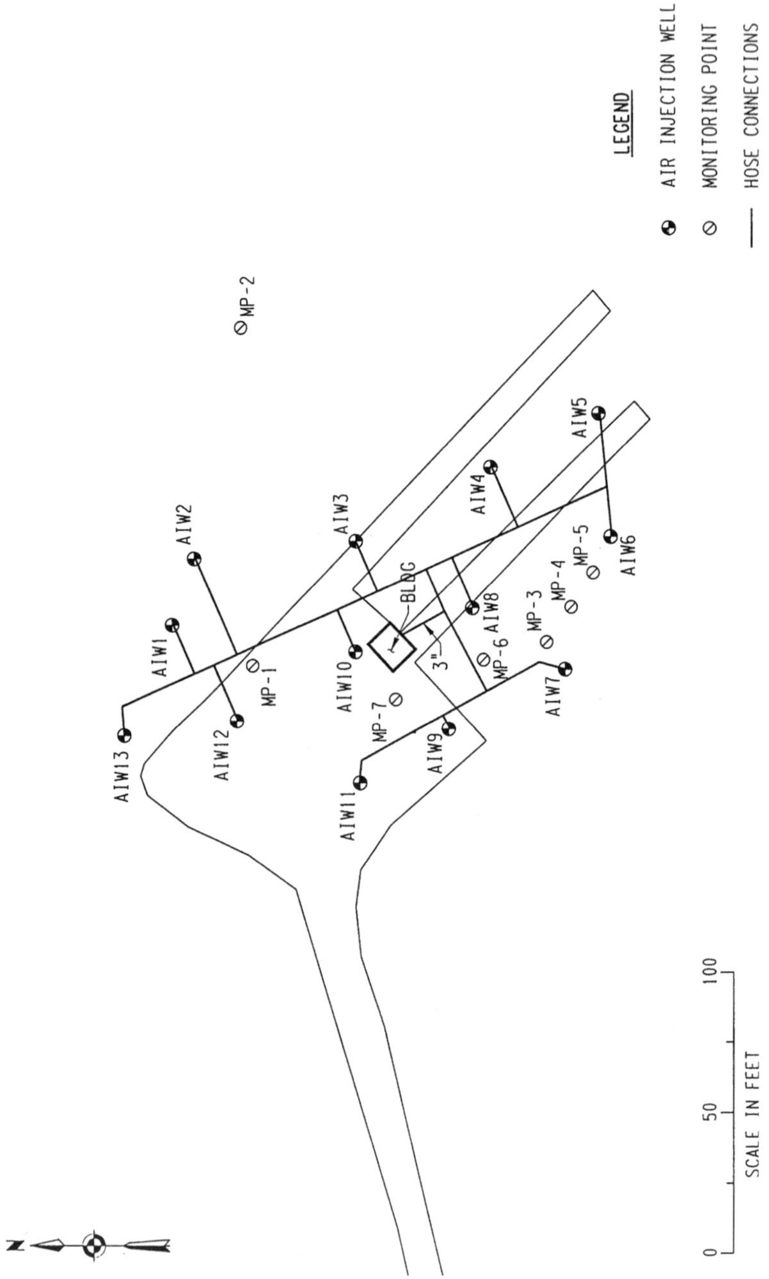

Figure 4. **Biovent system layout** (fire training area).

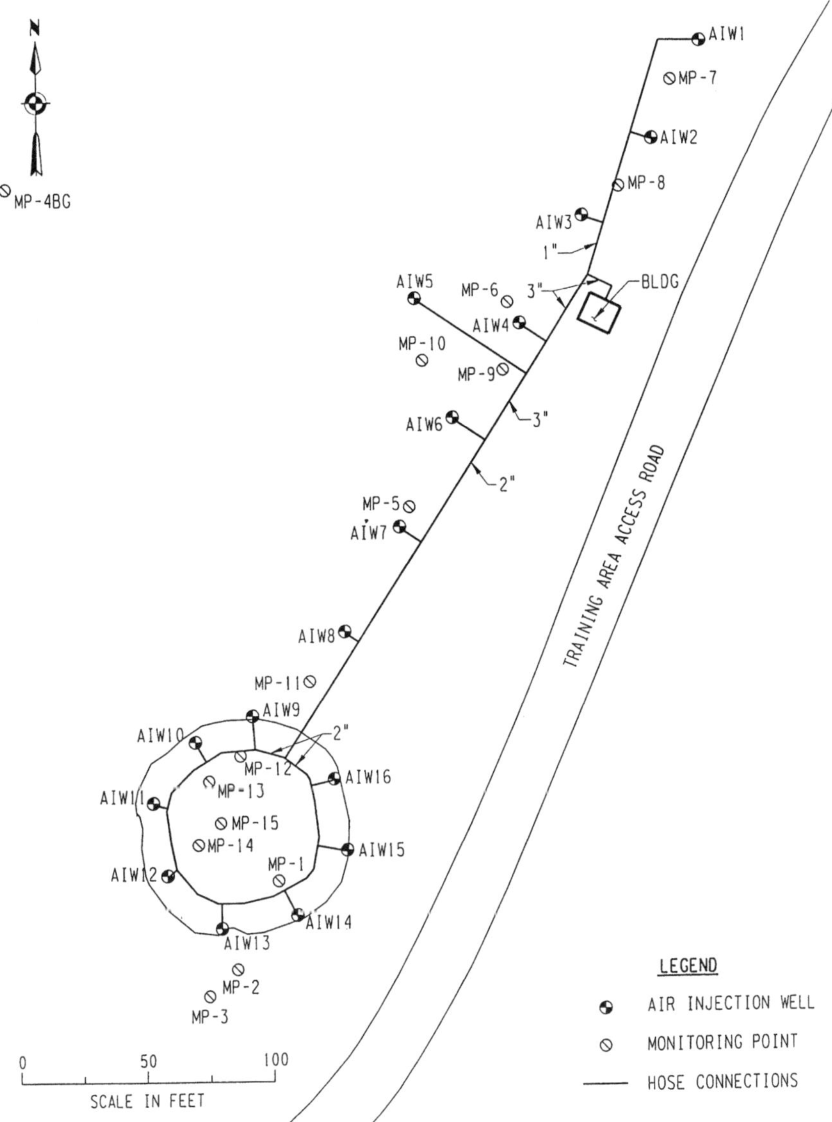

Figure 5. **Biovent system layout** (power plant drainage pipe).

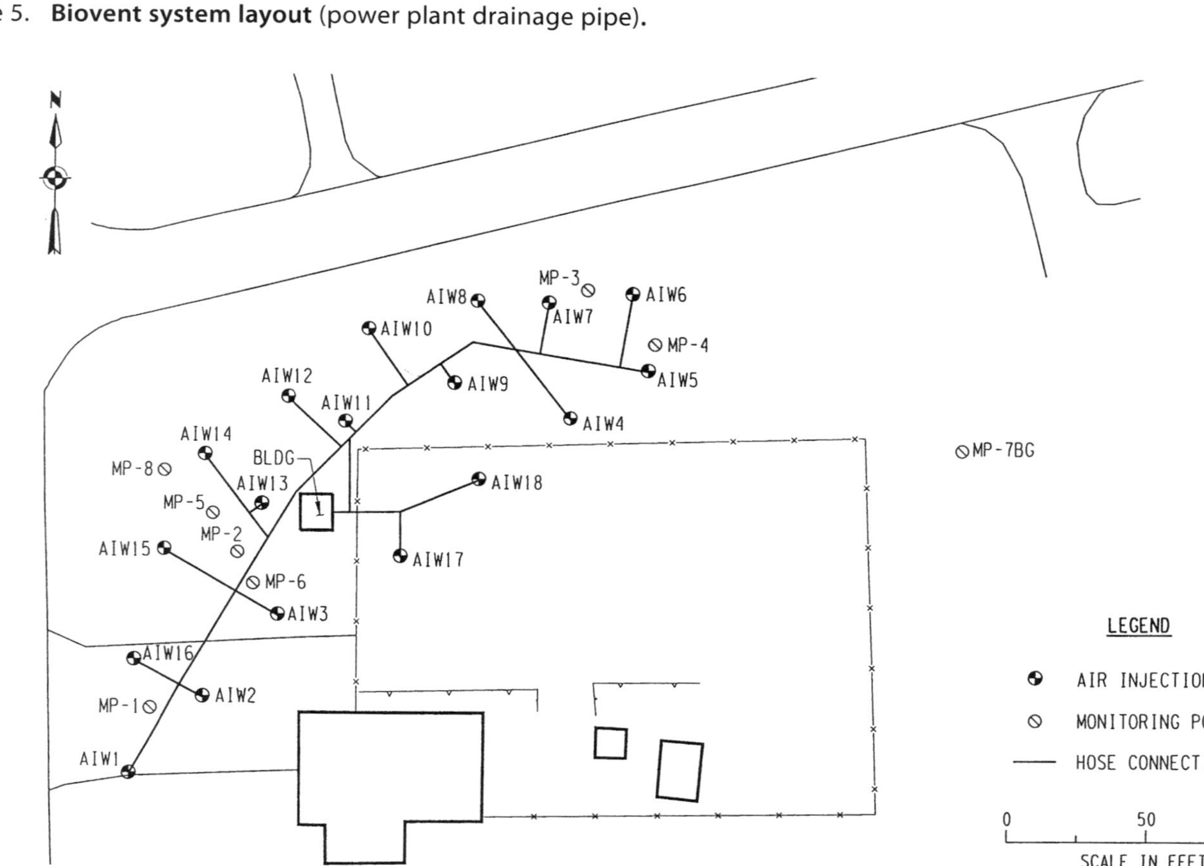

Figure 6. **Biovent system layout** (vehicle maintenance building).

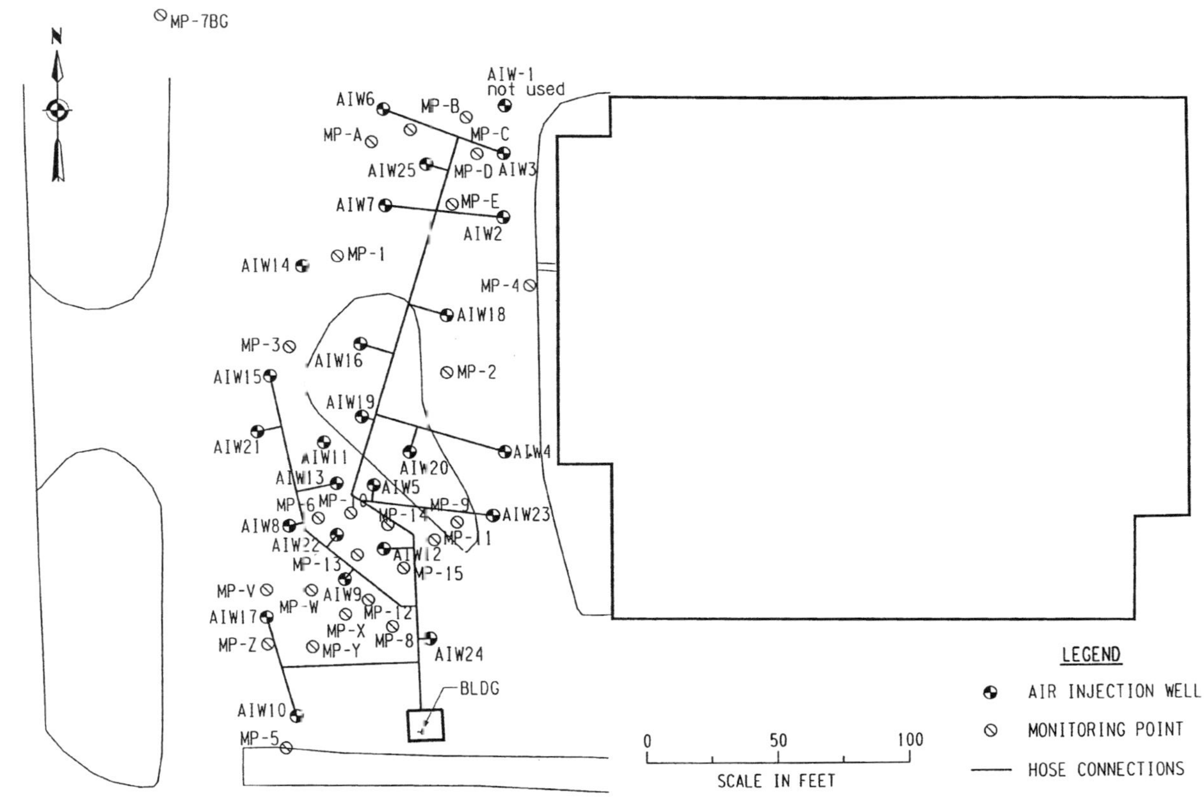

Law Environmental, Inc. 1993; ABB Environmental Services, Inc. (ABB-ES) 1991–1988; R.F. Weston 1985; and CH2M Hill 1983.

- RIs for OUs 8, 9, 10, and 11 by ABB-ES summarized in 1994 and 1995. ABB-ES also drafted an initial screening of alternatives in 1994 and a feasibility study (FS) for OUs 5 and 11 in 1995.
- Pilot-scale bioventing treatability study report for the Base Exchange Service Station (BXSS) area by Earth Tech in 1995.
- DAR, design specifications, and EE/CA reports (90% draft and 100% final) by URS Consultants, Inc. in 1995.

Detailed descriptions of site activities and the results of these previous studies are provided in the referenced documents for each respective OU. Site-specific removal areas and proposed removal actions are described in Sections 5.0 through 11.0 of the EE/CAs (URS, 1995b; URS, 1995h).

All work covered under this removal action was performed in accordance with the appropriate work plans submitted to the AFCEE, AFBCA, MDEP, and USEPA.

Appendix A of the EE/CA shows how site-specific cleanup criteria were developed. Applicable or relevant and appropriate requirements are detailed in the RI, FS, and EE/CA documents by URS and others.

CERCLA activities such as those being proposed for OUs 5, 8, 9, 10, and 11 must comply with all substantive environmental requirements and regulations. The substantive requirements of all federal, state, and local air, water, solid waste, toxic substance, wetland, endangered species, and similar regulations have been evaluated for applicability in the EE/CA, RI, and FS for Loring AFB, and applicable requirements were incorporated into the response action.

CERCLA Section 121 (e) provides an exemption from obtaining any permits for response activities conducted on-site. The DAR and EE/CA documents specify the applicable regulatory requirements and assumptions made with regard to substantive permit requirements and other specific environmental approvals (URS, 1995). For example, no air permits or specific air pollution control devices are required, due to the low anticipated volumes of volatile organic compounds and dust that might be generated. No special wastewater permits are required other than coordination and approval to discharge to the base wastewater treatment plant.

A wetlands mitigation plan has been issued for the base. Wetland delineation maps (Rev. 8, dated 6/27/95) were consulted from the OU 13 Mitigation Process Plan, Wetlands Management Program (HAZWRAP, 1995). No disturbance of wetlands occurred in the installation of the bioventing systems.

CHRONOLOGY OF EVENTS

This section summarizes the major events associated with the bioventing removal action at OUs 5, 8, 9, 10, and 11. The major events are listed below.

March 24, 1995	Public comment period on OUs 5, 9, 10, and 11 EE/CA begins
April 24, 1995	Public comment period on OUs 5, 9, 10, and 11 EE/CA ends
May 24, 1995	Action memorandum signed for OUs 5, 9, 10, and 11 EE/CA
September 5, 1995	Final bioventing work plan issued
September 11, 1995	Public comment period on OU 8 EE/CA begins
September 19, 1995	Drilling for bioventing well installation started
October 11, 1995	Public comment period on OU 8 EE/CA ends
December 6, 1995	Action memorandum signed for OU 8 EE/CA
December 7, 1995	Drilling for bioventing well installation completed
December 11, 1995	All biovent buildings and equipment installation completed
December 16, 1995	1st bioventing system started up (PPDP) at OU 9
December 19, 1995	2nd bioventing system started up (FTA) at OU 8
January 5, 1996	3rd bioventing system started up (FJETC) at OU 5
January 9, 1996	4th bioventing system started up (VMB) at OU 11
January, 18, 1996	Official "ribbon cutting" for bioventing installations
January 31, 1996	Bioventing systems 1–4 inspected and accepted by Air Force
February 1, 1996	Operation and maintenance period for bioventing begins

Table 2 shows the bioventing equipment installation schedule.

Table 2.   Bioventing Equipment Installation Schedule

Site	Building installed	Equipment installed	Utilities installed	Startup
FJETC	11-29-95	12-5-95	12-7-95	1-5-96
VMB	11-31-95	12-6-95	11-21-95	1-9-96
PPDP	11-31-95	12-6-95	11-21-95	12-16-95
FTA	12-4-95	12-8-95	11-30-95	12-19-95
ES	(Installation planned in September of 1996)			

PERFORMANCE STANDARDS AND CONSTRUCTION QUALITY CONTROL

This section discusses the performance standards and construction QC of the bioventing systems.

Performance Standards

The removal action objectives for the bioventing areas are described in detail in the EE/CA (URS, 1995b) and the draft initial screening of alternatives (ABB, 1994). Site-specific target compounds and remediation goals have been selected to achieve the overall goals of protecting human health and the environment.

As outlined in the DAR, a preliminary remedial goal of 870 mg/kg of total petroleum hydrocarbons (TPH) has been determined. This goal will be used for any soil confirmation sampling. The design basis for the extent of the removal actions was based on a more conservative 500 mg/kg goal to ensure all source areas are adequately removed. The level of 870 mg/kg will be used in conjunction with compound-specific goals for determining when the removal action objectives have been met. These goals are included in the investigation and design documents previously reviewed and approved by MEDEP and USEPA for this project.

The PRGs, established using the Loring Air Force Base Risk Assessment Methodology (HAZWRAP, 1994), are presented in the final DAR (URS, 1995a). The PRGs for each OU are chemical-specific concentrations that correspond to a level of risk that is within acceptable limits (i.e., 10^{-6} excess cancer risk for carcinogenic effects and a hazard index of one [1] for noncarcinogenic effects). The PRGs were developed based on specific scenarios to estimate human health risks, certain anticipated plant receptors, and ecological receptors, and to protect from contaminants leaching into groundwater. A detailed discussion on the development of the PRGs is presented in the appendixes of the EE/CAs (URS, 1995b; URS, 1995h). The OU-specific PRGs are also provided in the design drawings prepared by URS Consultants, Inc. (URS) and the removal action work plan prepared by Bechtel.

Air Permeability and Startup Testing

The air permeability tests were reviewed by the designer and approved by AFCEE prior to completing system installation. Some design changes were made by the designer, and installation followed these recommendations (see section titled "Injection Wells and Monitoring Points," below). Subsequent system checks during startup were all performed based on the design specifications, work plan, or startup plans submitted to the Air Force for review and approval. Checklists were used to monitor the completeness of the test programs. Final inspections of the installed equipment included a review of these checklists prior to approval.

Other testing of the bioventing equipment was done at the fabrication shop prior to delivery. These test results were submitted to AFCEE and URS for review before installation and startup.

Final inspection of the bioventing installations was performed and recorded on prefinal check lists, with a punch list included for items to be resolved.

Operation and Maintenance

Additional monitoring and testing of the bioventing systems will be conducted during the operation and maintenance (O&M) phase of the removal action. Regular reporting, monthly and semiannually, and evaluation of results will be included in this effort. Respiration test report results will be submitted as completed, twice a year. Quality control (QC) details are more completely described in the bioventing O&M plan.

Confirmation Sampling

Confirmation sampling has not yet been done for any of the bioventing areas. This sampling is anticipated to occur within three years. These samples will be taken in accordance with the bioventing work plan (Bechtel, 1995) when respiration tests indicate that this step is justified. Confirmation sampling at each site will commence when the biodegradation rate measured within the treatment zone is within 5% of the background biodegradation rate for two consecutive tests. Biodegradation rates are determined by respiration testing (URS Technical Specification 02020). The number of samples collected and their locations at OUs 5, 8, 9, 10, and 11 will be in accordance the DAR and preliminary design specification (URS, 1995a).

The bioventing work plan provides the rationale for the number of samples established for each OU. In summary, the number of samples were determined using the statistical evaluation method discussed in Volume 1 of the USEPA's guidance document, Methods for Evaluating the Attainment of Cleanup Standards (USEPA, 1989). The hot spot methodology defined the actual sampling patterns and the grid spacing required to detect hot spots of varying sizes with differing levels of confidence. The criteria used to develop the sampling scheme represent a 90% confidence level for the detection of a hot spot.

CONSTRUCTION QUALITY CONTROL

Construction work associated with the removal actions at OUs 5, 8, 9, 10, and 11 was executed in strict adherence to the AFCEE-approved construction QC plan and the project-specific construction QC plan. The QC process was implemented from the start of each definable activity. The implementation of a specified work activity observed a sequence of steps to be followed and tasks to be completed before starting new ones and before completing the overall work. This construction QC process involved a three-phase control approach that included the preparatory inspection phase, initial inspection phase, and the follow-up inspection phase.

Bechtel planned and executed the preparatory inspection phase, which included a discussion of construction activities that would be part of and would influence the actual construction work. This phase began with a meeting relative to the specific definable feature of work that involved the drilling subcontractor (Maine Test Boring), the site engineer, the construction supervisor, the safety and health representative, and the on-site geologist. Representatives of AFCEE, AFBCA, and MDEP also attended.

Items discussed at the preparatory phase meetings included:

- Specific construction approaches such as well depths, equipment to be used, duration of work activities, and in-situ testing to be performed.
- Safety practices for worker protection including air monitoring, selection of personal protective equipment, and protection of the worker population and unintentional intruders from site hazards.
- Lines of authority and responsibilities within Bechtel and between Bechtel and its subcontractor.
- The importance of having the entire construction team aware of changing conditions at the site.

The initial phase of the control process began with a site preconstruction and safety and health review. The review emphasized all the critical concerns identified and discussed in the preparatory phase, including the steps to be implemented during actual construction. Agreements were reached on specific site issues such as the exact locations to be drilled, underground utilities and interferences, actions to take due to changing site conditions, extent of the scope of the defined work, and safety and health requirements for the worker population and occasional visitors.

During the follow-up phase, Bechtel provided oversight and inspection of the field work, and kept an on-site QC representative to inspect and record daily all the activities related to the removal action. Additionally, the safety and health representative performed real-time monitoring of air emissions and oversight of overall safety conditions at each construction site, assigning and implementing the appropriate measures and procedures to protect the worker population and the environment.

CONSTRUCTION ACTIVITIES

The construction activities associated with the installation of the bioventing systems at OUs 5, 8, 9, 10, and 11 are outlined in the following section.

REMOVAL ACTION CONTRACT

The construction activities described in this report were performed under a removal action contract awarded to Bechtel with AFCEE as the contracting agency. In compliance with the work plan, all work performed under this delivery order was overseen and inspected by Bechtel. Maine Test Boring of Caribou, Maine, served as the drilling subcontractor under Bechtel and performed all activities associated with installation of air injection wells, monitoring points, and related tasks. Environmental Instruments, Inc. of Oakdale, California, fabricated, assembled, tested, and provided delivery and installation support of the skid-mounted bioventing equipment. Local area vendors supplied the bioventing buildings, power, heating and ventilation, and connecting pipe and wellhead fittings used at each site. Bechtel performed all air permeability testing, equipment installation, and startup activities.

Under this structure, Bechtel was the contractor to AFCEE, and the various subcontractors reported directly to Bechtel's construction superintendent. Bechtel was under the authority of the AFCEE field engineer, who was located on-site. URS Consultants, Inc. was the design engineer and approved design and field changes through AFCEE as needed.

SITE ACTIVITIES

The majority of the site activities were related to the drilling of wells, installation of monitoring points, and installation of buildings and equipment. Locally hired Bechtel laborers assembled and tested all of the wellhead piping and fixtures as well as the assembly and installation of the connecting air piping. The buildings were fabricated using a local small Maine business and were delivered fully assembled to the site.

Injection Wells and Monitoring Points

Air injection wells and monitoring points were installed as the first step in construction of the bioventing system. A drilling subcontractor was hired to install the wells and monitoring points. Wellhead completions were assembled and installed by Bechtel field staff. A Bechtel on-site geologist supervised the work of the drilling subcontractor. The on-site geologist recorded the drilling activities and subsurface information in a logbook. There were a large number of changes and additions. Table 3 shows where injection wells and monitoring points were installed.

Table 3. **Locations of Injection Well and Monitoring Point Installation**

OU	Site	Number of Injection Wells		Number of Monitoring Points	
		Design	Installed	Design	Installed
OU 5	FJETC	12	13	7	8
OU 8	FTA (option only)	16	16	7	37
OU 9	PPDP	18	18	15	21
OU 10	ES	7	7	9	9
OU 11	VMB	18	25	11	32

Initial plans were for a single in-situ air permeability test at each site. One well and multilevel monitoring points at three locations were to be placed for each permeability test. The initial results led to multiple air permeability tests and some significant changes in the test well locations in order to verify air permeability. Groundwater was found at shallower than expected depths in some areas and air permeability was not as uniform as previously assumed in the design. (ES was previously evaluated as being too small an area to warrant pilot testing. Currently a proposed in-situ test is being considered by the designer for this area.) Once the air permeability design basis was confirmed, and design changes approved by AFCEE, the remaining wells and monitoring points were installed.

The air injection wells were installed as specified by the design engineer (URS, 1995c). Each well was completed, packed with sand around the screened intervals, sealed with a bentonite plug, and grouted as specified. Above-grade well completions were installed using the design specification as a guideline for construction.

The monitoring points were installed as specified in the design specifications. Below-grade point completions were also installed.

Process Equipment

The DAR discusses the background, design basis, technical justification, and pilot testing that provides the rationale for the bioventing process equipment specified. The bioventing system processing equipment consists of a positive displacement air blower, air dryer, air filter, and other miscellaneous valves and fittings. As outlined in the DAR and pilot-testing report from the BXSS area, the bioventing system is designed to operate at an air injection rate of 2 to 10 standard ft^3/min per well. The air permeability assumptions are detailed in the DAR. Air permeability tests at each site confirmed or revised these assumptions.

The bioventing process equipment was purchased, delivered, received, checked, and assembled at the site as planned. Skid-mounted equipment assemblies were used to reduce on-site construction costs. The control systems were modified to include a remote terminal unit to allow remote monitoring of basic elements. These elements include pressure, level, and power alarms, as well as total operating hours and status of each motor. Power and utilities were provided as shown on the utility plan drawings. Equipment buildings were provided, with some minor modifications to the specifications, as approved by AFCEE.

Piping

Once the air injection wells, monitoring points, and process equipment were installed, the connecting piping was constructed. The piping design air flow rate is 30 ft/second.

Flexible rubber air hose was used to manifold air to the injection wells from the blower assemblies. Valving has been installed to allow regulation of the amount of air delivered to each well. All piping was installed above grade at all sites.

STARTUP TESTING

The bioventing equipment was installed and started up with a sequence approved by AFCEE. Mechanical and electrical completeness checks were done and motors were started under a closely monitored process. When the equipment was up and running, the air piping was connected to the wells. A 30-day testing period was then initiated to demonstrate the operating ability of the equipment. Inspections were completed and approved by AFCEE, leading to acceptance of the equipment on January 31, 1996.

In-situ respiration soil gas monitoring was performed as described in the DAR and design specifications. Due to freezing conditions during startup, no additional respiration testing or soil gas monitoring has been performed, outside of the initial conditions reported during the permeability testing. These tests will be done as soon as conditions permit. The results of the soil gas testing will also be reported in the semiannual O&M report with a simple summary of the data.

Instrumentation such as the flame ionization detector hydrocarbon analyzer will be calibrated and used in the field for startup and continued bioventing system operations. Quality assurance (QA)/QC requirements will be followed for instrument calibration and results reporting.

Lessons Learned

Challenges encountered during the removal actions associated with this report were caused by the differences in field conditions and the subsurface conditions found during testing, and resulted in changes in the number of monitoring wells installed. Direct communication between all parties involved (AFCEE, AFBCA, Bechtel, URS, and others) allowed most of these changes to be resolved quickly and efficiently.

Other lessons learned include the following:

1. Subsurface conditions were greatly different in the VMB and other areas than anticipated. The two most significant were groundwater conditions and numerous utility conditions, which interfered with air flow paths. Better up front information about groundwater and utility corridors should have been developed prior to design of this well system, and should be done at all other locations where biovent or SVE systems are to be installed.

2. Prefabricated skid-mounted bioventing equipment reduced the cost and effort required to install the systems in the field. Preconstructed buildings also reduced the overall cost.

3. The total cost of the O&M tasks required to maintain and monitor the bioventing systems was not completely recognized. Design changes could have been made to better automate the system. Remote terminal unit controls could have included more sensors that would have reduced the manual labor at the site used for monitoring and routine measurements. Available sensors are being evaluated for possible addition to presently installed systems and future systems.

4. The subsurface completions for monitoring wells froze and made access impossible during startup. Subsurface access became blocks of ice. Drawing air samples through the frozen ground caused the moisture in the air to freeze, plugging the sample tubing below the ground. Surface completions would have worked better, preventing access freeze-ups. Additional evaluation of monitoring requirements, methods, and winter operations is needed. (The monitoring points are currently frozen and have not been sampled.) In-situ gas detection equipment needs to be researched.

5. Adjustment of flow at individual wells is extremely important for the correct operation of the bioventing systems. The designed method of measuring flow at wellheads does not work at low flow rates. A replacement is being considered by the designer at URS.

6. Well seals (pure bentonite) are found to be compromised during extreme freeze-thaw conditions, allowing air to escape at the well riser-seal interface, through the seal, and/or at the seal-borehole wall interface. A well seal test using different polymer-modified bentonite seals has been completed at the base, and a technical report with results and recommendations will be issued.

7. The biovent systems become very noisy when greater than 50% of capacity is blown through the bypass system. Silencers are recommended for the systems where noise could be considered a problem. For presently installed systems, this includes VMB and PPDP. Silencers have been ordered for these systems.

8. The success of the QC program indicates that an effective program using a three-phase control approach for each definable feature of the project should be continued. These three phases include:

   - The preparatory phase that outlines actions in advance of construction.
   - The initial phase that outlines actions at the start of operations.
   - The follow-up phase that outlines the actions involving inspections and testing.

OPERATION AND MAINTENANCE

The O&M of the bioventing removal action described in this report is described in detail in the bioventing O&M plan (Bechtel, 1995a). Weekly, monthly, and semi-annual activities are described that involve checking operations, routine maintenance, periodic monitoring, and testing programs. Regular system evaluation and reporting of results are also planned.

A full-time staff, including a site O&M supervisor, is engaged in bioventing O&M activities. These activities have been contracted for the next 24 months, and will follow the approved work plan submitted to the Air Force and regulatory agencies for review.

REFERENCES

ABB-ES (ABB Environmental Services, Inc.). 1994. Final Remedial Investigation Report, North Flightline Operable Unit (OU 10). Volumes I through IX. Installation Restoration Program. Loring Air Force Base. Prepared for HAZWRAP. Portland, Maine. August.

ABB-ES, 1995. Final Remedial Investigation Report, OUs 8, 9, 10, and 11. Volumes I through IX. Installation Restoration Program. Loring Air Force Base. Prepared for HAZWRAP. Portland, Maine. September.

Bechtel (Bechtel Environmental, Inc.). 1995. Work Plan for Bioventing at OUs 5, 8, 9, 10, and 11, Final. September.

Bechtel, 1996a. Operation and Maintenance Plan for Bioventing at OUs 5, 8, 9, 10, and 11, Final. January.

Bechtel, 1996b. Removal Action Report for Excavations in OUs 5, 8, 9, 10, and 11, Final.

Earth Technology and HAZWRAP, 1995. Base Exchange Service Station Pilot-Scale Bioventing Treatability Study Report.

HAZWRAP (Hazardous Waste Remedial Actions Program), 1994. Loring Air Force Base Risk Assessment Methodology.

HAZWRAP, 1995. OU 13 Mitigation Process Plan, Wetland Management Program. ABB-ES, Woodlot Alternatives, Inc. June.

URS (URS Consultants, Inc.), 1995a. Final Design Analysis Report, OUs 5, 9, 10, and 11. (Prefinal, March.) Final, July.

URS, 1995b. Final Engineering Evaluation/Cost Analysis Operable Units 5, 9, 10, and 11. (Draft, March.) Final, June.

URS, 1995c. Bioventing and Excavation Specifications for Former Jet Engine Test Cell, Vehicle Maintenance Building, Power Plant Drainage Pipe and Entomology Shop. (Prefinal. March). Final, June.

URS, 1995 d, e, f.
Design Drawings, Removal Action - Power Plant Drainage Pipe; Sheet Nos. 3, C-1A, and C-1B.
Design Drawings, Removal Action-Entomology Shop; Sheet Nos. 3–4, and C-1A–C-1D.   Design Drawings, Removal Action - Vehicle Maintenance Building; Sheet Nos. 3, C-1A, and C-1B, and C-2A. June.

URS, 1995g. Preliminary design drawings for OU 8, Fire Training Area. August.

URS, 1995h. Final Engineering Evaluation/Cost Analysis Operable Unit 8. September.

USEPA (U.S. Environmental Protection Agency), 1989. Methods for Evaluating the Attainment of Cleanup Standards, Volume I.

USEPA, 1994. Superfund Removal Procedures Removal Response Reporting: POLREPs and OSC Reports (p. 1-1).

USEPA, 1995. Close Out Procedures for National Priorities List Sites. EPA/540/R-95/062. Washington, D.C. August.

CHAPTER 29

Electrokinetic-Enhanced Bioremediation in Clayey Soils

Sabanayagam Thevanayagam, Saddanathapillai Nesarajah, and **Vasoki Thevanayagam**, Department of Civil Engineering, State University of New York at Buffalo, New York

INTRODUCTION

There are approximately two million tanks storing gasoline in the U.S. leaking (Hutchins et al., 1991a). The movement, transformation, and fate of these contaminants in the environment are of particular interest because many of these contaminants are suspected carcinogens or mutagens and are toxic to humans. Several techniques are still being developed for the cleanup of these contaminants. Bioremediation is one technique that shows great promise and is fast becoming one of the most important forms of remediation since the mid-1980s.

During the bioremediation process, microorganisms consume organic contaminants as an energy source and oxidize them under favorable pH (5.5 to 8.5) and temperature (10-30°C) (Alexander and Scow, 1989; Battermann, 1986). Biological oxidation of organic substances requires the transfer of electrons from the organic molecules to some acceptor molecules. These acceptors are called terminal electron acceptors (TEAs). Typically oxygen, nitrate, sulfate, and some inorganic compounds (Mn^{2+}, Fe^{3+}, and CO_2) serve as electron acceptors. In addition to energy sources, microorganisms also require nutrients for growth and metabolic activities. It is essential to maintain a minimum nutrient level to support desired microbe growth. Nitrogen and phosphorus provide nutrients necessary for the growth and activities of the microorganisms.

The type and population of microorganisms naturally present in the subsurface are usually sufficient to initiate the biodegradation process (Ward, 1995; Hutchins et al., 1991b). However, the TEAs, which critically affect the growth of microorganisms and their activities, are not sufficiently available in the subsurface. Further, the nutrients such as nitrogen and phosphorus, which are also critical for the biodegradation process, are also not naturally available in sufficient quantity. Therefore, an external supply of TEAs (oxygen for aerobic, and nitrate, sulfate, etc., for anaerobic) and the above nutrients to the subsurface contaminant locations becomes essential to sustain effective biodegradation. Although it is possible to introduce such ingredients to contaminant zones via traditional pumping techniques in the high permeable sandy soils, in-situ bioremediation is less successful in low permeability soils due to the difficulties in uniform delivery of the above treatment reagents.

This Chapter presents the results of a feasibility study of injecting the above ingredients using electrokinetics to enhance bioremediation in clayey soils. The rationale for this strategy and the underlying principles are explained. This study specifically focused on the experimental evaluation of the injection of nitrate using electrokinetics to the contaminated zones in a soil mixed with naphthalene. Nitrate thus injected is used as a TEA as well as nutrient for the bioremediation process. Bench-scale biodegradation experimental data are presented.

ELECTROKINETICS-AIDED BIOREMEDIATION

Electrokinetics technology has been used in geotechnical engineering for many years for various ground improvement applications (Casagrande, 1983). Recently its use has been expanded into hazardous waste site remediation applications (Mitchell and Yeung, 1990; Acar, et al., 1993a); barriers for advective diffusive transport of ionic contaminants through compacted clay liners (Mitchell and Yeung, 1990); diversion schemes for waste plumes (Mitchell and Yeung, 1990); in-situ electrolysis and removal of metal contaminants (Acar et al., 1992; 1994; Lageman, 1993; Hicks and Tondorf, 1994); and to move soluble contaminants from the contaminated zone to a treatment zone (Shapiro and Probstein, 1993).

The flow of ions in the soil due to a d.c.-voltage gradient applied across a clayey soil would be influenced by (i) electromigration, (ii) electroosmosis, and (iii) diffusion. Transport of negatively charged ions by electromigration will be toward anode and vice versa for positively charged ions. Typically, transport of ions due to advection by electroosmosis will be toward the cathode. Transport due to diffusion could be either toward the anode or cathode, depending on the prevailing concentration gradient of the ion. The transport rate by diffusion is very small compared to the transport rates by electromigration and electroosmosis. Typically, the rate of transport by electromigration toward the anode is about one or more orders of magnitude higher than the transport rate toward the cathode by advection. Therefore a negative ion, if introduced at the cathode, could effectively travel toward the anode against the electroosmotic flow.

Except for oxygen (for aerobic process), the primary ingredients that are required for effective bioremediation are ionic molecules (nitrate, phosphate, sulfate, etc.) that are highly soluble in water. Nitrate is a negatively charged ion. If supplied at the cathode, nitrate could move from the cathode side to anode when a d.c. electric potential is applied across the soil. If one could reduce the potential for any chemical interaction of nitrate with other ionic substances in the soil, it would be available as nitrate within the soil mass. Hence, it could be used as a TEA as well as a nutrient for the bioremediation process. In a similar manner, one could inject other ingredients (phosphate, sulfate, etc.) as well. However, during the application of d.c. current, H^+ ions and OH^- ions are also produced near the anode and cathode, respectively. These ions would also migrate into the soil and alter the soil pH. Therefore, care must be taken to neutralize such pH during the electrokinetic injection process. Such pH changes could be controlled, by introducing $CaCO_3$ for example, at the anode to neutralize the H^+ ions. Figure 1 schematically shows a layout for field implementation of this technique.

EXPERIMENTAL EVALUATION

Injection of Nitrate

Clay soil specimens were prepared by mixing Georgia kaolin (Thiele Kaolin Company, Georgia) with water and allowing it to consolidate for seven days under the stress of 35 kPa. Following consolidation, the soil was placed in a 5-cm diameter and 27-cm long cylindrical tube with graphite electrodes at the ends. Nitrate solution at a concentration of 1200 ppm (mg/L) was placed in a tank and connected to the specimen via a tube at cathode (Figure 2). The concentration of the nitrate solution at the anode was maintained constant. A thin layer of calcium carbonate was placed at the anode to neutralize H^+ ions produced at the anode and hence control the potential for changes in the pH in the soil. A voltage gradient of 1 V/cm was applied across the soil for seven days.

Figure 1.   **Electrokinetic aided bioremediation - schematic.**

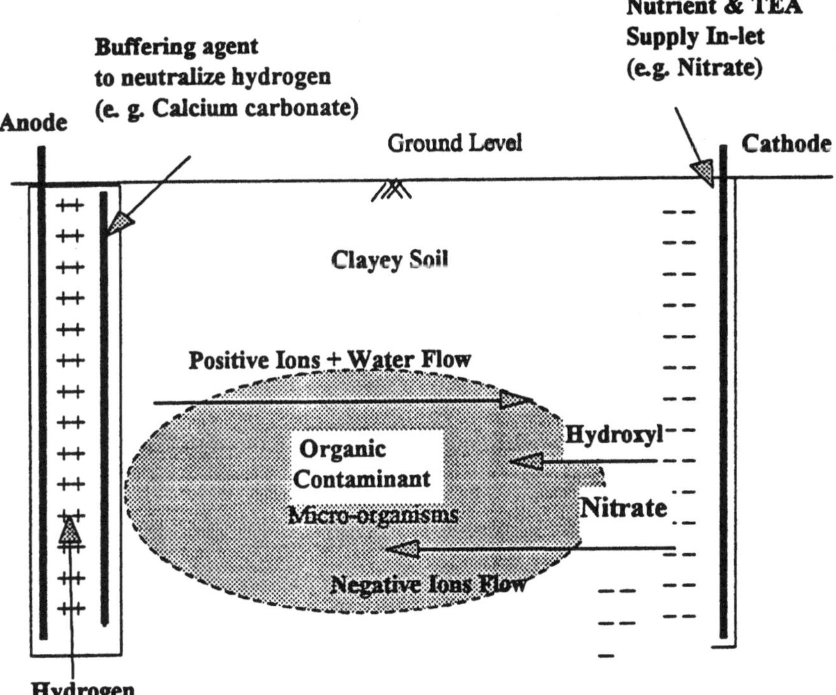

Figure 2.   Experimental setup for nitrate injection.

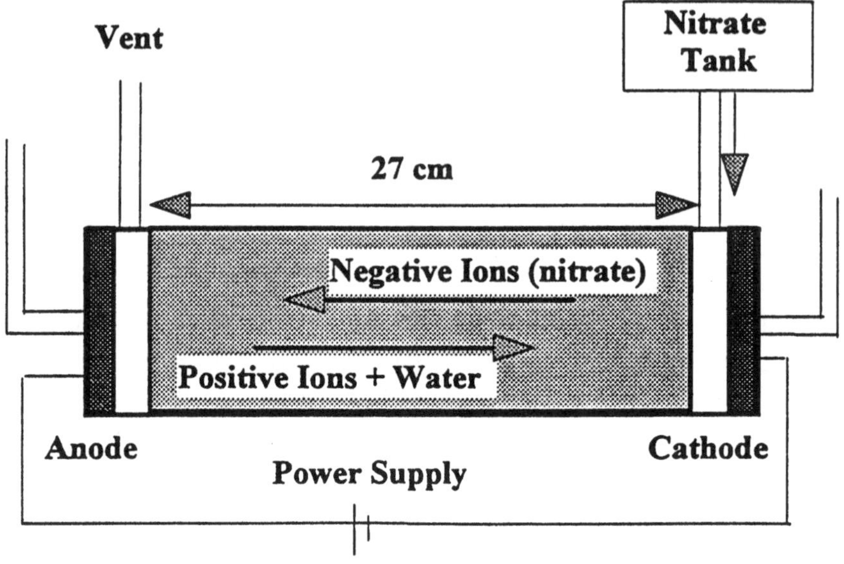

Figure 3.   Nitrate concentrations before and after electrokinetic treatment.

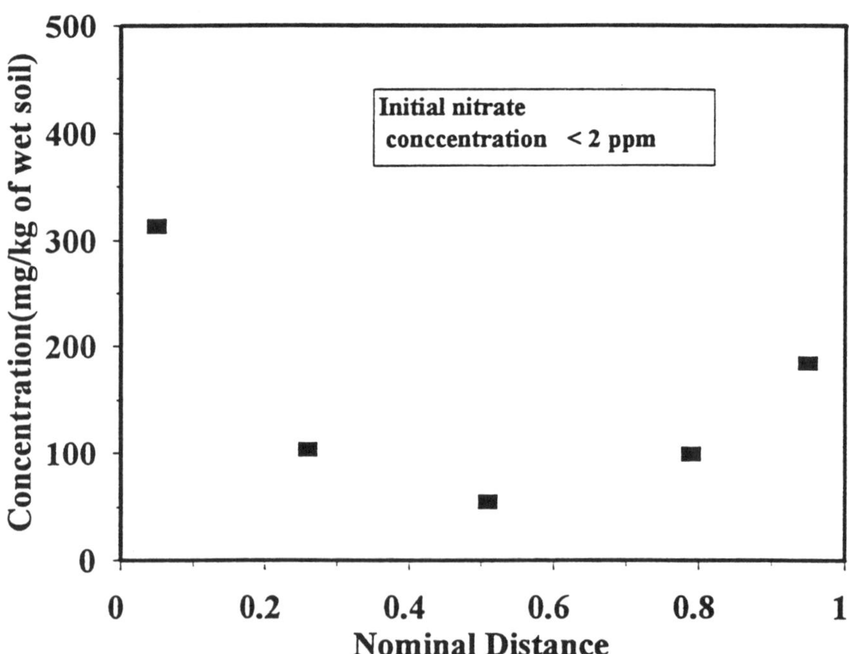

After seven days, soil samples were taken at different locations of the specimen for measurement of pH, water content, and concentration of nitrate. The in-situ pH was measured using a pH paper to the accuracy of 0.25. Ten grams of wet soil was taken from five different locations from the specimen and tested for nitrate concentration. To extract the nitrate, the soil samples were mixed with 100 mL deionized water and stirred thoroughly using A magnetic stirrer for 30 minutes. The supernatant was collected and tested for nitrate using nitrate electrode (Orion Model 93-07) (Thottan et al., 1994).

The nitrate concentration data before and after electrokinetic treatment are shown in Figure 3. The initial concentration of nitrate was about 2 ppm (mg/kg of wet soil). Following the electrokinetic treatment this concentration has increased to about 60 ppm in the middle portion of the soil column. Concentration close to the anode was 463 ppm. The change in the concentration of nitrate in the soil following electrokinetic treatment demonstrates the feasibility of injection of nitrate into fine-grained soils using electrokinetics. The high concentration near the anode compared to the values in the middle portion is due to accumulation of the nitrate near the anode. The pH in the specimen ranged from 6.5 to 8.5 for all specimens (Figure 4).

Biodegradation of Naphthalene

Following the success in injecting nitrate into clayey soils the study focused on feasibility evaluation of electrokinetic aided biodegradation. Table 1 presents a brief summary of the experimental program. In this study, naphthalene, one of the PAHs (polyaromatic hydrocarbons), was selected as the contaminant. A commercially

Figure 4.   **The pH profiles: before and after the bioremediation.**

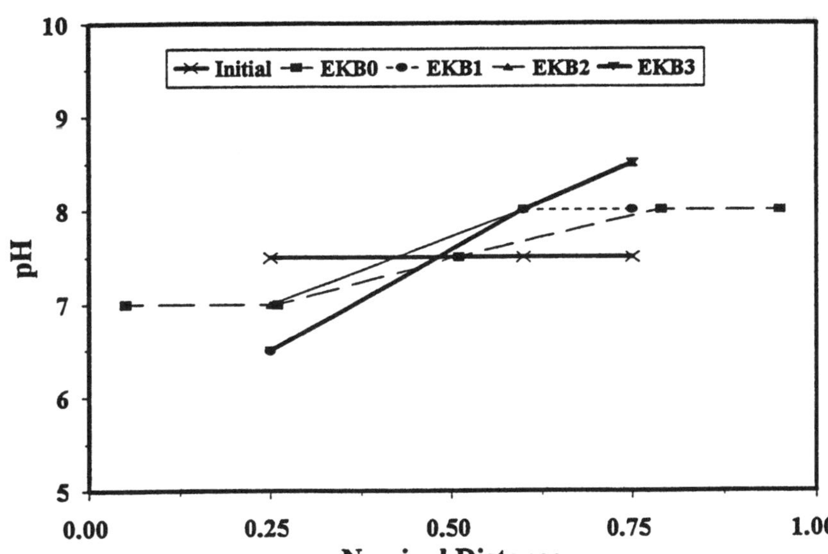

available top soil was mixed with kaolin at a ratio of 3:1 kaolin:top soil by dry weight. After mixing these two soils thoroughly, 300 mg of naphthalene powder was mixed with this soil mixture. Deionized water containing 100 mg/L of N in the form of NH_4^+ and 20 mg/L of P in the form of PO_4^{3-}, was added to this soil mix and blended thoroughly. The specimens were consolidated under a stress of 35 kPa for seven days. After consolidation, the contaminated soil was placed in the middle portion of a 27-cm length and 5-cm diameter cylindrical tube. The end portions of the tube were filled with a "clean" kaolin + top soil mix as shown in Figure 5. Three different specimens (EKB1 through EKB3) were prepared identically in this manner. The initial water content of each specimen ranged from 58 to 62%. The initial concentration of naphthalene in the contaminated soil was 200 ppm (mg/kg of wet soil). The initial nitrate concentration in the soil was 170 ppm. The specimens were subjected to 1V/cm of voltage gradient over a period of seven days with a constant supply of 2000 ppm (mg/L) of nitrate solution at the cathode. The nitrate concentration in the soil following electrokinetics was measured. The data are shown in Figure 6.

Following the electrokinetics treatment (nitrate injection), the specimens were sealed and kept in room temperature conditions in the same tube under anaerobic conditions. Soil samples were collected periodically from the contaminated soil (at the middle of the tube) and tested for nitrate, naphthalene, pH, and water content. For tests EKB1, EKB2, and EKB3 samples were collected after a duration of 35, 85, 122 days, respectively and tested. In the case of test EKB2 samples were taken for test a second time after a total duration of 140 days. The latter test data are summarized in Table 1 under the name EKB2A. The in-situ pH and nitrate concentration were measured as described earlier. For naphthalene, soil samples were

Figure 5. **Experimental setup for Tests EKB1-EKB3.**

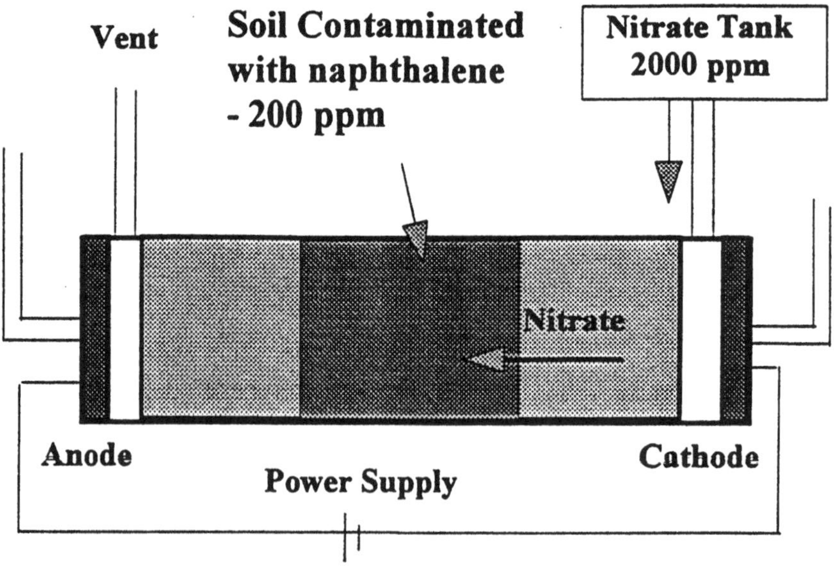

Table 1.   **Summary of Experiments**

Soil	Test		Before Bioremediation					After Bioremediation			
			Initial Conc. of	Conc. of Nitrate (mg/kg)		pH	EK duration	BR duration	Conc. of	Conc. of Nit.	pH
			Naph. (mg/kg)	Before Injection	After Injection		(days)	(days)	Naph. (mg/kg)	(mg/kg)	
Kaolin	Nitrate injection feasibility		0	2	60	5.5	7	N/A	N/A	N/A	6
Kaolin	Bioremed	EKB1	200	170	595	7.5	7	35	99	495	8
75% +	--iation	EKB2						85	99	455	
Top	feasibility	EKB3						122	70	396	
Soil 25%		EKB2A						140	49	351	

Note: N/A = Not applicable; EK = Electrokinetics; BR = Bioremediation

Figure 6.   **Final nitrate concentrations in the soil after electrokinetics.**

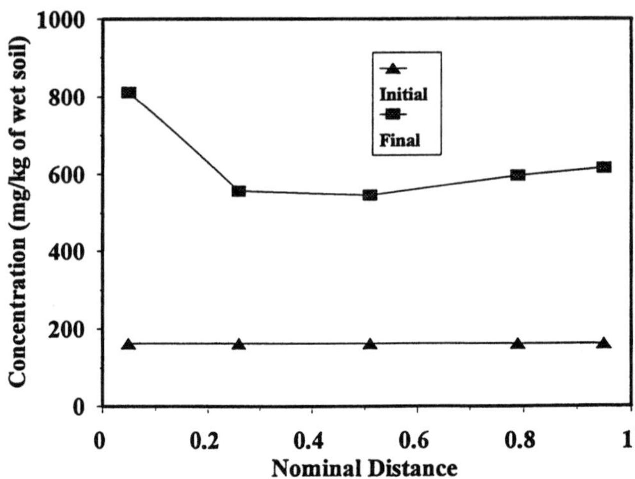

thoroughly mixed with methanol and were centrifuged. The supernatant was collected and tested for naphthalene using UV spectrophotometer.

Figure 7 shows the remaining naphthalene concentration in the soil at the end of each duration of bioremediation. Figure 8 presents the concentration of nitrate remaining in the soil. Figure 9 presents the data on the reduction (biodegraded) in the concentration of naphthalene versus the nitrate consumed. Naphthalene concentration has been reduced from 200 ppm (mg/kg of wet soil) to 50 ppm over 140 days, with a concurrent reduction in nitrate concentration from the initial value of about 595 ppm (mg/kg of wet soil) to 350 ppm over the same period. The consistent loss of nitrate with the loss of naphthalene in each test indicates that the loss of naphthalene is due to biodegradation under denitrifying conditions.

CONCLUSION

Feasibility of degrading organic contaminants using electrokinetic coupled bioremediation technique was studied. In this technique, ionic nutrients (e.g., nitrate) are introduced into the soil using electrokinetics. Bench-scale experiments on naphthalene contaminated kaolin soil show that the naphthalene concentration decreased from an initial concentration of 200 ppm to 50 ppm in less than 150 days, with a concurrent reduction in nitrate from an initial value of about 595 ppm to 350 ppm over the same period. These results show that the electrokinetic aided bioremediation technique can be a viable means for degradation of organic contaminants in clayey soils. However, further tests are needed to evaluate the effectiveness of this technique for different organic compounds.

Figure 7. **Remaining naphthalene concentrations vs. duration.**

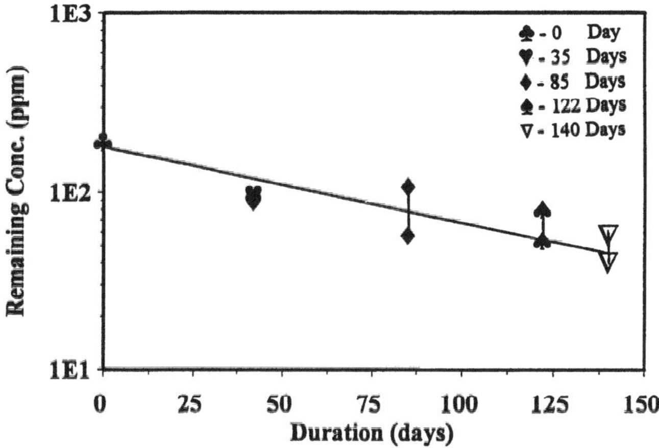

Figure 8. **Remaining nitrate concentration vs. duration.**

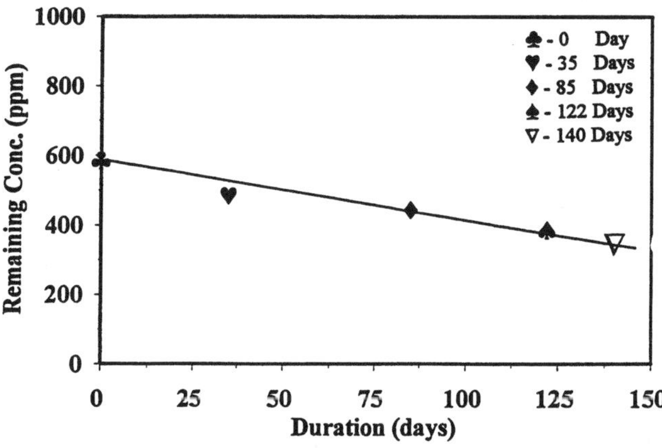

Figure 9. **Amount of naphthalene degraded vs. nitrate consumed.**

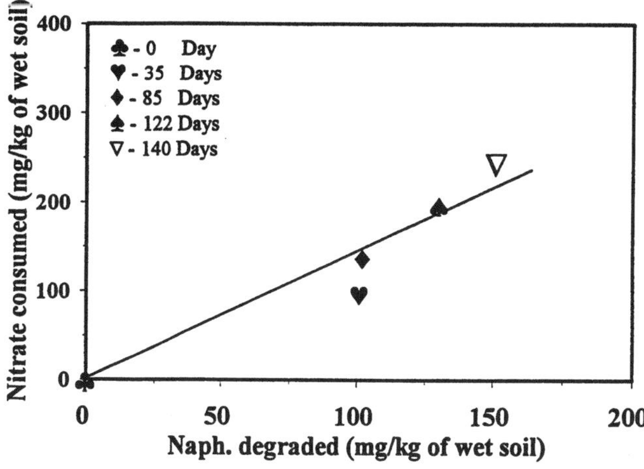

ACKNOWLEDGMENTS

The authors wish to thank Mr. T. Rishindran, former graduate student at Polytechnic University, who assisted in conducting laboratory tests reported herein.

REFERENCES

Acar, Y.B., Alshawabkeh, A.N., and Gale, R.J. (1993). "Fundementals of Extracting Species from Soils by Electrokinetics." Waste Management, 13, 141-151.

Acar, Y.B., Gale, R.J., Hamed, G.J., and Putnam, G. 1992. Acid/Base Distributions in Electrokinetic Soil Processing. *Transp. Res. Rec.*, 1288, 23-34.

Acar, Y.B., Hamed, J.T., Alshawabkeh, A.N., and Gale, R.J. 1994. Removal of Cadmium (II) from Saturated Kaolinite by the Application of Electrical Current. *Geotechnique* 44 (2), 239-254.

Alexander, M. and Scow, K.M., 1989. Kinetics of Biodegradation in Soil; In Reaction and Movement of Organic Chemicals in Soils, *Am. Soc. Agron.* Madison, WI. pp. 243-269.

Battermann, G. 1986. Decontamination of Polluted Aquifers by Biodegradation. In: *1985 International TNO Conference on Contaminated Soil*, pp. 711-722. (Assink, J.W. and Van den Brink, W.J., Eds.) Dordrecht, Nijhoff.

Casagrande, L. (1983). "Stabilization of Soils by means of Electroosmosis: State of Art." J. Boston Soc. Civ. Engrs., 69, 255-302.

Hicks, R.E. and Tondorf, S. 1994. Elecrorestoration of Metals Contaminated Soils. *Environ. Sci. Technol.* 28, 2203-2210.

Hutchins, S.R., Downs, W.C., Wilson, J.T, and Smith, G.B., Kovacs, D.A., Fine, D.D., Douglass, R.H., and Henddrix, D.J. 1991a. Effect of Nitrate Addition on Biorestoration of Fuel-Contaminated Aquifer: Field Demonstration. *Ground Water*, 29(4), 571-580.

Hutchins, S.R., Sewell, G.W., Kovacs, D.A, and Smith, G.A. 1991b. Biodegradation of Aromatic Hydrocarbons by Aquifer Microorganisms Under Denitrifying Conditions. *Environ. Sci. Technol.* 25, 68-76.

Lageman, R. 1993. Electroreclamation. *Environ. Sci. Technol.*, 27(13).

Mitchell, J.K. and Yeung, A.T. 1990. Electro-Kinetic Flow Barriers in Compacted Clay. *Transp. Res. Rec.*, 1288, 1-9.

Shapiro, A.P. and Probstein, R.F. 1993. Removal of Contaminants from Saturated Clay by Electroosmosis. *Environ. Sci. Technol.*, 27(2), 283-291.

Thottan, J., Adsett, J.F., Sibley, K.J., and MacLeod, C.M. 1994. Laboratory Evaluation of the Ion Selective Electrode for Use in an Automated Soil Nitrate Monitoring System. *Commun. Soil Sci. Plant Anal.*, 25 (17&18), 3025-3034

Ward, C.H. 1995. In: *Bioremediation: Innovative Site Remediation Technology*, by Anderson, W.C., Annapolis, MD, American Academy of Environmental Engineers.

CHAPTER 30

In Situ Bioremediation Using the UVB Technology: A Case Study

Susanne M. Borchert, Fayaz S. Lakhwala and **James G. Mueller**, SBP Technologies, Inc., Pensacola, Florida

Richard J. Desrosiers, SBP Technologies, Inc., White Plains, New York

INTRODUCTION AND SITE BACKGROUND

Under a Multi-Vendor Biotreatability Demonstration Program, SBP Technologies, Inc. (SBP) conducted a field demonstration of a microbiologically enhanced *in situ* groundwater and soil treatment technology at the Sweden-3 Chapman Superfund Site, Brockport, New York. This demonstration was performed under the United States Environmental Protection Agency's (USEPA's), Superfund Innovative Technology Evaluation (SITE) Program, with participation from the New York State Department of Environmental Conservation (NYSDEC) and the New York State Center for Hazardous Waste Management (NYSCHWM).

The Sweden-3 Chapman site is an inactive landfill used to dispose of construction/demolition debris and hazardous wastes between the years of 1970 and 1978. Site investigations conducted in 1985 by NYSDEC indicated that drums were buried throughout the landfill. These drums and on-site soils were sampled in 1987, indicating elevated levels of trichloroethylene (TCE), methylene chloride, tetrachloroethylene (PCE), and acetone, among other volatile organics (VOCs) and semivolatile organics (SVOCs). In 1992, a remedial investigation identified source areas and determined the extent of vertical and horizontal migration of contaminants at the site. Information and data generated during these investigations set the foundation for conducting the multi-vendor biotreatability demonstration.

DEMONSTRATION GOALS AND OBJECTIVES

The primary objective of the field study was to determine the effectiveness of the technology in reducing concentrations of eight target VOCs (acetone, MEK, MIBK, toluene, *cis*-1,2 DCE, *trans*-1,2 DCE, TCE, and PCE) in soil, and 10 target VOCs (those above plus TCA and VC) in water. Specifically, the remediation goals were to: (1) meet NYSDEC cleanup criteria for 90% of the target compounds in soil within a 50' x 50' plot, (2) demonstrate that groundwater circulation can be developed to effectively reduce the concentration of monitored constituents in groundwater, and (3) show that at least 51% of observed mass reduction in target organics was due to microbiological removal processes (i.e., biodegradation). This paper focuses on the results of the groundwater treatment component of this field demonstration.

TECHNOLOGY DESCRIPTION

The groundwater and soil treatment technology employed in these studies was based on the UVB system (Vacuum Vaporized Well) patented by IEG mbH-Germany (Figure 1). This system consists of a specially constructed double-screened well to simultaneously mobilize and treat contaminants from the vadose zone, capillary fringe, and the saturated zone. Treatment in the vadose zone is achieved by vacuum extraction through the partially exposed upper screen, while treatment in the saturated zone is achieved by a combination of soil flushing, air stripping, and bioremediation.

Vertical groundwater circulation in the saturated zone is established by creating a pressure differential (by a pump and vacuum blower) across the two screens in the well. In a standard circulation mode of operation, groundwater enters the well through the lower screen and leaves through the upper screen. In a reverse circulation mode, groundwater enters the well through the upper screen and leaves through the lower screen. While traveling through the remediation well, the groundwater passes through one or more in-well treatment systems which, for example, may include an air stripper/aerator, an *in situ* bioreactor, etc., depending on the type of chemicals being treated. Groundwater leaving the treatment well is supplemented with dissolved oxygen as a result of the stripping reactor. In addition, co-substrates, such as inorganic nutrients, may be added to further facilitate *in situ* biodegradation of contaminants in the aquifer. In this process, circulating groundwater is used as a medium to distribute reagents throughout the saturated zone.

The reduced pressure inside the well casing (created by a vacuum blower) causes ambient surface air to be drawn passively down to the stripping reactor. The negative pressure environment increases the rate of expansion of the air bubbles rising within the stripping zone. Air bubbles burst when they reach the air/water interface inside the well casing. The volatilized VOCs are subsequently transported through the well shaft and up to the off-gas treatment unit, or atmosphere, by means of the vacuum blower. After the bubbles burst, the groundwater falls along the walls of the well, thus producing a hydraulic pressure forcing the water horizontally into the aquifer through the upper screen section.

The dimension of the circulation cell created by the groundwater circulation well (GCW) is dependent on site-specific parameters such as hydraulic conductivities, hydraulic gradients, the saturated thickness being treated, and groundwater recirculation rates, among others. Part of the groundwater flow entering the well casing represents upstream influent ground water that enter through the up-

Figure 1.   Diagram of the UVB system (Vacuum Vaporizor Well) patented by IEG mbH-Germany.

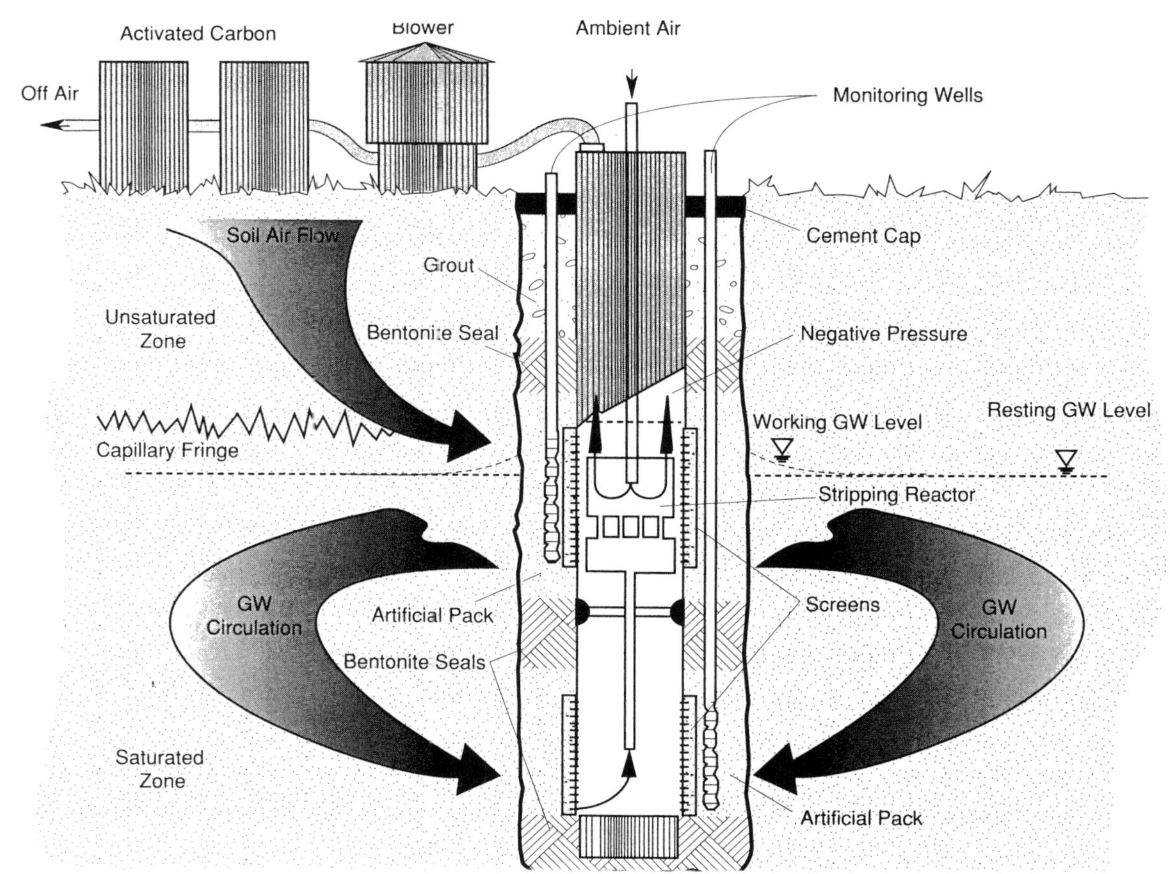

stream capture zone. An equal portion of groundwater leaving the well casing exits the circulation cell through the downstream release zone. The groundwater flow and the dimensions of the capture zone, circulation cell, and release zone are calculated using design aids based on numerical simulations of the groundwater hydraulics.

WELL CONSTRUCTION AND SYSTEM DESIGN

The GCW installed at the Sweden-3 Chapman site consisted of a 16-inch inner diameter (I.D.) steel casing set at a depth of 25.8 feet below ground surface (bgs). The lower screen (bridge slot) was 3 feet in length, and the upper screen (double-cased) was 7 feet in length. An 18-inch fixed packer was installed between the two screens at a depth of 22 feet bgs. A submersible pump was connected to the bottom of the packer to draw water from the lower screen for recirculation. The top of the packer was connected to a 7 foot, 12-inch outer diameter (O.D.) bioreactor. The bioreactor contained granular activated carbon (GAC) as a biosupport medium for native microorganisms. The top of the bioreactor was connected to an aerator/stripper with an aboveground ambient air intake pipe. The entire well was completed 2 feet aboveground and was sealed airtight with a flange. The top of the well was connected to a blower to create a vacuum in the well, passively pulling ambient air through the aerator/stripper. The pressure side of the blower was connected to gas-phase bioreactors followed by GAC drums for off-gas treatment.

OPERATION AND MONITORING

Groundwater entering the lower screen was first pumped through the *in situ* bioreactor. Untreated VOCs leaving the bioreactor were air-stripped as the water flowed through the stripper/aerator. The treated, aerated and oxygen-enriched groundwater was then discharged from the GCW through the upper screen at the water table level. As the circulation cell developed, groundwater circulated through the formation, to ultimately return to the GCW carrying newly-dissolved VOCs. To some extent, groundwater returned to the lower screen of the GCW provided dissolved oxygen to the bioreactor for aerobic degradation.

The system operated in the field for 14 months. During this time, the concentration of target VOCs in groundwater was monitored in 15 deep and shallow monitoring wells positioned strategically around the treatment system (Figure 2). Influent and effluent sampling ports were installed across the in situ bioreactor to evaluate its performance. GAC samples from the bioreactor were analyzed for microbial plate counts before, during, and after the demonstration. Groundwater samples from these wells were also analyzed for nitrate-N, nitrite-N, ammonia-N, phosphate, heterotrophic plate counts, and TCE degraders. Other physical and chemical groundwater parameters monitored included water levels, DO, pH, and temperature. The effects of two groundwater treatment flow rates (8 gpm and 22 gpm) were investigated. Air flow through the aerator/stripper varied between 70 and 250 cfm.

Figure 2. **Potentiometric head data collected from the shallow and deep monitoring wells.**

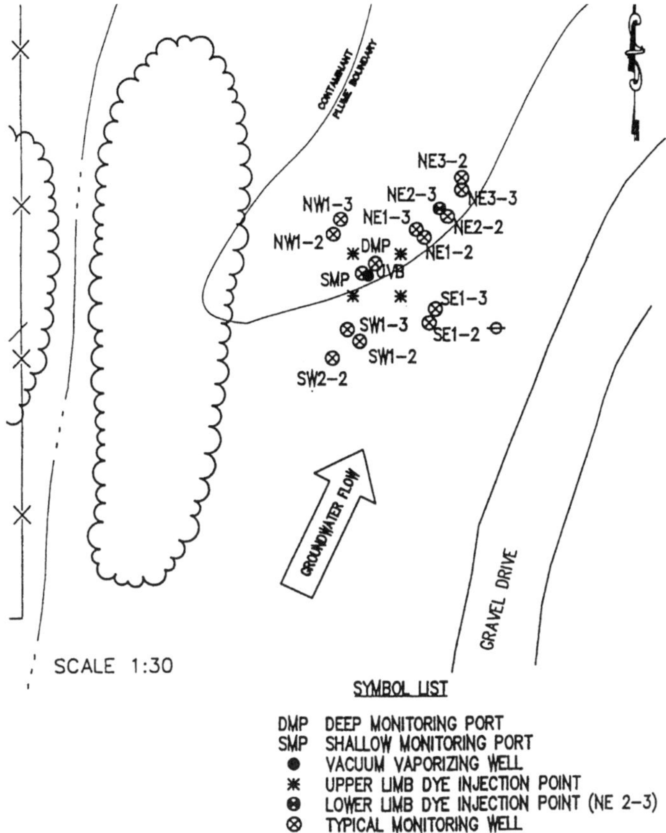

Field validation of the circulation cell formation and its dimensions was conducted using dye tracer and evaluating head changes in the aquifer. Florescent dyes (rhodamine WT and pyranine) were injected in selected deep and shallow wells and their movement toward and away from the treatment system was monitored in surrounding wells.

RESULTS AND DISCUSSIONS

Groundwater Circulation Cell

Mathematical modeling of the system for this site indicated that the effective radius of influence ranged from 40 to 50 feet. Potentiometric head data collected from the shallow and deep monitoring wells showed that vertical head differentials were established for a standard circulation cell. The field data showed that the extent of circulation was at least between 30 and 40 feet from the GCW.

A dye-tracer study utilized Rhodamine WT to evaluate the convergent flow of circulation, and pyranine to evaluate the upper divergent flow of the circulation cell. Rhodamine WT was injected into a deep monitoring well 30 feet upgradient from the GCW. Tracer dye was detected in the shallow and deep wells at the GCW, at other monitoring well locations, and in the soil profile 13 to 15 feet below grade, 10 to 15 feet away from the GCW. The Rhodamine WT data indicated that groundwater in the lower limb of the circulation cell moved toward the GCW, as expected. Pyranine was injected into the shallow aquifer at four radial locations 10 feet from the GCW. This tracer dye was detected in the deep GCW well within the first six hours. It was detected at several shallow and deep monitoring wells 30 feet from the GCW during the three-month study, and was found in soil samples 13 to 15 feet below grade, at least 28 feet radially from the GCW. The data indicated that groundwater was pulled vertically downward and toward the GCW as a function of standard circulation, and that groundwater movement was away from the GCW in the upper zone.

Whereas dye tracer data verified that convergent and divergent groundwater flow developed as a result of the GCW, potentiometric head data confirmed the effective radius of the circulation cell to be 30 to 40 feet (data not shown).

In Situ Bioreactor Performance

Table 1 summarizes the performance of the *in situ* bioreactor over the 14-month operation. After one month of operation, to allow for breakthrough of the GAC and for the biofilm to develop, bioreactor influent and effluent samples were collected periodically. Influent concentrations to the bioreactor were high initially, and gradually decreased over time. Influent and effluent concentrations of target VOCs across the bioreactor were averaged over block days of operating periods for mass balance purposes. In all, 31 kg (68 lb) of target VOCs were biologically degraded by the in situ bioreactor. Biofilm development on the GAC was confirmed visually as well as by microbial plate counts as shown in Table 2. On average, a two- and three-log increase in plate counts was observed after 10 and 15 months of operation, respectively. A ten-fold increase in the plate count was reported from the top to the bottom of the bioreactor, presumably because the bottom of the in situ reactor continuously received dissolved oxygen-enriched (3-4 ppm DO) water.

Table 1.   In Situ Bioreactor Mass Balance: Sweden-3 Chapman Site

Operating Period	Days	Concentration (ppb) Influent	Effluent	Flow (gpm)	Mass (g) Influent	Effluent	Biodegraded
8/15/94 to 8/31/94	17	1,187	792	8	900	601	300
9/15/94 to 9/26/94	12	592	434	8	317	233	84
10/6/95 to 3/10/95	115	5,731	1,098	8	29,420	5,636	23,785
4/13/95 to 6/15/95	54	403	20	22	2,672	133	2,539
6/16/95 to 7/30/95	45	281	42	22	1,553	232	1,321
8/1/95 to 9/30/95	61	527	152	22	3,946	1,138	2,808
10/1/95 to 10/23/95	23	177	153	22	500	432	68
TOTAL	327				39,309	8,405	30,904

Table 2. Total Heterotrophic Plate Counts from the In Situ Bioreactor Activated Carbon

Sample	Control	After 10 months	After 15 months
		(CFU/g dry weight)	
Top of the Bioreactor	3.50×10^3	1.13×10^5	5.78×10^5
Bottom of the Bioreactor	3.50×10^3	1.01×10^6	5.03×10^6
Average Bioreactor	3.50×10^3	5.62×10^5	2.80×10^6

Contaminants in Soil

Possible ways for the GCW to induce a decrease in soil chemical concentrations included soil vapor extraction, soil flushing, and biodegradation. The limited vadose zone present during the demonstration reduced the effectiveness of the vapor extraction because the screen was submerged for most of the demonstration. The chemical data indicated good reductions (74%) in the ketones (acetone, MEK, MIBK) which were easily mobilized. The data for the chlorinated solvents indicated a smaller reduction (less mobile) in total mass during the 15 months. Toluene had a very high removal rate of 87% and was easily mobilized.

Table 3 summarizes the mass of contaminant present in the plot with time, and Table 4 summarizes the results versus the NYSDEC compliance criteria. Approximately 70% of the soil samples collected meet the New York State Cleanup Criteria. However, due to the high concentrations of toluene present, the detection limits for the critical analytes (i.e., acetone and MEK) were higher than the NYSDEC criterion. Consequently, many data points were rejected and were not used in the analysis, affecting the percent of soil samples which meet the cleanup criteria.

Table 3. Total Soil Massa of Contaminants (g) in 50 x 50 Foot Plot at Various Times

Chemical Monitored	Background (0)	Months of Operation (Sampling Event)			
		3 (1)	5.5 (2)	10 (3)	14 (4)
Acetone	3700	4200	2700	1900	960
2-Butanone (MEK)	6300	8800	5300	4400	2100
cis-1,2-Dichloroethene (DCE)1900	3900	1500	1400	1200	
4-Methyl-2-pentanone (MIBK)2200	3500	2300	970	440	
Trichloroethene (TCE)	1500	8200	4500	660	3200
Tetrachloroethene (PCE)	380	3500	680	890	350
Toluene	58000	100000	55000	20000	7400

A factor affecting field data was the heterogeneous nature of the soil matrix. Visual contamination was observed along the micro fractures of the soil structure. Samples collected from individual split spoons had a high degree of chemical variability. The data collected three months after the baseline sampling reported an increase in all target analytes. This increase can be attributed to the mobilization of the contaminants from the heterogeneous soil matrix via soil flushing, and may be more representative of the initial contaminants present in soil versus the baseline data. This would account for a greater mass reduction in the plot than reported.

Contaminants in Groundwater

Over the 15-month operation, both a decrease and an increase in concentration of contaminants was observed. The processes that could cause a decrease in concentration of contaminants in groundwater include dilution, stripping, biodegradation across the *in situ* bioreactor, biodegradation in the aquifer, and convective movement away from the circulation cell. The processes that could cause an increase in concentration include convective movement into the cell from a source area (typically from highly contaminated vadose and capillary fringe zones).

Figure 3 shows a concentration profile of target chlorinated hydrocarbons (CHCs) and nonchlorinated hydrocarbons (nCHCs) within the 30-foot zone of influence of the GCW over the 15-month operating period. The concentrations of CHCs decreased by 48% and that of nCHCs decreased by 72% over the first six months. Heavy rains over the next three months elevated that water table by four feet and transported large quantities of contaminants from the vadose zone into the saturated zone. A four-foot increase in the saturated zone also increased the size of the circulation cell by 20 feet, thus drawing contaminants from other

Figure 3. **Concentration profile of target chlorinated hydrocarbons and nonchlorinated hydrocarbons within the 30-foot zone of influence of the GCW over the 15-month operating period.**

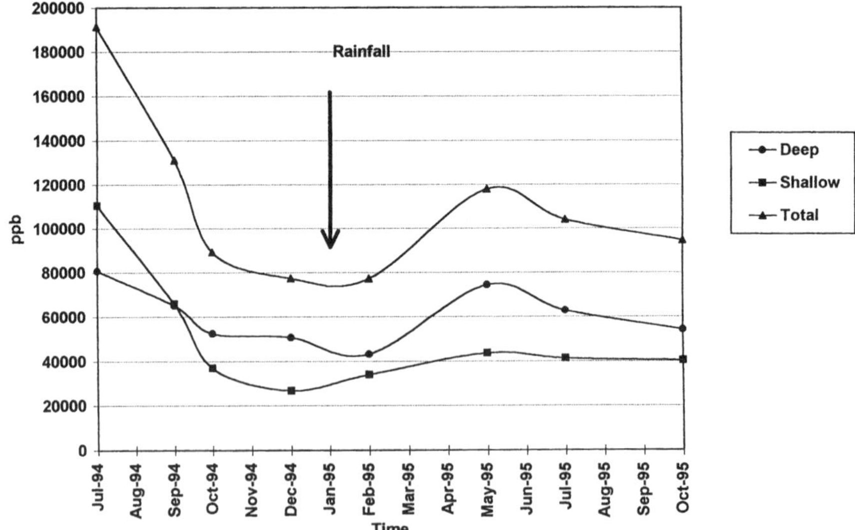

Table 4.   Compliance with NYSDEC Cleanup Criteria

VOC	Criterion (ppb)	Usable Data Points (#)	Data Points Meeting Criterion (#)	Data Points Meeting Criterion (%)
Acetone	200	19	0	0
2-Butanone (MEK)	600	25	4	16
cis-1,2-Dichloroethene (DCE)	600	46	22	48
4-Methyl-2-pentanone (MIBK)	2000	46	45	98
Trichloroethene (TCE)	1500	46	45	98
Tetrachloroethene (PCE)	2500	46	44	96
Results after 14 months				

nearby source areas into the cell. Consequently, the concentrations of CHCs and nCHCs increased by 79% and 15%, respectively. The increase in flow rate from 8 to 22 gpm also increased the effective capture zone, thus increasing the volume of contaminated water introduced to the treatment area. At the end of the study, the concentration of CHCs had decreased by 26%, and that of nCHCs by 78% (from their baseline levels).

Table 5 shows reductions in total CHCs and nCHCs as a function of distance from the GCW over the 15-month operation. The treatment efficiency for CHCs decreased almost exponentially with an increasing distance from the GCW. On the other hand, the treatment efficiency for the nCHCs decreased only within 20-feet from the GCW, and remained constant at 30- and 40-foot distances. This indicated that the mobilization and treatment of CHCs was probably influenced by other physical/chemical factors, while the nCHCs were more easily mobilized and treated.

Nutrients in Groundwater

Dissolved oxygen monitoring indicated that aerobic conditions were established in a 40-foot diameter zone surrounding the GCW. Concomitantly, significant decreases in nitrate-N and ammonia-N were observed in all the monitoring wells (data not shown). Combined, these data suggested that resident microflora was active (i.e., biodegradation) within the 40-foot diameter zone of influence. Outside this zone of treatment, anoxic to anaerobic conditions were likely present. Thus, decreases observed were likely the result of anaerobic biotransformation and dilution.

CONCLUSIONS

Results from this demonstration indicate that the UVB groundwater circulation technology offers a technically and economically feasible alternative to pump-and-treat technology for the treatment of groundwater contaminated with VOCs. Over the 15-month period of operation, the average concentration of target compounds

Table 5.   **Percent Reduction of Analyzed Contaminants in Groundwater, July 1994 to October 1995**

Area	CHCs	nCHCs	Total
UVB-400	99.40%	98.50%	99.20%
20 ft. away	50.90%	51.70%	48.90%
30 ft. away	36.10%	76.60%	53.50%
40 ft. away	26.30%	77.60%	48.90%

was reduced by 49% in monitoring wells within the GCW zone of influence. Moreover, 70% of the soil samples collected met NYSDEC cleanup criteria. Monitoring of the *in situ* bioreactor, off-gases and other physical, chemical, and microbiological parameters indicated that significant reductions were probably due to microbiological processes. However, the complex and varying test environment, coupled with budgetary constraints, made it difficult to quantify each removal pathway. The dye tracer studies indicated that an effective radius of greater than 30 feet was created.

ACKNOWLEDGMENTS

This work was performed under the United States Environmental Protection Agency's (USEPA's) Superfund Innovative Technology Evaluation (SITE) Program with participation from the New York State Department of Environmental Conservation (NYSDEC), and the New York State Center for Hazardous Waste Management (NYSCHWM). Research was performed, in part, in association with the USEPA's National Health and Environmental Effects Research Laboratory (NHEERL), Gulf Breeze, FL, under the terms of the Cooperative Research and Development Agreement between SBP Technologies, Inc. and the USEPA (FTTA-003 CRADA 0016-B-92). The bacterial strain *Burkholderia cepacia* was obtained from Dr. Malcolm Shields, Center for Environmental Diagnostics and Bioremediation, University of West Florida, Pensacola, FL. The authors would like to acknowledge Herb Skovroneck (SAIC), Scott Weber (NYSCHWM), Annette Gatchett (USEPA), Nick Kolak (NYSDEC), Eric Klingel (IEG), Bill Langley (IEG), and Kelley Loftus (Rochester Institute of Technology) for their participation and assistance during this project.

Mention of trade names and technologies does not constitute endorsement by the USEPA, NYSDEC, or NYSCHWM.

CHAPTER 31

Bioremediation of PCB- and PAH-Containing Sludges and Sediments in Land Treatment Units to Achieve Risk-Based Endpoints

John R. Smith and **Margaret E. Egbe**, Aluminum Company of America, Pittsburgh, Pennsylvania

INTRODUCTION

When it is determined that the presence of organics at a site creates a situation of unacceptable risk to public health and/or the environment, remedial action is required to reduce the respective chemical concentrations to acceptable levels and/or to minimize exposure to receptors. Bioremediation serves to address both of these goals, often at a lower cost compared to other more energy intensive remedial actions. As defined by the USEPA (1993a):

> "Bioremediation is an engineered process that uses microorganisms to decompose toxic, hazardous compounds to improve environmental quality and reduce human health risks. As a low energy natural process, it is an attractive alternative to conventional clean-up methods. The process residues are typically non-toxic and are easily reintroduced into the earth's biogeochemical cycles. Bioremediation technologies are potentially less disruptive to the environment and less expensive than other treatment options;" here, other treatment options refer to excavation followed by incineration and/or landfilling.

Mounting scientific evidence is beginning to support that bioremediation can be applied in a technically sound manner to provide for organics stabilization in conjunction with ultimate conversion to carbon dioxide, water, and bacterial cells (i.e., biostabilization). Based on a review of 123 articles and documents citing laboratory and field bioremediation projects, Loehr and Webster (1997) summarize weight-of-evidence data that serve to support the concept of biostabilization as being a scientifically sound and cost-effective remedial approach for soils, sludges, and sediments containing organic constituents. Here, initial active (i.e., engineered) bioremediation serves to provide relatively rapid reductions of the more soluble, more desorbable, and potentially more mobile lower molecular weight compounds due to their greater bioavailability. Subsequent passive (i.e., intrinsic) bioremediation - in conjunction with adequate engineering controls to provide for no direct contact by receptors - serves to ensure no chemical migration away from the actively treated material. This is due to the fact that the less soluble, less desorbable, and less mobile organics remaining after active bioremediation are retained in place and are biodegraded as they slowly desorb off the soil and become bioavailable. Here, the rate of desorption (i.e., bioavailability) does not exceed the rate of intrinsic biodegradation, thereby ensuring long-term protection of human health and the environment. In this context, active bioremediation refers to a time frame of months, and passive bioremediation refers to a time frame of years. Thus, subsurface migration of chemicals is prevented and groundwater is protected by the fact that the compounds that continue to slowly desorb off the soil will be biodegraded before they can travel any measurable distance and impact groundwater. Here it is to be stressed that under the proper environmental conditions, the remaining organics are not just retained within the soil matrix, but continue to biodegrade with ultimate conversion to carbon dioxide and water over a 5 to 10 year period, or longer.

While bioremediation of polycyclic aromatic hydrocarbon (PAH) containing soil/sludge material is scientifically documented and utilized on a full-scale basis (Smith et al., 1995; Loehr and Webster, 1997; USEPA, 1995a; USEPA, 1995b), the work presented in this chapter is significant in that it also focuses on biostabilization of polychlorinated biphenyls (PCBs). To date, PCB bioremediation has primarily been studied via laboratory bench-scale testing (Electric Power Research Institute, 1985; Unterman et al., 1988), with some limited pilot-scale projects (General Electric, 1992; Harkness et al., 1993). To date, there have been no full-scale bioremediation projects specifically focusing on PCB-contaminated media (USEPA, 1995b). This is primarily due to the fact that even though it is generally recognized that both aerobic and anaerobic biodegradation of PCB compounds does occur, the rate and extent of PCB biodegradation is not sufficient to achieve specific cleanup goals in a more conventional treatment time frame of weeks and months. As such, the present remedial options being most utilized for PCB-containing soils, sludges, and sediments is excavation followed by some high energy-intensive destruction process such as incineration, thermal desorption, or chemical oxidation, or the material is landfilled (USEPA, 1995b). For the cases of thermal desorption, chemical oxidation, or incineration, treatment costs are generally in excess of $500 per ton for soils, sludges, and sediments (USEPA, 1993b).

Alternatively, the approach of biostabilization via engineered land treatment represents a scientifically sound, low cost, and permanent remedial alternative for sludge/sediment materials containing PCBs and PAHs, with treatment costs generally ranging between $50 and $100 per ton (Colghazier et al., 1991). This contention is based on the fact that recent research has shown that within an engineered land treatment process, biodegradation rates of highly sorptive compounds like PCBs and PAHs are related to their bioavailability (i.e., desorption from solids) and that slow biodegradation rates are related to low bioavailability (Alexander, 1995). Thus, given a sufficient time period of years, the potential exists for biodegradation to achieve reductions (i.e., 90% or greater) as the compounds slowly desorb off the solid matrix and become bioavailable, assuming proper environmental conditions. From a full-scale perspective, it is to be noted that adequate institutional and/or engineering controls would need to be implemented as part of any full-scale biostabilization approach to ensure no exposure of target organics to environmental and/or human receptors. However, as the organic levels continue to reduce with time, the potential exists for such controls to be lessened as determined by site-specific risk-based factors.

For the purpose of demonstrating biostabilization of PCBs and PAHs in soil/sludge/sediment materials as being a technically sound, environmentally protective, permanent, and low-cost bioremediation approach, Aluminum Company of America (Alcoa) is in the midst of a multi-year program to demonstrate the efficacy of a sequential active-passive bioremediation process. This work is intended to build upon already existing weight-of-evidence data supporting biostabilization of PAHs (Loehr and Webster, 1997; Smith et al., 1995) and to expand the database to PCBs. As part of this program, three pilot-scale engineered land treatment units (LTUs) have been set up, operated, and continue to be monitored at Alcoa's Massena, New York, facility. These LTUs were loaded with sludge/sediment materials from two different lagoons (identified in this paper as A and B) at the Massena site. Due to site-specific conditions, both PCBs and PAHs exist together because these lagoons were historically used for settling of various plant process and sanitary wastewater flows. In 1994, different combinations of the lagoon sludge/sediment materials were tilled into respective LTU soil beds with two months of active bioremediation achieving reductions of the lower molecular weight, more desorbable and bioavailable PCB and PAH compounds. The LTUs are now in the passive (i.e., not actively tilled) bioremediation phase where 3-5 years of monitoring is being carried out to document no subsurface migration and the continued biodegradation of the remaining compounds as they become bioavailable. Thus, with passive bioremediation continuing over a 5 to 10 year period, it is expected that significant PCB and PAH destruction will be achieved, while at the same time ensuring no measurable subsurface migration. Alcoa's program is also intended to generate critical scientific data needed to better understand the environmental fate mechanisms controlling biostabilization from a more thermodynamic-based modeling perspective.

In support of the concept of engineered land treatment-based biostabilization serving to achieve risk-based endpoints for soil/sludge/sediment materials containing PCBs and PAHs, procedures and results from both the active and one year of passive bioremediation phases for the three pilot-scale LTUs are presented. The data presented and the conclusions made are strengthened by the fact that similar trends were observed for both PCBs and PAHs.

THEORETICAL CONSIDERATIONS OF PCB AND PAH BIODEGRADATION IN A SOIL/SLUDGE/SEDIMENT MATRIX

Conceptual Model Overview

It is hypothesized that the bioremediation of organics associated with solid phase materials (i.e., soils, sludges, sediments) containing organics occurs via various complex physical/chemical mechanisms and biological interactions associated with the solid-phase matrix components of nonaqueous phase liquid (NAPL), sand, and fine-grain aggregates (Smith et al., 1995). In this regard, Figure 1 schematically depicts the primary media characteristics and the primary physical/chemical and biological processes that are known to influence biodegradation. Here, the three potentially competing and concurrently occurring environmental fate mechanisms of sorption, volatilization, and biodegradation are proposed as the three primary processes which serve to influence both the rate and extent of organic reductions observed during bioremediation.

As illustrated in Figure 1, soil and/or sediment material impacted by an industrial sludge is depicted as a mixture of NAPLs, sand particles, and fine-grain aggregates. *NAPL* is defined as any free-phase hydrocarbon with a density either greater than, near, or less than that of water. *Sand* refers to noncohesive soils and can also include gravels. *Fine-grain aggregates* refer to cohesive soil particles comprised of individual silt and clay particles. Also present in soil/sludge media are natural organics, such as humics and other remnants of decayed vegetation.

Related to desorption and bioavailability, there are several specific subregions of the soil/sludge/sediment matrix where chemicals may exist, with each subregion having very different tendencies for organic release into the macropore water, as required for biodegradation to occur. It is further hypothesized, due to size

Figure 1. **Schematic of processes affecting biodegradation of organics in soil/sludge/sediment matrix.**

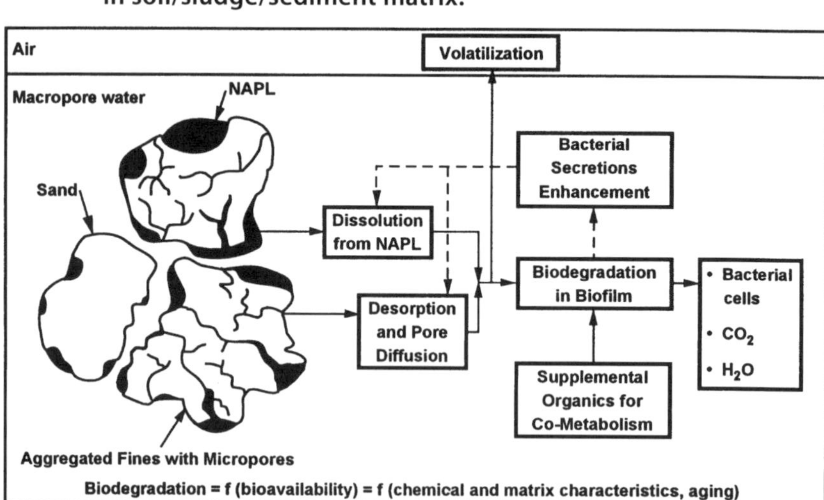

and mass transfer restrictions, that microorganisms exist in biofilms present on macropore surfaces or in bulk water (Rittman et al., 1990). These particular subregions include:

A: The external (macropore accessible) particle surfaces that may release sorbed organics relatively quickly;

B: The internal (micropore accessible) micropore particle surfaces not accessible by bacteria; and

C: The interior portions of NAPL droplets in which organics are dissolved.

Release of an organic chemical to the macropore water (i.e., desorption), is relatively rapid for subregion (A) defined above, and significantly slower for subregions (B) and (C). Specific to sparingly soluble organics such as PCBs and PAHs, this hindered desorption can be so significant that it often appears, over a time period of days and weeks, as if they were permanently fixed or bound to the solid matrix. However, this process of apparent sequestration is generally considered to be reversible, given sufficient time (Pignatello and Xing, 1996). Such time scales may involve years to decades (Alexander, 1997). Presently, the process of organic chemical sequestration is not well understood at the molecular scale and may involve more than simple sorption to solid surfaces and diffusion into subregions. Other mechanisms could involve the formation of oxidized coats and films, changes in the initial organic chemical or NAPL, and changes in the nature of the solid matrix (Pignatello and Xing, 1996). It is also an observed fact that the extent of sequestration increases with the amount of time a particular soil/sludge material is exposed to an organic (i.e., aging effect) (Alexander, 1997).

Figure 1, illustrates that the more soluble and biodegradable organics may serve to enhance the biodegradation of sparingly soluble organics, such as PCBs and PAHs via the process of co-metabolism. Here, the conditions can exist where the relatively low soluble concentration (i.e., ppb level) of a target organic substrate or substrates may not be sufficient to supply either a source of carbon or energy, or both, for the growth of the microorganisms. As such, the microorganism requires a second, more soluble, compound as its carbon and energy source. Schmidt and Alexander (1985) present a thorough literature review, as well as experimental data, which support that the rate and extent of biodegradation of low organic concentrations in a soil/sludge material may be controlled by the presence of higher concentrations of more soluble organics that support microbial growth and synthesis. They also note that with more than one type of organic compound present, biodegradation can occur simultaneously or in sequence, based upon the specific microorganisms that are present and the relative solubility and bioavailability of the organic substrates.

For the model presented, it is conceptualized that bioremediation of associated organic compounds is primarily a water-based process influenced by sorption mechanisms which control the accessibility of organics to bacteria present in attached biofilms within the macropore water, referred to in this Chapter as bioavailability. It is this bioavailability that is a primary controlling factor influencing both the rate and extent of organic biodegradation within a certain time frame (Alexander, 1997; Loehr and Webster, 1997). Biodegradation is thus a sequential process. Chemicals must first desorb and diffuse from micropores to macropores, and dissolve from NAPL and transport to the bulk macropore water phase. Once soluble in the macropore water, volatilization can also compete with biodegradation.

It is to be noted that this conceptual illustration of biodegradation being primarily a function of bioavailability is not meant to ignore the influences of chemical structure and specific environmental factors, and that organic biodegradation is also the result of complex biochemical microbial interactions that take place within and among bacterial cells. Rather, its intent is to show that interactive sorption/dissolution mechanisms of an organic chemical with a solid matrix are also highly influenced by chemical structure, and as such, sorption processes also serve to highly influence bioavailability and thus biodegradation. Furthermore, data presented in this Chapter, and reported by others (Alexander, 1997; Loehr and Webster, 1997; Pignatello and Xing, 1996; Smith et al., 1995) serve to support the hypothesis that within a time frame of months and years, as is the case for an active/passive bioremediation process, chemical dissolution and desorption can be predominant factors significantly influencing the overall rate and extent of organic biodegradation. Correspondingly, the influence of chemical structure on biochemical microbial type interactions is measured more in terms of hours, days, and weeks.

Bacterial secretions may also serve to enhance the rate and extent of organic dissolution and desorption and consequently, the rate and extent of biodegradation due to enhanced bioavailability (Guerin and Boyd, 1992; Stucki and Alexander, 1987; Thomas and Alexander, 1987; Rynaarts et al., 1990).

Relating the cited conceptual model to an engineered land treatment scenario, as active biodegradation proceeds over a weekly and monthly time frame, organic concentrations in soil/sludge/sediment material are reduced until an apparent plateau soil concentration is reached (Smith et al., 1995; Alexander, 1997; Loehr and Webster, 1997). This rate of reduction generally follows an empirical first-order rate expression (Middleton et al., 1991). The plateau level represents a condition where the organics are much less bioavailable than they were before active biodegradation. However, under the proper environmental conditions and given a sufficient time period of months and years, it is hypothesized that organic reductions will continue to biodegrade as they slowly become bioavailable. Such a two-step organic desorption rate process is cited and described by Pignatello and Xing (1996). While the observed soil plateau concentrations often exceed regulatory guidelines, the real question that needs to be addressed relative to potential risk from contaminant migration under these conditions is whether the desorbable aqueous phase organic concentrations have been reduced to where the bioremediated soil/sludge/sediment matrix does not represent a potential threat to groundwater. Furthermore, if the rates of organic desorption and subsurface migration have been reduced to where they are much less than the rate of biodegradation, then subsurface migration will not occur and the solid matrix organic concentrations will continue to decrease; thus the media is said to be biostabilized.

Influence of PCB and PAH Characteristics on Biodegradation

Referring to the above discussion and previously cited Figure 1, bioavailability, and thus biodegradation, is not only influenced by the physical/chemical characteristics of the particular soil/sludge/sediment matrix and aging, but also by the physical/chemical characteristics of the respective organic chemical(s) themselves.

Based on PCB and PAH physical/chemical data and fugacity modeling (Mackay et al., 1992a), Figure 2 illustrates that with increasing molecular weight, the relative effect of apparent adsorption increases due to the influence of increasing values in the octanol water partition coefficients (K_{ow}) of the respective PCB and PAH compounds. Correspondingly, the fact that the higher molecular weight compounds tend to adsorb more to the solid matrix results in the observed effect of increasing molecular weight also serving to reduce relative volatilization, desorption, subsurface transport, bioavailability, and thus the apparent rate and extent of biodegradation. From a pure NAPL-type chemical component standpoint, increasing molecular weight also results in less aqueous solubilities. Specifically related to PCBs and PAHs, the general trends cited in Figure 2 are also supported by numerous researchers.

Figure 2. **Influence of PCB and PAH physical/chemical characteristics on environmental fate mechanisms.**

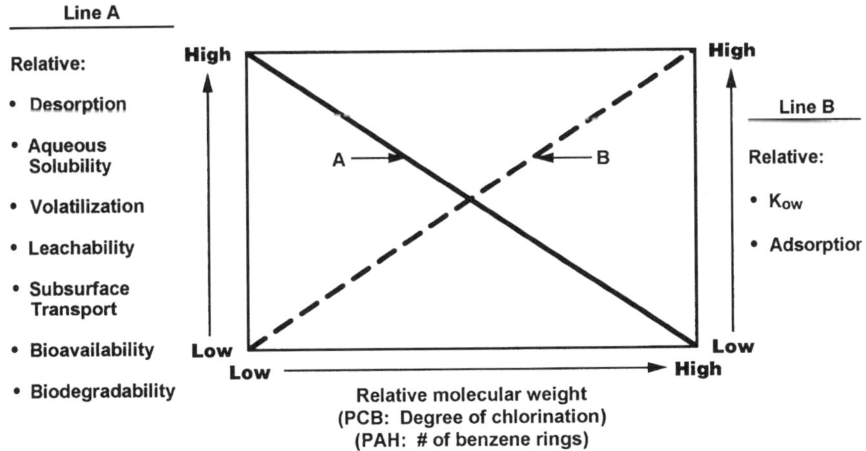

Polychlorinated Biphenyls

Chemical Characteristics

PCBs are defined as a class of organic compounds in which one or more chlorine atoms are attached to a biphenyl molecule. With ten available sites for an attached chlorine atom, there are 209 distinct, individual isomeric PCB compounds possible; these are commonly referred to as congeners. The general molecular formula for a PCB congener is $C_{12}H_{10}Cl_n$ with n ranging from 1 to 10. The resulting nomenclature is defined by this range, from monochlorobiphenyl (n=1) through decachlorobiphenyl (n=10). PCBs with the same total number of chlorines, but at different positions on the biphenyl molecule, are defined as congener homolog groups.

PCBs were used in the United States from the 1920s until about 1977, and were primarily manufactured and sold as Aroclor™ mixtures by the Monsanto Company, Inc. These mixtures were comprised of a variety of individual PCB congeners totaling 50 or so. Because of their suspected human carcinogenicity, PCB manufacturing in the United States was halted in 1977, and with a few exceptions, their use is now banned.

PCB Sorption/Dissolution

Numerous researchers have evaluated various mechanistic sorption processes with PCBs and soils (Chou and Griffin, 1986; Rifkin and Bouwer, 1994; Rifkin and LaKind, 1991; Achman et al., 1993; DiToro and Horzempa, 1982; Carroll et al., 1994; Witkowski et al., 1988; Sedlak and Andren, 1994; McLaughlin et al., 1994; DiGiano et al., 1993; Girvin et al., 1990) and dissolution from pure Aroclor and Aroclor/NAPL mixtures (Luthy et al., 1996; Adeel et al., 1996). These reported studies serve to support the relative trends cited in Figure 2 related to the influence of molecular weight. For example, Achman et al. (1993) report the water phase concentrations of individual PCB congeners to range from 0.46 to 8.0 ng/L, with the solid phase concentration ranging from 730 to 1,300 mg/kg; here, the majority of the water phase PCBs are the lower molecular weight mono-, di-, and tri-chlorobiphenyls.

Related to rates of desorption from soil/sediment, Carroll et al. (1994), Witkowski et al. (1988), and Girvin et al. (1990) observed that PCB desorption is a two-step process. First, there is an initial relatively rapid release on the order of hours and days, followed by a slower release on the order of months and years. They also cite that the initial rapid desorption step involves predominantly the less chlorinated, lower molecular weight compounds, and the slower step begins to desorb more of the higher chlorinated, higher molecular weight compounds.

Specific to PCB dissolution, Luthy et al. (1997) conducted a series of experiments in which the dissolution of respective PCB congeners from pure Aroclor 1242 and a mixture of pure Aroclor 1242 diluted in a hydraulic oil (i.e., Fyrquel 220) were studied. For these experiments, apparent equilibrium was achieved at

20 weeks, with maximum measured aqueous solubility indicated by the aqueous PCB water phase concentration leveling off. For the pure Aroclor in water experiments, a total aqueous phase PCB concentration of approximately 100 µg/L was achieved, as comprised of approximately 30% di-, 50% tri-, 10% tetra-, and 3% penta-chlorobiphenyls, compared to Aroclor 1242 itself, which is comprised of approximately 15% di, 47% tri-, 30% tetra-, and 8% penta-chlorobiphenyls. Here, the water phase concentration contained a relatively higher percentage of di- and lesser percentages of tetra- and penta-chlorobiphenyls. For the Aroclor 1242 diluted in hydraulic oil experiments, lower total PCB aqueous phase concentrations were achieved corresponding to lower Aroclor 1242 concentration in the hydraulic oil, with approximate agreement to Raoult's law. Thus, under the scenario that organics in a soil/sludge or sediment/sludge mixture could be present both as a sorbed phase and a separate NAPL phase (Mackay and Cherry, 1989; Feenstra, 1992), then for the particular case of Aroclor oil or Aroclor in hydraulic oil mixtures, it is supported that the PCB congener concentrations in waters associated with the mixtures would be diminished due to dilution in the NAPL phase and be skewed toward the lower chlorinated compounds.

PCB Volatilization

Related to the potential for PCB volatilization, Achman et al. (1993) cite that *volatilization is mediated by the partitioning of PCBs between water and aquatic particles suspended in solution because only truly dissolved PCBs are available for transfer across the air-water interface.* Thus, PCB volatilization is very congener-specific and reflective of the sorption processes that contribute to aqueous concentrations. This theoretical hypothesis is supported by work performed by Hornbuckle et al. (1993) where they simultaneously sampled air, water, and sediments of Green Bay in Lake Michigan and determined the more chlorinated congeners to be associated with the sediments, with only the less chlorinated congeners existing as dissolved compounds and being susceptible to volatilization. Thus, it is supported that loss of PCBs from soil/sludge material via volatilization is only significant for the lower chlorinated compounds. However, as cited in the following discussion, these are also the PCB congeners most readily biodegradable. Therefore, the potential does exist to minimize PCB volatilization and maximize biodegradation via a properly designed and operated bioremediation system.

PCB Biodegradation

Biodegradation of PCBs is reported to occur via both anaerobic (Anid et al., 1991; Bedard and Quensen, 1995; Williams, 1994; Fish and Principe, 1994; Brown et al., 1987a; Brown et al., 1987b; Quensen et al., 1988; and Mohn and Tiedje, 1992), and aerobic processes (Electric Power Research Institute, 1985; Harkness et al., 1993; Clark et al., 1979; Niaki, 1987; Furukawa, 1982; Rochkind-Dubinsky et al., 1986; and General Electric Company, 1992). However, the following discussion will focus only on aerobic PCB biodegradation because this is the process most applicable to the environmental conditions associated with the engineered land treatment units cited in this paper.

For the previously cited aerobic PCB biodegradation references, both laboratory and field studies support a general trend in that both the rate and extent of PCB biodegradation appears to decrease with an increasing degree of chlorination, assuming all other environmental factors are equal. This serves to support the concept that bioavailability has a major influence on biodegradation (see previously cited Figure 2) in that an increase in the degree of chlorination of specific congeners also generally corresponds to a decrease in maximum solubility (i.e., molecular weight) and an increase in respective K_{ow} values that correlate with greater partitioning to the solid phase of the respective congeners; i.e., less bioavailability.

Based on earlier PCB biodegradation research, primarily dealing with Aroclor mixtures, it was originally thought that penta- and higher congeners were generally resistant to biodegradation (Furukawa, 1982; Niaki, 1987; Clark et al., 1979). However, more recent work performed which focused on specific PCB congeners rather than Aroclor mixtures has determined that naturally occurring bacteria exposed to PCBs in the environment have developed various mechanisms for metabolizing PCBs and thus have much greater abilities for degrading a broader spectrum of PCB congeners (i.e., more chlorinated) than originally thought (Unterman et al., 1988; Bopp, 1986; Unterman et al., 1987; Bedard et al., 1986; McDermott et al., 1989; Bedard et al., 1987; Alcoa, 1994; Alcoa 1995a; Alcoa, 1995b; and Jansen, 1995). Such findings also agree with an Electric Power Research Institute (1985) report which cites PCB biodegradation at all levels of chlorination. This more recent work has begun to suggest that, depending on specific conditions, different and varied enzymatic processes may be involved based on the observation that PCB congeners, representative of all major structural classes, are biodegradable by certain bacterial strains. While the exact mechanisms by which many of the more chlorinated congeners are biodegraded are not fully understood, enzymatic dechlorination via mono-oxygenases or dioxygenases are now cited as possible mechanisms. For the case of mono-oxygenases (or hydroxylases), such oxidation tends to initiate biodegradative attack most probably by direct insertion of an oxygen atom into a carbon-hydrogen bond (Bedard et al., 1987).

It is also well reported that aerobic PCB biodegradation primarily occurs via co-metabolic mechanisms. As cited by Focht and Brunner (1985), PCB congeners with two or more chlorines per molecule are not known to be biodegraded where they serve as the only growth substrate. However, based on both laboratory and field studies, it is well documented that the addition of a second, more soluble and biodegradable substrate serves to enhance both the rate and extent of PCB biodegradation, especially related to the higher chlorinated congeners. Clark et al. (1979) cites this for the addition of acetate, and Focht and Brunner (1985), General Electric Company (1992), and Harkness et al. (1993) cite this for biphenyl addition to PCB-contaminated soils/sediments. While such co-metabolic mechanisms are not fully understood, PCB physical/chemical characteristics do serve to support the hypothesis that the relatively hydrophobic PCB congeners cannot supply sufficient energy for cell growth and synthesis. Such a contention is supported by various researchers who cite threshold concentrations for bacterial growth being between 2 µg/L to 10 µg/L, depending on the specific organic and environmental conditions (Law and Button, 1977; Shekata and Marr, 1971; van der Kooij et al., 1982; Schmidt and Alexander, 1985). Thus, it is quite possible that the growth on the more soluble substrate simply raises the total biomass in the system to the point where uptake and transformation of the less soluble PCB substrate is concurrently achieved. Here, the addition of such substrates does serve to increase the total population, which then results in enhanced PCB biodegradation via enhanced microbial interactions. Such microbial interactions also serve to support that a mixed microbial population is critical for PCB biodegradation in the field.

PCB Bioavailability

From the previously cited literature, two different mechanisms are cited as being the primary factors responsible for the observed recalcitrance of certain PCB congeners to biodegradation. These are: (1) the complexity of the PCB molecule related to the number and position of chlorine substituents that serve to impact steric and electronic hindrances to enzymatic catalysis; and (2) increasing PCB hydrophobicity with increasing molecular weight that serves to limit PCB bioavailability. Both of these factors are directly related to PCB physical/chemical characteristics.

Many earlier PCB studies that were originally performed attempted to establish general rules on the relationship between chemical structure and biological recalcitrance. Here, recalcitrance to biodegradation was cited as being primarily due to electronic and steric properties imparted by the types, numbers, and locations of bonds and substituents (Electric Power Research Institute, 1985; Pitter, 1985). However, based on more recent studies, (Jansen, 1995; Alcoa, 1995a; Alcoa, 1995b; General Electric Company, 1992; Harkness et al., 1993), many of these earlier observations have been disproved with mixed bacterial populations exhibiting various mechanisms for metabolizing all classes of PCBs, and thus bacteria have a greater ability than originally thought.

Based on the premise that there are varied and novel metabolic mechanisms potentially associated with PCB biodegradation, this more recent data is also beginning to support the hypothesis that bioavailability is also a critical factor, along with molecular structure, contributing to apparent PCB recalcitrance. Again, please note that the contention is made in the context of soil/sludge/sediment bioremediation processes over a time frame of months and years. For time frames of hours, days, and weeks, electronic and steric hindrances based on chlorination can play a significant role in terms of influencing PCB biodegradation. However, chemical structure related to degree of chlorination also has a significant effect on respective congener solubilities and partition coefficients which in turn influence bioavailability.

Based on pilot-scale bioremediation of Hudson River sediments, General Electric researchers determined that the PCBs associated with the sediments consisted of both a labile (i.e., readily bioavailable) fraction and a resistant fraction (General Electric Company, 1992; Harkness et al., 1993). Here, the labile fraction consisted of the lower molecular weight, less chlorinated PCBs that readily desorbed from the sediments. The resistant fraction is cited as consisting of the higher molecular weight PCBs that were dissolved in the natural organic matrix of the sediments, thus controlling desorption. During these same studies, biphenyl additional along with repeated inoculations with a purified PCB-degrading bacterial strain, possessing a known ability for higher chlorinated PCB congeners, failed to improve either the rate or extent of PCBC sediment reduction compared to using only the naturally occurring strains already present in the sediments.

To better elucidate the influence of sorption processes on biodegradation, Jansen (1995) studied the aerobic biodegradation of PCBs originating from Aroclor 1242 sorbed onto 0.5 micron polystyrene beads as surrogate soil particles. The studies were conducted in sealed sacrificial 126-mL reaction bottles augmented with biphenyl and a PCB-degrading innoculum. In the absence of particulates, PCB biodegradation was rapid and nearly complete for the more soluble mono-, di-, tri-, and tetra-PCB congeners. Increasing particle concentration reduced the congener biodegradation rate and extent. This effect was more pronounced for higher chlorinated congeners. In all cases, even where total biodegradation efficiency was significantly reduced, soluble phase PCB concentrations were negligible, suggesting that biodegradation rates of all soluble congeners were higher than desorption rates.

In somewhat related work, but without particulates, Envirogen, Inc. and Alcoa collaborated on laboratory studies to measure the biodegradation of only water-soluble PCB congeners; all experimental runs were performed in duplicate (Alcoa, 1995b). The soluble PCB congener solution was generated by equilibrating Aroclor 1242 with water over a 20-week period. The solutions containing only soluble PCB congeners were inoculated with a known PCB-degrading bacterial population. Measurements were taken after 1 and 14 days of incubation. Parallel runs with and without biphenyl addition were also carried out. The major finding from this work is that after 14 days of incubation, more than 99% of the solubilized PCB concentration was biodegraded; reduced from an average of 98.5 µg/L to an average of 0.7 µg/L total PCB, compared to microbial-killed control reactors which measured total PCB concentrations near 100 µg/L at Day 14. This total PCB reduction corresponds to over 99% reduction of the mono-, di-, tri- and tetrachloropiphenyls and 90% or more of the total penta- and hexa-chlorobiphenyls. It is also important to note that the homolog distribution in the PCB-equilibrated water was comprised of more than 90% mono-, di-, tri- and tetra-chlorinated congeners, with no hepta-chlorinated congeners or higher present. The work also demonstrated that greater than 97% of the total PCB degradation was achieved after only 1 day, with 14 days of incubation showing increasingly greater reductions of the tetra-, penta- and hexa-chlorinated biphenyls. Also, with the exception of the penta- and hexa-chlorinated congeners, the addition of biphenyl did not seem to have a major influence on PCB biodegradation compared to those reactors with no biphenyl addition with the same added inoculum concentration. It is also concluded from this work that there does appear to be some influence of chlorine position on the rate of biodegradation for specific di-ortho substituted PCB congeners, and that this effect also increases with the degree of chlorination. Regardless, substantial biodegradation was measured for all soluble PCB congeners, thus supporting that sorption processes are one of the predominant influences on PCB biodegradation in soil/sludge/sediment systems, and that all leachable and bioavailable PCB congeners should biodegrade, given the proper environmental conditions and a time frame of months and years.

Further supporting the concept of long-term biostabilization, Gan and Berthouex (1994) monitored PCB-contaminated sewage sludge that was applied to soil and monitored over a seven year period. Here, the range of quantifiable PCB half-lives was from 7 to 11 months for dichlorobiphenyls, 5 to 17 months for trichlorobiphenyls, and 11 to 58 months for tetrachlorobiphenyls and higher. Total PCBs had an average half-life of 19 months. In a similar long-term monitoring study, Alcock et al. (1995) also cite continued slow reduction of PCB soil/sludge concentrations over a 15- to 20-year period; total PCB levels were reduced from approximately 0.8 mg/g to approximately 0.05 mg/g. Here, the greater percent reductions were seen for the lower chlorinated congeners, and proportionally decreased with increasing degree of chlorination. Thus, for these two longer-term monitoring studies, it is hypothesized that PCB biodegradation was maintained and continued over an extended yearly time frame as the specific PCB congeners became bioavailable.

Polynuclear Aromatic Hydrocarbons

Chemical Characteristics

PAHs are defined as a class of neutral, nonpolar organic compounds consisting of two or more fused benzene rings in linear, angular, or cluster arrangements. PAHs generated as a result of fossil fuel-related production processes are associated with gas manufacturing, creosote wood treating, coal tar refining, coke manufacturing, petroleum refining, and aluminum smelting. Many of these processes still continue to operate and generate PAHs.

Due to the potentially acute and chronic toxicity primarily associated with the lower molecular weight PAHs, and the potential carcinogenicity associated with the higher molecular weight PAHs, the United States Environmental Protection Agency has designated 16 PAHs as being environmentally important and representative of PAHs as a class of compounds. These are the same PAH compounds which are included in EPA's Priority Pollutant List (USEPA, 1979). These 16 PAHs can be grouped into five different categories defined by the particular number of benzene rings comprising a particular compound; as such, the five groups are 2-, 3-, 4-, 5-, and 6-ring PAH compounds.

As given by Mackay et al. (1992b), PAHs are similar to PCBs in that they generally exhibit both low volatility and low aqueous solubility. Both of these properties decrease with increasing ring number (i.e., increasing molecular weight). It is these characteristics which strongly influence the removal of PAHs from soil/sludge/sediment materials via bioremediation. While many other PAH isomers exist, the 16 identified by the USEPA are usually given focused attention by both state and federal regulatory agencies in the United States and Canada.

PAH Sorption/Dissolution

Referring to previously cited Figure 2, very similar analogies can be made between PCBs and PAHs related to the environmental fate mechanisms which influence their overall bioavailability in a soil/sludge/sediment matrix. Here too, the different PAH compounds will behave differently with respect to the three environmental fate mechanisms as strongly influenced by sorption processes. As dictated by their physical/chemical characteristics, PAHs will exist as sparingly soluble compounds in solution and be subject to volatilization and/or biodegradation, with the rate and extent of these two processes decreasing with increasing molecular weight; i.e., increasing number of benzene rings.

PAH thermodynamic data (Mackay et al., 1992b) support that the lower molecular weight 2-, 3-, and 4-ring compounds will desorb off a soil/sludge/sediment matrix and be more bioavailable than 5- and 6-ring compounds. Here both the respective maximum aqueous solubilities and the log K_{ow} values vary over 4 to 5 orders of magnitude. Such a premise has been confirmed by Brusseau and Rao (1989), where they cite an empirical correlation which indicates that both the rate and equilibrium extent of PAH desorption for soil/sludge material into water decrease with increasing molecular weight.

PAH Volatilization

Once desorbed, volatilization is only expected to be significant for the 2- and 3-ring PAHs because they have the relatively higher values of Henry's law constants. Furthermore, accounting for the fact that sorption, diffusion, and dissolution processes control the rate and extent by which PAHs are present in macropore fluid, it is further expected that while volatilization may occur for the 2- and 3-ring PAHs, biodegradation will be the primary fate mechanism in a well-designed and maintained soil/sludge bioremediation system.

PAH Biodegradation

Numerous researchers have studied and documented the biodegradation of individual PAHs compounds. Under co-metabolic growth conditions, partial mineralization to CO_2 using radio-labeled PAHs has been demonstrated by Herbes and Schwall (1978) and Heitkamp and Cerniglia (1988, 1989) for 3-ring PAHs; by Mahaffey et al. (1988) for 4-ring PAHs; and by Heitkamp and Cerniglia (1988), Sikka (1991), and the Gas Research Institute (1992) for 5-ring PAHs. With a particular PAH serving as the sole source of cellular carbon and energy, Heitkamp, et al. (1988) cites radio-labeled mineralization to CO_2 for the 4-ring PAH pyrene; Kelley, et al. (1991) cites similar results for the 4-ring PAH fluoranthene. Using standard analytical and microbial characterization techniques on laboratory controlled mixtures, Mueller et al. (1990) also suggest the microbial utilization of fluoranthene (4-ring PAH) as the sole source of cellular carbon and energy for growth synthesis. For the radio-labeled PAH studies, several researchers document both CO_2 production as well as incorporation of the PAH organic carbon into cellular mass via synthesis.

Related to PAH co-metabolism, it is cited above that all 2- and 3-ring PAHs and the 4-ring PAHs of pyrene and fluoranthene, can be mineralized to CO_2 and H_2O under conditions where the individual compounds serve as the sole carbon source. The other 4-ring and all 5- and all 6-ring PAHs will only biodegrade under co-metabolic conditions. The PAHs that can be biodegraded without co-metabolism have respective maximum aqueous solubilities ranging from approximately 31 mg/L for naphthalene to 0.26 mg/L for fluoranthene (Mackay et al., 1992b). Those PAHs that can only be biodegraded under co-metabolic conditions have respective maximum aqueous solubilities ranging from 0.01 mg/L for benzo(a)anthracene (4-ring PAH) to 0.00026 mg/L for the 6-ring PAH benzo(g,h,i)perylene (Mackay et al., 1992b). Thus, as with PCBs, it can be hypothesized that the less soluble PAHs do not supply sufficient energy for cell growth and synthesis.

PAH Bioavailability

In addition to the well-defined research work cited above that focused on biodegradation of individual PAH compounds in somewhat pure systems, numerous bench-, pilot-, and full-scale applied research studies have been performed on a number of actual contaminated samples (Gas Research Institute, 1990; Guerin and Boyd, 1992; Mueller et al.; 1991; Howard et al., 1988; Sims and Overcash, 1982; and Loehr and Webster, 1996). These soil, sediment, sludge, groundwater, and wastewater related studies demonstrated that biodegradation of all 16 of the EPA's cited PAHs can occur. These data serve to support the concept that biodegradation of the 2-, 3, and 4-ring PAHs will almost always occur, and at a faster rate and to a greater extent than the relatively less soluble, less desorbable 5- and 6-ring PAHs. On this basis, the apparent recalcitrance sometimes reported for the 5- and 6-ring PAHs associated with soil/sludge material is not due to the fact that they are nonbiodegradable relative to the 2-, 3-, and 4-ring PAHs, but that they are less bioavailable due to adsorption/desorption, diffusion, and dissolution processes. PAHs have a molecular weight range of 128 for the 2-ring PAH naphthalene through 276 for the two 6-ring PAHs indeno(1,2,3-cd) pyrene and benzo(g,h,i) perylene compared to PCBs, which have a range of 154 for monochlorobiphenyls through 498 for decachlorobiphenyl (Mackay et al., 1992a; 1992b). Thus, with the thermodynamic properties influencing biodegradation (see Figure 2), it is expected that PAHs will be slightly more bioavailable, and thus more biodegradable, in a soil/sludge/sediment material.

As with PCBs, there are a few long-term PAH biodegradation studies that serve to support that PAH biodegradation in a soil/sludge/sediment matrix will continue over a yearly time frame as controlled by the rate at which the PAHs desorb and become bioavailable, assuming that the proper environmental conditions are maintained. As reported by Smith et al. (1995), two months of active full-scale land treatment in four different lifts of a creosote contaminated soil achieved total PAH reductions from approximately 3,000 mg/kg-dry weight (DW) to an average of approximately 1,000 mg/kg-DW (~66% reduction). The majority of this reduction was due to reductions of the more bioavailable 2-, 3-, and 4-ring PAHs. The lifts were left dormant with an additional five years of passive remediation achieving an average total PAH level for all four lifts of approximately 200 mg/kg-DW (~ 93% reduction); passive bioremediation refers to establishing a vegetative cover directly on the treated soil with no active tilling, watering, or nutrient addition. Active bioremediation half-lives were computed between 12 to 21 days for the 2-ring PAH naphthalene and proportionally increasing to between 108 to 178 days for 6-ring PAH compounds. However, half-lives of between 3 to 4 years were reported for the longer-term passive bioremediation phase, with the higher ring PAHs generally having the higher half-lives. Lastly, no subsurface migration of PAHs occurred based on none being detected in the subsurface soils beneath the treatment area. There were also corresponding reductions in PAH-TCLP soil extract levels as treatment proceeded through the active and yearly passive bioremediation phases. Other such similar long-term PAH biodegradation studies are reported by Loehr and Webster (1996).

STUDY APPROACH

Engineered land treatment is a generally accepted and proven process for treating soils, sludges, and sediments containing PAHs and other hydrocarbon based chemicals associated with creosote wood treating and oil refining operations (Smith et al., 1995; Ryan et al., 1991; Carlieo et al., 1992; USEPA, 1995a). However, land treatment has yet to be used for full-scale treatment of PCB-contaminated material due to the perception that PCBs are not as readily biodegradable compared to other compounds. In this regard, one of the primary objectives of an ongoing R&D effort by Alcoa is to better define how the rate and extent of PCB and PAH biodegradation for solid phase material is influenced by the rate and extent of chemical desorption (i.e., bioavailability). Here, it is hypothesized that as these chemicals slowly leach out of the solid matrix (i.e., become bioavailable), they will be biodegraded before migrating any measurable distance in the subsurface.

To better understand and to document this hypothesis, Alcoa conducted and directed laboratory-scale biodegradation treatability testing of lagoon sediment/sludge materials in 1993 and 1994 (Alcoa, 1994). Based on positive results, subsequent on-site LTU pilot-scale testing was begun in 1994. As detailed by Alcoa (1995a), part of this work entailed the design, startup, and active bioremediation operation of three engineered land treatment units (LTUs). Each LTU received varying combinations of sludges and sediments from two different water/wastewater lagoons at Alcoa's Messena, New York facility.

As schematically illustrated in Figure 3, each LTU is approximately 4 feet wide by 22 feet long, with a 6-inch tilled surface layer overlaying a compacted subsoil layer. The subsoil layer is comprised of compacted soil, geotextile, geonet, and a high-density polyethylene liner. At the time of application, contaminated lagoon sludges and sediments were mixed with clean soil to provide a 6-inch layer of a tillable soil/sludge/sediment mixture. LTU-1 received material from Lagoon A at a 9:1 sand:sludge ratio. LTU-2 received approximately a 1:1 blend of Lagoons A and B materials, with the blended material applied at a 14:1 sand:sludge ratio. LTU-3 received Lagoon B sludge/sediment material at a 20:1 sand:sludge ratio. All ratios are on a dry weight basis. The lagoon materials were mixed with uncontaminated coarser sand in the LTUs at the ratios cited in order to achieve proper physical consistency required for tilling and to help maintain conditions needed for stimulating and maintaining aerobic land treatment processes.

In 1994, each LTU was subjected to 60 days of active aerobic bioremediation treatment. Here, each treatment bed was tilled daily, kept moist via irrigation, and nitrogen and phosphorus (nutrients) were added as needed for bacterial growth and synthesis. No supplemental bacteria (i.e., bioaugmentation) or supplemental organics were added. With analytical procedures and results detailed elsewhere (Alcoa, 1995a), extensive PCB (via congener analysis) and PAH monitoring were performed to assess mass balance reductions while accounting for any volatilization (via air monitoring), runoff, and subsurface transport. In addition to PCBs and PAHs, extensive monitoring was also performed related to various metals, cyanide, fluoride, and volatile and semivolatile organics. Primarily, only PCB and PAH results are given in this paper.

Following two months of active bioremediation treatment (September and October, 1994), the LTUs were covered with plastic during the winter months. In the spring of 1995, the plastic was removed and the LTUs were covered with 3-6 inches of a soil, and a grass cover was established. In August and November of 1995, the actively treated soil layer, underneath the applied soil cover, was sampled and analyzed for PCB and PAH compounds to assess continued reduction via passive (i.e., more intrinsic type) bioremediation processes. Plans are to continue such yearly monitoring of passive bioremediation through the year 2000 to document the rate and extent of continued PCB and PAH reductions, reduced PCB/PAH water leachability, and low or nonmeasurable PCB/PAH subsurface migration.

RESULTS AND DISCUSSION

PCB and PAH Reductions

For the active bioremediation phase, Table 1 gives the PCB and PAH mass balance summary results for LTUs 1 and 2. As given, the soil reductions measured are primarily attributable to biodegradation, via mass difference, with minimum volatilization and transport to the subsoil and sump water. All results cited are based on average statistically significant values at a 90% confidence level.

Table 1.   Mass Balance Summary for LTUs During Active Bioremediation

| Pathway and Media | Percent of Initial Mass (°/0) | | | |
	LTU-1		LTU-2	
	PCB-C	PAH	PCB-C	PAH
Initial concentration (mg/kg-DW)	1·2	1660	8	970
Total measured reduction	39	61	27	67
Volatilized to air	0.003	0.013	0.02	0.012
Transport to subsoil	0.95	0.054	0.77	0.014
Transport to sump water [a]	0.0011	0.0008	0.0029	0.0011
Biodegraded (computed)	38	61	26	67

[a] Values are for unfiltered sump water. PCBs (<0.01 ppb) and PAHs (<20 ppb) were not detected in filtered water.

Figure 3. **Schematic diagram of pilot-scale land treatment unit.**

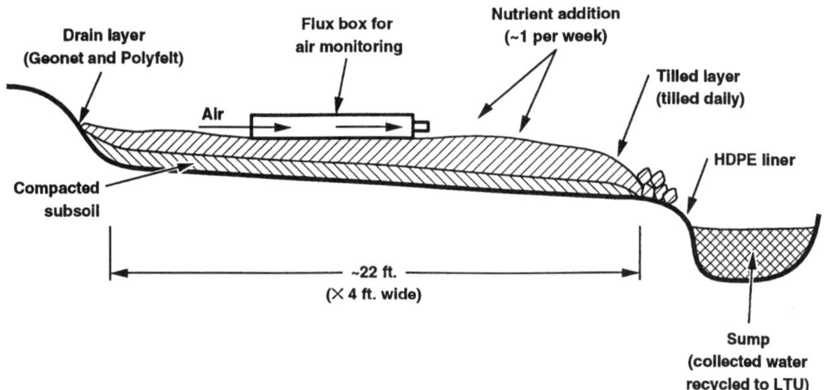

- **Untreated sludge added to clean sand at ~50:50 volume ratio (~10-20 sand:1 sludge on DWB)**
- **10 week operation**
- **Initial, midpoint and final sampling of air tilled layer**
- **Final sampling of compacted subsoil and sump water**
- **Extensive air monitoring**

Related to the expected low percent of PCB and PAH mass volatilized, it is noteworthy that for PCBs, only di-, tri-, tetra- and penta-chlorobiphenyls were detected initially in air above the LTUs at concentrations of approximately 0.8 µg/m³, 0.65 µg/m³, 0.2 µg/m³, and 0.05 µg/m³, respectively. After two months of active bioremediation, only a few trichlorobiphenyls were detected in air at a level of approximately 0.05 µg/m³, with all other PCBs not detected at a 0.01 µg/m³ detection level. Similar to PCBs, only 3-ring PAH compounds were initially detected in air above the LTUs at a level of approximately 160 µg/m³; no PAHs were detected in air at the end of two months of active bioremediation at a detection limit of 30 µg/m³. Thus, these results agree with expected bioavailability concepts in that volatilization of PCB and PAH compounds should be very low in terms of a total mass basis, and that there should be a preference for the lower molecular compounds having the relatively higher air concentrations. Also, only the lower molecular weight PCBs and PAHs should volatilize, and that any measured concentrations should proportionally decrease with increasing molecular weight (see previously cited Figure 2).

In terms of the mass transport of PCBs and PAHs to the compacted subsoil below the actively treated soil layer, it is noteworthy that for PCBs, the low percent of the total initial PCB mass measured corresponds to a total PCB concentration ranging between 0.06 to 0.1 mg/g-DW in the two LTUs. For PAHs, subsurface soil concentrations correspond to levels between 0.1 to 0.9 mg/g-dry weight.

For both the active (AB) and intrinsic (IB) bioremediation phases, Figures 4 and 5, respectively, give total PCB and total PAH soil concentrations with time. These results for the different time period are based on average statistically significant values at the 90% confidence level. For these two sets of plots, the time period of approximately Day 100 through 250 represents dormant winter conditions from December through the end of April.

Specific to PCBs, Figure 6 shows that during active bioremediation, the majority of the total PCB reductions were due to reductions of the more desorbable and bioavailable less chlorinated PCBs. Furthermore, the combination of active and passive bioremediation served to achieve further reductions of the 2 , 3 , 4 chlorinated PCB congener homolog groups, with no reduction of the 5- chlorinated and higher PCB congener homolog groups.

Similar trends were also observed with PAHs in that active bioremediation achieved the following PAH reductions: 90% of 3-ring, 84% of 4-ring, 27% of 5-ring, and 24% of 6-ring. Similar to PCBs, passive bioremediation served to result in greater percent reductions beyond these levels, especially for the 5- and 6-ring PAHs.

Thus, both the PCB and the PAH results serve to support the previously cited theoretical considerations predicting that the rate and extent of biodegradation will be greater for the more desorbable for the lower molecular weight compounds, and will proportionally decrease with increasing molecular weight.

Figure 4. Total PCB congener concentrations for LTUs during active and passive bioremediation.

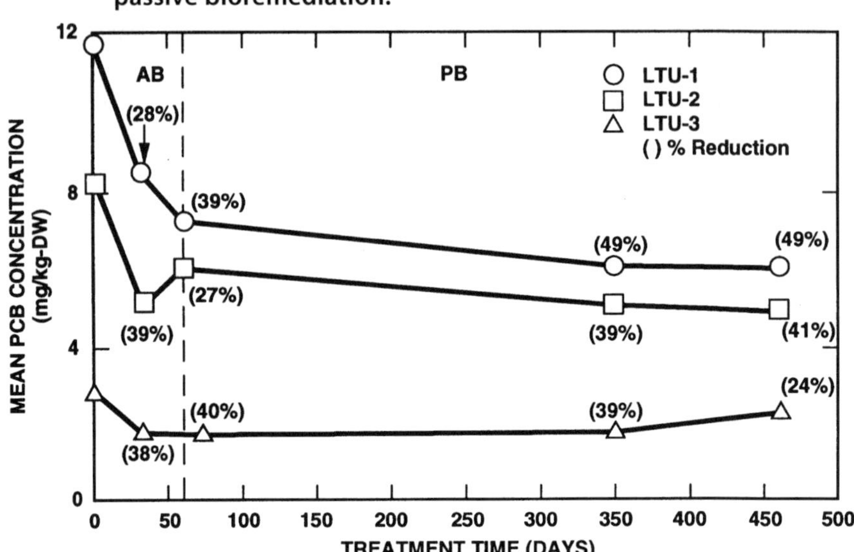

Figure 5. Total PAH concentrations for LTUs 1 and 2 during active and passive bioremediation.

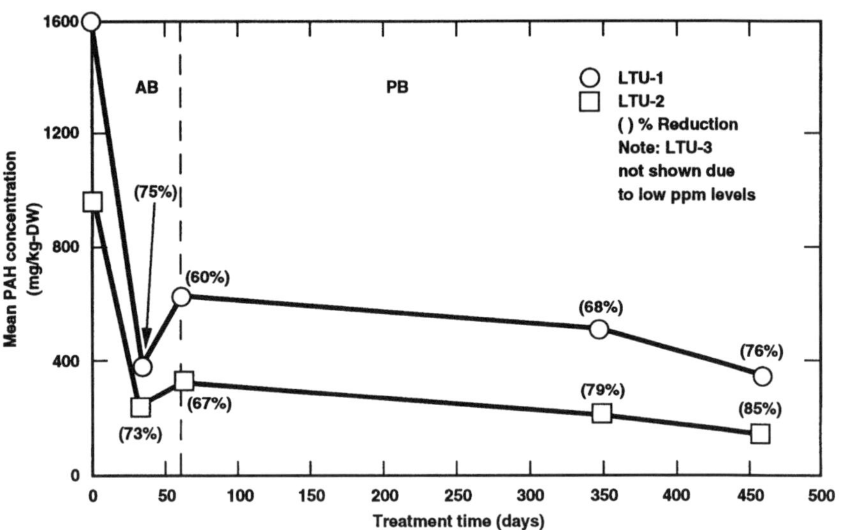

REDUCTIONS IN LEACHABILITY AND BIOAVAILABILITY

Results of Toxicity Characteristic Leaching Procedure (TCLP) soil extracts through various stages of land treatment serve to support the concept that biodegradation results in reduced desorption of PCBs and PAHs and thus reduced leachability

Figure 6. **Percent reductions of PCB homolog groups for LTUs 1 and 2 combined.**

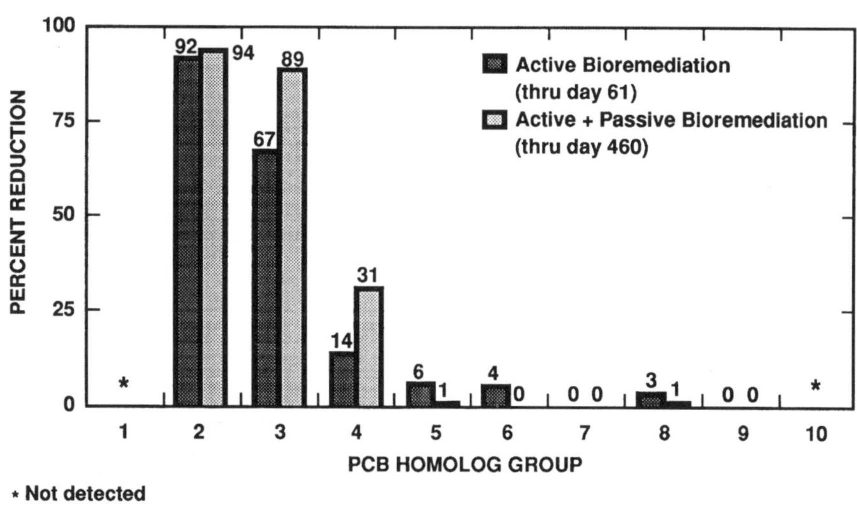

and bioavailability. Figures 7 and 8 show these TCLP reductions with time for total PCB and total PAH, respectively. The data represents single data points in time.

For PAHs, the majority of the reduction occurred during active bioremediation, while for PCBs, the TCLP reductions continued to proportionally increase through Day 450, well into the passive bioremediation phase.

Here, it is noteworthy that initial PAH-TCLP concentrations were an order of magnitude higher than PCB-TCLP concentrations. This is expected because PAHs were present in the soil/sludge/sediment matrix at order of magnitude higher concentrations. With no supplemental organics added, it is also theorized that these more leachable PAHs also served as co-metabolic substrates for the less soluble higher molecular PCB and PAH compounds.

The PCB homolog group plots in Figure 9 further serve to support the concept of bioavailability as being significantly affected by the desorbable leachable fraction. Here, the extent of PCB desorption decreases with increasing molecular weight (i.e., increasing degree of chlorination). These proportional reductions also agree with reductions in the PCB soil concentrations with time.

For PAHs, TCLP data showed that the majority of the bioavailable fraction consisted of 4-ring PAHs, which were reduced to very low ppb levels after active bioremediation.

Finally, water extracts of untreated soil/sludge/sediment material were initially toxic in terms of Microtox toxicity. After active and one season of passive bioremediation, Microtox toxicity was reduced to zero.

Figure 7. Total PCB congener concentrations in TCLP extracts for LTUs during bioremediation.

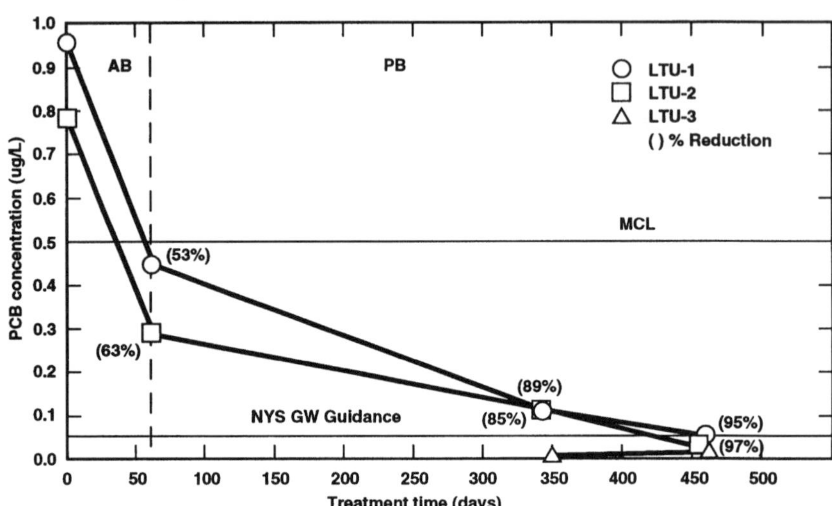

Figure 8. Total PAH concentrations in TCLP extracts for LTUs during bioremediation.

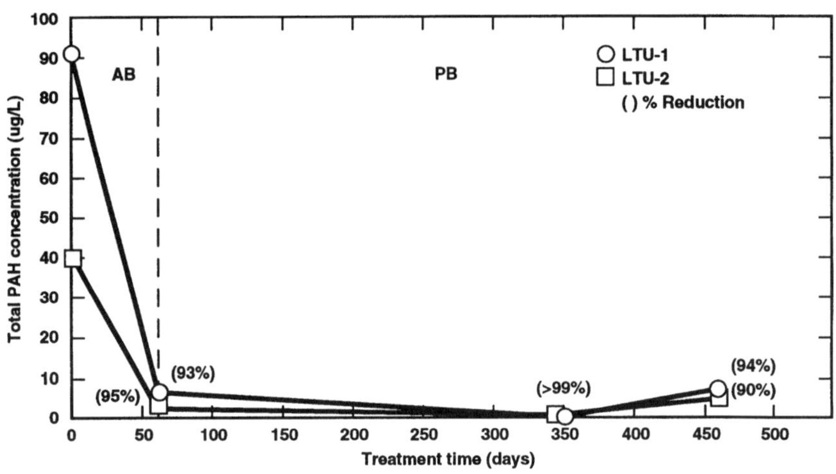

Supplemental Data Supporting Aerobic PCB and PAH Biodegradation

In addition to PCB and PAH analysis, the LTUs had elevated microbial counts. Total microbial populations ranged between 10^7 and 10^8 colony-forming units per gram of soil-dry weight (CFU/gm-dw). Correspondingly, both PCB degraders, identified as growing on biphenyl, and PAH degraders, identified as growing on phenanthrene, ranged from 10^2 CFU/gm-dw initially to 10^7 CFU/gm-dw after active bioremediation.

Also, there was very measurable nutrient (i.e., nitrogen and phosphorus) consumption during both active and passive bioremediation phases.

Finally, soil gas monitoring of the treated soil layer during the passive bioremediation time period showed no methane present and carbon dioxide and oxygen levels were near atmospheric conditions; i.e., approximately 0.5% CO_2 and approximately 20% O_2. Thus, aerobic conditions were present in the treated soil layer below the applied soil cap and vegetative cover.

SUMMARY AND CONCLUSIONS

For the three pilot-scale LTUs operated and monitored for PCB and PAH compounds, the following are noteworthy:

1. Greater than 40% PCB and greater than 90% PAH reductions were achieved via 2 months of active and one 6-month season of passive bioremediation in LTUs 1 and 2.
2. Both the rates and extents of degradation decreased with increasing molecular weight.
3. There was negligible volatilization, surface runoff, and subsurface migration of PCBs and PAHs.
4. Average TCLP-total PCB reductions ranged from approximately 0.86 μg/L to 0.03 μg/L, and TCLP-total PAH reductions ranged from approximately 65 μg/L to 4 μg/L.
5. PAHs served as co-metabolic substrates for PCBs and the higher molecular weight PAHs.
6. Reductions were achieved via biostimulation, with no bioaugmentation required.
7. Biodegradation achieved reduced bioavailability and Microtox toxicity.
8. Aerobic conditions existed during both the active and passive bioremediation phases.

Both the PCB and PAH data complement one another and independently serve to support theoretical predictions related to bioremediation of the soil/sludge/sediment contaminated materials. Specifics include:

1. Bioavailability is a function of desorption; this relates to both desorption from soils and sediments and NAPL dissolution.
2. Bioavailability decreases with increasing molecular weight. For PCBs, higher molecular weight relates to increasing extent of chlorination; for PAHs, it relates to increasing number of benzene rings.
3. The rate and extent of soil/sediment desorption and NAPL dissolution has a major influence on the rate and extent of biodegradation; influence of bioavailability.

Figure 9. PCB homolog concentrations in TCLP extracts for LTUs 1 and 2.

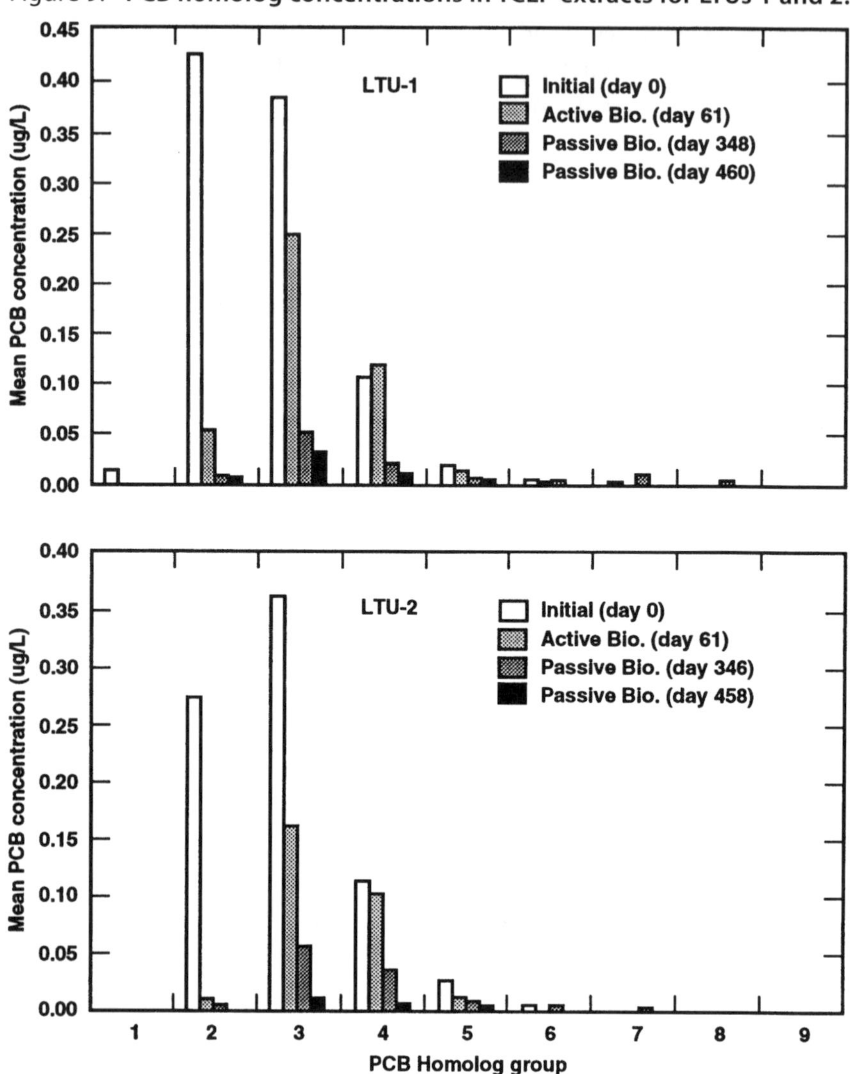

4. Related to PCBs and PAH, volatilization should be negligible and only be proportionally evident for the lower molecular weight compounds as controlled by desorption.

Engineered land treatment of PCB and PAH contaminated soils, sludges, and sediments can be designed and operated to provide a *technically sound, environmentally protective and cost-effective* remediation approach. Here, the concept of biostabilization is defined as:

1. *Active bioremediation* achieving reductions of more desorbable, leachable, bioavailable and more potentially mobile compounds; i.e., significant source reduction.

2. Subsequent *passive bioremediation* serves to ensure continued reductions of the remaining less desorbable and less mobile compounds as they slowly become bioavailable.

3. Measurable PCB and PAH chemical migration is reduced to zero, while passive bioremediation is achieving continuing reductions in PCB and PAH mass, bioavailability, and toxicity.

4. Engineered and/or institutional controls may be initially required to help ensure no or low environmental and human health risks; such controls can be reduced as PCB and PAH biodegradation proceeds, achieving reduced bioavailability and risk-based endpoints.

5. The entire biostabilization approach requires a 5- to 10-year time frame.

Long-Term Monitoring Plan

Under the proper environmental conditions, the data presented in this paper begin to support that all PCBs and PAHs are biodegradable; however, the rate and extent of biodegradation is highly influenced by their bioavailability.

To help support this premise, Alcoa plans to perform long-term passive bioremediation monitoring of the LTUs over a five-year time period. Such monitoring will include:

1. Yearly analysis of the treated soil layers. Such analyses will include:
   - replicate soil sampling for PCBs and PAHs;
   - nutrients, TOC, O&G, and pH;
   - microbial enumeration for total, PAH and PCB degraders;
   - soil gas analysis for oxygen, carbon dioxide, and methane; and
   - modified TCLP extract analysis for PCBs and PAHs.
2. Yearly analysis of storm runoff waters for PCBs and PAHs.
3. Yearly analysis of subsurface soil layers for PCBs and PAHs to measure any subsurface migration.
4. Related to developing a better scientific understanding of the controlling biological and physical/chemical mechanisms, additional terrestrial toxicity testing, and desorption/dissolution, mobility and biodegradation modeling work will also be concurrently carried out.

ACKNOWLEDGMENTS

While the work presented in this paper was directed and financially supported by Alcoa, the efforts of Warren Lyman of Camp, Dresser & McKee, and Ron Unterman and Randi Rothmel of Envirogen, Inc., especially during the early phases of the work, are to be recognized.

REFERENCES

Achman, D.R., Hornbuckle, K.C., and Elsenreich, S.J. 1993. Volatilization of Polychlorinated Biphenyl from Green Bay, Lake Michigan. *Environ. Sci. Technol.* 27(1), 75-87.

Adeel, Z., Luthy, R.G., Dzombak, D.A., and Smith, J.R. 1996. Leaching of PCB Compounds from Untreated and Biotreated Sludge-Soil Mixtures, Accepted for publication in *J. Contaminant Hydrology.*

Alcoa. 1994. Evaluation of In Situ Bioremediation Treatment of 60-Acre Lagoon Sludge Material for PCB Reduction/Immobilization at Alcoa's Massena, NY Facility. Internal report. EHS Technology Center, Alcoa Technical Center, Alcoa Center, PA.

Alcoa. 1995a. Bioremediation Tests for Massena Lagoon Sludges/Sediments - Program Summary, Volumes I and II. Internal report. EHS Technology Center, Alcoa Technical Center, Alcoa Center, PA

Alcoa. 1995b. Bioremediation Tests for Massena Lagoon Sludges/Sediments - Program Summary, Volumes I and II. Internal report. EHS Technology Center, Alcoa Technical Center, Alcoa Center, PA.

Alcock, R.E., McGrath, S.P., and Jones, K.C. 1995. The Influence of Multiple Sewage Sludge Amendments on the PCB Content of an Agricultural Soil Over Time. *Environ. Toxicol. Chem.*, 14 (4), 553-560.

Alexander, M. 1997. Sequestration and Bioavailability of Organic Compounds in Soil. In: *Environmentally Acceptable Endpoints in Soils.* (Linz, D. and Nakles, D., Eds.). Annapolis, MD, American Academy of Environmental Engineers.

Alexander, M. 1995. How Toxic are Chemicals in Soil? *Environ. Sci. Technol.* 29(11), 2713-2717

Anid, P.J., Nies, L., and Vogel, T.M. 1991. Sequential Anaerobic-Aerobic Biodegradation of PCBs in the River Model. In: *Onsite Bioreclamation.* (R.E. Hinchee and R.F. Olfenbuttel, Eds.).

Bedard, D.L. and Quensen, J.R. 1995. Microbial Reduction Cechlorination of Polychlorinated Biphenyls. In: *Microbial Transformation and Degradation of Toxic Organic Chemicals.* pp. 127-216. Wiley-Liss, Inc.

Bedard, D.L., Haberl, M.L., May, R.J., and Brennan, M.J. 1987. Evidence for Novel Mechanisms of Polychlorinated Biphenyl Metabolism in *Alcaligenes eutrophus* H850. *Appl. Environ. Microbiol.* 53(5), 1103-1112.

Bedard, D.L., Unterman, R., Bopp, L.H., Brennan, M.J., Haberl, M.L., and Johnson, C. 1986. Rapid Assay for Screening and Characterizing Microorganisms for the Ability to Degrade Polychlorinated Biphenyls. *Applied and Environmental Microbiol*ogy. 53(5), 1103-1112.

Bopp, L.H. 1986. Degradation of Highly Chlorinated PCBs by *Pseudomonas* strain LB400. *J. Indus. Microbiol.* 23-29.

Brown, J.F., Bedard, D.L., Brennan, M.J., Carnakan, J.C., Feng, H., and Wagner, R.E. 1987a. Polychlorinated Biphenyl Dechlorination in Aquatic Sediments. *Science* 236, 709-712.

Brown, J.F., Wagner, R.E., Feng, H., Bedard, D.L., and Brennan, M.J. 1987b. Environmental Dechlorination of PCBs. *Environ. Toxicol. Chem.* 579-593.

Brusseau, M.L. and Rao, P.S.C. 1989. Sorption Nonideality During Organic Contaminant Transport in Porous Media. In: *CRC Critical Reviews in Environmental Control.* Boca Raton, FL, CRC Press.

Carlieo, C.J., Schanke, C.A., and Consentini, C.C. 1992. Case Study of the Biological Treatment of Creosote Contaminated Soil. Paper presented at the AICHE 1992 Summer National Meeting, Minneapolis, MN.

Carroll, K.M., Harkness, M.R., Bracco, A.A., and Balcarcel, R.R. 1994. Application of a Permanent/Polymer Diffusional Model to the Desorption of Polychlorinated Biphenyls from Hudson River Sediments. *Environ. Sci. Technol.* 28, 253-258.

Chou, S.F.J. and Griffin, R.A. 1986. Solubility and Soil Mobility of Polychlorinated Biphenyls. In: *PCBs and the Environment*, Volume I. (J.D. Waid, Ed.). Boca Raton, FL, CRC Press.

Clark, R.R., Chian, E.S.K., and Griffin, R.A. 1979. Degradation of Polychlorinated Biphenyls by Mixed Microbial Cultures. *Appl. Environ. Microbiol.* 680-685.

Colghazier, E.W., Cox, T., and Davis, K. 1991. *Estimation of Resource Requirements for NPL Sites.* One of six unnumbered volumes of a University of Tennessee Report of Hazardous Waste Remediation. Knoxville, TN, University of Tennessee.

DiToro, D.M. and Horzempa, L.M. 1982. Reversible and Resistant Components of PCB Adsorption-Desorption Isotherms. *Environ. Sci. Technol.* 16, 594-602.

DiGiano, F.A., Miller, C.T., and Yoon, J. 1993. Predicting Release of PCBs at Point of Dredging. *ASCE J. Environ. Eng.* 119(1), 72-89.

Electric Power Research Institute. 1985. Application of Biotechnology to PCB Disposal Problems. Report No. CS-3807. Project No. 1263-16. Palo Alto, CA.

Feenstra, S. 1992. Evaluation of Multi-Component DNAPL Source by Monitoring of Dissolved-Phase Concentrations. In: *Subsurface Contamination by Immiscible Fluids.* (Weyer, K.U. and Balkan, A.A., Eds.). Proceedings of the International Conference of Subsurface Contamination of Immiscible Fluids, Calgary, Canada, April 18-20, 1990.

Fish, K.M. and Principe, J.M. 1994. Biotransformation of Aroclor 1242 in Hudson River Test Tube Microcosms. *Appl. Environ. Microbiol.* 60(12), 4289-4296.

Focht, D.D. and Brunner, W.W. 1985. Kinetics of Biphenyl and Polychlorinated Biphenyl Metabolism in Soil. *Appl. Environ. Microbiol.* 50(10), 1058-1063.

Furukawa, K. 1982. Microbial Degradation of Polychlorinated Biphenyls. In: *Biodegradation and Detoxification of Environmental Pollutants*, pp. 33-57. (Chakrabarty, A.M., Ed.). Boca Raton, FL, CRC Press, Inc.

Gan, D.R. and Berthouex, P.M. 1994. Disappearance and Crop Uptake of PCBs from Sludge-Amended Farmland. *Water Environ. Res.* 66(1), 54-60.

Gas Research Institute. 1990. Biological Treatment of Soils Containing Gas Plant Residues. Report No. GRI-90/01117, Chicago, IL.

Gas Research Institute. 1992. The GRI Accelerated Biotreatability Protocol for Assessing Conventional Biological Treatment of Soils: Development and Evaluation Using Soils from Manufactured Gas Plant Sites. Report No. GRI-92/0499, Chicago, IL.

General Electric Company. 1992. 1991 In Situ Hudson River Study: A Field Study on Biodegradation of PCBs in Hudson River Sediments. Corporate Research and Development, Schenectady, NY.

Girvin, D.C., Sklarew, D.S., Scott, A.J., and Zipperer, J.P. 1990. Release and Attenuation of PCB Congeners: Measurement of Desorption Kinetics and Equilibrium Sorption Partition Coefficients. Electric Power Research Institute Report GS-6875. Palo Alto, CA.

Guerin, W.F. and Boyd, S.A. 1992. Differential Bioavailability of Soil-Sorbed Naphthalene to Two Bacterial Species. *Appl. Environ. Microbiol.* 58(4), 1142-1152.

Harkness, M.R., McDermott, J.B., Abramowicz, D.A., Salvo, J.J., Flanagan, W.P., Stephens, M.L., Mondello, F.J., May, R.J., Lobos, J.H., Carroll, K.M., Brennan, M.J., Bracco, A.A., Fish, K.M., Warner, G.L., Wilson, P.R., Kietrich, D.K., Lin, D.T., Morgan, C.B., and Gatelay, W.L. January, 1993. In Situ Stimulation of Aerobic PCB Biodegradation in Hudson River Sediments. *Science* 259, 503-507.

Heitkamp, M.A. and Cerniglia, C.E. 1989. Polycyclic Aromatic Hydrocarbon Degradation by a Mycobacterium in Microcosms Containing Sediment and Water from a Pristine Ecosystem. *Appl. Environ. Microbiol.* 55(8), 1968-1973.

Heitkamp, M.A., Franklin, W., and Cerniglia, C.E. 1988. Microbial Metabolism of Polycyclic Aromatic Hydrocarbons: Isolation and Characterization of a Pyrene-Degrading Bacterium. *Appl. Environ. Microbiol.* 54(10), 2549-2555.

Heitkamp, M.A. and Cerniglia, C.E. 1988. Mineralization of Polycyclic Aromatic Hydrocarbons by a Bacterium Isolated from Sediment Below and Oil Field. *Appl. Environ. Microbiol.* 54(6), 1612-1614.

Herbes, S.E. and Schwall, L.R. 1978. Microbial Transformation of Polycyclic Aromatic Hydrocarbons in Pristine and Petroleum-Contaminated Sediments. *Appl. Environ. Microbiol.* 35(2), 306-316.

Hornbuckle, K.C., Achman, D.R., and Elsenreigh, S.J. 1993. Over-Water and Over-Land Polychlorinated Biphenyls in Green Bay, Lake Michigan. *Environ. Sci. Technol.* 27(1), 87-98.

Howard, P.H., Crisman, J., and Hueber, J. 1988. Review of Soil Degradation Rates for Polycyclic Aromatic Hydrocarbons, (Final Draft) Syracuse Research Corporation Report No. SRC-TR-88-093. Syracuse, NY, April, 1988.

Jansen, L. 1995. The Effects of Particulates on Aerobic PCB Biodegradation. M.S. Thesis, Civil Engineering Department, State University of New York at Buffalo, Buffalo, NY.

Kelley, I., Freeman, J.P., Evans, F.E., and Cerniglia, C.E. 1991. Identification of a Carboxylic Acid Metabolite from the Catabolism of Fluoranthene by a Mycobacterium. *Appl. Environ. Microbiol.* 57(3), 636-641.

Law, A.T. and Button, D.K. 1977. Multiple-Carbon-Source-Limited Growth Kinetics of a Marine Coryneform Bacterium. *J. Bacteriol.* 129, 115-123.

Loehr, R.C., and Webster, M.T. 1997. Effect of Treatment on Contaminant Availability, Mobility and Toxicity. In: *Environmentally Acceptable Endpoints in Soils.* Chapter 2. (Linz, D. and Nakles, D., Ed.). Annapolis, MD, American Academy of Environmental Engineers.

Luthy, R.G., Dzombak, D.A., Shannon, M., Unterman, R., and Smith, J.R. 1997. Dissolution of PCB Congeners from an Aroclor and an Aroclor/oil Mixture. *Water Research.* 31(3), 561-573.

Mackay, D., Shiu, D.Y., and Ma, K.C. 1992a. *Illustrated Handbook of Physical-Chemical Properties and Environmental Fate for Organic Chemicals; Vol. I - Monoaromatic Hydrocarbons, Chlorobenzenes and PCBs.* Boca Raton, FL, Lewis Publishers.

Mackay, D., Shiu, W.Y., and Ma, K.C. 1992b. *Illustrated Handbook of Physical-Chemical Properties and Environmental Fate for Organic Chemicals; Vol. II - Polynuclear Aromatic Hydrocarbons, Polychlorinated Dioxins and Dibenzofurans.* Boca Raton, FL, Lewis Publishers.

Mackay, D.M. and Cherry, J.A. 1989. Groundwater Contamination: Pump-and-Treat Remediation. *Environ. Sci. Technol.* 23(6), 630-636.

Mahaffey, W.R., Gibson, D.T., and Cerniglia, C.E. 1988. Bacteria Oxidation of Chemical Carcinogens: Formation of Polycyclic Aromatic Acids from Benzo(a)anthracene. *Appl. Environ. Microbiol.* 54(10), 2415-2423.

McDermott, J.B., Unterman, R., Brennan, M.J., Brooks, R.E., Mobley, D.P., Schwartz, C.C., and Dretrich, D.K. 1989. Two Strategies for PCB Soil Bioremediation: Biodegradation and Surfactant Extraction. *Environ. Prog.* 8(1), 46-51.

McLaughlin, D.B., Armstrong, D.E., and Andren, A.W. 1994. Oxidation of Polychlorinated Biphenyl Congeners Sorbed to Particles. In: *Proceedings of the 48th Industrial Waste Conference,* pp. 349-353. West Lafayette, IN, Purdue University.

Middleton, A.C., Nakels, D.V., and Linz, D.G. 1991. The Influence of Soil Composition on Bioremediation of PAH-Contaminated Soil. *Remediation,* 391-406.

Mohn, W.M. and Tiedje, J.M. 1992. Microbial Reductive Dehalogenation. *Microbiol. Rev.* 56 (3), 482-507.

Mueller, J.G., Chapman, P.J., Blattman, B.O., and Pritchard, P.H. 1990. Isolation and Characterization of a Fluoranthene - Utilizing Strain of *Pseudomonas paucimobilis. Appl. Environ. Microbiol.* 56(4), 1079-1086.

Mueller, J.G., Middaugh, D.P., Lantz, S.E., and Chapman, P.J. 1991. Biodegradation of Creosote and Pentachlorophenol in Contaminated Groundwater: Chemical and Biological Assessment. *Appl. Environ. Microbiol.* 57(5), 1277.

Niaki, S. 1987. Treatment Technologies for PCB-Contaminated Soils. Conference Proceedings -Haztech International Conference. St. Louis, MO, August 26-28.

Pignatello, J.J. and Xing, B. 1996. Mechanisms of Slow Sorption of Organic Chemicals to Natural Particles. *Environ. Sci. Technol.* 30(1), 1-11.

Pitter, P. 1985. Correlation of Microbial Degradation Rates with the Chemical Structure. *Acta hydrochim, hydrobiol* 13(4), 453-460.

Quensen, J.F., Tiedje, J.M., and Boyd, S.A. 1988. Reductive Dechlorination of Polychlorinated Biphenyls by Anaerobic Microorganisms from Sediments. *Science,* 752-754.

Rifkin, E. and LaKind, J. 1991. Dioxin Bioaccumulation: Key to a Sound Risk Assessment Methodology. *J. Toxicol. Environ. Health* 33, 103-112.

Rifkin, E. and Bouwer, E.J. 1994. A Proposed Approach for Deriving National Sediment Criteria for Dioxin. *Environ. Sci. Technol.* 28, 441-443.

Rittman, B.E., Balocchi, A.J., and Bareye, P. 1990. Fundamental Quantitative Analysis of Microbial Activity in Aquifer Bioremediation. U.S. DOE Progress Report No. DOE/ER/60-733-1. U.S. Gov. Print. Office, Washington, DC.

Rochkind-Dubinsky, M.L., Layler, G.S., and Blackburn, J.W. 1986. *Microbiological Decomposition of Chlorinated Aromatic Compounds,* pp. 142-145. New York, Marcel Dekker, Inc.

Ryan, J.R., Loehr, R.C., and Rucker, E. 1991. Bioremediation of Organic Contaminated Soils. *J. Hazardous Mater.* 28, 159-169.

Rynaarts, H.H.M., Bachmann, A., Jumelet, J.C., and Zehnder, J.B. 1990. Effect of Desorption and Intraparticle Mass Transfer on the Aerobic Biomineralization of a-Hexachlorocyclohexane in a Contaminated Calcareous Soil. *Environ. Sci. Technol.* 24(9), 1349-1354.

Schmidt, S.K. and Alexander, M. 1985. Effects of Dissolve Organic Carbon and Second Substrates on the Biodegradation of Organic Compounds at Low Concentrations. *Appl. Environ. Microbiol.* 49(4), 822-827.

Sedlak, D.L. and Andren, A.W. 1994. The Effect of Sorption on the Oxidation of Polychlorinated Biphenyls (PCBs) by Hydroxyl Radical. *Water Res.* 28, 1207-1215.

Shekata, T.E. and Marr, A.G. 1971. Effect of Nutrient Concentration on the Growth of Escherichia Coli. *J. Bacteriol.* 107, 210-216.

Sikka, H.C. 1991. Microbial Degradation of Polynuclear Aromatic Hydrocarbons Associated with Coal Gasification Sites. Progress Report to New York State Center for Hazardous Waste Management, Great Lakes Laboratory, SUNY College at Buffalo, Buffalo, NY.

Sims, R.C. and Overcash, M.R. 1982. Fate of Polynuclear Aromatic Compounds in Soil Plant System. *Residue Rev.* 88, 1.

Smith, J.R., Tomicek, R.M., Swallow, P.V., Weightman, R.L., Nakles, D.V., and Hebling, M. 1995. Definition of Biodegradation Endpoints for PAH Contaminated Soils Using a Risk-Based Approach. In: *Hydrocarbon Contaminated Soils.* Volume V, pp. 531-572. (Kostecki, P.T., Calabrese, E.J., and Bonazountas, M. Eds.). Amherst, MA, Amherst Scientific Publishers.

Stucki, G. and Alexander, M.. 1987. Role of Dissolution Rate and Solubility in Biodegradation of Aromatic Compounds. *Appl. Environ. Microbiol.* 53(2), 292-297.

Thomas, J.M. and Alexander, M. 1987. Colonization and Mineralization of Palmitic Acid by *Pseudomonas pseudoflava.* In: *Microbial Ecology,* pp. 75-80. New York, Springer-Verlag, Inc.

Unterman, R., Mondello, F.J., Brennan, J.J., Brooks, R.E., Mobley, D.P., McDermott, J.B., and Schwartz, C.E. 1987. Bacterial Treatment of PCB-Contaminated Soils: Prospects for the Application of Recobinant DNA Technology. *Proceedings— 2nd international Conference on New Frontiers for Hazardous Waste Management.* EPA Report No. EPA/600/9-87/018F. Pittsburg, PA. pp. 259-264.

Unterman, R., Dedard, D.L., Brennan, J.J., Bopp, L.H., Mondello, F.J., Brooks, R.E., Mobley, D.P., McDermott, J.B., Schwartz, C.C., and Kietrick, D.K. 1988. Biological Approaches for Polychlorinated Biphenyl Degradation. In: *Environmental Biotechnology.* p. 2. (Omenn, G.S., Ed.). New York, Plenum Publishing Corporation.

U.S. Environmental Protection Agency. August, 1995a. Bioremediation in the Field. U.S. EPA Rep. EPA/540/N-95-500 - No. 12. U.S. Gov. Print. Office, Washington, DC.

U.S. Environmental Protection Agency, Office of Solid Waste and Emergency Response. September, 1995b. Innovative Treatment Technologies: Annual Status Report (Seventh Edition). U.S. EPA Rep. EPA-542-R-95-008. U.S. Gov. Print. Office, Washington, DC.

U.S. Environmental Protection Agency, Office of Research and Development. February, 1993a. Bioremediation Issue Plan - Issue 26. U.S. Gov. Print. Office, Washington, DC.

U.S. Environmental Protection Agency. 1993b. The Assessment and Remediation of Contaminated Sediment, ARCS Program, Remediation Guidance Document. USEPA Great Lakes National Program Office. Contract No. 68-62-0134. U.S. Gov. Print. Office, Washington, DC.

U.S. Environmental Protection Agency. 1979. Water Related Environmental Fate of 129 Priority Pollutants. NTIS No. PB80-204381. U. S. EPA Report 440/4-70-0296. U.S. Gov. Print. Office, Washington, DC.

van der Kooij, D., Visser, A., and Kijen, W.A.M. 1982. Growth of Aeromonas hydrophila at Low Concentrations of Substrates Added to Tap Water. *Appl. Environ. Microbiol.* 43, 1139-1150.

Williams, W.A. 1994. Microbial Reductive Dechlorination of Trichlorobiphenyls in Anaerobic Sediment Slurries. *Environ. Sci. Technol.* 28(4), 630-635.

Witkowski, P.J., Jaffe, P.R., and Ferrara, R.A. 1988. Sorption and Desorption Dynamics of Aroclor to Natural Sediments. *J. Contaminant Hydrology* (2), 249-269.

CHAPTER 32

Bioremediation of Hydrocarbon Contaminated Soils by Using Pulp and Paper Residues: A Resource Recovery Management

Mohammad H. Golabi, M.L. Cabrera, W.P. Miller, L.A. Morris, and **M.E. Sumner,** Department of Soil and Environmental Science, The University of Georgia, Athens, Georgia

BACKGROUND

The release of petroleum products into the environment has become a growing concern over the years. Of particular concern is contaminant movement into the subsurface, where valuable groundwater and connected surface water resources are at risk (Lyman et al., 1990). When petroleum products are released into the subsurface, they move primarily by gravity down through the unsaturated zone, with only minor horizontal movement due to dispersion and capillary forces (Lyman et al., 1990).

Awareness of the problem has resulted in the development and use of a variety of remediation techniques to remove petroleum products from the subsurface. Among the variety of techniques for the cleanup of the petroleum contaminated sites is the in-situ biorestoration of the site. Biorestoration relies on naturally occurring or genetically altered microorganisms to transform contaminants to less hazardous compounds (Lyman et al., 1990).

Petroleum products are used extensively in chemical (chemical manufacturing, solvent) and transportation (gasoline, diesel, and jet fuels) industries. All these materials are handled multiple times from production to the final market. During transport and use these materials are occasionally released accidentally to soils and groundwater and can represent a significant threat to environmental quality. On the other hand, tremendous amounts of sludge and waste by-products are produced by paper and pulp industries around the nation, with a significant portion being dumped in landfills while it could be used as a viable resource to clean up contaminated soils via biodegradation.

Remediation of petroleum contaminated sites is a subject almost without limits. Sites can be as small and simple as a corner service station with little contamination, to large refinery contaminated sites with hundreds of compounds over every square inch of the ground, to a highway site with a large spill from a tanker turnover, to petroleum contamination offshore from a ruptured vessel. The field has reached the point where it is no longer a matter of digging out to property line and backfilling the excavation.

Most petroleum compounds are chronic toxic hazards to humans, especially the aromatic compounds routinely found in the most common contaminant, gasoline, since gasoline contains a substantial percentage of benzene (Cole, 1994). The good news, however, is that many petroleum products are readily consumed by soil bacteria, with the result that a petroleum contaminated site may, in effect, remediate itself, particularly if the bacterial activity is enhanced by an organic medium such as primary and secondary sludge produced by pulp and paper industries.

Bioremediation is not a new concept. Microbiologists have studied the process since the 1940s; however, bioremediation became known to a broader public in the United States only in the late 1980s as a technology for cleanup of shorelines contaminated with spilled oil (Hoff, 1993). Using sludges produced by pulp and paper industries for biodegradation, however, has not been looked at and deserves to be studied vigorously.

Bioremediation has broad application in terrestrial and freshwater environments for treating soils and sediments contaminated with oil and other substances, as well as for coastal environments impacted by oil spills. Bioremediation can be as simple as applying a garden fertilizer to an oil-contaminated beach, or as complex as an engineered treatment "cell" where soils or other media are manipulated, aerated, heated, and treated with various chemical compounds to promote degradation (Hildebrandt and Wilson, 1991).

As it is referred to by R.Z. Hoff (1993), bioremediation can provide a low-impact cleanup option for use in sensitive environments where other techniques are inadvisable or inappropriate. In such sensitive environments, intrusive cleanup activities can cause more overall detrimental impact than gain. Examples include marshes, soft intertidal substrates, and eelgrass beds (Hoff, 1993). Bioremediation also is a potential cleanup technique for use in tropical habitats such as mangroves (Scherrer and Mille, 1989). Though the research base is not extensive for evaluating the effectiveness of bioremediation in tropical areas, mangroves are highly sensitive to oil, and bioremediation could provide an effective way to treat residual oiling in these habitats (Hoff, 1993).

In many ways, bioremediation is now much more well established as an oil spill cleanup technique than it was a few years ago. Oil spill cleanup strategies for other parts of the world are now being designed to include measures to enhance natural degradation processes, or "bioremediation" (Kerry, 1991). Among these, nutrient enhancement as referred by Kerry (1991), has been used successfully at many sites including beaches contaminated by oil from the *Exxon Valdez* in Alaska. Ongoing research efforts continue to help define the limiting variables of the degradation process in various shoreline and other environments (Hoff, 1993). By applying appropriate techniques and proper medium, bioremediation can offer some very substantial benefits for the marine as well as contaminated inland soils and sediments.

As reported by Kerry (1991), the response to added nutrients indicated that organisms capable of the degradation of SAB (Special Antarctic Blend distillate) were present in Antarctic mineral soils and that nutrient applications enhanced this process. In-situ biorestoration of polluted sandy soils that is investigated by RIVM (National Institute of Public Health and Environmental Protection in Netherlands) is based on the estimation of the natural degradation processes in the soil itself (Visscher and Brinkman, 1989). In this investigation it was observed that by addition of nutrients (N,P) and oxygen source (peroxide or air), a 2900-6400 mg gasoline/kg soil was reduced to 10-600 mg gasoline/kg soil (depending on depth) after about six months (Visscher and Brinkman, 1989).

Research on response of bacterial populations to petroleum hydrocarbon contamination in tropical soils by Robert and Israel (1994) indicated that hydrocarbon-degrading populations of bacteria in contaminated soils were stimulated as a result of the addition of N and P fertilizers. Pometto and co-workers (1994) studied the bioremediation potential of soybean hulls for cleaning up petrochemical spills in soil and found that in every amendment with soybean hulls, biological activity was stimulated in contaminated gravel by 14 days. They also stated that a GC analysis showed 60% reduction in diesel fuel within the four week period. In a land treatment of contaminated soil from a former bulk fuel storage facility, Berger and Schwartz (1994) evaluated the effect of wood chips and dewatered sewage sludge or fresh cow manure for bioremediation of soils contaminated with diesel fuel. In their studies they found that after 17 days the diesel range organic contamination in all land treatment cells declined below the practical quantitative limits (PQL) of 10 mg/kg. Other workers in the field (Wilson and Portier, 1994; Barton and Portier, 1994; and Pitonzo and Amy, 1994) have reported that in-situ bioremediation of petroleum contamination had decreased considerably within a short period of time, and named the in-situ bioremediation as a successful process for cleaning up the contaminated sites.

In-situ treatment through chemical or biological transformation of the contaminant has the advantage of dealing with all aspects of the contamination problem and, by destroying the contaminant, of providing a permanent solution (Riser-Roberts, 1992). Pump-and-treat methods (aboveground) treat only the groundwater, while the contaminated soil continues to recontaminate the groundwater. In contrast, in-situ treatment can reach organics trapped or sorbed by the soil matrix (Lee and Ward, 1985). In addition, since microorganisms are capable of degrading a wide range of organic compounds (Riser-Roberts, 1992), the addition of sludge can enhance their activities for in-situ remediation of contaminated sites.

DEFINITIONS

It is important to define bioremediation within the context of biodegradation, a naturally occurring process. Biodegradation is a large component of oil weathering and is a natural process whereby bacteria or other microorganisms alter and break down organic molecules into other substances, eventually producing fatty acids and carbon dioxide (Hoff, 1993). The term "biodegradation" is often used to describe a variety of different microbial processes that occur in the natural ecosystem, such as mineralization, detoxication, co-metabolism, or activation (Alexander, 1980). Biodegradation can also be defined as the breakdown of organic compounds in nature by microorganisms, such as bacteria, actinomycetes, and fungi (Sims and Bass, 1984). Bioremediation is the acceleration of this process through the addition of nutrients or other materials to contaminated environments or through manipulation of the contaminated media using techniques such as aeration or temperature control (Hoff, 1993). The microorganisms derive energy and may increase in biomass from the degradation processes (Lee and Ward, 1985).

Mineralization occurs when there is complete biodegradation of an organic molecule to inorganic compounds; i.e., carbon dioxide, water, and mineral ions such as nitrogen, phosphorus, and sulfur (Sims and Bass, 1984). Under anaerobic conditions, methane may be produced, and nitrate nitrogen may be lost as N_2 or N_2O gas through denitrification. Microorganisms can also transform hazardous organic compounds into innocuous or less toxic organic metabolic products (Riser-Roberts, 1992). Chemical alteration can also be the result of co-metabolism (co-oxidation); i.e., growth on another substrate while the organic molecule is degraded coincidentally (Alexander, 1980). This process may be promoted by enzymes that catalyze reactions of chemically related substrates (Riser-Roberts, 1992).

Contaminants in solution in groundwater, as well as vapors in the unsaturated zone, can be completely biodegraded or transformed to new compounds (Wilson et al., 1986). Biodegradation converts petroleum components to compounds of lower molecular weight, while biotransformation can convert them to more polar compounds of carbon number equal to the parent compound (Atlas, 1977). Although the compositions of refined oils, crude oils, and oily wastes are quite different, approaches toward their stimulated biodegradation have been similar (Riser-Roberts, 1992).

Organisms that occur naturally in almost every soil system (the indigenous microbial population) appear to be the chief agents involved in the metabolism of chemicals in waters and soils (Riser-Roberts, 1992). Heterotrophic bacteria and fungi are responsible for most of the chemical transformations. Aerobic bacteria, on the other hand, including the actinomycetes, cyanobacteria, anaerobic bacteria, fungi, and some true algae have all been shown capable of degrading many classes of organic chemicals (Amdurer et al., 1985).

Many aerobic bacteria found in soil and water utilize biologically produced substances, such as the sugars, protein, fats, and hydrocarbons of plant and animal wastes, and can also metabolize petroleum hydrocarbons, converting them to carbon dioxide and water (Brown et al., 1985). Anaerobic bacteria are important for the biodegradation of many chlorinated pesticides and heavily halogenated organics (e.g., trichloroethylene, pentachlorophenol) (Riser-Roberts, 1992).

The ability of microorganisms to degrade or transform toxic substances has been applied to the treatment of environmental contamination by hazardous organic compounds (Riser-Roberts, 1992). Biodegradation has been used, with reasonable success, to treat aquifers contaminated with petroleum hydrocarbons (Lee and Ward, 1985). Biorestoration is useful for hydrocarbons, especially water-soluble compounds and low levels of other compounds that would be difficult to remove by other means.

As referred to by Robert (1992), Buckingham (1981) stated that natural soil bacteria may be present in dormant or slow-growing state, but when stimulated by a specific set of environmental conditions and/or by the addition of nutrients, may multiply rapidly and subsequently adapt to the new environment (Buckingham, 1981). Some of the more common genera of bacteria involved in biodegradation of oil products include *Nocardia, Pseudomonas, Acinetobacter, Flavobacterium, Micrococcus, Arthrobacter,* and *Corynebacterium* (Riser-Roberts, 1992). Tests have revealed that cultures containing more than one genus appear to have greater hydrocarbon-utilizing capabilities than many of the individual culture isolates (Riser-Roberts, 1992).

Degradation in contaminated soil and other aquifers may be affected by environmental constraints, such as dissolved oxygen; pH; temperature; toxicants; oxidation-reduction potential; availability of inorganic nutrients, including nitrogen, phosphorus, and other necessary nutrients; salinity; and the concentration and the nature of the organic material. Dissolved oxygen may promote biodegradation of wastes, such as aromatics, while some compounds, such as highly chlorinated compounds, can be degraded more readily without oxygen (Lee and Ward, 1985). The number and type of the organisms present in the environment will also play an important role in the degradation processes (Riser-Roberts, 1992). Optimizing the conditions of pH, temperature, soil moisture content, soil oxygen content, and nutrient concentration and availability, therefore, will stimulate growth of the organisms that will metabolize the particular contaminants present (Sims and Bass, 1984). Some hazardous compounds may be degradated more readily under aerobic condition, and some, under anaerobic conditions. Anaerobic conditions are always present at soil microsites (Riser-Roberts, 1992).

Biodegradation is a viable option for treatment of hazardous chemical spills. Its feasibility depends upon the availability of necessary equipment and manpower, as well as upon decontamination time restrictions (Buckingham, 1981). Once it is established, however, biodegradation is potentially less expensive than any other approach to neutralizing toxic wastes (Nicholas, 1987). As referred to by Riser-Roberts (1992), Ghisalba (1983) stated that such systems involve a low capital investment, have a low energy consumption, and are often self-sustaining operations because biological means of decomposition requires less energy than physiochemical processes and can be a competitive option under certain circumstances. Bioremediation, on the other hand, is the acceleration of this process through addition of nutrients or other materials to contaminated environments, or through manipulation of the contaminated media such as aeration and temperature control.

The use of sludge for bioremediation is a cost-effective method which enhances the microbial degradation of hydrocarbon by-products to CO_2, energy, and H_2O, and results in clean soil and water after a relatively short period of time. Moreover, biodegradation is environmentally sound, since it destroys organic contaminants and, in most cases, does not generate problems as with waste products (Riser-Roberts, 1992). However, using municipal sludges in biodegradation may be problematic because of the high concentration of heavy metal in them (Riser-Roberts, 1992). In contrast, sludges produced by pulp and paper mills are generally exclusive of heavy metal (Morris, 1995). On the other hand, due to the low level of N and P and other micronutrients, the application of fertilizers to the contaminated soil will certainly enhance the biodegradation by these sludges.

LABORATORY EXPERIMENTATION

An initial laboratory study was conducted to evaluate CO_2 evolution from a hydrocarbon contaminated soil after pulp and paper by-products were added to the soil. Four different treatment were used in this initial study:

1. Soil (500 gr) + primary sludge + $NH_4 NO_3$ + 50 mL of gasoline mixture (gas + oil).
2. Soil (500 gr) + primary sludge + $NH_4 NO_3$ + 50 mL H_2O
3. Soil (500 gr) + $NH_4 NO_3$ + 50 mL H_2O
4. Soil (500 gr) + $NH_4 NO_3$ + 50 mL of gasoline mixture

The amount of the sludge used was 1/3 of the soil volume used in each treatment.

Treated samples were placed in 1-liter mason jars with airtight lids. The lids were fitted with a rubber septum to sample the headspace. The experimental mason jars were placed in a laboratory hood for incubation.

At 1, 3, 6, 8, 10, 15, and 20 days after incubation, headspace samples were collected from each jar and injected into 1-mL glass vials. The mason jars were subsequently aerated after each sampling event. A varian 3600 standard chromatograph with a TCD detector was used to measure CO_2 concentration in the sample.

Ammonium nitrate was added to all treatments to provide necessary nutrients to enhance the bidegradation processes. According to Riser-Roberts (1992), addition of nutrients to hydrocarbon contaminated soil has proved useful in increasing microbial degradation. Since nitrogen and phosphate are the nutrients most frequently present in limiting concentrations in soil (EPA, 1985), research work in this area (Westlake, Jobson, and Cook, 1978) has shown that the application of fertilizer stimulates greater microbial growth and utilization of some components of oil.

RESULTS AND DISCUSSION

As shown in Figure 1, the results indicated that Treatment 1 in which both sludge and the gasoline mixture were added had the highest level of CO_2 concentration at the time of each measurement. The next highest CO_2 evolution was from Treatment 2, where only the sludge was added to the soil. Treatments 3 and 4 showed very low levels of CO_2 evolution. This suggested that the addition of primary sludge enhanced the biological activity of the soil, which in turn enhanced hydrocarbon decomposition and produced higher amount of CO_2 in the process. As can be seen from Table 1, the cumulative amount of CO_2 evolved is highest when both primary sludge and the gasoline mixture were added to the soil. However, when only the gasoline was added without sludge application, the cumulative CO_2 evolution was the lowest among all treatments. In fact, the difference between CO_2 evolved

Figure 1. **Showing the CO_2 evolution for each treatment at different incubation times.**

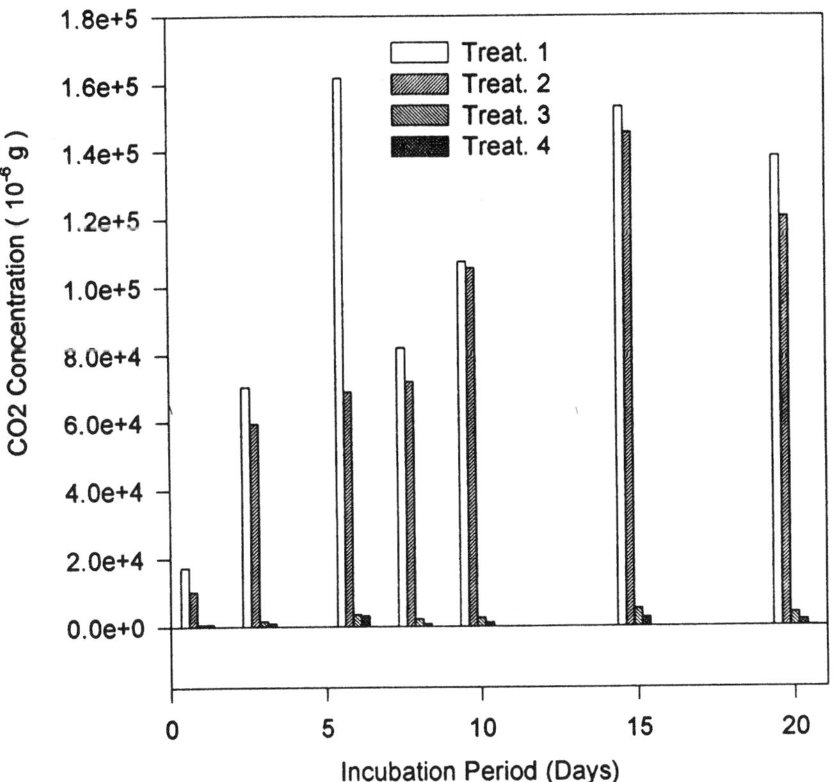

Table 1.   Cumulative CO_2-C Evolved (µg) from Each Treatment During the Incubation Process.

Incubation (days)	Treatments			
	Sludge Applied	Gasoline Added		
		YES	NO	Difference
1	YES	17,422	10,280	7,142
	NO	545	568	-23
3	YES	87,680	69,687	17,993
	NO	1,451	2,096	-644
6	YES	24,9535	13,8567	11,0968
	NO	4,605	5,586	-981
8	YES	33,1425	21,0428	12,0996
	NO	5,343	7,726	-2,383
10	YES	43,8895	31,5963	12,2931
	NO	6,534	10,292	-3,758
15	YES	59,1941	46,1469	13,0472
	NO	9,124	15,555	-6,431
20	YES	73,0160	58,1973	14,8187
	NO	10,952	19,438	-8,485

from Treatment 4 and 3 was negative at all times. This was an indication that the addition of gasoline mixture alone might have been detrimental to soil microorganisms, decreasing the microbial activities during the incubation period.

This experimentation is being extended to a larger-scale laboratory study to include other mill by-products in various other treatments. Gas chromatography (GC), which is currently being used to evaluate the CO_2 evolution will also be employed later to quantify remaining hydrocarbon in the soil in these later tests.

At this stage of the experimentation our conclusion is that the application of the primary sludge as energy source can enhance the microbial activity of the soil and may break down the hydrocarbon chain into carbon dioxide and water, hence improving the biodegradation rate of hazardous organic compounds such as petroleum products. Addition of vital nutrients such as nitrogen source is essential for this process. Application of primary sludge for biodegradation purposes would not only be effective and inexpensive but would also add a new spectrum to the waste management process of the pulp and paper industries: to use their mill by-products as a resource recovery entity rather than landfilling these valuable resources.

REFERENCES

Alexander, M. 1980. Biodegradation of Toxic Chemicals in Water and Soil. In: *Dynamics, Exposure and Hazard Assessment of Toxic Chemicals*, pp. 179-190. (Haque, R., Ed.). Ann Arbor, MI, Ann Arbor Science.

Amdurer, M., Fellman, R., and Abdelhamid, S. 1985. In Situ Treatment Technologies and Superfund. In: Proc. Intl. Conf. on New Frontiers for Hazardous Waste Management, Sept. 1985. EPA Report No. EPA-600/9-85/025. Hazardous Waste Eng. Res.. Lab., Cincinnati, OH.

Atlas, R.M. 1977. Stimulated Petroleum Biodegradation. *CRC Crit. Rev. Microbiol.* 5, 371-386.

Barton, K. and Portier, P. 1994. In-Situ Bioremediation of Petroleum Contaminated Soil, p. 397. Abstracts: The 4th General Meeting of ASM in Las Vegas, NV, May 23-27, 1994.

Berger, M. and Schwartz, L. 1994. Lab-Scale Composting and Land Treatment of Contaminated Soil from a Former Bulk Fuel Storage Facility, p. 396. Abstracts: The 4th General Meeting of ASM in Las Vegas, NV, May 23-27, 1994.

Brown, R.A., Norris, R.D., and Brubaker, G.R. 1985. Aquifer Restoration with Enhanced Bioreclamation. *Poll. Eng.* 17, 25-28.

Buckingham, B. 1981. Studies in Biodegradation. Assoc. Amer. Railroads. Res. and Test Dept., 1920 L St., N.W., Washington, DC.

Cole, G.M. 1994. In Situ Remediation Technologies. In: *Assessment and Remediation of Petroleum Contaminated Sites*, pp. 6-7,197-236. Boca Raton, FL, Lewis Publishers.

Environmental Protection Agency. 1985. Handbook Remedial Action at Waste Disposal Sites (Revised). EPA Report No 625/6-85/006.

Ghisalba, O. 1983. II. Microbial Degradation of Chemical Waste, an Alternative to Physical Methods of Waste Disposal. *Experientia* 39, 1247-1257.

Hildebrant, W.W. ahd S.B. Wilson. 1991. On-site remediation of organic impacted soils on field properties. Proceedings; California Regional Meeting of the Society of Petroleum Engineers. Ventura, California.

Hoff, R.Z. 1993. Bioremediation: An Overview of Its Development and Use for Oil Spill Cleanup. *Mar. Pollut. Bull.* 26(9), 476-481.

Kerry, E. 1991. Bioremediation of Experimental Petroleum Spills on Mineral Soils in the Vestfold Hills, Antarctica. *Polar Biol.* 13, 163-170.

Lee, M.D. and Ward, C.H. 1985. Environmental and Biological Methods for the Restoration of Contaminated Aquifers. *Environ. Toxic. Chem.* 4, 743-750.

Lyman, W.J., Noonan, D.C., and Reidy, P.J. 1990. *Cleanup of Petroleum Contaminated Soils at Underground Storage Tanks*, pp. 99-109. Park Ridge, NJ, Noyes Data Corporation.

Morris, L.A., W.L. Nutter, J.I. Sanders, E.A. Ogden, M.H. Golabi, W.P. Miller, M.E. Summer, M. Saunders, and L. Shuman. 1995. Mill Residue and By-products Utilization, Consortium Mid-Year Progress Report.

Nicholas, R.B. 1987. Biotechnology in Hazardous-Waste Disposal: An Unfulfilled Promise. *Am. Soc. Microbiol. News* 53, 138-142.

Pitonzo, B. and Amy, P. 1994. Bioremediation Potential of Diesel Fuel in a Nevada Desert Soil, p. 397. Abstracts: The 4th ASM General Meeting. University of Nevada, Las Vegas, NV, May 23-27, 1994.

Pometto, A.L., DiSpirito, A.A., Oulman, G.S., and Johnson, K.E. 1994. Bioremediation Potential of Soybean Hulls to Cleanup Petrochemical Spills in Soil, p. 395. Abstracts: The 4th ASM General Meeting. University of Nevada, Las Vegas, NV, May 23-27, 1994.

Riser-Roberts, E. 1992. *Bioremediation of Petroleum Contaminated Sites,* pp. 15-22;173. Boca Raton, FL, C.K. Smoley.

Robert, F. and Israel, S. 1994. Response of Bacterial Populations to Petroleum Hydrocarbon Contamination in Tropical Soils, p. 395. Abstracts: The 4th ASM General Meeting. University of Nevada, Las Vegas, NV, May 23-27, 1994.

Scherrer, P. and Mille, G. 1989. Biodegradation of Crude Oil in an Experimentally Polluted Peaty Mangrove Soil. *Mar. Pollut. Bull.* 20, 430-432.

Sims, R. and Bass, J. 1984. Review of In-Place Treatment Techniques for Contaminated Surface Soils. Vol. 1: Technical Evaluation. EPA Report No. EPA-540/2-82-003a.

Visscher, K. and Brinkman, J. 1989. Biological Degradation of Xenobiotics in Waste Management. In: *Hazardous Waste and Hazardous Materials,* Volume 6, No. 2, 1989. Mary Ann Liebert, Inc. Publishers.

Westlake, D.W.S., Jobson, A.M., and Cook, F.D. 1978. In-Situ Degradation of Oil in a Soil of the Boreal Region of the Northwest Territories. *Can. J. Microbiol.* 24,254-260.

Wilson, J.T., Leach, L.E., Henson, M., and Jones, J.N. 1986. In Situ Biorestoration as a Ground Water Remediation Technique. *Ground Water Mon. Rev.* 56-64.

Wilson, M.B., and Portier, R. 1994. Bioremediation of Petroleum Contaminated Soil Using Land Treatment Technologies, p. 396. Abstracts: The 4th ASM General Meeting, University of Nevada, Las Vegas, NV, May 23-27, 1994.

CHAPTER 33

Micronutrients and the Bioremediation of PHC-Contaminated Coral Sand

Paul E. Flathman, Mary L. Laski, Jason R. Trausch, and **John H. Carson, Jr.**, OHM Remediation Services Corporation, Findlay, Ohio

Charles K. So, OHM Remediation Services Corporation, Pleasanton, California

Patrick M. Woodhull, Douglas E. Jerger, and **Paul R. Lear**, OHM Remediation Services Corporation, Findlay, Ohio

INTRODUCTION

The objective of this study was to determine the benefit of micronutrient addition to the coral sand for enhanced treatment of PHCs. Micronutrient (i.e., trace element and vitamin) concentrations in the coral sand could be low and limit microbial growth on PHCs. The benefit of micronutrient addition was determined in electrolytic respirometer treatment vessels at a 1.5% (by-weight) solids concentration. Yeast extract addition was evaluated at 10 and 100 mg/L concentrations. Yeast extract is an inexpensive and readily available source of trace elements and vitamins.

SAMPLE COLLECTION AND CHARACTERIZATION OF CORAL SAND

Six 2.5 ft x 3.0 in. (76 cm x 7.6 cm) Shelby tubes containing brown-colored coral sand, which were collected at the site on February 2 and 3, 1995, were shipped to OHM's Treatability Laboratory in Findlay, Ohio. The six discrete samples were collected from a depth approximately 30 in. (76 cm) above groundwater to groundwater at the site. Upon receipt, those samples were removed from the Shelby tubes, mixed, and the coral sand was chemically, biologically, and physically characterized on a sample of the mixed composite.

Chemical analyses were performed according to USEPA (USEPA, 1992 and 1986) and American Society of Agronomy/Soil Science Society of America (ASA/ SSSA) (Page, 1982) methods. Bacterial population densities were determined according to the American Public Health Association's (APHA's) Standard Method 9215 B using spread plate techniques, and Method 9221 using most probable number (MPN) techniques (APHA, 1985). ASA/SSSA (Klute, 1986), U.S. Department of Agriculture (USDA) (Richards, 1954), and American Society for Testing and Materials (ASTM, 1987) procedures were used to physically characterize the coral sand. The objectives for performing those analyses were to characterize the coral sand and to make an initial assessment of the potential to utilize enhanced biological treatment for remediation of the PHC-contaminated coral sand at the site.

Analytical results for characterization of the coral sand are presented in Table 1. Total petroleum hydrocarbon (PHC) concentration in the grab sample used to characterize the coral sand was 17,000 mg/kg with 13%, 78%, and 9% of the PHCs, respectively, comprised of gasoline- (n-C_2 - n-C_{10}), diesel fuel- (n-C_{10} - n-C_{21}), and lubricating oil- (n-C_{21} - n-C_{40}) range PHCs. Analytical results for quantification of the PHCs in the coral sand by infrared analysis and gas chromatography were comparable. Ethylbenzene (11.2 mg/kg) and 2-methylnaphthalene (87.0 mg/kg) were the only other organics quantified. Iron and magnesium, which are required for energy-yielding metabolism by bacteria, were present in the coral sand.

Bacterial population density in the coral sand was very high as determined by spread plate and MPN analyses. Aerobic heterotrophic and diesel fuel-degrading bacterial population densities ranged from 10^6 to 10^8 cfu/g. The addition of nitrogen and phosphorus mineral nutrients would be required for full-scale biological treatment since they were either not detected or at the analytical limit of detection in the sample analyzed. The pH of the coral sand was 8.3, which is at the upper limit of the range acceptable for enhanced microbial growth on available organics, and for full-scale treatment, would be closely monitored.

EFFECT OF MICRONUTRIENTS ON THE BIODEGRADATION OF THE PHCs

Materials and Methods

Biological treatment was performed in 1.0-L laboratory-scale, electrolytic respirometer treatment vessels (Bioscience, Inc., Bethlehem, PA). Those vessels provided a direct and continuous measurement of oxygen uptake by microbial populations with oxygen generated within the closed system (Figure 1).

The experimental design for the study is presented in Table 2. Coral sand was added to replicate (n = 4) treatment vessels for each of the three treatments to a final solids concentration of 1.5% (by-weight). Diammonium phosphate (DAP), $(NH_4)_2HPO_4$, was added to all treatment vessels as a readily available source of the nitrogen and phosphorus required to support bacterial growth. All treatment vessels were phosphate-buffered (0.20 M) at a pH of 7.8. Treatment vessels were maintained at ambient laboratory temperature (~23°C).

Treatment 1, which was amended with only water-soluble mineral nutrients, served as a control. Treatments 2 and 3, respectively, evaluated the benefit of 10 and 100 mg/L additions of yeast extract to treatment vessels for enhanced biological treatment of the PHCs. Yeast extract contained a mixture of trace elements

Table 1. Characterization of coral sand.

Parameter	Concentration
Chemical	
PHCs, GC (mg/kg)	
Gasoline range, n-C_2 - n-C_{10}	2,140
Diesel fuel range, n-C_{10} - n-C_{21}	13,300
Lubricating oil range, n-C_{21} - n-C_{40}	1,560
Total, n-C_2 - n-C_{40}	17,000
Total PHCs, IR (mg/kg)	16,400
BETX (mg/kg)	
Benzene	<0.577
Ethylbenzene	8.58
Toluene	<0.577
Total xylenes	<0.577
pH	8.3
Alkalinity (mg/kg as CaCO3)	750,000
Available mineral nutrients (mg/kg)	
Ammonium-nitrogen (NH4-N)	<1.8
Nitrate-nitrite-nitrogen (NO3-NO2-N)	<6.1
Phosphate-phosphorus (PO4-P)	0.6
Metals, Total (mg/kg)	
Calcium	390,000
Iron	16.5
Magnesium	14,400
Biological	
Aerobic heterotrophic bacterial	1.8[a] X 106, 2.0 X 107 (22°C/96 h)
Population density (cfu/g)	1.4[b] X 108 (28°C/48 h)
	1.0[c] X 108 (28°C/48 h)
Diesel fuel-degrading bacterial	1.0[d] X 108 (28°C/48 h)
Population density (cfu/g)	
Physical	
Moisture content (%):	
wet-weight basis (as-received)	17.1
dry-weight basis	20.7
Field capacity (%)	
Wet-weight basis (as-received)	15.2
Dry-weight basis	18.0
Porosity (%)	15.1
Particle size (%)	
Coarse gravel, >4.75 mm	6.06
Medium gravel, 2.00 mm - 4.75 mm	3.88
Fine gravel, 1.70 mm - 2.00 mm	0.96
Coarse sand, 600 μm - 1.70 mm	19.47
Medium sand, 300 μm - 600 μm	23.24
Fine sand, 150 μm - 300 μm	22.03
Very fine sand, 75 μm - 150 μm	2.30
Coarse silt, 38 μm - 75 μm	1.81
Silt and clay, <38 μm	20.25

[a] R2A Agar Medium, Spread Plate Techniques

[b] Trypticase Soy Agar Medium, Spread Plate Techniques

[c] Trypticase Soy Broth, MPN Techniques

[d] Diesel Fuel-Basal Salts Medium, MPN Techniques

Figure 1. Schematic of an electrolytic respirometer treatment vessel.

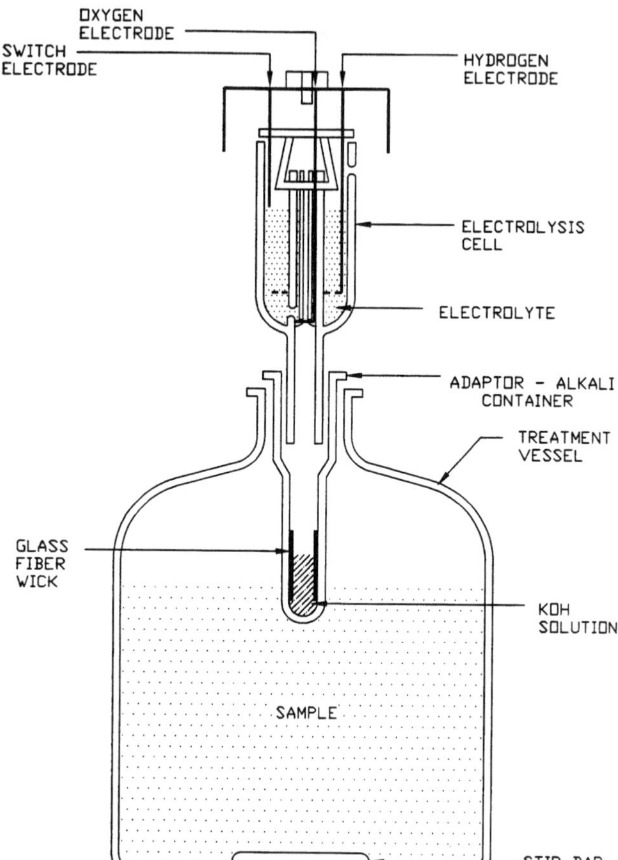

Table 2. Experimental design for the effect of micronutrient addition on biological treatment of the PHCs.

Treatment	Treatment[a] Vessel Number	Coral Sand (DWB)[b]	Mineral Nutrient Amendment	Phosphate Buffer (0.20 M, pH 7.8)	Trace Element Amendment (Yeast Extract)
1	1-4	15.00 g	0.80 g $(NH_4)_2HPO_4$	50.53 g K_2HPO_4 7.66 g KH_2PO_4	---
2	5-8	15.00 g	0.80 g $(NH_4)_2HPO_4$	50.53 g K_2HPO_4 7.66 g KH_2PO_4	10 mg
3	9-12	15.00 g	0.80 g $(NH_4)_2HPO_4$	50.53 g K_2HPO_4 7.66 g KH_2PO_4	100 mg

and vitamins which might not have been present in the coral sand in the quantities required to support enhanced biological treatment of the PHCs. Food-grade yeast extract is inexpensive and readily available. Final volume in each treatment vessel was 1.0 L. The pH in treatment vessels was monitored several times per week, and available nitrogen and phosphorus mineral nutrients were monitored weekly using field test kits (Hach Company, Loveland, CO). PHCs remaining in treatment vessels were quantified by gas chromatography on Days 18, 26, 38, and 56.

Ultimate biochemical oxygen demand (BOD_u) in treatment vessels, i.e., the total amount of oxygen required to biodegrade the immediately available organic matter present in a sample (Mitchell, 1974), was determined by fitting oxygen uptake data to a modified crescent curve (Shammas, 1984):

$$BOD = A + Be^{-Kt}$$

where

BOD = the amount of BOD expressed at time t (mg/L),

$A = BOD_u$, the ultimate amount of oxygen uptake to be expressed (mg/L),

B = a lag period parameter (mg/L) (Note: $\Omega A\Omega = \Omega B\Omega$ if the fitted crescent curve intersects the origin),

K = the first-order rate constant (hour^{-1}), and

t = time (hours).

When biological treatment of the readily available organics appeared to be completed, as determined by oxygen uptake data, a treatment vessel for each of the three treatments was sacrificed and the aqueous and solid fractions were analyzed for PHCs. In addition to an initial analysis of the coral sand for PHCs on Day 0, one treatment vessel for each of the three treatments was sacrificed on Days 18, 26, 38, and 56, and the solid and aqueous fractions within those vessels were similarly analyzed for PHCs.

Results and Discussion

Analytical results for analysis of the PHCs in the aqueous and solid fractions of treatment vessels for Treatments 1, 2, and 3 on Days 0, 18, 26, 38, and 56 are presented in Table 3. The total mass of the PHCs and of the DROs in treatment vessels over time are presented in Table 4.

As indicated by oxygen uptake data, by Day 18 (431 h), biodegradation of most of the readily available organics in treatment vessels appeared to be complete (Figure 2). A BOD of 197 mg/L for Treatment 1 on Day 18 was determined by fitting oxygen uptake data from the electrolytic respirometer treatment vessel to a modified first-order crescent curve. For Treatment 1, 89% of the 222 mg/L BOD_u was met, and 88% of the PHC mass initially present in the treatment vessel was biologically treated (Table 4). In addition, 97% of the PHCs initially present in the solid fraction of the treatment vessel were biologically treated (Figure 3, Table 3).

Table 3. PHCs in the solid (S) and aqueous (A) fractions of electrolytic respirometer treatment vessels for Treatment 1, 2, and 3 over time.

Treatment	Analysis	PHC Concentration								
		Day 0	Day 18		Day 26		Day 38		Day 56	
		S	S	A	S	A	S	A	S	A
		mg/kg	mg/kg	mg/L	mg/kg	mg/L	mg/kg	mg/L	mg/kg	mg/L
Treatment 1	Diesel Fuel Range, $n\text{-}C_{10}$ - $n\text{-}C_{21}$	6,400	156	7.19	823	3.15	464	2.42	759	1.45
	Lubricating Oil Range, $n\text{-}C_{21}$ - $n\text{-}C_{40}$	1,360	83	3.35	<719	2.27	271	0.77	540	0.93
	Total PHCs	7,760	239	10.54	<1,542	5.42	735	3.19	1,299	2.38
Treatment 2	Diesel Fuel Range, $n\text{-}C_{10}$ - $n\text{-}C_{21}$	6,400	544	4.94	566	2.71	512	2.76	559	1.48
	Lubricating Oil Range, $n\text{-}C_{21}$ - $n\text{-}C_{40}$	1,360	370	3.97	<503	1.36	318	2.94	<475	0.70
	Total PHCs	7,760	914	8.91	<1,069	4.07	830	5.70	<1,034	2.18
Treatment 3	Diesel Fuel Range, $n\text{-}C_{10}$ - $n\text{-}C_{21}$	6,400	466	6.09	699	2.51	676	2.99	441	2.00
	Lubricating Oil Range, $n\text{-}C_{21}$ - $n\text{-}C_{40}$	1,360	274	3.44	847	<1.10	355	1.02	325	0.89
	Total PHCs	7,760	740	9.53	1,546	<3.61	1,031	4.01	766	2.89

Table 4.　Total PHC and DRO mass in electrolytic respirometer treatment vessels over time.

Treatment No	Day 0	Day 18		Day 26		Day 38		Day 56	
	mg	mg	Reduction (%)	mg	Reduction (%)	mg	Reduction (%)	mg	Reduction (%)
Total PHCs									
1	116	14.1	88	1ND	ND	14.2	88	21.9	81
2	116	22.6	80	ND	ND	18.2	84	ND	ND
3	116	20.6	82	ND	ND	19.5	83	14.4	88
DROs									
1	95.9	9.53	90	15.5	84	9.38	90	12.8	87
2	95.9	13.1	86	11.2	88	10.4	89	9.87	90
3	95.9	13.1	86	13.0	86	13.1	86	8.62	91

Figure 2.   Oxygen uptake over time for biodegradation of the PHCs in an electrolytic respirometer treatment vessel containing mineral nutrients (Treatment 1).

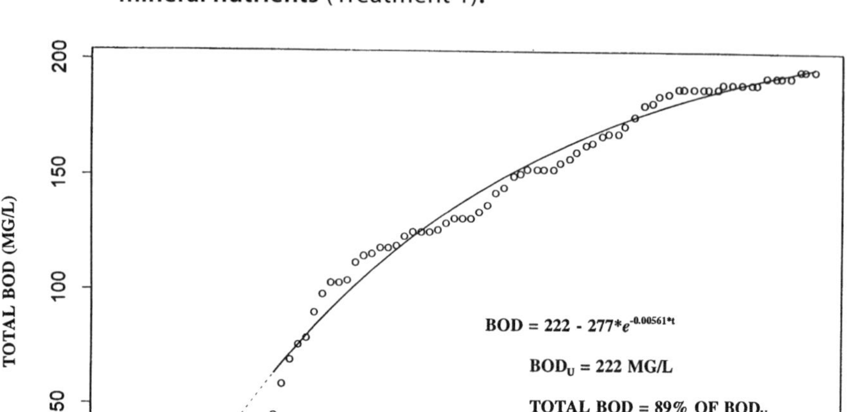

The concentration of the PHCs in the solid fraction was reduced from 7,760 mg/kg on Day 0 to 239 mg/kg by Day 18.

The gas chromatograms presented in Figure 3 for analysis of the PHCs on Day 18 established the fact that biological treatment of the PHCs was extensive and that, based on the fact that resolved peaks of assumed readily biodegradable compounds remained, additional biological treatment of those PHCs was anticipated. Treatment 1 was amended with only water-soluble mineral nutrients. Similar results were observed for the other treatments, and little change was observed in the extent of PHC degradation by the end of the laboratory. Based on those results, there does not appear to be an advantage for the addition of yeast extract to the coral sand for enhanced biological treatment of the PHCs.

A theoretical oxygen demand (ThOD) for complete oxidation of the PHCs was calculated to support oxygen uptake data for biological treatment of the PHCs. ThOD for aerobic mineralization of n-hexadecane, a PHC in the middle of the diesel fuel range, was calculated from the following relationship:

$$2\,C_{16}H_{34} + 49\,O_2 \rightarrow 32\,CO_2 + 34\,H_2O$$

The ThOD is 3.45 mg O_2 per mg of n-hexadecane or any other n-alkane. Since petroleum products contain PHCs in addition to n-alkanes, the calculated ThOD was considered to be an approximate value.

Figure 3. **Gas chromatograms of PHCs in the coral sand on Day 0 and in the solid fraction of the electrolytic respirometer treatment vessel for Treatment 1 on Day 18.**

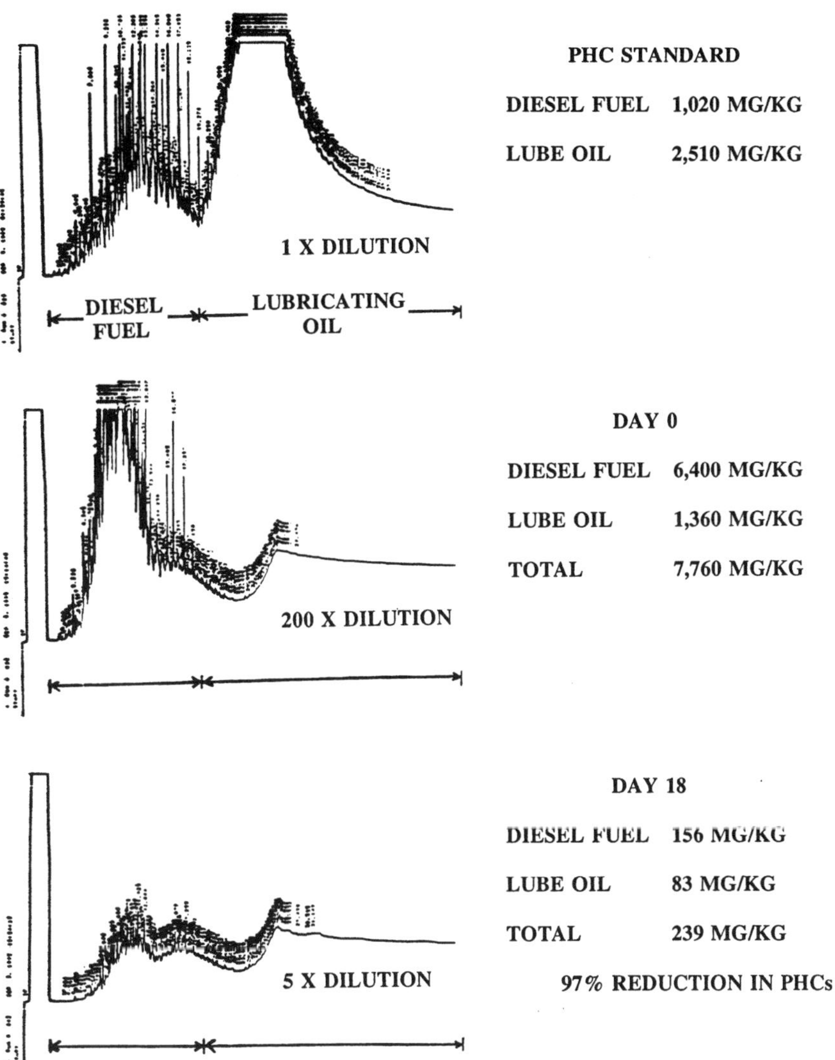

On Day 0, all treatment vessels contained 116 mg of PHCs (Table 4). By Day 18 for Treatment 1, only 14 mg of PHCs remained; i.e., 102 mg of PHCs were biologically treated. Based on the above equation for mineralization of n-alkanes, ~352 mg O_2 would be required for complete oxidation of the PHCs. Using the following relationship, it was calculated that 63% of the ThOD for PHC biodegradation was met:

$$(BOD_u / ThOD)100$$

A BOD_u of 63% of the ThOD was acceptable since not all of the PHCs would have been completely oxidized with microbial growth on the PHCs (Ramalho, 1983).

A BOD_u of 222 mg/L in a treatment vessel containing 15 g coral sand (dry-weight) equated to 14,800 mg O_2/kg coral sand. Based on a BOD_u:NH_4-N:PO_4-P ratio of 100:10:1, 1,480 and 148 mg/kg of NH_4-N and PO_4-P would be required, respectively, for biological treatment of the PHCs. Since nitrogen and phosphorus are naturally recycled in the environment, much lesser quantities of diammonium phosphate (DAP), $(NH_4)_2HPO_4$, would be added to the coral sand to provide those needed nutrients.

CONCLUSIONS

Based on the results of this study, the following conclusions can be stated:

- By Day 18, in a electrolytic respirometer treatment vessel amended with only water-soluble mineral nutrients, 88% of the PHC mass initially present in the treatment vessel was biologically treated, and a 97% reduction in the concentration of the PHCs in the solid fraction of a treatment vessel was observed.

  Similar results were observed for the remaining electrolytic respirometer treatment vessels, and little change was observed in the extent of PHC degradation by the end of the laboratory study.

- There does not appear to be an advantage for the addition of yeast extract to the coral sand for enhanced biological treatment of the PHCs.

  Overall, 84 ± 3% (x ± s, n = 8) of the PHC mass and 88 ± 2% (x ± s, n = 12) of the mass of the DROs initially present in electrolytic respirometer treatment vessels were biologically treated.

REFERENCES

American Public Health Association, American Water Works Association, and Water Pollution Control Federation. 1985. *Standard Methods for the Examination of Water and Wastewater*, 16th Edition. Washington, DC, American Public Health Association.

American Society for Testing and Materials. 1987. D422, Standard Method for Particle-Size Analysis of Soils. Philadelphia, PA, American Society for Testing and Materials.

Klute, A., Ed. 1986. *Methods of Soil Analysis, Part 1, Physical and Mineralogical Methods*, 2nd Edition. Madison, WI, American Society of Agronomy and Soil Science Society of America.

Mitchell, R. 1974. *Introduction to Environmental Microbiology*. Englewood Cliffs, NJ, Prentice-Hall.

Page, A.L., Ed. 1982. *Methods of Soil Analysis, Part 2, Chemical and Microbiological Properties*, 2nd Edition. Madison, WI, American Society of Agronomy and Soil Science Society of America.

Ramalho, R.S. 1983. *Introduction to Wastewater Treatment Processes*, 2nd Edition. New York, Academic Press.

Richards, L.A. 1954. Diagnosis and Improvement of Saline and Alkali Soils, United States Department of Agriculture Handbook No. 60. Washington, DC, Superintendent of Documents, U.S. Government Printing Office.

Shammas, N.C. 1984. Modified Crescent Curve Fitting, Program Number 341C, Users' Library. Corvallis, OR, Hewlett-Packard Company.

U.S. Environmental Protection Agency. 1992. Test Methods for Evaluating Solid Waste, Physical/Chemical Methods, 3rd Edition, SW-846, Update 1. Washington, DC, Office of Solid Waste and Emergency Response.

U.S. Environmental Protection Agency. 1986. Test Methods for Evaluating Solid Waste, Physical/Chemical Methods, 3rd Edition, SW-846. Washington, DC, Office of Solid Waste and Emergency Response.

CHAPTER 34

Sorption, Diffusion, and Biodegradation of Toluene in Fine-Grained Soils

Sankar N. Venkatraman, Bioremediation Applications Group, McLaren/Hart Inc., Warren, New Jersey

John R. Schuring, Department of Civil and Environmental Engineering, New Jersey Institute of Technology, Newark, New Jersey

David S. Kosson, Department of Chemical and Biochemical Engineering, Rutgers University, Piscataway, New Jersey

INTRODUCTION

Fine-grained soils contaminated with volatile constituents of petroleum and other hydrocarbons are ubiquitous throughout the United States (Siegrist and Van Ee, 1994). The movement of volatile organic compounds (VOCs) from these geological strata tend to be a chronic problem, with the contaminants migrating over many years along preferential pathways. Because of the inherent low permeability of these formations, conventional treatment technologies are not easily applicable to remediate the contaminated soils. In order to assist in plume delineation, assessment of ecological and human health risks, evaluation of intrinsic bioremediation, and in the selection and design of appropriate remedial alternatives, a detailed comprehension of the VOC fate and transport in low permeability soils is essential.

Considerable attention has been paid to understanding the individual contributions of primary subsurface mechanisms such as sorption, diffusion, and biodegradation on the movement of solutes in the vadose zone and groundwater (Arocha et al., 1996; Allen-King et al., 1994; Barone et al., 1992; Chen et al., 1992; Jin et al., 1994). Contaminant transport mechanisms have also been studied through the formulation of mathematical models and by examining the influence of different physical processes such as advection and dispersion in the solid, liquid, and vapor phases on the fate and transport of VOCs (Jury et al., 1983; Baehr, 1987; Priesack, 1991; Gierke et al., 1992). However, the information available from laboratory studies on the simultaneous sorption, diffusion, and biodegradation of VOCs, especially in fine-grained soils, is limited (West et al., 1995).

This Chapter presents the findings from experimental and computational studies conducted to describe the movement of toluene in fine-grained soils. Column experiments were conducted to examine toluene diffusion and biodegradation in clayey soils under aerobic and denitrifying environments. The experimental results were compared with a mathematical model developed to describe VOC transport in low permeability soil formations.

THEORETICAL DEVELOPMENT

The physical problem to be modeled is presented in Figure 1. A constant source of volatile, biodegradable contaminant (solute) at the bottom of a low permeable soil bed diffuses through the formation. Partitioning occurs between vapor, solid, and liquid phases and the contaminant is consumed as it comes in contact with degrading microorganisms. The following assumptions are made in the formulation of the model:

- Bed density, porosity, moisture content, and diffusion coefficient remain constant.
- Concentration gradients exist only in the z-direction
- Solute exists in the liquid, solid, or gas phase only; the bed is composed of a continuous gas phase and an immobile liquid phase
- Adsorption phenomena follow linear local equilibrium
- Biodegradation occurs only on the surface of soil particles in the solid-liquid interface, is assumed to follow first-order reaction kinetics, and occurs only in the liquid phase (Jin et al., 1994).
- Vapor phase diffusion is much faster than liquid or solid phase diffusion (Cussler, 1991)

Based on the above assumptions, the governing equation can be given as:

$$R_V \frac{\partial C_V}{\partial t} = D_{eff} \frac{\partial^2 C_V}{\partial z^2} - B_{eff} C_V \qquad (1)$$

The first term in the above equation is the transient term, the second term accounts for the diffusive flux, and the third term represents the biodegradation in the reactive bed. C_V is the vapor phase solute concentration; R_V is the retardation factor that accounts for the phase partitioning between the solid, liquid, and vapor phases and is given as:

$$R_V = \varepsilon \left(1 - \theta_w\right) + \frac{\varepsilon \theta_w}{H} + \frac{\rho_b K}{H} \qquad (2)$$

H in the above equation is the Henry's law constant relating liquid and vapor phase concentrations, K is the linear equilibrium partition coefficient between the solid and liquid phase concentrations, and ρ_b is the bed density. The D_{eff} term used in

Figure 1.   **Schematic of the system studied.**

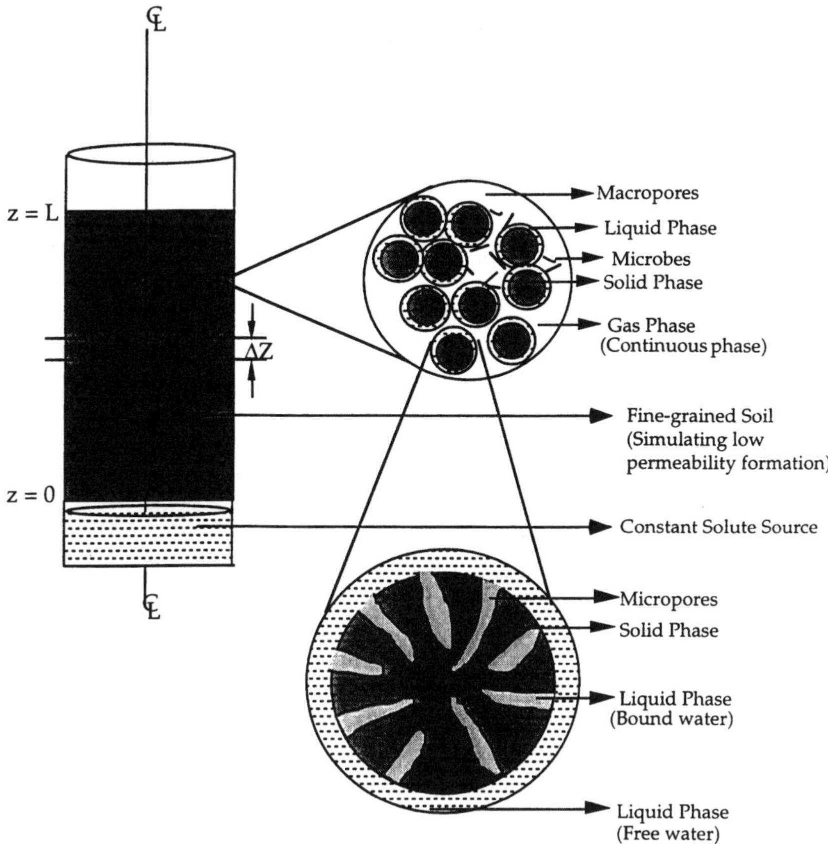

Equation 1 denotes the effective diffusivity that accounts for the tortuous path of the vapors in the bed and is determined from the free diffusion term D_V^o and the tortuosity value, τ_V:

$$D_{eff} = \frac{\varepsilon\left(1 - \theta_w\right)D_V^o}{\tau_V}$$

(3)

B_{eff} is the effective biodegradation term given as:

$$B_{eff} = \frac{\varepsilon\theta_w\mu_L}{H}$$

(4)

where μ_L is the first-order biodegradation rate constant in the liquid phase, ε is the bed porosity and θ_W is the relative saturation. The governing Equation 1 is subject to the following initial and boundary conditions:

Initial Condition: At time zero, the soil bed is solute-free

$$t = 0 \qquad\qquad C_V = 0 \qquad\qquad \text{for all } z \qquad\qquad (5a)$$

Boundary Condition 1: At the bottom layer of the bed, there is a constant solute source

$$z = 0 \qquad\qquad C_V = C_{Vo} \qquad\qquad \text{for all } t \qquad\qquad (5b)$$

Boundary Condition 2: At the top of the bed, the solute gradient is zero

$$z = L \qquad\qquad \partial C_V/\partial z = 0 \qquad\qquad \text{for all } t \qquad\qquad (5c)$$

The analytical solution to the governing equation (Equation 1) with the corresponding initial and boundary conditions (Equations 5a-5c) is given as (Carslaw and Jaegar, 1989):

$$C_V = \frac{C_{Vo}}{\cosh\left[l\left(\dfrac{B_{eff}}{D_{eff}}\right)^{1/2}\right]} - \frac{4C_{Vo}}{\pi}\sum_{n=0}^{\infty}\frac{(-1)^n\,exp\left[-B_{eff}\,t-\left(\dfrac{D_{eff}\,(2n+1)^2\,\pi^2\,t}{4l^2}\right)\right]}{(2n+1)\left[1+\left(\dfrac{4B_{eff}\,l^2}{(2n+1)^2\pi^2 D_{eff}}\right)\right]} \qquad (6)$$

The above analytical solution can be used to generate contaminant profiles once the soil/solute-specific parameters such as B_{eff} and D_{eff} are determined. In this study, batch and column experiments were conducted to determine the adsorption and diffusion parameters. Upon obtaining these values, theoretical time versus concentration curves were generated to model the experimental solute elution profiles.

MATERIALS AND METHODS

To simulate the physical situation described above, column experiments were designed and constructed and solute transport and degradation were then examined using toluene as the model solute. Soil samples used for the laboratory batch and column experiments were obtained from a confidential gasoline-contaminated site. The soils at the site had been previously exposed to toluene but did not contain any contamination at the time of the study.

Soil Characterization

Soil preparation entailed homogenization by the method of coning and quartering. The soil was then sieved through a 2 mm (#20) opening mesh and all the materials retained on the sieve were discarded. A section of the homogenized soil sample was sterilized using the gamma-irradiation technique to minimize biological solute losses during diffusion studies. Physical and chemical characterization of the soil was conducted at the Rutgers University Soil Testing Laboratory and the results of the analyses are presented in Table 1.

Table 1. **Characteristics of Soil used for the Studies**

Soil Property	Value
Moisture content	0.16-0.18 g/g dry soil
pH	6.72
Particle size	
Sand	36%
Silt	38%
Clay	26%
Organic matter	0.42%
CEC	10.8 meq/100 g
Intraparticle volume	0.025 mL/g
Intraparticle area	14.2 m^2/g
Specific surface area	15.5 m^2/g
Microbial enumeration	Mean values of triplicate enumeration
Aerobic degraders	3×10^6 cfu/g fresh soil
Denitrifiers	7×10^5 cfu/g fresh soil

Column Design and Construction

A schematic of the experimental apparatus set up to serve as a microcosm of a low-permeability geologic formation is presented in Figure 2. The apparatus shown is one of four columns that were simultaneously set up at the start of each experimental study. The principal components of each apparatus were the same; the difference was in the atmosphere provided in each column to facilitate the study of biodegradation rates under different microenvironments. Each experimental apparatus consisted of a solvent reservoir, a saturated vapor zone above the reservoir, the soil bed and vapor sampling ports at the top and bottom of the soil bed. The soil samples were analyzed for nitrogen and phosphorous content prior to the study. Care was taken to ensure that the soil was not limiting in nutrient concentrations and that the soil moisture content and density were representative of field conditions.

Figure 2.   Schematic of the experimental setup.

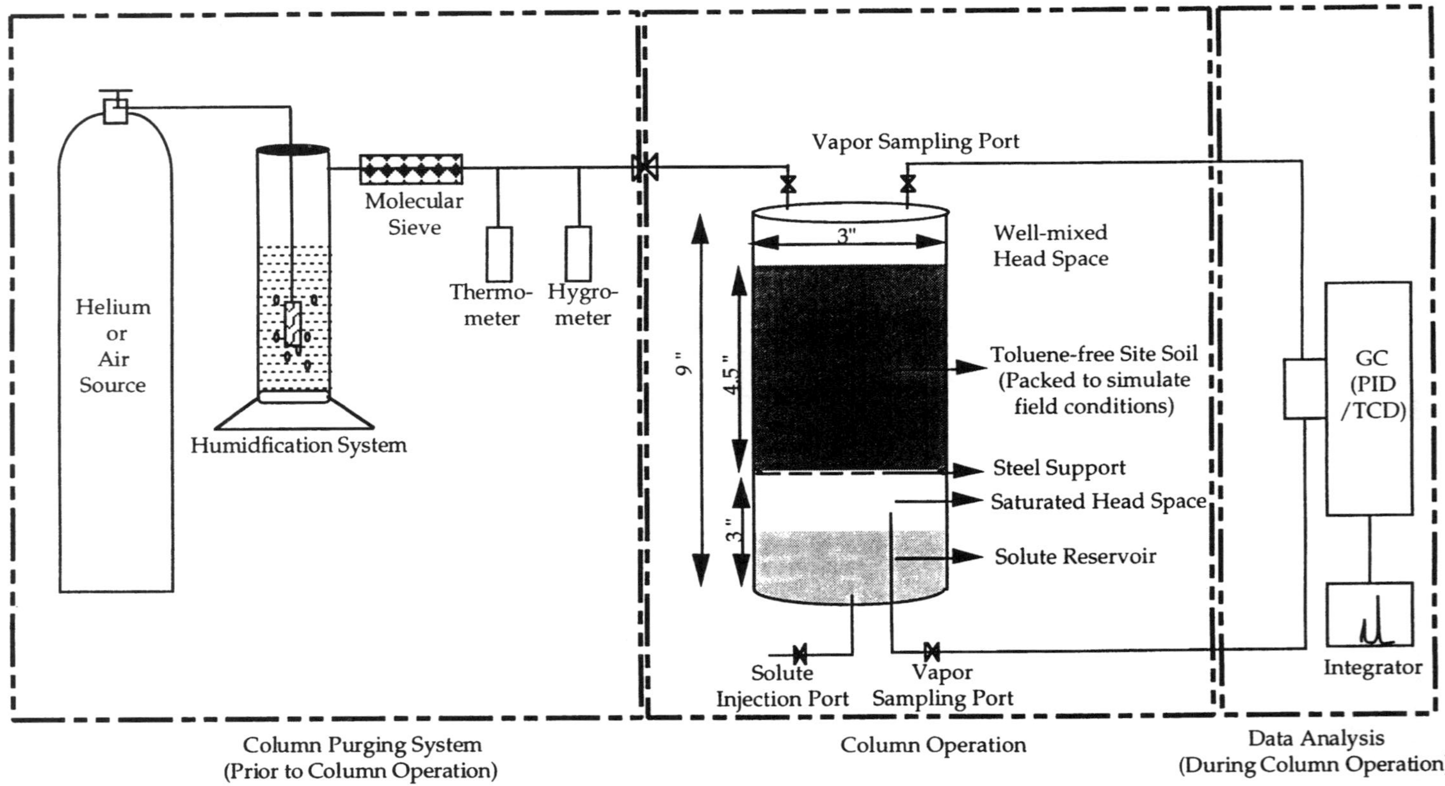

The columns used for the study were custom-made 7.5 cm (3 inch) diameter and 22.5 cm (9 inch) long glass pipes with fritted ends to provide for tight fittings (Kontes Glass Shop, Vineland, NJ). The column dimensions ensured that the wall effects were minimal and reasonable breakthrough times (less than 3 days for sterile column) could be obtained. The glass pipe was indented 0.15 cm, at a distance of 7.5 cm (3 inches) from one end of the column. A circular stainless steel screen mesh measuring 7.3 cm (2.875 inches) in diameter was placed on top of the indented portion to support the soil packing. The screen had a uniform pore size of 0.01 cm (100 microns) and was 0.16 mm thick. All the columns were packed with site soil to simulate low permeability conditions observed in the field. A compactive effort test was performed on the soil to determine the optimum soil density packing that would closely resemble field conditions. Further details on the compaction tests are presented elsewhere (Venkatraman et al., 1995). Following the soil column packing, the ends of the column were fitted with stainless steel plates provided with multiple ports for vapor sampling and solute injection.

Each experimental study entailed the setting up of four soil columns. The first column was used to study diffusion in conjunction with aerobic biodegradation. Following the packing of this column with biologically active site soil, it was purged with humid air. The second column was used to examine diffusive transport under anaerobic biodegradation conditions. This column was packed with exactly the same kind of soil used for Column 1, with the exception that a known amount of nitrate in the form of calcium ammonium nitrate decahydrate salt was solubilized in less than 5 mL of water and was mixed with the soil prior to packing. Following packing, the column was purged with either humid helium or humid argon to create an anaerobic atmosphere. The third column, packed with gamma-irradiated sterile soil, was the negative control column which provided information on reaction-free diffusion through soil beds. The fourth column, a positive control, was set up similar to the other columns and was not purged.

Following the column packing and purging, liquid solute was introduced into the bottom of the bed. The solute used in the studies was toluene, mixed with a low-volatile solvent, n dodecane. The presence of the solute source in the same column as the soil provided a uniform, constant VOC source at the bottom of the soil bed and resulted in minimal concentration fluctuations. The vapor concentration was maintained at 7.68 x 10 $^{-6}$ g cc^{-1} (2000 ppmv) toluene.

Analytical Methods

Vapor samples were obtained from the ports provided at the center of the top and bottom plates of the soil columns. A gas-tight luer-lock syringe equipped with an on-off valve was used for drawing the vapor samples. Sample volumes of less than 600 µL ensured that the extraction of the samples from the bed did not significantly affect the volume of the contaminant diffusing through the bed. Vapor samples were analyzed for toluene on a HP 5890 (Hewlett Packard, Wilmington, DE) gas chromatograph equipped with a packed SPB-1500, 10' x 1/8" column

(Supelco, Bellefonte, PA) and a photoionization detector (PID). Samples were analyzed for oxygen, nitrogen, methane, and carbon dioxide concentrations on a HP 5880 gas chromatograph (Hewlett Packard, Wilmington, DE) equipped with a packed Carboxen (15' x 1/8") column (Supelco, Bellefonte, PA) and a thermal conductivity detector (TCD). The analysis of the vapor samples provided cumulative gas concentrations at the top of the bed.

RESULTS AND DISCUSSION

Parameter Estimation

The list of parameters used in the model simulations is presented in Table 2. From the table it can be noted that several parameters were obtained from independent experiments. Details of these experiments are discussed in Venkatraman (1995). This approach reduced the number of floating parameters required to generate theoretical contaminant profiles and yielded site-specific values.

Solute Transport Studies: Reaction-Free Diffusion

For highly volatile compounds such as the monoaromatics, vapor diffusion is the dominant mode of transport through soil beds (Taylor and Ashcroft, 1972; Kreamer et al., 1988). The presence of tortuous channels in the matrix causes the diffusion of the contaminant vapors in packed beds such as soils, to be much slower

Table 2. **Summary of Model Parameters and Estimation Methods**

Parameter	Units	Typical Value	Mode of Determination
l : Length of the bed	cm	11.43	Measured value
D° : Solute free diffusivity	cm^2/second	0.0877	Literature value[a]
MC : moisture content	g water/ g dry soil	0.15	Measured value
ρ_b : Bed density	g dry soil/ mL bed	1.67	Measured value
ρ_s : Particle density	g solid/ mL solid	2.6	Measured value
ε_t : Total Bed porosity	mL tot. voids/mL bed	0.35	$1-(\rho_b / \rho_s)$
ε_{ip} : Intraparticle porosity	mL intraparticle voids/mL bed	0.025	Measure value (BET analysis)
τ_v : Vapor-phase tortuosity	Dimensionless	9.03	SASK model[b]
ε : Interparticle porosity	mL voids /mL bed	0.32	$\varepsilon_t - [(\rho_b)\ (\varepsilon_{ip})]$
θ_w : Relative saturation	mL water/mL voids	0.68	$[(MC)\ (\rho_b)]/ \varepsilon$
H : Henry's law constant	mL water/mL vapor	0.27	Literature value[c]
K : Solid-Liquid part. coeff.	mL in water/g in soil	0.015	Estimated
μ_L : Liquid phase biodeg. rate	g-solute consumed/ g-soil/second	1e-07	Regressed value

[a] Riddick et al., 1986.

[b] Schaefer et al., 1997.

[c] Turner et al., 1996.

than in free air. To account for the irregular channels in the soils, an effective diffusivity term is incorporated to describe the tortuous path of the vapor movement. The effective soil diffusion coefficient is influenced by overall porosity, soil skeletal density, bed density, and soil moisture content. Several correlations such as the Penman relation (Penman, 1940), Millington-Quirk relation (Millington and Quirk, 1961), and Van Bavel relation (Van Bavel, 1952), have been developed for the determination of soil tortuosities and effective diffusivities. The Millington-Quirk (M-Q) equation, which includes the bed porosity and the relative saturation terms to define the tortuosity factor, is the most frequently used relationship for estimating effective diffusivity. The general form of the M-Q equation to determine tortuosity is given as (Karimi et al., 1987):

$$\tau_V = \frac{\varepsilon^2}{\left[\varepsilon\,(1-\theta_W)\right]^{10/3}} \tag{7}$$

where θ_W is the relative saturation, defined as the ratio of volume of water to pore volume of voids, and ε is the interparticle porosity. Frequently, the exponential in the equation is a fitted parameter, yielding case-specific values. Although the M-Q relationship has been used for a long time in describing contaminant diffusion through porous beds, it most often underpredicts the diffusivity values and hence overpredicts the breakthrough times of contaminants (Jin et al., 1994; Lam, 1994). Further, the deviation in the actual and predicted values increases with increase in soil moisture content.

To overcome the drawbacks of the M-Q equations, the SASK model developed by Schaefer et al. (1997) was used in this study to determine the tortuosity values. The SASK model is based on the premise that to define the effective diffusivity more accurately, both the pore size distribution and the water distribution need to be considered. The soil bed is considered to be divided into two diffusional regimes - an interparticle pore volume regime that accounts for the pores in the interstitial spaces between particles, and the macropore volume regime which accounts for the larger interconnected pores. The tortuosity in each regime is described by the M-Q equation. Thus, with known values of bed density, moisture content, macropore volume, and interparticle pore volume, the tortuosity of the bed can be determined.

To illustrate the variability in the tortuosity values obtained from the two models, plots of experimental concentration and the model diffusion profiles obtained by using SASK and M-Q models for computing the tortuosity values are presented in Figure 3. A close match between the experimental and the predicted values (using SASK model for determining tortuosity) can be observed, demonstrating the effectiveness of the mathematical model to describe solute fate and transport. The model's over-prediction of the time required for the contaminant to reach steady state (by using the M-Q equation for tortuosity estimation) can also be noted from the plot.

Figure 3.   **Comparison of experimental results with model prediction.**

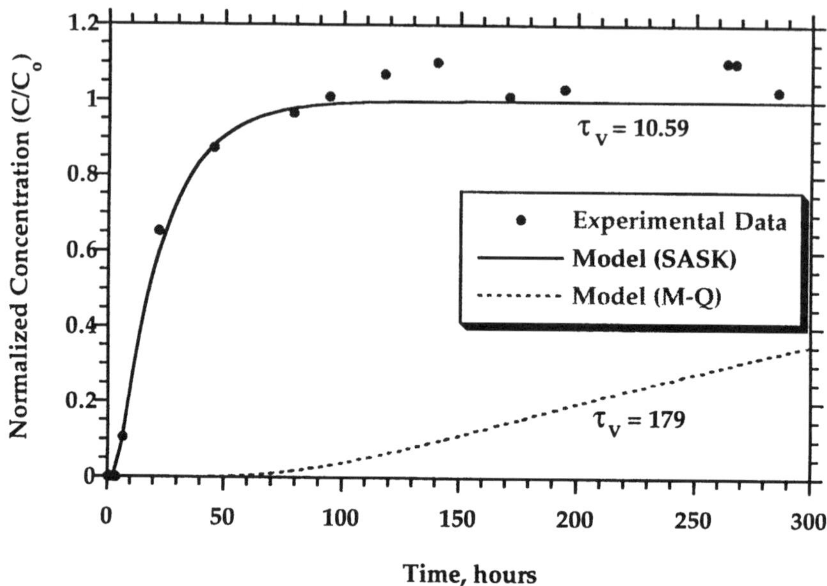

Solute Transport Studies: Anaerobic Denitrifying Column

The nonavailability of oxygen due to high diffusional resistance, especially in low permeability formations can potentially convert any aerobic environment to an anaerobic system. The high solubility of nitrates and easy assimilation of nitrate by denitrifying and facultative microorganisms make denitrification an effective alternative to aerobiosis in tight soil formations. Column studies were conducted to investigate the simultaneous sorption, diffusion, and biodegradation of toluene under anaerobic denitrifying environments. The concentration profile for solute diffusion in an anaerobic denitrifying column is presented in Figure 4a. The solid line in the figure represents the model prediction with the biodegradation rate obtained by the best-fit technique. This rate constant was estimated to be 0.017 day^{-1}. This rate is consistent with the rates of 0.007-0.020 day^{-1} observed during the field degradation of toluene via denitrification process (Mester, 1995). The observed value also agrees well with the first-order rate constant of 0.0224-0.035 reported by Hutchins (1991). Figure 4b presents the carbon dioxide and nitrogen generation profiles in a similar column, demonstrating the metabolic oxidative conversion of toluene to carbon dioxide and the reduction of nitrate to nitrogen gas.

Solute Transport Studies: Aerobic Column

Oxygen and carbon dioxide concentration profiles were monitored in the column head space of the aerobic column (Figure 5). After 2 days of column operation, carbon dioxide production was observed. However, after 1500 hours (over 62 days),

Figure 4a. **Toluene transport profiles in denitrifying column: comparison of experimental results with model prediction.**

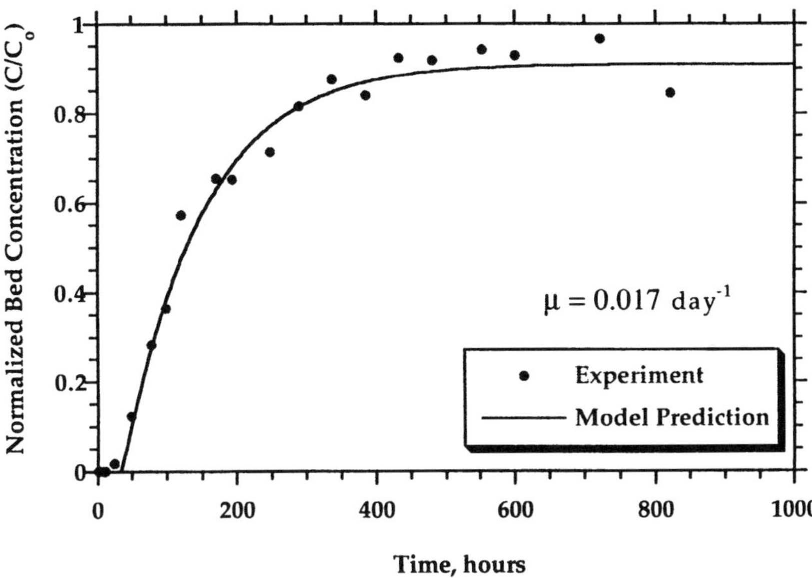

Figure 4b. **Nitrogen and carbon dioxide profiles in a denitrifying column.**

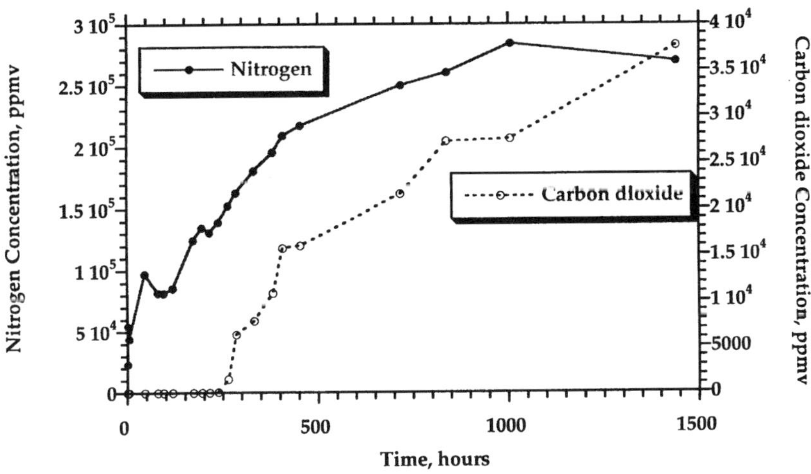

the oxygen concentration in the head space had decreased to less than 3% volume and consequently no further increase in carbon dioxide generation was observed, signifying a cessation in the toluene degradation activity.

Figure 5. **Oxygen and carbon dioxide profiles in an aerobic column.**

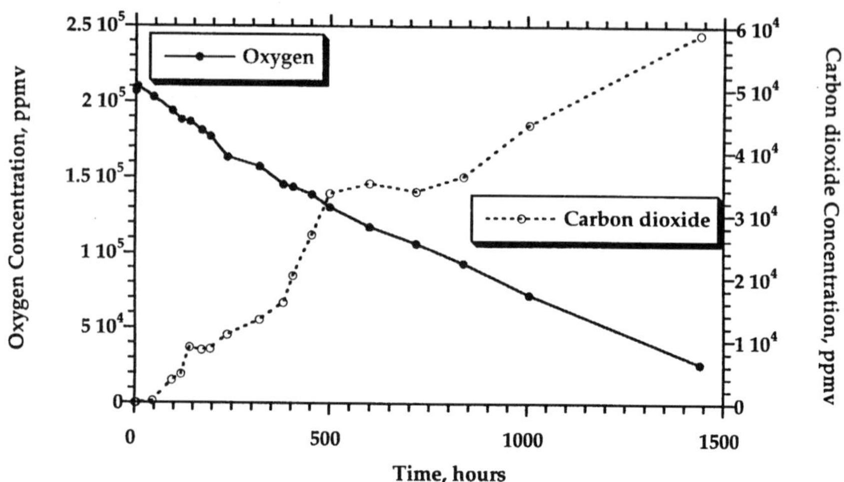

Solute Transport Modeling: Comparison of Column Results

Studies were also conducted to examine the effect of additional nitrate in the biodegradation of toluene under aerobic environments in fine-grained soils. Nitrate was introduced in the soils by mixing the soil with known amounts of calcium ammonium nitrate decahydrate prior to column packing. This was in addition to the nitrate and phosphate included with all soils, to serve as nutrients. Care was taken to ensure that the soils were not limiting in any of these nutrients. Two other columns - one without additional nitrate and one with sterile soil - were packed and assembled in a similar manner. All three columns were then purged with oxygen. The carbon dioxide generation profiles for the three columns are presented in Figure 6. As can be noted from the figure, the rate of carbon dioxide generation (assuming complete mineralization of the substrate) was highest in the aerobic column with additional nitrate. From this study it is evident that the nitrate serves as an effective alternate electron acceptor in low permeability formations even in the presence of oxygen. This is due to the fact that low permeability formations have pockets of anaerobic zones. The presence of nitrates in these areas provides the facultative anaerobes an effective alternate electron acceptor for the degradation. These findings indicate that in low permeability formations, addition of labile electron acceptors, such as nitrates and sulfates, in addition to oxygen can enhance the biodegradation rates of aromatic hydrocarbons.

CONCLUSIONS

An overall transport model incorporating diffusion, partitioning, and biodegradation processes in low permeability soils was developed. This was followed up with laboratory column experiments to study the simultaneous adsorption, diffusion, aerobic and denitrifying biodegradation of toluene in fine-grained soils.

Figure 6. **Effect of excess nitrate in aerobic columns: comparison of carbon dioxide profiles**

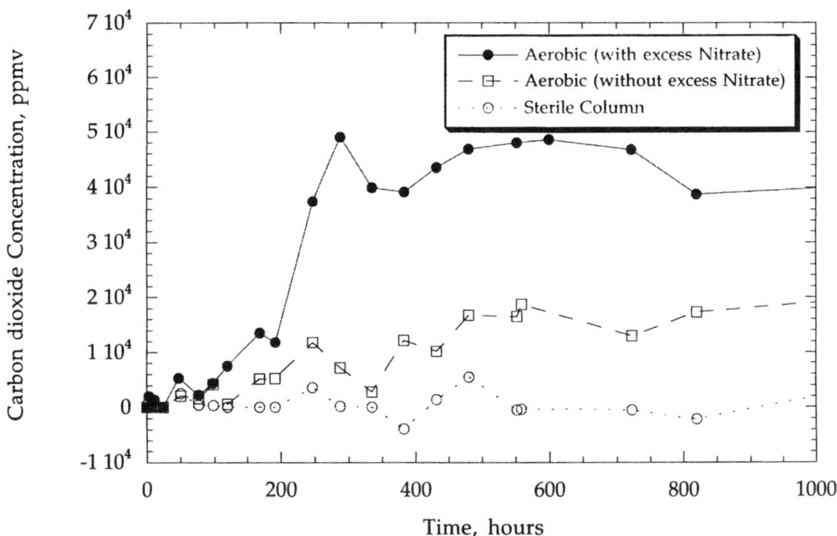

Reaction-free diffusion of toluene in low permeability formations was studied by conducting column experiments using sterile soils. The tortuosity values for estimating effective diffusivities were determined from the SASK model, and excellent matches between experimental toluene concentration profiles and model predictions were observed. Anaerobic denitrifying column experiments were conducted by mixing nitrate salt with the soil prior to the start of the studies. The toluene degradation rate constant under denitrifying conditions was determined to be 0.017 day^{-1}. This value agrees well with the field degradation rates of toluene under denitrifying conditions, as reported in the literature. Studies were also conducted to examine the effect of the presence of excess nitrate in the aerobic biodegradation of toluene in low permeability soils. Results indicated that introducing dissolved nitrate into the soils yielded higher toluene degradation as compared to the rates obtained from an aerobic column without nitrate addition. These findings indicate that bioremediation of aromatic hydrocarbons in fine-grained soils can be enhanced by introducing labile alternate electron acceptors together with oxygen to intercept regions of poor oxygen availability.

ACKNOWLEDGMENTS

This work was conducted when Dr. Venkatraman was a doctoral student at Rutgers University. This research was supported by the U.S. Department of Defense (DOD), Advanced Research Projects Agency (ARPA), Office of Naval Research (ONR) under the University Research Initiative (URI) Program (Grant N00014-92-J-1888, R&T a4Or41ruri), and by Hazardous Substance Management Research Center Project # 9-95215 (Site-30).

NOMENCLATURE

<u>English</u>

B_{eff} Effective biodegradation rate [sec^{-1}]
C_v Concentration of the solute [g cc^{-1}]
C_{vo} Initial soil vapor concentration [g cc^{-1}]
D°_V Free diffusivity of solute in the bed [cm^2 sec^{-1}]
D_{eff} Effective diffusivity of solute in the bed [cm^2 sec^{-1}]
H   Henry's law constant [-]
K   Solid-liquid partition coefficient [mL g^{-1}]
L   Length of the soil bed [cm]
R_V Retardation factor [-]
t   Temporal coordinate [sec]
z   Axial coordinate [cm]

<u>Greek</u>

τ_V Tortuosity in the vapor phase [-]
ε Bed porosity [-]
θ_W Relative saturation [-]
ρ_b Bed Density [g cc^{-1}]
μ_L Biodegradation rate in the soil bed [sec^{-1}]

REFERENCES

Allen-King, R.M., Barker, J.F., Gillham, R.W., and Jensen, B.K., 1994. Substrate and Nutrient-Limited Toluene Biotransformation in Sandy Soil. *Envir. Tox. Chem.* 13, 93-705.

Arocha, M.A., Jackman, A.P.,and McCou, B.H., 1996 Adsorption Kinetics of Toluene on Soil Agglomerates: Soil as a Biporous Sorbent. *Env. Sci Tech.* 30, 1500-1507.

Baehr, A.L. 1987. Selective Transport of Hydrocarbons in the Unsaturated Zone Due to Aqueous and Vapor Phase Partitioning. *Water Resour. Res.*, 23, 1926-1938.

Barone, F.S., Rowe, R.K., and Quigley, R.M. 1992. A Laboratory Estimation of Diffusion and Adsorption Coefficients for Several Volatile Organics in a Natural Clayey Soil. *J. Contam. Hyd.*, 10, 225-250.

Carslaw and Jaegar, 1989. *Conduction of Heat in Solids*, Second Edition, Oxford, England, Oxford Science Publication.

Chen, Y-M., Abriola, L.M., Alvarez, P.J.J., Anid, P.J., and Vogel, T.M., 1992. Modeling Transport and Biodegradation of Benzene and Toluene in Sandy Aquifer Material: Comparisons with Experimental Measurements. *Water Resour. Res.*, 27, 3189-3199.

Cussler, E.L. 1991. *Diffusion: Mass Transfer in Fluid Systems.* New York, Cambridge University Press.

Gierke, J.S., Hutzler, N.J., and McKenzie, D.B. 1992. Vapor Transport in Unsaturated Soil Columns: Implications for Vapor Extraction. *Water Resour. Res.,* 28, 323-325.

Hutchins, S.R. 1991. Biodegradation of Monoaromatic by Aquifer Microorganisms Using Oxygen, Nitrate or Nitrous as the Terminal Electron Acceptor. *Appl. Env. Microbiol.,* 57, 2403-2407.

Jin, Y., Streck, T., and Jury, W.A. 1994. Transport and Biodegradation of Toluene in Unsaturated Soil, *J. Contam. Hyd.,* 17, 111-127.

Jury, W.A., Spencer, W.F. and Farmer, W.J. 1983. Behavior Assessment Model for Trace Organics in Soil, I. Model Description. *J. Environ. Qual.,* 12, 558-563.

Karimi, A.A., Farmer, W.J., and Cliath, M.M. 1987. Vapor-Phase Diffusion of Benzene in Soil. *J. Environ. Qual.,* 16, 38-43.

Kreamer, D.K., Weeks, E.P., and Thompson, G.M. 1988. A Field Technique to Measure the Tortuosity and Sorption-Effected Porosity for Gaseous Diffusion of Materials in the Unsaturated Zone with Experimental Results from Near Barnwell, SC. *Water Resour. Res.,* 24, 331-341.

Lam, T.T. 1994. Adsorption and Diffusive Transport of Chlorinated Aliphatic Solvents in Unsaturated Soil. Ph.D. Dissertation, Rutgers, The State University of New Jersey, Piscataway, NJ.

Mester, K.G. 1995. Field Implementation of Anaerobic Denitrification for Toluene Degradation. Ph.D. Dissertation, Rutgers, The State University of New Jersey, Piscataway, NJ.

Millington, R.J. and Quirk, J.P. 1961. Permeability of Porous Solids, *Trans. Faraday Soc.,* 57, 1200-1207.

Penman, H.L. 1940. Gas and Vapor Movements in the Soil. I. The Diffusion of Vapors Through Porous Solids. *J. Agric. Sci.,* 3, 437-462.

Priesack, E. 1991. Analytical Solution of Solute Diffusion and Biodegradation in Spherical Aggregates. *Soil. Sci. Soc. Am. J.,* 55, 1227-1230.

Riddick, R.A., Bunger, W.B., and Sakano, T.K. 1986. *Organic Solvents: Physical Properties and Methods of Purification.* New York, John Wiley & Sons.

Schaefer, C.E., Arands, R.R., van der Sloot, H.A., and Kosson, D.S., 1997, Modeling Gaseous Diffusion Through Unsaturated Soil Systems. *J. Contam. Hydrol.,* In press.

Siegrist, R.L. and Van Ee, J.J. 1994. Measurement and Interpretation of VOCs in Soil: State of the Art and Research Needs. EPA/540/R-94/506; Office of Research and Development, U.S. EPA, Washington DC.

Taylor, S.A. and Ashcroft, G.L. 1972. *Physical Edaphology. The Physics of Irrigated and Nonirrigated Soils.* San Francisco, CA, W.H. Freeman and Company.

Turner, L.H., Chiew, Y., Ahlert, R., and Kosson, D.S., 1996. Measuring Vapor-liquid Equilibrium for Aqueous Organic Systems: Review and a New Technique., *AIChE Journal*, 42, 1772-1788.

Van Bavel, C.H.M. 1952. Gaseous Diffusion in Porous Media. *Soil Sci.*, 73, 91-104

Venkatraman, S. 1995. Enhancement of In-situ Bioremediation by Pneumatic Fracturing: Field Demonstration, Laboratory Studies and System Modeling., Ph.D. Dissertation, Rutgers, The State University of New Jersey, Piscataway, NJ.

West, O.R. Siegrist, R.L. and Jennings, H.L. 1995. In Situ Mixed Region Vapor Stripping in Low-Permeability Media. 1. Process Features and Laboratory Experiments. *Environ. Sci. Technol.*, 29, 2191-2197.

PART VI
STATE REGULATORY

CHAPTER 35

A Practical Approach for Assessing the Extent of NAPL in Gasoline-Impacted Soil

Peter Peuron and **Seth J. Daugherty**, Orange County Health Care Agency, Santa Ana, California

INTRODUCTION

Organic compounds and mixtures that are generally immiscible but sparingly soluble in water can exist in four basic phases in the subsurface: the separate liquid or oil phase, often referred to as nonaqueous phase liquid or NAPL; dissolved in water as a true aqueous phase solution; mixed with air as vapor phase in soil gas; or adsorbed to the surfaces of granular particles. Determining the phase of hydrocarbons at a contaminated site is one of the most important yet often neglected aspects of site characterization. This is especially true with respect to determining if immobile, trapped liquid phase hydrocarbon fuels are present, because they may act as a source of dissolved phase fuel constituents for long periods of time.

Knowing the primary contaminant phases at a site can lead to more effective remediation, more realistic cleanup criteria, and a better understanding of the length of time required for completion of cleanup. For example, if liquid phase NAPL is trapped in pore spaces, a pump-and-treat system used alone may take many years to clean up the contamination. Dissolved phase contamination, however, may be readily removable by pump-and-treat technology. Another example may be that a cleanup goal of removing all the hydrocarbon source material may be unrealistic if hydrocarbons have spread as a thin zone of liquid phase over an extensive area of the capillary zone. Moreover, the extent of liquid phase contamination will also provide a clearer understanding of time needed for remediation by natural attenuation. Finally, knowing the phase distribution is important for fate and transport evaluations used in risk assessments.

This Chapter presents a practical method to distinguish between liquid and dissolved phase hydrocarbons, based on the principles of physical chemistry as described in Feenstra, MacKay, and Cherry (1991). A more sophisticated approach is presented in Mott (1995).

BACKGROUND

Description of Liquid (NAPL), Dissolved (Aqueous), and Adsorbed Hydrocarbon Phases

Although subsurface hydrocarbons may exist in four basic chemical phases in the subsurface (liquid, dissolved, adsorbed and vapor), Lyman et al. (1992) recognized 13 "phase-loci" of contaminants in the subsurface, mainly based on further division of liquid and sorbed phases in different substrates. For example, they distinguished between liquid contaminants adhering to water-dry particles, liquid contaminants in pore spaces of the saturated zone, liquid contaminants in pore spaces of the unsaturated zone, liquid contaminants in fractured rock, and liquid contaminants floating on the water table. To avoid confusion, terms used to describe hydrocarbon phases will be more precisely defined.

Liquid phase hydrocarbons may saturate the pore spaces in the upper zone of the capillary fringe to the extent that the hydrocarbon is mobile under the force of gravity. If this free flowing liquid phase is a petroleum fuel, this mobile liquid phase is often called floating "free product." When this mobile free-flowing hydrocarbon liquid spreads to the point where it becomes discontinuous and is held by capillary forces to the extent that it is no longer mobile, it is properly considered as nonmobile trapped liquid, or as trapped NAPL, although it is sometimes incorrectly called "adsorbed" hydrocarbon, or less commonly "absorbed" hydrocarbon. Both free-flowing and trapped liquid phase hydrocarbons are often referred to as NAPL, or nonaqueous phase liquid. Trapped liquid phase hydrocarbons, immobile as separate phase, are often referred to as being at "residual saturation." Residual saturation for unweathered gasoline may range from 35,000 to 120,000 mg/kg (Hoag and Marley, 1986). Trapped liquid phase hydrocarbons are the primary source of groundwater contamination because the more soluble fuel constituents can dissolve out of nonmobile trapped NAPL fuels for decades (Hunt et al., 1988).

Soluble organic compounds may dissolve into water and form a true solution. If these compounds are constituents of complex mixtures such as fuels, their concentrations in water will be a function of both pure component solubility and mole fraction in the mixture. The more soluble hydrocarbon fuel constituents, such as benzene and methyl tertiary butyl ether (MTBE) may readily dissolve out of liquid phase fuels to form aqueous solutions concentrated enough to be of public health and environmental concern. Gaseous phase hydrocarbons can be considered as a solution of hydrocarbon vapor in air.

Adsorption is the process by which dissolved chemicals in a fluid attach to solid particles; in the case of organic compounds, most notably on the surface of solid organic matter (Devinny et al., 1990). Adsorbed organic compounds are sometimes referred to as the "solid" phase. The term "sorption" is considered to include adsorption, absorption, chemisorption, ion exchange, and desorption (Lyman et al., 1992). Absorption is generally taken to mean uptake of chemicals into (rather than on) another body such as a solvent (e.g., vapor into water, see Lyman et al., 1992), incorporation into the interior of a solid (National Research Council, 1990), or into microorganisms (Lyman et al, 1992)

In this Chapter, liquid phase hydrocarbons are usually referred to as NAPL. In practice, this often means relatively immobile liquid fuel trapped in the pore spaces of the subsoil, as the free-flowing "free product" is often removed. This trapped NAPL may be present in the vadose zone, the capillary fringe zone, or beneath the water table, and may reappear again as free product with changes in level of the water table. Nonliquid phase contaminants are referred to as "non-NAPL," a term which includes fuel constituents dissolved in water (aqueous phase), in the vapor phase, or adsorbed to the surface of organic matter.

THE METHOD DESCRIBED BY FEENSTRA, MACKAY, AND CHERRY (1991)

Feenstra et al. (1991) proposed a simplified method that can be used to determine whether soil samples contain NAPL. The method assumes equilibrium partitioning between organic contaminant phases that are present in complex mixtures. The maximum amount of a chemical that can be present as a dissolved species in water is expressed as the effective solubility of the chemical. The effective solubility, S^e_i, is stated as

$$S^e_i = X_i\, S_i \tag{1}$$

where:

X_i = Mole fraction of compound i in mixture of NAPL (fraction expressed as a decimal)

S_i = Solubility of compound i (mg/L).

When water is in direct contact with NAPL, individual constituents of the gasoline mixture will partition out of the mixture and into the water at a rate that should not exceed this equilibrium concentration in water. Contaminant concentrations in soil samples which exceed the capacity of the soil pore water to retain contaminants in the dissolved phase suggest the presence of NAPL in the soil sample. Soil samples that contain contaminant concentrations which are within the capacity of the soil water to contain dissolved contaminants imply that gasoline constituents are present as dissolved in the pore water, or possibly as some other non-NAPL phase of contamination.

When NAPL is not present in the soil, there is a theoretical limit to the mass of contamination which can be contained in the sorbed, dissolved, and gas phases (Feenstra et al., 1991). These three primary non-NAPL contaminant phases can be related to the actual amount of contamination present in a soil sample as

$$Ct = (Cs\, \rho b + Cw\, \theta w + Ca\, \theta a)/\, \rho b \tag{2}$$

where:

Ct = Contaminant concentration in soil (mg/kg dry weight)

Cs = Contaminant concentration adsorbed to organic carbon (mg/kg dry weight)

ρb = Soil bulk density (g/cm^3)

Cw = Contaminant concentration in pore water (mg/L)

θw = Water filled porosity (cm^3/cm^3)

Ca = Contaminant concentration in soil gas (mg/cm^3)

θa = Air-filled porosity (cm^3/cm^3).

The distribution coefficients for sorption (Kd) and for volatilization from the dissolved phase (the dimensionless Henry's law constant, Hc) can be substituted into this equation since,

Kd = Cs/Cw

Hc = Ca/Cw

By rearrangement:

Cs = Kd Cw

Ca = Cw Hc

Substituting these partitioning coefficients into (2)

$$Ct = (Kd\ Cw\ \rho b + Cw\ \theta w + Cw\ Hc\ \theta a)/\ \rho b \tag{3}$$

For petroleum hydrocarbons, sorption is assumed to occur predominantly onto the organic fraction of the soil. Therefore,

$$Kd = Koc\ foc \tag{4}$$

where:

Koc = Organic carbon to water partitioning coefficient (L/kg)

foc = Fraction of organic carbon in soil (fraction expressed as a decimal).

Equation 3 can be rearranged to express the soil contaminant concentration in terms of the contaminant concentration of the soil pore water as

$$Cw = (Ct\ \rho b)/(Kd\ \rho b + \theta W + Hc\ \theta a) \tag{5}$$

Equation 5 can be used to indicate whether NAPL is present in soil samples by representing the dry weight contaminant concentration as Ct and supplying known or estimated values for the other parameters. If the computed contaminant concentration exceeds the effective solubility of the contaminant given in Equation 1, it can be concluded that NAPL is present in the soil sample. Feenstra et al. note that the vapor phase can often be neglected since it may be present in soil samples in low quantities owing to sample handling practices, wet soil (i.e., low air-filled porosity) and low Henry's law constants. When this term is neglected, (5) reduces to

$$Cw = (Ct\ \rho b)/(Kd\ \rho b + \theta w) \tag{6}$$

When the adsorbed phase is negligible, the Kd ρb term may be dropped, leaving

$$Cw = (Ct\ \rho b)/\ \theta w \tag{7}$$

Simplifications for a Practical Approach for NAPL Identification

The method described below is an approach for data analysis. The purpose of this analysis is to determine the phase status of contaminated soil samples based on relationships of chemical properties that are expressed as ratios. A proportionality coefficient called the Aqueous Phase Saturation Ratio (APSR) is used to express the degree to which contamination in soil samples approaches or tends to exceed the capacity of the soil to retain the contaminants in pore water. It should be possible to apply the proposed method either to: (1) screen sites for the maximum possible extent of NAPL, or (2) delineate the extent of NAPL when organic carbon content is known to be negligible. In the method outlined below, the vapor and adsorbed phases are neglected. An alternative, simplified form of Equation 7 is presented which should be useful at most underground storage tank cleanup sites, at least as a screening level approach.

APSR Approach and Procedure for Calculating APSR

By assuming that contaminants in the soil are either in the liquid phase (NAPL) or in the aqueous (dissolved) phase, it is possible to calculate a "carrying capacity" of the soil pore water which, if exceeded, indicates the presence of NAPL (Feenstra et al., 1991). The assumption is conservative because neglecting organic carbon as a storage compartment for NAPL will result in positive identifications of NAPL at lower soil contaminant concentrations. As discussed earlier, the maximum concentration of a constituent of gasoline that can be present in water which is in contact with the gasoline is determined by multiplying the solubility of the constituent by its mole fraction in gasoline (Si MF). The product of this calculation is the effective solubility (S^ei) of the constituent of the gasoline mixture. It is possible to express the degree to which contamination in water tends to saturate the capacity of the water to retain the contaminants by dividing the pore water concentration (Cw, obtained using Equations 5, 6, or 7) by the effective solubility of the contaminant. This aqueous phase saturation ratio (APSR) may be expressed as

$$APSR = Cw/(Si\ MF) \tag{8}$$

The APSR represents a proportion between the known amount of a contaminant present in water and the maximum amount that can be held in the water. The degree of saturation will either be a decimal fraction below unity, indicating a contaminant concentration that does not exceed the point where NAPL must be present, or the ratio will indicate NAPL when its value is greater than 1. Note that for soil samples, calculation of Cw requires inputs for bulk density, water-filled porosity, and contaminant concentration by dry weight. A simpler expression of Equation 7 can be derived by using the relationship

$$Cw = Ctw/W \qquad (9)$$

where:

    Cw = Contaminant concentration in pore water (mg/L)

    Ctw = Concentration of contaminant in the soil (mg/kg wet weight)

    W = Soil water content (L/kg wet weight)

When Equation 9 is substituted into Equation 8, the Aqueous Phase Saturation Ratio is expressed as

$$APSR = Ctw/(Si\ MF\ W) \qquad (10)$$

The above equation represents a proportion between the known amount of a contaminant present in the soil and the maximum amount that can be held in the pore water. Actual contaminant concentrations of benzene, toluene, ethylbenzene, and xylene are represented as Ctw. Using this equation, NAPL status can be evaluated using actual contaminant concentrations reported from field data, rather than converting the data to a dry weight basis. Also, it is not necessary to account for bulk density and porosity in this form of the equation, since soil water content by wet weight (a directly measurable, or easily estimated parameter) accounts for these properties.

An APSR value near 1 is indeterminate (may or may not be indicative of NAPL). Marginal APSR values are considered to occur between 0.25 and 4. This range should adequately account for variations in soil water content and mole fraction, parameters which it was assumed would not vary from their average values by more than a factor of 2. Therefore, errors due to variability in these parameters should not vary from 1 by more than a factor of 4, either above or below 1 (0.25 to 4). In order for a soil sample to be considered as confirmed for NAPL, the APSR must exceed 4. Samples were, in general, confirmed as not being NAPL if the APSR value was less than 0.25.

Xylene and Ethylbenzene as Surrogates for Gasoline

Xylene and ethylbenzene are likely candidate constituents as gasoline surrogates because their solubilities are about the same as gasoline. Aromatic contaminant concentrations that correlate with TPH concentrations would tend to indicate that a given aromatic species behaves in a manner consistent with the partitioning characteristics of the gasoline mixture (or the "average" gasoline molecule). Such consistent behavior would imply that a particular species has physical and chemical properties that are similar to the net properties of gasoline. These aromatic constituents were chosen as surrogates for the phase status of gasoline, based on the results of regression analysis, as discussed later in this document. Further, xylene and ethylbenzene appear to exhibit reasonably ideal behavior with respect to dissolving in water in proportion to their mole fractions. The data reported by Cline et al. (1991), based on experimental studies of the partitioning of aromatic constituents into water from gasoline, showed that ethylbenzene and xylene isomers exhibit fairly ideal behavior with respect to dissolution. By contrast, concentrations of benzene and toluene were considerably higher than would be expected based solely on mole fraction and solubility.

Water Content

A wet weight soil water content of 0.15 (g/g or cc/g or L/kg) was assumed for soil samples obtained from the vadose zone. Samples which were obtained from the capillary fringe or below groundwater were assumed to be saturated. A wet weight soil water content of 0.25 l/kg was assumed for these samples. The soil water content values were calculated based on average volumetric water content values and average porosity values which were obtained from the literature (Rawls and Brakensiek, 1989).

Example Calculation

An example calculation using a soil sample with 170 mg/kg of xylene is given as follows:

$$\text{Ctw} = 170 \text{ mg/kg}$$
$$\text{MF} = 0.075$$
$$\text{S} = 180 \text{ mg/L}$$
$$\text{W} = 0.15 \text{ l/kg}$$
$$\text{APSR} = (170 \text{ mg/kg})/(180 \text{ mg/L} * 0.075 * 0.15 \text{ l/kg})$$
$$\text{APSR} = 170 \text{ mg/kg}/2.025 \text{ mg/kg}$$
$$\text{APSR} = 84$$

This result indicates that there is about 84 times more xylene in the sample than the theoretical maximum amount of xylene that could be present in the pore water as dissolved phase. This value is well above the value of 4 that confirms the presence of NAPL. Therefore, it is very likely that NAPL exists in this sample.

METHODS

A total of 411 APSR values were calculated for each of the aromatic petroleum hydrocarbons being considered (a total of 1644 APSR values for 411 soil samples). The samples were obtained from 20 UST cleanup sites, most of which were located in Huntington Beach, California. The data were obtained from site assessment reports that were provided by a variety of environmental consultants. Laboratory analyses were performed by a number of different state certified laboratories.

Samples were analyzed for Total Petroleum Hydrocarbons by a modified EPA Method 8015 analysis, and for benzene, toluene, ethylbenzene, and xylene by EPA Method 8020. Depth of sample data was also included in the database which was evaluated. It was necessary to estimate the values of other parameters used in APSR calculations because site assessment reports did not provide this information. Assumed solubilities, mole fractions, water contents and the net effective solubility in soil for each of the aromatic constituents are shown on Table 1. Solubilities (USEPA, 1988) and mole fractions (Lyman et al., 1992) were obtained from the literature. After performing the APSR calculations, soil samples were classified as confirmed NAPL, confirmed non-NAPL, or indeterminate, based on the previously discussed criteria.

Table 1.   **Contaminant Properties Used in Determining EffectiveSolubilities**

Contaminant	Solubility	Mole Fraction	Water Content	Effective Solubility
	(mg/l)	(dimensionless)	(l/kg)	(mg/kg)
Benzene	1780	0.015	0.15	4.005
Toluene	520	0.070	0.15	5.46
Ethylbenzene	170	0.013	0.15	0.3315
Xylene	180	0.075	0.15	2.025

Xylene to ethylbenzene ratios were evaluated in order to assess the reliability of these gasoline components as indicators of phase status. Since the properties of both of these constituents are similar, it is proposed that the partitioning behavior of each of these constituents would be similar, and therefore that the ratio of xylene to ethylbenzene concentrations should be somewhat constant. Correlation analysis was performed to verify that these constituents are typically present in soil samples in proportions which are equal to the ratio of their respective mole fractions (about 0.075/0.013 or 6/1).

RESULTS AND DISCUSSION

Correlation of Aromatic Hydrocarbons to TPH

The ability of each aromatic species to mimic the gasoline mixture was evaluated by comparing each aromatic petroleum hydrocarbon concentration against the TPH concentrations from the same soil samples. This analysis was performed for the 411 soil samples under consideration. Figures 1 through 4 show the scatter plots for this data. The coefficients of determination for xylene and ethylbenzene (0.762 and 0.771, respectively) were considerably higher than those for benzene and toluene (0.516 and 0.607, respectively). The toluene/TPH correlation was better than the benzene/TPH correlation. This result is consistent with the solubility of gasoline being much more like the solubility of toluene than that of benzene. Further, the average solubility of the gasoline mixture can be expected to range from 120 to 200 mg/L (Lyman et al., 1992), while the solubilities of ethylbenzene and xylene are approximately in the middle of this range (see Table 1). It appears that the similar solubilities of xylene, ethylbenzene, and the gasoline mixture play a significant role in these correlations. These results tend to validate the use of ethylbenzene and xylene as surrogates for the gasoline mixture. Based on these considerations, the APSR values for xylene and ethylbenzene were considered to be more reliable indicators of NAPL status.

Evaluation of APSR Values

In order to illustrate use of this method, results are first discussed for an individual cleanup site. Results for all 411 soil samples used in this study are then discussed separately. The illustrative site is located in Huntington Beach, California. Analytical results for 18 soil samples from the site are shown on Table 2. The APSR values for xylene indicate that there were seven samples confirmed as NAPL (APSR

Figure 1.   **Relationship between benzene and TPH concentrations in soil.**

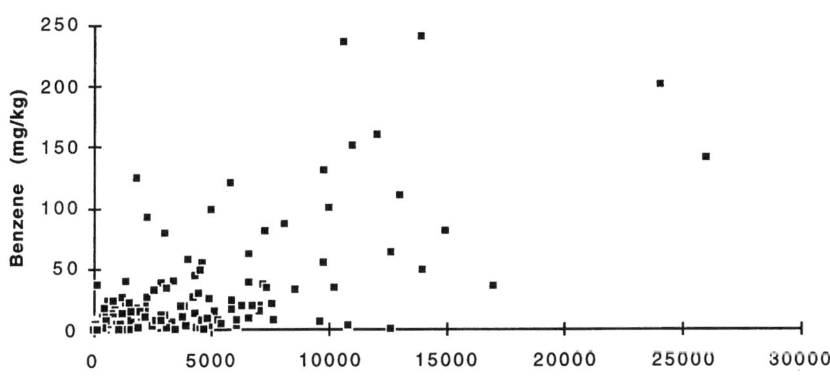

Figure 2.   **Relationship between toluene and TPH concentrations in soil.**

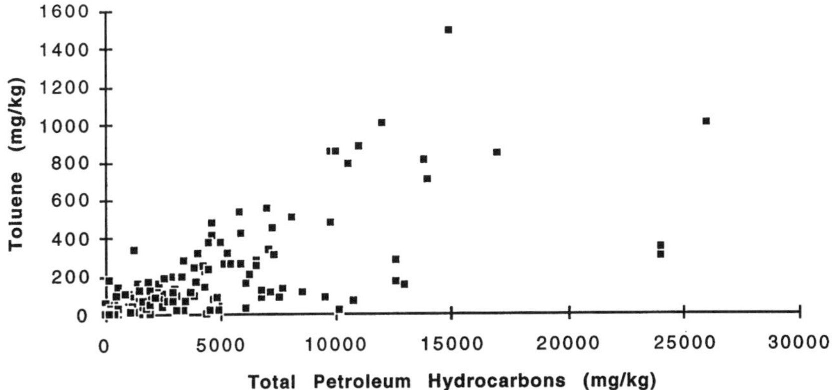

Figure 3. **Relationship between ethylbenzene and TPH concentrations in soil.**

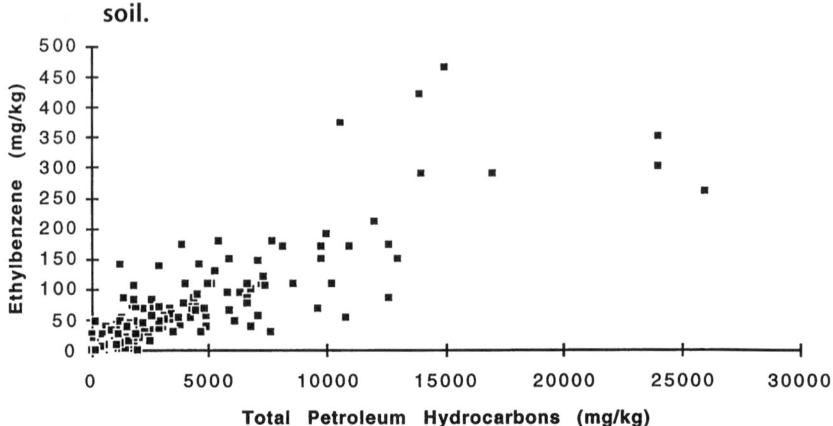

Figure 4. **Relationship between xylene and TPH concentrations in soil.**

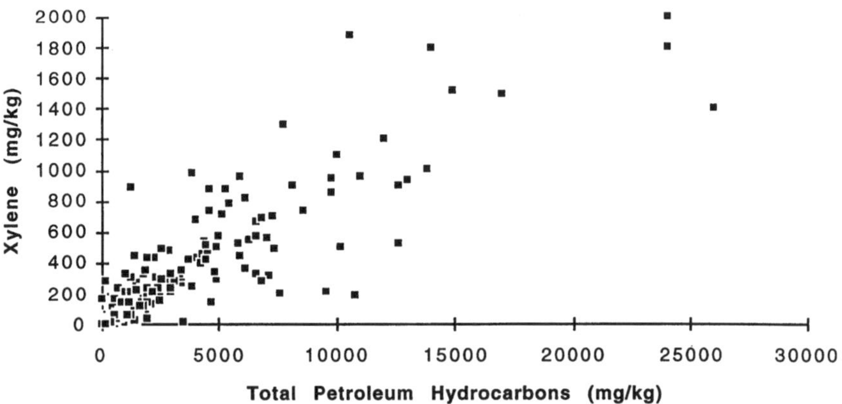

greater than 4), and eight samples confirmed as non-NAPL samples (APSR less than 0.25). Three samples were found to have xylene concentrations that fell in the indeterminate range (APSR less than 4, but greater than 0.25). The ethylbenzene APSR values agree with all of the determinations based on xylene data except for sample number 4 where the APSR for ethylbenzene slightly exceeds the 0.25 level. Several of the benzene and toluene APSR values indicate either a non-NAPL status or indeterminate status for samples that are confirmed to contain NAPL by the xylene and ethylbenzene APSR values. Note that all of the high xylene and ethylbenzene APSR values are associated with the high TPH values, and all high TPH values are associated with high APSR values. The highest confirmed non-NAPL TPH concentration is 5 ppm. The indeterminate APSR values are associated with TPH values of 7.7 ppm, 8.4 ppm, and 12.5 ppm. It could be argued that sample number 4 is also indeterminate, but the ethylbenzene APSR value (0.26) barely qualifies this sample as being in the "indeterminate" range. Other

Table 2. **Soil Sample Data and APSR Values for a Site in Huntington Beach, California**

No.	Depth (feet)	TPH	Benzene	Toluene	Ethyl-benzene	Xylene	APSR Benzene	APSR Toluene	APSR eth-ylbenzene	APSR Xylene
1	20	1.5	0.12	0.055	0.038	0.22	0.030	0.010	0.11	0.11
2	30	2.1	0.3	0.016	0.06	0.31	0.075	0.003	0.18	0.15
3	20	0.69	0.068	0.014	0.02	0.063	0.017	0.003	0.060	0.031
4	30	2	0.46	0.0025[a]	0.086	0.13	0.11	0.0005	0.26	0.064
5	35	1.2	0.33	0.0025[a]	0.025	0.11	0.08	0.0005	0.08	0.05
6	20	7.7	0.39	0.79	0.14	0.95	0.097	0.15	0.42	0.47
7	30	8.4	0.66	0.83	0.22	1.2	0.17	0.15	0.66	0.59
8	35	2	0.26	0.11	0.045	0.29	0.07	0.02	0.14	0.14
9	45	0.25[a]	0.068	0.0025[a]	0.011	0.052	0.017	0.0005	0.033	0.026
10	3	3800	9.8	97	42	250	2.5	17.8	127	123
11	5	180	0.25	4	2.4	17	0.06	0.73	7.2	8.4
12	10	7100	19	340	55	320	4.7	62	166	158
13	15	320	0.25	5.5	4	28	0.06	1.0	12.0	13.8
14	15	4900	25	55	56	500	6.2	10.0	168	246
15	30	*0.25[a]	0.066	0.0025[a]	0.016	0.025	0.017	0.0005	0.048	0.012
16	10	3400	2.5	32	33	270	0.62	5.9	99.5	133
17	17	4900	2.5	20	40	300	0.62	3.7	120	148
18	35	12.5	0.022	0.041	0.14	1.2	0.005	0.008	0.42	0.59

Contaminant concentrations in mg/kg.

[a] No detectable contamination, detection limit was 0.005 mg/kg for toluene and 0.5 mg/kg for TPH.

Contaminant concentration is assumed to be half of the detection limit.

data, including the TPH result (2 ppm) weigh in favor of a conclusion of non-NAPL status for this sample. An overview of this table indicates that there are three soil samples that fall into the transition range and that the other 15 samples can be classified somewhat definitively into NAPL or non-NAPL categories. This data set suggests that benzene and toluene APSRs do not tend to indicate false positive values for NAPL, but are prone to false negative indications. Two benzene APSR results appear to be false negatives (samples 11 and 13) while both benzene and toluene results fall into the indeterminate range repeatedly when NAPL is indicated by the xylene and ethylbenzene APSR results. In general, conflicting NAPL and non-NAPL indications should be resolved in favor of an NAPL conclusion because NAPL levels of one hydrocarbon can coexist with a concentration of another hydrocarbon that is not indicative of NAPL, but NAPL should be present if any one hydrocarbon species exceeds its net effective solubility (assuming organic carbon content is not anomalously high). The agreement between the xylene and ethylbenzene APSR values on Table 2 is important since these species have similar properties. Any significant degree of variation in the behavior of these parameters would tend to suggest a lack of reliability in the proposed method.

Table 3 shows the results of the APSR determinations for all 411 soil samples. The consistency between xylene and ethylbenzene results is again apparent. Of the 242 samples in which the xylene APSR indicated NAPL, the ethylbenzene APSR value concurred 240 times. The total number of ethylbenzene confirmations of NAPL was 253. Therefore, there were 11 cases in which ethylbenzene data confirmed NAPL while xylene results did not. In all of these 11 cases, the xylene APSR result classified the samples as indeterminate. There were no soil samples for which either xylene or ethylbenzene confirmed NAPL while the other constituent confirmed a non-NAPL result. There were 57 cases in which the ethylbenzene APSR indicated a non-NAPL status. The xylene result agreed with 54 of these. However, there were 25 samples for which ethylbenzene values indicated non-NAPL results, while the xylene APSRs were indeterminate. Overall the NAPL determinations based on xylene and ethylbenzene APSR values show reasonable consistency. It is noteworthy that there were no outright contradictions between the APSR results for xylene and ethylbenzene (i.e., a NAPL and non-NAPL result for the same soil sample). APSR values for benzene and toluene confirmed NAPL 64 times and 152 times, respectively. The failure of these constituents to confirm NAPL status at a more consistent rate is attributable to the higher solubilities of these species. The lower rate of NAPL confirmation for the more soluble gasoline constituents (benzene and toluene) is consistent with the fact that these species are leached out of the gasoline matrix more easily than xylene and ethylbenzene. Moreover, the similar solubilities of ethylbenzene, xylene, and the gasoline mixture imply that xylene and ethylbenzene should behave in a manner consistent with the gasoline mixture as a whole.

Table 3. Summary of APSR Results for 411 Soil Samples

Contaminant	NAPL	Indeterminate	Non-NAPL
Benzene	64	163	184
Toluene	152	121	138
Ethylbenzene	253	101	57
Xylene	242	90	79

Table 4 lists the average TPH values for each of the categories of NAPL status (confirmed NAPL, confirmed non-NAPL, and indeterminate). These results suggest that the transitional range from non-NAPL levels of contamination to NAPL may occur around 40 to 70 ppm TPH (since the average indeterminate TPH values for ethylbenzene and xylene are 46 ppm and 67 ppm, respectively). However, it should be noted that the standard deviations for TPH values in the indeterminate range were 78 ppm for ethylbenzene and 97 ppm for xylene. It appears that it is difficult to identify a narrow range of TPH values where the transition to NAPL will occur, due to variability between conditions at different sites (e.g., variation in the degree to which contaminants have weathered). In any site-specific uses of the approach it would be necessary to consider all available data and integrate relevant information into the assessment of the extent of NAPL. For example, the depth data shown on Table 2 indicates that all samples which were confirmed for NAPL were found within 17 feet of ground surface. While all non-NAPL samples were found at a depth of at least 20 feet below the surface. This suggests a continuous zone of NAPL which overlies a zone where leached contaminants have migrated. Note that sample number 18 falls into the indeterminate category. However, based on its depth (35 feet below the ground surface) and relatively low TPH (12.5 ppm), it is probably reasonable to include this sample with the non-NAPL group.

Table 4. Average TPH Values for NAPL Groups Determined by APSR

Contaminant	NAPL	Indeterminate	Non-NAPL
Benzene	7,381	1,384	248
Toluene	4,234	651	102
Ethylbenzene	2,221	46	4
Xylene	3,021	67	7

Xylene to Ethylbenzene Ratios

APSR values for xylene and ethylbenzene shown on Table 2 tend to be similar in magnitude. Actually, these species should always produce like APSR values. It is apparent from Equation 10 and Table 1 that the only significantly different numbers used in calculating the APSR of xylene and that of ethylbenzene are the contaminant concentrations (Ctw) and the mole fraction (MF). A basic premise of the proposed method is that the fate and transport properties of the species under consideration can be predicted based on the chemical and physical properties of each individual petroleum hydrocarbon constituent. Xylene and ethylbenzene have essentially the same physical and chemical properties. Therefore, the concentrations of each of these constituents in a soil sample (Ctw) should be in proportion to the relative concentration of each contaminant present in gasoline. Regardless of the phase that either xylene or ethylbenzene is in, the APSR values should be approximately equal, since each of the species tends to remain in, and leach out from, the liquid phase in equal proportion. This means that the contaminant concentration (Ctw, in the numerator of Equation 1) should somewhat consistently reflect a six to one ratio between xylene and ethylbenzene (the ratio of their respective mole fractions). As long as xylene Ctw values are consistently six times higher than ethylbenzene Ctw values, APSR values will continue to be similar in magnitude. The average xylene to ethylbenzene ratio for all 411 soil samples considered was about 5.4. This is the "batch" average where the sum of all xylene results were added together and this sum was divided by the sum of all ethylbenzene results. A more meaningful indicator of the consistency of this ratio can be found by calculating each xylene to ethylbenzene ratio separately and then finding the average. When this was done, an average xylene to ethylbenzene ratio of 5.7 was found for the 411 soil samples considered. Figure 5 shows the scatter plot of xylene to ethylbenzene concentrations for the 411 soil samples considered. The coefficient of determination for this relationship is 0.913, once again indicating a strong similarity in the behavior of these constituents. The slope of this graph is 5.4, which agrees with the average X/E ratios quoted above.

Sources of Variability

There are four primary sources of variability which must be taken into consideration when applying this method on a site-specific basis. All concern the simplifying assumptions which have been made in order to make NAPL assessments using data typically available at UST sites. On any given site, it may be necessary to modify one or more of the default assumptions concerning organic carbon content, significance of the vapor phase, soil water content, and contaminant mole fractions. The assumption of no organic carbon content tends to favor a conclusion that NAPL is present. In order to assess potential sensitivity to the adsorbed phase, calculations of NAPL status were performed using Equations 6 and 8. Table 5 shows differences in the total numbers of each APSR derived category of NAPL status when the organic carbon content is varied from 0 to 0.01%, to 0.1%, to 1%. At 0.01% organic carbon results were virtually identical with the results obtained when organic carbon was assumed to be absent. The APSR values were reduced by about 13%, however, when 0.01% organic carbon was factored in. At 0.1% organic carbon, pore water concentrations were about one-half of the concentrations obtained when the adsorbed phase was not included. This resulted in 14%

Figure 5.   **Relationship between xylene and ethylbenzene concentrations in soil.**

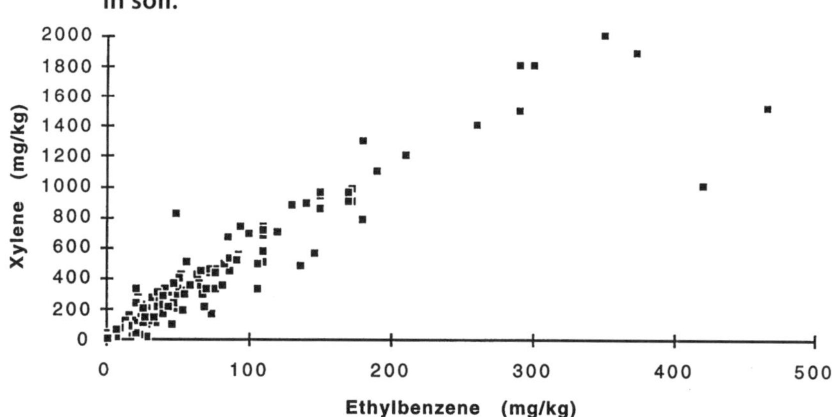

Table 5.   **Summary of APSR Results for Xylene When Fraction of Organic Carbon (foc) Varies**

% Organic Carbon	NAPL	Indeterminate	Non-NAPL
foc = 1 %	123	124	164
foc = 0.1%	208	100	103
foc = 0.01%	242	88	81
foc = 0%	242	90	79

fewer NAPL confirmations and 30% more confirmed non-NAPL results. At 1% organic carbon the results were affected to a significant degree. Only about 53% of the xylene and ethylbenzene NAPL determinations that had been confirmed without sorption were confirmed as NAPL. Pore water concentrations were about 13 times lower at this level of organic carbon. The average TPH for the indeterminate region was 548 ppm when 1% organic carbon was included. Table 6 lists the average TPH values for xylene APSR determinations when organic carbon was varied. Again, the potential impact of inclusion of the adsorbed phase is evident. As mentioned earlier, experience has shown that high organic carbon levels in deep soil are uncommon in the study area. Of course, this assumption will not be valid for all areas. Tests for organic carbon should be conducted when necessary. Note, however, that total organic carbon concentrations do not necessarily indicate accurate organic carbon values that can be applied in equations using the distribution coefficient for adsorption to organic carbon (Kd). This is because total organic carbon determinations typically include organic fractions in the soil which have very low adsorptive efficiency. Chunks of organic materials (e.g., leaf pieces, sticks, lignite) have a low capacity to affect adsorption relative to their weight,

Table 6.   Average TPH Values for NAPL Groups Determined by APSR for Xylene When Fraction of Organic Carbon (foc) Varies

% Organic Carbon	NAPL	Indeterminate	Non-NAPL
foc = 1 %	5,468	548	37
foc = 0.1%	3,297	170	10
foc = 0.01%	3,056	72	7
foc = 0%	3,021	67	7

whereas significant adsorption can be expected when humus content is high. Another confounding aspect of attempting to include this phase in calculations is that the constants which must be used in calculations are themselves uncertain. Experimentally derived Koc values for petroleum hydrocarbons have not been established; therefore, the Koc values for the constituents of concern must be estimated based on either octanol-water partition coefficients or aqueous solubilities (Lyman et al., 1992). Available estimates of these values vary by a factor of three, adding another source of error in attempting to assess the significance of the adsorbed phase. Subsurface heterogeneity constitutes another significant problem when attempting to apply organic carbon data to specific soil samples. Since both Total Organic Carbon (TOC) tests and tests for aromatic petroleum hydrocarbons are destructive in nature, the same sample cannot be tested by both methods. Given the rather extreme heterogeneity of the subsurface, TOC tests can be expected to provide only rough values for estimating adsorption to organic carbon. Considering the sources of error inherent in the standard approach for accounting for this phase, it is apparent that the contribution of organic carbon cannot be assessed with a high degree of accuracy on a site-specific basis. Note also that errors caused by overestimation of the adsorbed phase result in underestimation of the occurrence of NAPL. Since errors caused by underestimation of organic carbon content lead to overestimation of NAPL, this approach is conservative. Therefore, it can be applied as a screening methodology. Data from site assessment reports can be evaluated to identify zones where NAPL is clearly not present, and zones with possible NAPL can be located. More intensive analysis (including, perhaps, organic carbon, moisture content, and mole fraction determinations) can be applied in the suspect zones.

As discussed earlier, the influence of vapor phase contaminants on phase status determinations is considered to be insignificant. Calculations of APSRs were performed which included the vapor phase (using Equations 5 and 8). The results indicated that the potential mass of contaminant that can be present in the vapor phase has a minor effect on APSRs. Conclusions regarding NAPL status did not tend to be altered because the concentrations at which the vapor phase may be predominant (i.e., very low concentrations) always produce a non-NAPL result. Therefore, false positives (indicating NAPL when there is no NAPL) should not occur. At high contaminant concentrations (APSR near or above 4) the potential influence of the small mass of vapor phase contaminants in the soil sample is usually negligible. At least, the error is small compared to the variability of other factors which have already been accounted for by specifying an indeterminate range between the NAPL and non-NAPL limits. When organic carbon content is assumed to be zero, inclusion of the vapor phase reduced contaminant concentrations in pore water by a factor of about 10 per cent. Five of the 242 confirmed NAPL samples were classified as being indeterminate when this phase was included. The number of non-NAPL determinations did not change when vapor phase contaminants were included. The vapor phase had even less effect on the results when an organic carbon level of 1% was considered. In this case, pore water concentrations were reduced by about 1% when the vapor phase was included. It should be noted that inclusion of this phase is not conservative since fewer positive identifications of NAPL will be indicated. Yet, as Feenstra et al. (1991) point out, laboratory analyses may not properly account for the vapor phase due to sample handling practices that result in the loss of volatile components. Inclusion of the vapor phase may therefore be unnecessary. Based on the results obtained in this study, the impact of this phase is small even if it is captured in the laboratory. Both practical and theoretical uncertainties suggest that the simplifying assumption of no vapor phase contaminants is often appropriate.

By allowing for a variance of a factor of 2 from the assumed mole fraction values, actual mole fractions are considered to fall within a range of one-half to two times the assumed mole fractions. Therefore, ethylbenzene is assumed to comprise between 0.65% (0.0065) and 2.6% (0.026) of the total quantity of petroleum hydrocarbon molecules. Xylene is assumed to comprise 3.75% (0.0375) to 15% (0.15) of the TPH molecules. The validity of these estimates was checked by dividing the ethylbenzene concentration by the TPH concentration and the xylene concentration by the TPH concentration for all 411 soil samples. Each of these weight fractions were then multiplied by a molecular weight correction factor (the assumed molecular weight of gasoline over the molecular weights of xylene and ethylbenzene, or 82/106). The average mole fraction of ethylbenzene for all soil samples was 0.0189, with a median of 0.0128. The assumed ethylbenzene mole fraction was 0.013. The average xylene mole fraction according to the field data was 0.082, with a median value of 0.0624. The assumed xylene mole fraction was 0.075. These empirical estimates of mole fraction appear to be quite close to the assumed values; however, the variation in these empirical mole fractions was somewhat significant. Only about 62% of the empirical xylene/TPH mole fractions fell within the range of 0.0375 to 0.15. About 86% of the ethylbenzene/TPH empirically determined mole fractions fell within the specified range of 0.0065 to 0.026. It appears that the assumed values were accurate as averages, but it may be necessary to expand the range of allowable mole fraction variation, depending on the amount of conservatism required at specific sites. In a site-specific application of this method, empirical mole fraction data could be derived as discussed above, and mole fraction estimates could be based on these values. It is probably more accurate to use an estimated molecular weight for weathered gasoline (e.g., 95 grams per mole as per Lyman et al., 1992) instead of the lower value that was employed in this study.

The estimates of water content (0.15 by wet weight in the vadose zone and 0.25 by wet weight in the saturated zone) considered with the variability factor of 2 for this parameter imply that moisture content can vary from 0.075 to 0.30 in the vadose zone and from 0.125 to 0.50 (by wet weight) in saturated soil without substantial error. Actually, it should be possible to make more accurate estimates of soil water content by taking account of the soil type described in the boring logs of the samples under consideration. The literature provides values which are specific to soil type for this parameter (Rawls and Brakensiek, 1989). This approach is recommended for historical site data where soil water content values are typically not available. Of course, site-specific moisture content values can be obtained which would justify reducing the margin of error that is assumed in this study. In fact, all four of the parameters which were estimated can be measured site-specifically. However, it is usually not practical to collect this data at an average UST site. Therefore, the proposed method can provide reasonably accurate means for NAPL estimation that can be used at virtually any site.

SUMMARY AND CONCLUSIONS

An approach for distinguishing NAPL (liquid phase hydrocarbons) from non-NAPL, based on the ratio of soil contaminant concentrations to the capacity of the soil pore water to retain contaminants as dissolved phase, is presented. The approach is based on a method originally proposed by Feenstra, MacKay, and Cherry (1991). The degree to which soil contaminant concentrations tend to exceed the capacity of soil pore water to retain them in the dissolved phase is expressed as an aqueous phase saturation ratio (APSR). The approach was applied using a database consisting of 411 soil samples from 20 UST cleanup sites in Orange County, California. The original method was modified to allow for direct inputs of contaminant concentrations on a wet weight basis and soil water content (which replace bulk density, volumetric water content, and contaminant concentration by dry weight). Simplifying assumptions of no contaminant adsorption and no vapor phase contaminants were made in order to accommodate the methodology to data that is typically available in reports from UST cleanup sites. When APSR values for benzene, toluene, ethylbenzene, and xylene were evaluated along with TPH data, the following conclusions were made:

1. Xylene and ethylbenzene appear to be good surrogates for gasoline contamination in soil. The similarity of the solubilities of xylene, ethylbenzene, and gasoline support this conclusion. Correlations between concentrations of xylene and ethylbenzene with gasoline TPH levels were significantly better than correlations between benzene and toluene with gasoline TPH.

2. When APSR results were used to categorize samples into NAPL, non-NAPL, and indeterminate classes, the xylene and ethylbenzene values showed a high rate of agreement (over 90%) with each other. Notably, there were no cases in which one of these species confirmed NAPL while the other categorized the sample into a non-NAPL status.

3. A significant correlation was found between xylene and ethylbenzene concentrations in soil. Xylene concentrations were consistently about six times greater than ethylbenzene concentrations. Since this ratio is about equal to the ratio of the mole fraction of xylene to the mole fraction of ethylbenzene in gasoline, it appears that these compounds behave similarly in the soil environment.

4. The APSR was very sensitive to soil organic carbon content. APSR results were reduced by a factor of 13 when 1% organic carbon content was assumed. At 0.1% organic carbon, APSR values were about one-half of the values obtained when organic carbon was neglected. At 0.01% organic carbon, APSR results were 13% lower, a difference which appeared to have a negligible effect on conclusions pertaining to NAPL status of soil samples.

REFERENCES

Cline, P.A., Delfino, J.J., and Rao, S.C. 1991. Partitioning of Aromatic Constituents into Water from Gasoline and Other Complex Mixtures. *Environ. Sci. Technol.* 25(5), 914-920.

Devinny, J.S. 1990. Chemical and Physical Alteration of Wastes and Leachates. In: *Subsurface Migration of Hazardous Wastes.* New York, Van Nostrand Reinhold.

Feenstra, S., MacKay, D.M., and Cherry, J.A. 1991. A Method for Assessing Residual NAPL Based on Organic Concentrations in Soil Samples. *Ground Water Monitoring Review,* 11(2), 128-136.

Hoag, G.E. and Marley, M.C. 1986. Gasoline Residual Saturation in Unsaturated Uniform Aquifer Materials. *J. Environ. Eng.* 112(3), 586-604.

Hunt, J.R., Sitar, N., and Udell, K.S. 1988. Nonaqueous Phase Liquid Transport and Cleanup 1. Analysis of Mechanisms. *Water Resour. Res.* 24(8), 1247-1258.

Lyman J., Reidy, J.R., and Levy, B. 1992. *Mobility and Degradation of Organic Contaminants in Subsurface Environments,* Chelsea, MI, C. K. Smoley, Inc.

Mott, H.V. 1995. A Model for Determination of the Phase Distribution of Petroleum. *Ground Water Monitoring and Remediation.* 15(3), 157-167.

National Research Council. 1990. Ground Water Models: Scientific and Regulatory Applications. Washington, DC, National Academy Press.

Rawls, W.J. and Brakensiek, D.L. 1989. Estimation of Soil Water Retention and Hydraulic Properties. In: *Unsaturated Flow in Hydrologic Modeling,* (Morel-Seytoux, H.J., Ed.). Boston, MA, Kluwer Academic Publishers.

USEPA. 1988. Cleanup of Releases from Petroleum USTs: Selected Technologies. Washington, DC. Office of Underground Storage Tanks. EPA/530/UST-88/001.

CHAPTER 36
The Rhode Island Brownfields Program

Gregory S. Fine, Rhode Island Department of Environmental Management, Office of Waste Management, Providence, Rhode Island

Timothy M. O'Connor, Environmental Resources Management Group, Providence, Rhode Island (formerly of the Rhode Island Department of Environmental Management)

INTRODUCTION

Brownfields initiatives are a relatively new approach to three old problems: the continuing decay of urban areas, the loss of rural open space, and the adversarial relationship between the business community and environmental regulatory agencies.

From 1835 until 1939, Rhode Island cities were a hotbed of the industrial age. Huge mill complexes were used to produce products which were sold worldwide. These complexes have been gradually abandoned by their original industries (which moved to the suburbs) and left to compete for secondary uses such as small businesses, office space, and apartments. Today, many of these facilities remain either entirely abandoned or vastly underutilized; they have become symbolic of the decay of urban quality of life.

To make matters worse, many of these abandoned properties pose a risk to their neighbors. These risks are the result of the attractiveness of these buildings to transient populations and children, and more universally by the environmental contamination left behind by the former tenants.

Concurrently, industry continues to pay higher and higher prices for land and infrastructure in rural areas. This trend has two significant negative aspects: the loss of valuable, undeveloped land, and the continued decay of urban areas. The rising costs of developing rural properties is the result of many things, including the scarcity of prime land free of encumbrances such as wetlands, steep slopes, or zoning complications.

As these development costs continue to increase for undeveloped rural properties, it has started to become cost-effective for developers to consider the acquisition of abandoned urban property. From a redevelopment standpoint, however, the environmental contamination present at these sites represents a deal-breaker to the business community. Most if not all potential redevelopment options for these abandoned industrial sites have been stymied by the liability mandate of the federal Superfund program, the perceived unpredictability of regulatory cleanup requirements, and the huge costs associated with hazardous materials remediations. Thus, an approach to environmental cleanups which is acceptable to the business and environmental regulatory community is necessary.

In response to these circumstances, state and federal environmental regulatory agencies, along with state and local economic development groups, have started what are commonly referred to as Brownfields initiatives. These programs are aimed at attracting business back to these abandoned properties (the "Brownfields") and bringing them back to productive use, creating jobs, increasing urban tax bases, cleaning up the environment, and preserving open space (the "Greenfields") all at the same time. This is accomplished with financial incentives and by providing the means by which a potential developer can quantify and limit the environmental liability associated with a project.

In Rhode Island, the Department of Environmental Management (RIDEM), with the support of the Rhode Island Economic Development Corporation (RIEDC), has established a program designed to streamline the site remediation process in order to efficiently clean up underutilized, contaminated properties to allow these sites to be beneficially reused. This program, known as the Rhode Island Brownfields Program, accomplishes these goals by (1) developing and implementing generic, risk-based soil cleanup standards which are based on current and reasonably foreseeable future use of a property, and (2) establishing a liability system which provides a process for relief from environmental liability for innocent parties, including prospective purchasers, financial institutions, and voluntary performers. The relief is documented in a Settlement Agreement that can include a covenant not to sue issued by the state.

REGULATORY HISTORY

In order to fully appreciate the motivation behind Brownfields initiatives it is necessary to understand the evolution of hazardous site regulation. To address abandoned or uncontrolled hazardous waste or hazardous materials sites, the Comprehensive Environmental Response, Compensation and Liability Act (CERCLA) (also referred to as the Superfund program) was created in 1980. This program calls for a series of environmental site assessments to determine the highest priority abandoned hazardous materials sites. The assessment phase of the Superfund program concludes with a site scoring system which is used to determine if a site warrants a federally overseen cleanup. Sites which score above the established criteria are added to the National Priorities List (NPL) of hazardous material sites. These NPL sites then undergo a series of federally overseen investigations in order to determine an appropriate site remedy.

One of the main themes of CERCLA is that whenever possible, those parties responsible for a site should pay to clean it up. Accordingly, the law establishes a strict, joint and several liability system which makes each responsible party tied to a site individually responsible for a site's entire cost of investigation and cleanup. The law also makes liability retroactive; thus a party could be held liable for an act which was legal at the time disposal occurred. The assignment of liability, along with its overall slow pace, are Superfund's most criticized aspects. These are also the issues which are the principal impediments to potential redevelopment of contaminated sites, and which lead prospective purchasers toward properties not associated with the Superfund program or with any other form of potential environmental liability.

As a result of the slow progress made by the Superfund program, many states have developed their own hazardous sites programs. By 1990 (10 years after the original passage of CERCLA), 48 states had passed separate statutes which enabled the creation of a parallel program. This trend was borne out of dissatisfaction with the Superfund program by both the state environmental regulators and the regulated community. Since the establishment of the federal Superfund program, the states have been looking to clean up more sites and have a larger role in the process. Correspondingly, the regulated community often feels more comfortable dealing with a state program because it is less bureaucratic, more flexible, and has learned from Superfund's shortcomings.

In general, the state programs are also willing to regulate voluntary site cleanups where the EPA must address sites by priority, regardless of the cooperative nature of the responsible parties. Thus, sites which are not a high priority in the federal program can sit for years in their inventory of backlogged sites waiting to rise to the top. This inventory, known as the EPA's Comprehensive Environmental Response, Compensation and Liability Information System (CERCLIS), is a list of sites eligible for the NPL. These sites in the backlog are left with the stigma of potential Superfund liability as well as the undefined timetable with respect to whether a site is even eligible for a Superfund cleanup. The potential Superfund liability and its undefined schedule make sites caught anywhere in the program virtually undevelopable. Thus, the regulated community (site owners, operators, or developers) often attempt to stay free of the Superfund quagmire by volunteering to work with a state program.

The problems encountered with the backlogged CERCLIS sites has led the EPA to allow states which operate an effective hazardous sites program to take the lead role in providing guidance, assistance, and oversight to voluntary performers who conduct investigations and remediations at these sites. In February 1997 RIDEM formally documented this relationship through a Memorandum of Agreement (MOA) with the EPA. The MOA documents the EPA's commitment that, at active state response actions, the EPA will not plan or initiate federal action under Superfund law, unless exceptional circumstances exist.

By entering into the MOA, the EPA agrees not to list any sites on CERCLIS, or propose any site for listing on the NPL without a written request to do so from the state. The EPA has developed Comfort Letters which are sent to voluntary performers who complete site remedies under the state program clarifying that the EPA is not likely to list the particular site on the NPL. Comfort Letters are issued by the EPA only upon written request from the state. Comfort Letters serve to help define the limit of the environmental liability by eliminating the stigma of double jeopardy with regard to the federal government. For industry, a state covenant not to sue in conjunction with the EPA's commitment not to pursue federal action at the site provides a definitive liability endpoint, which is critical to making Brownfields redevelopment an attractive business venture.

RHODE ISLAND BROWNFIELDS PROGRAM

In March of 1993, RIDEM's Division of Site Remediation promulgated the Rules and Regulations for the Investigation and Remediation of Hazardous Material Releases (the *Remediation Regulations*). These regulations were aimed at creating a more efficient version of the federal Superfund program. The authority for these regulations was existing state laws regarding hazardous waste management, solid waste management, groundwater quality, surface water quality, and air quality.

In 1995, in an attempt to address Rhode Island Brownfields sites and stir economic redevelopment, the State of Rhode Island passed the Industrial Property Remediation and Reuse Act (hereinafter, the Brownfields Act). This legislation was drafted with significant input from state personnel from both RIDEM and RIEDC. The goals of this legislation are to (1) require the publication of numerical soil cleanup objectives which are protective of human health and the environment based on the current and reasonably foreseeable land use of a property and its surrounding natural resources; (2) allow for voluntary actions by parties other than the responsible parties (such as prospective purchasers) and insulating them from the liability normally associated with performing any work at a contaminated site; and (3) increase public involvement in remediation decisions.

In August 1996, RIDEM amended the *Remediation Regulations* to address the mandates of the Brownfields Act. This amendment process was a major undertaking, but it successfully addressed the first three directives of the legislation.

The Rhode Island Brownfields program operates under the jurisdiction of the *Remediation Regulations*, which require all performing parties to conduct consistent and effective investigations of and responses to releases of hazardous materials. The regulations establish a process designed to provide a stepped approach to solving these environmental problems, with the degree of investigation and remedial response proportional to site-specific circumstances. The regulations utilize a modified Risk Based Corrective Action (RBCA) process for responding to hazardous material releases through notification, investigation, and, if necessary, remediation. The elements of the modified RBCA approach are outlined in the

Risk Management section of the regulations, which establishes the requirements for response actions such that they are consistently protective of human health and the environment.

A jurisdictional release of hazardous materials occurs when analytical results indicate an exceedance of the appropriate reportable concentrations defined in these regulations (see Figure 1). If the Division determines that the reported release requires a response action, the area impacted by the release is considered to constitute a source area of contamination. A site with one or more source areas is considered to be a contaminated site. A contaminated site is the focus of the regulatory framework described in the *Remediation Regulations.*

The *Remediation Regulations* are also used as guidelines for voluntary cleanups, such as those performed under the Brownfields program. Voluntary parties bear no responsibility for the contaminated site, but may realize some benefit, economic or otherwise, from remediation. Such performing parties will not proceed under an enforcement mode, but instead will be informed of the necessary procedural steps in order to meet the requirements of these regulations through the issuance of a Voluntary Procedure Letter (see Figure 2).

Remediation of the contaminated site under the regulations shall be performed with the goal of providing permanent protection to human health and the environment. Contaminated sites are likely to enter the site management process during a phase of the Site Investigation (see Figure 3). The regulatory requirement to comprehensively characterize the nature and extent of contamination at a contaminated site ensures that enough information is collected to support the consideration of remedial alternatives, and thus represents the first major step toward

Figure 1. **Reporting a Release**

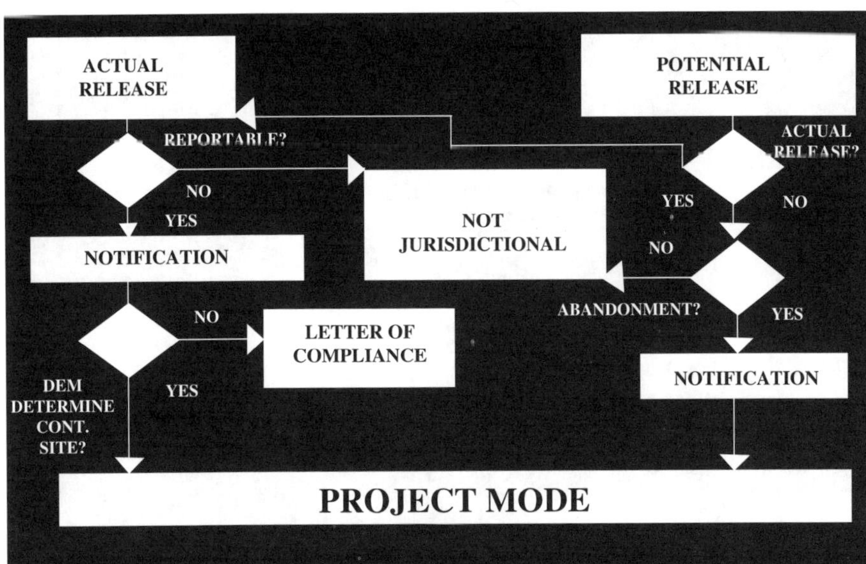

long-term protection. In order to further ensure ecological protection, the regulations specifically require the evaluation of any potential impacts to any environmentally sensitive areas at or in the vicinity of the contaminated site.

The Site Investigation process concludes with the selection of a site remedy or issuance of a Letter of Compliance if remedial action is not necessary. For sites requiring remedial action, the performing party must propose a remedy at the conclusion of the Site Investigation and must provide a demonstration that the preferred alternative is protective with respect to the specific requirements of the Risk Management section of the regulations.

The Risk Management provisions of the *Remediation Regulations* contains the cleanup standards for soil and groundwater. The cleanup standards streamline the remediation process because the tedious analysis to determine protective levels through site-specific risk assessments is no longer required, making the road to remedy selection easier and more predictable. These generic standards serve to economize cleanups as well because the elimination of the risk assessment phase of the process translates into large financial savings by the performing parties.

Settlement Agreements are another important component of the overall performance of the Rhode Island Brownfields Program because they represent a contract which legally codifies the commitment of the performing party to complete the approved remedy and RIDEM's commitment to an end point in the regulatory process. RIDEM may enter into Settlement Agreements with performing parties to perform response actions if the proposed response actions are appropriate and entering the agreement is in the public interest. Settlement Agreements, which may include a covenant not to sue, have an economic utility when used in

Figure 2. **Project Mode**

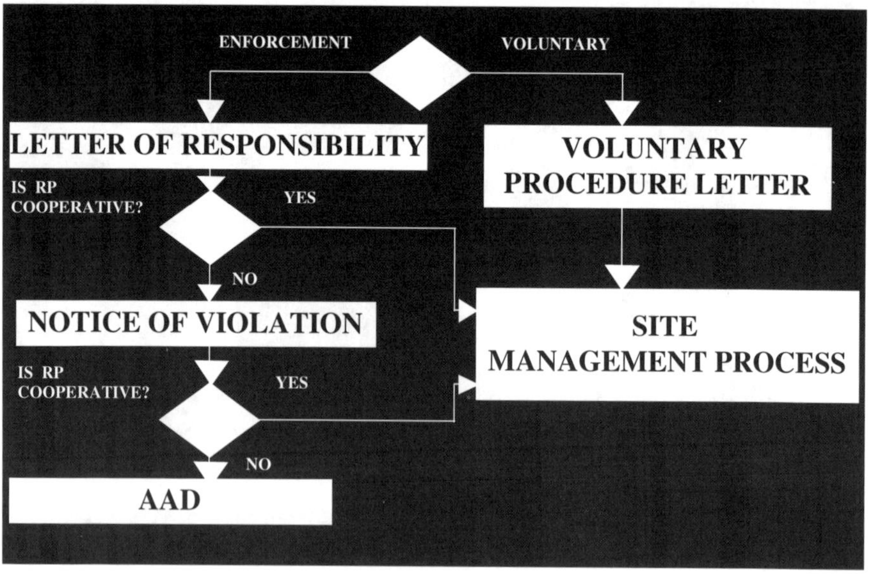

Figure 3. **The Site Management Process**

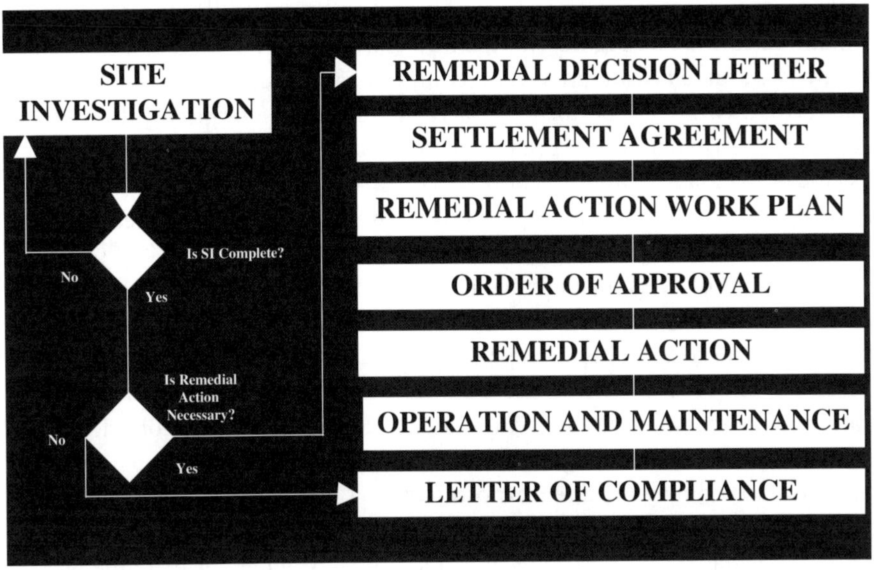

<table>
<tr><td>SITE INVESTIGATION</td><td>REMEDIAL DECISION LETTER</td></tr>
<tr><td>Is SI Complete? No / Yes</td><td>SETTLEMENT AGREEMENT</td></tr>
<tr><td>Is Remedial Action Necessary? No / Yes</td><td>REMEDIAL ACTION WORK PLAN</td></tr>
<tr><td></td><td>ORDER OF APPROVAL</td></tr>
<tr><td></td><td>REMEDIAL ACTION</td></tr>
<tr><td></td><td>OPERATION AND MAINTENANCE</td></tr>
<tr><td></td><td>LETTER OF COMPLIANCE</td></tr>
</table>

tandem with the soil standards, because together they quantify the full extent of environmental liability up front, allowing for sound business decisions regarding the redevelopment of a property. Thus, these two regulatory tools allow RIDEM to oversee expedited and cost-effective cleanups at sites where environmental contamination was previously a barrier to economic investment.

Additionally, in August 1996, the State of Rhode Island passed the Mill Building and Economic Revitalization Act. This statute provides economic incentives for the beneficial reuse of industrial buildings built prior to 1950. These incentives include property tax, inventory tax, and other tax relief to owners and operators which meet specific criteria.

BROWNFIELDS SITE REMEDIES

Under the Rhode Island Brownfields Program, a contaminated site remedy is the result of two processes: risk management and public involvement.

Risk Management

The Risk Management section of the *Remediation Regulations* has facilitated the remedial process by establishing a modified RBCA approach to site management. This section ensures that voluntary response actions continue to be risk-based by providing a tiered approach to remedial objective development.

The remedial objectives component establishes performance and concentration-based standards that are protective, given the current and reasonably expected future use of the property and its underlying groundwater. The Risk Management section provides three methods for determining protective remedial objectives for the hazardous substances found to exist in soil and/or groundwater at any given contaminated site (see Figure 4). Method 1 is a series of tables establishing conservative risk-based cleanup levels for common hazardous substances. Method 2 is a process by which the performing party can supplement or modify the Method 1 cleanup levels to reflect site-specific circumstances. Method 3 corresponds to site-specific human health and/or ecological risk assessments which may be used for assessing baseline risk and subsequently determining appropriate remedial objectives for all impacted media.

Regardless of method, all remedial objectives shall limit carcinogenic risk to a level associated with 1 excess cancer case in a million people, and shall limit noncancer risks to a level associated with a Hazard Index of 1, shall not adversely impact any environmentally-sensitive area, and shall not adversely impact groundwater or any other natural resource of the state. Soil cleanup levels consider direct exposure to humans (see Figure 5) and leachability to groundwater issues (see Figure 6). Accordingly, the direct exposure criteria are divided into residential and industrial/commercial land use categories, and leachability criteria are divided into GA/GAA (suitable for drinking water) and GB (unsuitable for drinking water) groundwater classifications.

Figure 4. **Remedial Objectives Schematic**

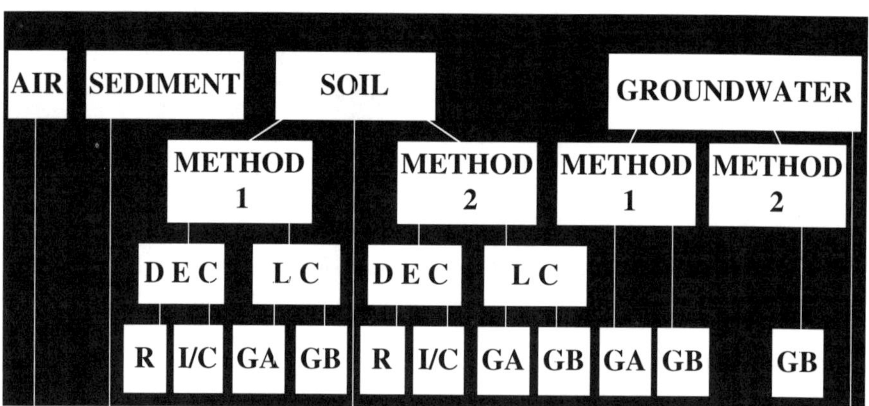

Public Involvement

When RIDEM enters a Settlement Agreement, each party's liability for the response actions (including any future liability to the state, relating to the release or threatened release that is the subject of the agreement) shall be limited as provided in the agreement pursuant to a covenant not to sue. The covenant not to sue may, at the discretion of the Department, be transferred to successors or assigns who are not otherwise found to be a responsible party under these regulations. The covenant not to sue may provide that future liability to the Department of a settling party under the agreement may be limited to the same proportion as that established in the original Settlement Agreement.

Before the finalization of any Settlement Agreement, RIDEM shall provide an opportunity for public comment for a period of 14 days after the date of the notice of the proposed agreement. RIDEM shall consider any written comments, views, or allegations relating to the proposed agreement. The proposed agreement shall be considered final when all substantive public comments have been addressed.

Public Notice is also required at two points during the Site Investigation. Prior to the implementation of field activities, the performing party must notify all abutting property owners and tenants that an investigation is about to occur. Additionally, prior to the formal RIDEM approval of the Site Investigation Report, the performing party must notify all abutting property owners, tenants, and local community well suppliers that the investigation is complete, and provide them with the findings of the investigation and any proposed remedial alternative which includes on-site treatment and/or containment of hazardous materials as part of the final remedy.

Figure 5. **Direct Exposure Criteria Pathway**

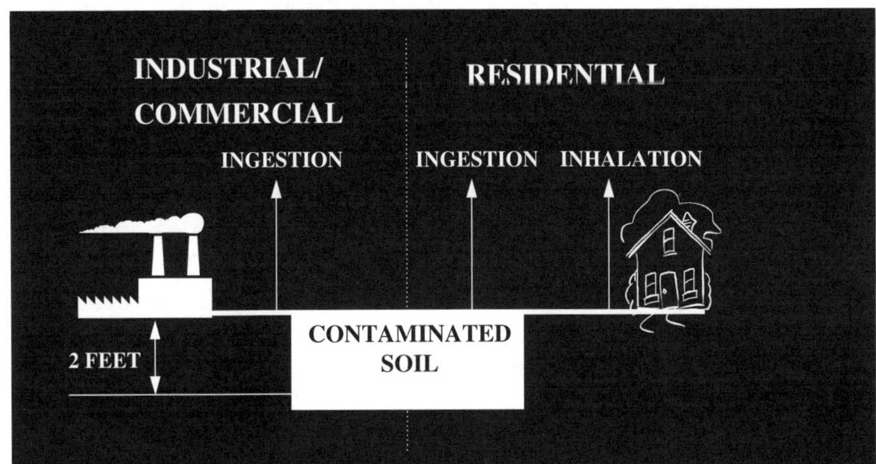

Figure 6. **Leachability Criteria Pathway**

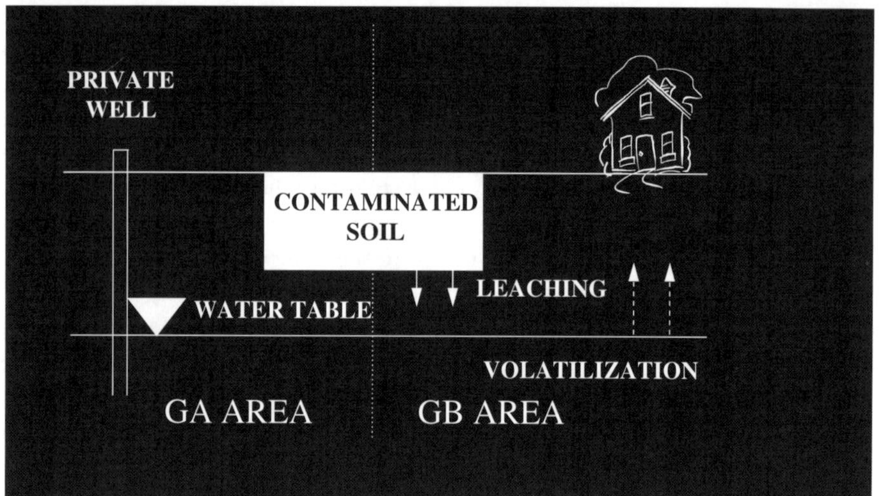

RHODE ISLAND BROWNFIELDS CASE STUDY

One site which capitalized on the Rhode Island Brownfields Program is the Long Wharf Area located in Newport, Rhode Island. The property which was the focus of this Brownfields project is a vacant lot centrally located on the Newport waterfront. The entire Long Wharf Area is listed on CERCLIS. The property is located in an area which was formerly a cove, and encompasses approximately 2.44 acres of land. The cove was filled in gradually during the 1800s. By the early 1900s, the entire cove was filled and used as a railroad yard. The property has been the subject of several proposed business ventures which were abandoned.

The property underwent site investigations in 1988, 1993, 1994, and 1995 consisting of the completion of soil borings and test pits, and sampling and analysis of soils for petroleum hydrocarbons and lead. Results of these analyses indicated that elevated levels of lead exist in soils throughout the property. The elevated lead levels were primarily located at depths greater than four feet below the ground surface.

A stigma of potential environmental liability remained with the property due to the presence of the contaminated soil and CERCLIS listing. A prospective purchaser, Eastern Resorts Company, L.L.C. (ERC), proposed to purchase the property and develop it into luxury timeshare condominium units. This proposal was based on resolving the environmental issues in an expedited manner consistent with the proposed use of the property. Furthermore, as a prospective purchaser they expected to be held harmless for existing conditions at the site.

RIDEM and ERC entered a Settlement Agreement which included a covenant not to sue pursuant to the Industrial Property Remediation and Reuse Act allowing ERC to develop the property. RIDEM approved recommendations, as presented by ERC's consultant, designed to address the elevated lead levels at the property which, when implemented, will protect human health and the environment pursuant to the standards set forth by the state. The consultants establish three general surficial areas at the property: paved parking and access areas, proposed and existing buildings, and proposed landscaped areas. Measures proposed to prevent exposure to the contaminated soil include dust control, installation of soil cover material, and engineering design considerations and controls to limit the need for soil excavation during future site development. Some of the smaller buildings associated with the project have already been built; however, the designs for the main building have not been finalized. RIDEM agreed in the Settlement Agreement to request a Letter of Comfort from the EPA when the site remedy (construction of the building) is complete.

CONCLUSION

The Superfund program is now 16 years old and the developing Brownfields Programs are clear examples of the lessons learned by the program and by the individual state regulatory programs. Specifically, the Brownfields movement is directly related to the concept of hazardous site cleanups based on the reasonably intended land usage, thus paving the way for less complicated hazardous waste site cleanups.

Rhode Island is fertile ground for Brownfields initiatives. Mill complexes are common along the banks of larger rivers which flow through urban areas such as: the Blackstone, Woonasquatucket, and Pawtuxet, and also common in more rural areas such as Burrillville and Hopkinton. Rhode Island was also hit hard by the recession of the nineties and thus there are also younger sites awaiting redevelopment.

The Brownfields program is working in Rhode Island because of the cooperative relationship between RIDEM, RIEDC, and the regulated community. This relationship is the result of the cooperative effort in writing the Rhode Island Brownfields Act, a Rhode Island Brownfields conference held in October 1995, brainstorming work shops held regarding the amendments of the Remediation Regulations, and RIDEM's overall commitment to use the command and control approach to site cleanups only when all voluntary options fail.

ACKNOWLEDGMENT

To Timothy Regan, Senior Engineer with RIDEM, for his review of drafts of this paper.

REFERENCES

Bartsch, C. and Collaton, E. 1994. Contamination and Industrial Site Reuse, *Northeast Midwest Economic Review*, December.

Harr, C.M. and Kayden, J.S. 1989. *Zoning and the American Dream.*

(RIDEM) Rhode Island Department of Environmental Management. 1995. Long Wharf Area Fact Sheet.

(RIDEM) Rhode Island Department of Environmental Management. 1996. Rules and Regulations for the Investigation and Remediation of Hazardous Materials Releases.

(RIGL) Rhode Island General Laws. 1995. Chapter 23-19.1 The Rhode Island Industrial Property Remediation and Reuse Act.

(RIGL) Rhode Island General Laws. 1996. Chapter 64.6 The Rhode Island Mill Building and Economic Revitalization Act.

Rhode Island Historic Preservation Commission, 1981. Providence

Industrial Sites-Statewide Historic Preservation Report P-P-6.

(USEPA) U.S. Environmental Protection Agency. 1991. Code of Federal Regulations Title 40.

(USEPA). 1992. Guidance for Performing Site Inspections under

CERCLA.

PART VII
FEDERAL/MILITARY

CHAPTER 37

Expedited CERCLA Removal Actions at Loring AFB, Maine

T. Wood and **S. Underhill**, Bechtel Environmental, Inc., Oak Ridge, Tennessee

D. St. Peter and **D. Hopkins**, Air Force Base Conversion Agency, Loring Air Force Base, Maine

J. Mueller, Air Force Center for Environmental Excellence, Loring Air Force Base, Maine

INTRODUCTION

Bechtel Environmental, Inc. (Bechtel), contracted (Prime Contract No. F41624-94-D-8072) by the Air Force Center for Environmental Excellence (AFCEE) as the removal/remedial action contractor for Loring Air Force Base (AFB), performed removal actions for contaminated surface and subsurface soil at Loring AFB in Limestone, Maine. This summary report describes the removal action work (excavations in Operable Units [OUs] 5, 8, 9, 10, and 11) performed during the 1995 construction season by Bechtel and its subcontractors.

SCOPE AND PURPOSE

The objectives of the removal action report are as follows.
- Documenting removal activities performed at OUs 5, 8, 9, 10, and 11 and providing verification that performance standards were met at these OUs, if applicable.
- Informing the U.S. Environmental Protection Agency (EPA) and the Maine Department of Environmental Protection (MDEP) that the removal actions at OUs 5, 8, 9, 10, and 11 were completed in accordance with:
  - Final Design Analysis Report, OUs 5, 9, 10, and 11
  - Final Engineering Evaluation/Cost Analysis Operable Units 5, 9, 10, and 11
  - Design drawings and technical specifications for OUs 5, 9, 10, and 11
  - Preliminary design drawings for OU 8, Fire Training Area excavations
  - Final Engineering Evaluation/Cost Analysis Operable Unit 8

The report was developed in accordance with EPA guidance documents Close Out Procedures for National Priorities List Sites and Superfund Removal Procedures Removal Response Reporting: POLREPs and OSC Reports.

The work items addressed in this report can be summarized as follows.

- Excavation of contaminated material (OUs 5, 8, 9, 10, and 11)
- Sampling and analysis (field screening and confirmation sampling)
- Demolition of existing structures
- Transportation of contaminated material
    - Site restoration (seeding, pavement repair, etc.)

Concurrent with the activities described above, other related and necessary tasks were executed to support the removal actions. Major tasks included:

- Placement of contaminated soil and demolition debris at Landfill (LF)-3
- Underground pipe cleaning at the Power Plant Drainage Pipe (PPDP)
- Transporting and off-site disposal of some hazardous liquids and soils
- Dispositional and investigative (e.g., toxicity characteristic leaching procedure [TCLP]) sampling

Except as detailed in this report, Bechtel has not independently verified historical information or prior reports upon which this report is based. Information available as of the issue date was used.

Site Description and Background

Loring AFB is in Aroostook County in the northeastern corner of Maine, about 3 miles from the Canadian border. The base occupies about 9,000 acres, mostly in the town of Limestone, Maine. The base was constructed between 1946 and 1953, and improvements were made throughout its operational life. Most recently, the base was part of the Air Combat Command. On September 30, 1994, it was officially closed and is now the responsibility of the Air Force Base Conversion Agency (AFBCA). Loring AFB is designated as Operational Location M by AFBCA. Figures 1 and 2 show the location of the base and the major OUs.

Specific information regarding the site characterization for OUs 5, 8, 9, 10, and 11 (e.g., site description, surface waters, ecological characteristics, geology, and hydrogeology) is contained in the design analysis report (DAR), the removal action work plan, and various remedial investigation reports cited in these documents

Summary Description of Operable Units 5, 8, 9, 10, and 11

Individual OUs included in this removal action are briefly outlined below. (Note: Not all sites within OUs 5, 8, 9, 10, and 11 are included in this report.)

- OU 5: Excavation in OU 5 involved two areas identified as Nose Dock Areas 1 and 3.

Figure 1. **Loring Air Force Base (Limestone, Maine) location map.**

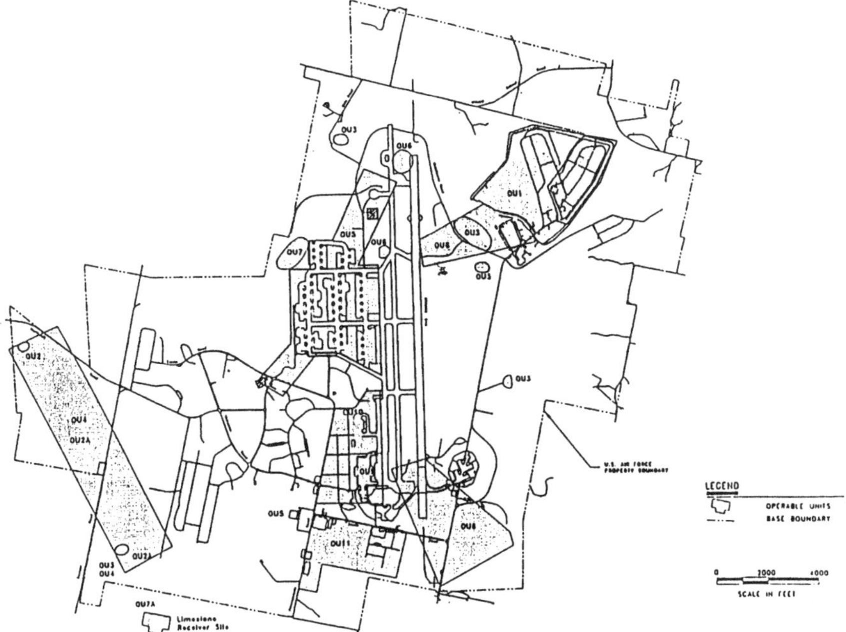

- OU 8: Excavation in OU 8 involved several areas at the Fire Training Area (FTA), including a circular excavation, a pipeline extending to the pit, a drain swale leading from the fire pit area, and three rectangular areas.
- OU 9: Excavation in OU 9 involved the Snow Barn and the PPDP. The excavations at the Snow Barn area were remedial actions taken under the State of Maine regulations for underground oil storage facilities, as specified under the Code of Maine Rules, Chapter 691. Although the Snow Barn excavations were not regulated under the Comprehensive Environmental Response, Compensation, and Liability Act (CERCLA), they were performed consistently with CERCLA removal action guidance and are presented as such in this removal action report.
- OU 10: Excavation in OU 10 involved one area at the Former Solvent Storage Building (FSSB) and the Entomology Shop (ES) building and associated drainlines.
- OU 11: Excavation in OU 11 involved one area at the Vehicle Maintenance Building (VMB), two areas at the Refueling Maintenance Shop Area (RMSA), and five areas at the Fuel Tank Farm (FTF).

Figure 2.   **Operable Units, Loring Air Force Base, Limestone, Maine.**

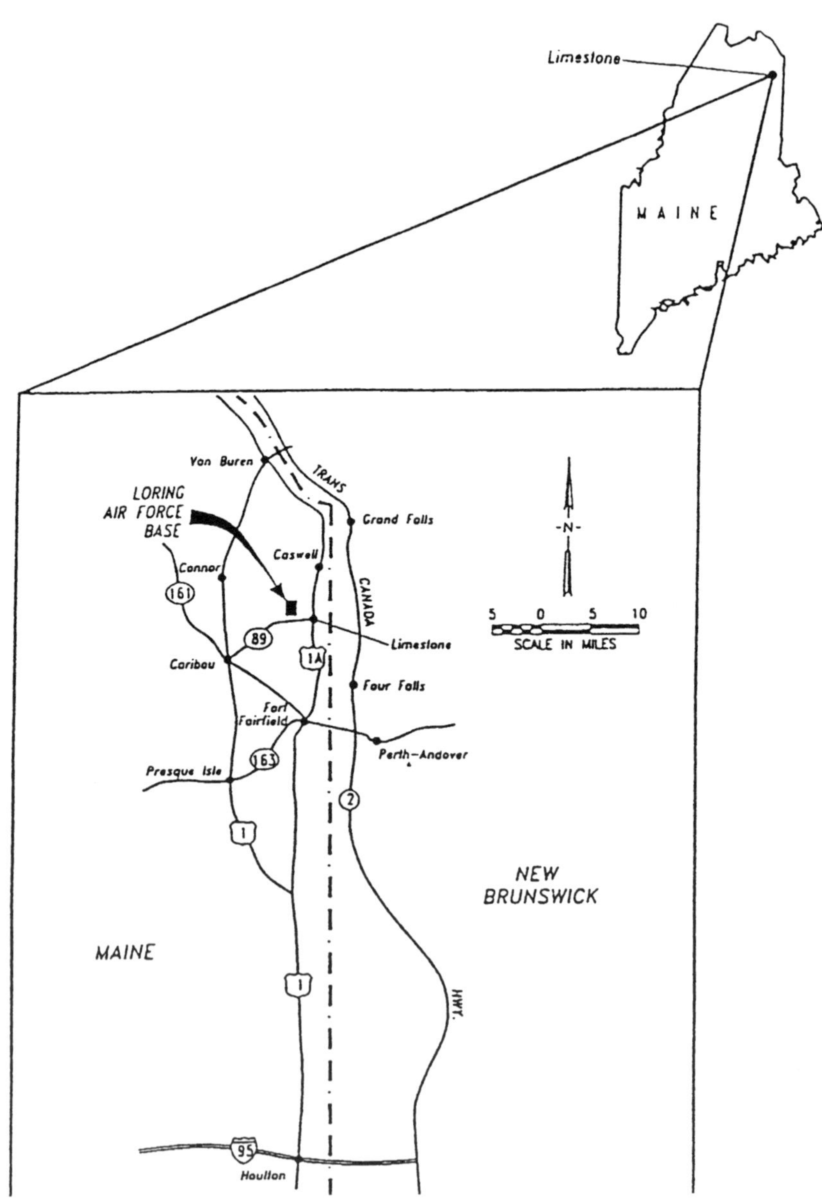

Regulatory Setting

The removal actions at OUs 5, 8, 9, 10, and 11 were performed in compliance with CERCLA Section 104. The removal actions were developed through an engineering evaluation/cost analysis (EE/CA), required by 40 CFR 300.415 (b)(4), which evaluated the applicability of all pertinent environmental requirements and regulations and were designed under 40 CFR 300.415 (c) and (d) removal actions.

The removal action at the Snow Barn, however, was covered under the State of Maine regulations specified under the code of Maine rules, Chapter 691 (Bureau of Oil and Hazardous Materials Control).

All work covered under this removal action was performed in accordance with the appropriate work plans submitted to MDEP and EPA. Any deviations from the work plan were discussed with AFCEE, AFBCA, and MDEP before enactment.

CHRONOLOGY OF EVENTS

This section summarizes the major events associated with the removal action at OUs 5, 8, 9, 10, and 11. The major events are listed below.

Public comment period on OUs 5, 9, 10, and 11 EE/CA	3/24–4/24/95
Action memorandum signed for OUs 5, 9, 10, and 11 EE/CA	5/24/95
Draft final work plan for removal action submitted	6/95
Field activities (see below)	8/95–12/95
Final work plan for removal action submitted	9/95
Public comment period on OU 8 EE/CA	9/11–10/11/95
Action memorandum signed for OU 8 EE/CA	12/6/95

PERFORMANCE STANDARDS AND CONSTRUCTION QUALITY CONTROL

The intent of the removal action at OUs 5, 8, 9, 10, and 11 was to remove the contaminated soil above preliminary remediation goals (PRGs) and dispose of it on-site at LF-3. In addition, confirmatory samples were collected from the base of the excavations, with the analytical results compared to applicable PRGs to verify the success of the removal action.

Throughout the actual removal of the contaminated soil, project-specific performance standards and quality control (QC) requirements were followed to monitor and control the removal operations and related support activities including field screening and testing, confirmatory sampling and analysis, reexcavation and reconfirmation (if necessary), and final site restoration.

Overall Summary of Performance Standards and Construction Quality Control

Performance Standards

The removal action or excavation for each particular area within each OU was executed to achieve levels equal to or less than the PRG levels. The PRGs, established using the Loring Air Force Base Risk Assessment Methodology, are presented in the final DAR.

The PRGs for each OU are chemical-specific concentrations that correspond to a level of risk that is within acceptable limits (i.e., 10^{-6} excess cancer risk for carcinogenic effects and a hazard index of 1 for noncarcinogenic effects). The PRGs were developed based on specific scenarios to estimate human health risks, certain anticipated plant receptors, ecological receptors, and to be protective of contaminants leaching to groundwater. A detailed discussion on the development of the PRGs is presented in the appendixes of the EE/CAs. The OU-specific PRGs are also provided in the design drawings prepared by URS Consultants, Inc. (URS) and the removal action work plan prepared by Bechtel.

In most instances, the removal actions performed at OUs 5, 8, 9, 10, and 11 accomplished the applicable performance standards defined by the PRGs. However, in a few instances where concentrations of certain compounds slightly exceeded the PRGs, residual risks were assessed to evaluate the appropriateness of further soil removal. The issues associated with residual risk assessments are discussed further in the Confirmation Test Results portion of this report. The residual risk assessments were calculated by HAZWRAP as requested by AFBCA.

Construction Quality Control

Construction work associated with the removal action at OUs 5, 8, 9, 10, and 11 was executed in strict adherence to the approved project construction QC plan. The QC process was implemented from the start of each definable activity. The implementation of a specified work activity observed a sequence of steps to be followed and tasks to be completed before starting new ones and before completing the overall work. This construction QC process involved a three-phase control approach that included the preparatory inspection phase, initial inspection phase, and the follow-up inspection phase.

Bechtel planned and executed the preparatory inspection phase, which included a discussion of construction activities that would be part of and would influence the actual construction work. This phase began with a meeting relative to the definable feature that involved the subcontractor (Soderberg), the site engineer, the construction supervisor, the safety and health representative, and the sampling team representative. Representatives of AFCEE, AFBCA, and MDEP also attended.

Items discussed at the preparatory phase meetings included:

- Specific construction approaches such as excavation depths, equipment to be used, duration of work activities, and field screening and analytical sampling.
- Steps to minimize cross-contamination and recontamination during removal operations.
- Safety practices for worker protection including air monitoring, selection of personal protection equipment, and protection of the worker population and unintentional intruders from site hazards.
- Lines of authority and responsibilities within Bechtel and between Bechtel and its subcontractor.
- The importance of having the entire construction team aware of changing conditions at the site.

The initial phase of the control process began with a site preconstruction and safety and health review. The review emphasized all the critical concerns identified and discussed in the preparatory phase, including the steps to be implemented during the actual construction. Agreements were reached on specific site issues such as the appropriate route for loaded trucks to LF-3, frequency of water application for dust abatement at the excavation and traffic areas, type of excavator buckets to be used to prevent/minimize spread of contamination, actions to take due to changing site conditions, extent of the scope of the defined work, and safety and health requirements for the worker population and occasional visitors.

During the follow-up phase, Bechtel provided oversight and inspection of the field work and kept an on-site QC representative to inspect and record daily all the activities related to the removal action. Additionally, the safety and health representative performed real-time monitoring of air emissions and oversight of overall safety conditions at each construction site, assigning and implementing the appropriate measures and procedures to protect the worker population and the environment.

Confirmation Sampling

The number of samples collected and their locations at OUs 5, 8, 9, 10 (with the exception of the Entomology Shop basement), and 11 were in accordance with the grid system presented on the site-specific URS drawings, Bechtel's work plan, and in the DAR. Sample locations were surveyed based on the URS drawings; however, in some instances, locations were varied slightly due to physical obstructions. At the Entomology Shop basement area, the sample locations were established in accordance with field conditions.

The DAR provides the rationale for the grid system and the sampling locations established for each OU. In summary, the sampling grids were determined using the statistical evaluation method discussed in Volume 1 of EPA's guidance document, Methods for Evaluating the Attainment of Cleanup Standards. The hot spot methodology defined the actual sampling patterns and the grid spacing required to detect hot spots of varying sizes with differing levels of confidence. The criteria used to develop the sampling scheme represent a 90% confidence level for the detection of a hot spot. In some instances, the URS-design confirmatory sample grids and locations were recalculated to meet field conditions. These locations (i.e., FTA and RMSA) were recalculated using the EPA Hot Spot methodology, and are presented in the individual site discussion.

During excavation, field screening/immunoassay testing was performed as specified in the respective work plans. When field screening and testing showed undetectable or relatively low contaminant concentrations, confirmation samples were taken for off-site analysis. Chain-of-custody forms were prepared and maintained for all confirmation samples.

Confirmation Test Results

Following excavation and field screening, confirmation sampling and analysis (testing) was performed for each site undergoing removal action. The sequence of events leading to the analytical results included:

- Determination and verification of areas to be excavated by review of the latest URS drawings and by field layout of each area.
- Excavation to the depths and areal extent shown on the drawings and as established in the field.
- Field screening of the excavations by visual monitoring and immunoassay testing.
- Field layout of the confirmation sampling points.
- Collection of confirmation samples in accordance with the URS specifications and the procedural requirements of the site work plan and shipment for off-site analysis.
- Receipt, review, and validation of confirmation test results.

Following these events, the confirmation test results were evaluated for compliance with the OU-specific PRGs. If the area was determined to be compliant, it was backfilled and restored after review by MDEP and authorization by AFCEE. If the area or some of the sample locations did not comply with PRGs, the area was reexcavated and resampled for the noncompliant analytes in accordance with requirements stated in the site-specific work plan.

Confirmation Test Results at OU 5

The removal action for OU 5 included two sites: Nose Dock 1 and Nose Dock 3.

Nose Dock 1. The contaminated soil at Nose Dock 1 was excavated 5 ft below surface grade; no reexcavation or reconfirmation sampling was required.

Thirty-three confirmation samples taken at Nose Dock 1 were tested for the compounds in the PRG list. The analytical data were reviewed against the method QC criteria and found to be acceptable for their intended use. The analytical results for all 33 samples showed that concentrations of the compounds in question were below the PRG levels. **Based on this information, the removal action at Nose Dock 1 has met performance standards.**

Nose Dock 3. The contaminated soil at Nose Dock 3 was excavated 2 ft below surface grade; no reexcavation or reconfirmation sampling was required.

Twenty-three samples taken at Nose Dock 3 were tested for the PRG compounds. The analytical data were reviewed against the method QC criteria and found to be acceptable for their intended use. The analytical results for all 23 samples showed that concentrations of the compounds in question were below the PRG limits. **Based on this information, the removal action at Nose Dock 3 has met performance standards.**

Confirmation Test Results at OU 8

The removal action for OU 8 (FTA) included several subareas :

- Circular Excavation: A circular area west of the fuselage area.
- Subareas 2 and 3: Two small rectangular areas northeast of Subarea 1.
- Subarea 4: A rectangular area west of the drainage ditch subarea.
- Drainage Ditch: A drainage ditch area northeast of Subarea 1 and along existing Access Road and Oklahoma Road.

Circular Excavation. The circular subarea was initially excavated to approximately 14 ft below surface grade and with side slopes as shown on the URS drawings. The bedrock layer encountered at the base of the excavation did not permit confirmatory soil samples to be collected because all soil was removed from the bedrock surface. Seven confirmation soil samples were taken at the side slopes of this subarea as shown on the URS drawings. Also, two additional soil samples were taken at the western quadrant of the circle at locations just above the bedrock layer.

Confirmation test results for eight samples showed concentrations less than PRG limits. Confirmation test results for one sample showed concentrations for ethylbenzene, total xylene, and total petroleum hydrocarbon (TPH) greater than PRG limits. The respective area was reexcavated and sampled at four points for reconfirmation testing; reconfirmation results showed concentrations less than PRG limits.

Subareas 2 and 3. Subareas 2 and 3 were both excavated 2 ft below surface grade. Each excavation had an area of approximately 18 ft by 18 ft. The total original number of confirmation samples for each area was 11, which meant a spacing of 4.5 ft between most sample points. Based on this configuration, any reexcavation required would have four resample points at about 2.25-ft spacing, which would be difficult to excavate. To provide more reasonable spacing between sample points and thus reduce the number of actual samples, new grid spacing was calculated using EPA's Hot Spot Location techniques, assuming that the hot spots represented 20% of the area to be sampled. The revised, total number of confirmation samples was eight for Subareas 2 and 3. These recalculations were reviewed and approved by both EPA and MDEP. The analytical results showed that the concentrations of the compounds in question were less than the PRG concentrations for all eight samples.

Subarea 4. Subarea 4 was excavated 8 ft below surface grade as shown on the URS drawings. The total original number of confirmation samples for this sub-area was 14. Using the same criteria as for Subareas 2 and 3 (see previous section), the total number of confirmation samples was revised to six. The analytical results showed that the concentrations of the compounds in question were less than the PRG concentrations for all six samples.

Drainage Ditch. The drainage ditch itself was excavated 2 ft below surface grade for its entire extent, as shown on the URS drawings. The total number of confirmation samples taken for the drainage ditch itself was 23 ; analytical results showed that concentrations of the compounds in question were less than the PRG concentrations for all 23 samples.

A rectangular area adjacent to the ditch (part of a designated "wetland area") was excavated a minimum of 4 ft below surface grad.). The total original number of confirmation samples for this rectangular area was 13. Using the same criteria as for Subareas 2 and 3, the total number was revised to four. Analytical results showed that concentrations of the compounds in question were less than the PRG concentrations for all four samples. The wetland area will be evaluated in the spring of 1996 for appropriate mitigation efforts. **Based on the above information, the removal action at the FTA has met performance standards for all subareas.**

Confirmation Test Results at OU 9

The removal action at OU 9 included two sites: the PPDP and the Snow Barn. *PPDP.* The removal action for the PPDP included three subareas.

- West: Area north of the Auto Hobby Shop and east of Pennsylvania Road
- Central: Rectangular area north of the Self Help Store
- East: Rectangular area west of the power plant

<u>West</u>. The contaminated soil at the west area was excavated to depths ranging from 2 ft to 12 ft. Reexcavation was required over the entire area, based on visual observations and air monitoring using an organic vapor analyzer prior to confirmatory sampling. The excavation depths generally decreased from the southeast corner (approximately 12 ft in depth) to the northwest corner (2 ft). Excavation depths along the southern wall were between 8 and 12 ft.

The total number of original confirmation samples was 17. The analytical results revealed five locations with PRG exceedances.

One location was reexcavated with four reconfirmation samples taken. Two reconfirmation sample results had TPH levels above 870 mg/kg and were subsequently excavated down to groundwater (approximately 2 ft below the water table). A recovery well was subsequently installed at this location to capture any potential free product on the groundwater surface. (A small sheen was noted on the groundwater surface at the time of excavation.)

Two locations showed elevated levels of polychlorinated biphenyls (PCBs) (Arochlor 1260 at 0.94 and 0.73 mg/kg, respectively). Since the levels were only slightly above the PRG (0.2 mg/kg), a residual risk analysis was performed. This analysis showed that the remaining levels were still below acceptable risks.

Based on the data and the residual risk assessment for the two locations, the performance standards for the west area of the PPDP have been met for polyaromatic hydrocarbons (PAHs) and PCBs. Two other locations had PRG exceedances for TPH; however, the biovent system adjacent to the Auto Hobby Shop will be installed in this area. It is reasonable to anticipate that the biovent system will remediate the TPH concentrations in the soils below PRG levels. Based on these excavation results, no further excavation is required at the PPDP west.

Central. The contaminated soil at the PPDP central area was excavated to the horizontal design limits and to a depth of 5 ft. An additional foot of soil was then removed over the entire excavation, based on the results of the field screening. A second field screening effort indicated that the northeast corner again required excavation. During this second reexcavation, an additional 2 ft of soil was removed.

Fourteen confirmation samples were then taken at the designed locations. PRGs for PCBs and PAHs were met at all locations, and therefore the performance standards were met for these analytes. Six of the 14 locations, however, had confirmation sample result PRG exceedances for TPH, ranging from 930 to 16,000 mg/kg. These elevated levels of TPH are anticipated to be remediated by the biovent system. The biovent system (installed in 1995 and covered under a separate removal action report) extends inclusively over the PPDP central area that contains the remaining petroleum-contaminated soil. Based on these excavation results, no further excavation is required at PPDP central.

East. The contaminated soil at the east area was excavated to a minimum depth of 8 ft. Only one location was reexcavated based on field screening. Following this, 14 confirmation samples were collected and analyzed for the chemicals of concern. The results showed that there were no exceedances of the PRGs. **Based on this, the performance standards have been met for all three PPDP subareas.**

Snow Barn. The removal action for the Snow Barn included three subareas :

- West: 4-ft trapezoidal excavation
- East: 2-ft rectangular excavation
- South: 2-ft rectangular excavation

West. The contaminated soil at the west area was excavated to a minimum depth of 4 ft. The entire area was reexcavated based on field screening. Sixteen confirmation samples were subsequently taken following this reexcavation. Results show that six locations had PRG exceedances for PCBs, and one location had a TPH PRG exceedance. These seven areas were reexcavated, and reconfirmation sample results showed that six locations met performance standards. Only one location had PRG exceedances, with two reconfirmation samples having PCB levels greater than 10 mg/kg. (Note: The performance standard for Arochlor 1260 was calculated to be 10 mg/kg for soil depths greater than 2 ft, since the construction worker scenario became the driving risk.)

Reconfirmation locations were reexcavated and sampled. The reconfirmation analyses for Arochlor 1260 revealed that seven of the eight reconfirmation samples were below the PRG level of 10 mg/kg. The remaining sample location had a Arochlor 1260 concentration of 13.5 mg/kg, which corresponded to a risk of 7.7 x 10^{-6}, which is below the MDEP target risk of 10^{-5}. During the reconfirmation sampling, PCB levels were found to exceed the Toxic Substances Control Act (TSCA) regulatory levels of 50 mg/kg (40 CFR, Part 761) and these soils were therefore restricted from disposal in LF-3. Seven field screening samples delineated the extent (lateral and vertical) of areas with PCB concentrations exceeding the TSCA limit . One cubic yard of soil was then excavated from this area, containerized, and disposed of off-site at an EPA-approved TSCA landfill.

East. The contaminated soil at the east area was excavated to a depth of 2 ft; no reexcavation or reconfirmation was needed. One location had a benzo(a)pyrene result equal to the PRG (1.8 mg/kg); however, no confirmation sample results had PRG exceedances.

South. The contaminated soil at the south area was excavated to a depth of 2 ft. One location was reexcavated based on field screening. Following the reexcavation, 14 confirmation samples were taken. Only two locations had analytical results above the PRGs. One location with a Arochlor 1260 level of 0.28, was reexcavated and sampled, with results indicating that this location met performance standards. The second location with a benzo(a)pyrene level of 1.9 mg/kg, was only slightly above the PRG of 1.8 mg/kg (corresponding to a 10^{-6} risk) and was not reexcavated, since the risk associated with this concentration was below the MDEP target risk of 10^{-5}. **Based on this information, the removal actions at the Snow Barn have met the performance standards.**

Confirmation Test Results at OU 10

The removal action for OU 10 included the following two sites: The Entomology Shop and the FSSB.

Entomology Shop. The Entomology Shop area included a basement and a series of drainlines.

Basement. The following specific activities were performed at the Entomology Shop basement:

1. Contaminated soil directly beneath the bottom foundation slab was excavated to a minimum of 14 ft below surface grade. Soils from the side slopes were also excavated as needed to remove the concrete rubble. Thirteen samples were taken at the basement area and tested for the PRG compounds. The analytical data were reviewed against the method QC criteria and found to be of acceptable quality for their intended use, with the exception of any rejected data. At the time of excavation, wet site conditions

did not allow safely taking a fourteenth sample per URS drawings. Concentrations of the compounds in question in eight soil samples were less than the area PRG limits; concentrations in soil samples exceeded the PRG concentrations for TPH.

2. The basement's concrete foundation and walls were demolished and removed in their entirety in accordance with the URS drawings. Concrete rubble from a chamber (cell) that contained water and sludge contaminated with chlordane was tested for TCLP pesticides and found to meet the land disposal restrictions for on-site disposal at LF-3.

3. Contaminated soil previously identified as hazardous waste at a hot spot at the southwest corner of the Entomology Shop was removed and disposed of off-site at a permitted disposal facility.

4. Sediment/sludge from a manhole at the northeast corner of the Entomology Shop was sampled and tested and determined to be hazardous. This material was removed and disposed of off-site at a permitted disposal facility.

Five soil confirmation results showed TPH concentrations greater than the PRG list (see Item 1 above). One sample was taken at the southwest quadrant of the basement floor at an approximate depth of 14 ft bgs. Four samples were taken at the side slopes at an approximate depth of 10 ft bgs. A biovent system will be installed in 1996 to remove TPH from the area soils above the groundwater. The current design of the biovent system encloses the eastern third of the Entomology Shop. No modifications to the biovent system are anticipated at this time.

Northeast drainline. The soil and drainline at the northeast drainline area was excavated to depths, widths, and side slopes in accordance with URS drawings and actual site conditions. The depth of the excavation varied from approximately 10 to 13 ft, and no reexcavation or reconfirmation sampling was required.

Seven confirmation soil samples were taken at the northeast drainline trench and tested for the PRG compounds. The analytical data were reviewed against the method QC criteria and found to be acceptable for their intended use. The analytical results showed that the concentrations of the compounds in question were less than the PRG levels in all seven samples.

Corrugated pipe. The contaminated soil and corrugated pipe (east drainline) was excavated to depths, widths, and side slopes in accordance with URS drawings and actual site conditions. The area was excavated a minimum of 2 ft below the prevailing surface grade, and no reexcavation or reconfirmation sampling was required.

Six confirmation soil samples were tested for the PRG compounds. The analytical data were reviewed against the method QC criteria and found to be acceptable for their intended use. The analytical results showed that concentrations of the compounds in question were less than the PRG limits in all six samples.

<u>South drainline</u>. The removal of the drainline included:

- Excavation of uncontaminated overburden soil that was stockpiled next to the trench and used as backfill material.
- Removal of the drainline and adjacent soils (approximately 1 to 2 ft around the pipe), which were disposed of in LF-3.

The total excavation for removal of the south drainline was to a depth varying from 10 to 14 ft bgs. Six confirmation soil samples were taken at the south drainline trench and tested for the PRG compounds. The analytical data were reviewed against the method QC criteria and found to be acceptable for their intended use. The analytical results showed that concentrations of the PRG compounds in question were less than the PRG limits in five samples.

The area encompassing sample point S-3 was reexcavated an additional 1 to 2 ft in depth and resampled for TPH. However, due to the imminent onset of winter, it was decided to backfill the area before the confirmation test results were received. The TPH reconfirmation results showed that concentrations still exceeded the PRG limits.

Reconfirmation test results from the sampling point showed that the TPH levels did not meet the PRG requirements. However, the risks posed to humans from direct human exposure to a TPH-contaminated single point would be minimal. The SR-3 sampling point is at 11 to 12 ft bgs, which is below the depth of 10 ft used in risk assessments for Loring AFB
and therefore is not considered a risk. Based on unlikely risks, direct human exposure from a single point TPH above the PRG is not likely, and therefore no further action is recommended for the Entomology Shop south drainline.

FSSB. The FSSB was excavated to a depth of 10 ft. However, the lateral limits of excavation were altered from the design. This change included removing the FSSB foundation from the excavation limits, because further review of the remedial investigation data indicated that there were no chemicals of concern with levels above the PRGs inside the foundation. Excluding this area from the excavation, only 13 of 16 confirmatory locations were subsequently sampled. Only one location had results that exceeded the PRG limit for both total organic solvents [TOSs] and TPH levels.

Following the first excavation, a small sheen was detected below the FSSB foundation leading to the center of the excavation. This sheen was removed (along with 0.5 ft of soil below the sheen) along with the portions of the foundation wall. This allowed the remaining three samples to be collected along with the reconfirmation samples. MDEP also took a sample from the base of the excavation where the former sheen was detected.

Based on the analytical results, four locations did not meet the performance. Since the performance criteria were not met, further actions are warranted. These actions, while not finalized, may consist of one or a combination of the following:

- In situ remediation
- Excavating and treating with soils from the Base Laundry area
- Disposing of off-site

A finalized remedial system, which will be selected based on further review of cost, technical feasibility, and applicable regulatory requirements, will be implemented in the spring of 1996 to comply with the PRG requirements at the site.

At the close of the 1995 construction season, the open excavation was marked and covered with a geotextile to show where contaminated soils still exist, and then temporarily backfilled. A recovery well was also placed prior to backfilling. This well will be used to observe groundwater levels and determine when water levels are low enough to begin reexcavating. **The performance standards, as described above, have not yet been met at the FSSB.**

Confirmation Test Results at OU 11

The removal action at OU 11 included three sites: FTF, RMSA, and VMB.

FTF. Five bermed areas at the FTF were associated with the removal action: 7810, 7811, 7812, 7820, and 7830. The excavation required removing the top 0.5 ft of soil. In total, 194 confirmation samples were taken and analyzed for lead. **Based on these data, the performance standards for the removal action within the five bermed areas in the FTF have been met.**

A sixth area, outside the bermed tank cells, includes a drainage ditch. This area, according to design reports and work plans, was scheduled to be incorporated into this removal action. However, due to free product encountered on the surface water in this ditch, an alternative removal action was incorporated (i.e., removing the free product via absorptive pads). To avoid cross-contamination, the drainage ditch soil will not be removed until all free product has been removed.

RMSA. The RMSA was divided into three areas:

- Ditch - area east of the Pest Management Laboratory
- East - area immediately east of the Refueling Maintenance Shop
- 5 ft by 5 ft - small excavation northeast of the Refueling Maintenance Shop

<u>Ditch</u>. The ditch area behind Building 7610 (Pest Management Laboratory) was excavated to a design depth of 2 ft. Results from field screening indicated that only Location needed to be reexcavated. After this area was reexcavated, 14 confirmation samples were taken. Confirmation results indicated that none of the contaminants of concern exceeded the PRG levels.

<u>East</u>. The excavation at the east area encompassed the limits from the URS design; however, some soil was not removed to protect the fire hydrant/water mainline and the sanitary sewer manhole. Originally, 24 confirmatory sample locations were designated, but due to the changes in field conditions, 20 samples were actually taken. These 20 samples all had concentrations below the PRGs.

The areas where the samples were not taken had been excavated below the water table (this area was formerly the location of a dry well), and therefore no vadose zone samples were possible. However, at the request of the Air Force (AFCEE 1995), four investigative samples were taken: two under the foundation, one in the soil adjacent to the manhole (1x), and one adjacent to the fire hydrant (2x). The analytical results indicated that of the contaminants of concern, only TPH had levels above the PRGs. Based on these residual TPH levels, additional excavations will be required in the spring of 1996.

<u>5 ft by 5 ft excavation</u>. The removal area design at the RMSA originally included excavating the drainage ditch immediately north of the Refueling Maintenance Shop. Prior to excavation, the remedial investigation, EE/CA, design, and area topography were reviewed. The results of this review did not indicate any PRG exceedances or any reason to believe that contamination existed in this area. The only exception is one sample that exceeded the PRGs for TPH and PAH. This is the 5 ft by 5 ft excavation. This review prevented about 200 yd^3 of soil from being excavated and disposed of in LF-3.

This area was initially excavated to a depth of 4 ft. Screening sampling consisted of one sample from each sidewall and one sample from the base of the excavation. Based on these results, one additional foot of soil was removed from the base (the four sidewall sample results were below PRG levels). Using EPA's Hot Spot location technique, one confirmatory sample was calculated to be needed. The results from this sample indicated that performance criteria were met. **Based on the excavation and sampling at the RMSA, the Ditch and 5 ft by 5 ft excavation have met performance standards; however, the East excavation has not met performance standards and further removal of contaminated soil will be done during the 1996 construction season.**

VMB. The VMB area consisted of a 2-ft excavation along the ditch and a 4-ft excavation south of the building. The southern portion of the drainage ditch and the southern portion of the 4-ft excavation were reexcavated based on field screening results. The reexcavation also extended approximately 100 ft past the design limits (based on visual observations and organic vapor analyzer readings). After these reexcavations, confirmatory samples were taken. Four additional confirmation samples were collected at the southern (downgradient) end of the ditch to account for increase in the excavation limits. Results for the 81 confirmatory samples from the ditch excavation and 19 from the 4-ft excavation showed two locations in the ditch area that had levels exceeding those of the PRGs. A residual risk assessment was done for these locations; results indicated that the risk was below acceptable limits, and reexcavation was not necessary.

Results for the 19 samples from the 4-ft excavation were below the PRGs for the contaminants of concern. **Based on this information, the performance standards for the removal action at the VMB have been met.**

CONSTRUCTION ACTIVITIES

Removal Action Contract

The construction activities described in this report were performed under a removal action contract awarded to Bechtel with AFCEE as the contracting agency. In compliance with the work plan, all work performed under this delivery order was overseen and inspected by Bechtel. Soderberg Construction Company of Caribou, Maine, served as the primary subcontractor under Bechtel and performed all activities associated with excavation, backfilling, grading, erosion control, and transport of excavated soils and debris. Bechtel performed all field screening activities (e.g., immunoassay testing) and confirmatory sampling. Confirmatory analyses were subcontracted to PACE Laboratories (now Katahdin Analytical Services), Ceimic Laboratories, and Alpha Laboratories. Clean Harbors Inc. was subcontractor to Soderberg for handling, transporting, and disposing of hazardous materials.

Under this structure, Bechtel was the contractor to AFCEE, and the removal action subcontractor, Soderberg, reported directly to Bechtel's construction superintendent. The analytical laboratories and a local survey contractor (Doody, Blackstone and Bubar) reported directly to Bechtel's resident engineer. Bechtel was under the authority of the AFCEE field engineer, who was located on-site. The AFCEE field engineer reported to AFBCA and to project representatives from EPA and MDEP.

Site Activities

The majority of the site activities were related to the excavation of contaminated soil from OUs 5, 8, 9, 10, and 11 and the subsequent disposal of this soil in LF-3. General activities associated with the excavations also included clearing and grubbing, backfilling, grading, establishing turf, and seeding and mulching. The total quantities associated with this removal action are summarized in Table 1.

Most activities were accomplished through the use of heavy earthmoving equipment including: excavators and backhoes for excavation work; dump trucks for soil transport; bulldozers, front end loaders, and compactors (vibratory and sheep foot) for backfill operations; heavy duty mulch blower for seeding and mulching; and pumps and storage tanks for the handling of water. Various types of equipment were also used for specialty tasks (e.g., hydrovac truck for the PPDP pipe cleaning, water truck for dust suppressant).

Contaminated soils and other miscellaneous demolition debris (e.g., Entomology Shop building) associated with these removal actions were disposed of at LF-3. All material placed in LF-3 was disposed of in accordance with the LF-3 subgrade construction work plan, design criteria, and regulatory framework governing land disposal. Soils were placed in a maximum of 1-ft layers, compacted, and surveyed. Erosion and sediment controls were also enacted and maintained throughout the construction season at LF-3 as described in the work plan.

Table 1. **Construction Activities Summary**

Operable Unit	Site	Actual Volume Excavated	Clear/Grub (Acres)	Seed/Mulch (Acres)	Pavement Restored (ft²)	Misc. Construction Activities
5	Nose Dock #1	474	0.5	0.2	•	• Recovery well added of FJETC area
	Nose Dock #3	261			•	• 2,747 ft² of pavement restored
8	FTA	4,510	0.2	-	•	• 1,000-gal underground storage tank removed • Underground oil lines removed • Fuselage removed • Wetland partially mitigated
9	PPDP	4,610	0.3	1.2	•	• Power plant drainline cleaned (400 ft) • 42 ft of drainline repaired • 0.1 acres of wetlands restored • Recovery well installed
	Snow Barn	2,518		0.1	• 1,211	• 1 yd³ of PCB TSCA hazardous soil removed/disposed
10	Entomology Shop	9,177		-	•	• 5,713 yd³ of uncontaminated soil excavated and stockpiled • Five hazardous waste drums (TCE sediment from manhole) removed/disposed • Two hazardous waste drums (Chlordane from basement sediment) removed/disposed • 1,600 gal of hazardous liquids (basement) removed/disposed • Entomology Shop demolished • 1,700 ft of drainline removed/disposed
	FSSB	417		-	•	• 22 ft of drainline replaced
11	FTF	9,710			•	• Free product collected
	RMSA	5,030	0.9	1.0	•	• Dry well removed • Oil interceptor, oil/water separator, and 600-gal storage tank removed/disposed • Sanitary sewer line replaced between building and manhole • Recovery well installed • Removal and disposal of 9,000 gal water from dry well and misc. underground structures • Wetland partially mitigated
	VMB	4,960	0.7	1.7	726	
	TOTALS	41,677	2.9	12.4	1,937	

Backfilling was done in accordance with specifications. In areas where pavement was to be restored, compaction was at a minimum of 95% as measured by in situ nuclear methods (ASTM D-2922). In the remaining areas, compaction was at a minimum of 90%, which was accomplished with four passes of the compaction roller. In some areas, the quantity of backfill was minimized to save on time, costs, and materials. For example, the FTF was not backfilled, but simply regraded to allow for drainage to the outlet structures, and the ditch at the VMB was only partially backfilled to improve drainage.

As part of the removal action at the PPDP, the drainage pipe (from the Power Plant to the discharge locations immediately west of Pennsylvania Road) was cleaned using a hydrovac system. The pipe cleaning encompassed all sections of the drainline including the drainlines leading from the Power Plant roof and the storm drain located in the fenced transformer area adjacent to the west end of the Power Plant. After the cleaning, the sections of line that were removed from the excavations were replaced. Water samples will be taken from the drainlines in the spring of 1996 during storm events to determine whether secondary contaminant sources still exist.

Lessons Learned

Most of the challenges encountered during the removal actions associated with this report were caused by the shortened construction season, which greatly affected the pace of work. Other challenges included differences between design drawings and field conditions, and the quick turnaround time needed for the TPH analytical results. Direct communication between all parties involved (AFCEE, AFBCA, Bechtel, MDEP, EPA, Soderberg) allowed most of the challenges to be solved quickly and efficiently.

Other lessons learned include the following.

1. During excavation in new areas, the construction crew and oversight engineers must be alert to changing field conditions that could require reassessment and revision of field operations and possible additional handling of the wastes (e.g., Entomology Shop).

2. A water management plan must be in place before work begins because water, either from natural conditions (rain, surface runoff, groundwater) or from past activities (e.g., liquids from tanks), will likely be encountered during excavation and will be of unknown quality. The water management plan will allow for efficient handling of all liquids, as long as all parties are aware and in agreement with this plan. This would include developing a set protocol for handling, sampling, containerizing, and disposing of waters of any origin and any quality. Development and agreement of the water management plan should be by the contractor, AFCEE, AFBCA, Loring Development Agency, MDEP, and other parties associated with water handling.

3. Immunoassay tests kits used for initial field sampling should be evaluated to determine whether an on-site lab would be a cost-effective alternative. In many instances, the immunoassay kits were too conservative to consistently satisfy performance standards, and many decisions had to be based on extrapolated data. For example, the immunoassay kit for PAHs only gave gross results for *total* PAHs, but individual compound analyses were needed (i.e., PRGs were compound-specific). In terms of correlation between immunoassay and analytical results, the following generalizations were developed: (1) PCB immunoassay kits had good correlation to analytical results; (2) PAH immunoassay kits provided average correlation with not a great amount of confidence; and (3) TCE/PCE immunoassay kits provided very poor correlation.

4. Materials must be evaluated for hazardous compounds before removal. Advanced planning is needed to establish segregated areas for holding suspect materials. Materials that cannot be confidently identified must be retained in a controlled area for analytical testing to comply with disposal requirements. These disposal requirements must be approved by AFCEE prior to any action taken by the contractor.

5. The success of the QC program indicates that an effective program using a three-phase control approach for each definable feature of the project should be continued. These three phases include:

    - The preparatory phase that outlines actions in advance of construction.
    - The initial phase that outlines actions at the start of operations.
    - The follow-up phase that outlines the actions involving inspections and testing.

6. Field activities should also be scheduled, if possible, to limit large quantities of samples being sent for off-site analysis. If large quantities of environmental testing are required, contingency laboratories must be available when the main laboratory becomes overwhelmed with analytical requests.

CHAPTER 38

Application of Immunoassay Field Analysis at a Field Site

Bradley A. Call, U.S. Army Corps of Engineers, Sacramento, California

INTRODUCTION

This chapter presents the analysis of immunoassay data collected during remediation of a site in the Presidio of San Francisco, California. The analysis was performed to access the usefulness of immunoassay field analysis and to determine its comparability to laboratory methods. Because laboratory analysis is the yardstick by which environmental cleanup is judged, the degree to which immunoassay can produce comparable results is fundamental to its utility.

Currently immunoassay field analysis can be used in two similar ways. The first is to compare a sample to a benchmark concentration. This qualitative measure (often referred to as semiquantitative) is useful when evaluating ongoing remedial efforts to meet a stipulated clean-up concentration. A second use is to actually quantify contaminant concentrations, analogous to the data provided by laboratory analysis. This capability is still maturing, but it has the promise to significantly improve the cost-effectiveness and timeliness of environmental cleanup work.

Quality control parameters often associated with immunoassay include correlation with laboratory results, evaluation of potential cross-reactivity problems, determining false negative/false positive rates, and comparing duplicate samples. With the exception of cross-reactivity, these parameters will be reviewed in this paper.

Where possible, the immunoassay data will be compared to both whole product concentrations (measured as total petroleum hydrocarbon [TPH]) and polycyclic aromatic hydrocarbon (PAH) compounds, quantified using standard Environmental Protection Agency (EPA) laboratory analytical methods. TPH was measured with the California LUFT modified version of EPA Method 8015, quantified against a diesel standard, and PAHs were measured with EPA 8270. It should be noted that the data presented in this report were not collected for the purpose of evaluating immunoassay field analysis. These data were collected in conjunction with remedial activities, and as such were not optimized to support the analyses presented in this chapter.

BACKGROUND

The data evaluated for this report were collected from the Building 1349 site which is located near the coastal bluffs on the western side of the Presidio. The facilities referred to as Building 1349 were constructed in the late 1800s (perhaps as late as 1906) as the central distribution point for a gravity feed fuel oil system that supplied most of the Presidio. The original facility included a large aboveground storage tank (AST) and distribution piping. The size of the original tank is unknown; however, it was located north of, and adjacent to, a replacement AST that was constructed in the 1950s (see Figure 1). The replacement AST had a capacity of 100,000 gallons, and was removed in 1995 as part of the environmental cleanup of the site (Montgomery Watson, 1995a).

Figure 1.   **Site map for Building 1349, Presidio of San Francisco, California (after Montgomery Watson, 1996).**

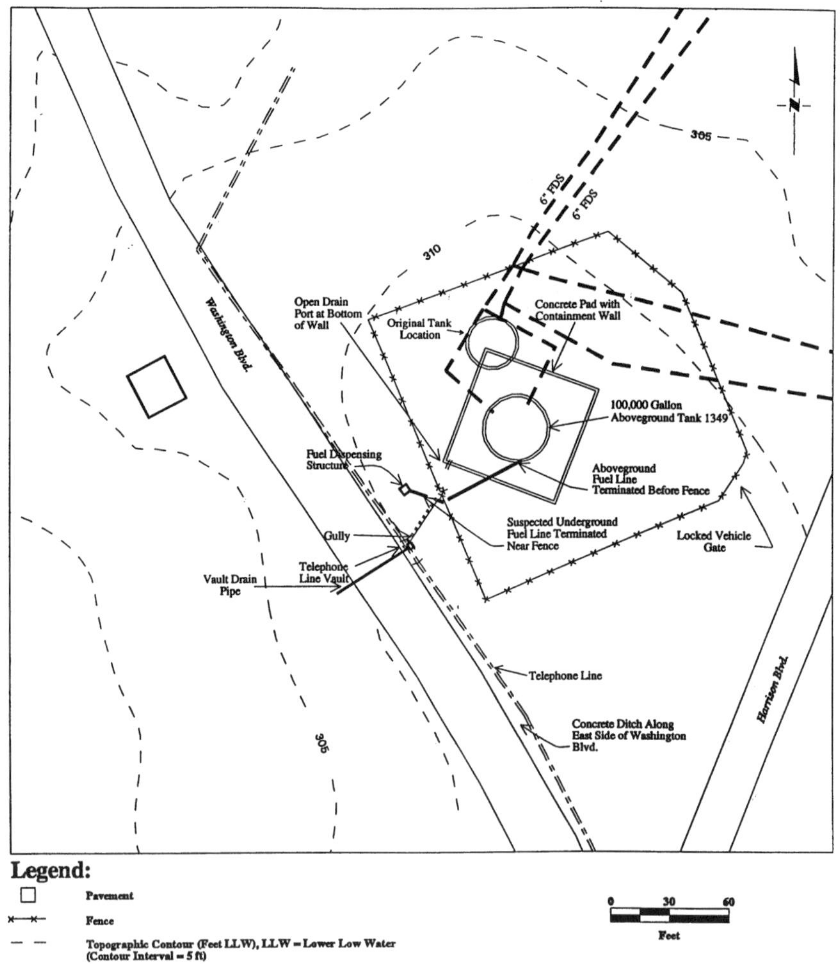

The fuel oil distribution system was taken out of service beginning in the 1940s and continuing through the early 1960s, as the Presidio was converted to natural gas heating. The 100,000-gallon replacement AST was converted to storage of diesel fuel during this time. In September 1992, fuel contamination was noted in the soil and an investigation was initiated.

In August 1993, Montgomery Watson began a three-phase investigation of the site. The investigation found that the contamination was located to the west of the replacement AST, in the vicinity of a fuel dispensing pump. The contamination was identified as a middle-range petroleum distillate (identified analytically as an "extactable range" petroleum hydrocarbon), consistent with the historical use of fuel oil and diesel at the site. Whole product concentrations averaged 3590 mg/kg, with a maximum concentration of 170,000 mg/kg. The highest concentrations were found at a depth of approximately 10 feet (Montgomery Watson, 1995a, 1996).

GEOLOGY

The shallow subsurface in the vicinity of Building 1349 consists of dry clayey silts with interbedded lenses of clay and varying amounts of fine gravels. Extremely weathered bedrock (siltstone and shale) is encountered at an average depth of 10 feet below the ground surface. This overlies more competent serpentinite and sandstone bedrock that is found at an average depth of 30 feet below the ground surface. The exact depth to bedrock can vary significantly due to vertical displacement along faults (Montgomery Watson, 1996).

REMEDIAL ACTIVITIES

The cleanup of the site was conducted by International Technology Corporation, and included removal of the AST, piping, and contaminated soil. During cleanup work, contaminated soil was identified beyond that shown in the investigation reports. The additional contamination was found to the east of the AST using a combination of visual observation and immunoassay analysis. Two separate excavations were conducted at this site. The eastern excavation was referred to as the "upper excavation," and the original as the "main excavation." To most observers, the upper excavation appeared to be associated with fuel oil, while the main excavation with diesel. Both excavations were generally limited to a depth of 10 feet or less. The data evaluated in this report was collected during this work.

Cleanup levels were established with the regulatory agencies prior to excavating soil. These concentrations consider the risk posed to human and ecological receptors, as well as impacts to groundwater resources. At a depth of 10 feet, cleanup concentrations for diesel and fuel oil were 1380 and 1900 mg/kg, respectively. The summation of carcinogenic PAHs was not to exceed 5.6 mg/kg (Montgomery Watson, 1995b).

DESCRIPTION OF IMMUNOASSAY ANALYSIS

Immunoassay, or more formally, enzyme linked immunosorbent assay (ELISA), is an analytical technique that uses the ability of antibodies to selectively bind to low concentrations of target compounds. This technology has been in use in the medical field for approximately 30 years but has only recently been adapted for environmental cleanup work. Antibodies are proteins produced by the immune systems of living organisms in response to foreign substances (referred to as antigens). During immunoassay analysis, both the target compound (antigen) and an enzyme conjugate compete for the antibody binding sites. This competition is governed by the relative concentrations of the target compound and the enzyme conjugate. Depending on the manufacturer, a variety of techniques are then used to decant the reagents and measure the concentration of the target compound. This is usually done by measuring a color change that is inversely proportional to the amount of the enzyme conjugate bound to the antibodies.

At this site, immunoassay analysis was done with the Ohmicron PAHs RaPID Assay(R) kit. Its use involves the following steps:

- Measurement of the soil sample quantity
- Extraction of the target compounds
- Immunoassay
- Measurement of results

Methanol is used for the extraction. The target compound recovery is affected by the type of soil, the vigor of the extraction agitation, and presence of potentially interfering substances in the soil. The samples are normally run in batches to maximize efficiency. Quality control (standard and control) samples are an integral part of the system. Various detection ranges are possible by varying the amount of sample diluent. Raw results are adjusted based on the volumes of extractant and diluent used. The calculated value is reported as a total PAH concentration, which in turn can be expressed as a TPH diesel concentration by using a linear conversion plot.

The target antigen for the antibodies in the Ohmicron PAH kit is phenanthrene. Because antibodies recognize chemical structures, they bind with differing affinity to the other PAHs with similar structures. This cross-reactivity is desirable when the goal is to determine the presence of a group of molecularly similar contaminants; however, it does complicate efforts to quantify specific types of PAH compounds, as discussed below. Pyrene, fluoranthene, chrysene, and anthracene react more strongly in the system, while benzo(k)fluoranthene and fluorene give a lesser response.

This kit is reputed to have a detection range of 100 to 5000 μg/kg (or parts per billion [ppb]) in soil. Table 1 shows the constituent specificity (expressed as a least detectable dose (LDD) as reported by Ohmicron. As shown in Table 1, the kit is sensitive to both carcinogenic and noncarcinogenic PAHs, as well as other petroleum constituents of concern at this site.

Table 1.   **Ohmicron PAHs RaPID Assay (R) Sensitivity in Soil (µg/kg)**

Compound	LDD	Molecular wt	#Rings
Phenanthrene	70	178	3
Anthracene	54	178	3
Fluoranthene	32	202	4
Pyrene	20	202	4
Fluorene	165	166	3
Naphthalene	6500	128	2
Benzo(a)pyrene	50	252	5
Benzo(k)fluoranthene	77	252	5
Chrysene	40	228	4
Benzo(b)fluoranthene	91	252	5
Indeno (1,2,3-c,d)pyrene	78	276	6
Heating oil	1280	NA	NA
Diesel fuel	1960	NA	NA

NA = Not Applicable
LDD = Least Detectable Dose

The immunoassay test kits were used by skilled environmental professionals, but with no special experience in performing such analyses. This is an important point when considering the data quality described below. Another factor to consider is that the field analyses were not corrected from wet to dry weight, as were the laboratory results. This potentially gives the immunoassay results a negative bias; however, the required adjustment (resulting from an estimated average soil moisture of 12 to 16%) is not considered significant when compared to other sources of error as discussed below.

DATA QUALITY

For each batch of samples, immunoassay data quality indicators included blank, control, and standard samples. Data quality was also evaluated later, through the use of duplicate comparison. This duplicate comparison includes analysis of relative percent differences between immunoassay samples, as well as checks for false negative and positive results, using laboratory values as a benchmark.

Ohmicron provides standard and control samples for process calibration and monitoring. It is recommended that such samples be run with each batch of primary samples. During the analysis of these samples, it was noted by field personnel that the kit is sensitive to both the ambient temperature and the time taken to conduct the procedure. Difficulties with standard preparation may have resulted in an undetermined bias to some results. Increased familiarity with the kit procedures provided more consistent and reliable results.

The precision of the immunoassay results were evaluated by measuring the reproduciblity of total PAH concentrations in duplicate samples. The duplicates were initially collected as two separate contiguous samples. Later, a single sample was collected and the material was partially homogenized by breaking up soil clumps and shaking the sample. This homogenized sample was then spilt. These procedures are consistent with typical duplicate collection procedures conducted during site remedial activities, but they are not optimum for comparison purposes. The resulting duplicate data can be expected to differ somewhat due to the heterogeneous nature of soil and uneven contaminant distribution. Under these circumstances, agreement within one order of magnitude is generally acceptable. Twenty pairs of duplicates were collected. Of these, six pairs were censored when either one or both values of the sample pair were reported to be zero. The resulting analysis on the 14 remaining pairs found a mean relative percent difference (RPD) of 34.8, which shows good reproducibility, and therefore precision, of the data.

Another measure of data quality is the rate of reported false negatives and positives. When possible, this evaluation was applied to both the immunoassay TPH and total PAH results. Difficulties with the TPH data arise because the dilution used most often provided quantitation from 350 to 5500 mg/kg, which complicates comparison to laboratory results outside that range. With these restraints in mind, no false positives, and one false negative (immunoassay; <540 mg/kg and laboratory; 3500 mg/kg) were noted for the immunoassay TPH data. For the evaluation of the total PAH data, the difference in process detection limits (immunoassay versus laboratory) is a complication. The laboratory PAH analysis detection limits were generally higher than that for the immunoassay total PAH. One false positive result (immunoassay result indicating significant [greater than 10 mg/kg] total PAH concentrations in the absence of laboratory detection of PAHs), was found in the data set. Once again, it should be mentioned that this kit has been designed with a wide cross-reactivity to facilitate its use as a PAH screening tool. Two false negative results (immunoassay total PAH results of zero with companion laboratory detection of PAHs) were found. This corresponds to a 1.72% false negative rate, which is well within the acceptable SW-846 criteria for immunoassay analysis.

A summary of all of the immunoassay data (to include that without companion laboratory data) is shown in Table 2. It shows the characteristics often associated with data from contaminated sites; it is nonnormal and positively skewed.

Table 2.   **Immunoassay TPH & Total PAH Summary Statistics (mg/kg)**

Parameter	TPH	Total PAH
Total number of samples	216	216
Number of samples w/ quantitation	69	197
Percent w/ quantitation	31.9%	91.2%
Average	2230	34.5
Median	1950	0.95
Std. Deviation	1400	61.0
Range	5410	290
Skewness	0.034	0.96

SEMIQUANTITATIVE COMPARISON

In evaluating immunoassay performance as a whole product (TPH) threshold predictor, success was judged as agreement between the immunoassay and laboratory sample pairs. Agreement is defined as the following:

- The immunoassay lower bound result is less than or equal to the laboratory result
- The immunoassay quantified result is greater than or equal to the laboratory result
- The immunoassay upper bound result is greater than or equal to the laboratory result

Many immunoassay results were not quantitated (i.e., expressed as greater than, etc.) because the calibration range is finite (for example, values less than 350 mg/kg or greater than 5500 mg/kg). A dilution change is necessary to return to the calibrated range. While this is always possible, in practice this adds both time and expense to the process and is not generally necessary. Only 69 of the 215 immunoassay samples resulted in quantitated TPH concentrations.

The soil samples for immunoassay and laboratory analysis were collected as two separate contiguous samples. This introduces the soil heterogeneity and contaminant distribution problems discussed above. Despite this, the agreement was over 80% (Table 3).

Table 3.   **TPH Threshold Comparison**

Parameter	Upper Excavation	Main Excavation	Combined
sample pairs	22	110	132
agreement	21	86	107
percent agreement	95.5	78.2	81.1

Establishing PAH criteria for defining agreement is much more complex. This is due to the difficulty in comparing a total PAH immunoassay measurement to individual laboratory PAH concentrations. Additional complexity is introduced by the fact that the laboratory detection limits are normally higher. To make a meaningful assessment of agreement, the first step was to evaluate all available laboratory results. It was found that naphthalene, phenanthrene, and fluorene (in this relative frequency) were the most commonly identified PAHs. Benzo(a)pyrene was found only twice and always in conjunction with higher concentrations of the other PAHs. It was also noted that when PAHs were reported by the laboratory, the associated immunoassay result was normally four to ten times greater. With these facts in mind it was decided to approach the problem by comparing the immunoassay total PAH value to either the laboratory naphthalene or total PAH result. Agreement was defined as the following:

- For samples with no laboratory PAH detection, sample pairs agree if the immunoassay result is approximately within a factor of 10 of the laboratory naphthalene detection limit
- For samples with a laboratory PAH detection, the sample pair agree if the immunoassay quantified result is greater than or equal to the total PAH laboratory result, or the immunoassay upperbound result is greater than the sum of the values reported by the laboratory

Using these criteria, 96.3% of the pairs were in agreement. In one of the four pairs that did not agree, a significant immunoassay result was reported with no corresponding laboratory detection. In two pairs there was no immunoassay detection with a corresponding laboratory detection, and in the last pair the immunoassay result was approximately half that detected by the laboratory. But generally, as mentioned above, the immunoassay results were much higher than the corresponding laboratory results. This may indicate the presence of PAHs not analyzed for by the laboratory. Another possibility is that other fuel non-PAH constituents are being detected by the kit. As shown in Table 4, the pairs that did not agree were collected in the diesel contaminated main excavation.

Table 4.   **PAH Threshold Comparison**

Parameter	Upper Excavation	Main Excavation	Combined
sample pairs	22	85	107
agreement	22	81	103
percent agreement	100	95.3	96.3

QUANTITATIVE COMPARISON

Quantitative results were obtained for 91% of the immunoassay total PAH tests. Thirty-two percent of the test results fell within the calibration range of the linear conversion plot, and thus yielded quantitated TPH equivalent data. To evaluate the accuracy of the quantitated immunoassay data, four comparisons were made:

- TPH equivalent with the whole product concentration reported by the laboratory
- Total PAH with the summation of laboratory reported PAH results
- Total PAH with naphthalene as reported by the laboratory
- Total PAH with phenanthrene as reported by the laboratory

In the first comparison, the immunoassay TPH equivalent value was compared with the EPA 8015 result. In the others, the immunoassay total PAH results were compared to the sum of the EPA 8270 detections, as well as with the two PAHs that had the highest frequency of laboratory detections, naphthalene and phenanthrene. Naphthalene was the most frequently detected (44 times), but is the PAH that is most dissimilar to phenanthrene in regard to molecular weight and ring structure. Phenanthrene was detected 35 times and is the benchmark for the immunoassay test kit.

When considering the entire data set, the immunoassay TPH values were found to correlate only moderately with EPA 8015 results. Using linear regression, the calculated correlation coefficient (R), was 0.59. The strength of the correlation coefficient is defined as follows (Devore, 1987):

- Strong, R 0.8
- Moderate, 0.5 > R < 0.8
- Weak, R 0.5

As mentioned above, the site cleanup concentrations for diesel and fuel oil were 1380 and 1900 mg/kg, respectively. To evaluate immunoassay comparability in the concentration range of interest, EPA 8015 results of 2000 mg/kg or less were compared to the associated TPH equivalent values. The resulting correlation coefficient was 0.83, which indicates that the kit performs well within this range. Figure 2 shows a scatter plot of the EPA 8015 and associated immunoassay data in the range of 0 to 5,000 mg/kg. The figure shows that the accuracy of the immunoassay result decreased with increasing concentration. Various lines were fit to the data in Figure 2 to further evaluate the comparability. The best fit (R value of 0.91) was obtained with a power curve. The trend, as shown in the figure, was for the kit to report a lower concentration than the EPA 8015 analysis. As estimated from this curve, the expected difference varied from approximately 30% at a concentration of 2000 mg/kg, to 37% at 4000 mg/kg.

To assess the immunoassay PAH results, they were compared to three values: the sum of laboratory-detected PAHs, naphthalene, and phenanthrene. The respective correlation coefficients were; 0.45, 0.27, and 0.52. These coefficients indicate a weak immunoassay PAH association with total PAH and naphthalene, and a moderate one with phenanthrene. The correlation coefficients are summarized below:

- Immunoassay TPH versus EPA 8015 (all values) - R = 0.59, moderate
- Immunoassay TPH versus EPA 8015 (2000 mg/kg) - R = 0.83, strong
- Immunoassay total PAH versus total PAH - R = 0.45, weak
- Immunoassay total PAH versus naphthalene, R = 0.27, weak
- Immunoassay total PAH versus phenanthrene, R = 0.52, moderate

CONCLUSIONS

At this site, the immunoassay kit was used for measuring concentrations of diesel and fuel oil, as well as the associated PAH constituents. Overall, the kit performed very well. The kit provided timely data which was useful for both controlling soil excavation activities as well as for site characterization. Quality control measures, such as reproducibility and false negatives, were generally within acceptable limits.

The kit was very successful when used in the semiquantitative mode. Over 81% of the whole product (TPH) sample pairs were found to agree. When considering only fuel oil, this agreement was excellent at over 95%. The immunoassay kit performance as a threshold predicator for total PAH was also good; however, it is more difficult to make a direct comparison. Site-wide, total PAH agreement was 96.3%. The kit is very sensitive to PAHs and was found to overestimate by a factor of four to ten. Overall, the kit was judged to be a very effective as a threshold predictor, although the PAH results would benefit from a site-specific adjustment factor.

Figure 2. **Comparison of laboratory and immunoassay data.**

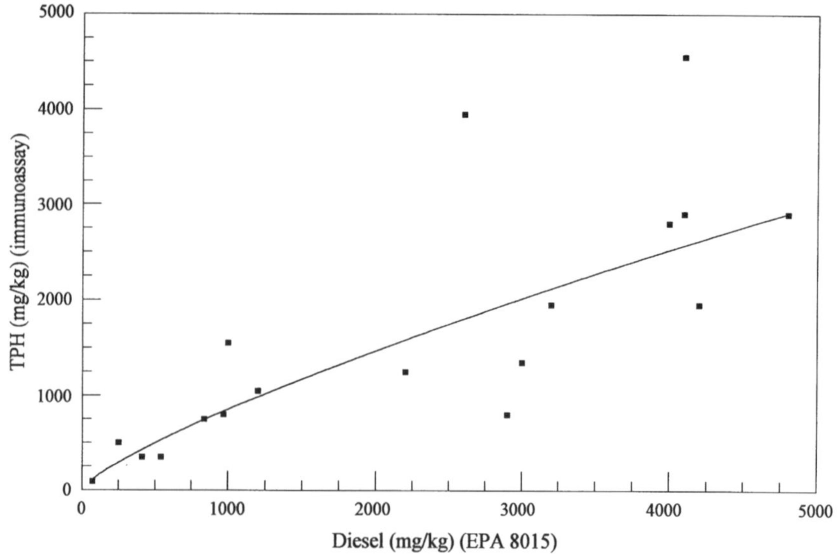

When used to quantify whole product (once again, measured as TPH), the immunoassay results were found to correlate strongly with laboratory data in the concentration range of interest and to correlate only moderately across the entire data set. In association with this, it was found that the kit tends to underestimate at the higher concentrations. A weaker correlation was found with laboratory results for total measured PAHs, naphthalene, and phenanthrene, which is ironic considering the kit's intended application.

While it is not clear why the accuracy of the immunoassay kit fell off with increasing concentration, the relative extraction efficiency of methanol versus methylene chloride (used by the laboratory) may play a role. Extraction efficiency is an aspect of this technology that should be further evaluated. The immunoassay results should also be corrected for moisture content, as is done in the laboratory.

The explanation behind the poor correlation of the immunoassay total PAH results to the laboratory values is due to a number of factors. First, the antibodies are sensitive to a wide variety of molecularly similar compounds. Undoubtedly this includes numerous compounds that are present in the fuel mixture but were not the subject of laboratory analysis. A second factor is the extraction efficiency problem noted above. This is exacerbated with very insoluble compounds, such as PAHs. Increased attention to sample pair preparation would also improve the correlation. But despite these problems, the PAH correlation was highest for the target antigen, phenanthrene, as it should have been. Taken as a whole, the kit is judged to be an excellent screening tool for the presence of PAH and similar compounds.

It should be noted that the good agreement of the immunoassay results with the laboratory TPH values was obtained with the conversion plot provided by Ohmicron. This plot presumably originates from analysis of a "fresh" product. At this site, both the diesel and fuel oil had "weathered," which raises the possibility that even better agreement would be possible by preparing a site-specific conversion plot.

Implications of this technology for site characterization and remedial operations, where analytical costs often represent a large percentage of the project budget, are evident. It can also be anticipated that by better understanding the limitations, immunoassay analysis will be better applied in the future.

ACKNOWLEDGMENTS

The conclusions of this paper are solely those of the author and do not reflect official policies of the U.S. Government. Preparation of this paper would not have been possible without the careful field work of David Korfus and Mary Plessas of International Technology Corporation. The author would also like to thank all of those, especially Lynn DeGeorge of Montgomery Watson, who reviewed an earlier version of this paper.

REFERENCES

Devore, J.L. 1987. Probability and Statistics for Engineering and the Sciences, Brooks/Cole Publishing Company, Monterey, California.

International Technology Corporation. 1996. Aboveground Storage Tank Closure Report, Building 1349, Presidio of San Francisco, May.

Montgomery Watson. 1995a. Building 1349 Site Investigation Report, Presidio of San Francisco, California, January.

Montgomery Watson. 1995b. Fuel Product Action Level Development Report (FPALDR), Presidio of San Francisco, California, May.

Montgomery Watson. 1996. Building 1349 Additional Site Investigation Report, Presidio of San Francisco, California, May.

Ohmicron, Product Information, PAHs RaPID Assay(R) Test Kit.

CHAPTER 39

Technical Requirements for On-Site Low Temperature Thermal Treatment of Nonhazardous Soils Contaminated with Petroleum/Coal Tar/Gas Plant Wastes

James B. Harrington, New York State Department of Environmental Conservation, Albany, New York

Chris Renda, Environmental Services Network, Denver, Colorado

BACKGROUND

The Interstate Technology and Regulatory Cooperation Working Group (ITRC) is a national coalition of state environmental regulatory agencies working cooperatively with federal agencies and other stakeholders to improve acceptance and interstate deployment of innovative environmental technologies. It is made up of 26 states, three federal agencies, and public stakeholders whose mission is to facilitate cooperation in the common effort to test, demonstrate, evaluate, verify, and deploy innovative environmental technologies. The ITRC, co-chaired by Jim Allen of California and Nancy Worst of Texas began in March 1995 as part of the Western Governors Association's Demonstration of Innovative Technology (DOIT) project.

The ITRC was organized into various work groups that address different areas and projects in support of the overall mission. Each subgroup is responsible for development of products that relate to the specific technology or subject area. The focus of this presentation is the Low Temperature Thermal Desorption (LTTD) task group. Other work groups included: In situ Bioremediation, Expedited Site Characterization, Case Studies, and Electronic Bulletin Board. In the first one and a half years of operation most of these work groups completed products that can be used to meet our objective of reducing impediments to the transfer of environmental technology from state to state. A protocol of minimum requirements for the on-site low temperature thermal treatment of nonhazardous soils contaminated with petroleum/coal tar/gas plant wastes is the first product completed by the LTTD task group. Details of that protocol follow as the main body of this paper. These initial efforts have been so successful that the Vice President's National Performance Review has presented the "Hammer Award" for breaking down barriers in the reinvention of government.

In June 1996 the DOIT project ended, and ITRC reorganized as an independent entity. The reorganized ITRC is made up of the member states, the federal partners, as well as other private and public stakeholders, and is associated with the Western Governors' Association and the Southern States' Energy Board. The co-chairs have become the managing directors and the management team is made up of the task group leaders as well as the federal partners and the stakeholder representatives. An advisory group made up of Commissioners of some of the member states is being assembled. The reorganized ITRC has also expanded in scope by adding several new work groups addressing implementation, policy, and communication issues as well as two new technology classes: metals in soil and permeable treatment walls.

The Technology Specific Task Group

The LTTD group was one of the first groups organized under ITRC. The mission of this group is to create a format and protocol that can be used to make the permitting and operational requirements of environmental technology more consistent from state to state. The group selected low temperature thermal desorption as the technology for development of this model because it was a technology with which many states had some experience, and it was initially considered an easy task that could be used for format development. Unfortunately, the different state regulations that apply to this class of technologies contain several issues that are controversial and need to be addressed on a larger than state-by-state basis.

As a first step, the group took on the challenge of creating a protocol of minimum requirements for the approval of low temperature thermal treatment of soils contaminated with petroleum and manufactured gas plant wastes. While it seemed that such a protocol could be quickly developed and agreed to because the use of this technology was fairly widespread, at least in some states, achieving a consensus among work group members first, and the entire ITRC second, was challenging. However, the objective was achieved and a final protocol was adopted on March 12, 1996. This protocol is a consensus document that was developed by members of the group which include representatives from state members, industry, the EPA, and public stakeholders. It has been accepted for use (a concurrence letter received) by 13 states, with several more expected. A list of the states who have submitted a concurrence letter is included as Appendix B of this paper.

The LTTD work group is currently developing a protocol specifying minimum technical requirements for the low temperature thermal treatment of soils contaminated with hazardous and low level mixed waste containing chlorinated compounds and/or mercury. This protocol is expected to be completed by spring of 1997. In addition, an evaluation of the potential use of this format for different types of environmental technology is expected to be completed by early summer of 1997.

The Purpose

The purpose of this chapter is to present the initial product of the LTTD work group. The easiest way to do that is to present the protocol in its entirety, rather than attempting to explain it and perhaps create confusion. Further, presenting the document in its entirety provides a resource that can be used in the future.

Acknowledgments

The protocol itself contains an acknowledgment section that details the agencies and individuals who participated in this effort. The contribution of those agencies and individuals was essential in completing this document. As Chair of this Task Group, I would like to provide special recognition to Brian Sogorka from New Jersey, who led the subgroup during the creation of this first product, and to Chris Renda from Environmental Services Network who provided an essential role in compiling the work of the subgroup and facilitating the group's progress through many issues. From a larger perspective, I would like to acknowledge the ITRC co-chairs, Jim Allen from California and Nancy Worst from Texas, who provided support and direction when it was needed but left the group to complete its work.

PREFACE

The Interstate Technology and Regulatory Cooperation Work Group (ITRC) is exploring mechanisms for interstate cooperation which may decrease the amount of time it takes for new technologies to become widely accepted and integrated into the site cleanup process. ITRC Technology Specific Task Groups (TSTGs) are focusing on several technologies, one of which is Low Temperature Thermal Desorption (LTTD). In preparing this document, the LTTD Task Group used the following "basic assumptions":

- For purposes of this document, the term "nonhazardous" takes the federal definition as defined in 40 CFR.
- The LTTD group has elected to produce technical requirements which should be followed for all LTTD applications, as opposed to a protocol or detailed test plan for a site-specific application. Because of the wide diversity of thermal treatment technologies, the group feels it is not feasible to establish a generic protocol appropriate for all sites.
- These technical requirements were developed to provide stakeholders (including vendors) with some degree of predictability and consistency of requirements from state to state. However, states reserve the right to go beyond these requirements, but should have a rationale for doing so.
- Alternatives to these requirements may also be acceptable, on a case-specific basis, but there should be a technical basis for the alternative.
- Because of the wide variability among states, the technical requirements do not include any emission criteria for air, or cleanup criteria for soil or water.

INTRODUCTION TO PROTOCOL

Background

The legal and regulatory uncertainties surrounding the cleanup of waste sites discourages the testing and use of innovative technologies and innovative applications of accepted technologies. Technology developers have difficulty gaining regulatory approval for the use of new technologies. Their difficulties are compounded by the requirement for developers to demonstrate a technology's performance in each state targeted for technology deployment.

In response to this concern, the Western Governors' Association convened a meeting of western regional regulators during the summer of 1994 to discuss ways to increase cooperation among states on the review, permitting, and evaluation of promising new remediation technologies. This group, now called the Interstate Technology and Regulatory Cooperation (ITRC) Working Group, has been expanded to states outside the region and includes federal, industry, tribal and public advisors as well.

Under direction from western governors, participating regulatory agencies which are cooperating with the ITRC will report to those governors in June 1996. The agencies will recommend mechanisms to be incorporated into state policy to facilitate interstate cooperation in order to shorten the time it takes technologies to go from demonstration to widespread application. One of the mechanisms under review is the development of baseline regulatory requirements and standardized protocols for verifying a technology's cost and performance.

During subsequent meetings of the ITRC Working Group, initial areas of technical focus were chosen and subgroups began work on establishing reporting and demonstration protocols for specific technologies. The Low Temperature Thermal Desorption (LTTD) Task Group (regulators from California, Florida, Illinois, New Jersey, New York and EPA) developed a preliminary plan to provide a set of composite LTTD regulatory requirements for all participating states. However, early work confirmed a suspected limitation — most states have not yet established generic regulatory requirements, but rather have established requirements on a site-specific basis. The objective of the LTTD Task Group therefore became the development of a baseline of technical requirements which might be acceptable to several state regulatory agencies. If successful, this reduction of regulatory impediments should help to lower the cost of permanent remedies for contaminated sites.

Status of LTTD Use/Acceptance

- present several perspectives (states, EPA, consultants, public, environmental groups, industry, vendors, etc.)
- trends towards fixed vs. mobile facilities.

Scope of Document

The LTTD Subgroup elected to begin with the relatively "straightforward" case of requirements for nonhazardous soils contaminated with petroleum hydrocarbon contaminants, coal tar, and other manufactured gas plant (MGP) contaminants and defer discussion on more problematic contaminants such as chlorinated compounds. This document deals with contaminants including gasoline, mineral spirits, kerosene, jet fuel, fuel oil, crude oil and cutting oil, coal tars, tar soils, purifier box waste, purifier box waste contaminated soil, and a combination of all of these contaminants. A future version is planned which will also address soils contaminated with hazardous wastes such as polychlorinated biphenyls (PCBs), chlorinated solvents, and pesticides.

Because of the wide range of variations from state to state, this document does not address media cleanup criteria (soil, water, air) or waste classification sampling requirements. This document does not attempt to address whether any particular LTTD unit or afterburner is classified as an incinerator. That determination, along with associated requirements, will be made by individual states.

In addressing areas of agreement among states, the LTTD Task Group has chosen to lay out technical requirements, as opposed to guidance or recommendations, for implementation of LTTD because it is a fairly well-developed technology. In keeping with the objective of providing requirements, the word "shall" is used throughout this document, rather than softer words such as "should."

The Need for Flexibility and Variances

The LTTD subgroup recognizes that on some sites, states may choose to go beyond this set of requirements. It is incumbent upon operators to find out from regulators whether there are additional or alternate requirements applicable; and it is in the states' best interest to allow variances from these technical requirements based on specific technology applications. Variances also should be provided to allow for the use of appropriate alternative sampling or analytical methods.

In order to provide flexibility in the technical requirements, variances for alternate sampling, analytical, waste processing or monitoring methods may be used if:

1. The method has previously been used successfully under similar site conditions, as documented by a regulatory agency; or
2. The method has been tested successfully by independent, nonregulatory verification entity; or
3. The method is approved by the agency, based upon site-specific conditions or technology modifications; the following criteria should be considered:
   a. waste stream homogeneity (e.g., verification sample frequency could be decreased for a highly homogeneous waste stream, and increased for a heterogenous waste stream);

b. contaminant concentration in waste stream (e.g., verification sample frequency could be decreased for a waste stream that is only moderately contaminated, and increased for a heavily contaminated waste stream);

c. automatic shut-down conditions (e.g., shut-down condition based on soil exit temperature could be eliminated based on a higher verification sample frequency);

d. receptor proximity (e.g., fugitive dust control requirements could be relaxed based on receptor proximity).

The Need for Public Involvement

The LTTD Task Group recognizes the need for stakeholder involvement when selecting new technologies for the cleanup of contaminated sites. In keeping with the full ITRC, they have adopted the concepts, in principle, put forward in "A Guide to Tribal and Community Involvement in Innovative Technology Assessment." This guide clearly points out the desire and need for "meaningful community involvement" at the site implementation level.

Although emphasis is placed on public and tribal involvement at the site-specific level, technology developers need to be aware of the types of information the community will require for their decision making process. The guide can be used as a checklist by technology developers and regulators. Examples of concerns which can be considered in a generic sense include noise levels, air emissions, risk to the public, permanence of the remedy, and cost.

Cost and Performance Reporting Requirements

The ITRC has adopted the "Guide to Documenting Cost and Performance for Remediation Projects" as a model to standardize cost and performance reporting. The LTTD group further recommends that the data elements found in the Cost and Performance Report for the TH Agriculture & Nutrition Company Superfund Site (TH Ag Report) be used. An outline of cost and performance reporting elements taken from the TH Ag Report is provided in Appendix A of this report.

PRETREATMENT SOIL SAMPLING

Sample Parameters

For purposes of this document, the objective of pretreatment sampling is to identify the range of soil types and contaminant concentrations expected on the site. This information is necessary in order to select the appropriate soil for the thermal treatment test runs and to ensure that the most heavily contaminated samples are selected for the test run. It is assumed that the site has been adequately characterized during a remedial investigation. Therefore, sample frequency requirements are not addressed in this document.

Pretreatment soil sampling for petroleum contaminated or coal tar/MGP wastes and contaminated soils shall include the parameters for the contaminant source outlined in Tables 1 and 2, respectively. Pretreatment soil sampling parameters shall also include any additional contaminants of concern associated with the soil. (See Section entitled Feed Soil Limitations.) Recommended methods for the various sampling parameters are presented in the following section. Sample data collected during an investigation of the site may be substituted for the following requirements, as appropriate.

Analytical Methods

EPA/ASTM methodologies shall be utilized for all parameters. The specific methodologies are presented in Table 3.

Sample Quality Assurance/Quality Control (QA/QC)

All QA/QC required by the analytical method shall be completed. Lab QA/QC summary documentation (including nonconformance summary report and chain-of-custody) shall be submitted with analytical results. Full USEPA Contract Laboratory Program (CLP) deliverables or equivalent shall be maintained and shall be available upon request for at least three years. Ultimate responsibility for QA/QC documentation belongs with the responsible party of a site or the vendor conducting a demonstration. However, the responsibility party contracts with another entity, such as an analytical laboratory, to house the actual QA/QC data.

FEED SOIL LIMITATIONS

Soil contaminated with elevated levels of heavy metals shall not be treated unless the air permit specifically allows treatment of the material. The generator of the soil shall certify, based upon site history or previous sampling/characterization, that halogenated organic compounds (including PCBs) are not contained in the soil to be treated. As an added precaution, to prevent inadvertent treatment of chlorinated contaminants, soil shall be pretested for total organic halogen (TOX), using EPA SW846 Method 9020. TOX analysis is not required if site soils have been analyzed for chlorinated volatile organics, pesticides and PCBs. (The LTTD Task Group is drafting a separate document which specifies requirements for treating soils contaminated with chlorinated constituents.) If there is any doubt as to the nature of constituents, sampling is required.

The following soil conditions require pretreatment or a test run to ensure the technology will be effective:

1. soil moisture >35%
2. material > 2" diameter

Table 1.   **Soil Sampling Parameters for LTTD Treatment of Petroleum Contaminated Soils**

Petroleum Contaminant	Analytical Parameters
Gasoline, Mineral Spirits	VO+10[a], Lead[b]
Kerosene, Jet Fuel	VO+10, Naphthalenes[c]
Fuel Oil No. 2, Diesel Fue	TPHC, PAH[d]
Fuel Oil Nos. 4 & 6, Hydraulic Oils, Cutting Oil, Crude Oil, Lubricating Oil	TPHC, PAH[d]

[a] Environmental Protection Agency (EPA) target compound list volatile organic (VO) or priority pollutant VO scans including xylene with a gas chromatograph/ mass spectrometer (GC/MS) library search for the ten highest peaks.

[b] Lead analysis required for leaded gasoline sources. Soil may have elevated metals prior to petroleum spills occurring. However, metals other than lead, are not typically parameters of concern for petroleum spills. Operating temperatures are usually low enough to prevent significant volatilization of metals.

[c] Naphthalenes, including naphthalene, methyl naphthalenes, di-methyl naphthalenes; may be analyzed in base/neutral+15 (B/N+15) fraction or in VO fractions; if analyzed in VO fraction, instrument shall be calibrated for these analytes. Quantitation of all isomers found shall be performed against at least one methyl naphthalene standard and at least one di-methyl naphthalene standard.

[d] Polynuclear Aromatic Hydrocarbons (PAH) as per EPA Priority Pollutant List.

Table 2.   **Soil Sampling Parameters for LTTD Treatment of Coal Tar/Gas Plant Wastes and Contaminated Soils**

Contaminant	Analytical Parameters
Coal Tar and Coal Tar Contaminated Soils	BTEX,[a] PAH, TPHC, Metals,[b] BNA[c]
Purifier Box Waste and Box Waste Contaminated Soils	Metals,[b] Cyanide, Total and Reduced Sulfur
Combined Coal Tar and Box Wastes or Contaminated Soils	BTEX, PAH, TPHC, Metals,[b] BNA, Cyanide, Total and Reduced Sulfur

[a] BTEX compounds consist of benzenes, toluenes, ethyl benzenes and xylenes.

[b] Metals: At MGP sites, certain metals (e.g. arsenic, cadmium, chromium, copper, lead and nickel) could be present at elevated levels, typically up to several hundred parts per million. However, operating temperatures are usually low enough to prevent significant volatilization of metals.

[c] BNA compounds are base/neutral/acid extractables.

Table 3.　Methods of Analysis for LTTD Sites

Parameter	Method of Analysis
BTEX	SW846 8240 (Packed Column) or 8260 (Capillary Column)
Polynuclear Aromatic Hydrocarbons (PAH)	SW846 8270
TPHC	SW846 8015B
Metals	SW846 6010
BNA	SW846 8270
Cyanide	SW846 9010 (manual) or 9012 (automated)
Sulfur	ASTM 3176, 3177 methods 427C, 428A

3. soil has high plasticity[1]
4. soil has high humus content[1]
5. for petroleum contaminated media only - either soil TPHC >20,000 ppm or greater than 25% LEL in gas in desorption chamber [2]
6. for coal tar contaminated media only - coal tar product > 2% [2]

SOIL TREATMENT VERIFICATION SAMPLING

Sample Parameters

Soil treatment verification sampling for petroleum or coal tar/MGP contaminated soils shall include the parameters outlined in the Tables 1 and 2. In addition, any other site-specific contaminants of concern for the treated soil shall be included in the parameter list. Verification sampling is not required for any contaminants which will be unaffected by thermal treatment, including metals.

Sample Frequency

Post-treatment soil sampling will require 1 composite sample for each 100 cubic yards or 140 tons of treated soil, using method ASTMC702-87. Each composite shall be comprised of 5 discrete samples. Samples shall be biased to include the most heavily contaminated soils.

As an alternative to composite samples, 5 discrete samples for each 100 cubic yards or 140 tons of treated soil may be collected. On a case-by-case basis, based upon documented efficiency of the treatment system, the post-treatment sample frequency may be reduced.

---

[1]The value will be regarded as "high" if the plasticity or humus content is significant enough to impact the efficiency of the treatment unit.

[2]Limitation is not applicable for indirect heat units. Reference 3 in the Reference List indicates level of concern is 100,000 ppm for total organic carbon.

Special consideration is required for volatile organics sampling. Samples for volatiles shall be collected using specialized sampling techniques to minimize loss of volatile contaminants.

Analytical Methods

EPA/ASTM methodologies presented in Table 3 shall used. For verification sampling, gas chromatography methods with a mass spectrometer detector system are required for analysis of volatile/semivolatile contaminants. Mass spectrometer methods are not required if:

1. Contaminant identity is known;
2. The contaminant chromatographic peak is adequately resolved from any other peak; and
3. At least 10% of the sample analyses (minimum of one sample) are confirmed using the appropriate chromatograph/mass spectrometer detection system.

Sample QA/QC

All QA/QC required by the analytical method shall be completed. Lab QA/QC summary documentation (including nonconformance summary report and chain-of-custody) shall be submitted with analytical results. Full USEPA Contract Laboratory Program (CLP) deliverables shall be maintained and shall be available upon request for at least three years. Ultimate responsibility for QA/QC documentation belongs with the responsible party of a site or the vendor conducting a demonstration. However, the responsible party may contract with another entity, such as an analytical laboratory, to house the actual QA/QC data.

SOIL HANDLING AND STOCKPILING

Pretreatment soil stockpiles shall be stored on a surface such as concrete or an impermeable liner of appropriate thickness. The stockpile shall be covered by a secured plastic cover of appropriate thickness or stored within the confines of a building. At a minimum, the staging area for the stockpiles shall be constructed to prevent surface water and precipitation from entering the area, and to collect leachate. All soil stockpiles shall remain covered to prevent the generation of dust. Water spray or equivalent shall be utilized as necessary to prevent dust generation. Monitoring shall be provided to ensure that unacceptable levels of dust generated from the movement and handling of soil do not migrate from the site.

Post-treatment soil shall be stored in the same manner as pretreated soil until analytical testing has confirmed that the soil has successfully been treated. A physical barrier, such as a curb or a wall, shall be maintained to separate the pretreatment from the post-treatment stockpiles. All areas shall be restored, to the extent practicable, to preremediation conditions with respect to topography, hydrology, and vegetation, unless an alternate restoration plan is approved by the governing agency.

SYSTEM OPERATING REQUIREMENTS

Primary Unit Operations

Unit shall be operated within the operating envelope created during the test run, conducted to optimize system performance. Operating conditions such as minimum temperature range, residence time and airflow in primary units and afterburners shall be determined during the test.

If conditions warrant (e.g., wide variation in soil type on site), this test shall include separate runs for treatment of coarse and fine soil contaminated with one of the specific petroleum products or coal tar/MGP wastes. For example, if treating diesel contamination, two runs would be required: one with coarse soil and one with fine soil. If any adverse feed soil conditions as listed in the Section entitled Feed Soil Limitations (e.g., high TPHC, high plasticity, humus) exist, soils exhibiting these conditions shall be treated during an appropriate number of test runs. The maximum soil processing rate shall be based on amount of contaminant reduction (contaminant concentration times soil feed rate) demonstrated during the test. The thermal desorption unit shall be equipped with a baghouse and afterburner or equally effective air pollution control device.

Air Emission Control Unit Operations

An afterburner or equally effective air pollution control device is required in order to ensure adequate hydrocarbon control. Operating conditions (temperature and duration) will be determined during the test run, subject to individual state approval.

Monitoring Parameters

The following parameters shall be monitored and recorded during operation of the unit:

1. exit soil temperature
2. baghouse pressure drop
3. soil processing rate
4. afterburner temperature (if applicable)
5. exit air temperature from the desorption chamber

Automatic Shutdown Provisions

The following shall trigger automatic shutdown of contaminated soil feed:

1. Primary burner failure — Instantaneous shutdown
2. Outlet soil temperature below setpoint which is based on type and amount of contamination, soil type, and test run. — 10 minute delay
3. Afterburner temperature (if applicable) below set point used in test run. — 30 second to 2 minute delay

<table>
<tr><td>4. Blower failure or loss of negative pressure at the desorber.</td><td>Instantaneous shutdown</td></tr>
<tr><td>5. Baghouse pressure drop(if applicable) outside the operating envelope determined during test run.</td><td>Instantaneous shutdown</td></tr>
</table>

Fugitive Emissions Control

Fugitive emissions control is required. Fugitive emissions control shall be accomplished by maintaining negative pressure in equipment designed to operate at negative pressure. Controls to limit fugitive dust emissions at the treated soil outlet shall be in place. Treated soil shall be moisturized within the enclosed soil discharge conveyor to minimize dust generation.

AIR EMISSIONS MONITORING REQUIREMENTS

From a state's point of view, air emissions levels are a major factor in determining whether a process can be permitted. This section will focus on emission monitoring requirements, frequency of monitoring, and parameters. Air emissions criteria are not addressed because these are determined by individual states.

Emission Monitoring - Stack Testing

Stack testing is required for a new unit or if new equipment is added to a previously tested unit; however, it is not needed each time an approved unit is set up. Stack testing is required each time a new type of soil contamination is being treated.

Initial stack testing parameters shall include:
- total hydrocarbons
- particulates
- carbon monoxide (CO)
- oxygen
- sulfur oxides (SOx) - if box waste is treated
- PAHs as sampled with EPA modified Method 5 (coal tar only)
- applicable metals

Sites with soils having elevated background or metals from other sources shall screen for metals emission and baghouse efficiency. Samples representing the highest concentration of metals shall be collected from the site.

Emission Monitoring - Continuous Emission Monitors (CEM)

CEMs shall include oxygen and carbon monoxide (CO).

SAMPLING AND ANALYTICAL METHODS

EPA methodologies shall be as specified in 40 CFR Part 60, Appendix B.

Sample QA/QC

All QA/QC required by the analytical method shall be completed. Lab QA/QC summary documentation (including nonconformance summary report and chain-of-custody) shall be submitted with analytical results. Ultimate responsibility for QA/QC documentation belongs with the responsible party. However, the responsible party may contract with another entity, such as an analytical laboratory, to house the actual QA/QC data.

WATER DISCHARGE REQUIREMENTS

The operation of some treatment equipment may generate various types of water. Possible sources of water generation include condensate from the treatment system, storm water runoff, noncontact cooling water, and soil stockpile leachate. All such water shall be collected and treated, recycled, or discharged in accordance with applicable regulations. Water generated from the treatment system shall be treated on-site as necessary and used to moisturize the processed soils. If process water is used to remoisturize soil, treatment verification sampling shall occur after remoisturization. Any excess water which is generated shall be disposed in accordance with individual state requirements. In general, water can be disposed at a permitted off-site commercial facility, a publicly owned treatment works (POTW) or on-site in accordance with a National Pollution Discharge Elimination System (NPDES) permit.

OPERATIONS RECORD KEEPING

The following records shall be maintained on-site or at another approved location:
- Summary of confirmation sample results
- Operating logs, including:
  CEM records or logs
  Shutdown events
  Monitoring parameters
- Failed batches

GENERAL QA/QC

An independent certified laboratory is required for all analytical testing for environmental media including air, soil, and water. An in-house certified laboratory may be used if at least 10% of the samples are verified by an independent certified laboratory. These provisions apply to both mobile and fixed laboratories.

HEALTH AND SAFETY

A written Health and Safety Plan shall be developed and implemented in accordance with Occupation Safety and Health Administration (OSHA) regulations 20

CFR 1910.120, the Hazardous Waste Operations and Emergency Response Rule. The plan shall address the following elements:

Key Personnel
Health and Safety Risks
Training
Protective Equipment
Medical Surveillance
Spill Containment
System Maintenance Safety
Air Monitoring
Site Control
Decontamination
Emergency Response
Confined Space Entry
System Operation Safety

REFERENCES

1. Federal Remediation Technologies Roundtable, March 1995. Guide to Documenting Cost and Performance for Remediation Projects: EPA-542-B-95-002.
2. Participants of the DOIT Tribal and Public Forum on Technology and Public Acceptance, May 1995. A Guide to Tribal and Community Involvement in Innovative Technology Assessment.
3. U.S. Environmental Protection Agency, November 1993. Innovative Site Remediation Technology, Thermal Desorption, Volume 6: EPA 542-B-93-001.
4. U.S. Environmental Protection Agency, October 1994. How to Evaluate Alternative Clean-up Technologies for Underground Storage Tank Sites (A Guide for Corrective Action Plan Reviewers.): EPA 510-B-94-003.
5. U.S. Environmental Protection Agency, January 27, 1995. Cost and Performance Report, Thermal Desorption at the T H Agriculture and Nutrition Company Superfund Site, Albany, GA.
6. Lee, B. 1995. Draft Engineering Bulletin: Treatment of Manufactured Gas Plant (MGP) Soils, California Environmental Protection Agency Department of Toxic Substances Control, Alternative Technology Division.

ACKNOWLEDGMENTS

The members of the Interstate Technology and Regulatory Cooperation (ITRC) Low Temperature Thermal Desorption (LTTD) Task Group wish to acknowledge the individuals, organizations, and agencies that contributed to this technical requirements document.

The LTTD effort, as part of the broader ITRC effort, is funded primarily by the U.S. Department of Energy and U.S. Department of Defense. The U.S. Environmental Protection Agency and Association of State and Territorial Solid Waste

Management Officials are providing technical support and the Western Governors Association is staffing the working group.

The Task Group also wishes to recognize the efforts of its participating members. State regulatory representatives who developed this document included Mr. Brian Sogorka (NJ) who chaired the group, Mr. Tom Conrardy (FL), Mr. Tom Douglas (FL), Mr. Ted Dragovich (IL), Mr. Jim Harrington (NY), Mr. Bal Lee (CA), and Mr. Matt Turner (NJ). Representatives from EPA, Mr. Jim Cummings and Mr. Paul dePercin, provided a valuable federal perspective. Stakeholder participation and review was provided by Ms. Anne Callison of Lowry AFB RAB. Ongoing group facilitation and technical support was provided to the group by Ms. Chris Renda of Environmental Services Network. A beneficial industry perspective was provided by Mr. Jim Cudahy of Focus Environmental, Inc.

In addition the group would like to thank representatives of ITRC member states and federal agencies who provided thoughtful comments on the various draft documents. Written responses were received from the states of California, Florida, Illinois, Louisiana, Massachusetts, New Jersey, New York, Pennsylvania, Texas, and the U.S. Department of Energy Headquarters (EM-50). The LTTD group also wishes to thank the technology vendors whose comments provided a needed industry perspective: Electric Power Research Institute (EPRI), Maxymillian Technologies, RUST Geotech of the DOE Grand Junction Project Office, and Kaiser-Hill of the Rocky Flats Environmental Technology Site.

Special appreciation is extended to Mr. Chris McKinnon of the Western Governors Association and Ms. Ginger Swartz of Swartz and Associates for their guidance throughout the writing of this document.

APPENDIX A: OUTLINE OF COST AND PERFORMANCE REPORTING ELEMENTS

1. Executive summary
2. Site information
3. Background
    a. Contaminant Location and Geologic Profileb.
    b. Contaminant Characterization
    c. Soil/waste characteristics affecting treatment cost or performance
4. Treatment System Description
    a. Thermal desorption system description and operation
        - Detailed Description
        - Automatic Feed-Cutoff Conditions
    b. Operating parameters affecting treatment cost or performance
    c. Project timeline
5. Treatment System Performance
    a. Cleanup Goals/Standards
    b. Treatment Performance Data

  - Test Run Data Summary
  - Full-scale Sustained Run Data Summary
  c. Performance Data Assessment
  d. Performance Data Completeness
  e. Performance Data Quality
6. Treatment System Costs
  a. Procurement Process
  b. Cost Data Quality
  c. Treatment Cost Elements
  d. Before Treatment Cost Elements
  e. Post Treatment Cost Element
7. Observations and Lessons Learned
  a. Cost Observations and Lessons Learned
  b. Performance Observations and Lessons Learned
8. References
9. Appendix
  a. Treatability Study Results (if applicable)
    - Objectives
    - Test Description
    - Performance Data
    - Lessons Learned
    - Full Scale Treatment Activity (soil data)
  b. Test Run Data
  c. Full Scale Treatment Activity Soil Data

APPENDIX B: SUMMARY OF VERBAL/WRITTEN INDICATIONS OF STATE CONCURRENCE AS OF SEPTEMBER 1, 1996

(a) We agree that the requirements are appropriate and commit to using them to the maximum extent feasible;

(b) We agree that the requirements are appropriate; however, have an organizational, regulatory, policy, or statutory conflict. (Please indicate what the conflict is).

(c) We agree conceptually with the requirements and will use and evaluate them in a test mode; or

(d) We do not believe the requirements are appropriate. (Please indicate the reasons why.)

STATE	Total Document "Level A"	Air Quality - "Level B" All Others -" Level A"	Mixed for specified sections	Total Document "Level B"	Total Document "Level C"	Nonconcurrence for one or more sections - "Level D"	Status of concurrence process
AZ			X	X			Letter Submitted
CA		X		X			Letters Submitted
CO	X		X				Letter Submitted
CO (UST)		X		X			Letter Submitted
DE							Just starting
FL		X		X			Letter Submitted
ID							
IL			X	X			Letter Submitted
KS							Just starting
KY			X		X		Letter Submitted
LA		X		X			Letter Submitted
MA				X			Letter Submitted
NE							In mid-process
NJ			X	X			Letter Submitted
NM							In mid-process
NV							
NY			X			X	Letter Submitted
OH							
OR					X		In mid-process
PA			X	X		X	Letter Submitted
SD							
TN							Just starting
TX							In mid-process
UT (UST)	X						Letter Submitted
UT (AIR)			X	X			Letter Submitted
WA							Just starting

CHAPTER 40

A Sensible and Successful Remedial Investigation, Risk Assessment, and Feasibility Study Approach for BRAC Sites AOC 41, 43G, and 43J
Fort Devens, Massachusetts

John C. Snowden, ABB Environmental Services, Inc., Portland, Maine

Charles George, U.S. Army Environmental Center, Aberdeen Proving Ground, Maryland

INTRODUCTION

Government agencies dealing with Base Realignment and Closure (BRAC), and active military installations, have the task of completing the Remedial Investigation/Feasibility Study (RI/FS) process in the most sensible, technically sound, and cost-effective manner possible. The situation is complicated when the contaminated sites are within DOD property, and the contamination at the sites is not anticipated to impact potential commercial/industrial or residential receptors. Should these sites be dealt with in the conventional RI/FS process, or should a more pragmatic approach be used? An RI/FS approach was put forth to the U.S. Environmental Protection Agency-New England (USEPA) and the Massachusetts Department of Environmental Protection (MADEP) by the U.S. Army Environmental Center (USAEC) and ABB Environmental Services, Inc. (ABB-ES) to deal with the existing contamination at three RI/FS sites (Area of Contamination [AOC] 41, 43G, and 43J) at Fort Devens, Massachusetts. Fort Devens was officially closed in March 1996. A BRAC Reuse Plan was completed for Fort Devens in December 1994. In this Reuse Plan the areas around each of these AOCs is designated to remain Army property, which allows for the Army to restrict future land use of these properties. This designation will allow for a more realistic and cost-effective approach to remediation of existing site-related contaminants.

BACKGROUND

Fort Devens is located approximately 35 miles northwest of Boston, Massachusetts, in the counties of Middlesex and Worcester, and the installation covers approximately 9,280 acres. The installation is divided into three geographic areas, the North Post, the Main Post, and the South Post. The BRAC Reuse Plan for Fort Devens calls for the Army to retain the entire South Post (approximately 3,500 acres) and small portions of the North and Main Posts for an Army Reserve Enclave. AOC 41 is located in the east-central portion of the South Post, which has been and will continue to be used by military personnel for small arms and small caliber artillery firing ranges. AOC 43G and 43J are located within the Main Post Army Reserve Enclave. The future use for both AOC 43G and 43J is planned to be vehicle storage and/or maintenance for Army Reserve units.

Each of the AOCs were originally identified as Study Areas (SA) in the Fort Devens Master Environmental Plan (MEP) in 1991. A Site Investigation (SI) was conducted at each of the three AOCs in 1992 to determine the presence or absence of environmental contaminants in the different environmental media found at each SA. Supplemental SI programs were conducted in 1993 at each of the three AOCs, and an RI was completed at each AOC in 1994/1995.

RESULTS OF FIELD INVESTIGATIONS

AOC 41

The findings of the SI at SA 41 indicated that surface soil contamination was present on a small (less than 1 acre) area of waste material, and that chlorinated solvents (trichloroethylene (TCE) at 200 part per billion (ppb), tetrachloroethylene (PCE) at 10 ppb, and 1,1,2,2-tetrachloroethane (1,1,2,2-TCA) at 170 ppb) were present in groundwater above USEPA and MADEP drinking water guidelines. Based on these findings, a Supplemental SI (SSI) was recommended to further investigate the nature and distribution of chlorinated solvents in groundwater.

The SSI was conducted in 1993 and concentrated on determining the potential source of the chlorinated solvents detected in the groundwater and potential site impacts on surface water and sediment quality in nearby New Cranberry Pond. The results of the SSI indicated that the distribution of chlorinated solvent contamination was distributed across the center of the AOC, and that a potential source area was not within the portion of the site investigated during the SSI. Based on the results of the SSI, an RI/FS was recommended. During this period the site designation was changed from SA 41 to AOC 41.

The RI at AOC 41 was completed in 1994 and focused on refinement of the hydrogeologic conditions, the nature and distribution of the groundwater contamination, and the location of a potential groundwater contaminant source area(s). The results of the RI indicated several site conditions that are controlling the groundwater contamination.

1. The contaminants detected in the groundwater were distributed into two areas of the site and appeared to be limited to the soil at the water table.
2. The horizontal and vertical permeabilities of the fine grained soils found at the water table were limiting the downgradient migration of the contaminants.

3. A man-made culvert in New Cranberry Pond was causing the surface water of the pond to recharge the groundwater, which in turn causes the groundwater to flow north-northwest away from the pond.

Based on these findings it appeared that the contaminants detected at AOC 41 have not adversely impacted the surface water and sediment quality in nearby New Cranberry Pond. These findings also indicated that if contamination was to migrate off-site that it would be forced to flow, via regional hydrogeologic conditions, approximately 3,000 feet to the northeast and eventually discharge into the Nashua River. In addition, it appears that the chlorinated solvent contaminants detected in the groundwater are not impacting the only present and planned future drinking water source in the South Post (Well D-1) located approximately 2,500 feet north of AOC 41.

A risk assessment was completed for the groundwater around Well D-1 to identify if the existing groundwater quality posed an unacceptable risk to troops using this water point. The risk assessment assumed that a soldier would use Well D-1 for potable water for 14 days per year. Two exposure durations were used: 10 years, which is probably longer then the individual exposure, and two years which appears more typical. The results of the risk assessment indicated that there are no unacceptable risk to human health from Well D-1 groundwater. Based on the risk assessment results, the findings of the RI, and the future land use of the South Post, it was concluded that an FS was not necessary at AOC 41. A groundwater monitoring record of decision (ROD) will be put in place for this and other AOCs in the South Post, and groundwater samples will be collected from Well D-1 to confirm that the water quality will not be impacted in the future.

AOC 43G

The SI at SA 43G was completed in 1992. The SI focused on a small historic gas station and whether the activities at this SA had impacted the soil and groundwater at this site. The findings of the SI indicated that no residual soil or groundwater contamination was present at this SA. However, due to the proximity of SA 43G to the then-active Army Air Force Exchange Service (AAFES) gas station, and based on the results of a past underground storage tank (UST) removal at this station, the AAFES gas station was added to SA 43G, and an SSI was recommended to better define the distribution of contaminants in soil sand groundwater.

Field investigation activities for the SSI at SA 43G were undertaken and completed in 1993 and were developed to assess the distribution of soil and groundwater contamination in two areas at the AAFES gas station. These two areas encompassed the three then-active 10,000-gallon gasoline USTs (Area 2) and the former waste oil UST (Area 3). The SSI included soil sampling and the installation of additional monitoring wells to supplement the existing groundwater monitoring network. The results of the SSI groundwater sampling indicated that high concentrations (up to 2,000 ppb) of benzene, toluene, ethylbenzene, and xylenes

(BTEX) were present in the groundwater downgradient of Areas 2 and 3. Based on the SSI results, it was evident that the existing monitoring well network was insufficient to fully define the distribution of groundwater contaminations downgradient of the gas station. Because of this, the SA was recommended for an RI/FS. During this period the site designation was changed from SA 43G to AOC 43G.

The RI was completed in 1994. This program was designed to assess the distribution of groundwater contamination downgradient of Area 2 and 3 and further define the concentrations of BTEX contaminants in soil adjacent to the three then-active 10,000-gallon gasoline USTs in Area 2. The results of the RI indicated the approximate distribution of benzene (the contaminant with the lowest drinking water standard) downgradient of the AOC. The distribution of the remaining site-related contaminants in groundwater were defined based on USEPA and MADEP drinking water standards.

A risk assessment of contaminants detected in soil and groundwater was completed as part of the RI. The risk assessment divided the AOC into a source area and perimeter (downgradient/crossgradient) area. The risks associated with the contaminant concentrations detected in soil and groundwater in each area were assessed for a potential future commercial/industrial worker. No unacceptable risk from exposure to subsurface soil was found. The results of the risk assessment indicated that contaminants in groundwater in the source area and perimeter areas would cause an unacceptable risk to potential future commercial/industrial workers via ingestion.

Based on the results of the RI and the risk assessment, an FS was completed to determine the implementability of a range of remedial alternatives. Details of the FS approach are presented in a section below.

AOC 43J

The SI at SA 43J was also completed in 1992. This program was similar to the SI completed at SA 43G with the exception that an abandoned UST was found at the historic gas station and removed from the SA during the field investigation. The results of the subsequent sampling indicated that the distribution of soil and potential groundwater contamination was not adequately defined and that a SSI was needed to better assess the site conditions.

The SSI at SA 43J was completed in 1993. This program concentrated on determining the distribution of the soil and groundwater contamination emanating from the former gasoline and waste oil USTs. The results of the SSI sampling indicated that high concentrations (up to 1,000 ppb) of BTEX were present in the groundwater downgradient of the former USTs. Based on the SSI results, it was evident that the existing monitoring well network was insufficient to fully define the distribution of groundwater contaminations downgradient of the former USTs. Because of this, the SA was recommended for an RI/FS. During this period the site designation was changed from SA 43J to AOC 43J.

The RI was completed in 1994. This program was designed to assess the distribution of soil and groundwater contamination downgradient of the former USTs. The results of the RI did define the approximate distribution of benzene (the contaminant with the lowest drinking water standard) downgradient of the AOC. The distribution of the remaining site-related contaminants in groundwater were defined based on USEPA and MADEP drinking water standards.

A risk assessment of contaminants detected in soil and groundwater was completed as part of the RI. The risk assessment divided the AOC into a source area and perimeter (downgradient/crossgradient) area. The risks associated with the contaminant concentrations detected in soil and groundwater in each area were assessed for a potential future commercial/industrial worker. No unacceptable risk from exposure to subsurface soil was found. The results of the risk assessment indicated that contaminants in groundwater in the source area and perimeter areas would cause an unacceptable risk to potential future commercial/industrial workers via ingestion.

Based on the results of the RI and the risk assessment, an FS was completed to determine the implementability of a range of remedial alternatives. Details of the FS approach are presented in the section below.

FEASIBILITY STUDY APPROACH AND RECOMMENDED REMEDIAL ALTERNATIVE

Based on the results of the RI and the risk assessments, the proposed future land use under the Fort Devens BRAC Reuse Plan, and the existing site conditions, an FS approach was assembled for AOC 43G and 43J. Because of the future land use of the South Post and the fact that no potential exposure pathway for detected groundwater contaminants exists, an FS was not prepared for AOC 41. Instead, a proposed plan will be prepared for this AOC, and a monitoring ROD will be established to monitor specific site monitoring wells and Well D-1.

Four to five remedial alternatives were assessed in the FS for both AOC 43G and 43J, they included the following:

Alternative 1: No Action
Alternative 2: Intrinsic Bioremediation
Alternative 3: Intrinsic Bioremediation/Passive In-Situ Bioremedial Containment
Alternative 4: Intrinsic Bioremediation/Hydraulic Containment
Alternative 5: Groundwater Collection and Treatment/Soil Treatment

Each of these alternatives was assessed in a screening of alternatives and a detailed analysis in accordance with USEPA guidance ("Guidance for Conducting Remedial Investigation and Feasibility Studies Under CERCLA"). This process ensured that alternatives met remedial response objectives which were formulated based on environmental impacts outlined in the RI contamination assessment, risk assessment, and applicable or relevant and appropriate requirements (ARARs) identified by the USEPA and the MADEP. This screening also compared potential alternatives to preliminary remediation goals (PRG), remedial action objectives, and general response objectives. In addition, the screening identified appropriate technologies, conducted a technical screening and evaluated potential process options.

The final step for the five alternatives was a comparative analysis of the remedial alternatives to highlight the advantages and disadvantages of each alternative. This comparison showed that Alternative 1 (No Action) would be sufficient to eliminate the site-related fuel contamination at both AOCs (43G and 43J). However, this alternative does not include long-term groundwater monitoring which would limit the effectiveness of this alternative. The comparison also showed that the remaining alternatives would be equally effective in mitigating the migration of the site-related contamination beyond the Army Reserve Enclave boundary.

Based on the analysis of the FS, the preferred remedial alternative for both AOC 43G and 43J is Alternative 2: Intrinsic Bioremediation. This alternative includes: initial groundwater sampling to better define the biodegradation rates at each AOC; groundwater modeling to refine the future downgradient distribution of the groundwater contaminant plume; annual long-term groundwater monitoring for up to 30 years to assess the effectiveness of the alternative; and a five year data review to determine if this alternative is continuing to mitigate the migration of site related contaminants.

Alternative 2 also includes a contingency process which ensures that if the intrinsic bioremediation does not mitigate the migration of the site-related contaminants and reduce the contaminant concentrations, then one or more of the additional remedial technologies (as presented in Alternatives 3 through 5), will be implemented. This feature ensures that the groundwater quality at and beyond the Army Reserve Enclave boundary will not be adversely impacted by contaminants emanating from these AOCs in the future.

CONCLUSIONS

Based on the results and interpretation of the physical and chemical data, and taking into account the future land and groundwater use of the South Post, the Army recommended that a No Action proposed plan and ROD be completed for AOC 41.

Similarly; the FS approach for AOCs 43G and 43J used the following site conditions to determine the best remedial alternative.

1. The AOC-related contaminants are readily biodegradable.
2. Existing AOC groundwater chemistry appears to support in- situ biodegradation.
3. Groundwater modeling indicates that the contaminant plume at each AOC appears to have reached steady state and that contaminant concentrations,

in excess of established PRGs, will not migrate off the Army Reserve Enclave boundary downgradient of each AOC.

4. The planned reuse of land at both AOCs is anticipated to remain similar to the current use and no future groundwater exposure scenario are expected.

5. The Army plans to retain these areas of Fort Devens and to control the future land and groundwater uses.

By using these site conditions and future use scenarios, the Army is able to save both time and money in developing and implementing the remedial actions for these AOCs. In addition, the Army team at Fort Devens has established a working partnership with both the USEPA and MADEP which helped in making these remedial actions possible. Currently, the Army team is working with the USEPA and MADEP on other AOCs at Fort Devens using this approach to determine the best economical and time-saving remedial action decisions.

CHAPTER 41

The Use of Innovative Technologies for Environmental Restoration by the U.S. Army Corps of Engineers

Donna R. Kuroda, Headquarters, U.S. Army Corps of Engineers

Jeffrey L. Breckenridge, Hazardous Toxic and Radioactive Center of Expertise, U.S. Army Corps of Engineers

Johnette C. Shockley, U.S. Army Corps of Engineers (U.S. Geological Survey)

INTRODUCTION

The U.S. Army Corps of Engineers (Corps) plays a significant role in the restoration of the nation's hazardous, toxic, and radioactive waste (HTRW) sites. In addition to addressing U.S. Army needs, the Corps' environmental restoration activities include support of the U.S. Environmental Protection Agency Superfund program and provision of remediation assistance to other federal agencies such as the Department of Defense and Department of Energy. The Corps is also responsible for restoration of property formerly owned or used by the U.S. Department of Defense. Contaminants commonly found on HTRW sites include explosive wastes, solvents, petroleum products, heavy metals, and mixtures. To achieve more effective HTRW cleanups, the Corps is turning to innovative technologies. Innovative technologies may be generally defined as those lacking full-scale cost and performance data. The Corps is moving forward with innovative technologies on four fronts: first, the Innovative Technology Advocate (ITA) Program was initiated to foster the use of innovative technologies; second, the research, development, and demonstration program is making new technologies available for environmental remediation; third, the Corps is developing guidance documents on the application and design requirements for innovative technologies; and fourth, Corps districts are using innovative technologies to remediate sites.

INNOVATIVE TECHNOLOGY ADVOCATE PROGRAM

In 1989, the Corps established ITA positions at Headquarters, the Missouri River Division, and the Kansas City and Omaha Districts. In 1994, an ITA position was established at the Tulsa District. Additional ITA positions were also established at the New England Division, Baltimore District, Sacramento District, and Alaska

District. The mission of the ITAs is to inform, encourage, and support the use of HTRW innovative technologies for restoration of sites administered by the Corps. They gather and disseminate information regarding innovative technologies to HTRW personnel and maintain their own Home Page on the World Wide Web. The ITAs hold workshops, seminars, and site visits to review significant topics such as remediation and site characterization technologies, treatability studies, and contracting barriers. They are leading the Corps' effort to gather cost and performance data on innovative and proven technologies so that accurate comparisons can be made between technologies. The ITAs work both separately within their own organization and together within USACE to accomplish their goals. To formalize many of the ITA activities, an HTRW Innovative Technology Action Plan has been developed to cover critical areas such as the formal process for innovative technology selection; training, education, and sharing lessons learned; standard format for collecting cost and performance data; appropriate contracting tools; regulatory flexibility through partnering; risk sharing and indemnification; and incentives to accelerate commercialization of R&D efforts. This comprehensive strategy will give the Corps a cohesive and proactive plan for using innovative technologies to perform HTRW cleanups.

RESEARCH, DEVELOPMENT, AND DEMONSTRATION

Research, development, and demonstration efforts associated with newly developed technologies is a joint effort between the Corps research and development laboratories (such as U.S. Army Engineer Waterways Experiment Station and U.S. Army Cold Regions Research and Engineering Laboratory) and the U.S. Army Environmental Center. The Corps laboratories are responsible for technology development from concept through large-scale pilot testing. The Army Environmental Center then typically takes responsibility for large-scale field demonstration and eventual fielding of the technology.

Several examples of technology development for HTRW applications follow. In the field of bioremediation, the laboratories are investigating variations such as bioslurry, composting, landfarming and in-situ techniques for treatment of explosive wastes. Additional cold climate techniques for bioremediation are also under evaluation. Natural restoration is another important area worthy of attention. Another kind of technology, the frozen wall barrier, which is a containment technology, is being studied for its application to HTRW sites.

In the area of site characterization, the Site Characterization and Analysis Penetrometer System (SCAPS) is being developed as a joint effort with several other federal agencies. There are SCAPS units at the Kansas City, Tulsa, and Savannah Districts. Microwells are also being developed as an alternative to conventional drilling techniques for environmental investigations. Other site characterization techniques developed include a field testing kit for explosives.

HTRW GUIDANCE DEVELOPMENT DOCUMENT PROGRAM

This program develops and maintains Corps' technical guidance needs of the many programs related to HTRW; e.g., Installation Restoration Program and Formerly Used Defense Sites. The objectives are to provide consistency in the investigation of HTRW sites and design of remedial actions, to minimize overall time and costs required for the design and construction process, and to ensure that these efforts are completed at an established, uniform level of quality. Moreover, repetitive application of the economical and efficient investigation procedures and construction materials and methods established under this program reduces lost design effort, change orders during construction, and long-term maintenance costs over the life of projects. The Corps produces several types of guidance documents which are available to the public. Only those guidance documents that apply to innovative technologies are discussed below.

- **Guide Specifications** (GS) are used to assist bidders on remedial action projects by assuring uniformity to the description of work to be performed. Guide Specs recently completed include Landfarming and Soil Washing. A guide specification to document Cost and Performance of Innovative Technologies, Composting and Low Temperature Thermal Desorption is currently under development. In the future, a guide specification for Advanced Oxidation Processes is planned.

- **Engineer Technical Letters** (ETL) contain "advance" information on design, engineering, and construction projects. They are considered intermediary publications that will eventually be republished in the more permanent media such as Engineering Regulations or Engineering Manuals. ETLs for Landfarming and Chemical (UV) Oxidation and Low Temperature Thermal Desorption are complete. A joint Reactive Barriers ETP/Technical Protocol is under development with the U.S. Air Force. An ETL on Nutrient and Microbial Testing for Soil Bioremediation Processes Safety and Health Aspects of HTRW Remediation Technologies is also being developed.

- **Engineering Manuals** (EM) contain technical guidance of a continuing nature concerned primarily with Engineering and Design Projects. EM on Soil Vapor Extraction are complete, and one on Soil Washing is planned.

REMEDIATION OF HTRW SITES

The major thrust of this chapter is the use of innovative treatment technologies for remediation of HTRW sites by Corps districts. In functional terms, all technologies exclusive of incineration and solidification/stabilization for source control and pumping with conventional treatment for groundwater are considered innovative. Modifications to proven technologies can be considered innovative.

Data calls were made within the Corps to obtain information about specific HTRW sites involving innovative technologies. As such, the information obtained should not be considered to be totally comprehensive. However, data obtained do provide significant information from which conclusions and trends can be discerned. Additionally, the knowledge base associated with historical information is useful for future applications.

The most commonly used innovative technologies for soils include bioremediation, soil vapor extraction, low temperature thermal desorption, and soil washing. Other technologies used to a lesser degree are macroencapsulation, dechlorination, soil flushing, and solvent extraction. For groundwater, the most common innovative technologies are UV oxidation, bioremediation, and air sparging. Horizontal well applications reflect an innovative method of injection or extraction. Moreover, technologies may be combined together in a treatment train for a more effective means to perform a restoration.

Bioremediation

Bioremediation technologies involve degradation of contaminants by microbial organisms. Nutrients, oxygen, or other amendments may be used to enhance the biodegradation process. Bioremediation is effective for organic contaminants, especially simple hydrocarbons. However, bioremediation of chlorinated chemicals is more difficult to perform and can require special techniques. There are several approaches to bioremediation, but all fall under two categories, ex-situ or in-situ. Ex-situ processes require the excavation or removal of contaminated media and transport to treatment facilities. In-situ processes involve treatment in place.

Ex-Situ Bioremediation

Ex-situ bioremediation is the most common category of bioremediation.
- **Landfarming** is a relatively simple technique in which contaminated soil is spread over a given area and periodically tilled to aerate. Usually indigenous bacteria are used. Collection of leachate and/or volatiles may be required. This versatile technique is used widely in warmer climates and is being tested and tried in colder climates.
    - The Corps provided technical assistance to Ft. Polk, LA, to develop a plot of eight acres for remediating soils from petroleum product spills on the installation. Once the soil is remediated, it is removed to allow more contaminated soil to be treated. This technique has saved Ft. Polk more than $1 M as an alternative to off-site dumping.
    - Ft. Ord, CA, is developing a similar type of facility to be known as the Fort Ord Soil Treatment Area.
    - Landfarming projects have been completed at Ft. Jackson, SC; Matagorda Island, TX; Davis Monthan Air Force Base, AZ; and Ft. Ord, CA. Landfarming projects are underway at Ft. Ord, CA, (groundwater treatment train); Ft. Bragg, NC; and former Stead Air Force Base, NV.

Landfarming projects are being designed at Bethel Bank Project, AK, and former Glasgow Air Force Base, MT. Landfarming is an option at a site at Ft. Richardson, AK. Landfarming is planned for the former Chennault Air Force Base, LA.

- Cold climate techniques for landfarming are being studied in Farmers Loop, Fairbanks, AK, and Fairbanks Airport, AK.

- **Composting** is a technique that involves combining contaminated soil with amendments such as horse manure or potato waste. The mixture is turned or aerated periodically until remediation is complete.
  - Composting is underway at the Umatilla Army Depot Activity, OR, which is on the National Priorities List. The contaminants are explosive wastes (TNT, RDX, and DNT).

- **Soil pile** is a general technique similar to composting in which contaminated soil and bacteria are heaped in a pile that is turned occasionally until the remediation has been completed.
  - This technique has been used successfully at Sheppard Air Force Base, TX, and Williams Air Force Base, AZ. It is in the design stage at Ft. Wainwright, AL.

- **Bioslurry/bioreactor systems** involve reactor vessels in which an aqueous slurry is created by combining soil or sludge with water and other additives (bioslurry) or contaminants in extracted groundwater are put in contact with microorganisms through attached or suspended biological systems (bioreactor). This process can be aerobic or anaerobic, depending on the conditions needed to meet the cleanup endpoint.
  - This technique is being used at Schilling Landfill, OH, at a Superfund project as part of a treatment train. Data from a demonstration project at Joliet Army Ammunition Plant, IL, showed that this approach will meet cleanup goals for TNT-contaminated soil. Further studies are planned to extend this process for other contaminants and to optimize results.
  - Anaerobic bioslurry treatment of TNT-contaminated soil was the subject of a demonstration effort at Weldon Spring, MO. Another anaerobic bioslurry study is being planned for the Iowa Army Ammunition Plant.

In-Situ Bioremediation

These bioremediation techniques offer the advantage of providing "in place" treatment, thereby avoiding soil excavation and groundwater extraction.

- A **phytoremediation** project is planned for the Iowa Army Ammunition Plant, IA. Phytoremediation was implemented at Lake Eufaula, OK, to mitigate metals contamination.
- **Bioventing** is the most common in-situ bioremediation technique and involves supplying air through wells into the contaminated soil so that the bacteria are stimulated by the oxygen. It works especially well for simple hydrocarbons and can be used where contamination is deep.

- Bioventing is underway at Ft. Carson, CO. This technique is being planned for Davis Monthan Air Force Base, AZ; Ft. Greeley, AK, and Kincheloe Air Force Base, MI. At Kelly Air Force Base, TX, the Corps set up a soil vapor extraction system that will later be used for bioventing to remove remaining contamination.
- **In-situ biodegradation** involves supplying oxygen and nutrients to stimulate naturally occurring bacteria. It can be used for soils and groundwater. Generally, this process includes conditioning of infiltration water with nutrients and an oxygen or other electron acceptor source.
  - This technique is being used to treat soils and groundwater at the Aua Fuel Farm in American Samoa for diesel fuel contamination.
- **Natural or intrinsic restoration** is an alternative which should be considered if immediate risk is low and off-site migration is a minimal concern. Natural subsurface processes are known to reduce soil or groundwater contaminant concentration to acceptable levels without external stimulation.
  - After collecting more than a year of data at a U.S. Army Reserve site at State College in PA, the results indicate that the contamination is being remediated by naturally present bacteria.
  - A Record of Decision was signed for a groundwater site contaminated with TCE at Hanford, WA. The Corps is placing monitoring wells to demonstrate that contamination is being remediated by naturally occurring organisms.
- **Soil Vapor Extraction** (SVE). This technology involves applying a vacuum through the use of extraction well to remove the more volatile contaminants from soil. The process typically includes a system for handling off-gases. It can be followed up by bioventing.
  - SVE remediations have been completed at the Rocky Mountain Arsenal, CO, (Motor Pool Area) and Sacramento Army Depot, CA, (Tank 2 OU)
  - SVE remediations are underway at sites on Commencement Bay (Superfund) South Tacoma Channel, Tacoma, WA; Langley Air Force Base, VA; Luke Air Force Base, Glendale, AZ; and Reese Air Force Base, Lubbock, TX.
  - SVE remediations are planned for Sacramento Army Depot, CA, (Burn Pits); ThermoChem Superfund, MI; Garden State Cleaners, Minotola, NJ; Holloman Air Force Base, NM (Bx Gas Station); and Marsh Run Park, New Cumberland, PA.
  - SVE is under consideration at Silresim Superfund, Lowell, MA; Union Chemical Company (Superfund), South Hope, MA; Vance Air Force Base (Site 12), Enid, OK; and George Air Force Base (Flightline Fuel Spill), Victorville, CA.
- **Low Temperature Thermal Desorption** (LTTD). This technology involves heating contaminated soils at temperatures high enough (200 to 600°F) to drive off

the volatile or semivolatile contaminants, but well below those of incineration. The volatilized components must be collected and further treated. LTTD is a physical separation process not designed to destroy organics. This technology is often combined with other technologies.

- An LTTD project is complete at Port Moller, AK. LTTD projects are underway at Ft. Wainwright (stockpile of petroleum contaminated soil), AK; Savanna Army Depot Activity (Fire Training Area), Savanna, IL; Ft. Campbell (UST removal and cleanup), KY; and a site at the O'Hare International Airport, Chicago, IL.
- LTTD projects are planned for American Thermostat (Superfund), South Cairo, NY; Caldwell Trucking (Superfund), Fairfield Township, NJ; Claremont Polychemical (Superfund), Old Bethpage, Nassau County, NY; Metaltec/Aerosystems (Superfund), Franklin Borough, NJ; Waldick Aerospace (Superfund), Wall Township, NJ; Ft. Greeley (Texas Tower), AK; and Wright Patterson Air Force Base, Dayton, OH.

- **Soil Washing**. Wash water separates small soil particles on which contaminants are sorbed from larger contaminant-free soil particles. The resultant reduction in contaminant volume reduces costs of further treatment. Soil washing is often used as part of a treatment train.

- Soil washing was used successfully at a Saginaw River, MI, sediment project for concentrating PCBs to 10 ppm from 1-2 ppm. The contaminated soil was then treated using an experimental ex-situ bioremediation technique.
- Soil washing is underway at Gould Battery (Superfund), Portland, OR; Sacramento Army Depot (Burn Pitts), CA; and Ft. Wainwright (petroleum contaminated soil stockpile), AK.
- Soil washing is planned for Savanna Army Depot Activity (Old Burning Grounds), Savanna, IL, and Sand Creek, Commerce City, CO.

- **UV Oxidation**. This technology is the most widely used innovative technology for groundwater technology. Ultraviolet (UV) radiation, ozone, and/or hydrogen peroxide are used to destroy organic contaminants as part of a pump-and-treat system.

- UV oxidation is underway at Bofors Nobel (Superfund), Muskegon, MI; Sacramento Army Depot, CA; Rocky Mountain Arsenal (CERCLA water treatment plant), CO; New Bedford Harbor (Superfund), New Bedford, MA; and Hanscom Air Force Base (for Sites 1, 2, & 3), MA.
- UV oxidation is planned at Milan Army Ammunition Plant, Milan, TN; Redstone Arsenal (three separate locations), Huntsville, AL; Groveland Wells 1 and 2 (Superfund), Groveland, MA; and Southern Maryland Wood Preservers (Superfund), Hollywood, MD.
- UV oxidation is under consideration at two sites on March Air Force Base, CA.

- **Air Sparging.** This technology involves injecting air into the groundwater to release volatile contaminants which then can be captured by a soil vapor extraction (SVE) system.
  - This approach is under consideration at Dover Air Force Base (WP-21), Dover, DE; Sacramento Army Depot (Parking Lot 3), CA; Nellis Air Force Base, NV, and Hastings East Industrial Park Groundwater, Hastings, NE.
- **Other Technologies.** The Corps has used, is planning to use, or is considering other innovative technologies for soil and groundwater remediation, but not as extensive as those covered above.
  - Dechlorination, an ex-situ technology, was used to treat soils containing PCBs at Wide Beach (Superfund), NY.
  - Soil flushing is being used at the Lipari Landfill (Superfund), Glouster County, Pitman, NJ.
  - Solvent extraction is planned for the Norwood PCB Site (Superfund), Norwood, MA.
  - Macro Encapsulation has been complete for a project at the Naval Ordnance Plating Shop, KY.
  - Modified Air Stripping using sieve plates was conducted to remove TCE and diesel contaminants at the Cold Regions Research and Engineering Laboratory, Hanover, NH.
  - Funnel and Gate reductive dechlorination is being installed at the Denver Federal Center, and planned at the Air Force Reserve site at Moody Air Force Base, GA, and Duluth Air Force Base, MN.
  - Electro-osmosis is under consideration at the Offutt Air Force Base in Omaha, NE.

CONCLUSIONS

Because of inherent risks associated with innovative technologies, fear of failure promotes "defensive engineering," resulting in a natural bias toward tried and true technologies; i.e., proven technologies. Inadequate demonstration data/information is considered the largest source of risk associated with innovative technologies. Available information is often based only on bench-scale efforts. As a result, full-scale cost and performance data have not been available to support increased application of innovative technologies.

As more HTRW remediations involving innovative technologies are completed, availability of full-scale cost and performance data will increase. As a result, there will be less reluctance associated with consideration and selection of innovative technologies. In particular, increased numbers of bioremediation applications are anticipated as understanding of potential applications grows. In-situ techniques will likely receive greater consideration as the knowledge base increases. Enhanced research and development efforts for in-situ treatment of contaminated groundwater (especially for metals-contaminated) will be reflected in future innovative technology applications. The authors conclude that within the Corps, the use of innovative technologies for HTRW remediation is expected to increase.

ACKNOWLEDGEMENTS

The authors thank their colleagues in the U.S. Army Corps of Engineers for their generous support and responsiveness needed to prepare this paper.

HOW TO OBTAIN USACE GUIDANCE

USACE publications, except guide specifications, are listed in EP 25-1-1 Index of Publications.

USACE documents, including Engineer Manuals (EMs), Engineer Technical Letters (ETLs), Engineer Pamphlets (EPs), Engineer Regulations (ERs), Engineer circulars (ECs) and guide specifications, military and civil, can be obtained from:

National Institute of Building Sciences (NIBS) Construction Criteria Base (CCB) CD-ROM disk system

-To subscribe ($500/yr) call NIBS @ (202) 289-7800

HND TECHINFO Web Site

The system is available on the Internet at

<http:/www.hnd.usace.army.mil/techinfor/tiindex.html>

All documents except guide specifications are also distributed in hard copy, usually to your library.

Hard copies are also available from:

USACE Publications Depot
2803 52nd Avenue
Hyattsville, MD 20781-1102

Request documents (except guide specifications) by signed memorandum which shows your complete mailing address and office symbol.

Your library can help you obtain Technical Manuals (TMs) and other documents. TMs are Army or Triservice documents.

U.S. Army Corps of Engineers Innovative Technology Home-Page is also available on the web at:

<http://www.mrd.usace.army.mil/mrded-h/itech.html>

PART VIII
MANUFACTURED GAS PLANT SITES

CHAPTER 42

Physiologically Available Cyanide (PAC) in Manufactured Gas Plant Waste and Soil Samples

Brian Magee, Ogden Environmental & Energy Services, Westford, Massachusetts

Alexander Taft, **Wanda Ratliff** and **Jack Rourke** Massachusetts Natural Gas Council

Joan Kelly, NET Atlantic, Windham, New Hampshire

Jim Sullivan and **Oscar Pancorbo**, Massachusetts Department of Environmental Protection, Wall Experiment Station, Lawrence, Massachusetts

INTRODUCTION

Earlier in this century, gasification of coal was the primary process for producing gas for cooking and heating. Cyanide-containing chemicals were removed from the gas before storage and distribution. Cyanide was removed by passing the manufactured gas stream through a bed of iron metal. Because cyanide and iron have strong affinity to bind to each other, the iron and the cyanide formed chemical bonds, and the cyanide was removed from the gas. Accordingly, the residues of these processes contain various iron-complexed compounds which are recognized by their deep blue color. The most common of these is ferric ferrocyanide, commonly known as Prussian Blue.

Prussian Blue is prevalent in areas where manufactured gas was purified and where process residues were disposed of. Technically, Prussian Blue is a cyanide-containing compound. However, Prussian Blue and other similar iron-complexed cyanide compounds are almost totally devoid of toxic effects, as shown herein. Accordingly, it is important when characterizing such sites to use an analytical method that is relevant to risk assessment needs.

As recently noted by Shifrin et al. (1996), no analytical method for cyanide currently exists that is able to distinguish its chemical forms that are relevant to evaluating potential adverse health effects. Three methods in wide use yield results that have questionable relevance to human health risk assessment. Each is summarized briefly here. Further details on these methods can be found in Shifrin et al. (1996) and Theis et al. (1994).

The EPA Total Cyanide Method uses a strong acid solution with a pH of less than 1.0 to release cyanide from waste samples and convert it to hydrogen cyanide (HCN), which is measured colorimetrically. Other methods of measuring released HCN include silver nitrate titrations and ion chromatography. The EPA method is not relevant to risk assessment because the extraction conditions are far more acidic than those in the human stomach. In addition, the samples are boiled under a vacuum to maximize the release and subsequent capture of HCN.

The Weak Acid Dissociable Cyanide Method has been used by many because it more closely mimics the conditions in the human stomach. In this method, the sample is incubated in a moderately acidic acetate buffer (pH 5-6) which releases cyanide and forms HCN. This method has been criticized by some risk assessors, however, because the pH of the medium is higher than the normal pH of the stomach (pH~2). Additionally, the method employs zinc acetate to precipitate certain iron cyanide species, which are then not measured. As with the total cyanide method, it also employs boiling of the samples under a vacuum.

The Amenable Cyanide Method, which was developed for wastewater samples, is a method in which one-half of the sample is analyzed for total cyanide. The other half is chlorinated to destroy any free cyanide or other cyanide species that are able to react with the chlorination agent, and then subjected to the total cyanide method. Amenable cyanide is defined as the difference between these two measurements. Because of the heterogeneity of waste site samples, erroneous positive and even negative amenable cyanide results can be reported that are purely artifactual. Plus, as with the other methods, the samples are boiled under vacuum for the designated time.

None of the above methods are relevant to human health risk assessment. Thus, there is a great need for a method that uses conditions that more closely mimic the human stomach. Such a method would yield cyanide concentrations that are directly useful in risk assessment. In addition, there is a need for a method that measures the amount of cyanide that might be absorbed in the gastrointestinal tract without knowledge of the exact composition of the cyanide-containing sample. Such a method would be equally applicable to MGP waste, to metal electroplating waste, and waste of unknown origin or from mixed sources.

As will be shown below, Prussian Blue is practically nontoxic to animals and humans. If it were known with certainty that manufactured gas plant (MGP) waste contained Prussian Blue as the only cyanide-containing compound, there would perhaps not be a great need for a physiologically available cyanide (PAC) method. However, other forms of cyanide are possibly present in such wastes, including ammonium ferric ferrocyanide, simple ferrocyanide ($Fe(CN)_6^{-4}$), free cyanide (CN^-), thiocyanate (SCN^-), and other, unidentified, species (Edison Electric Institute, 1984). Such species might have been initially formed during purification of the gas or they might have formed due to degradation of complex compounds, such as by the action of ultraviolet light.

Theis et al. (1994) recently demonstrated that most of the cyanide in samples of MGP waste was iron-complexed rather than free, but their methods were unable to distinguish between ferrocyanide, ferric ferrocyanide (insoluble Prussian Blue), potassium ferric ferrocyanide (colloidally soluble Prussian Blue), or ammonium ferric ferrocyanide. No methods are available that can do so, and the toxicity and gastrointestinal release of free cyanide from these species may differ. However, the PAC method is not dependent on the actual form of cyanide present. If the species can release free cyanide, forming HCN under the conditions of the assay, the available cyanide will be measured and will be useable in a human health risk assessment.

BACKGROUND

Toxicity of Prussian Blue

Because some forms of cyanide can be very toxic in certain circumstances, it is important to understand the differences between free cyanide compounds, which can be toxic, and iron-complexed cyanide compounds, such as Prussian Blue, which are not. Free cyanide is defined as hydrocyanic acid (HCN, or "cyanide gas"), the cyanide anion (CN^-) or simple salts of the cyanide anion (e.g., NaCN, KCN), which freely form CN^- in water.

Free cyanide when inhaled or ingested can cause adverse effects in humans if the amount taken in is greater than the amount that the body's detoxification system can handle. If the rate of exposure is low, even the free cyanide is not toxic. The body's enzyme system (called rhodenase) can detoxify the free cyanide and convert it to a nontoxic chemical that is rapidly eliminated by the kidneys (Ballantyne, 1987). For instance, many foods, such as lima beans, contain cyanide-containing chemicals. Small amounts of free cyanide can be formed in the body when people consume these foods, but the free cyanide is rapidly detoxified, and no toxic effects occur (Winkler, 1958; Montgomery, 1980).

Another example of a situation in which free cyanide does not cause adverse effects is use of the blood pressure reducing drug, nitroprusside. This chemical is a commonly prescribed drug that breaks down to free cyanide in the bloodstream. At the doses used by medical doctors, the body efficiently detoxifies the free cyanide, and no adverse effects occur (Vesey, 1987).

In summary, free, uncomplexed cyanide (HCN or the CN anion, CN^-) has the potential to cause adverse effects when present at high dose rates that exceed the body's detoxification enzymes. However, Prussian Blue and other iron-complexed cyanide chemicals have almost no potential to cause toxic effects, because the cyanide is bound so tightly to the iron in the complex that it cannot even be absorbed into the body, let alone cause effects, as is described below. In fact, Prussian Blue is used as an antidote for thallium poisoning and radionuclide poisoning.

Thallium is a toxic metal that has been used in the past as a pesticide. It is still used in some less-developed countries. Prussian Blue was discovered to speed the removal of thallium from the body by binding to the thallium in the gastrointestinal tract. In studying the mechanism of this beneficial use of Prussian Blue, it was found that the Prussian Blue was not appreciably absorbed into the systemic circulation from the gastrointestinal tract; it did not release significant amounts of free cyanide; and it did not cause any adverse effects. Prussian Blue is now considered the antidote of choice for thallium poisoning (Moeschlin, 1980; Thompson, 1981; Saddique and Peterson, 1983).

Because of its favorable properties, Prussian Blue is also considered the antidote of choice for radionuclide poisoning. Radionuclides such as cesium-137 are extremely toxic because they are only very slowly excreted from the body once incorporated into the systemic circulation. Prussian Blue was discovered to speed the removal of such radioactive metals from the body by binding to them in the gastrointestinal tract.

For instance, the International Atomic Energy Agency (Melo et al., 1994) recommends the use of Prussian Blue in such cases. Also, the National Committee on Radiation Protection (NCRP, 1979) has stated that the clinical dose is "1 gram given orally with water 3 times daily for up to 3 weeks or longer as required." The Prussian Blue is well tolerated because it is not absorbed from the gastrointestinal tract and has low toxicity. It was successfully used to treat many people and animals after the Russian nuclear power plant accident in Chernobyl. In addition, many people were treated with Prussian Blue after an accident several years ago in Goiania, Brazil caused 140 people to be exposed to radioactive cesium. In studying the mechanism of this clinical use of Prussian Blue it has been found that the Prussian Blue was not appreciably absorbed into the systemic circulation from the gastrointestinal tract; it did not release significant amounts of free cyanide; and it did not cause any adverse effects.

For instance, in 16 studies in humans, doses ranging from 43 mg/kg/day to 286 mg/kg/day were given for periods of 1 day to 126 days with no toxic effects reported (Heath et al., 1983; Meggs et al., 1994; Pai, 1987; Stevens et al., 1974; Villanueva et al., 1990; Kamerbeek, 1971; Madshus et al., 1966; Stromme, 1968; Madshus and Stromme, 1968; Simonovic et al., 1984; Simonovic et al., 1986; Kostial et al., 1987; Ming-hua et al., 1988; Farina et al., 1991; Lipszstein et al., 1991; and Melo et al., 1994.) These studies described the cases of 106 people given Prussian Blue therapeutically. Of that number, 76 people received large total doses of Prussian Blue over the course of 10 to 126 days. Total doses per person exceeded 2,000 mg/kg. The average total dose for these 76 people exceeded 4,000 mg/kg.

Prussian Blue has also been extensively studied in laboratory animals. In 15 studies with rats and mice, doses of Prussian Blue ranging from 100 mg/kg/day to 7,100 mg/kg/day were given orally for periods of time as great as 120 days with no toxic effects reported (Heydlauf, 1969; Lehmann and Favari, 1985; Rauws, 1974; Rios and Monroy-Noyola, 1992; Kamerbeek, 1971; Nigrovic, 1963, 1965; Nigrovic et al., 1966; Richmond and Bunde, 1966; Iinuma et al., 1971; Kostial et al., 1980, 1981, 1982; Kargacin et al., 1985.) Total doses of Prussian Blue ranged from 400 mg/kg to 192,000 mg/kg in the 15 studies, with the average total dose being 49,000 mg/kg.

In three studies with sheep, doses of Prussian Blue ranging from 6 mg/kg/day to 286 mg/kg/day were given orally for periods of 15 to 23 days with no toxic effects reported (Ioannides et al., 1991; Pearce, 1994; Giese, 1988). Total dose ranged from 91 mg/kg to 4,300 mg/kg, with the average dose being 2,000 mg/kg.

In three studies with cows, Prussian Blue doses of 5 mg/kg/day to 33 mg/kg/day were given orally for periods of 15 to 30 days with no toxic effects reported (Unsworth et al., 1989; Arnaud et al., 1988; Giese, 1988). Total dose ranged from 100 mg/kg to 500 mg/kg, with the average total dose of 250 mg/kg.

GASTROINTESTINAL ABSORPTION OF PRUSSIAN BLUE

The issue of gastrointestinal absorption of iron-complexed cyanide compounds is complicated, because several processes that are relevant to human health risk assessment could occur. If the intact complex is absorbed, it might dissociate systemically to hexacyanoferrate. The hexacyanoferrate might be excreted intact, or it might dissociate to form free cyanide. Alternatively, the complex might dissociate in the gastrointestinal tract to form hexacyanoferrate. This species could be absorbed. Once absorbed, it might be excreted intact or dissociated systemically to form free cyanide. Finally, any hexacyanoferrate that might be formed in the gastrointestinal tract might dissociate there to free cyanide. Once formed, the free cyanide could be absorbed.

To properly evaluate the various metabolic studies that have been performed on Prussian Blue salts, the pharmacokinetic disposition of certain cyanide containing species must be known. Absorbed free cyanide is excreted in the urine as thiocyanate (ATSDR, 1995). Absorbed hexacyanoferrate (ferrocyanide) has been shown to be excreted in the urine in rats, dogs, and humans (Dvorak et al., 1971; Kleeman et al., 1955; Miller and Winkler, 1936). Because of its stability and rapid excretion, ferrocyanide has been used clinically to determine pharmacokinetic parameters in humans and experimental animals. Fecal excretion of these chemicals is not quantitatively important, and any chemical found in the feces after oral dosing experiments can be assumed to be chemical that passed through the gastrointestinal tract unabsorbed.

For instance, Dvorak et al. (1971) administered ^{59}Fe-labeled potassium ferrocyanide intravenously to rats and found that 76% of the radiolabel was detected in the urine within 24 hours. Because unbound iron is not excreted in the urine (Dagg et al., 1966) and because the dissociation constant for ferrocyanide is small (10^{-35} M) (Nielsen et al., 1990), the result indicates that 76% of intact hexacyanoferrate was excreted in the urine.

Kleeman et al. (1955) also found that ferrocyanide was not degraded in the blood stream to available cyanide. Dogs were given 100 mg/kg of ferrocyanide intravenously. In five studies, the recovery of ferrocyanide in the urine at 24 hours was 94 to 99.7%. Urinary excretion was so rapid that at least 80% of the ferrocyanide was excreted in the first 3 hours.

Dvorak et al. (1971) also showed in rats that small amounts of hexacyanoferrate can be absorbed in the gastrointestinal tract. Four groups of rats, each with six animals, were administered ^{59}Fe-labeled potassium ferrocyanide orally in doses ranging from 0.5 to 50 mg/animal. On average, 2.06% of the radiolabel was found in the urine after 24 hours, and 0.06% of the dose was found in the tissues. This implies that 2.1% of the hexacyanoferrate was absorbed.

In addition, Dvorak et al. (1971) studied the absorption of colloidally soluble Prussian Blue given intravenously and orally in rats. In the intravenous experiments, very little of the ^{59}Fe label at the ferric position was detected in the urine after 24 hours (0.11%). However, 54.1% of the ^{59}Fe label in the hexacyanoferrate group was detected in the urine after 24 hours. These results indicate that once present in the systemic circulation, colloidally soluble Prussian Blue can dissociate to ferric ion and hexacyanoferrate ion.

When given orally to two groups of six rats, little absorption of colloidally soluble Prussian Blue (40 mg/animal) was demonstrated. After 24 hours, only 0.2% of the ^{59}Fe in the ferric position was absorbed. All of this amount was found in the tissues. After 24 hours, only 0.14% of the ^{59}Fe in the hexacyanoferrate group was absorbed, with 99% of this amount found in the urine. Although no evidence was found that hexacyanoferrate degraded to form free cyanide, estimates of the amount of free cyanide that could have been released in the body can be made. Formation of free cyanide was 0.14% if all of the hexacyanoferrate dissociated.

Intestinal absorption of Prussian Blue was studied in piglets by Nielsen et al. (1988). Colloidally soluble Prussian Blue ($KFe^{III}[Fe^{II}(CN)_6]$) and insoluble Prussian Blue ($Fe^{III}_4[Fe^{II}(CN)_6]_3$) radiolabeled with ^{59}Fe in the hexacyanoferrate group were given by gastric intubation to male and female piglets (6.9 kg) of the "Deutsche Landrasse" strain. Absorption was analyzed by measuring whole body retention of the iron label 14 days after dosing.

In 11 animals given colloidally soluble Prussian Blue and 11 animals given insoluble Prussian Blue, only 0.2% of the iron label was absorbed. It is not known if the hexacyanoferrate ion was absorbed or if it degraded to free iron, which was absorbed, and free cyanide. If the latter occurred, free cyanide could also have been absorbed. Thus, maximum absorption of cyanide is estimated as 0.2% in 22 animals.

In another experiment, four animals were given either the iron-labeled colloidally soluble or insoluble Prussian Blue, and urine and feces were analyzed for seven days in addition to whole body counting of the label. Similar results were seen. The fraction of the administered dose that was present in the urine or the body was 0.7% for the soluble Prussian Blue and 0.2% for the insoluble Prussian Blue. Maximum absorption of cyanide is estimated as 0.7% and 0.2%, respectively.

In addition, two animals received insoluble Prussian Blue labeled with ^{14}C. No ^{14}C was detected in CO_2 measured over three hours after dosing. Although of short duration, this experiment supports the conclusion that little free cyanide was released or formed.

Nielsen et al. (1990) studied the gastrointestinal absorption of colloidally soluble Prussian Blue (K Fe^{III}[Fe^{II}(CN)$_6$) in humans by monitoring radiolabeled iron in the whole body and in the urine and feces and by monitoring radiolabeled carbon in expired air and in urine. Three volunteers were given 500 mg of the soluble Prussian Blue labeled with ^{59}Fe in either the ferric or ferrous position and with ^{14}C in the cyanide group.

The radiolabeled iron was completely recovered after 7 days (99.5 to 99.9%). 99.7% of the $^{59}Fe^{II}$ in the hexacyanoferrate group was recovered in the feces within 7 days. 99.4% of the $^{59}Fe^{III}$ was recovered in the feces within 7 days. The amount recovered in the urine after 7 days and retained in the body after 7 days was assumed to be absorbed through the gastrointestinal tract. Also, 0.22% of the $^{59}Fe^{II}$ in the hexacyanoferrate group was absorbed, and 0.04% of the $^{59}Fe^{III}$ was absorbed.

Absorption of the ^{14}C was 0.25 to 0.42% as determined by the amount of radiolabel in the urine over 7 days. In addition, a small amount (0.01 to 0.02%) of the oral ^{14}C dose was exhaled in the breath as $^{14}CO_2$ within 2 hours of dosing. Because absorption of radiolabeled carbon was greater than absorption of radiolabeled iron in the hexacyanoferrate group, it is possible that a fraction of the absorbed Prussian Blue or Prussian Blue-derived hexacyanoferrate group dissociated releasing free cyanide, either in the gastrointestinal tract or in the systemic circulation.

Because the urinary excretion of unbound iron is negligible (Dagg et al., 1966), and Dvorak et al. (1971) and Nielsen et al. (1990) showed that radiolabeled iron in the ferric position was not excreted in the urine, urinary ferrous iron was presumed to represent intact hexacyanoferrate. Thus, the difference between the urinary ^{14}C (0.42%) and ^{59}Fe (0.15%) provides an estimate of urinary excretion of free cyanide (probably as thiocyanate) (0.27%). According to Nielsen et al. (1990), 70% of systemic free cyanide is excreted in urine as thiocyanate. Thus, an estimate of the amount of free cyanide formed from the colloidally soluble Prussian Blue either in the gastrointestinal tract or in the systemic circulation is 0.27%/0.70 = 0.39% of administered dose.

Arnaud et al. (1988) administered ^{14}C-labeled ammonium ferric ferrocyanide (AFCF) orally to two dairy cows. Radioactivity was measured in urine, feces, blood, milk, and expired air for nine days. No HCN was detected in expired air. Most of the radioactivity was recovered in the feces within three days (>90%). This indicates low gastrointestinal absorption, because absorbed free cyanide and ferrocyanide are principally excreted in the urine.

The total amount of radioactivity excreted in the urine and milk was 0.5% of administered dose for animal one and 0.3% of administered dose for animal two.

These amounts represent estimates of gastrointestinal absorption of cyanide either as AFCF, hexacyanoferrate that dissociated from the AFCF in the rumen, or free cyanide that formed from dissociation of hexacyanoferrate. The results of Dvorak et al. (1971) suggest that it was principally intact hexacyanoferrate. In fact, the authors also showed by chromatographic means that the urinary radioactivity was comprised of 89% hexacyanoferrate and 11% thiocyanate.

It can be assumed that the urinary thiocyanate (0.11 x 0.47% = 0.052 % of oral dose) was formed from free cyanide. According to Nielsen et al. (1990), 70% of systemic free cyanide is excreted in urine as thiocyanate. Thus, an estimate of the amount of free cyanide formed from the Prussian Blue either in the gastrointestinal tract or in the systemic circulation was 0.07% of administered dose. If the radioactivity in the milk is assumed to be 100% free cyanide or thiocyanate, then the estimate of free cyanide formed is 0.14%.

In conclusion, several in vivo metabolic experiments have been performed using radioactive iron as a tracer. These experiments demonstrate that little (<1%) of the iron in iron-complexed cyanide compounds is absorbed when the compounds are administered orally to rats, pigs, and humans. Because iron ions are nutritionally required, gastrointestinal absorption is efficient, and it is anticipated that iron released by the breakdown of an iron-complexed cyanide species would be quantitatively absorbed. Thus, the lack of significant absorption of iron suggests that cyanide complexes are not absorbed and do not dissociate in the gastrointestinal tract to form free cyanide which could then be absorbed. More compelling experimental results come from experiments in which radiolabeled carbon is used as a tracer. Radiocarbon experiments with cows and humans agree with the radioiron experiments, demonstrating that formation of free cyanide is <1% of administered dose.

MATERIALS AND METHODS

The Physiologically Available Cyanide (PAC) Method is designed to measure the amount of "available" cyanide, the amount of cyanide "available" to be absorbed in the stomach upon ingestion of cyanide containing soil or waste material. Ideally, an analytical method designed to mimic the conditions of the human stomach would use the temperature and pH of the stomach as well as the time that an ingested soil or waste sample would be expected to be present there.

Physiological temperature is 37° Celsius. Gastric pH varies over the course of a day, but it is approximately 2.0 before and after meals. Directly after meals, pH rises for a short period to approximately 5, but it stabilizes after 2-3 hours at approximately 2.0. (Dotevall, 1961; Ovesen, et al.; 1985; Malagelada et al., 1976; Hannibal and Rune, 1983; Dressman et al., 1990).

Site-derived soil or waste material, if ingested, would have a stomach residence time ranging from one to four hours (Malagelada et al., 1976, 1977; Granger et al., 1985; Youngberg et al., 1987). Several samples of various cyanide-containing media were analyzed at pH 2 for four hours as will be discussed, but it was determined that four hours was an excessive time period for a routine analytical method. Thus the PAC method was designed to be executed at elevated temperature for a shorter period of time.

Analyses were performed on sieved (#10), one-gram, wet weight samples of cyanide-containing soil from a former manufactured gas plant. The equipment setup was as required in the EPA total cyanide method (Method 335.3). The vacuum flow rate was maintained at 290 to 350 mL/minute. Temperature was controlled at 72 to 78°C for one hour by a thermostatically-controlled constant temperature bath. In several analyses, the temperature was controlled by voltage-regulated heating blocks.

The buffering medium was a phosphate buffer titrated to a pH of 2.0. Magnesium chloride was added as specified in the EPA total cyanide method. Cyanide was collected in sodium hydroxide in the absorber tube to which cadmium carbonate was added to prevent sulfide interference. Cyanide analyses were performed colorimetrically by the autoanalyzer method following the methodologies and requirements of EPA 335.3 and the CLP SOW 788 or the titrimetric method specified in Standard Method 4500-CN D (APHA, 1992). In each batch of samples, a laboratory control sample of aqueous potassium cyanide was shown to have recovery greater than 80%, and a laboratory control standard of Prussian Blue was shown to have a recovery of less than 10%.

EXPERIMENTAL RESULTS

Free Cyanide Recovery at 37° for Four Hours

Shown in Table 1 are the results of free cyanide released from various cyanide-containing samples in a pH 2 buffer over four hours at physiological temperature (37°C). Potassium cyanide (KCN) is a simple cyanide salt that is classified as "free" or "available" cyanide. It was mixed with 1 gram of laboratory-prepared soil consisting of 50% fuller's earth and 50% commercially available potting soil. Recovery of "free" cyanide was generally high compared to recovery of complexed cyanide, but it was not completely recovered as it would be if the samples had been boiled for one hour. Thus, even "free" cyanide may not be completely converted to HCN and absorbed in the human stomach.

Prussian Blue (77% purity) was insoluble ferric ferrocyanide $(Fe^{III}_4[Fe^{II}(CN)_6]_3)$ obtained from Aldrich Chemical Company. Less than 1% was recovered in all experiments, with the average recovery being 0.19% for 11 samples analyzed on four different days. Percent recovery was calculated based on the assumption that the standard compound was 77% pure and that the remainder of the weight did not contain any cyanide species. If, as is likely, the impurities contained simple ferrocyanide or even free cyanide, then the actual recoveries would be less than those reported in Table 1.

MGP-contaminated soil was obtained from a site in Massachusetts. Less than 0.05% of the total cyanide was recovered in all cases, with the average recovery

Table 1.   Free Cyanide Recovery After Four Hours At 37° from pH 2 Buffer

Sample	Range in the Total Amount of Cyanide Present (mg)	Mean Percent Recovery	Range of Percent Recovery	Number of Replicates
KCN in Soil	0.5 - 5.0	61.5	35 - 85	11
Prussian Blue[a]	0.7 - 7.5	0.19	0.01 - 0.80	11
MGP-Contaminated Soil	3.6 - 4.5 [b]	0.03	0.02 - 0.04	5

[a] Aldrich Chemical Company, 77% purity, assumes remainder contains no CN.

[b] After the four-hour incubation, the absorber tube was replaced and the sample was boiled for 60 minutes. Total cyanide was defined as the sum of the cyanide found in the two absorber tubes.

being 0.03% for five samples analyzed on two different days. Each sample was incubated for four hours at 37% in the pH 2 buffer. Then, the absorber tube was replaced, and the sample was boiled for 60 minutes. Total cyanide was defined as the amount recovered in the two absorber tubes. If more cyanide had been recovered from a total cyanide analysis using concentrated sulfuric acid in lieu of the pH 2 buffer, as is likely, then the actual recoveries over the four hour period would be less than those reported in Table 1.

The results presented here for Prussian Blue should be compared to the results from several similar experiments described in the literature with various iron-complexed cyanide species that are collectively known as Prussian Blue salts. First, Kamerbeek (1971) incubated colloidally soluble Prussian Blue (potassium ferrihexacyanoferrate: $KFe^{III}[Fe^{II}(CN)_6]$) for four hours with gastric juice (pH 2) and 0.1N HCl in the presence of oxygen. No release of free cyanide was detected. Because this was an unpublished experiment that was briefly cited by Stevens et al. (1974), no details of the study design are available, such as the minimum detection limit. Nonetheless, the results reported here for insoluble Prussian Blue are consistent with those found by Kamerbeek (1971) for colloidally soluble Prussian Blue.

In addition, several experiments were performed by Verzijl et al. (1993). These investigators measured the release of free cyanide in vitro in artificial gastric and intestinal fluids from three Prussian Blue salts that are used therapeutically as antidotes to radiocesium poisonings. The three salts were colloidally soluble Prussian Blue ($KFe^{III}[Fe^{II}(CN)_6]$), insoluble Prussian Blue ($Fe^{III}_4[Fe^{II}(CN)_6]_3$), and ammonium ferric hexacyanoferrate ($NH_4Fe^{III}[Fe^{II}(CN)_6]$).

The release of cyanide over a five-hour period at 37° was measured for the test compounds in distilled water, simulated gastric juice (pH 1.2) containing HCl, NaCl, and pepsinum, and simulated intestinal fluid (pH 6.8) containing tris-(hydroxymethyl)-aminomethane, HCl, and pancreatin.

Cyanide release was measured by the evolution of HCN from different Prussian Blue salts in a special device. The liberated HCN was quantitatively determined by its reaction with picric acid and formation of a stable colored complex that was analyzed spectrophotometrically. The method of standard addition using NaCN standards was used for quantitation.

The results of Verzijl et al. (1993) cannot be directly compared to the results presented here because the methods differ significantly. The results presented here for insoluble Prussian Blue are for four-hour incubations of 1.2 mg Prussian Blue at pH 2.0 under reduced pressure to allow for collection of released HCN(g) in the NaOH absorber tube. Verzijl et al. (1993) used samples of 100 mg, a passive collection device, and a pH of 1.2. Nonetheless, the fraction released over four hours under gastric conditions is 0.009% compared to the value of 0.19% presented here. In both cases, release of free cyanide is much less than 1%.

Verzijl et al. (1993) also measured the in vitro gastric release of free cyanide from soluble Prussian Blue (0.04% over four hours) and ammonium ferric ferrocyanide (0.06% over four hours).

The results presented in Table 1 estimate gastric release of free cyanide. However, it is possible that there is a measurable release of free cyanide in the intestine that is not quantitated by use of a pH 2 buffer. In fact, Verzijl et al. (1993) have measured the rate of release of free cyanide in simulated intestinal fluid. For all Prussian Blue salts, the rate of release of free cyanide under gastric conditions exceeded that under intestinal conditions by factors of 4- to 10-fold. The authors' estimates of release of free cyanide assuming 1 hour of residence in the stomach and 23 hours residence in the intestines was 0.02% for insoluble Prussian Blue, 0.03% for colloidally soluble Prussian Blue, and 0.07% for ammonium ferric ferrocyanide.

Free Cyanide Recovery with the Physiologically Available Cyanide (PAC) Method

The PAC method is a performance-based method. The exact temperature and vacuum flow conditions are suggested, but not specified, because of the differences in laboratory equipment. Accompanying each batch of samples, a KCN laboratory control standard and a Prussian Blue laboratory control standard must be run. Recovery of the KCN over one hour must exceed 80% and recovery of the Prussian Blue must be *less than* 10%. In one laboratory, this was achieved at a temperature of 72° and a flow rate of 300-350 mL/min. In another laboratory, this was achieved at a temperature of 78° and a flow rate of ~290 mL/min.

Shown in Table 2 are the results of PAC analyses of insoluble Prussian Blue standards in three laboratories. Mean recoveries ranged from 2.5% to 4.1%. Shown in Table 3 are the results of PAC analyses by three laboratories of soils containing MGP residuals from three sites. Mean PAC recoveries ranged from 4% to 14% of total cyanide.

Table 2.   Cyanide Recovery From Prussian Blue, Physiologically Available Cyanide (PAC) Method

Sample	Total Amount of Cyanide Present (mg)	Mean Percent Recovery by PAC	Range of Percent Recovery by PAC	Number of Replicates
Lab 1	0.5	2.7	1.3 - 8.9	9
Lab 2	5	4.1	3.5 - 4.8	3
Lab 3	0.01	2.5	2.5 - 3.0	4

Table 3.   Cyanide Recovery From MGP-Contaminated Soil

Sample	Range in Total Amount of Cyanide Present (mg)	Mean Percent Recovery by PAC	Range of Percent Recovery by PAC	Number of Replicates or Samples
Lab 1, Site 1	1.0 - 2.3 a	4	2 - 5	3c
Lab 2, Site 1	1.9 - 3.6 a	13	6 - 16	6c
Lab 2, Site 1	4.1 - 4.6 a	14	13 - 16	2c
Lab 3, Site 2	0.2 - 12 b	5	1 - 11	6d
Lab 3, Site 3	0.005 - 0.02 b	7	3 - 10	4d

[a] After PAC analyses the absorber tube was replaced and the sample was boiled for 60 minutes. Total cyanide was defined as the sum of the cyanide found in the two absorber tubes.

[b] A separate subsample was analyzed for total cyanide.

[c] Replicates of a single sample.

[d] Samples from different site areas.

By design, this method overestimates the amount of cyanide released under gastric pH conditions and physiological temperature. This is because the method was designed to be performed for one hour under conditions that would allow recovery of <80% of a free cyanide standard, potassium cyanide. However, as noted above, the recovery of potassium cyanide from a four-hour, pH 2 incubation at physiological temperature is typically less than 80%. The conditions of the PAC method were defined with a margin of safety so that one would not underestimate the presence of free cyanide in a sample at the expense of overestimating the amount of physiologically available cyanide from iron-complexed cyanide compounds. As shown in Table 4, release of free cyanide from insoluble Prussian Blue was less than 0.2% in in vitro and in vivo studies. However, release of free cyanide from insoluble Prussian Blue with the PAC method was greater than 2.5%. Release of free cyanide from all iron-complexed compounds was less than 0.7% in in vitro and in vivo studies. Thus, the PAC method may overestimate the amount of release of free cyanide from iron-complexed cyanide compounds by as much as a factor of ten. However, the use of the PAC method is health-protective and ensures that any free cyanide present in a waste sample will be properly detected and included in the human health risk assessment.

Table 4. Comparison on In Vitro and In Vivo Estimates of Free Cyanide Release

Cyanide Species	Conditions	Fraction Of Cyanide Released	Citation
Insoluble Prussian Blue	In vitro simulated gastric, 4 hours	0.19%	This report
MGP soils	In vitro simulated gastric, 4 hours	0.03%	This report
Colloidally soluble Prussian Blue	In vitro simulated gastric, 4 hours	none detected	Kamerbeek (1971)
Insoluble Prussian Blue	In vitro simulated gastric, 4 hours	0.01%	Verzijl et al. (1993)
Colloidally soluble Prussian Blue	In vitro simulated gastric, 4 hours	0.04%	Verzijl et al. (1993)
Ammonium ferric ferrocyanide	In vitro simulated gastric, 4 hours	0.06%	Verzijl et al. (1993)
Insoluble Prussian Blue	In vitro simulated gastrointestinal, 24 hours	0.02%	Verzijl et al. (1993)
Colloidally soluble Prussian Blue	In vitro simulated gastrointestinal, 24 hours	0.03%	Verzijl et al. (1993)
Ammonium ferric ferrocyanide	In vitro simulated gastrointestinal, 24 hours	0.07%	Verzijl et al. (1993)
Colloidally soluble Prussian Blue	In vivo absorption, humans	0.39%	Nielsen et al. (1990)
Ammonium ferric ferrocyanide	In vivo absorption, cows	0.14%	Arnaud et al. (1988)
Colloidally soluble Prussian Blue	In vivo absorption, rats	0.14%	Dvorak et al. (1971)
Colloidally soluble Prussian Blue	In vivo absorption, pigs	0.7%	Nielsen et al. (1988)
Insoluble Prussian Blue	In vivo absorption, pigs	0.2%	Nielsen et al. (1988)

CONCLUSION

The Physiologically Available Cyanide (PAC) method is a simple method that can be routinely applied to field samples containing cyanide. The method is preferred over traditional methods because it yields a measurement of the amount of cyanide in a soil or waste sample that is physiologically relevant. Method results can be directly used in human health risk assessment with toxicity benchmarks based on "free" or "available" cyanide.

The PAC method yields results for authentic field samples of manufactured gas plant waste that are, on average, 4% to 14% of the results of the Total Cyanide Method. Actual results vary from sample to sample and site to site, because the fraction of cyanide that is physiologically available varies due to differences in historical production methods, differential weathering, and other factors.

Use of the method allows risk assessors and regulatory agencies to consider the large database that demonstrates the low order of toxicity of iron-complexed cyanides that are commonly found at MGP sites and also allows them to ensure that site samples with free cyanide are properly identified and included in risk-based decision making.

REFERENCES

APHA. 1992. *Standard Methods for the Examination of Water and Wastewater*, 18th Edition. (American Public Health Association, American Waterworks Association, Water Environment Federation: Washington, DC).

Arnaud, M.J., Clement, C., Getaz, F., Tannhauser, F., Schoenegge, R., Blum, J., and Giese, W. 1988. Synthesis, Effectiveness, and Metabolic Fate in Cows of the Caesium Complexing Compound Ammonium Hexacyanoferrate Labelled with ^{14}C. *J. Dairy Sci. 55, 1-13.*

ATSDR. 1995. Toxicological Profile for Cyanide. (U.S. Department of Health and Human Services, Agency for Toxic Substances and Disease Registry: Atlanta, GA).

Ballantyne, B. 1987. Toxicology of Cyanides. In: *Clinical and Experimental Toxicology of Cyanides* (Ballantyne, B. and Marrs, T.C., Eds.). Wright Publishers, Bristol England.

Dagg, J.H., Smith, J.A., and Goldberg, A. 1966. Urinary Excretion of Iron. *Clinical Sci. 30, 495-503.*

Dotevall, G. 1961. Gastric Secretion of Acid in Diabetes Mellitus during Basal Conditions and after Maximal Histamine Stimulation. *Acta Medica Scandinavica 170, 59-69.*

Dressman, J.B., Bernardi, R.R., Dermentzoglou, L.C., Russel, T.L., Schmaltz, S.P., Barnett, J.L., and Jarvenpaa, K.M. 1990. Upper Gastrointestinal (GI) pH in Young, Healthy Men and Women. *Pharma. Res. 7, 759.*

Dvorak, P., Gunther, M., Zorn, U., and Catsch, A. 1971. Metabolisches Verhalten con Kolloidalem Ferrihexacyanoferrat(II). *Naunyn-Schmiedebergs Arch. Pharmak.* 269, 48-56.

Edison Electric Institute. 1984. Handbook on Manufactured Gas Plant Sites. Prepared for Utility Solid Waste Activities Group by Environmental Research & Technology, Inc. and Koppers Company.

Farina, R., Brandao-Mello, C.E., and Oliveira, A.R. 1991. Medical Aspects of ^{137}Cs Decorporation: The Goiania Radiological Accident. *Health Phys.* 60, 63-66.

Giese, W.W. 1988. Ammonium-Ferric-Cyano-Ferrate(II) (AFCF) as an Effective Antidote Against Radiocaesium Burdens in Domestic Animals and Animal Derived Foods. *Br. Vet. J.* 144, 363-369.

Granger, D.N., Barrowman, J.A., and Kvietys, P.R. 1985. *Clinical Gastrointestinal Physiology.* Philadelphia, W.B. Saunders Company.

Hannibal S. and Rune, S.J. 1983. Duodenal Bulb pH in Normal Subjects. *Eur. J. Clinical Invest.* 13, 455-460.

Heath, A., Ahlmen, J., Branegard, B., Lindstedt, S., Wickstrom, I., and Andersen, O. 1983. Thallium Poisoning- Toxin Elimination and Therapy in Three Cases. *J. Toxicol.-Clin. Toxicol.* 20, 451-463.

Heydlauf, H. 1969. Ferric-Cyanoferrrate(II): Am Effective Antidote in Thallium Poisoning. *Eur. J. Pharmacol.* 6, 340-344.

Iinuma, T.A., Izawa, M., Watari, K., Enomoto, Y., Matsusaka, N., Inaba, J., Kasuga, T., and Nagai, T. 1971. Application of Metal Ferrocyanide-Anion Exchange Resin to the Enhancement of Elimination of ^{137}Cs from Human Body. *Health Phys.* 20, 11-21.

Ioannides, K.G., Kamtzios, A.S., and Pappas, C.P. 1991. Influence of Prussian Blue in Reducing Transfer of Radiocesium into Ovine Milk. *Health Phys.* 60, 261-264.

Kamerbeek, H.H. 1971. Therapeutic Problems in Thallium Poisoning. *Proefschrift Utrecht,* Tilburg, Gianotten, as cited in Stevens et al., 1974.

Kargacin, B., Maljkovcic, T., Mlanusa, B., and Kostial, K. 1985. The Influence of a Composite Treatment for Internal Contamination by Several Radionuclided on Certain Health Parameters in Rats. *Arh. hig. rada toksikol.* 36, 165-172.

Kleeman, C.R., F.H. Epstein, M.E. Rubini and E. Lamdin. 1955. Initial Distribution and Fate of Ferrocyanide in Dogs. *Am. J. Physiol.* 182, 548-552.

Kostial, K., Vnucec, M., Tominac, C., and Simonovic, I. 1980. A Method for a Simultaneous Decrease of Strontium, Caesium and Iodine Retention After Oral Exposure in Rats. *Int. J. Radiat. Biol.* 37, 347-350.

Kostial, K., Kargacin, B., Rabar, I., Blanusa, M., Maljkovic, T., Matkovic, V., and Ciganovic, M. 1981. Simultaneous Reduction of Radioactive Strontium, Caesium and Iodine Retention by Single Treatment in Rats. *Sci. Total Environ.* 22, 1-10.

Kostial, K., Kargacin, B., and Rabar, I. 1982. The Effect of Chemotherapy for Mixed Fission Products on the Toxicokinetics of Cadmium and Mercury in Rats. *Sci. Total Environ.* 24, 287-289.

Kostial, K., Kargacin, B., and Simonovic, I. 1987. Reduced Radiostrontium Absorption in a Human Subject Treated with Composite Treatment for Mixed Fission Product Contamination. *Health Phys.* 52, 371-372.

Lehmann, P.A. and Favari, L. 1985. Acute Thallium Intoxication: Kinetic Study of the Relative Efficacy of Several Anitodal Treatments in Rats. *Arch. Toxicol.* 57, 56-60.

Lipsztein, J.L., Bertelli, L., Oliveira, C.A.N., and Dantas, B.M. 1991. Studies of Cs Retention in the Human Body Related to Body Parameters and Prussian Blue Administration. *Health Phys.* 60, 57-61.

Madshus, K. and Stromme, A. 1968. Increased Excretion of ^{137}Cs in Humans by Prussian Blue. *Zeitschrift fur Naturforschung* 23, 391-392.

Madshus, K., Stromme, A., Bohne, F., and Nigrovic, V. 1966. Diminution of Radiocaesium Body-Burden in Dogs and Human Beings by Prussian Blue. *Int. J. Rad. Biol.* 10, 519-520.

Malagelada, J.-R., Longstreth, G.F., Summerskill, W.H.J., and Go, V.L.W. 1976. Measurement of Gastric Functions During Digestion of Ordinary Solid Meals in Man. *Gastroenterol.* 70, 203-210.

Malagelada, J-R., Longstreth, G.F., Deering, T.B., Summerskill, W.H.J., and Go, V.L.W. 1977. Gastric Secretion and Emptying After Ordinary Means in Duodenal Ulcer. *Gastroenterol.* 73, 989-994.

Meggs, W.J., Hoffman, R.S., Shih, R.D., Wesiman, R.S., and Goldfrank, L.R. 1994. Thallium Poisoning from Maliciously Contaminated Food. *Clinical Toxicol.* 32, 723-730.

Melo, D.R., Lipsztein, J.L., de Oliveira, C.A.N., and Bertelli, L. 1994. ^{137}Cs Internal Contamination Involving a Brazilian Accident, and The Efficacy of Prussian Blue Treatment. *Health Phys.* 66, 245-252.

Miller, B.F. and A. Winkler. 1936. The Ferrocyanide Clearance in Man. *J. Clin. Invest.* 15, 489492.

Ming-hua, T., Yi-fen, G., Cheng-yao, S., Chang-ging, Y., and De-chang, W. 1988. Measurement of Internal Contamination with Radioactive Caesium Released from the Chernobyl Accident and Enhanced Elimination by Prussian Blue. *J. Radiol. Prot.* 8, 25-28.

Moeschlin, S. 1980. Thallium Poisoning. *Clinical Toxicol.* 17, 133-146.

Montgomery, R.D. 1980. Cyanogens. In: *Toxic Constituents of Plant Foodstuffs.* New York, Academic Press.

Muller, W.H. 1969. Cs^{137}-Dekorporation mit kolloidal-loslichem Berliner Blau bei der Ratte. *Strahlentherapie.* 137, 705-707.

NCRP. 1979. Management of Persons Accidentally Contaminated with Radionuclides: Recommendations of the National Council on Radiation Protection and Measurements. NCRP Report No.65. National Council on Radiation Protection and Measurements, Washington, DC.

Nielsen, P., Dresow, B., Fischer, R., Gabbe, E.E., Heinrich, H.C., and Pfau, A.A. 1988. Intestinal Absorption of Iron from ^{59}Fe-labelled Hexacyanoferrates(II) in Piglets. *Arzneim.-Forsch/Drug Res.* 38, 1469-1471.

Nielsen, P., B. Dresow, R. Fischer, and H.C. Heinrich. 1990. Bioavailability of iron and cyanide from oral potassium ferric hexacyanoferrate(II) in humans. *Arch. Toxicol.* 64, 420-422.

Nigrovic, V. 1963. Enhancement of the Excretion of Radiocaesium in Rats by Ferric Cyanoferrate (II). *Int. J. Rad. Biol.* 7, 307-309.

Nigrovic, V. 1965. Retention of Radiocesium by the Rat as Influenced by Prussian Blue and Other Compounds. *Phys. Med. Biol.* 10, 81-91.

Nigrovic, V., Bohnr, F., and Madshus, K. 1966. Dekorporation von Radionukliden (Untersuchungen an Radiocaesium). *Stralentherapie* 130, 413-419.

Ovesen, L., Bendtsen, F., Tage-Jensen, U., Pedersen, N.T., Gram, B.R., and Rune, S.J. 1985. Intraluminal pH in the Stomac, Duodenum, and Proximal Jejunum in Normal Subjects and Patients with Exocrine Pancreatic Insufficiency. *Gastroenterol.* 90, 958-962.

Pai, V. 1987. Acute Thallium Poisoning. *W.I. Med. J.* 36, 256-258.

Pearce, J., Unsworth, E.F., McMurray, C.H., Moss, B.W., Logam, E., Rice, D., and Hove, K. 1989. The Effects of Prussian Blue Provided by Indwelling Rumen Boli on the Tissue Retention of Dietary Radiocaesium by Sheep. *Sci. Total Environ.* 85, 349-355.

Pearce, J. 1994. Studies of Any Toxicological Effects of Prussian Blue Compounds in Mammals -A Review. *Fd. Chem. Toxic.* 32, 577-582.

Rauws, A.G. 1974. Thallium Pharmacokinetics and Its Modification by Prussian Blue. *Naunyn-Schmiedeberg's Arch. Pharmacol.* 284, 295-306.

Richmond, C.R. and Bunde, D.E. 1966. Enhancement of Cesium-137 Excretion by Rats Maintained Chronically on Ferric Ferrocyanide. *Proc. Soc. Expt. Biol. Med.* 121, 664-670.

Rios, C. and Monroy-Noyola, A. 1992. D-Penicillamine and Prussian Blue as antidotes against thallium intoxication in rats. *Toxicology* 74, 69-76.

Saddique, A. and Peterson, C.D. 1983. Thallium Poisoning: A Review. *Vet. Hum. Toxicol.* 25, 16-21.

Shifrin, N.S., Beck, B.D., Gauthier, T.D., Chapnick, S.D., and Goodman, G. 1996. Chemistry, Toxicology, and Human Health Risk of Cyanide Compounds in Soils at Former Manufactured Gas Plant Sites. *Regul. Toxicol. Pharmacol.* 23, 106-116.

Simonovic, I., Kargacin, B., and Kostial, K. 1986. The Effect of Composite Oral Treatment for Internal Contamination with Several Radionuclides on ^{131}I Thyroid Uptake in Humans. *J. Appl. Toxicol.* 6, 109-111.

Simonovic, I, Kostial, K., and Kargacin, B. 1984. [131]I Uptake in Human Thyroid after Antidote Treatment for Mixed Fission Products Contamination. *Int. J. Radiat. Biol.* 46, 459-462.

Stevens, W., van Peteghem, C., Heyndrickx, A., and Barbier, F. 1974. Eleven Cases of Thallium Intoxication Treated with Prussian Blue. *Int. J. Clin. Pharmacol.* 10, 1-22.

Stromme, A. 1968. Increased Excretion of ^{137}Cs in Humans by Prussian Blue. In: *Diagnosis and Treatment of Deposited Radionuclides.* (Kornberg, H.A. and Norwood, W.D., Eds.). Amsterdam, Excerpta Medica Foundation.

Theis, T.L., Young, T.C., Huang, M., and Knutsen, K.C. 1994. Leachate Characteristics and Composition of Cyanide-Bearing Wastes from Manufactured Gas Plants. *Environ. Sci. Technol.* 28, 99-106.

Thompson, D.F. 1981. Management of Thallium Poisoning. *Clinical Toxicol.* 18, 979-990.

Unsworth, E.F., Pearce, J., McMurray, C.H., Moss, B.W., Gordon, F.J., and Rice, D. 1989. Investigations of the Use of Clay Minerals and Prussian Blue in Reducing the Transfer of Dietary Radiocaesium to Milk. *Sci. Total Environ.* 85, 339-347.

USEPA. 1993. Wildlife Exposure Factors Handbook. U.S. Environmental Protection Agency. EPA/600/R-93/187a.

Verzijl, J.M., Joore, H.C.A., van Dijk, A., Wierckx, F.C.J., Savelhoul, T.J.F., and Glerum, J.H. 1993. *In Vitro* Cyanide Release of Four Prussian Blue Salts Used for the Treatment of Cesium Contaminated Persons. *Clinical Toxicol.* 31(4), 553-562.

Vesey, C.J. 1987. Nitroprusside Cyanogenesis. In: *Clinical and Experimental Toxicology of Cyanides,* (Ballantyne, B. and Marrs, T.C., Eds.). Wright, Bristol.

Villanueva, E., Hernandez-Cueto, C., Lachica, E., Rodrigo, M.D., and Ramos, V. 1990. Poisoning by Thallium: A Study of Five Cases. *Drug Saf.* 5, 384-389.

Winkler, W.O. 1958. Report on Methods for Glucosidal HCN in Lima Beans. *J.A.O.A.C.* 41, 282-287.

Youngberg, C.A., Berardi, R.R., Howatt, W.F., Hyneck, M.L., Amidon, G.L., Meyer, J.H., and Dressman, J.B. 1987. Comparison of Gastrointestinal pH in Cystic Fibrosis and Health Subjects. *Dig. Dis. Sci.* 32, 472-480.

CHAPTER 43

Probabilistic Model for the Estimation of Costs for the Environmental Management of MGP Sites

Alessandro Battaglia and **Brett Shamory**, Remediation Technologies, Inc., Pittsburgh, Pennsylvania

David Nakles, Remediation Technologies, Inc., Monroeville, Pennsylvania

Thomas Hayes, Gas Research Institute, Chicago, Illinois

INTRODUCTION

Utilities must often estimate the costs of managing their manufactured gas plant (MGP) sites in the absence of significant site data. Activities requiring such cost estimates include:

- utility rate cases;
- financial disclosures;
- insurance settlements;
- cost allocations between responsible parties;
- escrow determination for property transfers; and
- general financial and environmental planning.

These activities often required accurate and defensible cost estimates based on existing records and information, in advance of significant site characterization efforts.

The conventional approach to estimating costs in these cases is to develop a single management strategy, estimate its cost, and present the estimate as a single value or, at most, a range of costs. This approach provides little or no characterization of the inherent uncertainty in the cost estimate. In the absence of an explicit characterization of the uncertainty, decision-makers either intuitively assess the uncertainty or ignore the issue of uncertainty altogether. In either case, the decision-maker is put at risk if the uncharacterized uncertainty is significant.

To overcome the limitations inherent in traditional cost estimation techniques RETEC, on behalf of the Gas Research Institute (GRI), has developed a methodology for estimating the costs of managing MGP sites that utilizes probabilistic techniques and explicitly includes and characterizes the uncertainty associated with all aspects of the management process. A computer program was developed to implement this methodology. The computer program is referred to in the following as the *GRI MGP Cost Model.* The costs considered in this model are those incurred beginning with a preliminary assessment of the site and ending with the implementation of a site remediation strategy.

OVERVIEW OF THE MODEL

There are two basic sources of uncertainty in estimating costs:

- the uncertainty in the management actions which will be required (e.g., will remediation be needed at all? If it is, will it be excavation and disposal? Will it be pump-and-treat? Will it be in-situ bioremediation?); and
- the uncertainty in the cost of each management action, which is related to uncertainties in estimates of quantities and durations (e.g., How many cubic yards of soils? How many years of pump-and-treat?).

Uncertainty in the Sequence of Management Actions

To address the uncertainty in the sequence of management actions and decisions that can be undertaken at an MGP site, the probabilistic technique called decision tree analysis is used in the *GRI MGP Cost Model.* In decision tree analysis, potential sequences of actions for the environmental management of an MGP site are logically ordered and the probabilities and costs (or cost probability distributions) are assigned independently to each action. By appropriately combining the probabilities and costs of each independent action, each sequence of actions is assigned a probability and a cost. In standard decision tree analysis, each action is assigned a single cost value that represents the best available cost estimate for the action. However, no explicit quantification of the uncertainty in the cost value assigned to each action is specified in standard decision tree analysis.

Considering the sequence of environmental management actions relevant to the environmental management of MGP sites, a decision tree can be developed. Figure 1 presents the general structure of such a decision tree used for the spreadsheet model. The decision tree starts with a Preliminary Assessment (PA). Depending on site conditions, a Site Investigation (SI), Interim Response (IR), RI/FS, and Remedial Strategy (RS) (i.e., no action or one of four remediation options, RS 1 to RS 4) may or may not be implemented. The decision tree specifies the probabilities of the occurrence of each of these actions and the cost probability distribution (or a single value estimate of the cost) associated with each action. The probability of each pathway (i.e., each possible sequence of actions in the decision tree) is equal to the product of the probability of occurrence of each action along that pathway. The cost of each outcome is the sum of the costs associated with the implementation of each action along the pathway. It should be noted that, since the cost of each management action is specified as a probability distribution (i.e., as a random variable), the sum of the costs of management actions along

Figure 1.   **General decision tree for the environmental management of an MGP site.**

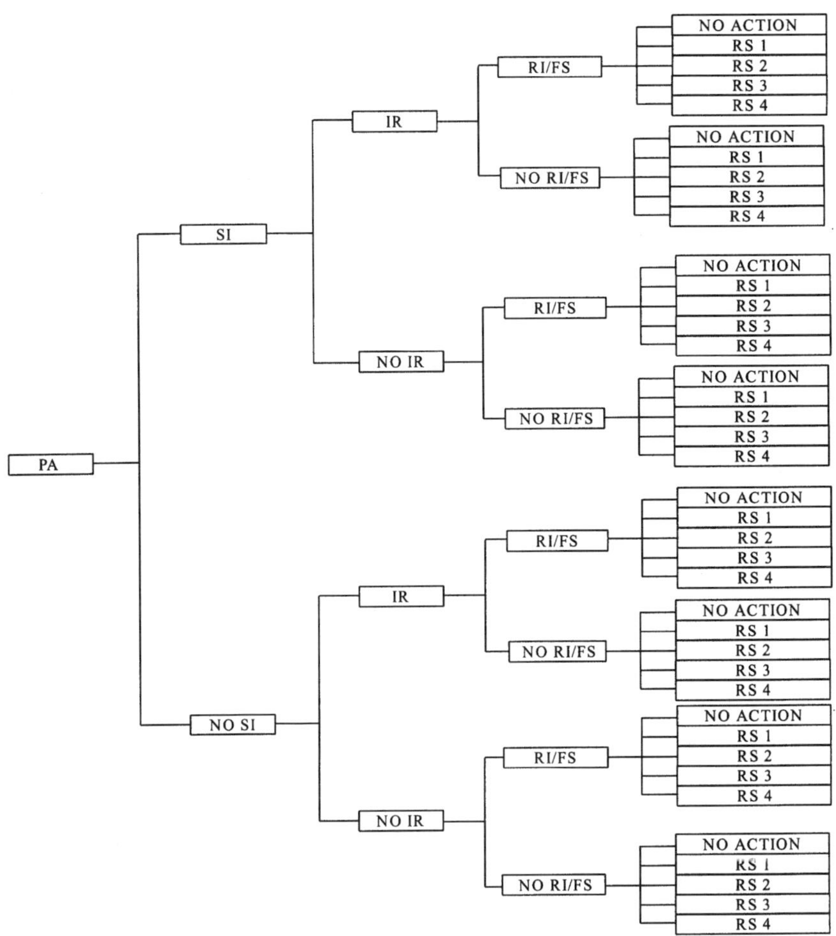

each sequence is performed according to the algebra of random variables. As can be seen in Figure 1, there are 40 possible pathways in the illustrative decision tree in the figure. In the *GRI MGP Cost Model*, Ten (10) remedial scenarios, plus the No Action scenario, can be evaluated, providing a total of 88 possible pathways. Each of these pathways has a cost probability distribution associated with it, which is calculated from the individual probabilities and cost estimates of the actions that comprise the pathway.

The probability distribution of costs for the overall management of an MGP site is calculated by appropriately combining the individual cost probability distributions of each of the possible pathways weighed by their probability of occurring.

Uncertainty in the Cost of Management Actions

To address the uncertainty in the estimation of costs of each management action in the *GRI MGP Cost Model*, the parameters relevant to the cost estimation are specified as probability distributions, rather than single values. This approach modifies the standard decision tree analysis to explicitly incorporate uncertainty into the cost estimates. In the *GRI MGP Cost Model*, probability distributions can be assigned to all parameters relevant to the estimate of costs of management actions. Cost equations have been programmed, whose inputs can be probability distributions, rather than single values. The computer program uses the programmed cost equations to generate cost probability distributions for all management actions. The probability distributions for the cost of each management action are combined and carried through the cost calculation. The result of this analysis is not a single value of the cost of each management action, but a cost probability distribution. The numerical technique known as Monte Carlo analysis is used to combine the probability distributions of uncertain actions and costs to yield cost distributions for various sequences of management actions at a site.

Summary

In summary, the *GRI MGP Cost Model* combines both decision tree and Monte Carlo analysis. The spreadsheet model allows the user to perform a decision tree analysis and specify parameters as probability distributions rather than single values. These probability distributions are carried through the cost calculation and result in each management action being characterized by a cost distribution rather than a single value of cost. The cost probability distributions for each management action are then combined by standard decision tree analysis techniques to obtain the probability distribution for the overall environmental management cost of an MGP site. As such, the uncertainty in the estimate of the cost of each management sequence of actions is quantified based on the uncertainty of all input assumptions.

ADVANTAGES

The benefits of GRI's probabilistic cost analysis are significant. First, GRI's framework allows the uncertainty in costs to be explicitly characterized. Conventional cost analyses typically generate one cost per remedial strategy or a range of costs (low, best estimate and high) for each strategy. While these cost analyses may provide useful information, they provide little insight into the underlying uncertainty because they typically span a very wide range of costs and provide no information on the likelihood of the occurrence of each cost. To put cost estimates in perspective, a probability distribution is needed to determine the likelihood that a particular cost (or a cost within a given range) will occur. Such distributions typically demonstrate that low and high costs, while possible, are extremely unlikely.

Second, the decision tree framework allows critical assumptions to be examined for their influence on the resulting cost distribution. For example, the effect of a changing regulatory environment on site management costs can be accounted for in the decision tree. In fact, more stringent regulations will result in higher probabilities being assigned to extensive (and costly) remedial scenarios. Conversely, regulations that are risk-based may result in lower cost remedial scenarios. GRI's model allows changes in the decision tree probabilities to be readily implemented and, therefore, cost differences associated with different regulatory scenarios to be quickly evaluated.

Third, understanding the probability that a certain cost, or cost range, will occur allows for more informed decisions. For example, in the event that the cost estimates are part of settlement negotiations between two parties, the two parties typically estimate their own costs, which are often quite divergent since they are based on vastly different sets of assumptions. Such an approach makes it difficult for the two sides to find common ground and often leads to a breakdown in negotiations. GRI's probabilistic cost analysis is conducive to bringing the parties together, by demonstrating that a wide range of costs are possible and characterizing the uncertainty introduced by the assumptions so that allocation of the uncertainty can be considered in the negotiation.

A CASE STUDY

To illustrate the use of the *GRI MGP Cost Model,* an illustrative example has been developed. In this example, the process used for estimating costs associated with the environmental management of an MGP site is presented explicitly. In the following, the site is referred to as the Illustrative MGP site. In this example, it is assumed that the only information available on the site is based on a Preliminary Assessment (PA).

Management Actions

The first task in the use of the *GRI MGP Cost Model* is the development of environmental management scenarios and the assignment of probabilities to them. The environmental management actions for the Illustrative MGP Site are summarized in the decision tree presented in Figure 2. This decision tree is a collapsed, site-specific, version of the general decision tree shown in Figure 1, where those pathways having zero probability of occurring have not been included.

The PA of the Illustrative MGP Site recommends a limited environmental sampling program at the site to evaluate the presence or extent of releases. As such, in estimating future costs for the environmental management of the site, it is assumed that a Phase I Site Investigation (SI) would be performed. Therefore, the PA and SI actions are assigned probabilities of 100% in the decision tree. Based on the absence of any immediate hazard, it is also assumed that no Interim Response (IR) action would be required at the site. As such, the IR pathway is assigned a 0% probability in the decision tree.

Figure 2.   Decision Tree for the Illustrative MGP site.

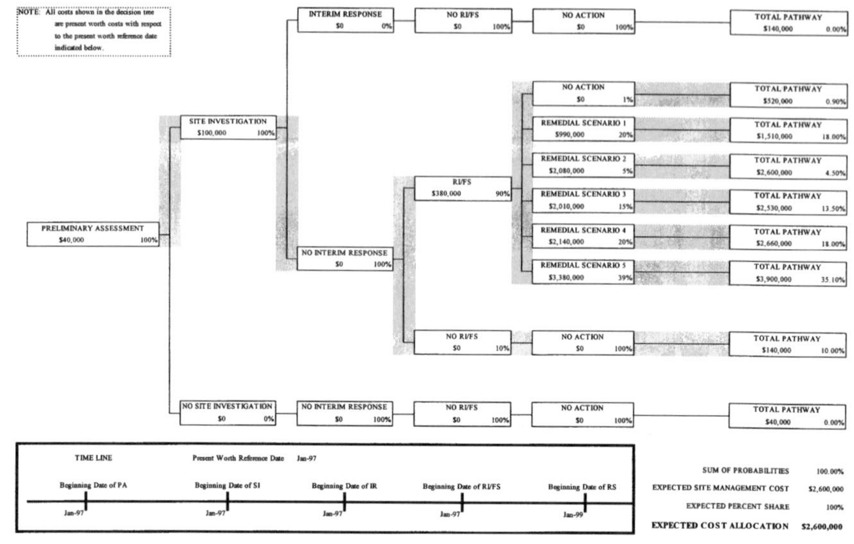

Given the potential for environmental releases to have occurred at the site, it is considered likely that an RI/FS would eventually be required for the site following the SI. However, it is believed that the possibility exists that, based on the SI and RI/FS results, no further action might be required at the site. As such, a probability of 90% is assumed in the decision tree for the RI/FS pathway and of 10% to the no RI/FS pathway.

If, based on the results of the SI, no RI/FS is required, it is assumed that no additional action would be required for the site. If, however, the SI were to indicate the presence of site impacts, a number of remedial scenarios are applicable to the Illustrative MGP Site. These scenarios and their estimated probabilities of occurrence are summarized in Table 1. This table also shows the expected value (i.e., mean or average) of the cost of each remedial scenario. The probability distribution whose average cost is shown in Table 1 is automatically generated by the model, based on the probability distributions of all input parameters, and is automatically copied in the decision tree shown in Figure 2.

Estimated Costs

The present value of the cost for the environmental management of the Illustrative MGP Site was estimated using the probabilistic model discussed in this paper. The results from the model represent the total cost of environmental management for the site. Table 2 presents the relevant statistics for the estimated probability distribution of costs for the Illustrative MGP Site. Figures 3 and 4 show the cost probability histogram and the cumulative cost probability distribution of the overall site management cost.

Table 1. **Remedial Scenarios - Illustrative MGP Site**

REMEDIAL ACTION	No Action	Remedial Scenario 1	Remedial Scenario 2	Remedial Scenario 3	Remedial Scenario 4	Remedial Scenario 5
No action	X					
Institutional controls		X	X	X	X	X
Groundwater monitoring		X	X	X	X	X
Paving and sealing		X	X			
Groundwater pump-and-treat for 30 years[a]						X
Air sparging for 30 years[a]			X			
Soil vapor extraction for 30 years(a)		X	X			
Air sparging for 5 years[a]				X	X	
Soil vapor extraction for 5 years[a]				X	X	
Targeted excavation of MGP structures			X	X	X	
Excavation of MGP process areas				X	X	
Excavation of areas where gas purification residuals may have been placed					X	
Above-ground structure demolition			X	X	X	
Excavation of unencumbered areas						X
PROBABILITY	1%	20%	5%	15%	20%	39%
EXPECTED VALUE OF COST [b]	$0	$990,000	$2,080,000	$2,010,000	$2,140,000	$3,380,000

[a] Expected value of the duration.

[b] This expected value refers to the total Remedial Scenario and represents the value of the cost.

INTERPRETATION OF THE RESULTS

The histogram shown in Figure 2 and the statistics shown in Table 2 indicate that a very wide range (between $90,000 and $7,060,000) of costs is considered possible for the Illustrative Site. This is due to the fact that essentially no information is available regarding environmental conditions at the site. If the cost analysis were to be repeated after completing a site investigation, the range of costs would likely shrink considerably.

Table 2 also indicates that the low and high tails of the distribution are very unlikely. For example, even though the minimum is $90,000, there is only a 15% probability that the cost will be less than $1,300,000. Conversely, there is a 95% probability that the cost will be less than $4,800,000, which means that there is only a 5% probability that the cost will be greater than $4,800,000. The minimum

Table 2. **Cost Statistics Summary Illustrative MGP Site**

SIMULATION DESCRIPTION		PROBABILITY OF COST <=	COST
Simulation Results for LEBACOST.WK4		5%	$140,000
Iterations= 2000		10%	$190,000
Simulations= 1		15%	$1,300,000
# Input Variables= 50		20%	$1,500,000
# Output Variables= 1		25%	$1,730,000
Sampling Type= Latin Hypercube		30%	$2,040,000
Runtime= 00:05:52		35%	$2,250,000
		40%	$2,410,000
		45%	$2,540,000
SIMULATION STATISTICS		50%	$2,640,000
		55%	$2,750,000
Minimum=	$90,000	60%	$2,860,000
Maximum=	$7,060,000	65%	$3,010,000
Mean =	$2,594,085	70%	$3,190,000
Std. Deviation =	$1,312,711	75%	$3,400,000
Variance =	1.723E+12	80%	$3,670,000
Skewness =	-0.036	85%	$3,960,000
Kurtosis =	2.795	90%	$4,330,000
Mode =	$150,000	95%	$4,800,000

cost refers to an optimal situation whereby, following a site investigation, essentially no impacts are discovered and no remediation is required. The maximum cost, instead, refers to a situation in which very significant impacts are discovered and extensive soil and groundwater remediation is required.

The bulk of the probability is for costs in the $2,000,000 to $4,000,000 range, which is consistent with expected management costs for a large number of MGP sites. The mean, or expected value, of the site management cost is approximately $2,600,000. This represents a weighted average of all potential costs, weighed according to their probability of occurrence. Lacking the information provided by the cost probability distribution, this cost should be considered the best estimate for this site.

Different costs values generated by the model have different meanings. For example, as previously discussed, the expected value represents the best estimate of site management costs. The 5% percentile cost ($140,000 - see Table 2) could be considered a reasonable minimum, while the 95% percentile cost ($4,800,000) could be considered a reasonable maximum. As might be expected, the actual minimum and maximum of the cost distribution are generally unlikely, and may not be considered realistic. Finally, another relevant cost value is the median or 50

Figure 3. **Cost probability histogram illustrative MGP site.**

Figure 4. **Cumulative cost probability distribution illustrative MGP site.**

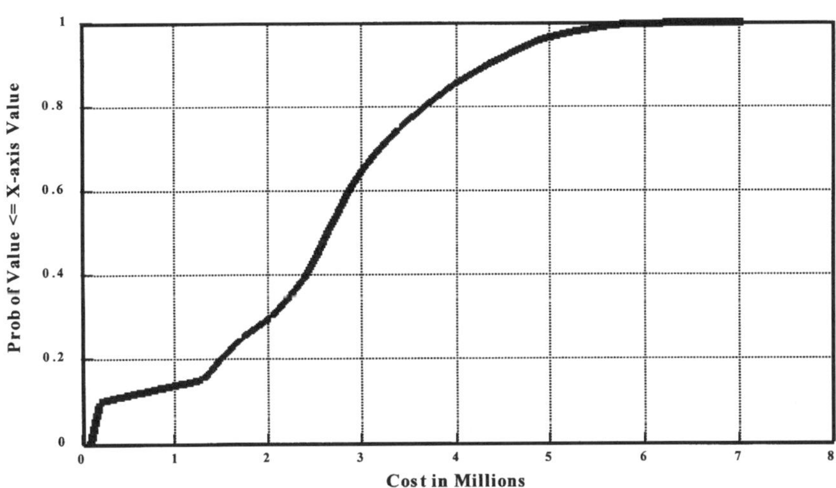

percentile cost ($2,640,000 - see Table 2). There is a 50% probability of incurring a cost above the median, and a 50% probability of incurring a cost below the median.

CONCLUSIONS

When utilities must estimate the costs of managing their MGP sites based on limited data, probabilistic cost analysis can be effectively used for structuring such an ill-defined problem and characterizing the uncertainty in potential costs. GRI has developed a spreadsheet-based framework for performing these analyses, which has been used on more than 100 MGP sites. The *GRI MGP Cost Model* software represents a useful MGP site management tool, providing utilities with an effective and defensible way of generating cost estimates for site remediation which explicitly include a quantification of the uncertainty involved in this process.

CHAPTER 44

Reaction Window Treatment of Ground Water at a Former MGP Site

Marco D. Boscardin, **Mark T. Mahoney**, **Mark L. Jones**, and **Michael P. Walker**, GEI Consultants, Winchester, Massachusetts

Joseph M. Callanan, Massachusetts Electric Company, Westborough, Massachusetts

INTRODUCTION

Site Setting

The site occupies approximately six acres adjacent to a small cove on Massachusetts' North Shore. A site plan is shown in Figure 1. Land adjacent to the site was formerly occupied by a manufactured gas plant (MGP) that operated from the mid-1800s to the 1960s.

Site Stratigraphy

A surface layer of miscellaneous fill covers the majority of the site. The miscellaneous fill consists primarily of silty sand, ranges from 0 to about 10 feet thick, and pinches out to the east, on the cove shoreline. Natural silty sand, from 2 feet to 16 feet thick was observed below the miscellaneous fill. Occasional coarse sand layers were observed in this stratum. DNAPL staining was observed in the coarse layers. Up to 12 feet of low permeability silt was observed below the silty sand. Dense nonaqueous phase liquid (DNAPL) staining was observed at the silty sand-silt interface. This interface dips toward the cove and act as a pathway for DNAPL migration. The silt grades into a lean clay with increasing depth. General site stratigraphy is shown on Figure 2.

Site Contaminants

Contaminants associated with historical MGP processes have been detected in site soil, groundwater, and marine sediment. Additionally, an intermittent sheen has been observed on the surface water near the time of low tide, and low pH groundwater has been detected in monitoring wells and discharging from an earthen slope adjacent to the cove. The sheen is apparently due to DNAPL migrating along the silty sand-silt interface, and "day lighting" on the cove shoreline. The low pH

Figure 1. **Plan view of site showing proposed remedial system - reaction windows are located near ends and center of subsurface barrier wall.**

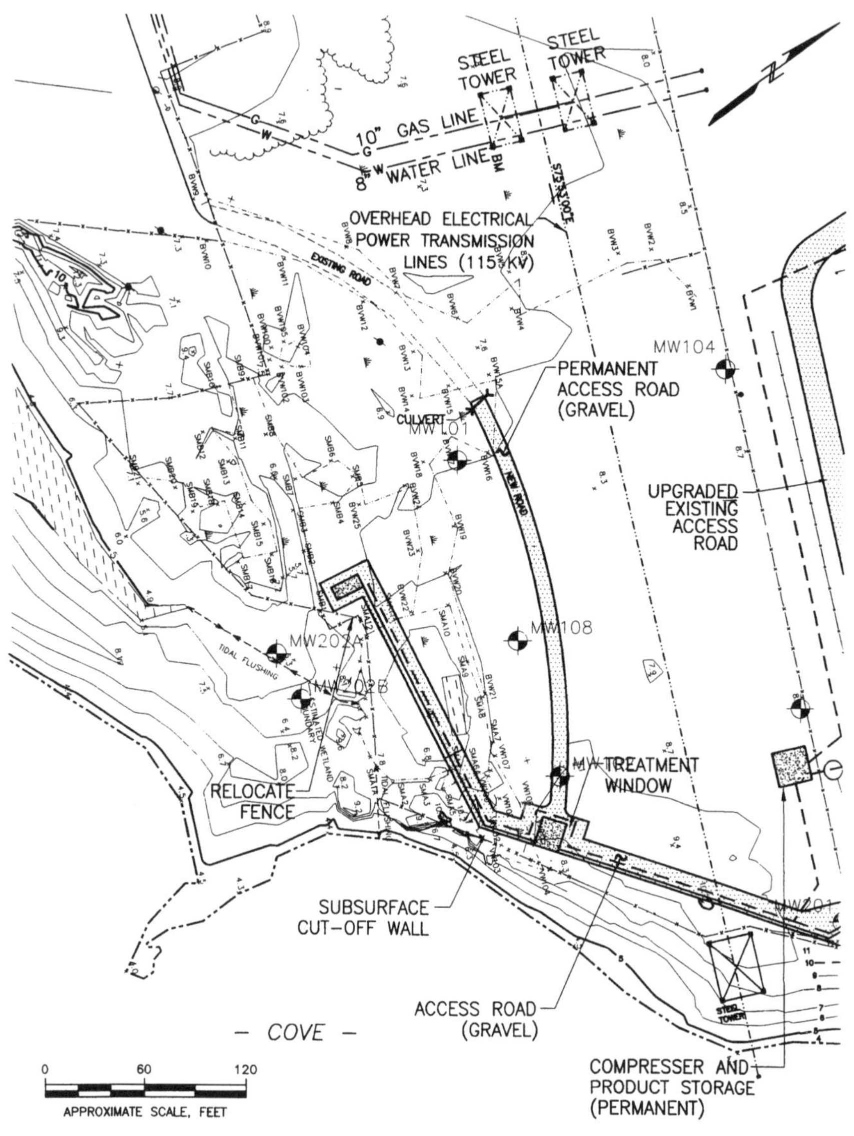

Figure 2. **Site shallow stratigraphy. Cross section is located near center reaction window and is oriented approximately perpendicular to subsurface barrier wall.**

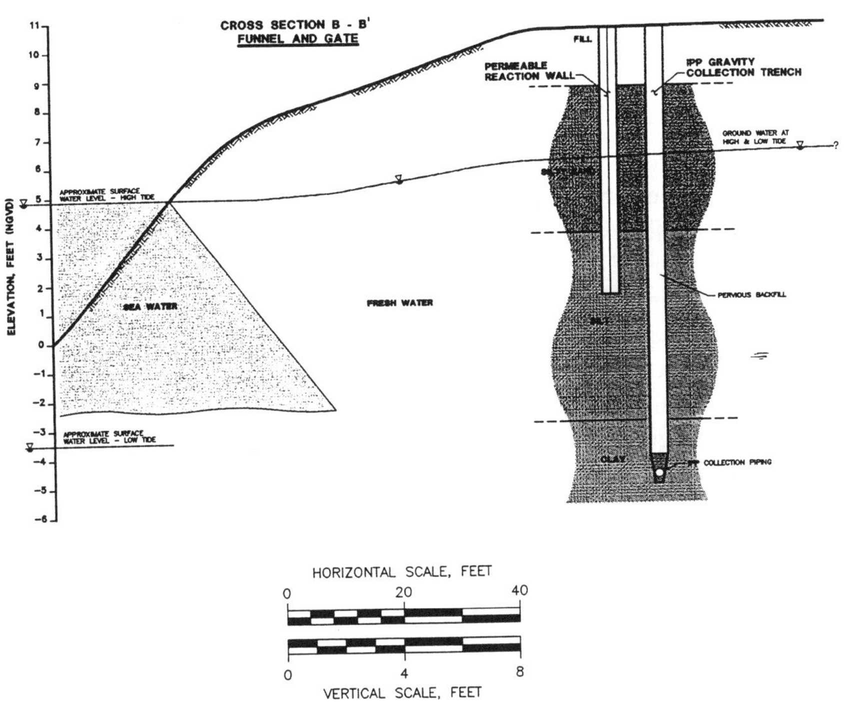

groundwater has been detected in portions of the site that were filled. Fill materials on MGP sites typically include sulfur-rich purifier box wastes. Reduction of sulfur in the wastes produces sulfuric acid and lowers groundwater pH. Groundwater pH measurements as low as 2 have been made in central portions of the site.

Due to the observed sheens, the potentially responsible parties (PRPs) were notified that a condition of Substantial Release Migration exists, and an Immediate Response Action (IRA) was required.

Proposed Remediation

A Focused Feasibility Study (FFS) was performed to evaluate remedial options for abating (1) the migrating DNAPL, and (2) low pH groundwater. Remedial action alternatives (RAAs) reviewed during the FFS included: (1) Encapsulation with wetlands restoration; (2) Excavation and disposal/treatment of approximately 18,500 cubic yards (c.y.) of contaminated soil with wetlands restoration; (3) In-situ groundwater treatment with DNAPL collection; and (4) No action (for comparison). RAA #1 was not selected due to impacts on site hydrology, and uncertainties associated with sufficiently abating migrating DNAPL to eliminate the sheen. RAA #2 was not selected due to significant costs associated with contaminated soil recycling/disposal. RAA #3 was selected based on expected effectiveness and cost savings.

A conceptual design for the selected RAA included a high density polyethylene (HDPE) subsurface barrier wall containing reaction windows to filter and treat groundwater passing through the barrier wall. A horizontal drain system with product pumps and a collection system will be installed behind the barrier wall for DNAPL collection.

A design support study (DSS) was performed to develop the information necessary for RAA final design, and drawings and specifications for construction. Issues that were resolved during the DSS included: extent of the area where groundwater that flows through the reaction windows originates (termed the "capture zone"); effectiveness of various limestone materials for elevating groundwater pH as it passes through the reaction window; and effectiveness of PAC/sand mixture for organic removal as the groundwater passes through the reaction window.

The final design included using the results of the DSS for planning barrier wall length and orientation, the number and dimensions of reaction windows, construction materials and procedures, and the treatment sequence in the reaction windows. These design requirements were incorporated in the general civil engineering design, i.e., general site work including project layout, planning construction access roadways, minimizing disturbance to wetland resource areas.

EVALUATION OF CAPTURE ZONE AND PRELIMINARY DESIGN FEATURES

The extent of the capture zone was evaluated using groundwater modeling methods described by Starr and Cherry (1994). The approach included development of a groundwater model for the site which was calibrated by comparing to the pre-remediation, or existing site conditions and then used to investigate various remedial configurations (i.e., different combinations of barrier wall lengths and orientation, and reaction window placement and dimensions). The groundwater model was used for selecting the design configuration, and for predicting the groundwater flow rate through the reaction windows in that configuration.

Groundwater modeling was performed using Flowpath v5.11. As a check, several replicate modeling runs were made using Visual MODFLOW v1.49. The output from these replicate runs closely matched the output from the Flowpath v5.11 runs. Calibration of the groundwater model indicated the model simulated measured seasonal high and low groundwater elevations satisfactorily. The groundwater model was used to investigate various recharge conditions, and was also used to simulate flooding conditions due to seawater inundation. Approximately 75 model simulations were run with the following factors varied in different combinations for each model run:

- The number of reaction window structures. Runs were made with 1 to 3 filter treatment structures inserted in the barrier wall;
- The location of the reaction windows. Reaction windows were located at the middle of the barrier wall, at the barrier wall ends, and various combinations thereof;
- The hydraulic conductivity of the reaction windows. The hydraulic conductivity of the materials in reaction windows was varied from 1 order of magnitude less than the hydraulic conductivity of the surrounding soil to 1 order of magnitude more than the hydraulic conductivity of the surrounding soil;
- The width of the filter treatment structures. Widths of the filter treatment structures were varied from 5 feet to 40 feet for the configurations described above;
- The length of the barrier wall. The barrier wall length was varied from approximately 100 feet to approximately 450 feet;
- The plan location and orientation of the barrier wall. Various barrier wall configurations were modeled, and modified to accommodate site constraints (barrier wall to be installed above the elevation of high tide, not to interfere with electric power transmission line tower foundations); and
- The presence of a narrow, high permeability drainage trench or net behind the barrier wall.

Groundwater modeling results indicated the barrier wall and reaction window shown in Figure 3 provided an appropriate capture zone for groundwater treatment for the range of conditions expected to be encountered.

FIELD AND LABORATORY DESIGN TASKS

Field Sampling

Field tasks included collecting groundwater samples from nine monitoring wells for laboratory analyses for inorganic and organic compounds. A summary of the analytes and mean, maximum, and minimum measured concentrations are shown in Table 1. The laboratory results were used to identify a "worst case" groundwater specimen that would be collected from the identified monitoring well with high dissolved organic and low pH. The worst-case groundwater sample was used for subsequent laboratory batch and column testing. The worst-case groundwater sample had 10.05 milligrams per liter (mg/L) polycyclic aromatic hydrocarbons (PAHs), 83.2 mg/L volatile organic compounds (VOCs), a pH of 2.1, and an acidity of 4,860 mg/L as $CaCO_3$. The groundwater from this well also had the highest iron concentration, 1,200 mg/L. Iron and other dissolved metals were of concern in the design due to the potential for precipitation of these metals as the pH was elevated. Precipitation of iron and other metals may result in reduced permeability of reaction window materials. Groundwater inorganic contaminants are similar to those encountered in coal mine-associated acid drainage (Zelmanowitz et al., 1995)

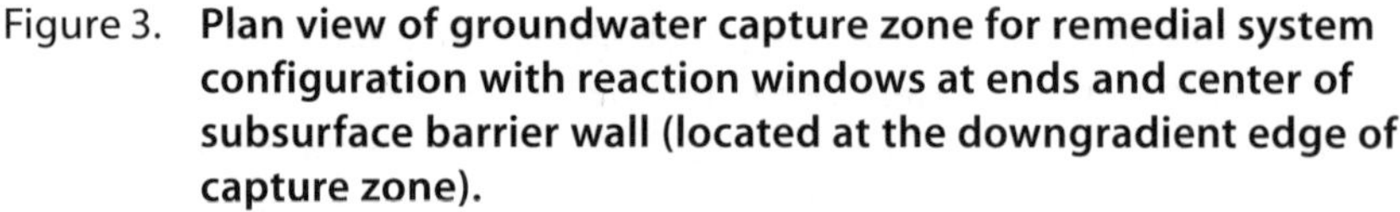

Figure 3. **Plan view of groundwater capture zone for remedial system configuration with reaction windows at ends and center of subsurface barrier wall (located at the downgradient edge of capture zone).**

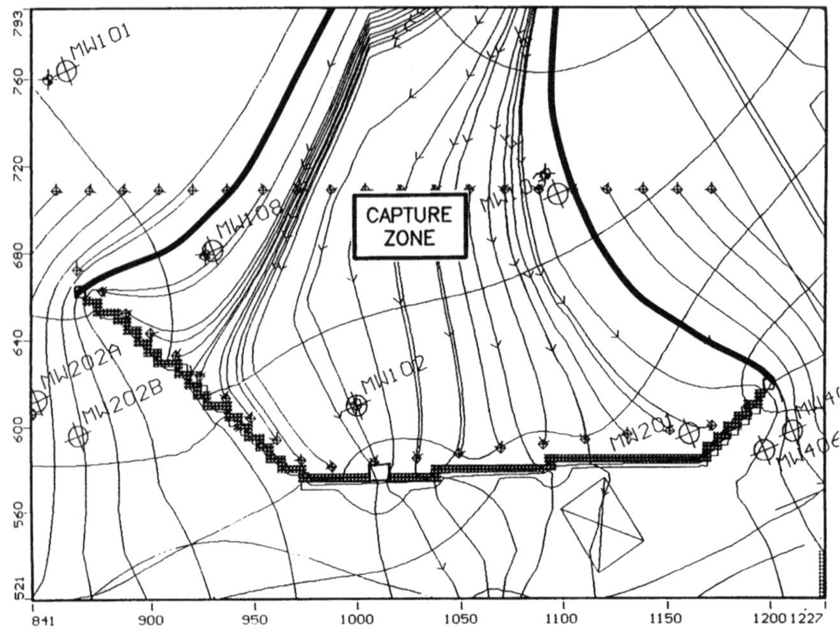

An "average" groundwater specimen was also identified for subsequent laboratory testing. The "average" groundwater specimen consisted of a mixture of equal volumes of groundwater collected from six of the nine monitoring wells tested. Elevated PAH and VOC concentrations, low pH, and high acidity were identified in the six monitoring wells.

Laboratory Batch Testing

Laboratory batch testing was performed to evaluate the effectiveness and required dosages for limestone pH treatment, and adsorptive capacity of powdered activated carbon (PAC) for organic compounds in site groundwater. Batch testing involved placing site groundwater and candidate reaction media in contact, and measuring contaminant reduction versus time.

Limestone Batch Testing

Limestone batch testing was conducted in several phases. The first phase involved determination of a suitable limestone to neutralize the acidity of the low pH groundwater. Specimens from five commercially available limestones were used: (1) limestone chips; (2) coarse crestite; (3) hi-calcium granules; (4) hi-calcium sand; and (5) crusher fines. Detailed geological, physical, and chemical properties

Table 1. Summary of Ground Water Laboratory Analyses Data

Compound	Minimum Concentration[a]	Maximum Concentration[a]	Mean Concentration[a]
Total PAHs	Not detected	13.9	3.2
Total VOCs	Not detected	105.2	29.6
Total cyanide	Not detected	266	30
Amenable cyanide	Not detected	2.2	0.4
Total/dissolved calcium	29.9/28.8	417/375	124/109
Total/dissolved iron	7.5/0.3	1,200/1,087	369/291
Total/dissolved manganese	Not detected	8.1/7.1	3.9/4.0
Total/dissolved sodium	Not detected/33.9	3,152/3,334	746/897
pH (standard units)	2.1	6.8	4.0
Hardness (mg-eqCaCO$_3$/L)	136	1,850	754
Alkalinity (mg/L CaCO$_3$)	[b]	89.6	19
Acidity (mg/L CaCO$_3$)	165	4,860	1,590
Chloride	45	5,176	1,058
Sulfide	0.6	1.8	1.0
Total suspended solids	30	1,554	415
Total dissolved solids	750	11,324	5,570
BOD	Not detected	39.6	9.6
COD	65	838	360
TOC	33	287	107

were supplied by the vendors. The specimens were tested using both "average" and "worst-case" groundwater under variable and constant pH conditions.

A plot of pH versus time for individual tests for crusher fines, hi-calcium sand, hi-calcium granules, and limestone chips test samples for "worst-case" groundwater is shown on Figure 4. The buffering capacities for the crusher fines, hi-calcium sand, and to a lesser degree hi-calcium granule test samples and limestone chips were acceptable. The buffering capacity of the crestite was poor and the crestite was eliminated from further consideration.

Limestone specimens were also investigated by immersion in the "average" groundwater for a period of two hours. Under similar test conditions, "average" groundwater samples exhibited a sharper rise in pH with time compared to the "worst" case samples. At the conclusion of the test period, final equilibrium pH values were similar for both ground water conditions.

Tests were also conducted for the worst-case groundwater samples under both open and closed to the atmosphere conditions to evaluate effects of atmospheric CO_2 exchange. Test results for crusher fines are shown in Figure 5, and based on inorganic carbon concentration as an indicator, the results for both open and closed to the atmosphere conditions are similar. Results indicate that atmospheric CO_2 exchange is not a significant factor in the buffering process.

During the second phase of the limestone batch testing, the four limestone specimens were immersed in the "worst-case" and "average" sample water for a period of two hours under: (1) variable pH; and (2) constant pH conditions. Comparison of test results in Figures 6 and 7 illustrates the differences between dissolution rates for the constant and variable pH test runs. These testing results indicate that raising pH creates an environment that more readily facilitates the dissolution of calcium carbonate.

Powdered Activated Carbon (PAC) Batch Testing

Two equilibrium adsorption isotherm batch tests were performed to determine the adsorptive capacity of the PAC and to evaluate its efficiency in removing dissolved organic from groundwater. The first round of batch tests investigated the adsorptive capacity of the PAC for "worst case" and "average" groundwater contaminant concentrations. The second round of batch tests investigated the adsorptive capacity of the PAC under variable pH conditions - test runs for pH values of 7.0 and 9.0 were conducted using "worst case" groundwater samples (pH was adjusted by addition of 6.0 N NaOH). An additional test run was conducted for

Figure 4.   **Plot of pH versus time for limestone batch testing. Crusher fines and hi-calcium sand provided most rapid pH response for "worst-case" site groundwater sample.**

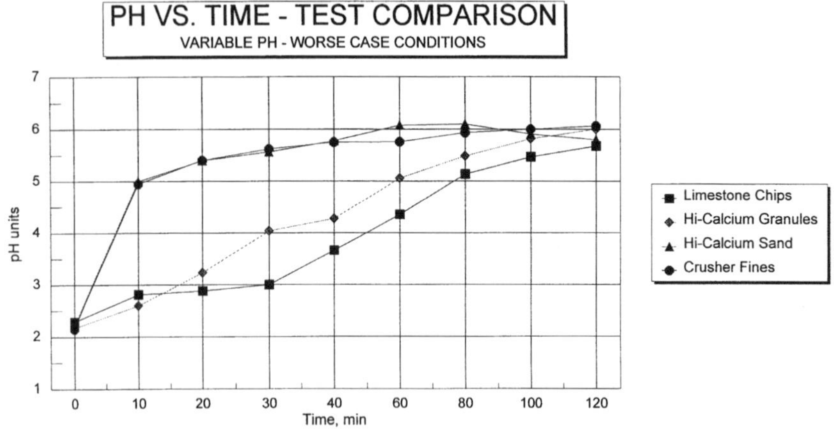

granular activated carbon (GAC) under existing groundwater pH conditions for "worst case" groundwater concentrations to compare GAC and PAC performance. Influent and effluent contaminant concentrations were monitored using total organic carbon (TOC) concentrations as an indicator.

Figure 8 presents a comparison of all the carbon isotherms generated during the isotherm batch testing study. The results show that all of the PAC test runs outperformed the GAC by a significant margin, for the range of influent concentrations tested. For influent TOC concentrations above 30 mg/L carbon adsorption capacity appears to increase as influent pH increases. However, results indicate only a small increase in carbon capacity occurs above pH 7.0 for the range of influent concentrations tested (Figure 8).

Laboratory Column Testing

Four columns were constructed from 2-inch i.d. Excelon tubing. The columns were approximately 3 feet long and were oriented vertically. "Worst case" groundwater was pumped upward through the columns from a reservoir through each column at a flow rate of about 2 mL/min. Effluents were collected in individual glass beakers. The experimental setup is illustrated on Figure 9. The flow rate through the columns was selected based on the predicted flow rate through the reaction windows from groundwater modeling. Influent and effluent contaminant concentrations were monitored using total organic carbon (TOC) concentrations. Total VOCs were measured as benzene using a portable gas chromatograph.

Figure 5. **Plot of inorganic carbon concentration versus time for batch testing using limestone crusher fines. Tests were performed for open and closed (to atmosphere) cases to test potential effects of atmospheric exchange of CO_2 on limestone dissolution rate, with inorganic carbon used as an indicator of dissolved CO_2. Tests indicated CO_2 exchange is not significant, and is not likely to impact buffering reaction for limestone crusher fines.**

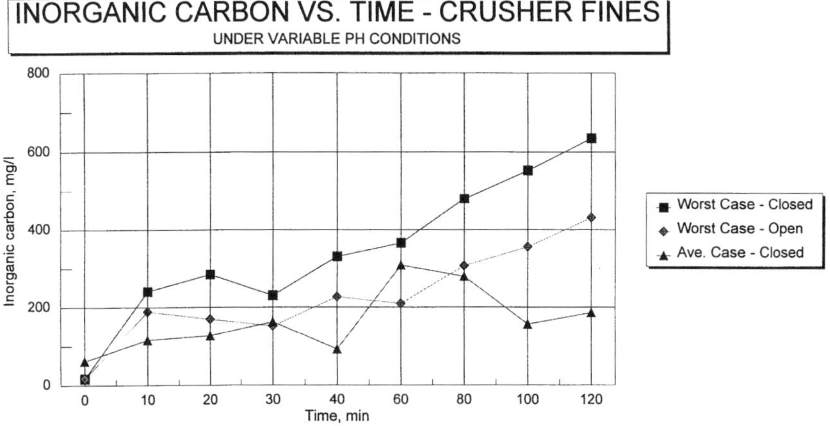

Figure 6.   Plot of inorganic carbon concentration versus time for batch testing of candidate limestone materials. Tests were performed at varying pH for closed to atmosphere cases to evaluate CO_2 dissolution rate. Tests indicated dissolution of limestone proceeds at a relatively constant rate for crusher fines than for other limestone materials.

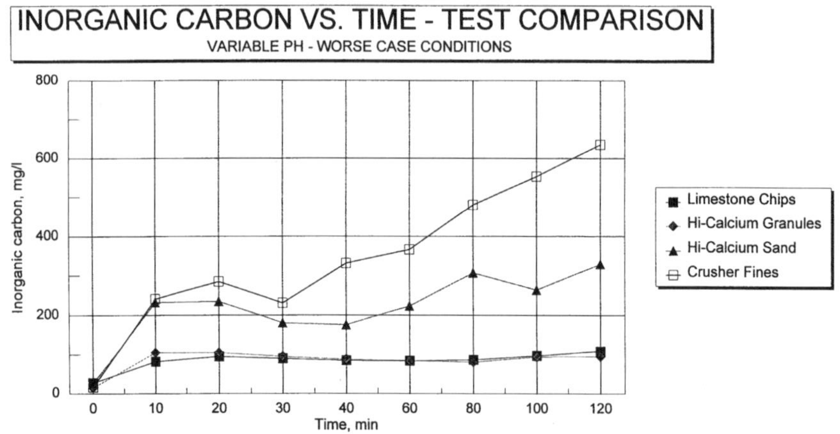

Figure 7.   Plot of inorganic carbon concentration versus time for batch testing of candidate limestone materials. Tests were performed at constant pH for closed to atmosphere cases to evaluate effect of constant pH on CO_2 dissolution rate. Tests indicated dissolution of limestone generally increases to a maximum point, followed by a dropoff as the system becomes saturated with dissolved CO_2.

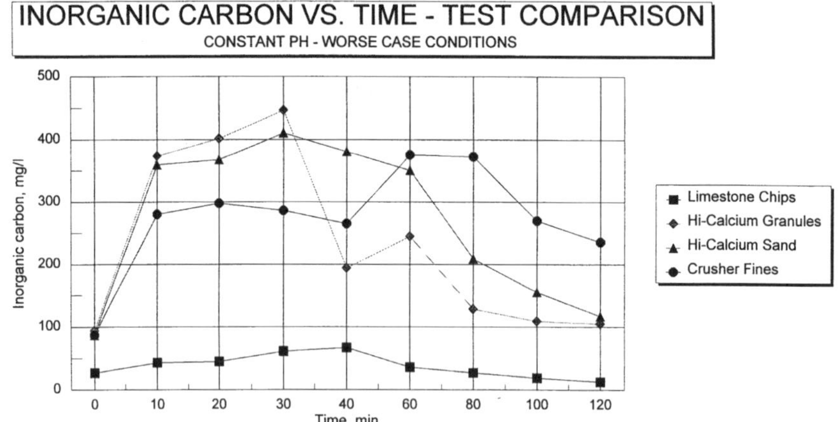

Figure 8. **Isotherms for batch testing of 5% powdered activated carbon (PAC) and silica sand, and 5% granular activated carbon (GAC) and silica sand mixtures using "worst case" site groundwater. Organic contaminants are VOCs and PAHs. The pH was varied from about 2 to 9 to evaluate the effects of pH on PAC and GAC adsorption capacity, using total organic carbon (TOC) measurements as an indicator of organic contaminant residual concentration.**

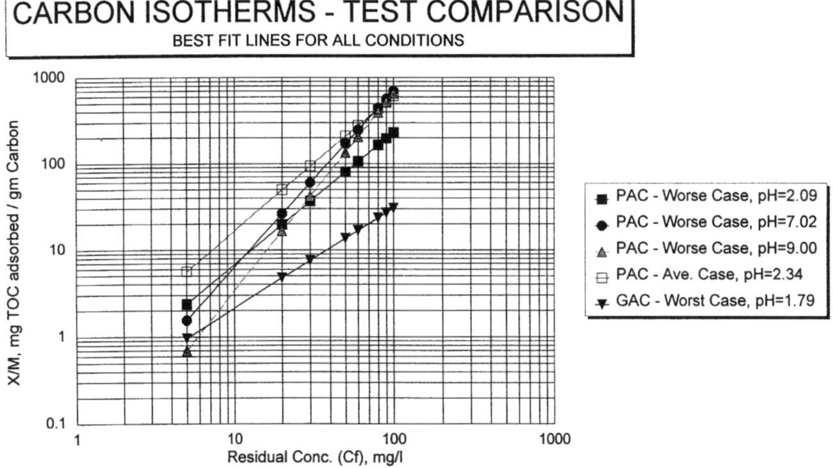

Column test setups are described below and are similar to those used for previous studies for column testing for inorganic contaminants (Choi and West, 1995) and organic contaminants (Rael et al., 1995):

- Column No. 1 contained a 60-cm crusher fines bed and was used to measure the effects of pH change for a limestone bed thickness that approximated the likely full-scale field bed thickness. (A primary concern was that pH not be raised above the target range of: $6 \leq pH \leq 8$). As expected, the effluent TOC and VOC concentrations remained relatively unchanged from the influent concentrations for this column setup.

- Column No. 2 contained a 60-cm layer of silica sand and was used: (1) as the control unit; and (2) to assess fine sand filtration as a potential pretreatment process. As expected, effluent pH was minimally affected and ranged from 2.0 to 2.4 over the duration of the test period. In addition, there was very little change in TOC and VOC concentrations during treatment of over 11 liters of groundwater. The most apparent benefit of the silica sand layer was its ability to filter or remove the suspended solids present in the influent groundwater. Column No. 2 produced the clearest effluent of all the columns.

Figure 9. **Column testing setup.**

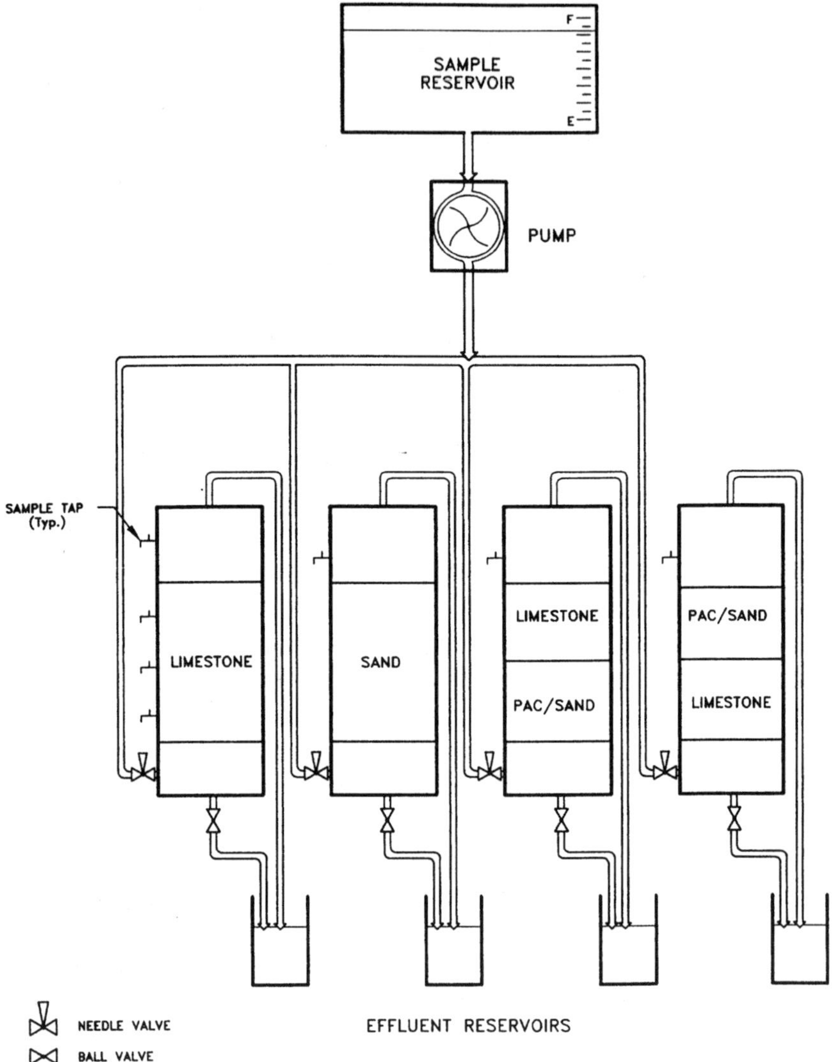

Columns No. 3 and No. 4 replicated the treatment process proposed for use in the field, with the PAC/sand and limestone treatment units placed in two different orders (Figure 9). The PAC/sand mixture used a 5% PAC to sand (mass/mass) ratio (Rael et al., 1995). The column testing was terminated after achieving a breakthrough concentration, selected as 30% of the influent contaminant concentration. Based on a worst-case influent TOC concentration of 157 mg/L, 50 mg/L was defined as the breakthrough concentration. Breakthrough occurred after 9 days of testing, or after about 15 L of groundwater volume throughput.

- Column No. 3 contained a 60-cm layer of 5% PAC/silica sand mixture followed by a 30-cm layer of crusher fines. Test results for Column No. 3 are shown in Figure 10. Approximately 18 L of groundwater passed through Column No. 3. The pH was raised to pH > 6 after 1 L was treated. TOC concentration reductions ranged from approximately 70% to 95%. Influent VOC concentrations were reduced to below detection throughout treatment of the total 18 L.

- Column No. 4 contained a 30-cm thick layer of crusher fines followed by a 60-cm layer of 5% PAC/silica sand mixture. Effluent pH ranged from an initial value •f pH 2 to a high of pH 7.6 after less than 1 L treated, pH 6 after 15 L treated, and ended with a drop to pH 4 after 19.5 L treated, possibly due to short-circuited flow along the limestone test column wall boundary. TOC concentration reductions ranged from approximately 70% to 95%. Effluent VOC concentrations were reduced to below detection levels through treatment •f 14.8 L of groundwater.

Testing results indicated Column No. 4 provided the better performance. The carbon capacity for Column No. 3 was calculated to be approximately 0.065 lb TOC per lb of 5% PAC/sand. The carbon capacity for Column No. 4 was calculated to be approximately 0.071 lb TOC per lb of 5% PAC/sand.

The ratio of breakthrough TOC concentration to influent worst-case TOC concentration (50 mg L^{-1}/157 mg L^{-1}) represents an approximately 68% reduction in dissolved organic, assuming TOC concentration is a representative indicator for individual organic compound response. Current site data indicate that up to 50% reductions may be required for some organic compounds in groundwater to achieve Massachusetts Contingency Plan (MCP) GW-3 standards. A change from about 68% to 50% reduction in dissolved organic means a greater carbon capacity may be utilized and still achieve treatment goals. Based on Column No. 4 test data, carbon capacity of approximately 0.092 lb TOC per pound PAC can be used in design. Based on these column test results, full-scale operation is expected to require approximately 525 lb PAC per reaction window. Based on predicted groundwater flow rates of approximately 282 gallons per day per reaction window, the estimated carbon capacity of approximately 0.092 lb TOC per lb PAC/sand mixture, and the worst-case influent concentration of 157 mg/L TOC, daily PAC consumption would be 5.92 lb. The design is conservative - "worst case" groundwater is not expected to flow through all reaction windows. Daily PAC consumption will be lower for "average" groundwater, which is probably more representative of groundwater at the majority of the reaction windows.

REMEDIAL DESIGN

After reviewing various subsurface barrier wall materials and installation methods, a proprietary system was selected that incorporates HDPE barrier wall con-

Figure 10. **Top: Column testing results for PAC/sand = limestone treatment sequence. Effluent pH maintained steady around 6. Influent TOC/Effluent TOC increased from < 1% to approximately 70%. Effluent VOCs less than 1 ppm detection limit. Bottom: Column testing results for limestone = PAC/sand treatment sequence. Effluent pH maintained steady around 6. Influent TOC/Effluent TOC increased from < 1% to approximately 40%. Effluent VOCs less than 1 ppm detection limit until about 19 L treated.**

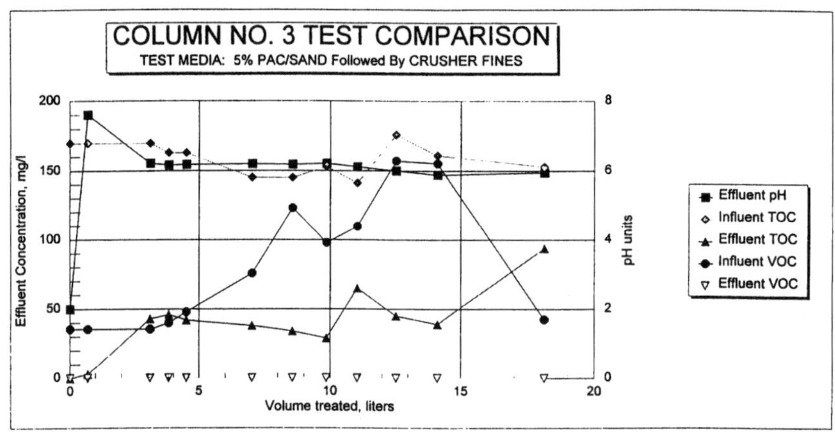

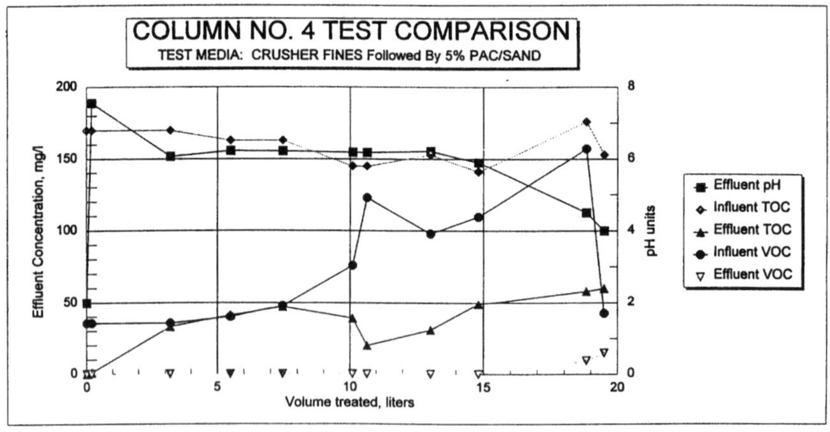

struction with horizontal drain installation. A plan view of the proposed remedial system was shown in Figure 11.

The reaction windows contained in the barrier wall will be constructed from formed-in-place concrete. A sketch of the reaction windows is shown in Figure 11. Prior to pouring the concrete, the forms will be lined with HDPE. After the forms are removed, the HDPE will remain in place as a protective coating for the concrete. The reaction windows will contain a silica sand filtration chamber, followed by limestone treatment, silica sand, and PAC/sand filter/treatment chambers. Monitoring wells will be placed in the silica sand layers and downstream from the PAC/sand chamber.

Operation and maintenance will include periodic monitoring of influent and effluent for pH and TOC with spot-checks for specific organic compounds. Reaction window materials will be changed on an as-needed basis, depending on the

Figure 11. **Reaction window, plan view.**

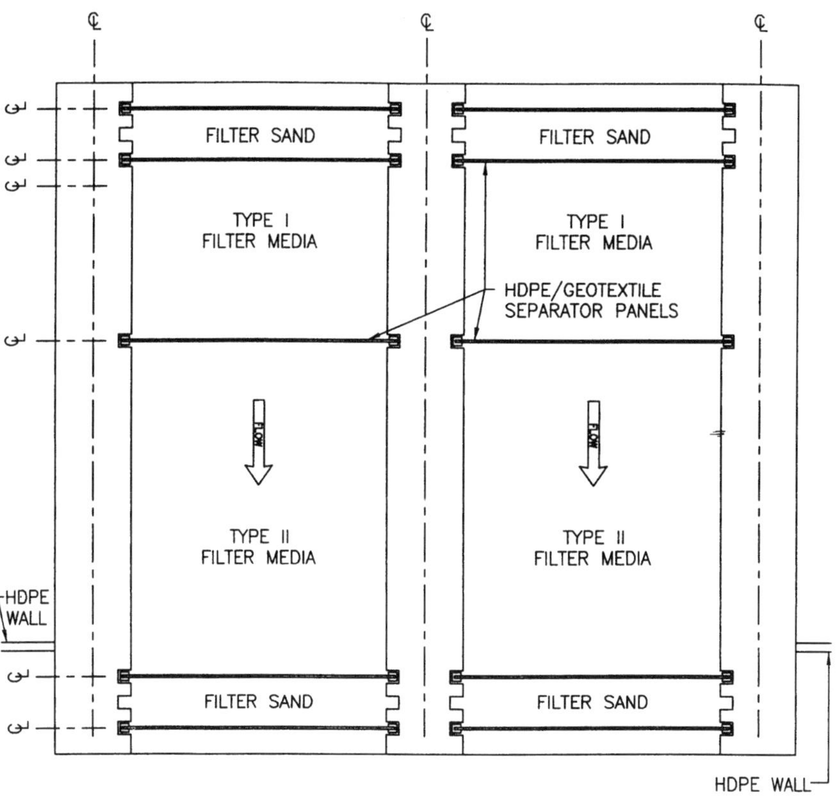

system effectiveness indicated by the monitoring results. Treatment materials will be removed from the reaction windows with a small backhoe and transported from the site for disposal. The ability to monitor, remove, and replace reaction materials provides flexibility - modifications (e.g., different reaction materials) can be made quickly and cost-effectively.

REFERENCES

Choi, J. and West, T.R. 1995. Evaluation of Phosphate Pebble as a Precipitant for Acid Mine Drainage Treatment. *Environ. Eng. Geoscience* 2, 163-171.

Rael, J., Shelton, S., and Dayaye, R. 1995. Permeable Barriers to Remove Benzene: Candidate Media Evaluation. *ASCE J. Environ. Eng.* 5, 411-415.

Starr, R.C. and Cherry, J.A. 1994. In Situ Remediation of Contaminated Ground Water: The Funnel and Gate System. *Ground Water.* 32, 465-476.

Zelmanowitz, S., Boyle, W.C., Armstrong, D.E., and Park, J.K. 1995. Ability of Subsoils to Buffer Extremely Acidic Simulated Coal Pile Leachates. *ASCE J. Environ. Eng.* 11, 411-415.

CHAPTER 45

Commercial Application of the GAC-FBR to the Cleanup of Contaminated Groundwater at MGP Sites

Robert F. Hickey, April Seybold, Dan Wagner, Veronica Groshko, and **Raj V. Rajan**, EFX Systems, Inc., Lansing, Michigan

Alfred Leuschner, Remediation Technologies, Concord, Massachusetts

Tom Hayes, Gas Research Institute, Chicago, Illinois

INTRODUCTION

Manufactured Gas Plants (MGPs) were employed in the United States between the early 1800s and the mid-1900s to produce gas from coal and oil. The residual tar or dense nonaqueous phase liquids (DNAPLS) from the manufacturing process were, in many cases, stored and/or disposed of on-site. Soils at these facilities typically contain coal tar residues, including BTEX and polynuclear aromatic hydrocarbons (PAH), phenolic compounds, and aliphatic hydrocarbons. In some cases, the coal tars have migrated into the subsurface, resulting in contamination of the surrounding groundwater. At these sites there is a need for treatment of the groundwater and/or removal of the coal tars in the subsurface. The major PAHs in groundwater are principally lower molecular weight, 2-4 ring, PAHs such as naphthalene, methylnaphthalenes, phenanthrene, and pyrene. Aqueous solubilities generally limit the 5-6 ring PAHs to low μg/L levels. Economical and reliable processes for the treatment of groundwater and effluent streams generated during subsurface DNAPL removal is essential to meet overall remediation objectives at these facilities.

THE GAC-FBR PROCESS

The fluidized bed reactor (FBR) is a high-rate, biological fixed-film treatment system that offers stable, efficient destruction of organic pollutants from dilute aqueous streams. FBR systems can be operated under conditions where control of biofilm thickness and mean cell residence times can be achieved (Jeris et al., 1977). The Granular Activated Carbon-Fluidized Bed Reactor (GAC-FBR) system modification of the process combines the advantages of biological and physical treatment in a single, robust unit operation (Hickey et al., 1991; Wagner et al., 1993).

Development of the biological FBR process dates back to the late 1960s (Jeris et al., 1974). Initial work focused on laboratory and pilot-scale denitrification of nitrified municipal wastewater (Jeris and Owens, 1975). Application of the process for aerobic degradation of organics, nitrification (Jeris et al., 1977; Sutton et al., 1979), anaerobic treatment of sewage (Switzenbaum and Jewell, 1980), and high-strength wastes (Hickey and Owens, 1981; Suidan et al., 1981) followed.

In the mid-1980s, it was recognized that the technology had the potential of substantially reducing the cost of treating contaminated groundwater and process effluents containing low levels of toxics. The key features of by the FBR process are:

- Large surface area for biomass attachment,
- High biomass concentrations,
- Ability to control and optimize biofilm thickness,
- Minimal plugging and channeling,
- High mass transfer properties through maximum contact between biomass and substrate,
- No off-gas produced; therefore, no air quality concerns, and
- Tailored biomass carrier to enhance system performance for certain classes of contaminants.

The key advantages to using GAC as the biomass carrier are:

- High removal efficiencies from time zero, because the GAC adsorbs contaminants during startup,
- The system performs as a biological reactor once the biofilm is established,
- Contaminants such as BTEX, PAHs, and phenolics adsorbed onto GAC during startup are, to a large extent, desorbed and degraded (bioregeneration) once an active biofilm is established, and
- Dual biological and adsorptive removal mechanisms ensure robust performance during perturbations in contaminant concentrations, interruption in feed, etc.

The GAC-FBR process has been successfully used to treat groundwater containing BTEX and other petroleum hydrocarbons at a number of sites (Hickey et al., 1993, 1991). The utility of GAC-FBR technology to treat waters from MGP sites, where PAHs are the primary contaminants of concern, was examined using a 550 gpd (2,080 liter per day) pilot-scale system. Based on results generated, the GAC-FBR process was installed at a Superfund site to process a 15 gpm effluent water stream containing BTEX and PAHs from a subsurface DNAPL removal process (CROW™ hot water extraction process), and for treatment of groundwater at a site with high concentrations of reduced iron.

MATERIALS

Pilot Reactor

A pilot-scale (550 gpd) GAC-FBR constructed from glass and Teflon™ components was used for this study (Figure 1). The reactor was fed a synthetic waste stream containing eight PAHs for over one year. The PAHs added were representative of the major components identified in samples of groundwater from an MGP site where subsurface DNAPL removal was scheduled. The PAHs and phenolic compounds in the synthetic feed included naphthalene, acenaphthylene, phenanthrene, fluoranthene, pyrene, benzo(b)fluoranthene, benzo(a)pyrene, benzo-(ghi)perylene, nitrophenol, dinitrophenol, and methyl dinitrophenol. This contaminated waste stream was amended with nitrogen, phosphorus (at a COD/N/P ratio of 100/5/1), trace nutrients, and oxygen. The influent water was maintained at room temperature (23°C). The reactor was operated as a one-pass system (no recycle used) at a hydraulic retention time HRT) of between 6 and 7 minutes. The GAC-FBR was inoculated with a consortium able to degrade 2- to 4-ring PAHs obtained from soil samples from several MGP sites.

COMMERCIAL SCALE REACTORS

Commercially available fluidized bed reactor systems (Envirex, Inc.) with maximum flow capacities of 30 and 190 gpm (110 to 720 L/min) were used for this work. These skid-mounted systems were equipped with an on-board high purity oxygen pressure swing adsorption (PSA) generator, biomass growth control system, and effluent oxygen control system.

Figure 1. **Schematic of pilot-scale GAC-FBR treating PAHs and nitrophenols.**

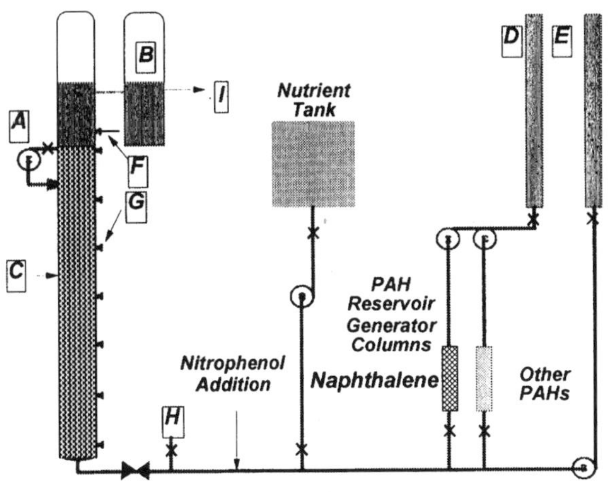

RESULTS

Pilot-Scale Testing

The GAC-FBR was biologically started in June 1992. The initial applied organic loading rate (OLR) to the system was approximately 1.6 kg COD/m^3-d. More than 80% of this applied OLR was naphthalene. Complete removal of naphthalene was observed from system startup. Effluent naphthalene concentrations were near or below detection limits (3 µg/L) over the first ten days, with removals of greater than 99%. Dissolved oxygen (DO) consumption through the reactor was initially much lower than required to oxidize the amount of naphthalene and other PAHs removed in the GAC-FBR. Removal during the initial startup was due primarily to adsorption. From Day 16 on, the measured DO consumption was commensurate with the mass of oxygen required to completely oxidize the mass of PAHs being added to the reactor, indicating removal was due to biological oxidation.

The results at the highest OLR examined, 4.6 kg COD/m^3-d are presented in Table 1. Even at the OLR of 4.6 kg COD/m^3-d, analysis of PAH concentrations through the profile of the FBR showed that the bulk of the removal was accomplished in the bottom half of the reactor. This indicates that the system was capable of handling higher PAH concentrations and OLRs than those applied during the pilot test.

Table 1.   **Performance data for a laboratory-pilot GAC-FBR treating a mixture of PAHs.**

Number of Rings	Influent Conc. (µg/L)	Effluent Conc. (µg/L)	Removal (%)
2	5600	<3.0	>99.9
3	516	<10	98.1
4	72	1.5	97.9
5	1.17	0.23	80.3
Total PAH	6189	<14.8	>99.7
Applied OLR - 4.6 kg COD/m3-d			
HRT = 6 min			

Batch kinetic assays using biomass from the GAC-FBR and 2-4 ring PAHs as carbon sources (conducted with appropriate abiotic and inhibited controls) verified that PAH disappearance was primarily due to biological oxidation and not adsorption. Acute toxicity, measured as EC_{50} in the Microtox® assay, indicated essentially complete reduction in toxicity. While the influent EC_{50} was around 10%, effluent samples did not exhibit any toxicity, indicating the absence of toxic intermediates or by-products from the biological oxidation process. Detailed analysis of results from the pilot-scale feasibility study are presented elsewhere (Hickey et al., 1995; Rajan et al., 1995; Sunday et al., 1994).

Field Demonstration

Based on the positive results obtained at the pilot-scale, field efforts were undertaken to examine the effectiveness of the GAC-FBR at different, but representative cases for the treatment of waters at MGP sites. These included treatment of an effluent stream generated from deep subsurface DNAPL removal (CROW™ process) at a Superfund site, and treatment of groundwater containing BTEX, PAH, and high concentrations of reduced iron (>100 mg/L).

Site 1: Treatment of Groundwaters and Process Waters from CROW™ Extraction of DNAPL from the Subsurface

Site Description. At this Superfund site, formerly an active MGP facility for the Pennsylvania Power and Light Company (PP&L) located in Stroudsburg, PA, dense nonaqueous phase liquid (DNAPL) coal tar was trapped in a stratigraphic depression 30 feet below the surface. Excavation was not a feasible option. A hot water (CROW™) extraction technology was the alternative selected for removal of the tar. Details of the CROW™ process option have been presented elsewhere (Leuschner et al. 1996). At this site, up to 100 gpm of water was heated to 160°F and injected into six 2-inch wells. These injection wells were placed in a ring around two 6-inch production wells from which a mixture of water and tar was extracted. To maintain hydraulic containment, approximately 15 gpm more than the water flow injected (115 gpm total) was pumped from the production wells. This excess water was treated using the GAC-FBR prior to surface discharge.

Treatment System Description. A schematic representation of the treatment train used at the PP&L site is presented in Figure 2. Subsequent to extraction, the water was acidified to aid in destabilizing any tar-water emulsion and to accelerate the tar-water separation process. Iron oxidation and removal, to prevent precipitation and plugging of the water heating systems reinjection wells with oxidized iron, followed gravity separation of the DNAPL. Fe(II) in the groundwater was oxidized using hydrogen peroxide at a basic pH. Oxidized iron particulates were settled in a holding tank. A portion of the extracted water (15 gpm) was continuously pumped to the GAC-FBR, to remove dissolved organic contaminants. The remaining water was reheated and reinjected, with appropriate pH control to ensure a Langelier Saturation Index (LSI) that would prevent scaling in the heater and precipitation of calcium and magnesium in the reinjection wells. Effluent from the GAC-FBR was passed through a filter and carbon polishing units prior to direct discharge into Brodhead Creek.

System Performance. Early in the project, problems with the heater resulted in low temperature water (essentially groundwater) being extracted and treated. The water had relatively low concentrations of organics present.

Weekly grab samples of the effluent from the FBR and the carbon polishing units were collected. Essentially complete removal (>99%) of BTEX, naphthalene, and the other PAH was achieved with this dilute concentration feed.

Figure 2.   **Process schematic - groundwater extraction and DNAPL removal from subsurface.**

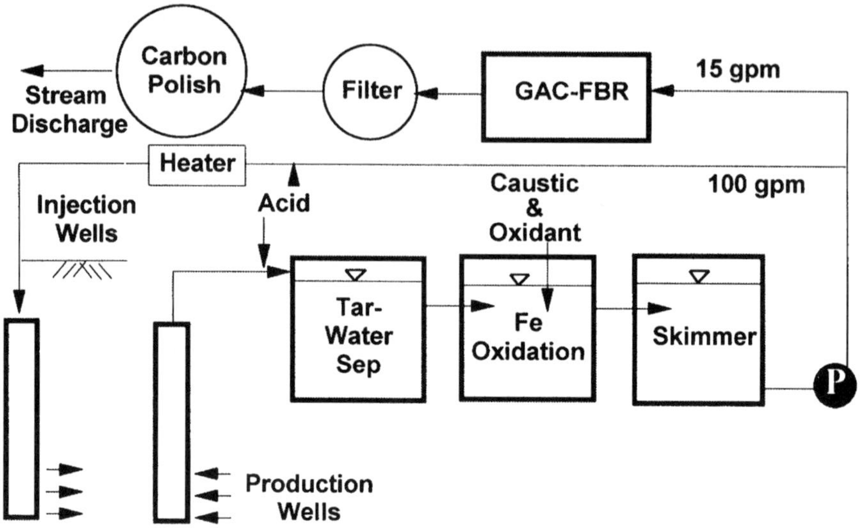

Once heat was applied to the system and DNAPL removal visually verified, the concentration of PAHs in the water increased.

Results over a six-month period are presented in Table 2. As observed during the pilot test, approximate 99% removal of BTEX and lower ring PAHs was achieved. The GAC polishing unit was observed to provide some additional removal of PAH, particularly for the 5- and 6-ring PAHs. The average PAH concentrations in the GAC-FBR effluent prior to filtration and GAC adsorption was 324 µg/L (95.3% removal efficiency). The concentration of PAHs in the GAC-FBR influent and effluent, as well as the effluent after GAC polishing during this six-month period are presented in Figure 3. Removal by the post treatment system was primarily due to capture of emulsified, high-ring PAHs. The average concentration of the higher-ring PAHs in the influent were well in excess of their solubility. A partial listing of influent concentrations of 4- through 6-ring PAHs and their solubility limits are presented in Table 3. Better removal of emulsified tars in the up-front treatment process would have reduced the need for any post-treatment. Nonetheless, the high removal of all PAHs (>95%) was achieved by the GAC-FBR. When coupled with post filtration and carbon polishing, greater than 99% removal was consistently achieved.

Site 2: Treatment of MGP Groundwaters High in Reduced Iron

Site Description. This field demonstration was conducted at a former MGP facility in New York. The primary objective of this study was to demonstrate the ability of the GAC-FBR technology to treat groundwaters containing BTEX and PAHs concurrent with elevated levels of iron (greater than 100 ppm). These high

Table 2. **Average performance data for BTEX and PAH by the GAC-FBR system.**

	Influent Conc. (µg/L)	Effluent Conc. (µg/L)	Removal (%)
Benzene	67	<1	>98.5
BTEX	500	5	99.0
PAH			
2-ring	3360	30.0	99.1
3-ring	2847	3.6	99.9
4-ring	553	27.8	95.0
5-ring	101	11.0	89.1
6-ring	33.2	6.3	81.0
Total PAH	6890	78	98.9

Average values of 15 sampling events from September 25, 1995 through February 14, 1996.

Figure 3. **Total PAH concentrations at PP&L.**

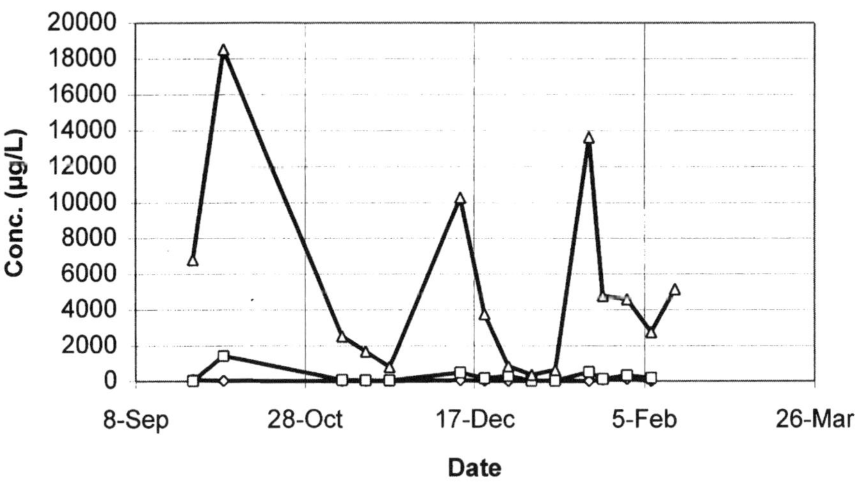

Table 3.   Comparison of influent concentration and solubility of selected 4- through 6-ring PAHs.

Compound	Influent Conc. (µg/L)	Solubility (µg/L)	Factor Above Solubility
Pyrene	282	140	2.0
Chrysene	59.3	6	9.9
Benzo(a)anthracene	74.5	10	7.5
Benzo(a)pyrene	76	3.8	20.0
Dibenzo(a,h)anthracene	6.25	0.5	12.5
Benzo(ghi)perylene	23.2	0.26	89.2
Indeno(1,2,3-c,d)pyrene	10	<0.5	>20.0

levels of reduced iron are present in the groundwater due to the presence of acid purifier box wastes at this site in the location where the groundwater was extracted from.

System Description. A schematic representation of the treatment train used at this site is presented in Figure 3. Iron removal was accomplished using permanganate oxidation, precipitation, flocculation, and settling operations prior to the treatment of organics in the GAC-FBR. Groundwater was pumped at 10 gpm, using a sump located in a storm sewer extension at the site. The storm sewer was installed in a granular fill bedding. This acted as a collection system for groundwater at the site. To operate the pilot treatment system under the site permit exclusion, effluent from the GAC-FBR was discharged to the existing wastewater treatment facility. The concentration of organics in the groundwater was quite low (400 µg/L benzene, 800 µg/L total naphthalenes and 100 µg/L or less of the other regulated PAHs). The resulting influent BODs averaged less than 20 mg/L. Effluent discharge criteria for organics were 10 µg/L for benzene, toluene, xylenes, and each of the 16 regulated PAHs. The entire treatment train was housed inside a temporary, heated, tension fabric structure.

Influent and effluent parameters that were measured on a regular basis included concentrations of total and dissolved iron and manganese, BTEX, VOCs, phenolics, PAHs, BOD and COD, pH, and DO. Standard EPA methods were used for all the analyses listed above.

System Performance. Three operational regimes were tested during the course of the study at this site. These were:

- Iron oxidation and settling in the pretreatment unit, followed by removal of any residual oxidized iron by filtration using a multimedia filter and subsequent treatment of the organics in the GAC-FBR,

- Iron oxidation in the pretreatment unit, followed by treatment of organics in the GAC-FBR and subsequent removal of any residual oxidized iron and suspended solids by filtration using a multimedia filter, and
- Augmentation of BTEX and naphthalene concentrations in the groundwater to examine performance capacity for treatment of groundwater with high concentrations of organics and, therefore, high applied OLRs.

Average results from these three periods are presented in Table 4. Details of operation during these periods are provided below.

Operational Period 1. During the first operational period, total PAHs in the raw influent averaged around 500 µg/L, over 75% of which was accounted for by naphthalene; BTEX in the influent averaged 250 µg/L. The concentrations of BTEX and naphthalene were reduced by approximately one-half in the pretreatment unit, resulting in a low applied OLR to the GAC-FBR. Effluent concentrations from the GAC-FBR of less than 5 µg/L BTEX (<1 µg/L benzene) and less than 10 µg/L total PAHs were consistently achieved. Accumulation of biomass in the GAC-FBR was limited during this period due to the low applied OLR.

During this first phase, before tight control of the iron oxidation system was achieved, the GAC-FBR system was exposed to shock loads of residual oxidant (permanganate), varying amounts of residual flocculant (Drewfloc 2270), and episodes with relatively high concentrations of reduced iron. The quality of effluent from the system was not adversely affected by any of these situations. Much of the DO consumption observed in the GAC-FBR during this period could be attributed to degradation of the residual Drewfloc flocculant (biodegradable polyacrylic acid) from the iron oxidation system. BTEX and PAHs were consistently reduced to below detection limits in the effluent from the GAC-FBR during this two-month period (Table 4).

Operational Period 2. In an attempt to eliminate losses of PAHs and BTEX in the pretreatment unit, the multimedia filter (thought to be biologically active) was moved downstream of the GAC-FBR. This change in configuration resulted in oxidized iron that carried over from the metals oxidation system entering the GAC-FBR system. Over the next few weeks, effluent quality remained high. Effluent BTEX and PAH concentrations remained below detection limits.

Table 4. **Performance of GAC-FBR during three different operating modes.**

Period	BTEX (µg/L)		PAH (µg/L)	
	Influent	Effluent	Influent	Effluent
1	271	<1	174	<5
2	272	<1	153	<5
3	11,640	<1	1331	<5

Operational Period 3. Following successful treatment of the groundwater for over 10 weeks, the ability of the system to handle higher applied OLRs was tested. This was achieved by supplementing BTEX and naphthalene to the groundwater as it entered the iron oxidation system. The resulting influent concentrations of BTEX and naphthalene ranged from 5,000-20,000 µg/L and 1,000-8,000 µg/L, respectively (Figure 4a). This corresponded to applied OLRs of 1.5 to 7 kg COD/m³-d. Rapid accumulation of biomass occurred, resulting in growth of the fluidized bed, up to the bed level control point. DO consumption in the system in-

Figure 4. **Concentrations of BTEX and naphthalene at a New York MGP site (a-influent; b-effluent).**

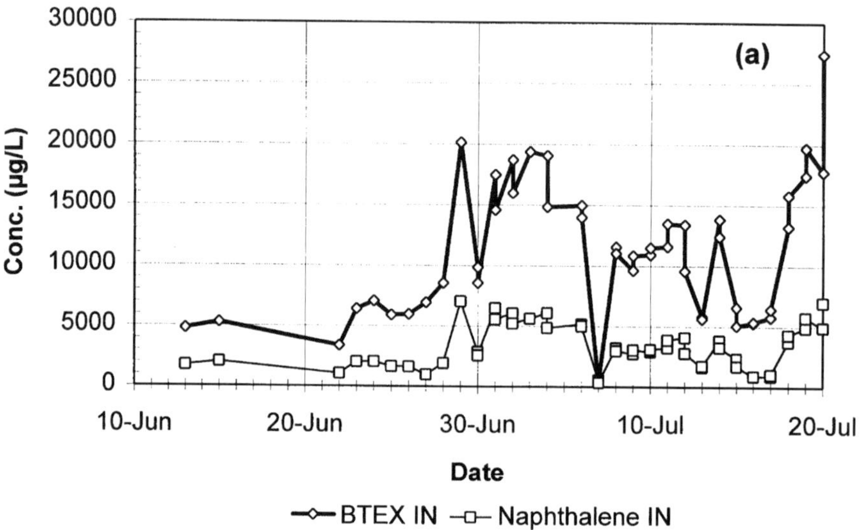

creased concurrently. Removal efficiencies for BTEX and naphthalene remained greater than 99%; effluent BTEX and naphthalene concentrations remained below detection limits (Figure 4b). Vertical concentration profiles of BTEX and naphthalene, as shown in Figure 5, indicated that the bulk of the removal was achieved in the bottom several feet of the reactor. Based on these data, if oxygen solubility was the only limiting factor to the maximum applied OLR that could be achieved, an OLR of approximately 15 kg COD/m³-d could be used before any deterioration in effluent quality would be observed.

Economic Evaluation. Based on the results of this demonstration and the expected concentrations of BTEX and PAHs in the groundwater at this site, comparative cost estimates of two alternative treatment systems to handle 200 gpm were prepared by Remediation Technologies, Inc. based on vendor quotes. The processes compared for removal of organics were the GAC-FBR and air stripping followed by carbon adsorption (liquid phase carbon for the semivolatiles [PAH] and other organics remaining after stripping) with vapor phase control. On an annualized basis (including both capital and operating costs), the GAC-FBR system was projected to cost less than 30% of the alternative process configuration, $72,200 versus $252,000 per year. Based on this, the GAC-FBR was recommended for full-scale application.

SUMMARY

The ability of the GAC-FBR to treat groundwater containing PAHs was shown to be feasible, using a 550 gpd pilot. Removal efficiencies of 2- to 4-ring PAHs were 99% for all OLRs applied. Based on results from concentration profiles of PAHs and DO and batch tests using biomass from the system, this removal was due primarily to biological oxidation once a mature biofilm developed (less than three

Figure 5. **Vertical Concentration Profile of BTEX and Naphthalene in the GAC-FBR at a New York MGP Site**

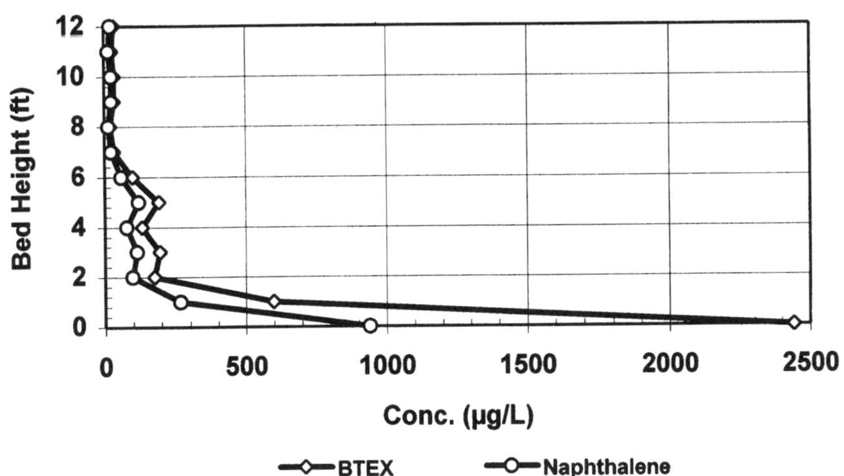

weeks). Only 1.2% of the PAHs fed to the GAC-FBR over a seven-month period was retained on the GAC. These high-removal efficiencies were replicated in two field demonstrations under the typical conditions in the field (i.e., variability in inlet concentration, the presence of emulsified tars, DNAPL, reduced iron, etc.).

Based on the results obtained from two field sites, the GAC-FBR appears to be a reliable, cost-effective treatment option for treating waters from MGP sites. At both sites, high effluent quality was continually achieved. Under normal operation, BTEX and lower molecular weight PAHs were generally reduced to below detection limits. High overall PAH removal efficiency was maintained even with significant amounts of emulsified PAHs. The presence of excess potassium permanganate, polymer, or iron in the influent at the NY site did not affect performance. Applied OLRs of up to 7.0 kg COD/m³-d were successfully treated.

Comparative economic analyses were performed by the site owner's consulting firm for the GAC-FBR and air stripping followed by carbon adsorption for the vapor phase and GAC treatment for the remaining PAHs and other semivolatiles in the liquid phase. Results demonstrated that the GAC-FBR was only 30% the total operating cost of the physical-chemical treatment alternative. Total annual costs (capital and operating costs) for the GAC-FBR were estimated to be just less than $0.69 per 1000 gallons treated.

REFERENCES

Hickey, R.F. and R.W. Owens. 1981. Methane Generation from High-Strength Industrial Wastes with the Anaerobic Biological Fluidized Bed. In: *Biotechnology and Bioengineering Symposium No. 11*, pp. 399-413. New York, John Wiley and Sons.

Hickey, R.F., et al. 1993. Application of the GAC-FBR for Treatment of Gas Industry Wastewaters. Presented at Institute of Gas Technologies 6th International Symposium on Gas, Oil and Environmental Biotechnology, Institute of Gas Technologies, Colorado Springs, CO, November.

Hickey, R.F., et al. 1994. Demonstration of Commercial Scale GAC-FBRs for Gas Industry Waste Treatment. Proceedings, 7th International IGT Symposium on Gas Oil and Environmental Biotechnology, December 12-14.

Hickey, R.F., et al. 1995. Treatment of PAH in Waters using the GAC-FBR Process. In: *Biological Unit Processes for Hazardous Waste Treatment*, pp. 21-28. (Hinchee, R.E. and Skeen, R.S., Eds.). Columbus, OH, Battelle Press.

Hickey, R.F., Wagner, D.J., and Mazewski, G. 1991. Treating Contaminated Groundwater using a Fluidized Bed Reactor. *Remediation*, 2, 447-460.

Jeris, J.S. and Owens, R.W. 1975. Pilot Plant, High-Rate Biological Denitrification. J. Water Pollut. *Control Fed.*, 47, 2043-2057.

Jeris, J.S., Beer, C. and Mueller, J.A. 1974. High Rate Biological Denitrification using a Granular Fluidized Bed. *J. Water Pollut. Control Fed.*, 46, 2118-2128.

Jeris, J.S., Owens, R.W., Hickey, R.F. and Flood, F. 1977. Biological Fluidized Bed Treatment for BOD and Nitrogen Removal. *J. Water Pollut. Control Fed.*, 49, 816-831.

Leuschner, A.P., et al. 1996. Full-scale Enhanced Recovery System for Coal Tar Removal from a Former MGP Site. Paper presented at the Conference on Hazardous Waste and Environmental Management in the Gas Industry, Albuquerque, NM, January 22-24.

Rajan, R.V., et al. 1995. Pilot and Full-scale Treatment of PAHs in Wastes from MGP Sites using the GAC-FBR Process. Proceedings, 2nd International Petroleum Environmental Conference, New Orleans, September 25-27.

Suidan, M.T., Cross, W.H., Fong, M., and Calvert, J.W., Jr. 1981. Anaerobic Carbon Filter for Degradation of Phenols. *J. Environ. Eng. Div., ASCE*, 107, EE3, 563-579.

Sunday, A.L., et al. 1994. Treatment of Manufactured Gas Plant Site Groundwaters in a Biological Fluidized Bed Reactor. Proceedings of the AWMA National Conference.

Sutton, P.M., Shiek, W.K., Woodcock, C.P., and Morton, R.W. 1979. Oxitron System Fluidized Bed Wastewater Treatment Process: Development and Demonstration Studies. Presented at the Joint Annual Conference on the Air Pollution Control Association and the Pollution Control Association of Ontario, Toronto, Canada, April.

Switzenbaum, M.S. and Jewell, W.J. 1980. Anaerobic Attached-Film Expanded Bed Reactor Treatment. *J. Water Pollut. Control Fed.*, 52, 1953-1965.

Wagner, D. J., Sunday, A. L. and Hickey, R. F., "Application of the Biological GACFBR Process for Gas Industry Wastes," GRI Topical Report, (1993).

CHAPTER 46

Case Study: MGP Site Remediation Using Enhanced DNAPL Recovery

Alfred P. Leuschner, Mark W. Moeller, and **Jason A. Gerrish**, Remediation Technologies, Inc., Concord, Massachusetts

Lyle A. Johnson, Western Research Institute, Laramie, Wyoming

INTRODUCTION

The Brodhead Creek Site is a former manufactured gas plant (MGP) facility located in Stroudsburg, Pennsylvania. The site has been remediated using an enhanced dense nonaqueous phase liquid (DNAPL) recovery technique termed Contained Recovery of Oily Wastes (CROW™). This paper summarizes the work performed since the start of the project, provides results of the remediation, and remediation costs.

The Brodhead Creek Site occupies a floodplain area of approximately 12 acres at the confluence of Brodhead Creek and McMichael Creek. The plant was in operation from 1988 to 1944. In 1955, Brodhead Creek overran its banks and flooded the site. Due to this flood, the Army Corps of Engineers designed and built a flood control levee for the local creeks. In 1980, as the flood control levee on Brodhead Creek was extended, a coal tar discharge was noted. Emergency response measures performed by Pennsylvania Power & Light Company (PP&L) included installation of a filter fence and underflow dams, and the installation of a coal tar recovery pit. In 1981, a cement and bentonite slurry wall was installed along Brodhead Creek. The slurry wall is 648 feet long and 17 feet deep and serves to protect Brodhead Creek from the discharge of free coal tar from the site. In 1982, the site was placed on the National Priorities List (NPL). From 1982 to 1983, a free coal tar recovery pumping system was installed and 8,000 gallons of coal tar were recovered from the subsurface. A Remedial Investigation (RI) was conducted from 1987 to 1989, and a Feasibility Study (FS) was conducted from 1989 to 1991.

According to the RI and subsequent investigations, free subsurface coal tar existed in two areas at the site: the area near the recovery well RCC and the area near monitoring well MW-2. These areas became known as the RCC and MW-2 areas, respectively. Figure 1 shows the locations of these areas within the site. The

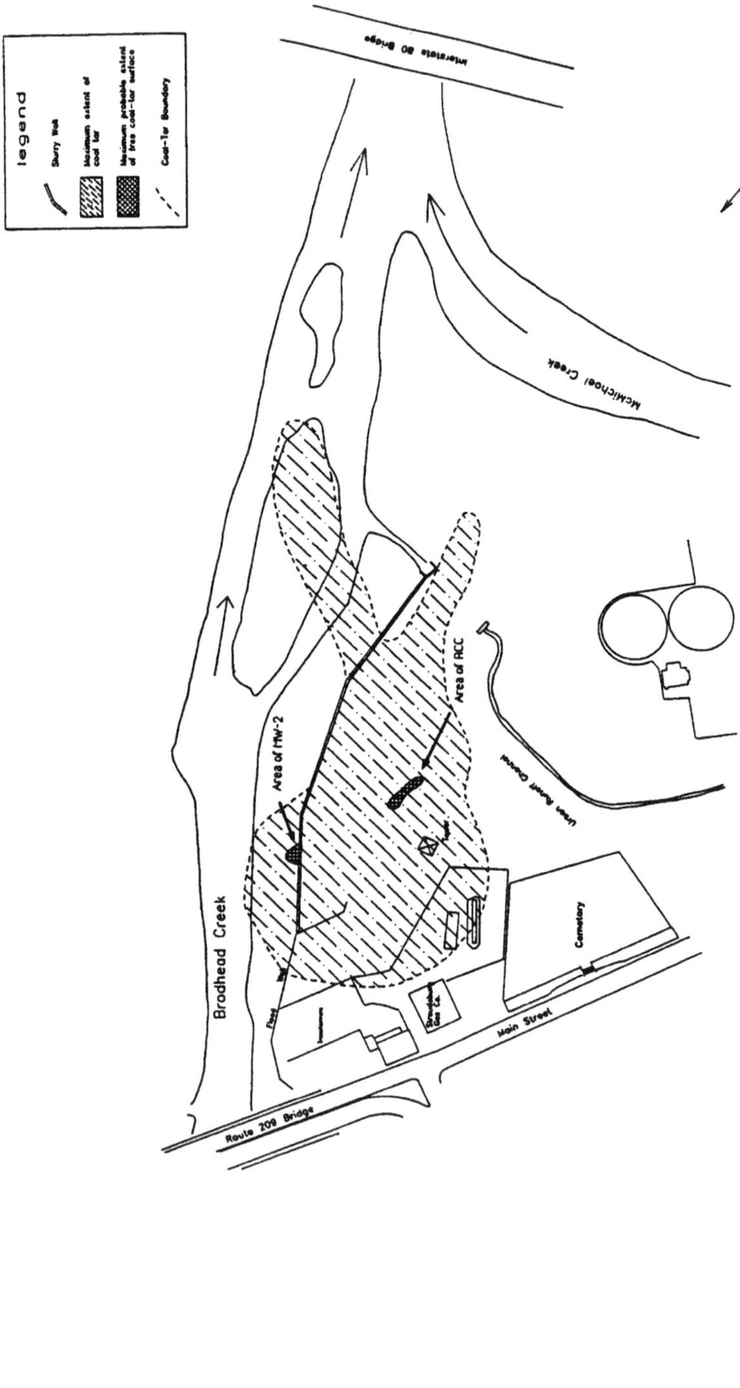

Figure 1. Locations of RCC and MW-2 areas.

coal tar appeared to have permeated the porous stream gravel layer near the surface, and accumulated on a silty sand layer 20 to 30 feet below the surface. Investigations revealed that the silty sand layer effectively contained further vertical migration of the coal tar. The two areas of free coal tar accumulation are stratigraphical depressions within the stream gravel layer. The RCC area is a natural depression area, but the MW-2 area is not. Coal tar appears to have accumulated in the MW-2 area due to the presence of the slurry wall. Figure 2 shows that dense coal tar may have accumulated near MW-2 only after the slurry wall was installed. The total volume of free coal tar in the RCC area was estimated at 6,000 gallons, and the total volume of free coal tar in the MW-2 area was estimated to be less than 300 gallons in the RI.

Figure 2.   **Vertical distribution of free coal tar.**

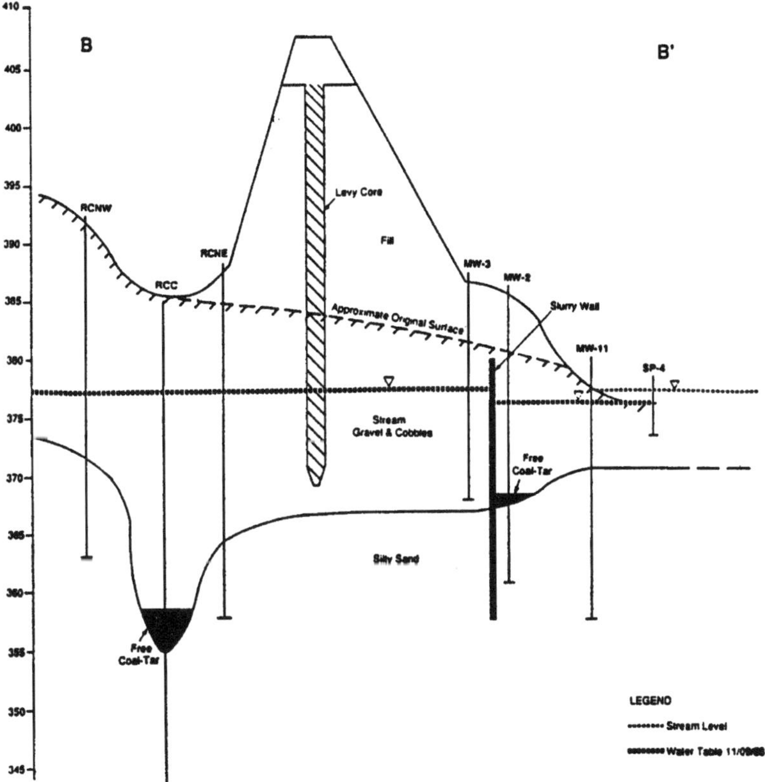

In March of 1991, the Record of Decision (ROD) was issued that targeted enhanced product recovery as the selected remedial technology for Operable Unit 1 (OU-1), the free, subsurface coal tar. The Consent Decree was negotiated and signed in December 1991 by the potentially responsible parties (PRPs). The Consent Decree was entered in the U.S. District Court for the Eastern District of Pennsylvania on September 2, 1992. EPA Region III agreed to allow coal tar removal in the RCC area first, before committing to a removal technology for the MW-2 area. According to the ROD, all recovered coal tar is to be incinerated. This paper discusses the remedial efforts for the RCC area.

The remediation entails the injection of hot water into the subsurface to mobilize the coal tar. The hot water is injected at the perimeter of the tar formation and withdrawn from the center of the formation. In this way, the site is hydraulically isolated. Tar is mobilized by the hot water, and will flow to the production wells, where along with the hot water, it is removed from the subsurface.

On the surface, an initial water treatment step separated the coal tar from the water. Next, dissolved inorganics, predominantly iron and manganese, were removed through oxygenation of the water and pH adjustment. The inorganics were subsequently separated by gravity. Next, the water was reheated and reinjected into the subsurface. A greater amount of water was produced than injected so the formation would be hydraulically isolated. This overdraw of water was treated in a fluidized bed reactor and discharged to Brodhead Creek.

Figure 3 provides a timeline for this entire project. The ROD was signed in March of 1991. From then until September 1992, the Consent Decree was negotiated and eventually signed. The design process required another 19 months to complete, and included preparation of the Remedial Design Work Plan, 30% Design, Prefinal Design, and Final Design. The design for the remedy was completed and approved by EPA Region III in March 1994. Construction was initiated in June 1994 and continued until November 1994. Startup required approximately 10 months of operation and necessitated several design modifications as will be discussed later. Actual operation began in July 1995 and continued through May 1996, when the system was shut down and decommissioning was started.

PREVIOUS STUDIES

An initial laboratory study was performed on material from the Brodhead Creek Site. This study was sponsored by the Gas Research Institute (GRI) and the EPA SITE Program. It was performed by the Western Research Institute (WRI) and Remediation Technologies, Inc. (RETEC). The purpose of this study was to demonstrate the applicability of this technology to material found at MGP sites as well as to develop some site-specific data for the Brodhead Creek Site. The results of this study showed that the tar content of the site soils could be substantially reduced as water flushing temperatures were increased to an optimum level of 156°F. Ultimately, over 60% of the coal tars in soil could be removed with 98% of the

Figure 3. Project timeline, Brodhead Creek Superfund Site, Stroudsburg, PA.

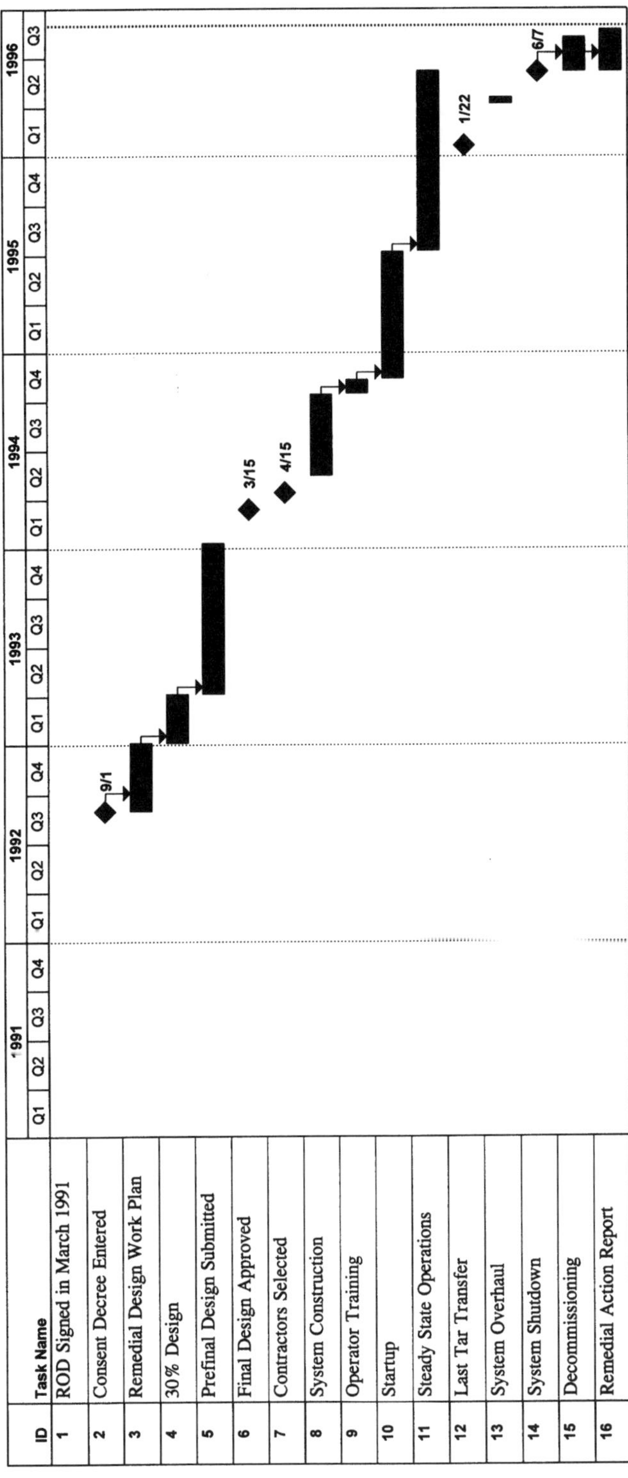

removable tar being recovered after 19 pore volumes flushed through the surface (Leuschner and Johnson, 1990). These data formed the basis for design of the system ultimately installed at the site.

A predesign investigation was performed by Atlantic Environmental Services, Inc. for the purpose of delineating the pool of coal tar in the RCC area of the site. Thirty-eight borings were installed in this area of the site and 23 were finished as piezometers. Over a four-month time period, tar level measurements were taken three times. Of the 23 peizometers, 11 consistently showed free tar levels. These levels varied from a few inches to over six feet depending on the piezometer location (Atlantic, 1993). The results of this study were then used to define the areal extent of the remedy.

SYSTEM DESIGN AND INSTALLATION

The CROW™ system for the RCC area at the Brodhead Creek Site was designed to recover approximately 6,000 gallons of subsurface coal tar. The CROW™ process relies on injection and recovery wells drilled in such a pattern that groundwater will be laterally contained within the area of concern. Hot water is injected into the aquifer at several points surrounding the tar accumulation in order to heat and mobilize the main accumulation. Heating the tar reduces both the density and viscosity of the tar to enable the tar to migrate with the groundwater flow.

Tars are contained vertically during the hot water flush by controlling temperatures during the hot water displacement. Downward migration of tar is reduced by thermal expansion of the tar to a lower density. Balancing the hot water injection and production rates controls the upper boundary of the contaminated area preventing fluid displacement into the overlying material. Volatilization of the lighter constituents of the tar can be controlled by maintaining a cold water cap over the pattern area. However, at the Brodhead Creek Site, approximately 15 feet of groundwater exists over the area to be swept. This acts as a natural cold water cap. Any volatile organics would subsequently condense once they reach the cold water and migrate back to the sweep area.

Mobilized tar and soluble constituents are contained laterally by groundwater isolation. Based on specific hydraulic measurements from the Brodhead Creek Site, WRI modeled several injection and production scenarios in order to design the appropriate well pattern for the RCC area. The well pattern and pumping rates were specifically designed to contain all the mobilized liquid phases by controlling the potentiometric surface around the affected area.

The CROW™ well pattern included six external injection wells and two internal production wells. The injection wells were installed by boring with a 10-inch hollow stem auger, placing a 20- to 25-foot segment of five-inch stainless steel well and well screen within the hole, backfilling the annulus with an 0-variety sand pack, grouting the sand pack to the surface, and sealing the well in place with concrete. One of the wells was installed through the flood control levee. In order to install this well, all well installation plans had to be approved by the Bureau of

Flood Protection and a soil ramp had to be constructed adjacent to the dike in order to access the well installation point. A special track-mounted rig was mobilized to install this injection well.

The production wells were installed by initially advancing a 24-inch borehole to the bottom of the stream gravel layer. However, some of the cobbles within this layer were eight inches in diameter, thus slowing the advancement of the hole. The first attempt utilized a bucket auger rig. After this rig advanced the hole approximately 14 feet, the surrounding earth caved into the hole, backfilling it to only eight feet. At this point, a 24-inch casing was advanced and the material within the casing was removed with the rig. However, the casing met refusal at 21 feet, four feet short of its final position. Without the ability to drive the casing, a Barber rig was mobilized to the site. Several of the existing monitoring wells were cut back and flush mounted in order to provide the clearance necessary to bring in the large equipment trucks that were mobilized with a Barber rig. The Barber rig was successful at installing the two 25-foot segments of 14-inch galvanized well and well screen. The annulus was backfilled with 0-variety sand, grouted to the surface, and sealed with a concrete cap. All soil cuttings were stored within a 20 cubic yard rolloff box and disposed of at an industrial waste landfill. The well installation activities produced approximately 30 cubic yards of soil waste materials.

The heated aquifer water altered the tar density and viscosity such that it can be recovered with the water phase. The in-well submersible pumps used can create a tar-water emulsion that is pumped into the separation tanks on the surface. The tar at the Brodhead Creek Site has a specific gravity of 1.06 at 70°F. However, when heated to 100°F, the tar has a specific gravity of approximately 1.00 due to thermal expansion. Therefore, in order to separate the tar from the water, any emulsion must be broken and the temperature must be cooled.

During the treatability testing, WRI attempted to break simulated emulsions by adjusting the water chemistry. WRI concluded that reducing the pH to 5.0 would sufficiently break any tar/water emulsion. Therefore, an acid tank was plumbed into the production line so the pH of the produced waters could be lowered to 5.0 before beginning the separation process. The remainder of the separation process relied on gravity separation. In order to accomplish this, the temperature of the liquid in the tanks would have to be reduced as much as possible. Because this system was designed to operate during the winter, all separation tanks were specified to be 0.25-inch, noninsulated carbon steel. This would allow the ambient air to sufficiently cool the liquids and promote gravity settling.

Each of the tar separation tanks was designed to provide 2.9 hours of retention time. Because the three separation tanks were connected with gravity lines with no pumps or mixers, the total retention time available was 8.7 hours. Each of these tanks were 20,000 gallons in volume. A fourth 20,000-gallon tank was used as a reservoir prior to recycling the water. Finally, a fifth tank, 10,000 gallons, was used to store tars generated by the recovery process prior to disposal.

In order to contain lateral movement of liquids, the injection/production system was operated in a manner to induce a gradient toward the production wells. The gradient was significant enough to overcome other natural gradients such that all injected water was captured by the production wells. This is accomplished by producing more water than is injected. This system was designed to continuously produce 115 gpm from the aquifer, while injecting only 100 gpm. After coal tar separation, 100 gpm of the remaining water was pumped to the water heater which raises the water temperature to 180°F. The heater water is then reinjected into the ground. The remaining 15 gpm was treated by an Envirex Model 190 GAC/FBR followed by four 200-pound carbon adsorption (CA) units (two parallel sets of two). The GAC/FBR provides a continuous reactor feed of 190 gpm. The forward flow from the tar separation system is 15 gpm and the reactor recycle flow is 175 gpm. The reactor tower is 15 feet high and 4 feet in diameter. The reactor is filled with water and granular-activated carbon which remains in fluidized suspension by the upward 190 gpm flow through the reactor. The fluidized carbon provides a tremendous amount of surface area in which to grow bacterial colonies. The Model 190 unit is equipped with nutrient feed and oxygen injection systems which are calibrated to provide optimal bacterial growth conditions. The CA units were installed at the request of the EPA to provide a polishing step, and to reduce the sampling, analytical, and water storing requirements prior to discharge.

OPERATIONS RESULTS

From November 1994 through July 1995, the CROW™ process was intermittently operated in a startup mode at the site. There was one significant problem that caused a series of operational problems which required system modifications. The addition of hot water in the subsurface at the site caused an increase in reduced (dissolved) iron in site groundwater from background levels of less than 2 ppm to over 8 ppm. When the groundwater was brought to the surface and exposed to the air, oxygen was introduced into the water. This caused the iron to oxidize and precipitate. The iron precipitate caused coal tar separation problems, fouling of the GAC/FBR, heater fouling, fouling of the CA units and plugging of the injection wells.

The problem was solved through a variety of changes. A schematic of the process with modifications is shown in Figure 4. To aid in the separation of coal tar and water, acid injection to the production water was used to lower the pH to 5.0. This assisted in cracking any emulsions formed and allowed the tar to settle. Subsequently, the pH was raised to 7.0 and peroxide was added to oxidize the iron in the produced water. These solids were then gravity separated in Tank 2. A series of bag filters were added throughout the process, before the heater, before the injection wells, and before the CA units, along with a sand filter, to remove any precipitated solids that carried over from the gravity separators. A well juttering system was installed on the injection wells to assist in keeping them clear of solids (Moeller et al., 1995).

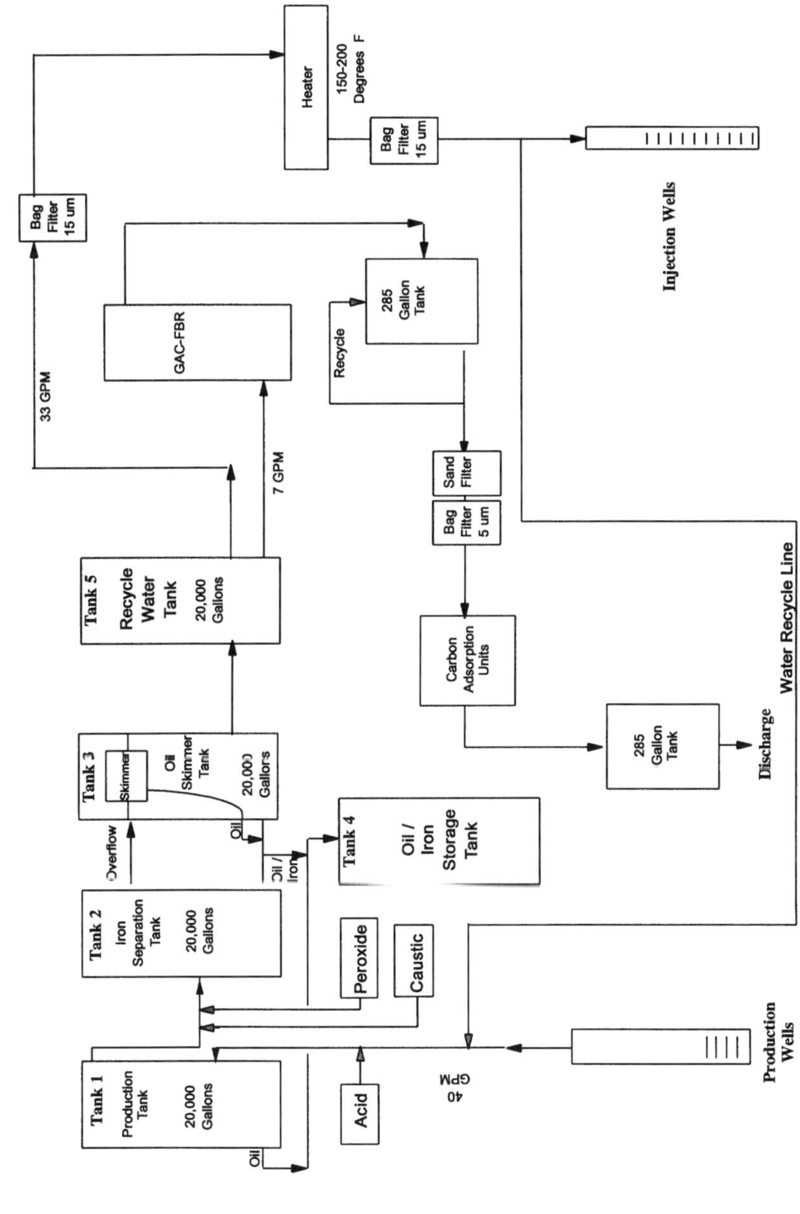

Figure 4. Process flow diagram, as-built.

From July 1995 to June 1996 the system was continuously operated and recovered coal tar from the subsurface. Figure 5 shows injection and production flow rates throughout operation. Design flows of 100 gpm injection and 115 gpm production were never reached predominantly due to the initial iron solids problem. Injection flow rates of 33 gpm and production of 40 gpm were typical of this operation. Temperature of injection and production water are shown in Figure 6. Injection temperatures varied from 160 °F to 200°F, and production temperatures varied from 100°F to 150°F. The system was shown to be effective in hydraulically isolating the area being swept by the CROW™ process. Figure 7 shows the temperature profile of the groundwater throughout the RCC area. Thermocouples were installed throughout the site to record these data. The figure shows that groundwater temperatures of 150°F were achieved around the injection wells (IW-1 through IW-6) and 130°F temperatures were achieved at the production wells (PW-1 and PW-2). However, within 20 feet outside of the 150°F isotherm, the groundwater was at the background temperature of 55°F. This shows that not only was the hot water being effectively directed at the area being swept, but also that mobilization of coal tar outside of the sweep area was not occurring.

Over the ten-month period of operation, a total of 29 pore volumes of hot water were swept through the RCC area. The hot water was effective in recovering the tar from the formation. Figure 8 presents the cumulative coal tar recovered with respect to pore volume flow. A total of 1,424 gallons of tar was recovered as pure tar (no water or organics). This was determined through chemical analyses of the material disposed of from the site. Along with the tar, water and inorganic solids (iron sludge) were removed by tankers. Table 1 shows the quantities removed. Along with the 1,424 gallons of tar, 6,064 gallons of water and 3,206 of inorganic solids were collected and disposed of.

Figure 5.   **Production and injection rates, CROW System, Brodhead Creek.**

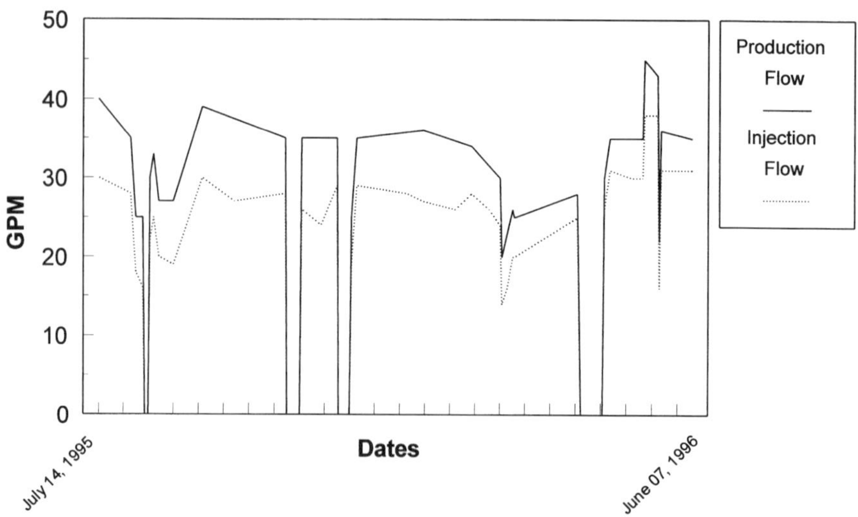

Figure 6. **Production and injection temperatures, CROW System, Brodhead Creek.**

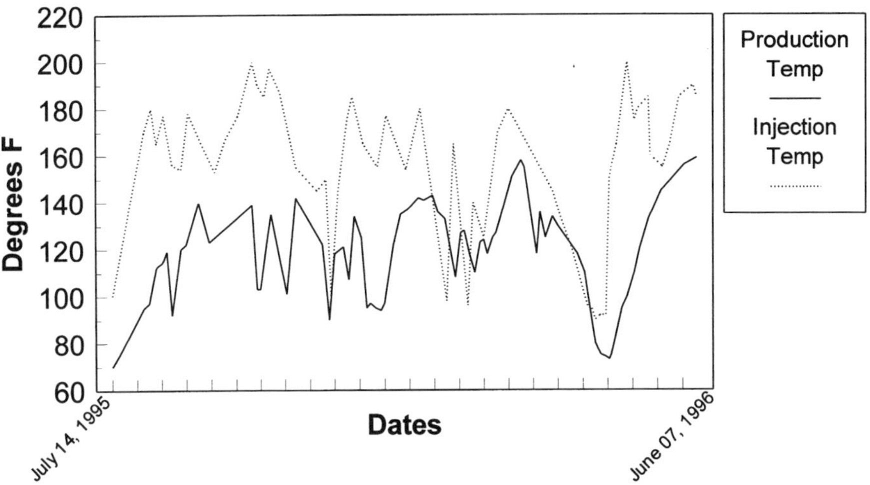

Figure 7. **Subsurface temperature contours, Brodhead Creek.**

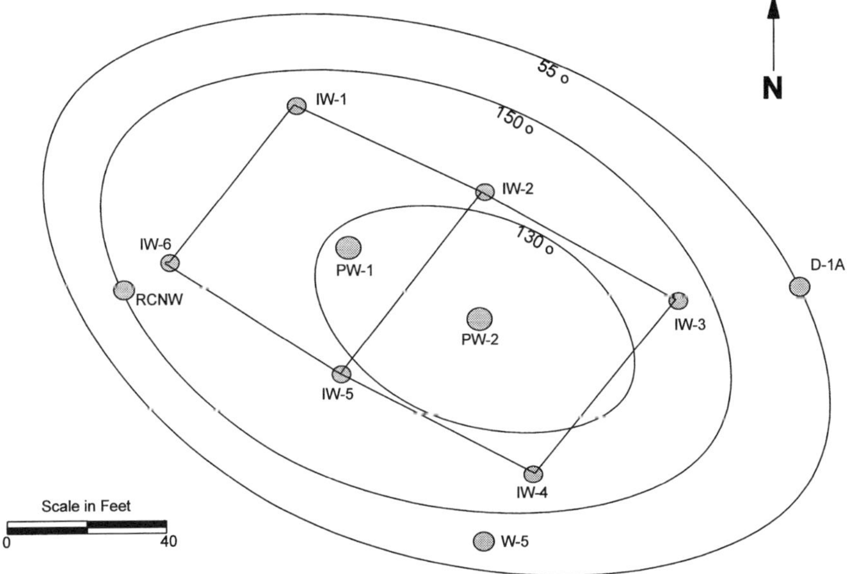

The GAC/FBR and CA units were found to be effective in treating the slip-stream of water generated from production. During startup, there were several exceedances of the daily discharge limit of total PAHs. These were due entirely to solids carryover and not soluble organics in the discharge water. After operations began, only one exceedance of the daily discharge limit for total PAHs occurred in the ten months of operation. This was again associated with solids carryover. Figure 9 presents discharge quality results throughout startup and operations.

Figure 8.   **Cumulative tar recovery vs. pore volume, Brodhead Creek DNAPL enhanced recovery process.**

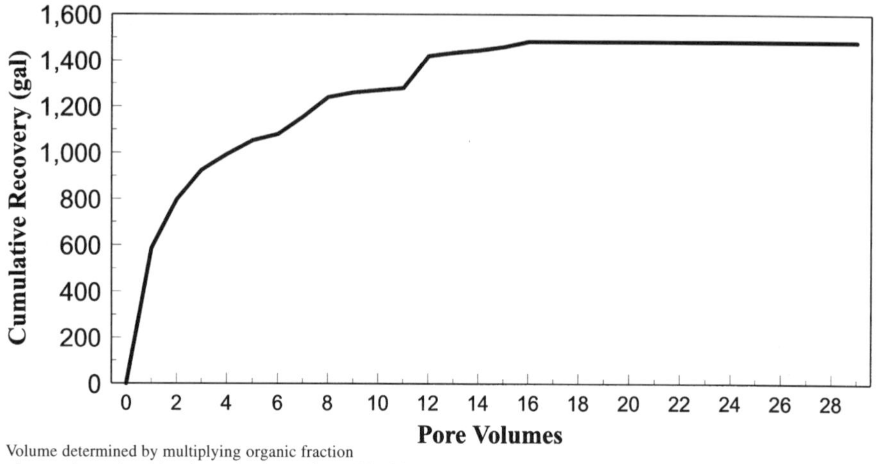

Volume determined by multiplying organic fraction
of separation tank residues by total volume of stored liquid.

Table 1.   **Material Disposed of After CROW™ Process Operation**

Constituent	Quantity
Tar	1,424 gallons
Water	6,064 gallons
Inorganics	3,206 pounds

Figure 9.   **Discharge results, total PAHs, Brodhead Creek Site.**

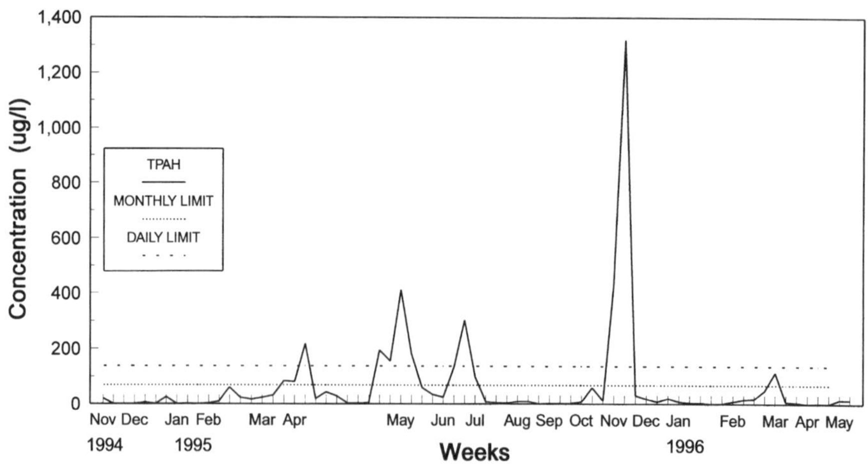

COSTS

This program was funded by the site owner, PP&L, the EPA SITE Program, Gas Research Institute (GRI), Electric Power Research Institute (EPRI), and Department of Energy (DOE). There were various components to the costs. These included treatability testing, the predesign investigation, design, construction, operation, and decommissioning of the site. In addition, the costs for these components were evaluated based on whether the work performed was associated with remediation of the site, costs associated with a Superfund site which would not have resulted if this site were not on the NPL, and costs associated with research and development (R&D) since this was the first time this technology was applied in the field. These costs are presented in Table 2. They are the actual costs incurred and not estimates.

Table 2.   **Cost Summary Brodhead Creek Site Remediation**

Cost Component	R & D Cost	Superfund Site Cost	Remediation Cost	Total Cost
Treatability	$208,000	-----	$ 15,000	$ 223,000
Predesign Investigation	$ 58,000	-----	$110,000	$ 168,000
Design	-----	$251,000	$175,000	$ 426,000
Construction	$180,000	-----	$925,000	$1,105,000
Operation	$400,000	$ 45,000	$455,000	$ 900,000
Decommissioning	-----	$ 15,000	$ 77,000	$ 92,000
TOTAL	$846,000	$311,000	$1,757,000	$2,914,000

A significant amount of initial treatability testing was performed in the laboratory by WRI to demonstrate that this technology was feasible for a MGP site. Therefore, the majority of the treatability costs have been allocated to R&D. The costs allocated to the remediation are reflective of subsequent treatability test costs since the R&D effort was performed.

The predesign investigation also had an R&D component. The work was predominantly performed by Atlantic Environmental Services, Inc. with EPRI sponsorship. A portion of the work performed involved an evaluation and comparison of a variety of sampling techniques for MGP sites. The remainder of the costs are associated with delineating the location of free coal tar. This was used in the design of the remedy.

The design component costs have been allocated between the remedy and additional costs associated with a full Superfund design. The full Superfund design required a 30% design, 60% design, 95% design, and a final design. This process took over a year. On a non-NPL site, this process would have been shortened and streamlined. The costs allocated to remediation are reflective of subsequent CROW™ designs.

The construction component was applied almost entirely to the cost of remediation. The only costs applied to R&D were those associated with the GAC/FBR water treatment system.

The operation costs are split between R&D, additional Superfund costs, and remediation. This was done because of the startup problems encountered, and the costs associated with retrofitting the system to remove iron from the production water were R&D associated costs. In subsequent designs, these issues have been accounted for, and therefore, should not be repeated. Superfund costs are associated with an additional month of operation to verify that the remedy was complete.

Lastly, decommissioning costs have been allocated to the remediation of the site, and additional Superfund site related costs. The additional Superfund related costs are associated with sampling which was required by the EPA above what was required for disposal of material.

REFERENCES

Atlantic Environmental Services, Inc., July 1993, Brodhead Creek NPL Site Supplemental Field Investigation Report, prepared for Pennsylvania Power & Light Company, Allentown, PA.

Leuschner, A.P. and Johnson, L.A., Jr. 1990. *In Situ* Physical and Biological Treatment of Coal Tar Contaminated Soil, presented at the Natural Water Well Association Conference, Petroleum Hydrocarbons and Organic Chemicals in Ground Water: Prevention, Detection, and Restoration, Houston, TX.

Moeller, M.W., Gerrish, J.A., Villaume, J.F., and Johnson, L.A., Jr. 1995. Implementation of a Full-Scale Enhanced Recovery System for Coal Tar at a Former MGP Superfund Site, presented at the Tenth Annual Conference on Contaminated Soils, University of Massachusetts, Amherst, MA.

CHAPTER 47

Risk Assessment and Risk Management Issues at Manufactured Gas Plant (MGP) Sites.

Charles A. Menzie, Menzie-Cura & Associates, Inc., Chelmsford, Massachusetts

INTRODUCTION

This paper provides an overview of risk assessment and risk management issues at Manufactured Gas Plant (MGP) sites. I first became involved in the assessment of such sites in 1985 and had the good fortune of leading the effort to write Volume III, *Risk Assessment*, of the Gas Research Institute's (GRI) compendium on Manufactured Gas Plant Sites published in 1987. In the past decade much has been learned about the characteristics of MGP sites, the health and ecological risks they pose, and methods for addressing these risks. Remediation of these sites has spanned a broad spectrum of technologies ranging from containment to sophisticated treatment of soils and groundwater. To a large extent, this was a period of learning how to implement risk management strategies at these sites.

This learning experience at MGP sites - with its successes and failures - is paralleled by similar experience at contaminated sites in general. Thus, evolving risk-management approaches can be found for petroleum hydrocarbons, PCBs, metals (e.g., lead), and many other classes of soil contaminants at specific sites or regions. While we have learned much about specific issues at MGP sites, we have also learned about risk management approaches for soil contamination. The emergence of risk management as an organizing framework for addressing soil contamination issues is perhaps the most critical development over the past decade. Approaches for addressing contamination at MGP sites can be best understood within this larger context.

THE EMERGENCE OF RISK MANAGEMENT

Risk assessment by itself is an orphan. A request - often heard from industrial clients or regulatory agencies - to "go do a risk assessment" is incomplete if it is separated from the overall risk management context. Risk assessment is a tool, not an end in itself. To understand how, or even when, to use risk assessment, the risk management context must be identified and the questions or information needs clearly defined. While this may appear obvious, there are many risk assessors and risk managers who fail to have the interactions necessary to develop and implement effective risk management approaches. To some degree this reflects historical misconceptions concerning the roles of risk assessment and risk management.

There is a perspective that risk assessment is best conducted and "kept pure" if it is carried out independently from risk management. This perception - that risk assessment and risk management must be kept apart - is a long-standing misinterpretation of the 1983 National Research Council (NRC) Red Book. Warner North, one of the authors of the Red Book, notes their report never called for a "separation" between risk assessment and risk management but only a "distinction" between the two (EPA Risk Policy Report, June 14, 1996, Volume 3, Number 8, p. 9).

The emergence of risk management as an organizing framework that incorporates risk assessment is underscored by recent reports from the Commission on Risk Assessment and Risk Management (1996) and the NRC (1996). The Commission has proposed a conceptual approach that emphasizes solutions to environmental risk reduction through an integrated process involving stakeholders (Figure 1). This theme is also present in the NRC's *Understanding Risk: Informing Decisions in a Democratic Society*. The report presents seven principles which will influence how risk assessment, risk characterization, and risk management are carried out in the future:

1. Risk characterization should be a decision-driven activity, directed toward informing and solving problems.

Figure 1. **Framework for risk management.**

2.   Coping with a risk situation requires a broad understanding of the relevant losses, harms, or consequences to the interested and affected parties.
3.   Risk characterization is the outcome of an analytic-deliberative process.
4.   The analytic-deliberative process leading to a risk characterization should explicitly deal with problem formulation early on in the process.
5.   The process should be mutual and recursive.
6.   Those responsible for a risk characterization should begin by developing a provisional diagnosis of the decision situation so that they can better match the analytic-deliberative process to the needs of the decision, particularly in terms of the level and intensity of effort and representation of parties.
7.   Each organization responsible for making risk decisions should work to build organizational capability to conform to the principles of sound risk characterization.

There are a number of examples of emerging risk management approaches for soil contamination problems. The first publication of the GRI Volume III report for MGP sites in 1987 - since updated and to be published this year - is one of the first to outline an integrated approach for site investigation, risk assessment, and site remediation (Figure 2). The risk assessment components were tiered; screening-level assessments led toward more sophisticated approaches depending on the information needs of the risk-management decisions. This integrated risk-management approach was presented in numerous fora during the late 1980s and became widely used by industry and regulatory agencies for MGP and other sites.

Figure 2.   **Risk assessment as part of site management.**

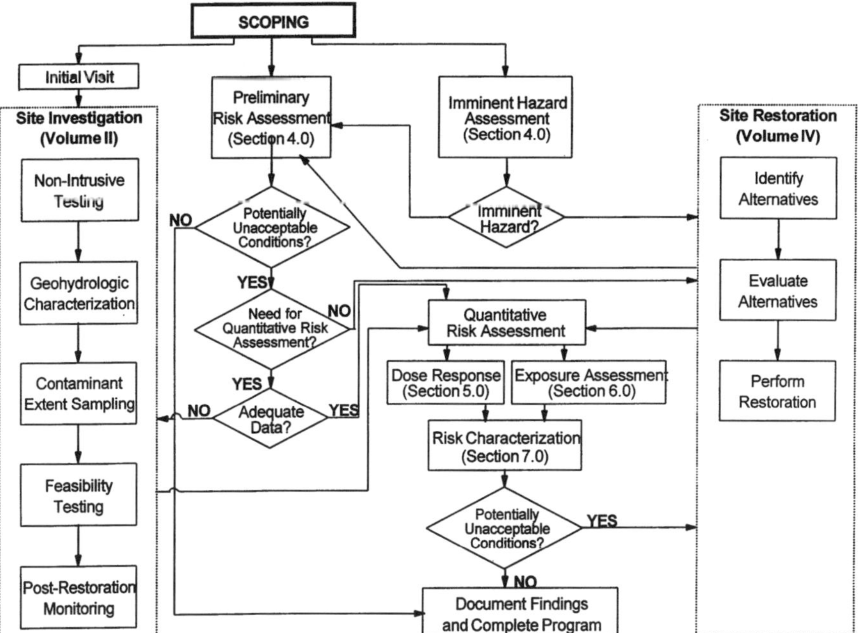

A similar concept is embodied in the ASTM (1995) Risk Based Corrective Action (RBCA) for petroleum release sites (Figure 3). RBCA involves a tiered assessment strategy wherein risk-related information, decisions, and remedial options are considered in a sequential manner. ASTM is currently working on a RBCA approach that can be applied to all sites.

RBCA promises to be a useful and understandable way to organize and interrelate risk management components for addressing contaminated soils. As a result, various organizations, groups, and agencies are using RBCA to help fashion their approaches. For example, the Total Petroleum Hydrocarbon (TPH) Criteria Group has adopted the RBCA strategy. Furthermore, a number of states have also moved toward risk-management approaches that use tiered strategies and which clearly link risk assessment activities to specific risk-based questions and remedial options. RBCA has served as an organizing framework for many of these states.

The 1994 to 1999 period will mark the emergence of risk management approaches to site assessment. This means more than simply using the results of risk assessments to guide decision making. As discussed by the Commission and NRC reports cited above, there are a number of factors and stakeholder issues that may enter into risk management decisions. Additionally, there will be an increasing emphasis on the goals of the risk management effort and on providing solutions to environmental risk reduction.

A RBCA approach is consistent with guidance developed for MGP sites by GRI (1987, 1996a) and the emergence of risk management strategies. I have been involved in the development of the generic RBCA standard and believe that it offers a unifying approach for risk-based decision making at MGP sites. Andy Middleton of RETEC is a RBCA instructor who has developed course materials for GRI to illustrate how RBCA could be applied to MGP sites. Also, Tom Hayes, who manages GRI's MGP effort, has expressed interest in developing this idea further. It seems appropriate to explore the use of the generic RBCA standard for MGP sites.

Risk assessment methods have continued to develop over the past decade as we have thought more about how to characterize exposure, effects, risk, and the uncertainties in these estimates. As a result, risk assessment techniques range from simple screening-level methods to sophisticated two-dimensional Monte Carlo analyses. Accompanying this range of methods has been a debate about the correct way to conduct risk assessments. This debate has often pitted risk assessors working for regulatory agencies, consulting companies, and industry against one another. This is unfortunate and has often led to frustration on all sides. From a risk management perspective, it should be clear that any of the methods might be appropriate, depending on the information needs of the decisions.

The ability of the parties to communicate with one another and with other stakeholders about the approach is also a critical consideration in the selection of

Figure 3.   RBCA flowchart.

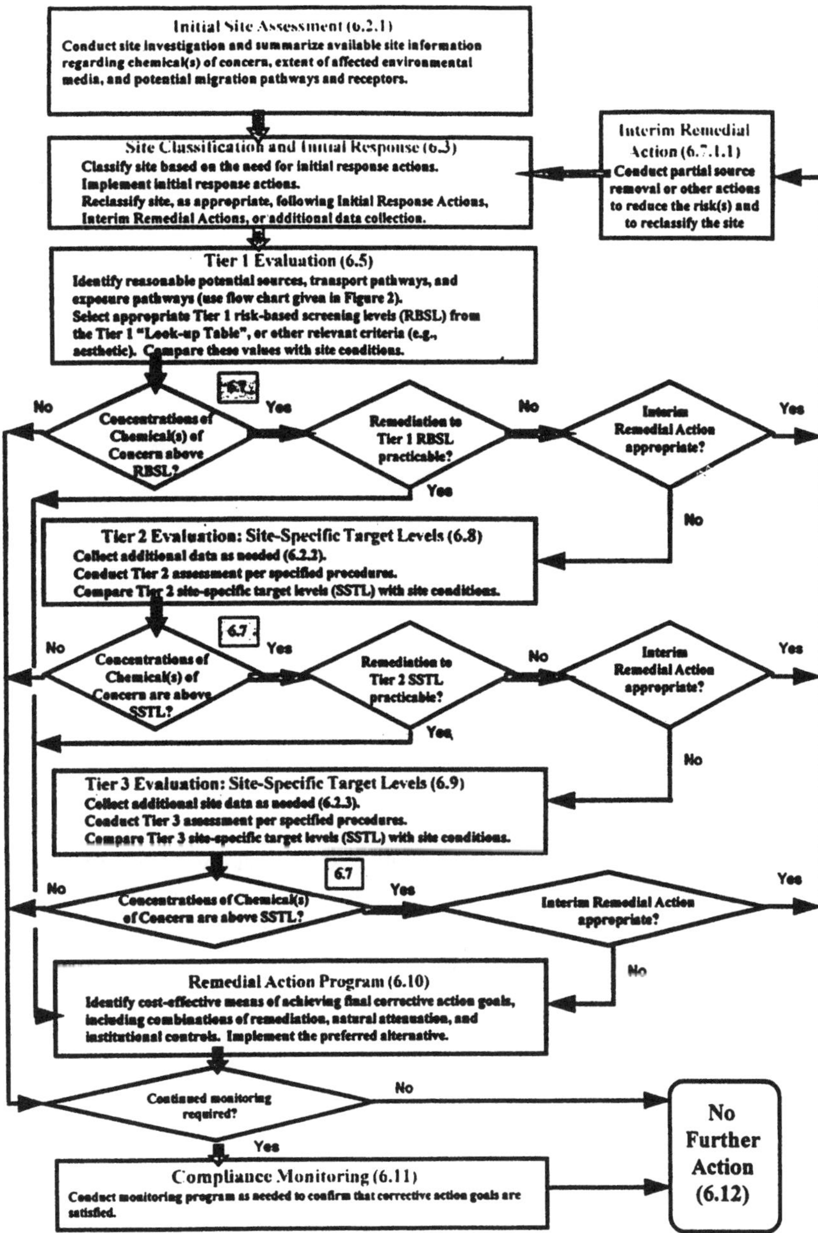

appropriate methods. The methods chosen should yield results that can be effectively used in a decision-making framework. Risk communication plays a significant role in risk management decisions.

These elements underscore the value of tiered assessment strategies that link risk assessment methods to decisions and that progress from simple to more sophisticated analyses. Successful decisions - short of litigation - are most likely to come about when all sides have a clear understanding and acceptance of the assessment process and how it will be used within the decision-making framework. This is perhaps one of the strongest messages of the recent Commission and NRC reports.

MAJOR RISK MANAGEMENT ISSUES AT MGP SITES

Over the past decade we have learned much about evaluating risks and implementing risk management decisions at MGP sites. Much of this experience is reflected in the updated GRI guidance (GRI, 1996a). I won't describe the contents of those documents in this chapter; rather, I will highlight some of the major risk management issues based on our experience and that of others. To this end, I contacted individuals within the utility industry, research groups, and consulting companies to learn what they believed to be the most pressing issues at MGP sites. What I found was that some issues are local, while others are common throughout the country. These issues are discussed below.

Recognizing and Characterizing Local Conditions

Many MGP sites are located in urban environments where contamination exists due to nonsite-related sources. Even in rural areas, activities can give rise to local conditions where contaminants are elevated in soils surrounding a site. How should such local conditions be taken into account within the risk management process? Is it possible to differentiate between the site and general conditions present in the region? The issue of how to account for local conditions is multifaceted and can cause frustration on all sides. Most often, the issue arises at MGP sites with respect to polycyclic aromatic hydrocarbons (PAHs), a class of chemicals present in coal tars that usually drives risks at these sites. The issue of local conditions has also arisen for lead at some sites. PAHs and other contaminants such as lead can be present in urban and some rural areas as a result of other anthropogenic and natural sources. Concentrations of PAH and lead in soils associated with local conditions can exceed risk-based screening levels and calculated risk thresholds. This phenomenon clouds risk management decisions with respect to the site and how it should be differentiated from surrounding areas.

The issue can be addressed, in part, by a process that recognizes that chemicals can be present due to off-site sources. The site can then be evaluated within the context of local conditions. There are a range of risk management considerations such as risk reduction objectives, liability, and feasibility that should be part of this evaluation.

PAHs as a class are usually present in soils as a result of nonsite-related local conditions. Menzie et al. (1992) found that the higher molecular weight carcinogenic PAH compounds were present in all soils, and highest in urban areas where concentrations ranged between 0.6 to 3.0 mg/kg. Slightly higher levels were reported in a study conducted in three New England cities by Bradley et al. (1994). These investigators report an arithmetic mean of 9.0 mg/kg for the three cities with a 95% upper confidence interval of 12.4 mg/kg. The corresponding values for total PAH were 18.4 and 24.8 mg/kg (Bradley et al., 1994). The two studies suggest that carcinogenic and total PAH in soils are typically in the mg/kg (ppm) range due to local conditions in urban environments.

Sediments near urban areas also typically contain PAH compounds due to local conditions. Hoffman et al. (1984) and Menzie et al. (in preparation) have identified urban runoff as a primary source of PAH in sediments of urban estuaries such as Narragansett Bay (RI) and Massachusetts Bay (MA). Menzie et al. (in preparation) estimated potential sediment enrichment factors to illustrate the degree to which urban runoff could influence local sediment conditions. This work indicated that PAH concentrations in runoff could be orders of magnitude greater than local sediments. Bradley et al. (1994) describe the nature and quantity of PAH loadings associated with runoff, rivers, and sewage treatment plants. This study showed that urban runoff carried the highest concentrations of persistent high molecular weight PAH. As discussed by Hoffman et al. (1984) and Menzie et al. (in preparation), most of these chemicals are due to combustion sources (e.g., motor vehicles) as well as drips and leaks of petroleum products (e.g., crankcase oils.)

Regional data collection efforts can be helpful in placing the results obtained at a site into perspective. In some cases, studies directed at establishing local background conditions can be especially beneficial. Unfortunately, because the collection of off-site data is often seen as a cost that is not yielding site-specific information, there is a tendency to perform such studies poorly in an effort to save money. This can limit the value of the assessment and result in misleading characterizations. As a general rule, a sound statistically-based sampling design is an important starting point for establishing local or regional conditions.

An individual's perspective on local conditions and acceptable concentration levels for soils is affected by a wide range of factors. These include an individual's perception of risk, the reliance placed on numbers (e.g., screening levels), their familiarity with the physical and chemical properties of the contaminants, the role that the individual has in the risk management process, and the breadth of their risk management experience. In such cases, constructive and frequent communication is needed to develop a common base of understanding on all sides of the issue.

Surprisingly, risk assessors whose experience is limited to the conduct of risk assessments, may have little risk management experience and a poor understanding of the needs of the risk manager. They can become focused on the process, in particular on the sets of equations that yield risk estimates and target levels (e.g., soil concentrations.) While these risk estimates and target level values are important and valuable for risk management, a strict calculation-based approach can preclude consideration of other factors. I refer to this as "risk assessment myopia," a phenomenon that can lead to outcomes that challenge common sense. Examples include the derivation <u>and reliance upon</u> target levels well below ambient levels (e.g., soil concentrations in the low parts per billion for some PAH compounds), as well as values at which soils would be almost saturated with a chemical (e.g., many noncarcinogenic organic chemicals such as xylene and toluene and inorganic compounds.) Such outcomes should be a flag that other factors should be considered within the risk management framework. Local conditions is one of these factors.

Understanding and Using Information on Exposure

Many of the decisions regarding assessment and remediation at MGP sites are based on exposure issues. Exposure distinguishes sites from one another, governs risk estimates, and can be effectively reduced through a number of remedial options. In cases where treatment alternatives designed to reduce toxicity are unavailable or infeasible, remedial alternatives that reduce or eliminate exposure offer the most promising opportunities for risk reduction.

Assessing exposure associated with soils and wastes at MGP sites typically involves understanding:

- vertical and horizontal distribution of chemicals;
- concentrations of chemicals;
- bioavailability of the chemicals;
- ability of chemicals to migrate in gaseous, aqueous, nonaqueous phases, or solid form;
- current and planned uses of the site, and how people and ecological receptors might be exposed via these uses;
- other pathways by which people or ecological receptors might be exposed to chemicals present at the site.

These are elements of a site conceptual model. The large knowledge base on the exposure characteristics of chemical contamination at MGP sites has made it possible to develop risk management guidance for these sites. However, sites can differ from one another, and planned uses for such sites may also differ.

Site-specific conditions and land use plans provide opportunities for creative and effective approaches for reducing risks and restoring land to a practical use. MGP site managers around the country are seeking solutions that involve eliminating exposure pathways at sites so that the site can be used for industrial/commercial use or even as recreational use areas. The value of creative solution-oriented approaches to restoring urban properties is becoming increasingly recognized. The concept has received much attention within Brownfields initiatives, and some MGP sites are being redeveloped in this context. Some utilities have entered into consent orders to evaluate and remediate (restore) sites to industrial land use. Others have proceeded with assessments and restoration projects consistent with state policies that allow for consideration of land use.

Restoration and risk reduction projects that rely upon eliminating or minimizing exposure pathways require knowledge concerning how exposures occur, the ability of chemicals to move through barriers, and methods that effectively eliminate sources or prevent migration from sources to locations where exposure could occur. Approaches may involve the removal of "hot spots" at or near the soil surface, combined with various containment or barrier methods including the footprint of buildings, capping with paving, and subsurface venting.

The ability to implement risk management solutions that involve controlling or eliminating exposure is dependent on state regulations. Some states - often those with many urban sites - are making an effort to develop flexible approaches for site restoration. Others have adopted more conservative approaches that explicitly or implicitly incorporate a "no degradation" policy. These positions reflect different points along a risk management spectrum. They are not necessarily inconsistent. Frequent and constructive communication among the stakeholders (the agencies, utilities, and others) is especially important when alternative solutions are sought. In some cases, legislation may need to be modified to allow for creative alternative approaches to be employed by the agencies.

Understanding and Using Information on Effects

Risk management decisions at MGP sites is most often driven by the presence of coal tar either as a nonaqueous phase or mixed with soil or other media. Coal tars are complex and variable mixtures of hydrocarbons and there is limited information on the toxicity of these mixtures. Most risk assessments have evaluated effects by relying on the limited information available for priority pollutant compounds associated with the tars, primarily PAHs. As a result, there is large uncertainty associated with evaluating the effects of these compounds. Other chemicals that have driven decisions at sites include volatile organic hydrocarbons such as benzene (in atmosphere and groundwater), iron cyanide compounds, and occasionally metals such as lead and arsenic.

The Electric Power Research Institute (EPRI) has taken the initiative to reduce the uncertainty associated with evaluating the toxic effects of coal tars and constituents within this waste material (GRI, 1996b). This program will improve our understanding of the adverse health effects of complex mixtures of coal tars and will improve the predictive ability of assessing the risks to mammals. The EPRI project is also evaluating the effects of the soil matrix in reducing the bioavailability of coal tar constituents. The primary study is now complete and data are being evaluated.

The presence of iron cyanides at MGP sites has caused concern because of the specter associated with the word "cyanide." For example, simple cyanide salts are well-known acute poisons and can be lethally toxic to people following short-term exposure to relatively high concentrations. This toxicity is associated with the physiological release of hydrogen cyanide. The iron cyanides present at MGP sites do not readily form or release hydrogen cyanide and are considered to have little or no toxicity. While most scientists within regulatory agencies, consulting companies, and the industry now recognize the large differences in toxicity among cyanide compounds, concern still remains that some exposure conditions might result in releases of at least small amounts of toxic cyanide. As a result, there have been several efforts underway to characterize these kinetics. As discussed in the update for Volume III, such releases are expected to be small if not negligible. Chemical analytical methods have been explored for deriving conservative estimates of cyanide exposure for selected exposure routes (e.g., incidental ingestion of the waste materials).

Iron cyanides can cause visible blue staining of materials, an aesthetic impact that can serve as a reminder that "cyanide" is present, and affect the perception of risk. This is a risk management issue in some parts of the country where sites are being redeveloped.

Ecological Issues

Ecological issues are being evaluated with increasing frequency at sites throughout the country. In the past five years there has been a concerted effort by state and federal agencies to develop guidance on conducting ecological assessments, and utilize the results in decision making. At MGP sites, the conduct of ecological assessments has usually been driven by issues related to releases of coal tars to surface water bodies and resultant sediment contamination. Because many MGP sites were located on lakes, coastal harbors, and rivers, some facilities have experienced such releases. The problem - when and if it arises - can be highly visible, persistent, and can cause concern for the health of the aquatic biota. The general issue of sediment contamination has taken center stage in the evaluation of aquatic ecological risks by the U.S. Environmental Protection Agency, National Oceanic and Atmospheric Administration, as well as many state agencies. Increased attention to this matter can, therefore, be expected. Guidance on how to conduct these assessments for hydrocarbons and other contaminants is being developed. The greatest risk management challenge is how to remediate these conditions.

In contrast to aquatic ecological risks, terrestrial ecological risks have not been a predominant issue. This is due, in part, to the locations of most sites within towns and cities where terrestrial habitats for wildlife are small and fragmented. In such locations, human health risks are typically judged to be more important from a risk management perspective.

Setting Priorities

Some utilities may have a number of MGP sites along with other environmental problems. Tackling all these possible problem areas simultaneously can be an overwhelming task from a technical as well as business perspective. It is also evident from experience that sites vary in the degree of risk they pose in the short and long term. They also differ with regard to a host of other factors that may bear on risk management decisions. Thus, some utilities are utilizing comparative risk management approaches to help prioritize their sites. These have included formal approaches including comparative site-assessment models as well as methods based on professional judgment. Some utilities and states have made progress on using comparative risk management approaches to address sites.

CONCLUSIONS

Risk management methods have proved effective for addressing environmental issues at MGP sites. Experience indicates that MGP sites share many common characteristics (GRI, 1995). These commonalities permit the development of generic risk management approaches which can simplify the conduct of assessments and facilitate communication of results. Site-specific factors such as land use, size, location, and degree of contamination govern exposure and can be considered when evaluating appropriate remedial methods.

Implementing effective risk management strategies generally requires: (1) a sound approach, and (2) acceptance of that approach by the parties involved in the decision. Because people have different technical backgrounds, responsibilities, and experiences, they may differ in their opinions regarding what constitutes a sound approach. Communication (constructive and frequent) is important in reaching a common base of understanding. The development and use of RBCA and other risk management methods should facilitate this process. Technical transfer workshops and seminars can be helpful for increasing awareness in these approaches.

REFERENCES

ASTM Standards: Designation: E 1739-95. 1995. Standard Guide for Risk-Based Corrective Action Applied at Petroleum Release Sites.

Bradley, L.J.N., Magee, B.H., and Allen. S.L. 1994 Background Levels of Polycyclic Aromatic Hydrocarbons (PAH) and Selected Metals in New England Urban Soils. *J. Soil Contam.* 3(4), 349-361.

Commission on Risk Assessment and Risk Management. June/1996. Risk Assessment and Risk Management in Regulatory Decision-Making.

Environmental Protection Agency. 1996. Risk Policy Report. 3:8, 9.

Gas Research Institute. 1987. *Management of Manufactured Gas Plant Sites, Volume III Risk Assessment.* Prepared by Atlantic Environmental for GRI.

Gas Research Institute. 1995. *Management of Manufactured Gas Sites, Volume I: Wastes and Chemicals of Interest.* Prepared by: Remediation Technologies, Inc., Atlantic Environmental Services, Inc., and Keystone Environmental Resources, Inc.

Gas Research Institute. 1996a. *Management of Manufactured Gas Plant Sites, Volume III Risk Assessment.* Prepared by Menzie-Cura & Associates and Remediation Technologies, Inc.

Gas Research Institute. 1996b. *Environmentally Acceptable Endpoints in Soil: Risk-Based Approach to Contaminated Site Management Based on Availability of Chemicals in Soil.* Published by American Academy of Engineers.

Hoffman, E.J., Mills, G.L., Tatimer, J.S., and Quinn, J.G. 1984. Urban Runoff as a Source of Polycyclic Aromatic Hydrocarbons to Coastal Waters. *Environ. Sci. Technol.* 18, 580-587.

Massachusetts Bays Program. 1995. Measurements and Loadings of Polycyclic Aromatic (PAH) in Storm Water, Combined Sewer Overflows, Rivers, and Publicly Owned Treatment Works (POTWs) Discharging to Massachusetts Bays. Prepared by Menzie-Cura & Associates, Inc., Chelmsford, MA.

Menzie, C.A., Cura, J., Freshman, J., LeFrey, E. (In preparation). Characteristics of PAHs in Massachusetts Stormwater and Implications for Coastal Areas. Menzie-Cura & Associates, One Courthouse Lane, Suite Two, Chelmsford, MA.

Menzie, C.A., Potocki, B.B., Santodonato, J. 1992. Exposure to Carcinogenic PAHs in the Environment. *Environ. Sci. Technol.*, 26, 1278-1284.

National Research Council (NRC). 1983. *Risk Assessment in the Federal Government: Managing the Process.* Washington, DC, National Academy Press.

National Research Council (NRC). 1996. *Understanding Risk: Informing Decisions in a Democratic Society.* (Stern, P.C., and Fineberg, H.V., Eds.). Washington, DC, National Academy Press.

CHAPTER 48

Site Investigation Considerations

Dennis F. Unites, Atlantic Environmental Services, Inc., Colchester, Connecticut

INTRODUCTION

Manufactured gas plants (MGPs) served as a source of gas for heating, cooking, and illumination from the early 1800s until the 1950s. Three processes were used to generate the gas: coal gasification, the water gas process, and oil gasification. While there were a multitude of variations, the nature of the processes and associated contaminants share many similarities among the sites. This "commonality" was recognized by the U.S. Environmental Protection Agency in its efforts to develop presumptive remedies for MGP sites. A review of reports from several hundred sites shows a consistent pattern in the types of contaminants found and their physical distribution.

THE SHARED CHARACTERISTICS OF MGP SITES

To understand the distribution of contaminants at a site, it is necessary to know a bit about the processes involved in gas manufacturing. In the coal gasification process, coal was heated in a sealed retort. The resulting gas passed through a water seal or "hydraulic main," then through a series of condensers or scrubbers to remove tar, ammonia, and hydrogen sulfide. The gas was then stored in a low-pressure gasometer or holder prior to distribution.

Coal gasification was a continuous process, with a number of retorts being heated and charged on a regular basis. The carburetted water gas process and the oil gas process, in contrast, were cyclical. In the water gas process, steam was injected into an incandescent bed of coal or coke. The resulting water gas was then carburetted with oil and passed through a superheater to "fix" the gas. The system was then "run" with air to reheat the bed of coals. In the oil gas process, the bed of coals is omitted, and an oil-air mixture is used to heat the checker brick of the generator. The gas in both processes passed through a hydraulic main as in the coal gas process. However, in order to deal with the cyclic nature of the process, the gas streams were generally subject to tar removal then storage in the relief holder, prior to scrubbing for sulfide removal.

The liquid stream from these processes consisted of tars, which were heavier than water, oils which were lighter than water, and water which was enriched in ammonia and organic chemicals. The oil gas and water gas processes tended to produce more tar and oil per unit of gas than the coal gas process. It is inaccurate to call this liquid a "waste stream" because it was truly a by-product that formed the basis of the early organic chemical industry. The tars and oils were separated from the water, using gravity separation in an overshot undershot weir, then stored in a "tar well" or holder bottom to allow further gravity separation. They were then either used as fuel as part of the operation, subjected to further on-site refining (generally water removal), or sold to tar refiners. Because of their value as a product, tars and oils were not disposed of on site. Rather, releases to the environment generally came from leaks or spills during operations or as the result of conditions occasioned by the decommissioning or demolition of the plant.

These process-related conditions result in the first commonality among the gas plant sites. That is, most contamination is found in the tar-handling areas or in the relief holders which may have been abandoned with several hundred gallons of tar or tarry sludge in place. The latter situation has resulted in the targeting of holder removal as a primary source to be removed during remediation.

The second commonality among the sites is the chemical and physical nature of the contaminants. Chemicals of concern include polycyclic aromatic hydrocarbons (PAHs), volatile aromatics such as benzene, and cyanide as a by-product of the sulfide removal system. Other chemicals of concern are presented in Table 1.

Metals may also be present in the coal or ash, paint, or in pressure control devices. Phenolics may also be of concern at some sites but it would appear that the time since closure and the ready biodegradation of these compounds has reduced concentrations to below levels of concern. Unless the sites have been used for other purposes, characterization based on PAHs, monocyclic aromatic hydrocarbons (MAHs), limited metals, and cyanide should be adequate.

The third commonality among the sites is a result of the dense nonaqueous phase liquid (DNAPL) nature of the tar. Because this liquid is heavier than water, given a suitably coarse or fractured media, it will sink until either residual saturation or an impermeable barrier is reached. In developing an investigation strategy, one has to determine the "bottom" of the contamination and be aware that a sloping confining layer or fracture system can conduct the DNAPL away from the direction predicted solely on the basis of the piezometric gradient. One also has to be concerned with penetrating confining layers or bottoms of structures and causing the spread of the DNAPL. Conversely, if the natural material is fine-grained and not fractured, migration of the viscous DNAPL will be significantly impeded.

The nature of the tars result in another similarity among the sites. Coal tar is a black to reddish-brown liquid with a distinctive odor. Chemical analysis is not required to determine that a heavily contaminated soil sample is indeed heavily contaminated. However, a little goes a long way, and samples with small amounts

Table 1.   Wastes and Chemicals of Interest at MGP Sites

WASTES	• Free tars, oils, and lampblack
	• Organic-contaminated soils
	  - Heavily contaminated
	  - Lightly contaminated
			•Organic-contaminated vessel, surface, and groundwaters
			•Purifier wastes
			•Mixed wastes and fill		

CHEMICALS	• Inorganics	•Metals	•Volatile	•Phenolics	•Polynuclear Aromatic Hydrocarbons
	Ammonia Cyanide Nitrate Sulfate Sulfide Thiocyanates	Aluminum Antimony Arsenic Barium Cadmium Chromium Copper Iron Lead Manganese Mercury Nickel Selenium Silver Vanadium Zinc	Aromatics Benzene Ethyl Benzene Toluene Total Xylenes	Phenol 2-Methylphenol 4-Methylphenol 2,4- Dimethylphenol	Acenaphthene Acenaphthylene Anthracene Benzo(a)anthracene Benzo(a)pyrene Benzo(b)fluoranthene Benzo(g,h,i)perylene Benzo(k)fluoranthene Chrysene Dibenzo(a,h)anthracene Dibenzofuran Fluoranthene Fluorene Naphthalene Phenanthrene Pyrene 2-Methylnaphthalene

Source: Remediation Technologies, Inc., et al., 1987.

of free NAPL can appear much worse than they are. Even samples with concentrations below one part per million will have the distinctive coal tar odor. Selection of samples for analysis should focus at determining the lower levels of contamination rather than the higher end.

The chemical nature of the tars provides the basis for a final similarity among the sites which must be addressed in the site investigation. Many of the PAH compounds present in tar are resistant to biodegradation. Because of this resistance, they may be present in surface soils and pose an exposure pathway which requires evaluation during a risk assessment. Difficulties arise, however, in separating the PAHs which are from other sources. In areas having considerable industrial activity and combustion of fossil fuels, it may be very difficult to determine true background conditions.

ISSUES IN SITE INVESTIGATION

The investigation of an MGP site employs the techniques used in the evaluation of any other contaminated site. If one keeps in mind the basic nature of the MGP contaminants and processes as well as the desired end use of the site, one can streamline the investigation process. The streamlining is a result of the development of a good site history, the application of areal screening methods, and the use of appropriate chemical analytical techniques.

SITE HISTORY

The development of a good site history is important in knowing where to look for contaminants. Site plans, insurance maps, and areal photographs should be used to determine the location of tar-handling equipment, tar wells, and relief holders. Aerial photographs may be available to determine disposal areas. In the absence of photographs, it should be kept in mind that most of the waste hauling was nonmechanized in the early days of operation, so materials would often go into the most convenient low spot. Information in company records may also provide information about demolition practices.

AREAL CHARACTERIZATION

Soil gas and geophysical surveys have been commonly used at MGP sites to provide areal characterization. Both techniques are highly dependent on site conditions and have had mixed results. Both active and passive soil gas techniques have been used at MGP sites. Active methods; that is, the collection of gas samples and analysis in the field or lab, appear to work best on sites where pavement provides a cap and moderately granular materials are present in the subsurface. Passive collectors may perform better on open sites. In either case, sample collection on a grid rather than random works best for interpretive purposes. One variation of active sampling, using a direct push sampler to collect a groundwater sample followed by analysis in the field for benzene or naphthalene, appears to show promise as an areal screening method.

Various electromagnetic and seismic geophysical techniques have been used at sites. Geophysical methods have not been useful in determining tar deposits. They are useful on some sites in determining subsurface structures which control migration and also can be used to locate the extent of conductive wastes like ash and purifier materials.

CHEMICAL ANALYSIS

There are numerous issues that center around the chemical analysis of samples collected at MGP sites:

- Are there indicator parameters that can be used to reduce analytical costs?
- Are lumped parameter methods such as TPH appropriate?
- Do test kits work?

The answer to all of the above is, "it depends on how you are going to use the data."

An initial site characterization should have the base of a number of analyses that can give a relatively accurate measure of the range of individual PAH, MAH, metal, and cyanide compounds in the various risk exposure pathways. This characterization could be determined using GC or GC/MS technique for the organics and other standard methods for the inorganic compounds. If the site history is in question, a full GC/MS scan for priority pollutants might be required. This specificity is required for use in the site risk assessment.

Once the preliminary assessment has been completed, the analytical program can be tuned to respond to particular needs. It may be necessary to determine the extent of contamination for design purposes or to determine whether contaminants have been removed during remediation. In those cases, lumped parameter tests or test kits can give information in a timely manner at a somewhat reduced cost.

CONCLUSION

Investigation experience at several hundred MGP sites has shown that a number of similarities exist among the sites. By recognizing these commonalities and focusing the investigation toward the proposed end use, significant cost savings can be realized.

REFERENCES

Remediation Technologies, Inc.; Atlantic Environmental Services, Inc.; and Keystone Environmental Resources, Inc., 1987. *Management of Manufactured Gas Plant Sites, Volume I, Wastes and Chemicals of Interest*, prepared for Gas Research Institute, October.

PART IX—RADIOACTIVITY

CHAPTER 49
Cost/Risk/Benefit Analysis of Alternative Cleanup Requirements for Plutonium Contaminated Soils

Barbara J. Deshler, IT Corporation, Las Vegas, Nevada

INTRODUCTION

The U.S. Environmental Protection Agency (EPA) recently published several documents bearing on the cleanup of federal facility sites contaminated with radioactive material. These include an issues paper (EPA, 1993a) and working draft radiation site cleanup regulations (EPA, 1994). The EPA's efforts potentially impact numerous U.S. Department of Energy (DOE) sites, including sites in Nevada contaminated with plutonium. This cost/risk/benefit analysis was undertaken to better understand the issues and potential impacts for the cleanup of these sites (DOE, 1995).

The analysis estimated the costs, risks, and benefits associated with applying alternative plutonium cleanup regulations to three southern Nevada sites: the Nevada Test Site (NTS), Nellis Air Force Range, and Tonopah Test Range (TTR). Plutonium contamination exists in surface soils at these sites as a result of (1) atmospheric detonations of nuclear devices and (2) safety-shot tests wherein nuclear devices were subjected to conventional explosives to determine whether the devices could attain criticality.

ALTERNATIVE CLEANUP REGULATIONS EVALUATED

Two types of proposed regulations were evaluated: (1) concentration-based regulations that specify maximum allowable plutonium concentrations in soil and (2) dose-based regulations of the form specified in the EPA's draft regulation. The EPA's draft proposal defines a two-tiered dose limit in terms of the maximum annual committed effective dose to the "reasonable maximally exposed individual" (RME) (EPA, 1994). Specifically, the proposal is that for locations to be released without restrictions, the maximum allowable RME dose is 15 millirems per year (mrem/yr) in excess of natural background radiation for 1,000 years after the completion of the cleanup. For locations to be released with active control, the proposed limit is 75 mrem/yr. The analysis evaluated the costs, risks, and benefits of applying alternative soil-concentration and dose limits at the NTS and related sites.

ASSUMED REMEDIAL APPROACH

For the purpose of estimating costs and risks, the analysis assumed a least-cost strategy for achieving regulations. Least-cost strategies were evaluated to maximize the potential for identifying regulations with benefits justifying costs. For dose-based regulations, the assumed strategy is to rely to the maximum extent possible on active control (rather than cleanup) to permit areas to be released in accordance with the regulation. Thus, plutonium concentrations are assumed to be reduced only to the levels necessary to achieve the upper limit of a two-tiered regulation (e.g., 75 mrem/yr rather than 15 mrem/yr). Concentration targets that allow a margin of error for meeting dose- or concentration-based regulations were also evaluated.

To achieve target concentration levels, topsoil is assumed to be removed from contaminated areas. Excavation is assumed to be accomplished with standard earthmoving equipment. Soil would be transported without containerization and disposed of as low-level waste. Rather than using existing waste disposal sites, a less costly approach is assumed wherein contaminated soil is disposed of in existing craters created by underground testing.

SOIL CONCENTRATION TARGETS FOR MEETING DOSE-BASED REGULATIONS

According to the EPA draft dose-based regulations, "The RME is defined as the individual receiving the radiation exposure experienced by the 95th percentile and above of the population at a released site (i.e., the upper five percent exposure level for individuals at the site)" (EPA, 1994). Accordingly, a dose-exposure model was developed to estimate the distribution of annual exposures that might be experienced by individuals residing in locations with specified plutonium soil concentrations. Given such distributions, RME exposures were obtained by reading off the 95th percentile values.

The dose-exposure model accounts for uncertainties and variabilities in (1) plutonium concentrations in indoor dust and in indoor/outdoor air, (2) quantities of soil and indoor dust ingested, (3) quantities of indoor/outdoor air inhaled, and (4) the relationship between ingestion and inhalation exposures and effective dose. For a fixed soil concentration, exposures were estimated to be lower at nuclear sites than at safety-test sites, based on data showing lower radionuclide resuspension rates at nuclear sites due to the incorporation of plutonium in glass created during nuclear blasts. To achieve 15 mrem/yr and 75 mrem/yr RME doses, concentration targets were estimated to be 169 picocuries per gram (pCi/g) and 845 pCi/g for nuclear sites, and 70 pCi/g and 349 pCi/g for safety-test sites, respectively.

AREAS REQUIRING REMEDIATION

Given soil concentration targets, areas requiring remediation were identified from contour maps showing estimated plutonium concentrations by location. The contour maps were developed using data from the NTS Radionuclide Inventory and Distribution Program (McArthur, 1991) together with assumptions to address data deficiencies. Due to data gaps and other limitations, considerable uncertainties were estimated, especially at the lowest concentration contours. For example, the area contaminated above 10 pCi/g was estimated to range between 17,000 and 220,000 hectares, with an expected value of 90,000 hectares. Out of the 90,000 hectares, compliance with the EPA's two-tiered 15/75 mrem/yr dose regulation would require remediating an expected 276 hectares (to achieve the maximum 75 mrem/yr dose), allow release with active control of an expected 1,772 hectares (which would produce doses between 15 and 75 mrem/yr), and permit unrestricted release of the remaining areas.

REMEDIATION COSTS

To account for major policy uncertainties, costs were estimated under alternative scenarios, including whether or not soil would be processed to reduce waste volume and whether or not TTR wastes could be disposed of on the TTR. Costs were then partitioned into fixed costs (cost components that show little or no dependency on the amount of remediation required, such as necessary road construction to permit waste transport), costs that depend mainly on the area requiring remediation (such as certification costs and revegetation costs), and costs that depend mainly on the volume of soil requiring remediation (such as excavation, transport, and disposal costs). Separate estimates were then generated for fixed costs, costs per unit area, and costs per unit volume.

Estimates of soil volume were obtained by multiplying the area requiring remediation by the estimated average depth of remediation. Average remediation depth was estimated to be between 5 and 12 cm and to be roughly independent of target cleanup level.

Remediation was found to be less costly with no treatment of soil to reduce waste volumes. Cost-benefit comparisons were, therefore, made assuming no soil processing. Figure 1 provides the expected values and uncertainty ranges (90% confidence intervals) for total costs, assuming no TTR disposal under various dose-based regulations. Estimates with TTR disposal are roughly $40 million lower due to the elimination of the need to construct a road for transporting TTR waste.

INDIVIDUAL RISKS

To calculate individual risks, the exposure model used to convert plutonium soil concentrations to doses was modified to account for exposure duration and coupled with a dose-response model. Uncertainty in exposure duration was quantified using data on residence time (mean of approximately 10 years and a 90%

Figure 1.   **Total costs of achieving dose-based regulations at the Nevada Test Site and related sites, assuming no disposal on the Tonopah Test Range.**

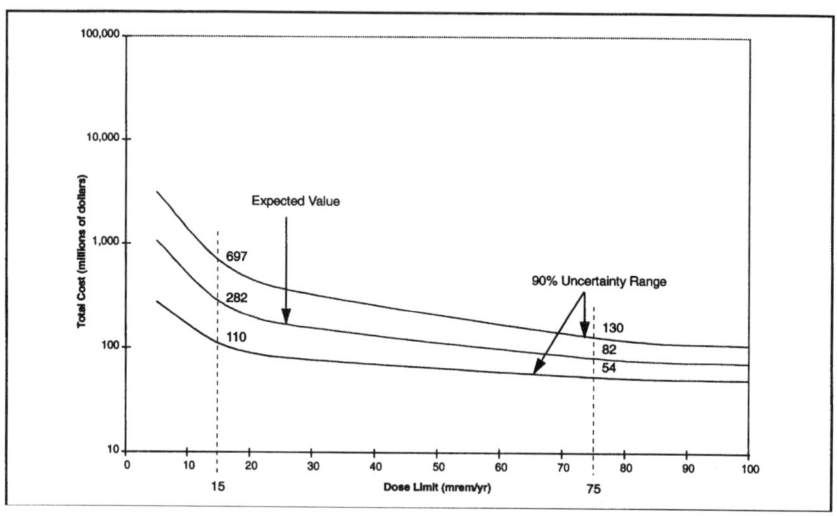

range between 0.4 and 36 years) (U.S. Bureau of the Census, 1991). EPA inhalation and ingestion risk factors were used after independent analyses produced similar results: $3.8 \cdot 10^{-8}$ per pCi inhaled and $2.3 \cdot 10^{-10}$ per pCi ingested (EPA, 1993b). Uncertainties were quantified using a Monte Carlo analysis based on variabilities in organ doses, cancer probabilities per unit dose, and correlations among organ doses derived from autopsied organs.

Figures 2 and 3 provide the resulting expected values and 90% uncertainty ranges for individual risk for nuclear and safety-test sites under dose-based regulations. The numbers indicate the estimated incremental probabilities of premature cancer fatality for a randomly selected individual residing at a location having the highest plutonium concentration allowed by the cleanup regulation. Under dose-based regulations, estimated individual risks are nearly identical for the two types of sites. However, this equivalency was not found under concentration-based regulations. If nuclear and safety-test sites are cleaned to achieve the same concentration limits, expected individual risks are roughly a factor of two higher at safety-test sites.

POPULATION RISKS

Population risks were estimated by modifying the model used to calculate individual risks, accounting for estimated future population sizes, and estimating risk as a function of time based on the decay rate of ^{239}Pu. The modifications to the individual risk model included replacing the residential land-use scenario with a combined scenario representing residential and commercial exposures. Also, the

Figure 2. **Individual risk for person at location having highest allowable plutonium concentration—nuclear site, dose-based regulations.**

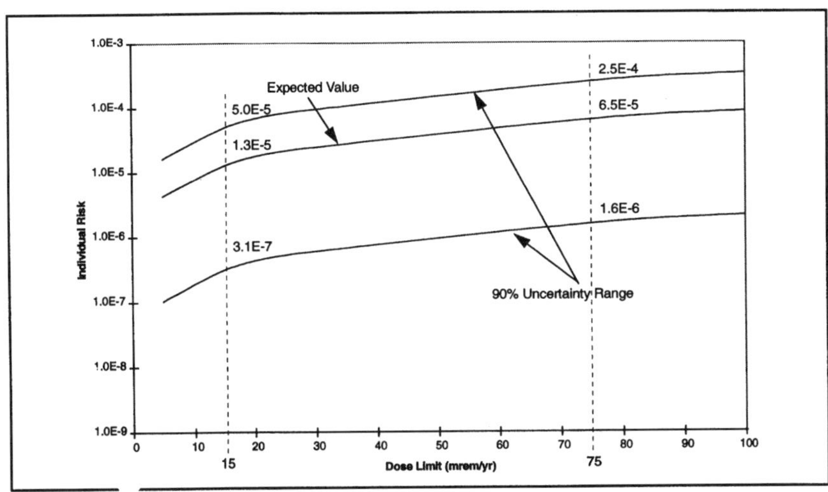

Figure 3. **Individual risk for person at location having highest allowable plutonium concentration—safety-test site, dose-based regulations.**

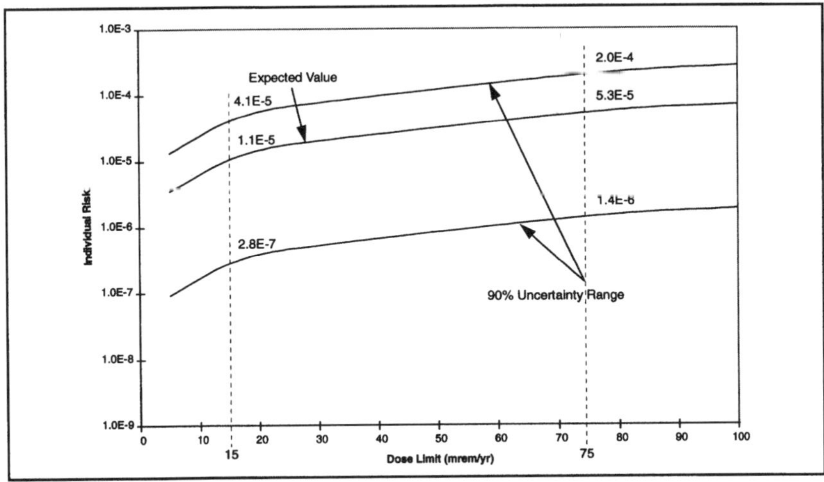

Monte Carlo analysis was altered so as to account only for variabilities in population averages, not individual-to-individual variabilities. For example, individual-to-individual variations in breathing rate used as input to the individual risk analysis were replaced with estimated population averages. Uncertainty in future population density was represented roughly by taking the current densities for Nye County (0.00386 persons/hectare), the state (0.03886 persons/hectare), and Nevada's most populated county (0.386 persons/hectare) as the 5, 50, and 95 percentiles of the probability distribution (U.S. Bureau of the Census, 1992). Population density was assumed constant over time, and the population at risk was assumed to be all individuals located on areas contaminated at 10 pCi/g or above. A model relating plutonium concentrations and areas was used to account for probabilistic dependencies between existing plutonium concentrations, post-remedial concentrations, and the sizes of contaminated areas.

Population risk was expressed in terms of estimated numbers of excess cancer fatalities to the population at risk and in terms of fatalities averted by the alternative regulations. Since the decay rate of plutonium is slow, risks potentially accumulate over a very long time duration (more than 100,000 years). Estimated fatalities were expressed in terms of total fatalities over all time, total discounted fatalities (by applying a discount rate to future fatalities), and fatalities over the first 1,000 years. The number of excess cancer fatalities occurring during the first 1,000 years was estimated to range between 0.034 and 10.1, with an expected value of 2.9. The expected total population at risk is 8,811 people, with an uncertainty range of 924 to 31,200.

Figure 4 provides the estimated numbers of fatalities averted during the first 1,000 years given cleanup to the alternative dose-based regulations. Fatalities averted are not strongly dependent on the level of the dose- or concentration-based regulation because post-remedial population risk is proportional to the average post-remedial plutonium concentration in the contaminated area, which is mostly determined by the large areas with relatively low concentrations near 10 pCi/g.

WORKER RISKS

Risks to workers engaged in remediation activities include fatal and nonfatal occupational accidents during waste handling and transportation and excess cancers associated with occupational exposures to radionuclides. Dominant risks to workers were estimated to be fatal accidents during soil transport and heavy equipment operation. Accident statistics from similar operations were used to estimate uncertainty in the rate of fatal accidents per unit volume of soil remediated.

Worker risks were estimated to be low (less than one expected fatality except under cleanup up to 10 pCi/g, where extremely high volumes would be remediated). Low worker risks were estimated due to the use of standard effective safety practices and procedures, and because transport of waste on private roads is likely to reduce the frequency of transport accidents.

Figure 4.   Public fatalities averted during the first 1,000 years under dose-based regulations.

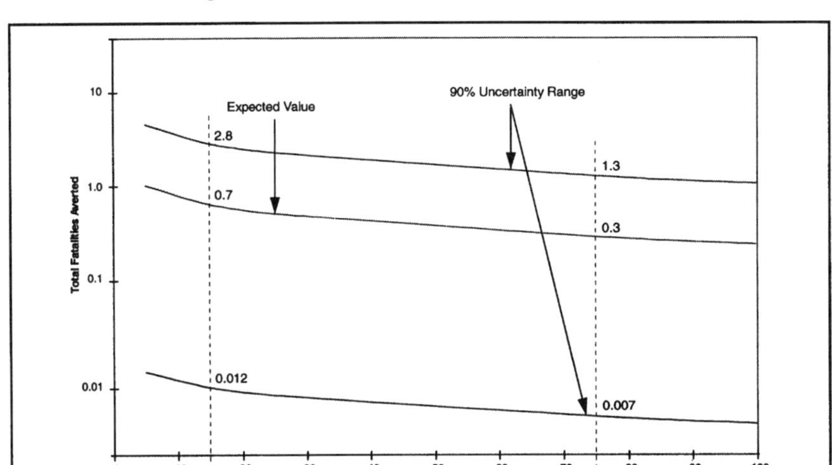

COST-BENEFIT COMPARISONS

If the estimated costs of compliance are divided by the estimated numbers of public cancer fatalities averted, the "value of life" necessary to justify application of a regulation on the basis of a cost-benefit comparison can be estimated. The values computed in this way depend strongly on the time period over which fatalities are accumulated and on whether fatalities estimated to occur in the distant future are discounted. If only those fatalities averted during the first few hundreds or thousands of years are counted, or if fatalities in the distant future are discounted, the values of life necessary to justify applications of the regulations are considerably higher than those typically recommended to justify other government risk-reducing investments. For example, if effects over only the first 1,000 years are counted, the value of life required to justify remediation is $240 million for a 75 mrem/yr standard and $390 million for a 15 mrem/yr standard. Values up to approximately $10 million have been cited as the maximum for concluding that risk-reducing investments are cost-effective.

LIMITATIONS AND CONCLUSIONS

Limitations of the analysis relate to the lack of hard data to validate the numerous judgments inherent in the analysis and limits in scope. Scope limitations include the omission from the analysis of potentially important impacts; for example, possible socioeconomic impacts of remediation (e.g., boom-bust cycles on local communities) and nonfatal risks. In addition, environmental impacts were not quantified (although a qualitative analysis of environmental impacts was conducted).

Despite these limitations, the results of the analysis provide a strong argument for postponing large-scale cleanup and release of contaminated lands pending development of more cost-effective cleanup technologies and/or more pressing demands for use of the contaminated lands.

There is little economic motivation, at this time, to release contaminated areas for public or private use. If contaminated land were to be cleaned and released, high population densities are unlikely to occur in the foreseeable future, due to limited water resources and lack of commercially valuable resources. Consequently, population risks would not be very large, and the risks averted through cleanup would not be large. Furthermore, because surface soil contamination exists over extremely large areas, meeting even moderate concentration- or dose-based standards would be extremely costly. Also, excavating surface soils over large areas would produce long-term, potentially irreversible damage to sensitive ecosystems. The cost-effective approach to risk management is the continuation of existing controls which ensure that individuals do not engage in activities in contaminated areas that may result in significant exposures.

A final conclusion relates to advantages and disadvantages of dose-based regulations of the form specified in the EPA draft. Dose-based regulations offer greater flexibility (compared to concentration-based regulations) for incorporating site-specific differences that affect exposures and risk. However, a relatively complex form of analysis requiring many assumptions is needed to determine the level of cleanup necessary to achieve such standards, particularly when expressed with reference to an RME. While justifiable for large sites such as the NTS, a simpler analysis using default assumptions may be more appropriate for sites that are less costly to clean up.

REFERENCES

(DOE) U.S. Department of Energy, 1995, Cost/Risk/Benefit Analysis of Alternative Cleanup Requirements for Plutonium-Contaminated Soils On and Near the Nevada Test Site, DOE/NV--399, Las Vegas, NV.

(EPA) U.S. Environmental Protection Agency, 1994, Preliminary staff working draft of Radiation Site Cleanup Regulations, Title 40 CFR Part 196, May 11, 1994.

(EPA) U.S. Environmental Protection Agency, 1993a, Issues Paper on Radiation Site Cleanup Regulations, EPA 402—R-93-084, September 1993, Washington, DC.

(EPA) U.S. Environmental Protection Agency, 1993b, Health Effects Summary Tables, Annual Update, EPA 540-R-93-058, Washington, DC.

McArthur, R.D., Desert Research Institute, 1991, Radionuclides in Surface Soil at the Nevada Test Site, Publication #45041, Las Vegas, NV.

U.S. Bureau of the Census, 1992, *Statistical Abstract of the United States 1992*, 112th Edition, U.S. Government Printing Office, Washington, DC.

U.S. Bureau of the Census, 1991, American Housing Survey for the United States in 1989, Current Housing Reports, H-150/89, U.S. Government Printing Office, Washington, DC.

CHAPTER 50

Probabilistic Evaluation of Radiological Exposures From Plutonium in Soils

Talaat Ijaz, McLaren/Hart ChemRisk, Cleveland, Ohio

Roy Eckart and **Eugene Rutz**, University of Cincinnati, College of Engineering, Cincinnati, Ohio

INTRODUCTION

The evaluation of radiological risks to either individuals or to Critical Population Groups (CPG), will have inherent uncertainties associated with the analysis result due to a number of independent sources including: the mathematical descriptions used to evaluate potential contaminant transport and intake; and the parametric variability of the inputs used by the models.

It is the last source that presents the most significant contribution to uncertainty; however, this source of dispersivity is quantifiable with a reasonable level of effort. Data typically used for radiological exposures are either measured data or are derived from a compiled database, each of which has inherent variabilities. The purpose of this study was to evaluate the parametric uncertainties associated with a Radiological Exposure Assessment (REA) for a specific site contaminated with plutonium-239,240 in soil. Using the resident farmer as the exposure scenario, models were compiled for radionuclide pathway analysis for the two most significant pathways: inhalation of resuspended dust, and ingestion of food products raised in contaminated soil.

UNCERTAINTY ANALYSIS TECHNIQUES

Four major techniques can be employed to assess uncertainty: (1) Differential Analysis, (2) Response Surfaces, (3) Analytical Methods, and (4) Sampling Methods. The first three methods are deterministic in nature, and will provide a limited interpretation of uncertainty. A detailed description and comparison of a number of solution methods is presented in IAEA Safety Series 100 (IAEA, 1989). Due to the varied nature of the input distributions, random sampling techniques are selected for an assessment of this kind. Such an approach also presents the uncertainty output as a probabilistic interpretation.

Random Sampling Techniques

Random sampling techniques are based on treating each model input parameter as random variable with an assigned probability distribution. This ensures that the fundamental characteristic of the parameter are included in the stochastic evaluation. Specific values for each parameter are selected based on a sampling procedure which generates an array of sampled values for each input. The two major techniques for random sampling that are commonly used are Simple Random Sampling (SRS) and Latin Hypercube Sampling (LHS).

Given the numerous models used in an REA, and their complexity, Latin Hypercube Sampling was selected as the most appropriate sampling technique. Simple Random Sampling does have the advantage of simplicity. In addition, SRS is a robust sampling technique and works well with many types of mathematical models. However, the main disadvantage is that it requires too many simulations for an effective evaluation of uncertainty (Andres, 1987). Latin Hypercube Sampling has a number of advantages as a random sampling strategy, including the fact that it generates unbiased estimators; most importantly, with a limited number of simulations, the underlying characteristics of the distributions are preserved (Ijaz, 1994).

DISTRIBUTION SELECTION

The most resource-intensive phase of uncertainty analysis is the determination and definition of the distributions to be used for each input parameter. This phase begins with data collection and compilation, and from the accumulated information a distribution is determined. However, the quantity and quality of data seldom meets expectations. Radioecological assessments of plutonium isotopes are usually based on minute concentrations in environmental media and this is reflected in the quality of transfer coefficients. In addition, such data are usually sparse, hence exacerbating data quantity and quality concerns. In most cases, only a range of values is available, and only minor inferences can be made with regard to distribution. Once a range has been established, various other sources of information, both objective and subjective are needed to provide characteristics to the data from which the distributions can be generated. The information requirement for the generation of input distributions is significantly larger than that used to determine parameter ranges. An indispensable component is engineering judgment; this allows determinations to be made with the benefit of rationalization.

Once the parameters have been determined, the random sampling generators can be employed to generate the input arrays. This analysis employed the Crystal Ball software program (Decisioneering, Inc., 1993) to generate and sample the input parameters distributions using Latin Hypercube Sampling (LHS). The dose assessment models were compiled into a spreadsheet format from within which Crystal Ball was executed. The generator was optimized to 250 LHS simulations for each input; this is less than would be required for SRS. The optimization ensures adequate distribution sampling and preserves the characteristics of the distribution. Increasing the number of simulations has a limited effect upon the output; beyond 500 simulations, tail sampling becomes exaggerated.

RADIOLOGICAL EXPOSURE ASSESSMENT

The Radiological Exposure Assessment (REA) establishes the models used for radionuclide transport, uptake, and dose assessment for ^{239,240}Pu in a semi-arid environment, within the framework of a residential farm scenario. This assessment was performed for Area 11 (Plutonium Valley), of the Nevada Test Site. This 2,200-acre site on the eastern extreme of the Nevada Test Site (NTS) was used for "safety shots" events, and as a result the soils are contaminated with plutonium and americium. Contaminant dispersion is generally limited to the blast centers and the immediate outlying regions. Although high concentrations exist at ground zero, the majority of the area has significantly lower concentrations of ^{239,240}Pu. The plutonium content of the soils of the area studied was characterized statistically using a stratified random sampling technique (Gilbert, 1987). Using the data from Gilbert and Eberhardt (1975) and Gilbert (1977), the weighted mean for Area 11 ^{239,240}Pu soil concentration is 0.153 ± 0.016 nCi/g.

For the purpose of this REA, the Resident Farmer scenario was used for dose evaluation. Although this scenario is hypothetical, it does activate all pathways, thus ensuring quantification of all uncertainties. Site occupancy is full-time and all consumables are produced on-site. Due to the characteristics of the site, no groundwater transport was considered; however, irrigation of the family farm from uncontaminated sources will continue. Inclusion of a groundwater model will introduce uncertainties associated with hydrogeological modeling; while this assessment of uncertainty is possible, the spatial and temporal variations of hydrogeological parameters may overshadow the fundamental assessment of uncertainties associated with plutonium in soils. All modeled exposure routes are shown in Figure 1. The two main mechanisms considered are the ingestion of food products raised on site and the inhalation of resuspended dust.

INGESTION PATHWAY

The ingestion pathway was considered as a compilation of six subpathways (ingestion of leafy and nonleafy vegetation, ingestion of meat, liver, milk, and inadvertent soil ingestion). Concentrations in plant media were determined using site-specific concentrations ratios and simplistic foliar deposition models. To account for plutonium accumulation in cattle, a series of submodels were employed. The submodels utilize soil to plant concentration ratios as inputs to the cattle biokinetic model. Due to the nature of plutonium transfer to plants, characterization of concentrations in plant media is highly imprecise. Reliance is upon limited measured concentrations of plutonium, which in most cases is at minute levels. As a consequence, large variabilities in the data were evident. Inclusion of foliar

Figure 1. Exposure pathways used for exposure assessment.

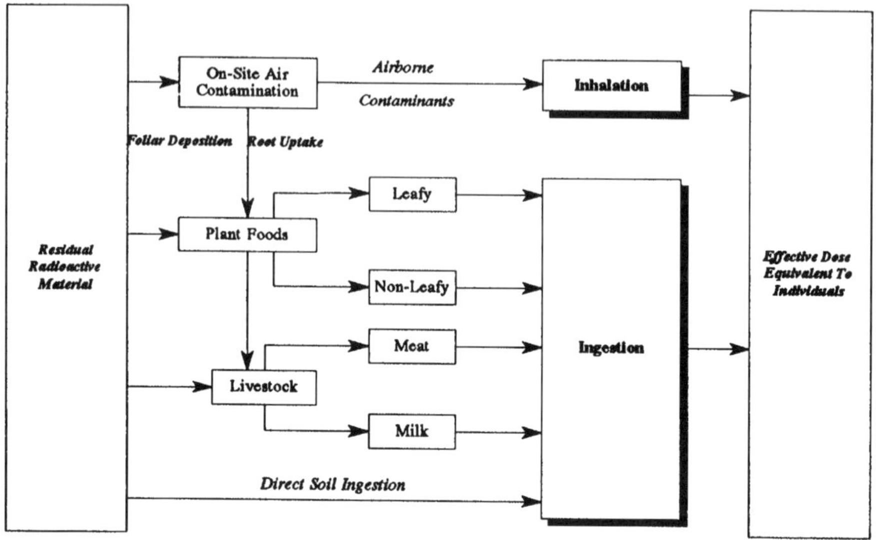

deposition models increased the data requirement, as all processes from particle interception to removal by weathering need to be included. Due to the native vegetation density, it is unlikely that forage will provide the entire vegetation requirement for cattle. Even though the restoration will return a significant proportion of the area to a condition similar to the current situation, it is unlikely that the standing biomass will be dense enough to support cattle. Thus, supplemental hay is considered the most likely feed source. Resilient strains of alfalfa can be cultivated in the local environment. The requirement for irrigation exists; however, an off-site water source is envisioned.

INHALATION PATHWAY

The inhalation pathway evaluates the dose received from the inhalation of contaminated dust. The dust becomes a respirable source when it is resuspended by an interaction with the top soil. In addition, there will be a background concentration of dust in the air due to wind generated natural resuspension. To evaluate the dose received from the inhalation of resuspended dust, there are essentially three components that are required. The aerosol source term generated by the dust resuspension component (μg/m^3), the breathing rate (m^3/hr) quantifies an intake rate, and the final component is the exposure duration (hr/day).

These components form a linear relationship to evaluate the dose from a given concentration of contaminant in the soil. There are also interrelationships that can be of concern when dealing with short-term exposures. To account for time-based variability in the evaluation of exposure, Time-Activity Scenarios were created. The concentration in air is a function of resuspension, which in turn is a function of activity. These interrelationships are outlined in Figure 2. The critical correlation is between breathing rate and activity. Activities with higher exertion levels will induce higher breathing rates. The breathing rate is also dependent upon a multitude of population descriptors including age, weight, sex, and physiology. Activity levels may also affect dust resuspension potentials. Activities that include interactions with the ground, such as soil tilling, walking, etc., will increase the dust resuspension potential.

Breathing Rates

The International Commission on Radiological Protection (ICRP, 1975) defines a daily breathing rate based on 8 hours of light activity, 8 hours of non-occupational light activity, and 8 hours of rest for a daily inhalation rate of $23m^3$/day. This approach assumes that only light activity is associated with the occupational environment. While this may be consistent with the activity lifestyle of average individuals, this does not hold for agricultural workers. Values of breathing rates presented in the Exposure Factor Handbook (USEPA, 1989) are compiled based upon age, sex, and activity levels. The activity levels are expressed qualitatively, and are not correlated to specific activities, i.e., digging, ranching, etc. Thus, breathing rates and activities need to be defined qualitatively based on a relative scale of resting, light, moderate, and heavy activity levels. The advantage of using the Exposure Factor Handbook categorization scheme is that it has been defined according to a criterion developed by the EPA for the air quality criteria document; in addition, values are available for various age and sex cohorts.

Figure 2.   **Interdependencies of the inhalation exposure pathway.**

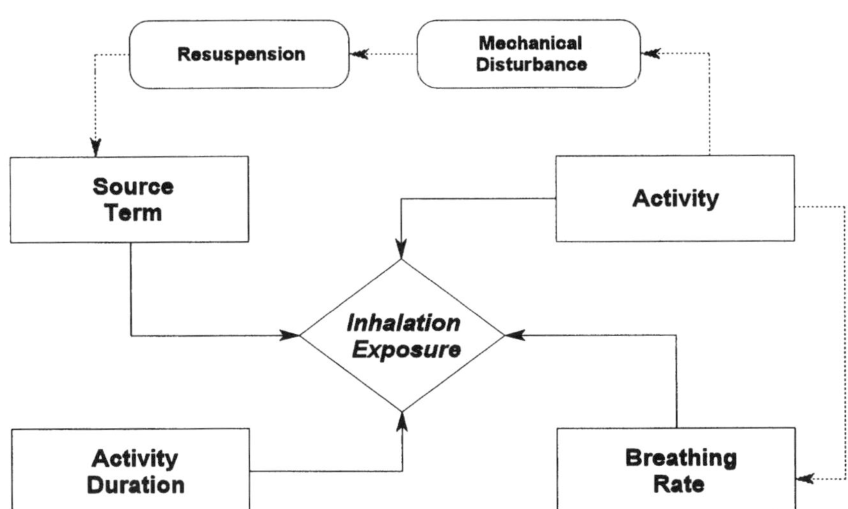

Daily Activity Patterns

Activity patterns are occupation dependent; for the resident farmer, activity profiles are not available due to seasonal and regional variability. Additional components include time indoors versus outdoors. It will be assumed that all activities categorized as being "heavy," "moderate," and "light" are conducted outside. This is consistent with the agricultural scenario. The remaining activity category, "resting," will be assumed to be pursued indoors. Since no specific data of this form is available, a compilation of sources was used.

Dust Resuspension

The driving force behind the significance of the dust resuspension potential is two-fold: first, the significance of inhalation as a dose-delivery mechanism for plutonium, and second, the potential for wind and mechanical resuspension in an arid environment. Actinides, in general, do not readily transfer from food chain pathways into humans; the absorption through the gastrointestinal tract is minimal. Consequently, the resuspension and inhalation of plutonium will be the controlling pathway for risk assessments of this kind.

There are two mechanisms by which contaminated soil can be resuspended into the atmosphere: wind-driven resuspension, primarily due to the motion and impact of particulates with diameters less than 1mm (Healy, 1980); and mechanical resuspension generated by disturbances in the top soil structure. Quantification of the resuspension potential can be characterized by Resuspension Factor, Resuspension Rate, or Mass-Loading. Given the limitation and availability of measured data for Area 11, values of dust mass loading are considered to be the most appropriate for inclusion in this REA. To account for the fact that small particles ($<$50 μm) have a higher resuspension potential, an Enhancement Factor (EF) was employed. The EF is the ratio of the activity density of the suspended particulates to the activity density in the soil (Shinn, 1991). Results from earlier studies conducted at Area 11 show the EF value to be close to unity, (EF=1.04, Shinn et al., 1986). Empirical studies (Shinn, 1991) also show that mechanical disturbances can significantly change the values of EF. Three mass loading values were designed (background, disturbed, and indoors), corresponding to the three scenarios envisioned, and two enhancement factors (disturbed and undisturbed) are required.

Normal Undisturbed: Background.

The reported background level measured for Area 11 (Shinn et al., 1989a) is 41 ± 12 μg/m^3. Since no other information is currently available to the contrary, it is assumed that these background dust loadings represent an acceptable level of dust resuspension the NTS.

Elevated Levels During Soil Disturbance

The only on-site data that are available are for very specific activities which are not compatible with the activities foreseen as part of a typical exposure assessment. Dust loadings for the activities envisioned may be available for predominantly agricultural areas which are contrary to the conditions at the NTS. The enhancement factors (EF) are site and contaminant-specific, as there is a strong dependency upon particle size distributions and as well as the form of mechanical disturbance to the soil. The unique characteristics of the safety shot tests determine the range of ^{239,240}Pu particle sizes and the association with the soil.

Various elevated dust loadings measured during specific activities have been reported in the literature. In an attempt to provide realistic ranges of dust loadings under differing disturbance conditions, a compilation of measurements was made. Selection criteria was based on conditions similar to those existing at the NTS. Table 1 lists the most appropriate values.

Table 1.   **Measured Dust Loading Values (µg/m^3)**

Disturbance Type	Dust Loading	Source
Area 11: CAT (Cleanup & Treatment)	88.0	Shinn et al., 1989a
Area 11: CAT, Post stabilization	41.0	Shinn et al., 1989a
Area 18: Raked and wet	15.4	Shinn and Homan, 1987
Area 18: Raked, wet, and dried	30.3	Shinn and Homan, 1987
Bikini: Bulldozing	136.3	Shinn et al., 1989b
Bikini: 1 week stabilization	69.5	Shinn et al., 1989b
Bikini: Vehicular traffic (median)	100.0	Shinn et al., 1989b
Bikini: Vehicular traffic (peak)	210.0	Shinn et al., 1989b
Bikini: Vehicular traffic (Time Ave.)	28.0	Shinn et al., 1989b
Bikini: Foot traffic (median)	26.0	Shinn et al., 1989b
NAAQS 24 hr maximum	260.0	Anspaugh, 1974
NAAQS Annual average	75.0	Anspaugh, 1974
Anspaugh: Generic value	100.0	Anspaugh & Phelps, 1974b

Disturbed source mass loadings range from a low of 26 (foot traffic) to the 260 µg/m^3 NAAQS 24 hour maximum. There is a large range associated with this data; the propensity of values in the 20-100µg/m^3 range indicating a strong central tendency about these points. The statistics of the compiled data produce a mean and standard deviation of 90.7 and 74.0 µg/m^3, respectively; the underlying distribution is positively skewed and shows characteristics of being lognormal.

Indoor Dust Concentrations

Using a control volume to represent a building, Layton et al., 1992, modeled the transport of contaminated dust into and out of a hypothetical dwelling by the infiltration of contaminated outdoor air, exfiltration of building air, resuspension of tracked particles from floors to indoor air, and the gravitational settling of particles. The parameter requirement for this model is extensive; most values are estimated from U. S. national averages for building structural compositions, infiltration values are taken from well documented sources. The resultant indoor mass loading for a typical farmhouse can be represented by a lognormal distribution with a geometric mean of 13 µg/m^3, and a geometric standard deviation of 1.97 µg/m^3.

Enhancement Factors

As discussed earlier, the use of non-site-specific data is not viable due to the dependency upon the original source and the particulate size distributions (Shinn, 1991). Enhancement factors vary widely from site to site, hence emphasizing the importance of site-specific parameters. The EF values evaluated by Shinn (Shinn et al., 1989), which are specific to Area 11, are utilized here as they are compatible with the default background mass loading. The two reported EF values are 1.04 for Background Conditions, and 3.71 for Disturbed Soil Conditions.

The distributions and values used for the inhalation parameters are listed in Table 2. Similar values used for the ingestion pathway are too numerous to be presented here because their significance to the final results are minor; these values are published in Ijaz, 1994.

Table 2.   Parameter Distributions for Inhalation Exposures

PARAMETER	Units	Distribution	Limits
Dust Resuspension: Mass Loading			
Background	µg/m^3	Normal	AM: 41, ASD: 12
Disturbed Soil	µg/m^3	Lognormal	AM: 90.7, ASD: 74.0
Indoor	µg/m^3	Lognormal	GM: 13.0, GSD: 1.97
Dust Resuspension: Enhancement Factor			
Background	-	Constant	1.04
Disturbed Soil	-	Constant	3.71
Indoor	-	Constant	1.04
Human Breathing Rates			
Resting	m^3/hr	Triangular	Max: 1.13 Mode 0.73 Min: 0.14
Light	m^3/hr	Triangular	Max: 1.8, Mode 0.83, Min: 0.14
Moderate	m^3/hr	Triangular	Max: 5.54, Mode 2.45, Min: 0.8
Heavy	m^3/hr	Triangular	Max: 11, Mode 4.8, Min: 2.0
Human Activity Patterns			
Resting	hr/day	Normal	AM: 14.33, ASD: 4.55
Light	hr/day	Normal	AM: 7.22, ASD: 1.86
Moderate	hr/day	Normal	AM: 1.22, ASD: 3.73
Heavy	hr/day	Normal	AM: 1.22, ASD: 2.48

Note: AM: Arithmetic Mean; ASD: Arithmetic Standard Deviation; GM: Geometric Mean; GSD: Geometric Standard Deviation

RESULTS

The stochastic evaluation using the LHS provides a quantitative measure of the dispersion of the DSR as well as generating bounds of confidence for the given result. The general characteristics of the output distribution conform to a lognormal distribution, with a significant degree of peakness about the central tendency and a distinctively positive skew (Figure 3). The supposition of a lognormal distribution is supported by the descriptive statistics generated from the output data.

Due to the presence of the extreme values, the arithmetic mean will be an overestimation of central tendency; hence, the median and geometric mean are used. The median value approximates the GM, confirming the measure of central tendency. The median DSR evaluated was 0.30 with a 95% confidence interval, based on lognormal statistics of 0.26 to 0.34 {(mrem/yr)/(pCi/g)}. The evaluated geometric mean (GM) and standard deviation (GSD) were evaluated to be 0.30 and 2.56 {(mrem/yr)/(pCi/g)}, respectively.

Uncertainty by Pathway Components

A deterministic analysis showed that 89% of the Dose to Source Ratio (DSR) is attributable to inhalation; the remaining 11% due to ingestion is not insignificant. In order to isolate the major source of uncertainty, the model was restructured to calculate the probabilistic DSR based on the parametric variability of each of the two pathways. This was achieved by maintaining constant input values for parameters significant to one pathway, while parametric sampling and stochastic gen-

Figure 3. **Frequency output data for ^{239,240}Pu DSR: {(mrem/yr)/pCi/g)}.**

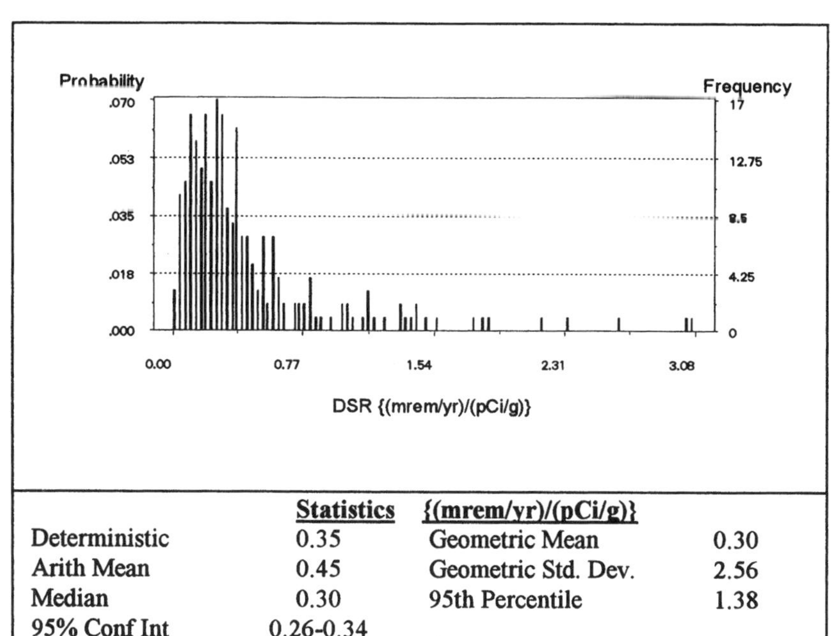

	Statistics	{(mrem/yr)/(pCi/g)}	
Deterministic	0.35	Geometric Mean	0.30
Arith Mean	0.45	Geometric Std. Dev.	2.56
Median	0.30	95th Percentile	1.38
95% Conf Int	0.26-0.34		

eration was used for the parameters of the other. Thus, two distinct outputs were generated, one that includes the variability of the inhalation pathway parameters, and the other that isolates the ingestion pathway parameters. Using distribution-free statistics, measures of the dispersivity were generated for each; percentile plots, including the values from the initial generation (TOTAL-DSR) are presented in Figure 4.

Evaluation of percentiles for each pathway DSR reveals a significantly larger dispersion of the inhalation DSR compared to the ingestion DSR. Both frequency plots for the inhalation and ingestion DSRs approximate lognormal distributions, indicating the cause of the high dispersion is primarily due to the high tail values. The higher dispersion of the inhalation DSR is a function of the lognormal distributions of resuspension and the high standard deviation of the activity duration values.

An additional investigation of the contributions to the dispersion of the inhalation pathway was conducted to isolate the sources of uncertainty. For the inhalation pathway the parameters were categorized by three classifications: breathing rates, activity duration, and mass loading data. The most dispersive evaluations were due to activity duration, and mass loading; the lowest dispersion was from the breathing rates (Figure 5).

Parameter Importance

Parameter importance was determined using rank correlation coefficients (RCC's) evaluated from the sampled values and the output DSR (Figure 6). Assessments using the variance propagation method were also made, and showed consistency with regard to the ranking of parameters, although there were discrepancies in the magnitude of the importance function. The RCC analysis also identified negative trends that were omitted by the variance propagation method.

Figure 4. Percentile plots by pathway (includes TOTAL-DSR).

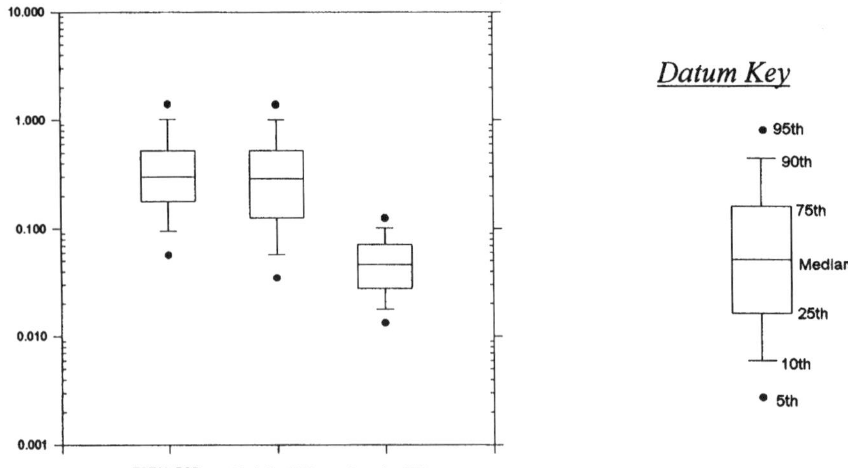

Figure 5. **Percentile plots for the three parameter classifications.**

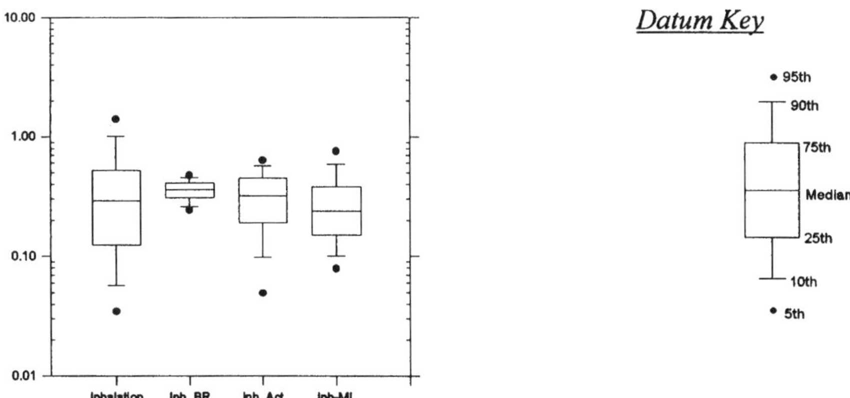

Figure 6. **Rank correlation coefficients for top 13 parameters.**

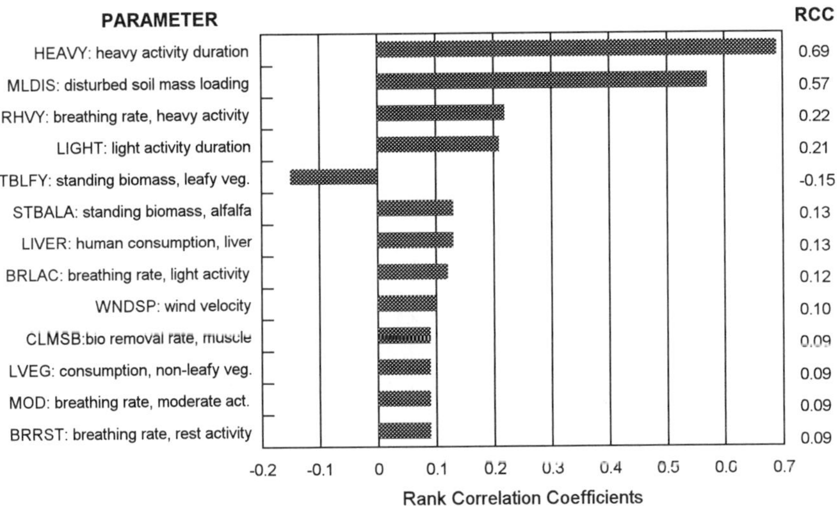

In all cases the parameters associated with the inhalation pathway were found to be the most significant contributors in terms of model sensitivity and dispersion. Of the 13 parameters with RCC > 0.09, seven are parameters affecting the inhalation DSR. All four breathing rates are highly correlated, as are the disturbed soil mass loading, and the light and heavy activity duration. The remaining parameters are major contributors to the vegetation ingestion dose, specifically the foliar deposition pathway and the nonleafy consumption rate.

The uncertainty associated with the inhalation parameters is evident from the distributions used. The heavy activity duration parameter is described by a normal distribution with a significant standard deviation. This dispersion is a direct consequence of the data quality; such inadequacies have been successfully conveyed by the uncertainty evaluation and this is evident from the results. The mass loading for disturbed soils is described by a lognormal distribution, also with a significant degree of dispersion. This dispersion, along with the tail characteristics of a lognormal distribution, generates the extreme values that have been noted throughout this analysis. This parameter is also weakly correlated to the heavy activity breathing rate, thus compounding the effect of high values, and ensuring the importance of both parameters.

CONCLUSIONS

Of the two principal means of random sampling, Latin Hypercube Sampling (LHS) is the most promising. The ability of LHS to sample discrete, probability-based intervals ensures the complete sampling of a given distribution. The most apparent distinction between the two types of generators is the preservation of the shape and characteristics of the distribution being sampled, regardless of the distribution type. Since the majority of the distributions selected are based on scientific judgment in the absence of data, the integrity of the distributions ensures the significance of the interpretation of data. The only limitation of the LHS scheme is its propensity to sample tail values from lognormal distributions. This effect increases with the number of simulations specified; and hence has relevance to the optimization of the generator. By increasing the number of simulations, the LHS generator produces an equivalent number of probability-based intervals. There is little or no evidence of the sensitivity correlations being dependent on the number of simulations. Marginal changes in the calculated correlation coefficients were noted, but no major redistribution of importance occurred. Hence the determination of generator optimization is governed by the limit of accuracy required by the uncertainty analysis, and not by any constraints imposed by the sensitivity analysis.

It is evident that the predominant sources of dispersion of the ^{239,240}Pu dose and DSR are the activity duration, mass loading, and breathing rate components of the heavy activity scenario. Using the data provided by an examination of output dispersion as a function of parameter classification, the activity duration has been identified as the major contributor to DSR dispersion about the measure of central tendency. The high standard deviations of this type of data are a property of the limited amount of data available. This uncertainty in the activity profiles of the resident farmer is due to the hypothetical definition of the exposure scenario. In essence, this uncertainty, although parametric, can also be classified as a Type 2 uncertainty. It is this uncertainty that determines the dispersion of the result. The propensity of extreme values, and the generation of the large tails of the resultant lognormal distribution is attributable to the dust resuspension data. This range of possible values is a function of the various soil disturbance mechanisms envisioned. The dispersion of values generated about the measure of central tendency is attributable to the large deviations noted in the activity duration data. This dispersion can be minimized by increasing the sample size used to generate the activity duration data. Conversely, the breathing rates, which are of equal importance in determining the inhalation DSR, do not significantly contribute to the output dispersion. These parameters are constrained about the mean value by the excellent quality of data. It is this quality of data that determines the dispersion, and hence the uncertainty of the dose evaluation.

REFERENCES

Andres, T.H. 1987. Statistical Sampling Strategies. In: Proceedings of an NEA Workshop on Uncertainty Analysis for Performance Assessments of Radioactive Waste Disposal Systems. OECD Seattle.

Anspaugh, L.R. and Phelps, P.L. 1974. Results and Data Analysis: Resuspension Element Status Report. In: *The Dynamics of Plutonium in Desert Environments*, pp. 265-298, (Dunaway, P.B. and White, M.G., Eds.), NVO-142.

Anspaugh, L. R. 1974. Appendix A, Resuspension Element Status Report: The Use of NTS Data and Experience to Predict Air Concentrations of Plutonium Due to Resuspension on the Enewatak Atoll. In: *The Dynamics of Plutonium in Desert Environments*, pp 298-310, (Dunaway, P.B. and White, M.G., Eds.), NV0-142.

Decisioneering, Inc. 1993. Crystal Ball Version 3.0: Forecasting and Risk Analysis for Spreadsheet Users.

Gilbert, R.O. 1987. *Statistical Methods For Environmental Pollution Monitoring*. New York, Van Nostrand Reinhold Publishers.

Gilbert, R.O. 1977. Revised Total Amounts of Pu-239,240 in Surface Soil at Safety Shot Sites. In: *Transuranics in Natural Environments*, pp. 423-429, (White, M.G., Dunaway, P.B., and Wireman, D.L., Eds.). NVO-181.

Gilbert, R.O. and Eberhardt, L.L. 1975. Statistical Analysis of Pu-239,240 and Am-241 Contamination of Soil and Vegetation on NAEG Study Sites. In: *The Radioecology of Plutonium and Other Transuranics in Desert Environments*, (White, M.G., and Dunaway, P.B., Eds.), NVO-153.

Healy, J.W. 1980. Review of Resuspension Models. In: Transuranic Elements in the Environment. W. C. Hanson, Ed., U.S. Department of Energy, USDOE Report, DOE/TIC-22800.

(IAEA) International Atomic Energy Agency. Evaluating the Reliability of Predictions Made Using Environmental Transfer Models: IAEA Safety Series 100. STI/PUB/835, 1989.

(ICRP) International Commission on Radiological Protection. 1975. Report of the Task Group on Reference Man. Publication 23. Pergamon, New York.

Ijaz, T. 1994. Implementation of Latin Hypercube Sampling to Evaluate Uncertainties Associated with Exposures from Plutonium in Soils. Ph.D. Thesis, University of Cincinnati.

Layton, D.W., Anspaugh, L.R., Bogen, K.T., and Straume, T. 1992. Risk Assessment of Soil Based Exposures to Plutonium at Safety Shot Sites Located at the Nevada Test Site and Adjoining Areas. LLNL-Draft.

Shinn, J.H. 1991. Enhancement Factors For Resuspended Aerosol Radioactivity: Effect of the Top Soil Disturbance. UCRL-JC-105257.

Shinn, J.H., Essington, E.H., Miller, F.L., O'Farrell, J.A., Orcutt, J.A., Romney, E.H., Shugart, J.W., and Sorom, E.R. 1989a. Results of Cleanup and Treatment Test at the NTS: Evaluation of Vacuum Removal of Pu-Contaminated Soil. *Health Physics*, 57(5), 771-779.

Shinn, J.H., Homan, D.N., and Robison, W.L. 1989b. Resuspension Studies at Bikini Atoll. UCID-18538.

Shinn, J.H. and Homan, D.N. 1987. Plutonium-Aerosol Emission Rates and Human Pulmonary Deposition Calculations for Nuclear Site 201, NTS. In: *The Dynamics of Transuranics and Other Radionuclides in Natural Environments*, pp. 261-278. (Howard, W.A. and Fuller, R.G., Eds.), NVO-272.

Shinn, J.H., Hofmann, C.B., and Homan, D.N. 1986. A Summary of Plutonium Aerosol Studies: Resuspension at the NTS, UCRL-90746.

(USEPA) U.S. Environmental Protection Agency. 1989. Exposure Factor Handbook. Office of Health and Environmental Assessment. EPA 600/8-89/043.

PART X—MTBE

CHAPTER 51

Significant Findings and Water-Quality Recommendations of the Interagency Oxygenated Fuel Assessment

John S. Zogorski, U.S. Geological Survey, Rapid City, South Dakota

Arthur L. Baehr, U.S. Geological Survey, West Trenton, New Jersey

Bruce J. Bauman, American Petroleum Institute, Washington, D.C.

Dwayne L. Conrad, Texaco EPTD, Bellair, Texas

Robert T. Drew, American Petroleum Institute, Washington, D.C.

Nic E. Korte, Oak Ridge National Laboratory, Grand Junction, Colorado

Wayne W. Lapham, U.S. Geological Survey, Reston, Virginia

Abraham Morduchowitz, Texaco Inc., Beacon, New York

James F. Pankow, Oregon Graduate Institute of Science and Technology, Beaverton, Oregon

Evelyn R. Washington, U.S. Environmental Protection Agency, Washington, D.C.

INTRODUCTION AND SCOPE

At the request of the U.S. Environmental Protection Agency (USEPA), the Office of Science and Technology Policy, Executive Office of the President, has coordinated an interagency assessment of the scientific basis and efficacy of the nation's winter oxygenated-gasoline program. This program mandates that compounds referred to as oxygenates be added to gasoline in select metropolitan areas across the United States (U.S.) to reduce the amount of carbon monoxide in the atmosphere in the winter. Many other areas of the U.S. have voluntarily chosen to use oxygenated fuels to abate air pollution and, in severe ozone nonattainment areas, reformulated gasoline is required. Fuel oxygenates also are used to enhance the octane of conventional gasolines - a practice that started in the late 1970s and continues today (1997).

Methyl *tert*-butyl ether (MTBE) is the most commonly used fuel oxygenate. The U.S. production was estimated to be 8.0 billion kilograms in 1995 (American Chemical Society, 1996). Essentially all of the MTBE produced is used for fuel oxygenation. Ethanol (EtOH) is the second most used oxygenate in gasoline blending. EtOH production in the U.S. was estimated to be 4.3 billion kilograms in 1994 (James, 1995). No data are available to estimate the portion of this production used in gasoline. Gasoline in carbon monoxide nonattainment areas must contain no less than 2.7% oxygen by weight. By volume, this requirement corresponds to 14.8% for MTBE and to 7.3% for EtOH (Conrad, 1995). Other oxygenates in limited commercial use include ethyl *tert*-butyl ether (ETBE), *tert*-amyl methyl ether (TAME), and diisopropyl ether (DIPE). Methanol (MeOH) is only being used as an oxygenate in a limited test program in California. *Tert*-butyl alcohol (TBA) has been added to gasoline in the past and is another potential fuel oxygenate, but is not currently produced for this purpose.

This chapter summarizes significant findings and water-quality recommendations of the Interagency Oxygenated Fuel Assessment reported in Zogorski et al., 1996, hereinafter referred to as the IOFA Water-Quality Report. The Assessment addresses water-quality issues arising from the production, distribution, storage, and use of fuel oxygenates and their movement in the hydrologic cycle. It summarizes the scientific literature, data, and agency information on the sources, concentrations, behavior, and fate of fuel oxygenates and their aqueous degradation products in groundwater and in surface water. It also assesses the implications for drinking-water quality and aquatic life. Recommendations for further database compilation, monitoring, reporting of information, follow-up assessments, reports, and research efforts also are indicated. The scope of the Assessment intended to include all oxygenates and their aqueous degradation products. However, except for MTBE, little to no data and few scientific publications are available on the occurrence and behavior of oxygenates and their degradation products in groundwater, surface water, and drinking water. Some monitoring data are available for MTBE; however, these data sets are limited in scope.

SOURCES AND RELEASES

Like hydrocarbon components of gasoline, fuel oxygenates are introduced to the environment during all phases of the petroleum fuel cycle, but the major sources of these compounds are likely associated with the distribution, storage, and use of oxygenated gasoline. Releases of gasoline containing oxygenates to the subsurface from, for example, underground storage tanks, pipelines, and refueling facilities are point sources for entry of oxygenates, as well as gasoline hydrocarbons, into the hydrologic cycle (Figure 1). Urban and industrial runoff and wastewater discharges also represent potential sources of oxygenates to the environment. Water in contact with the spilled gasoline at the water table or in the unsaturated zone dissolves oxygenates along with aromatic hydrocarbon constituents (for example, benzene, toluene, ethylbenzene, and xylenes, which are commonly referred to as BTEX). Such water can contain high oxygenate concentrations; for example, MTBE concentrations as high as 200,000 micrograms per liter (Davidson, 1995; Garrett et al., 1986) have been measured in groundwater at a site of gasoline leakage.

Petroleum storage tanks represent the largest population of potential point sources of alkyl ether oxygenates to natural waters. For example, in 1988 the USEPA estimated that there were about 2 million underground storage tanks (USTs) at more than 700,000 facilities (Lund, 1995). In the last several years, USTs have been removed at many facilities, and the current U.S. UST universe is an estimated 415,000 UST facilities (about 195,000 are service stations) (U.S. Environmental Protection Agency, 1995a) with about 1.2 million tanks. The USEPA's statistics indicate that slightly more than 300,000 sites have been identified as having contamination levels that require corrective action (U.S. Environmental Protection Agency, 1995a). Cleanup has been completed at about 130,000 of these sites. Because of the inherent difficulty in determining leak rates, leak durations, and in determining the use of oxygenates in the released gasoline, a reliable estimate of past or current annual release of fuel oxygenates from these sources is difficult to develop. According to USEPA and state requirements, USTs need to be fully upgraded to meet stringent release prevention and detection standards by December 1998. The concurrent improvement in the physical condition of USTs and in release-detection capabilities, coupled with a reduction in the population of tanks, should contribute to a considerable reduction in the annual volume of oxygenated gasoline released to natural waters from USTs.

Estimated releases of MTBE, MeOH, and TBA during industrial activity are mostly to the atmosphere (Table 1) and are reported to the USEPA in annual Toxics Release Inventory (TRI) (U.S. Environmental Protection Agency, 1993) submissions by manufacturers. The release of MTBE to the environment is almost entirely associated with its production, distribution, storage, and use as a fuel oxygenate, whereas releases of MeOH and TBA occur from various other industrial and commercial uses. Industrial releases of other fuel oxygenates ETBE, TAME, DIPE and, most notably, EtOH, are not included in the TRI.

Figure 1. **The movement of fuel oxygenates in the environment** (adapted from Squillace et al., 1995).

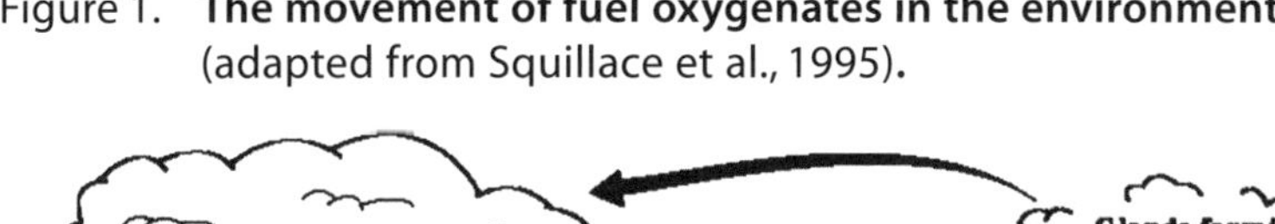
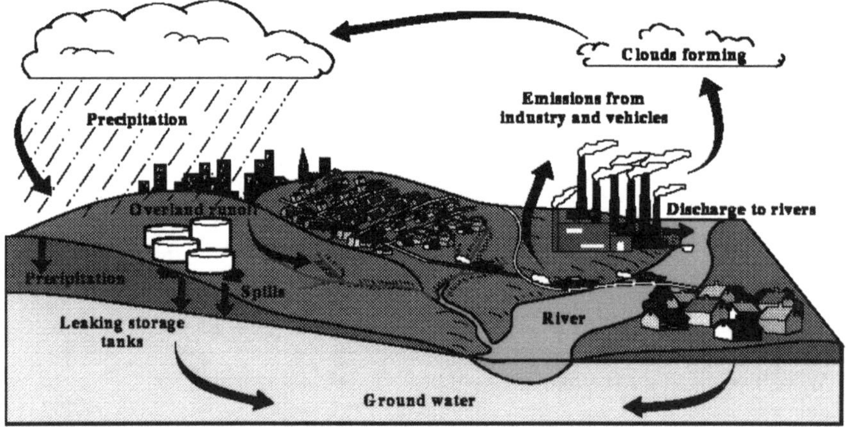

Annual estimates of exhaust emissions from vehicles (Calvert et al., 1993; Hoekman, 1992; Kirchstetter et al., 1996; and Stump et al., 1990), evaporative losses from gasoline stations and vehicles, and releases from storage tanks have not been reported in the scientific literature but are perceived to be important sources of oxygenate (and of other fuel components) releases to the environment. Fuel oxygenates in the atmosphere are considered a nonpoint source to the hydrologic cycle because of the dispersive effect of weather patterns and occurrence in precipitation.

Table 1.   Estimated releases of MTBE, TBA, and MeOH for the U.S. for report year 1993

| | Reported releases (kilograms per year) by: | | |
Releases to	Petroleum refineries	All other facilities	Total
Methyl *tert*-butyl ether			
Atmosphere	1,398,026	274,571	1,672,597
Underground injection	288	3,979	4,267
Land	184	2	186
Water	41,799	74	41,873
Total releases	1,440,297	278,626	1,718,923
Percent	83.8	16.2	100
Number of facilities	78	58	136
***tert*-butyl alcohol**			
Atmosphere	40,498	705,127	745,625
Underground injection	253	138,410	138,663
Land	0	158	158
Water	5	79,051	79,056
Total releases	40,756	922,746	963,502
Percent	4.2	95.8	100
Number of facilities	8	59	67
Methanol			
Atmosphere	369,895	77,544,790	77,914,685
Underground injection	9,670	12,645,520	12,655,190
Land	401	779,716	780,117
Water	17,134	4,523,137	4,540,271
Total releases	397,100	95,493,163	95,890,263
Percent	0.4	99.6	100
Number of facilities	67	2,198	2,265

(MTBE, methyl *tert*-butyl ether; TBA, *tert*-butyl alcohol; MeOH, methanol.
Source: U.S. Environmental Protection Agency, 1993)

OCCURRENCE IN AIR AND WATER AND SIGNIFICANCE TO DRINKING WATER AND AQUATIC LIFE

The Clean Air Act, Clean Water Act, and Safe Drinking Water Act do not require the monitoring of fuel oxygenates in air, groundwater, surface water, or drinking water. Therefore, comprehensive data to document the occurrence of fuel oxygenates in the major compartments of the hydrologic cycle across the nation are not available from these programs.

MTBE and other fuel oxygenates generally are not included in routine ambient air monitoring done by local and state agencies. An exception is the State of California, which recently began to monitor for MTBE and other volatile organic compounds (VOCs) in ambient air. Air-quality data for MTBE have also been collected in six cities as part of special studies (Allen and Grande, 1995; Kelly et al., 1993; Zweidinger, written communication, 1995) and these data are summarized in the IOFA Water-Quality Report (Table 6 in Zogorski et al., 1996). These data are not sufficient to provide a national perspective; however, MTBE was found in urban air, and the median concentrations in these urban areas ranged from 0.13 to 4.6 parts per billion by volume. Concentrations of MTBE in air near known sources of MTBE (gasoline stations, roadways, parking lots and garages, and so forth) (Johnson et al., 1995; McCoy and Johnson, 1995; NATLSCO, 1995) are higher, and in many cases considerably higher than in ambient urban air.

Little data exist on the occurrence of any of the fuel oxygenates in surface-water bodies, including streams, rivers, lakes, and reservoirs. Similarly, little data exist on oxygenate occurrence for drinking water derived from surface water.

Storm water was sampled and analyzed for MTBE in 16 U.S. cities during 1991-95 (Delzer et al., 1996). The studies in these 16 cities were completed to characterize storm-water runoff in cities with populations exceeding 100,000. These studies were not specifically designed to determine the occurrence and sources of MTBE, other oxygenates, or other VOCs originating from gasoline. The northeastern states and California, which are high-use areas of MTBE oxygenated-gasoline, were not sampled. The compilations, however, provide insight on the occurrence of MTBE in storm water in select cities. MTBE was detected in about 7% of 592 storm-water samples (Table 2). When detected, concentrations ranged from 0.2 to 8.7 micrograms per liter, with a median of 1.5 micrograms per liter. Eighty-three percent of the detectable concentrations occurred during the winter season. The median and mean concentrations of MTBE for all samples were shown to be statistically different during the October 1 to March 31 period, in comparison to the April 1 to September 30 period. The Wilcoxon rank sum test (one-sided p-value = 0.0000) and a modified t-test based on Tukey's biweight estimator (one-sided p-value = 0.0002) were used to establish the aforenoted statistical inferences. MTBE was detected both in cities using oxygenated gasoline to abate carbon monoxide nonattainment and in cities presumed to have used MTBE in gasoline for octane enhancement.

Table 2.   Occurrence of MTBE in storm water for 16 cities, 1991-95.

Time period	Number of samples	Number of detections	Percent occurrence	Percent of total detections in the indicated time period
Apr. 1 - Sept. 30	305	7	2.3	17
Oct. 1 - Mar. 31	287	34	11.8	83
Entire year	592	41	6.9	--

[MTBE, methyl *tert*-butyl ether]

Water-quality criteria for oxygenates to protect aquatic life have not been established. For the alkyl ether oxygenates, chronic aquatic-toxicity data are lacking, and only limited information exists on acute toxicity and bioconcentration. Considerably more acute-toxicity studies have been completed for EtOH and MeOH. The maximum concentration of MTBE detected in storm water reported in the previous paragraph is four to five orders of magnitude below the median lethal concentration for the most sensitive species investigated to date. More toxicity studies of aquatic animals and plants, bioaccumulation information, and ambient levels in surface water are needed before the significance of MTBE to aquatic life can be assessed.

Studies have been conducted to establish taste and odor thresholds for oxygenates in water. These studies provide useful information regarding the potential impacts on the palatability of drinking water and the ability of the public to detect MTBE-containing water before it is ingested. These studies indicate that the taste threshold for MTBE, ETBE, and TAME are 39 to 134, 47, and 128 micrograms per liter, respectively (TRC Environmental Corporation, 1994; Vetrano, 1993a, 1993b) . Similarly, the odor-detection threshold of MTBE, ETBE, and TAME in water are 45 to 95, 49, and 194 micrograms per liter, respectively (TRC Environmental Corporation, 1994; Vetrano, 1993a, 1993b). These taste and odor threshold values for MTBE are in the range of the USEPA's draft drinking-water lifetime health advisory (Figure 2).

Two data sets provide information on the occurrence of MTBE in groundwater and in drinking water derived from groundwater. Limited or no data are available for any other fuel oxygenate. The first data set was collected as part of the U.S. Geological Survey's (USGS) National Water-Quality Assessment Program (Gilliom et al., 1995). MTBE was included on a list of analytes for groundwater samples collected in 20 major basins across the U.S during 1993-94. In addition, retrospective efforts of that program have summarized MTBE occurrence data from a few state and regional groundwater assessment programs. The second data set was compiled during the preparation of the IOFA Water-Quality Report using information requested from states by USEPA, through its ten regions, on drinking-water programs that have analyzed for MTBE. In response to this request, data on the occurrence of MTBE in public drinking-water supplies derived from groundwater were provided by seven states and four states provided data from private wells. It is likely that some large water utilities monitor for constituents such as MTBE as part of routine scans for VOCs; therefore, additional information on MTBE occurrence may exist, but could not be compiled in the time frame of this assessment.

At least one detection of MTBE has occurred in groundwater in 14 of 33 states surveyed (Table 3). These 14 states are located throughout the United States. MTBE was detected in 5% of more than 1,500 wells sampled. Most of the detections were in shallow groundwater in urban areas. MTBE was not detected in approximately 99% of the samples from wells that were screened in shallow groundwater in agricultural areas (Squillace et al., 1995, 1996; Zogorski et al., 1996). MTBE was not detected in samples from approximately 98% of the wells that were screened in deep aquifers or deep parts of shallow aquifers; of the remaining 2% of wells, the highest MTBE concentration reported was 7.9 micrograms per liter. The mechanism for occurrence at depth is unknown. MTBE was reported in drinking water supplied by one or more utilities in six of the seven states that provided data for public water supplies derived from groundwater, and in all four of the states that provided data for private drinking-water wells (Table 3).

Because only a few states have information on MTBE in drinking water, it is not possible to describe MTBE concentrations in drinking water nationwide. A federal drinking-water standard has not been established for MTBE; however, the USEPA has issued a draft drinking-water lifetime health advisory of 20 to 200 micrograms per liter (U.S. Environmental Protection Agency, 1995b) (Figure 2). The draft health advisory is being revised by the USEPA and is expected to be available in the latter part of 1997. MTBE has been detected in samples from 51 public drinking-water systems as of 1996 based on limited monitoring (Table 4). When detected, however, the concentration of MTBE was generally low and nearly always below the lower limit of the current draft USEPA health advisory. This indicates that the consumption of drinking water was not a major route of exposure for these few systems. Additional MTBE monitoring data for drinking water are needed, both from the states that provided information for the preliminary assessment and from states that were unable to provide said information, before the significance of drinking water as a route of exposure can be assessed for the U.S.

Past and recent releases from USTs are perceived to be an important source for the entry of high concentrations of fuel oxygenates to groundwater. However, routine monitoring of groundwater at gasoline USTs has focused on BTEX compounds; as a result, little monitoring information is available for fuel oxygenates. A California Senate Advisory Committee has recently initiated a request to collect MTBE concentration data for groundwater at UST sites. This information should be helpful in understanding MTBE water-quality concerns from USTs. At least 10 states have established action levels or cleanup levels or both, for MTBE

Table 3. Summary information for MTBE in groundwater and in drinking water derived from groundwater for States surveyed.

	No. of States	No. of Programs	No. of Wells (w) or Drinking-Water Systems Sampled	No. of States Surveyed	No. of Programs Surveyed	No. of Wells (w) or Drinking-Water Systems with a Detection of MTBE	Percent of Wells (W) or Drinking-Water Systems with a Detection of MTBE
				In which there was at least one detection of MTBE			
Groundwater							
Groundwater resource assessment	33	90	1,516 (w)[a]	14	25	76 (w)[a]	5 (w)
Drinking water derived from groundwater							
Public	7	8	NK[b]	6	7	NK[b]	NK[b]
Private	4	4	NK[b]	4	4	NK[b]	NK[b]

[MTBE, methyl *tert*-butyl ether. NK, not known]

[a] Includes sampling of several springs.

[b] The total number of systems or wells sampled was not reported for each state.

Figure 2. Comparison of MTBE health advisories/guidelines, drinking-water standards, taste and odor thresholds, action levels, and cleanup levels.

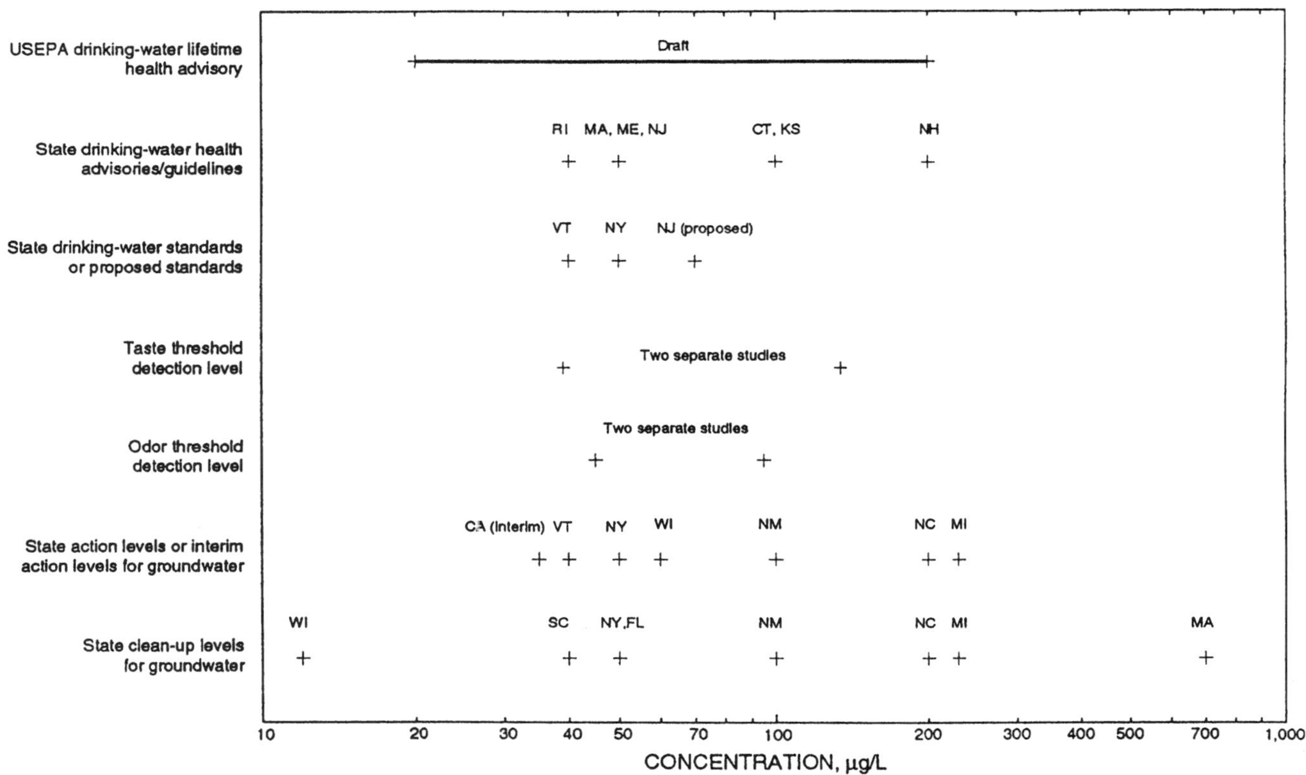

Table 4.   Occurrence and concentration ranges of MTBE in drinking-water samples from those programs, systems or wells in which there was at least one detection of MTBE.

State	No. of Systems or Wells With Detections[a]	Reporting Level (μg/L)	No. of Systems with Detections of MTBE Within Indicated Ranges of Concentration			
			> reporting level to 10μ g/L	>=10 to <20 μg/L	>=20 to <=200 μg/L	>200 μg/L
			Public water systems			
New Jersey	35	NK[b]	34	1	0	0
Iowa	1	NK	0	1	0	0
Illinois	7	NK	3	1	2	1
Texas	7	NK	5	1	1	0
Colorado	1	0.2	1	0	0	0
			Private wells			
Missouri	4	NK	0	0	0	4
Indiana	2	NK	0	0	1	1

[a] When multiple detections were reported for a single system or well, the most recent sample was used in this compilation. This table only includes those states where the number of unique public water systems or private wells could be identified. Public water-system data from the state of Rhode Island are not presented in this table, but appendix 7 of Zogorski et al.(1996) indicates MTBE has occurred in at least one public water system in that state. Private well data for the states of Texas and Rhode Island are also not presented in this table, but appendix 7 indicates MTBE has occurred in samples from at least one private well in each state.

[b] NK, unknown

at sites where gasoline releases have occurred and are being remediated. MTBE monitoring data may be available from these and other states; this information would provide insights on the potential significance of oxygenate releases from USTs, as well as MTBE concentrations in groundwater at these sites.

BEHAVIOR AND FATE OF FUEL OXYGENATES IN THE HYDROLOGIC CYCLE

Water in contact with spilled gasoline at the water table or in the unsaturated zone dissolves oxygenates along with aromatic hydrocarbon constituents. Water that is equilibrated with an oxygenated fuel can contain very high concentrations of the oxygenate. For example, for a gasoline that is 10% by weight MTBE, the concentration of MTBE in water in chemical equilibrium with the gasoline at room temperature would be on the order of 5,000,000 micrograms per liter, and the total BTEX concentration would be on the order of 120,000 micrograms per liter. Groundwater MTBE concentrations as high as 200,000 micrograms per liter have been measured in samples from wells near gasoline spills (Davidson, 1995; Garrett et al., 1986). The fact that measured MTBE concentrations are typically lower than solubility levels for gasoline oxygenated with MTBE is due to either dilution by uncontaminated water or depletion of MTBE from the gasoline source. Alcohol oxygenates could occur at high concentrations in water adjacent to spilled gasoline because of the high solubility of alcohols; however, there are no published studies that report alcohol concentrations in groundwater resulting from releases of oxygenated fuels.

Although oxygenated gasolines containing 10 to 15% by volume oxygenates can result in high concentrations of fuel oxygenates in water, i.e., 5,000,000 to 7,500,000 micrograms per liter MTBE, these concentrations are generally not high enough to increase either the water solubilities (cosolvent effect) (Barker et al., 1991; Pinal et al., 1990; Poulsen et al., 1992), or the transport rates of the BTEX compounds (Daniel, 1995; Davidson, 1995). Gasolines containing larger percent volumes of oxygenates are capable of causing this cosolvency effect, but such fuels are not in widespread use (Barker et al., 1991; Horan and Brown, 1995; Poulsen et al., 1992).

MTBE and other alkyl ether oxygenates are much less biodegradable than BTEX compounds in groundwater. Furthermore, the alkyl ether oxygenates sorb only weakly to soil and aquifer materials; therefore, their transport by groundwater will not be retarded to any significant extent (Daniel, 1995; Davidson, 1995). These factors explain why MTBE has been observed to persist at higher concentrations and to advance ahead of BTEX plumes in groundwater at gasoline spill sites. Conversely, EtOH would be expected to degrade much more rapidly than BTEX compounds; therefore, EtOH is not expected to persist much beyond the source area and the immediate contaminant plume at a gasoline spill site. However, no data exist on the occurrence of EtOH in groundwater to verify the hypothesis of EtOH's nonpersistence and nonmigration at spill sites.

Because of their occurrence in the atmosphere and their favorable and rapid partitioning to water, fuel oxygenates will occur in precipitation in direct proportion to their concentrations in air. For example, changes in the ambient air concentration of MTBE due to increased or decreased usage of oxygenated gasoline will affect the concentration of MTBE in precipitation. Cooler temperatures will cause larger concentration of MTBE in precipitation for a given atmospheric concentration. Assuming a concentration of 1 part per billion by volume MTBE in the atmosphere, the concentration of MTBE in precipitation would increase almost tenfold, from about 0.2 to 1.5 micrograms per liter, if the temperature decreases from 20° to 0° Celsius. Based on this range of proportionality and MTBE atmospheric concentration data collected in a few cities, precipitation for general urban atmospheres could contain submicrogram per liter to about 3 micrograms per liter of MTBE. Theory predicts that precipitation concentrations will be higher near roadways, parking lots and garages, and gasoline stations that consistently have higher air concentrations of MTBE. Available data suggest that the alkyl ether oxygenates have lifetimes in the atmosphere that range from 4 days to 2 weeks (Wallington et al., 1988). The main degradation pathway seems to be reaction with hydroxyl radical to form *tert*-butyl formate. The atmospheric source of alkyl ether oxygenates in urban areas is believed to be continual, and results in low concentrations of MTBE in water relative to point sources such as USTs. The atmospheric source is also areally extensive due to the dispersive effect of weather patterns.

Precipitation delivers oxygenates directly to streams, rivers, lakes, and reservoirs as it falls on these surface-water bodies or enters surface water through overland runoff and storm-water drainage. While they are volatile from water, the alkyl ether oxygenates, such as MTBE, in large rivers and some streams will not be lost quickly by volatilization (Pankow et al., 1996). Precipitation falling on land and recharging aquifers, together with diffusion of oxygenates from the atmosphere through the unsaturated zone, delivers oxygenates to shallow groundwater. From shallow groundwater it is possible that dissolved alkyl ether oxygenates will move deeper into an aquifer toward wells and surface-water discharge areas. The concentration of an oxygenate in groundwater along a particular flow path is related to the age of groundwater; that is, its residence time in the aquifer from recharge location, and the rates of any removal mechanisms. Dissolved MTBE and other alkyl ether oxygenates would advance deeper into the aquifer than BTEX compounds because they are less degradable than BTEX compounds; however, the significance of this deeper migration is uncertain because of the paucity of in-situ monitoring and degradation studies.

REMEDIATION

The chemical properties of oxygenates compared to the BTEX compounds provide insight about the anticipated performance and costs of remediation techniques typically applied to remove BTEX at gasoline spill sites. The presence of MTBE and other alkyl ether oxygenates does not prevent the application of conventional (active) remedial methods (air stripping, carbon adsorption, and soil-vapor extraction) for fuel spills, but the presence of these oxygenates raises the cost (Garrett et al., 1986; International Technology Corporation, 1991; Truong and Parmele, 1992). For example, the alkyl ether oxygenates can be removed from water using aeration, but only with much higher air/water ratios than required for BTEX removal. Similarly, vacuum extraction and air-sparging techniques will be much less efficient for alkyl ether oxygenates than for BTEX, and infeasible for alcohol oxygenate remediation. The application of intrinsic (passive) bioremediation, an emerging, relatively inexpensive approach to the management of conventional gasoline spills, may be limited because of the slowness with which alkyl ether oxygenates are biodegraded and the tendency of these compounds to migrate from release sites (Horan and Brown, 1995; Yeh and Novak, 1991). For contaminant plumes containing EtOH or MeOH, however, intrinsic bioremediation or enhancements to in-situ bioremediation will remain effective (Brusseau, 1993; Butler et al., 1992).

RECOMMENDATIONS FROM THE IOFA WATER-QUALITY REPORT

The IOFA Water-Quality Report provides an assessment of the water-quality consequences of using oxygenates in gasoline. The specific recommendations, as summarized below, will allow a more comprehensive understanding of those consequences. Specific recommendations are given for MTBE and the other alkyl ether oxygenates largely because of their extensive commercial use and their persistence in, and migration with groundwater. Ethanol, the only alcohol currently in wide-scale commercial use as a gasoline oxygenate, is expected to undergo rapid biodegradation in groundwater and in surface water, except when concentrations are high enough to be toxic to microorganisms.

At the present time, sufficient monitoring data are not available in the U.S. to characterize human exposure to the alkyl ether oxygenate compounds by the consumption of drinking water. Typical oxygenate concentrations are not known either for treated drinking water, or for the surface- and groundwater sources of that drinking water, except in select states. Similarly, little information is available on either the concentrations of oxygenates in surface water to which aquatic animals and plants are typically exposed, or the toxicities of those compounds to a broad range of aquatic species.

So that resources needed for implementing the recommendations can be used wisely and efficiently, it is recommended that an ad hoc panel representing the public and private sectors develop a comprehensive and interdisciplinary monitoring, research, and assessment plan to determine the significance of the use of alkyl ether fuel oxygenates on drinking-water quality and aquatic life. The plan should identify lead agencies and organizations for various topics, and provide the necessary details, which were beyond the scope of the assessment reported herein, such as defining specific goals; establishing the timing and relative priority of research and monitoring needs; developing coordination between industry and various local, state, and federal agencies; creating a national database for analytical determinations; and selecting data sets for entry into the database. Some components of the plan, such as monitoring and creating the national database should be piloted in select cities and states to further work out details, such as common field sampling and analytical methods, and to make appropriate modifications if deemed necessary. Selecting cities and states that are actively engaged in monitoring for alkyl ether oxygenates in air and water would allow for the timely completion of the suggested pilot phase, with minimal additional expenditure.

Three broad recommendations that are made in the IOFA Water-Quality Report are to:

1. Obtain more complete monitoring data and other information that would: (a) enable an exposure assessment for MTBE in drinking water, (b) characterize the relation of use of MTBE and other alkyl ether oxygenates in gasoline to water quality, and (c) identify and characterize major sources of MTBE to the environment.
2. Complete additional behavior and fate studies to expand current knowledge.
3. Complete aquatic-toxicity tests to define the threat posed to aquatic life and establish, if warranted, federal water-quality criteria.

Completion of the exposure assessment for MTBE in drinking water should be given priority. Monitoring of MTBE in drinking water for this purpose should initially be targeted to high MTBE-use areas and those environmental settings that are otherwise thought to be most susceptible to contamination. State and federal agencies and large public water utilities are encouraged to start voluntary monitoring of MTBE in drinking water immediately so that the drinking-water exposure assessment can be completed. Essential elements of each of the three recommendations are described below.

Better Characterize Ambient Concentrations, Oxygenated Gasoline Use, Sources to the Environment, and Relations Between Gasoline Use and Water Quality

- Add the alkyl ether oxygenates MTBE, ETBE, TAME, and DIPE to existing VOC analytical schedules and as USEPA routine target analytes for drinking water, wastewater, surfacewater, groundwater, and remediation studies. This inclusion will provide considerable additional information on the occurrence of these oxygenates in water at little additional expense.

- Create a national database of analytical determinations of the alkyl ether oxygenates in air, and in ground, surface, and drinking water. The database should also include, for example, information on sampling dates, locations, detection limits, analytical method, quality control, and the reporting agency. Carefully selected monitoring data obtained by federal, state, and local agencies should be reported for inclusion in the database. The difficulty and limited success in compiling monitoring data from the states for the assessment reported herein illustrates the need for the recommended database. The goal should be to assemble a database that allows exposure assessments for the alkyl ether oxygenates in drinking water and for aquatic life in surface water. The USEPA, in late 1970s and 1980s, completed similar drinking-water assessments for other VOCs.

- Characterize the seasonal volumes of gasoline containing the alkyl ether oxygenates that are sold in the major regions and cities of the U.S. Such usage information is presently unavailable in any systematic form and at the necessary scale, and this lack of data prevents attempts to correlate water-quality data with seasonal use. Establishing relations between the use of MTBE and the oxygenate's presence in groundwater and surface water will enable the identification of those hydrologic and geologic settings that are most susceptible to contamination.

- Determine the annual releases of the alkyl ether oxygenates to air, land, and water from all sources (for example, industrial releases, refueling losses, auto emissions, and storage-tank releases), both point and nonpoint in nature. Presently, estimates of this information are available only for MTBE, TBA, and MeOH, for some sources, and only for those industries that must file annual TRI reports. Precise estimates of annual releases will be difficult to make from currently available information but order-of-magnitude estimates should be made to identify major "data gaps" and to identify the primary source(s) of the alkyl ether oxygenates for air, surface water, groundwater, and drinking water. Once the releases are characterized, the feasibility of further reducing primary sources of fuel oxygenates to the environment should be determined.

- Annually review the information specified above and, when adequate information is available, prepare interpretive report(s) on the water-quality consequences of the use of MTBE and other alkyl ether oxygenates. At a minimum, the report(s) should include: (1) exposure assessment for drinking water and aquatic life; (2) relation between oxygenated fuel usage and the seasonal occurrence of alkyl ether oxygenates in surface water, for various climatic and hydrogeologic settings, and (3) the significance of various sources of fuel oxygenates to air, surface water, groundwater, and drinking water. The continued need to collect information on alkyl ether oxygenates should be reassessed after the above-noted report(s) is completed.

Expand the Current Understanding of the Environmental Behavior and Fate of Alkyl Ether Oxygenates

- Characterize, via field monitoring in select cities and regions of the nation, the occurrence, movement, and mass fluxes of alkyl ether oxygenates between ambient air, precipitation, surface water, the unsaturated zone, and shallow and deep groundwater. This monitoring is needed to: (1) determine the significance of urban atmospheres as nonpoint sources of contamination of alkyl ether oxygenates to surface and groundwater; (2) identify climatic and hydrogeologic settings where movement of alkyl ether oxygenates in the hydrologic cycle will be of concern; and (3) evaluate the influence of drainage and storm-water management programs on contaminant movement. These studies should also determine the importance of overland runoff in altering the levels of alkyl ether oxygenates in urban streams and rivers, and in water recharged to shallow aquifers.

- Conduct additional laboratory and field research on the degradation of the alkyl ether oxygenates in the different compartments of the hydrologic cycle, under a variety of environmental conditions. Information is needed on degradation half-lives and pathways. It is especially important to clarify how persistent the alkyl ether oxygenates can be in soil and groundwater under a range of redox conditions, and to identify all of the important degradation products for possible monitoring and behavior/fate research. The potential for plants to take in and otherwise alter the concentration of alkyl fuel oxygenates in the root zone should be determined.

- Conduct further theoretical modeling on the movement of the alkyl ether oxygenates from land surface to shallow groundwater. This research should: (1) test the hypothesis that the frequent occurrence of MTBE in shallow urban groundwater at low concentrations can be due to recharged precipitation or to overland runoff; (2) identify the associated predominant transport mechanisms; and (3) provide insights on the climatic and hydrogeologic factors that favor such movement of alkyl ether oxygenates in the unsaturated zone.

Expand the Current Understanding of the Aquatic Toxicity of Alkyl Ether Oxygenates

- Initiate studies of the chronic toxicity of alkyl ether oxygenates to a broad range of aquatic animals and plants indigenous to surface water. Additional acute-toxicity tests should also be completed for the same species. These studies are needed to: (1) define the extent of any threat posed by the alkyl ether oxygenates to aquatic life; and (2) collect sufficient information to form the basis of federal water-quality criteria, if warranted.

REFERENCES

Allen, M. and Grande, D. 1995. Reformulated Gasoline Air Monitoring Study. Pub. No. AM-175-95. Madison, Wisconsin, State of Wisconsin Department of Natural Resources Bureau of Air Management.

American Chemical Society. 1996. Facts and Figures for the Chemical Industry, 1. Production by the U.S. Chemical Industry. *Chem. Eng. News.* June 24, 40-46.

Barker, J.F., Gillham, R.W., Lemon, L., Mayfield, C.I., Poulsen, M., and Sudicky, E.A. 1991. Chemical Fate and Impact of Oxygenates in Groundwater: Solubility of BTEX from Gasoline-Oxygenate Compounds. Am. Pet. Inst. Pub. 4531. Washington, DC.

Brusseau, M.L. 1993. Complex Mixtures and Groundwater Quality. EPA/600/S-93/004. Washington, DC, U.S. Environ. Prot. Agency, Robert S. Kerr Environ. Res. Lab.

Butler, B.J., Vandergriendt, M., and Barker, J.F. 1992. Impact of Methanol on the Biodegradation Activity of Aquifer Microorganisms. Presented at SETAC, 13th Ann. Meet., Nov. 8-12, 1992, Cincinnati, OH.

Calvert, J.G., Heywood, J.B., Sawyer, R.F., and Seinfeld, J.H. 1993. Achieving Acceptable Air Quality: Some Reflections on Controlling Vehicle Emissions. *Sci.* 261, 37-45.

Conrad, D.L. 1995. Facts of Gasoline/Oxygenate Releases to the Environment - A Review of the Literature. Port Arthur, TX, Texaco Res. and Dev. Dept.

Daniel, R.A. 1995. Intrinsic Bioremediation of BTEX and MTBE: Field, Laboratory and Computer Modeling Studies. NC State Univ., Raleigh, NC, Dept. of Eng., Thesis.

Davidson, J.M. 1995. Fate and Transport of MTBE - The Latest Data. Proc. 1995 Pet. Hydrocarbons and Org. Chem. Ground Water: Prev., Det., and Remed. Conf., Nov. 1995, Houston, TX.

Delzer, G.C., Zogorski, J.S., Lopes, T.J., and Bosshart, R.L. 1996. Occurrence of the Gasoline Oxygenate MTBE and BTEX Compounds in Urban Stormwater in the United States, 1991-95. U.S. Geol. Surv. Water-Resources Invest. Rep. 96-4145.

Garrett, P., Moreau, M., and Lowry, J.D. 1986. MTBE as a Ground Water Contaminant. Paper presented at the 1986 NWWA/API Conf. Pet. Org. Chem. Ground Water, Houston, TX, 227-238.

Gilliom, R.J., Alley, W.M., and Gurtz, M.E. 1995. Design of the National Water-Quality Assessment Program: Occurrence and Distribution of Water-Quality Conditions. U.S. Geol. Surv. Circ. 1112.

Hoekman, S.K. 1992. Speciated Measurements and Calculated Reactivities of Vehicle Exhaust Emissions from Conventional and Reformulated Gasolines. *Env. Sci. Tech.* 26, 1206-1216.

Horan, C.M. and Brown, E.J. 1995. Biodegradation and Inhibitory Effects of Methyl-Tertiary-Butyl Ether (MTBE) Added to Microbial Consortia. Proc.

10th Ann. Conf. Hazardous Waste Res., May 23-25, Manhattan, KS., Great Plains/Rocky Mt. Haz. Sub. Res. Ctr.

International Technology Corporation. 1991. Cost-Effective, Alternative Treatment Technologies for Reducing the Concentrations of Methyl Tertiary Butyl Ether and Methanol in Ground Water. Am. Pet. Inst. Pub. 4491. Washington, DC.

James, F. 1995. Agribusiness Keeps Its Clout in GOP's House. *Chicago Tribune.* Nov. 5.

Johnson, T., McCoy, M., and Wisbith, T. 1995. A Study to Characterize Air Concentrations of Methyl Tertiary-Butyl Ether (MTBE) at Service Stations in the Northeast. Am. Pet. Inst. Pub. 4619, Washington, DC.

Kelly, T.J., Callahan, P.J., Plell, J., and Evans, G.F. 1993. Method Development and Field Measurements for Polar Volatile Organic Compounds in Ambient Air. *Env. Sci. Tech.* 27, 696-702.

Kirchstetter, T.W., Singer, B.C., Harley, R.A., Kendall, G.R., and Chan, W. 1996. Impact of Oxygenated Gasoline Use on California Light-Duty Vehicle Emissions. *Environ. Sci. Tech.* 30, 661-670.

Lund, L. 1995. Changes in UST and LUST: The Federal Perspective. *Tank Talk,* 10(2-3), 7. Lake Zurich, Ill., Steel Tank Institute.

McCoy, M., Jr., and Johnson, T. 1995. Petroleum Industry Data Characterizing Occupation Exposures to Methyl Tertiary-Butyl Ether (MTBE) 1983-1993. Am. Pet. Inst. Pub. 4622. Washington, DC.

NATLSCO. 1995. Service Station Personnel Exposures to Oxygenated Fuel Components - 1994. Am. Pet. Inst. Pub. 4625. Washington DC.

Pankow, J.F., Rathbun, R.E., and Zogorski, J.S. 1996. Calculated Volatilization Rates of Fuel Oxygenate Compounds and Other Gasoline-Related Compounds from Rivers and Streams. *Chemosphere.* 33, 921-937.

Pinal, R., Suresh, P., Lee, L.S., and Cline, P.V. 1990. Cosolvency of Partially Miscible Organic Solvents on the Solubility of Hydrophobic Organic Chemicals. *Env. Sci. Tech.* 24, 639-647.

Poulsen, M., Lemon, L., and Barker, J.F. 1992. Dissolution of Monoaromatic Hydrocarbons into Groundwater from Gasoline-Oxygenate Mixtures. *Env. Sci. Tech.* 26, 2483-2489.

Squillace, P.J., Pope, D.A., and Price, C.V. 1995. Occurrence of the Gasoline Additive MTBE in Shallow Ground Water in Urban and Agricultural Areas. U.S. Geol. Surv. Fact Sheet FS-114-95.

Squillace, P.J., Zogorski, J.S., Wilber, W.G., and Price, C.V. 1996. Preliminary Assessment of the Occurrence and Possible Sources of MTBE in Groundwater in the United States, 1993-1994. *Env. Sci. Tech.* 30, 1721-1730.

Stump, F.D., Knapp, K.T., and Ray, W.D. 1990. Seasonal Impact of Blending Oxygenated Organics with Gasoline on Motor Vehicle Tailpipe and Evaporative Emissions. *J. Air Waste Manage. Assoc.* 4, 872-880.

TRC Environmental Corporation. 1994. Odor Threshold Studies Performed with Gasoline and Gasoline Combined with MTBE, ETBE, and TAME. Am. Pet. Inst. Pub. 4592. Washington, DC.

Truong, K.N., and Parmele, C.S. 1992. Cost-Effective Alternative Treatment Technologies for Reducing the Concentrations of Methyl Tertiary Butyl Ether and Methanol in Groundwater. *Hydrocarbon Contam. Soils and Groundwater.* 2, 461-486.

U.S. Environmental Protection Agency. 1993. Toxic Release Inventory, 1987-1993. Washington, DC, Office of Pollution Prevention and Toxics Environmental Assistance Division (7408), (digital data files).

U.S. Environmental Protection Agency. 1995a. Drinking Water Regulations and Health Advisories. Washington, DC, Office of Water.

U.S. Environmental Protection Agency. 1995b. UST Corrective Action Summary for 4th Quarter FY 95, Quarterly Statistical Table, Oct. 30, 1995. Washington, DC, Office of UST.

Vetrano, K.M. 1993a. Final Report to ARCO Chemical Company on the Odor and Taste Threshold Studies Performed with Methyl Tertiary-Butyl Ether (MTBE) and Ethyl Tertiary-Butyl Ether (ETBE). TRC Proj. No. 13442-M31. Windsor, CT, TRC Environ. Corp.

Vetrano, K.M. 1993b. Odor and Taste Threshold Studies Performed with Tertiary-Amyl Methyl Ether (TAME). Am. Pet. Inst. Pub. 4591. Washington, DC.

Wallington, T.J., Dagaut, P., Liu, R., and Kurylo, M.J. 1988. Gas-Phase Reactions of Hydroxyl Radicals with the Fuel Additives Methyl *Tert*-Butyl Ether and *Tert*-Butyl Alcohol Over the Temperature Range 240-440 K. *Env. Sci. Tech.* 22, 842-844.

Yeh, C.K. and Novak, J.T. 1991. Anaerobic Biodegradation of Oxygenates in the Subsurface - Proc. 1991 Pet. Hydrocarbons and Org. Chem. Ground Water: Prev., Det., Rest., Nov. 20-22, 1991, Houston, TX.

Zogorski, J.S., Morduchowitz, A., Baehr, A.L., Bauman, B.J., Conrad, D.L., Drew, R.T., Korte, N.E., Lapham, W.W., Pankow, J.F., and Washington, E.R. 1996. Fuel Oxygenates and Water Quality: Current Understanding of Sources, Occurrence in Natural Waters, Environmental Behavior, Fate, and Significance. Final report submitted to Off. Sci. and Tech. Policy, Washington DC. Also published as chapter 2 in Interagency Assessment of Oxygenated Fuels, Off. Sci. and Tech. Policy, June,1997. Pages 2-1 to 2-80 with appendices.

Zweidinger, R. 1995. U.S. Environmental Protection Agency, written communication.

CHAPTER 52

Purge vs. No Purge for Sampling BTEX and MTBE in Groundwater

J. Patrick Byrnes, **Jeffrey E. Briglia**, and **L.J. Buddy Bealer**, EnviroTrac Ltd., Deer Park, New York

INTRODUCTION

Over the past several years numerous researchers have been evaluating what procedures are needed to obtain a "representative" groundwater sample from a monitoring well. It is widely accepted that in order to collect a groundwater sample truly representative of the aquifer surrounding a monitoring well, that the stagnant water in the well casing must be removed or purged before collecting a sample for water quality analysis (Barcelona et al., 1985; Keely and Boateng, 1987; Robin and Gillham, 1987; Puls and Powell, 1992; Herzog et al., 1988; Barcelona et al., 1994). However, recent efforts to conduct more detailed and reliable sampling at lower costs have produced a variety of well purging and sampling techniques. As a result, a growing number of researchers are questioning the need for, or validity of, traditional well purging practices (Robin and Gillham, 1987; Puls and Powell, 1992; Kearl et al., 1992; Kearl et al., 1994; Puls and Paul, 1995).

Previous studies have suggested that the number of well volumes to be purged to obtain representative groundwater samples ranged from less than one to up to 20 (Fenn et al., 1977; Humenick et al., 1980; Pettyjohn et al., 1981; Scalf et al., 1981; Gillham et al., 1983; Unwin and Huis, 1983; Wilson and Dworkin, 1984; Herzog et al., 1988). The term "well volume" or "bore volume" has been defined by researchers in several ways; however, for this study, one well volume is defined as the volume of water standing inside the well casing under static conditions. The most commonly used or traditional well purging protocols recommend that three to five well volumes be evacuated from a well (Scalf et al., 1981; Barcelona et al., 1985; EPA, 1991b; EPA, 1991; NJDEPE, 1992). Others have contested the "rules of thumb" applied to well purging by suggesting that the stabilization of water quality parameters (dissolved oxygen, temperature, pH, specific conductivity, etc.) may provide more representative samples (Barcelona et al., 1985, Keely and Boateng, 1987; Gibs and Imbrigiotta, 1990; Barcelona et al., 1994; Puls and Paul, 1995).

According to Gibs and Imbrigiotta (1990), none of the above-recommended criteria for well purging results in representative samples.

Recent studies by Kearl et al. (1992) and Puls and Powell (1992) have shown that flow in the well screen is horizontal and laminar, with minimal mixing with stagnant water in the overlying casing above the screen. In summary, these studies suggest that groundwater samples collected directly from the screened zone are representative of groundwater in the surrounding formation, and that purging well volumes prior to sampling is unnecessary.

Routine groundwater sampling using traditional purging is costly, and more cost-effective sampling methods need to be developed. The objective of this study was to determine if well purging is needed for monitoring wells that are routinely sampled for BTEX and MTBE compounds. Results of a groundwater sampling study comparing results from samples collected without well purging and samples collected following traditional well purging methods is presented. Monitoring wells in the study were constructed with screens that bridged or extended above the water table, which eliminates the existence of a stagnant water column. Comparative sampling was conducted using both sampling techniques at 13 sites in New York state exhibiting gasoline-contaminated groundwater, particularly benzene, toluene, ethylbenzene and xylenes (BTEX) and methyl-tertiary-butyl-ether (MTBE).

SITE CHARACTERISTICS, WELL CONSTRUCTION, AND HYDROGEOLOGY

Over a several-month period in 1995, a total of 13 retail gasoline sites (Sites #1-13) in New York were randomly selected for the study. The only criteria for selection was that the site is routinely sampled on a quarterly basis. Each site is an existing or former retail service station that has had a documented release of gasoline from the UST system. For ease of discussion, study sites are separated into two groups — Upstate and Long Island. The Upstate Group includes sites located in Putnam and Westchester County and the Long Island Group sites are located in Kings, Nassau, and Suffolk Counties, totaling three sites Upstate and 10 sites on Long Island (Figure 1).

All monitoring wells included in the study were installed using hollow-stem auger drilling methods. Wells are constructed of 4-inch diameter schedule 40 PVC casing and slotted well screen (typically 0.02 inch slot). The borehole annulus is sand packed (typically with Morie #2 washed sand) over the entire length of the well screen and extended approximately 2 feet above the screen. A bentonite seal with a thickness ranging from 1 to 2 feet was installed above the sand pack, followed by a cement/bentonite grout to the surface. Because the wells are constructed to monitor petroleum hydrocarbons, well screens are positioned to bridge the water table. Screen lengths typically range from 10 to 15 feet. A typical well construction diagram is shown in Figure 2.

Figure 1.   **General locations of study sites.**

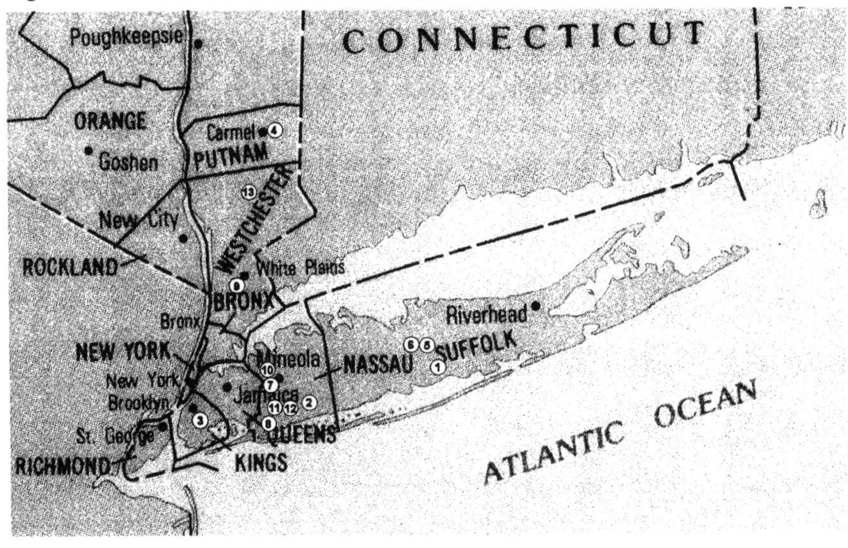

Figure 2.   **Typical monitoring well construction.**

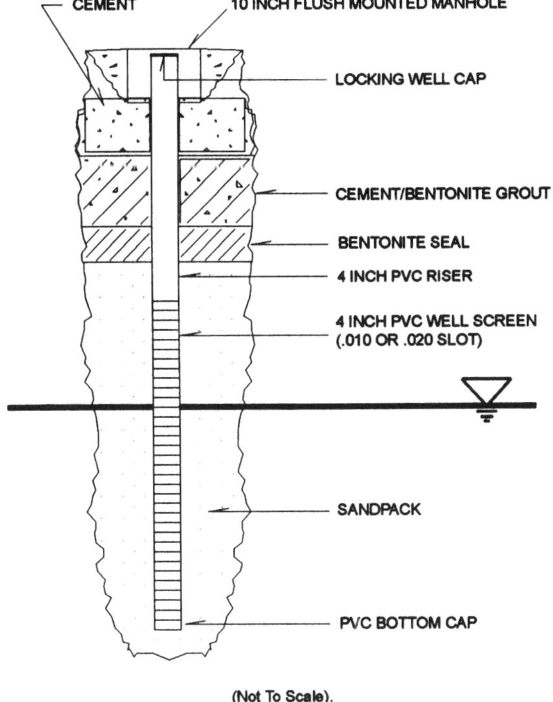

Upstate New York is dominated with bedrock covered with glacial deposits of till and stratified drift of variable thickness. The till, consisting of mixtures of clay, silt, sand, gravel, and boulders, mantles the uplands and small tributary valleys and typically is found beneath stratified drift deposits. Drift sediments (reworked by modern streams) are encountered along the floors of valleys and flat plains. These deposits include clay, silt, sand, and gravel which form the principal aquifer systems of Upstate New York (Heath, 1964). Long Island lies in the Coastal Plain province and is underlain by drift predominantly comprised of stratified sand and gravel. These deposits form the principal aquifer system. The shallow water-table aquifer or Upper Glacial aquifer was sampled in this study. The Upper Glacial aquifer consists of highly permeable sands and is continuous throughout Long Island.

The water table for all the study sites is present under unconfined conditions in unconsolidated sediments consisting of till deposits or stratified sands. Depth to water at the sites ranged from approximately 5 to 72 feet below land surface, with an average water depth of approximately 20 feet. Table 1 provides a general description of each site.

Table 1.   **General Site Characteristics**

Site #	Depth to Water, (ft)	Geology	Location
1	12	Medium-Coarse Sands	Long Island
2	8	Medium-Coarse Sands	Long Island
3	10	Fine- Silty Sands	Long Island
4	7	Fine Sands	Upstate
5	29	Medium-Coarse Sands	Long Island
6	24	Medium-Coarse Sands	Long Island
7	38	Medium-Coarse Sands	Long Island
8	14	Medium-Coarse Sands	Long Island
9	5	Fine Sands	Upstate
10	72	Medium-Coarse Sands	Long Island
11	11	Medium-Coarse Sands	Long Island
12	14	Medium Sands	Long Island
13	5	Fine- Silty Sands	Upstate

METHODOLOGY

Purging and Sampling Procedures

Over a several month period in 1995, a total of 84 wells were sampled before purging (no purge) and after purging at each of the 13 sites. Purging and sampling were conducted predominantly using hand bailers. For this study, all monitoring wells were sampled without purging and then purged three to five well volumes prior to collecting a second set of samples for laboratory analysis. Purging was performed typically with a 3-inch diameter PVC bailer. This technique consists of manually lowering the bailer in and out of the well casing until the sufficient volume of groundwater has been removed. All groundwater samples were collected using a Teflon® bailer with the exception of Site #1, in which purging and sampling was performed using dedicated pumps. In addition, due to the presence of a deep water table (approximately 72 feet) at Site #10, well purging was accomplished using a submersible pump, and samples collected using Teflon® bailers. Following sample collection before and after purging, water quality parameters consisting of dissolved oxygen (DO) and temperature were field-analyzed in situ.

To prevent cross contamination, the sampler wore vinyl gloves which were discarded after each sample was collected. Bailers and all other devices used for purging and sampling were thoroughly cleaned and decontaminated between use. Samples were collected in 40 mL laboratory supplied glass vials with open-top screw cap, Teflon® septa and HCL preservative. The vials were filled and tightly capped to exclude air. The sample vials were then labeled and immediately placed on ice in a cooler following strict chain-of-custody procedures. Samples were then shipped within 24 hours to the laboratory, where they were stored at 4°C until the analyses could be performed. Samples were analyzed by a New York state certified laboratory for BTEX and MTBE using EPA Method 624 or 602. Quality Assurance/Quality Control (QA/QC) field and trip blanks were also analyzed for each site sampled, to document sample integrity.

Statistical Analysis

Of the 84 wells that were sampled for BTEX and MTBE, 72 were tested for water quality parameters which included temperature and dissolved oxygen (DO). Note that samples from two monitoring wells were not used in the analysis. The wells in question, MW-4 from Site #1 and MW-1 from Site #4, were being hydraulically influenced as part of remedial activities at the time of the field testing.

Based on the fact that no assumption about the distribution of concentrations could be justified, a nonparametric test was necessary. The Mann-Whitney U-test (U-test) was selected as the nonparametric test used in the study. This test is used to determine if two sample sets come from the same population. For the purpose of this study, two sample sets were concentrations and field measurements before and after purging, and the population was the groundwater matrix. Statistically, if these two samples are shown to originate from the same population, then the matrix from which they come will also be the same, thereby supporting that purging of "bridged" wells is unnecessary.

Scatter plots of BTEX and MTBE concentrations after purging versus concentrations before purging were constructed and a linear regression analysis performed. The correlation coefficient (r) of the linear regression was also calculated in order to determine the relevance of the fit. A correlation coefficient equal to one implies a perfect fit, whereas a correlation coefficient of zero means that there is no linear data correlation.

Results and Discussion

BTEX and MTBE

Typical data generated for each site is shown for Site #2 in Figure 3. Statistical results are summarized in Table 2. Note that the statistical analysis was performed on sets of paired data which had at least one detectable concentration in order to not skew the analysis. This result shows that at the 0.01 (99%) significance level, there is no difference in the samples taken before and after purging. Scatter plots of BTEX and MTBE concentrations after purging versus concentrations before purging were constructed and are provided in Figure 4a and 4b, respectively. Theoretically, if there was no difference in "populations," then the points in the scatter plots would form a one-to-one (slope=1) relationship. In the case of BTEX, the best-fit line of the scatter plot yielded a slope of 0.99 with a correlation coefficient equal to 0.87. For MTBE, the resulting slope equaled 1.05 with a correlation coefficient equal to 0.98. From these results it can be surmised that BTEX and MTBE concentrations before purging are just as representative as after purging.

Table 2.   Summary of Statistical Analyses

	BTEX	MTBE	DO	Temperature
N	67	52	72	74
z	0.59	0.59	1.56	0.59
m	0.99	1.05	0.87	0.98
r	0.87	0.98	0.57	0.94

N - Number of paired samples.

z - z-score from Mann-Whitney U-test.

m - slope of best-fit line of after vs before purging.

r - correlation coefficient of the best-fit line.

According to many researchers, volatile organic compounds (VOCs) such as BTEX would be present in lower concentrations in well water prior to purging. In fact, stagnant water samples (i.e., before purging) have been found to be significantly lower (as much as 85%) in VOCs than those tested after purging in a wide range of hydrogeologic settings (Keely and Boateng, 1987; Robin and Gillham, 1987; Herzog et al., 1988; Gibs and Imbrigiotta, 1990; Barcelona et al., 1994). This is based on the belief that VOCs in stagnant well water off-gas or volatilize within a well. However, data presented in this study suggest that this is not a significant factor. Of the 67 wells that had detectable BTEX concentrations in groundwater, 36 wells (54%) had higher or equal concentrations before purging. Although off-gassing of VOCs is most likely occurring, data presented here supports the belief that traditional purging practices cause more volatilization to occur. The manner in which wells are purged and sampled is often inconsistent with the use of the resultant data. The volume of water purged, the rate at which it is withdrawn, and

Figure 3.   **Typical study site.**

Site #2

Monitor Well I.D.	Before Purging						Iter Purging					
	Benzene	Toluene	Ethyl Benzene	Xylenes	Total BTEX	MTBE	Benzene	Toluene	Ethyl Benzene	Xylenes	Total BTEX	MTBE
MW-1	1.4	ND	ND	ND	1.4	ND	1.1	ND	ND	ND	1.1	ND
MW-2	20	2.6	230	490	742.6	39	2.9	7.9	600	2200	2810.8	86
MW-3	66	9400	2400	10000	21866	220	99	12000	2400	9200	23699	230
MW-4	440	2000	3500	12000	17940	250	250	1200	2100	7400	10950	190
MW-5	83	3300	3500	16000	22883	280	64	1900	2800	11000	15764	270
MW-6	6.2	5	320	190	521.2	260	1.9	2.7	210	120	334.6	150
MW-7	280	790	460	610	2140	42	370	2400	390	1500	4660	ND
MW-9	ND	ND	ND	ND	ND	ND	ND	ND	ND	ND	ND	ND
MW-10	0.7	ND	ND	ND	0.7	2400	0.5	ND	ND	ND	0.5	730
MW-11	34	3.4	3.3	4.3	45	ND	94	14	52	40	200	16
MW-12	980	45	850	35	1910	130	5000	74	900	250	6224	460
MW-13	3800	390	2800	1100	8090	850	6500	1500	2800	2500	13300	1600
MW-14	1200	350	670	1400	3620	190	2600	1000	790	1500	5890	340
MW-15	1500	6.5	ND	ND	1506.5	270	3900	38	21	72	4031	390

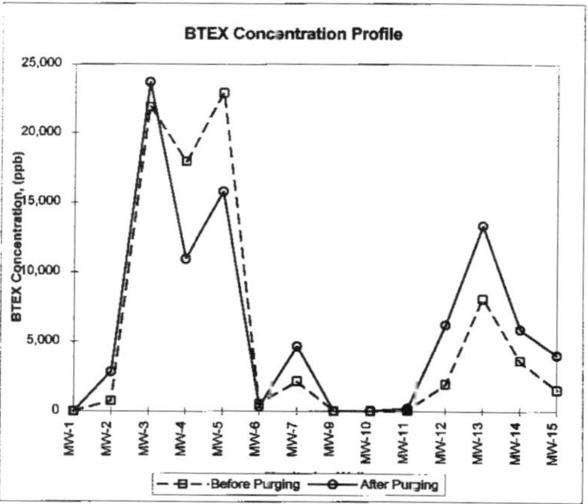

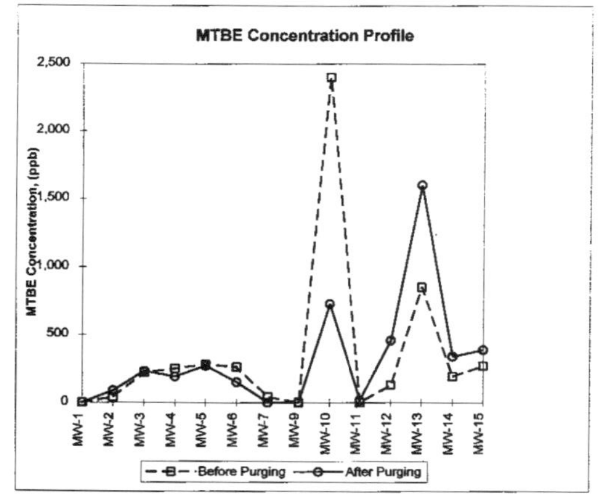

Figure 4.   After purge vs. before purge profiles.

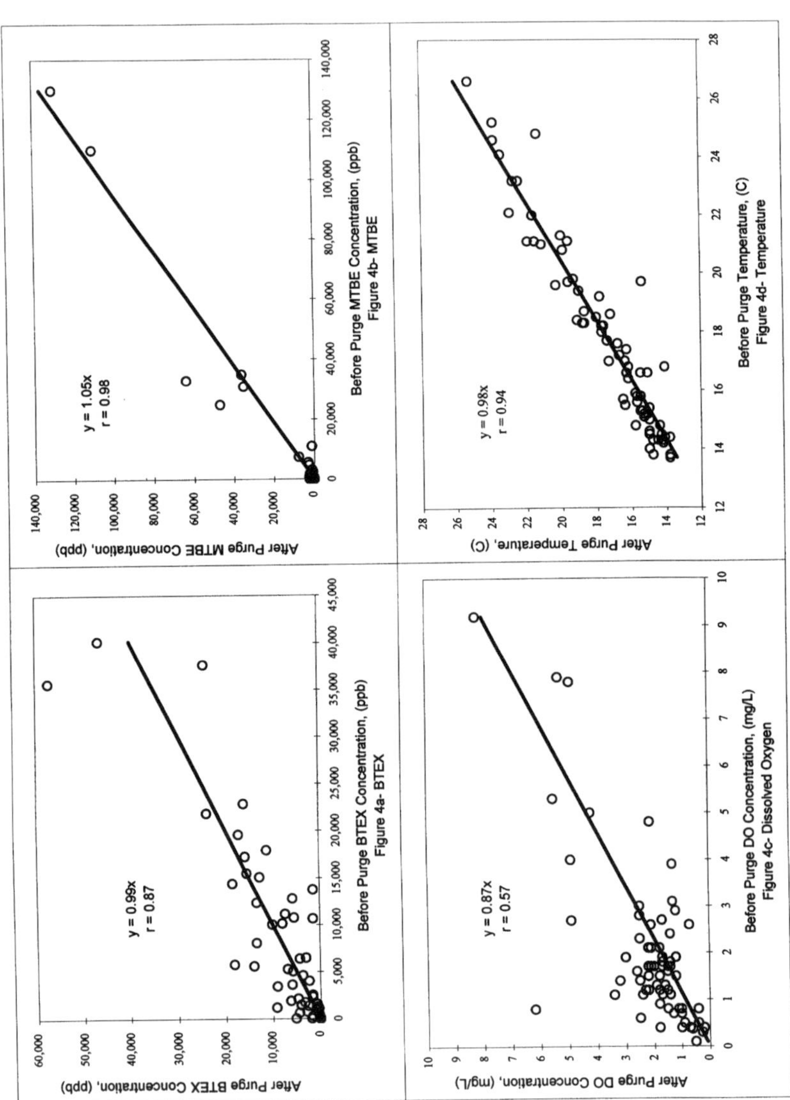

the location of the sampling device intake all contribute to how well the sample represents the adjacent water-bearing zone. It is important to emphasize that for this study a majority of the well purging was completed by hand bailing. Although many groundwater sampling protocols recommend that bailers not be used when sampling for VOCs, hand bailing is the most common. The major concern in using bailers is that they cause the most disturbance within a monitoring well. The raising and lowering of a bailer results in a similar action to that of a plunger, causing significant volatilization of VOCs. Another significant observation is that most studies have involved well screens below the water table. In this case one would expect that a stagnant water column would exist above the well screen. However, in this study well screens extend significantly above the water table, which would eliminate the presence of this stagnant water. Therefore, the entire screened interval would be renewed by natural flow from the adjacent formation.

Water Quality Parameters

The results from the statistical analysis of water quality parameters showed that groundwater was not significantly different before or after purging for both DO and temperature at a 99% confidence interval (Table 2). Scatter plots for DO and temperature are shown in Figures 4c and 4d, respectively. As shown in the figures, the correlation coefficient for temperature is greater than values for DO. This is most likely due to the nature of DO, which tends to be more sensitive and vary more than temperature measurements.

Previous studies have demonstrated that DO levels in stagnant or unpurged well water are greater than samples collected after purging. Barcelona et al. (1994) showed that DO levels were significantly higher (6 times) in unpurged water compared to purged well water. This higher DO level is said to be the direct result of the stagnant water being exposed to oxygen within the well casing headspace. However, findings presented in this study indicate just the opposite, with DO concentrations being greater after purging 63% of the time. A more detailed data analysis shows that higher DO levels after purging were not significant (only 1.9 times higher). In fact, on average, DO levels before and after purging were within 0.8 ppm of each other. These findings further support the opinion that water in the well is being renewed with the adjacent formation water. Once again it must be noted that wells in this study are constructed with screen zones that bridge the water table, whereas most other studies were conducted with wells that were screened below the water table, which produces a stagnant water column.

Hydrogeologic Setting - Long Island vs Upstate

To determine if purging results were affected by different hydrogeologic settings, Long Island and Upstate sites were compared. These comparisons are shown in Figures 5a and 5b. Based on these figures, results for both areas are very similar and indicate that hydrogeologically BTEX and MTBE concentrations before and after purging show a strong correlation. The figures also show that correlation coefficients for BTEX and MTBE for Upstate sites are better than for Long Island sites. This observation contradicts the general opinion. In the literature, little research has been performed in fine-grained or slowly recovering wells (i.e., Upstate) compared to coarse-grained and highly recovering wells (i.e., Long Island). According to Gillham et al. (1983), wells in fine-grained sediments that recharge slowly should not be purged because purging strips the sample of VOCs. McAlary and Barker (1987) reported significant volatilization losses shortly after purging wells in fine-grained sediments. These findings are similar in this study, which indicate higher BTEX concentrations for unpurged samples 60% of the time for Upstate sites. This observation supports the opinion that unpurged samples are more representative of adjacent formation water.

CONCLUSIONS AND RECOMMENDATIONS

Results presented in this study strongly support the elimination of purging monitoring wells that bridge the water table for groundwater sampling of petroleum hydrocarbons, particularly BTEX and MTBE. Comparisons of results showed no significant statistical differences between samples collected by traditional purging methods or samples collected without purging. Furthermore, the elimination of purging has several advantages over traditional purging: (1) it reduces potential cross-contamination between sample wells from purging equipment; 2) it eliminates concerns regarding purge water storage and/or disposal, and; 3) finally, it is important to note that considerable cost savings can be realized from this approach. By eliminating well purging prior to sampling, the savings of the time involved with purging can be substantial.

The primary goal of any sampling effort should be to obtain the most accurate and representative sample. Observations and data presented in this study strongly suggest that modifications to current sampling procedures are justified. From a hydrodynamic standpoint, horizontal laminar flow of groundwater occurs naturally within a monitoring well that is screened across the water table and is representative of groundwater in the adjacent formation. Therefore, it is recommended that wells should not be purged, and water samples analyzed by a laboratory be collected using Teflon® bailers. Bailers should be decontaminated between wells, and dedicated sampling line discarded after each use. Sampling technique should consist of lowering the sample bailer very slowly into groundwater until the entire bailer is full. The filled bailer is then retrieved and the appropriate sample containers filled. The remaining steps should follow standard sampling and chain-of-custody protocols.

Finally, it is important to stress the need for the development of standard groundwater sampling procedures for VOCs, particularly, BTEX and MTBE. These procedures must be consistent, scientifically defensible, and reduce wastes, manpower, and costs. A first step to realizing this goal is the elimination of well purging.

Figure 5.   **After vs. before purge profiles for Upstate and Long Island.**

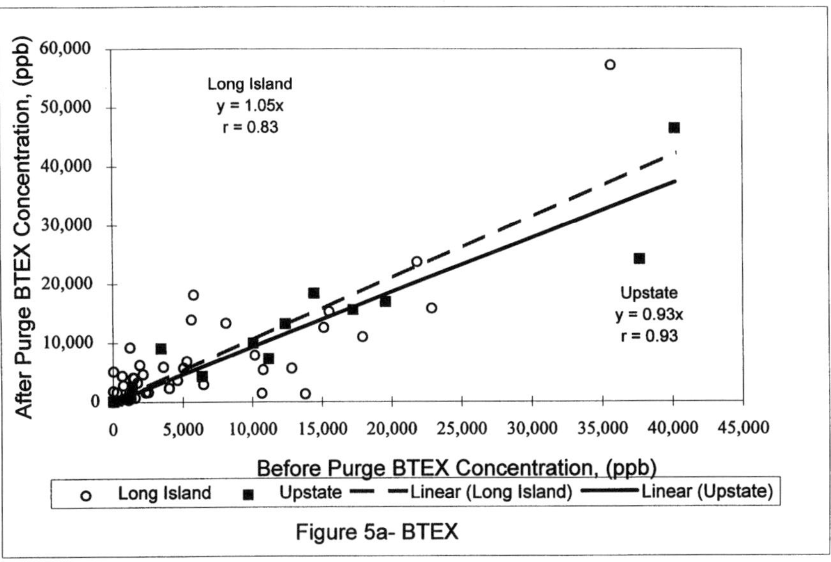

Figure 5a- BTEX

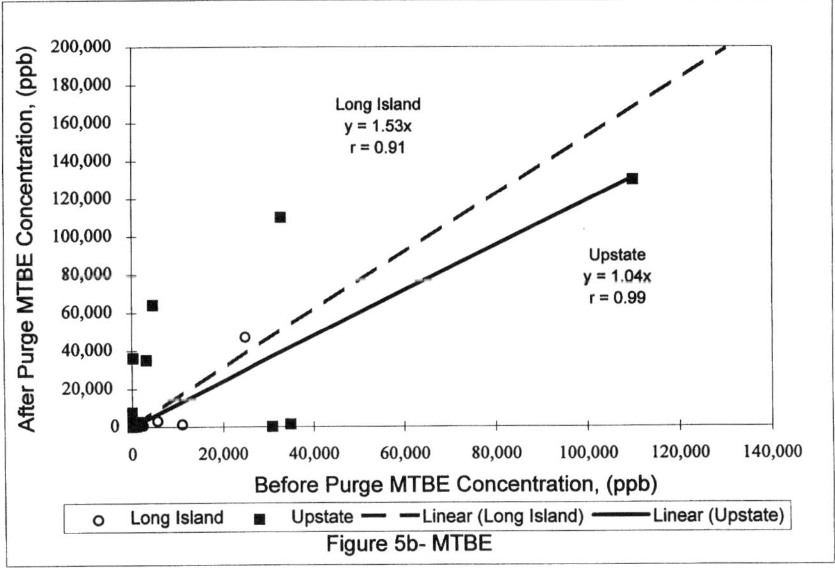

Figure 5b- MTBE

ACKNOWLEDGMENTS

The authors would like to thank Dan Farrier of Shell Oil, Karen Gomez and Chris O'Neil of the NYSDEC for their support and EnviroTrac's sampling gang Ron Kiely, Chris Helmes, Jim and Tim Byrnes for collecting all the groundwater samples.

REFERENCES

Barcelona, M.J., Wehrmann, H.A., and Varljen, M.D. 1994. Reproducible Well Purging Procedures and VOC Stabilization Criteria for Ground-Water Sampling. *Ground Water.* 32-1, 1, 12-22.

Gibs, J. and Imbrigiotta, T. 1990. Well-Purging Criteria for Sampling Purgable Organic Compounds. *Ground Water.* 28-1, 68-78.

Gillham, R.W., Robin, M.J.L., Barker, J.F., and Cherry, J.A. 1983. Groundwater Monitoring and Sample Bias. Environmental Affairs Department, American Petroleum Institute, Washington, DC.

Heath, R.C. 1964. Ground Water in New York: New York State Water Resources Commission Bulletin GW-5I.

Herzog, L., Chou, S.J., Vailkenburg, J.R., and Griffin, R.A. 1988. Changes in Volatile Organic Chemical Concentrations After Purging Slowly Recovering Wells. *Ground Water Monitoring Review,* 8-4, 93-99.

Humenick, M.J., Turk, L.J., and Colchin, M.P. 1980. Methodology for Monitoring Ground Water at Uranium Solution Mines. *Ground Water,* 18-3, 262-273.

Kearl, P.M., Korte, N.E., and Cronk,T.A. 1992. Suggested Modifications to Ground Water Sampling Procedures Based on Observations from Colloidal Borescope. *Ground Water Monitoring Review,* 12-2, 155-61.

Kearl, P.M., Korte, N.E., M. Stites, and J. Baker. 1994. Field Comparison of Micropurging vs. Traditional Ground Water Sampling. Ground Water Monitoring Review 14, no. 4: 183-90

Keely, J.F. and Boateng, K. 1987. Monitoring Well Installation, Purging, and Sampling Techniques - Part 1: Conceptualizations. *Ground Water,* 25-3, 300-13.

McAlary, T.A. and Barker, J.F. 1987. Volatilization Losses of Organics During Ground Water Sampling from Low-Permeability Materials. *Ground Water Monitoring Review,* 7-4, 63-68.

(NJDEPE) 1992. The NJDEPE Field Sampling Procedures Manual.

Pettyjohn, W.A., Dunlop, W.J., Cosby, R., and Keely, W.J. 1981. Sampling Ground Water for Organic Contaminants. *Ground Water,* 19-2, 180-189.

Puls, R.W. and Paul, C.J. 1995. Low-Flow Purging and Sampling of Ground Water Monitoring Wells with Dedicated Systems. *Ground Water Monitoring Review,* 15-1, 116-23.

Puls, R.W. and Powell, R.W. 1992. Acquisition of Representative Ground Water Quality Samples for Metals. *Ground Water Monitoring Review,* 12-3, 167-176

Robin, J.J.L., and R.W. Gillham. 1987. Field Evaluation of Well Purging Procedures. *Ground Water Monitoring Review,* 7-4, 85-93.

Scalf, M.R., McNabb, J.F., Dunlap, W.J., Cosby, R.L., and Fryberger, J.L. 1981. Manual of Ground-Water Quality Sampling Procedures, National Water Well Association, Dublin, OH.

Unwin, J.P. and D. Huis. 1983. A Laboratory Investigation of the Purging Behavior of Small-Diameter Monitoring Wells. In: *Proceedings of the Third Annual Symposium on Ground Water Monitoring and Aquifer Restoration*, pp. 257-262. National Water Well Association, Dublin, OH.

(USEPA) Barcelona, M.J., Gibb J.P., Helfrich J.A., and Garske E.E. 1985. Practical Guide for Ground Water Sampling. EPA/600/2-/85/104.

(USEPA) Fenn, D., Cocozza, E., Isbister, J., Braids, O., Yare, B., and Roux, P. 1977. Procedures Manual for Ground Water Monitoring at Solid Waste Disposal Facilities. EPA/530/SW-611.

(USEPA) 1991b. Handbook: Ground Water Volume II: Methodology. EPA/625/6-90/016b, 141 pp.

(USEPA) 1991. Compendium of ERT Groundwater Sampling Procedures. EPA 540/P-91/007, 63 pp.

Wilson, L.G. and Dworkin, J.M. 1984. Development of a Primer on Well Water Sampling for Volatile Organic Substances, U.S. Geological Survey, Reston, VA.

Biographies

Paul T. Kostecki, Associate Director, Northeast Regional Environmental Public Health Center, School of Public Health, University of Massachusetts at Amherst, received his Ph.D. from the School of Natural Resources at the University of Michigan in 1980. He has been involved with human and ecological risk assessment and risk management research for the last 12 years. Dr. Kostecki has co-authored and co-edited over 50 articles and 16 books on environmental assessment and cleanup including: *Remedial Technologies for Leaking Underground Storage Tanks; Soils Contaminated by Petroleum Products; Petroleum Contaminated Soils, Vols. 1, 2 and 3; Hydrocarbon Contaminated Soils and Groundwater, Vols. 1 2, 3 & 4; Hydrocarbon Contaminated Soils, Vols. 1, 2, 3, 4 & 5; Principles and Practices for Petroleum Contaminated Soils; Principles and Practices for Diesel Contaminated Soils, Vols. 1, 2, 3, 4 & 5; SESOIL in Environmental Fate and Risk Modeling, Contaminated Soils, Volume 1* and *Risk Assessment and Environmental Fate Methodologies*. Dr. Kostecki also serves as Associate Editor for the *Journal of Soil Contamination*, Chairman of the Scientific Advisory Board for *Soil and Groundwater Cleanup Magazine* as well as an editorial board member for the journal of *Human and Ecological Risk Assessment*.

In addition, Dr. Kostecki serves as Executive Director for the Association for the Environmental Health of Soils (AEHS). He is a member of the Navy's National Hydrocarbon Test Site Advisory Board and a member of the Steering Committee for the Total Petroleum Hydrocarbon Criteria Working Group and the Association of American Railroads Environmental Engineering and Operations Subcommittee.

Edward J. Calabrese is a board certified toxicologist who is professor of toxicology at the University of Massachusetts School of Public Health, Amherst. Dr. Calabrese has researched extensively in the area of host factors affecting susceptibility to pollutants, and is author of more than 300 papers in scholarly journals, as well as 24 books, including: *Principles of Animal Extrapolation; Nutrition and Environmental Health, Vols. 1 and 2: Ecogenetic: Safe Drinking Water Act: Amendments, Regulations and Standards; Soils Contaminated by Petroleum: Environmental & Public Health Effects: Petroleum Contaminated Soils, Vols. 1, 2 and 3; Ozone Risk Communication and Management; Hydrocarbon Contaminated Soils, Vols. 1, 2, 3, 4 and 5; Hydrocarbon Contaminated Soils and Groundwater, Vols. 1, 2, 3 and 4; Multiple Chemical Interactions; Air Toxics and Risk Assessment; Alcohol Interactions with Drugs and Chemicals; Regulating Drinking Water Quality; Biological Effects of Low Level Exposures to Chemicals and Radiation; Contaminated Soils; Diesel Fuel Contamination; Risk Assessment and Environmental Fate Methodologies; Principles and Practices for Petroleum Contaminated Soils, Vols. 1, 2, 3, 4 and 5, Contaminated Soils, Volume 1* and *Performing Ecological Risk Assessments*. He has been a member of the U.S. National Academy of Sciences and NATO Countries Safe Drinking Water Committees, and the Board of Scientific Counselors for the Agency for Toxic Substances and Disease Registry (ATSDR). Dr. Calabrese also serves as Director of the Northeast Regional Environmental Public Health Center at the University of Massachusetts and Chairman of the BELLE Advisory Committee.

Marc Bonazountas was born in 1945 in Athens, Greece. He is a civil and environmental scientist, specializing in mathematical modeling of all media: air, soil, water, biota; and in global environmental management procedures via advanced technologies such as geographic information systems (GIS) and remote sensing. Dr. Bonazountas received his engineering degree from the National Technical University of Athens, Greece (1969); his doctorate from the National Technical University of Munich, Germany (1973); his diploma in computer sciences from the Data Processing Institute of Athens (1969). He undertook further studies in water resources systems at the Massachusetts Institute of Technology (1975), and in public policy at Harvard University (1982).

He obtained his Professorship, Habilitation (1984), from the National Technical University of Athens, where he is currently tenured in the Civil Engineering Department, engaged in teaching and research in environmental systems. Prior to joining the National Technical University of Athens, he was with Arthur D. Little, Inc., Cambridge Massachusetts for ten years. He has been involved in numerous US EPA research and technology projects, has developed mathematical models such as SESOIL, and is currently engaged in environmental modeling and telematics-related programs of the European Commission, Brussels. He has authored more than 100 scientific publications and is on the editorial board of several scientific journals: *Journal of Soil Contamination, Journal of Mathematical Environmental Modeling, Journal of Coastal Zone Management,* and others. He is a member of CHESS, the Council for the Health and Environmental Safety of Soils, and a member of several European Commission evaluation and planning committees.

List of Contributors

Mohamed S. Abdel-Rahman, Pharmacology and Physiology Department, New Jersey Medical School, University of Medicine & Dentistry of New Jersey, 185 South Orange Avenue, Rm. 1-655, Newark, NJ 07103-2714

Priscilla Anderson Zieber, EMCON, 18912 North Creek Parkway, Suite 100, Bothell, WA 98011

David Ariail, EPA - Region 4, 345 Courtland St., NE, Atlanta, GA 30365

Mikhail V. Artsimovich, Institute for Nuclear Research, 47 Prospekt Nauki, Kiev-28, 252028, Ukraine

Arthur L. Baehr, U.S. Geological Survey, 810 Bear Tavern Rd., Suite 206, West Trenton, NJ 08628

Ralph S. Baker, ENSR Consulting and Engineering, 35 Nagog Park, Acton, MA 01720

Serge Barbeau, Environmental Engineering Division, City of Montreal, 999 de Louvain Street East, Montreal, Quebec, Canada H2M 1B3

Christopher C. Barker, Department of Civil and Environmental Engineering, University of Massachusetts Lowell, One University Avenue, Lowell, MA 01854

Alessandro Battaglia, Remediation Technologies, Inc., 3040 William Pitt Way, Pittsburgh, PA 15238

Bruce J. Bauman, American Petroleum Institute, 1220 L Street, N.W., Washington, DC 20005

L.J. Buddy Bealer, EnviroTrac Ltd., 561 P Acorn Street, Deer Park, NY 11729

Linda Blalock, NC Department of Environmental Management, Groundwater Section, P.O. Box 29578, Raleigh, NC 27626

Susanne M. Borchert, SBP Technologies, Inc., P.O. Box 76, Aberdeen Proving Ground, MD 21010

Marco D. Boscardin, GEI Consultants, Inc., 1021 Main Street, Winchester, MA 01890

Donald J. Bourgeois, EMCON, 6 Riverside Drive, Suite 101, Andover, MA 01810

Jeffrey L. Breckenridge, Hazardous Toxic and Radioactive Center of Expertise, U.S. Army Corps of Engineers

Jeffrey E. Briglia, EnviroTrac Ltd., 561 P Acorn Street, Deer Park, NY 11729

Clifford J. Bruell, Department of Civil and Environmental Engineering, University of Massachusetts Lowell, One University Avenue, Lowell, MA 01854

Robert A. Burke, Camp Dresser & McKee Inc., Riverside Technology Center, 840 Memorial Drive, Cambridge, MA 02139

J. Patrick Byrnes, EnviroTrac Ltd., 561 P Acorn Street, Deer Park, NY 11729

M.L. Cabrera, Department of Soil and Environmental Science, The University of Georgia, Athens, GA 30605

Bradley A. Call, U.S. Army Corps of Engineers, 1325 J Street, Sacramento, CA 95814-2922

Joseph M. Callanan, Massachusetts Electric Company, 25 Research Drive, Westborough, MA 01582

John H. Carson, Jr., OHM Remediation Services Corporation, 16406 U.S. Route 224 East, Findlay, OH 45839-0551

Dwayne L. Conrad, Texaco EPTD, 4800 Fournace Place, Bellair, TX 77401-2324

William J. Cooper, Drinking Water Research Center, Florida International University, Miami, FL 33193

Eric W. Daley, EMCON, 6 Riverside Drive, Suite 101, Andover, MA 01810

Andrew Dasinger, Sykorsky Aircraft, Connecticut

Seth J. Daugherty, Orange County Environmental Health, 2009 E. Edinger Avenue, Santa Ana, CA 92705

Richard Desbiens, D'Aragon, Desbiens, Halde associes Itee, 555 Rene-Levesque Boulevard West, 19th Floor, Montreal, Quebec, Canada H2Z 1B1

Barbara J. Deshler, IT Corporation, 4330 S. Valley View Blvd., Suite 114, Las Vegas, NV 89103

Richard Desrosiers, SBP Technologies, Inc., 106 Corporate Park Drive, Suite 106, White Plains, NY 10604

Robert T. Drew, American Petroleum Institute 1220 L Street, N.W., Washington, DC 20005

John W. Duggan, Massachusetts Department of Environmental Protection, Northeast Regional Office, 10 Commerce Way, Woburn, MA 01810

Roy Eckart, University of Cincinnati, College of Engineering, Cincinnati, OH 45221

Margaret E. Egbe, Alcoa Technical Center, 100 Technical Drive, Alcoa Center, PA 15069

Gregory S. Fine, Rhode Island Department of Environmental Management, Office of Waste Management, 235 Promenade Street, 3rd Floor, Providence, RI

Paul E. Flathman, OHM Remediation Services Corporation, 16406 U.S. Route 224 East, Findlay, OH 45839-0551

P. Forbes, Air Force Base Conversion Agency, Loring, ME

Thomas L. Foster, EMCON, 15055 SW Sequoia Parkway, Suite 140, Portland, OR 97224

Timothy D. Foster, Land Tech Remedial, Inc., 569 Main Street, Monroe, CT 06468

Steve J. Gagnon, U.S. Army Corps of Engineers, Waltham, MA

Vivek Galav, Department of Civil & Architectural Engineering, University of Miami, Coral Gables, FL 33124

Yakov Galperin, Global Geochemistry Corporation, 6919 Eton Avenue, Canoga Park, CA 91303-2194

Charles George, U.S. Army Environmental Center, Building E-4480, Edgwood Area, Aberdeen Proving Ground, MD 21010-5401

Jason A. Gerrish, Remediation Technologies, Inc., 9 Pond Lane, Concord, MA 01742-2851

John K. Ghirardi, GEI Consultants, Inc., 1021 Main Street, Winchester, MA 01890-1970

Gregory M. Gibbons, Parsons Engineering Science, Inc., 1000 Jorie Boulevard, Suite 250, Oak Brook, IL 60521

Tej Gidda, School of Engineering, University of Guelph, Ontario, Canada N1G 2W1

Ileen S. Gladstone, GEI Consultants, Inc., 1021 Main Street, Winchester, MA 01890-1970

Mohammad H. Golabi, Department of Soil and Environmental Science, The University of Georgia, Athens, GA 30605

David L. Gottshall, Camp Dresser & McKee Inc., Riverside Technology Center, 840 Memorial Drive, Cambridge, MA 02139

Veronica Groshko, EFX Systems, Inc., 3900 Collins Road, Suite 1011, Lansing, MI 48910

Jeffrey A. Hamel, EMCON, 6 Riverside Drive, Suite 101, Andover, MA 01810

James B. Harrington, Low Temperature Thermal Desorption Task Group, New York State Department of Environmental Conservation

Thomas Hayes, Gas Research Institute, 8600 West Bryn Mawr Avenue, Chicago, IL 60631

Robert F. Hickey, EFX Systems, Inc., 3900 Collins Road, Suite 1011, Lansing, MI 48910

Dalibor Hodko, Lynntech, Inc., 7610 Eastmark Drive, Suite 105, College Station, TX 77840.

James Holly, Geophex, Ltd., 605 Mercury Street, Raleigh, NC 27603

D. Hopkins, AFBCA, Loring Air Force Base, ME

John E. Hunt, Barr Engineering Company, 8300 Norman Center Drive, Suite 300, Minneapolis, MN 55437-1026

Lubna Hussain, School of Engineering, University of Guelph, Ontario, Canada N1G 2W1

Talaat Ijaz, McLaren/Hart ChemRisk, 29225 Chagrin Blvd., Cleveland, OH 44122

Douglas E. Jerger, OHM Remediation Services Corporation, 16406 U.S. Route 224 East, Findlay, OH 45839-0551

Evan T. Johnson, Tighe & Bond, Inc., 53 Southampton Road, Westfield, MA 01085

Lyle A. Johnson, Western Research Institute, Laramie, WY 82071

Mark Johnson, DAHL & Associates, Inc., 4390 McMenemy Road, St. Paul, MN 55127

Mark L. Jones, GEO Consultants, Inc., 1021 Main Street, Winchester, MA 01890

Stephen L. Kane, Photovac Monitoring Instruments, 25-B Jefryn Blvd. W., Deer Park, NY 11729

Jimmy Kao, Geophex, Ltd., 605 Mercury Street, Raleigh, NC 27603

Isaac R. Kaplan, Global Geochemistry Corporation, 6919 Eton Avenue, Canoga Park, CA 91303-2194

Raymond S. Kasevich, KAI Technologies, Inc. 170 West Road, Suite 4, Portsmouth, NH 03801

Joan Kelly, NET Atlantic, Windham, NH

Linda Kelley, ICF Kaiser Engineers, Inc., 1800 Harrison Street, Oakland, CA 94612

William R. Kerfoot, K-V Associates, Inc., 281 Main Street, Falmouth, MA 02540

Marvin Kirshner, Port Authority of New York and New Jersey, One World Trade Center, New York, NY 10048

Yefim I. Knizhnik, 2201 Strahle Street, Apt. D-304, Philadelphia, PA 19152

Nic E. Korte, Oak Ridge National Laboratory 2597 B + 3/4 Rd., Grand Junction, CO 81503

David S. Kosson, Department of Chemical and Biochemical Engineering, Rutgers University, Piscataway, NJ 08855-0909

Rajesh T. Kukreja, Lynntech, Inc., 7610 Eastmark Drive, Suite 105, College Station, TX 77840.

Donna R. Kuroda, Headquarters, U.S. Army Corps of Engineers

Charles N. Kurucz, Department of Management Science & Industrial Engineering, University of Miami, Coral Gables, FL 33124

Fayaz S. Lakhwala, SBP Technologies, Inc., One Sabine Island Drive, Gulf Breeze, FL 32561

Michael A. Lamarre, Mobil Business Resources Corporation, 464 Doughty Boulevard, 2nd Floor/Environmental, Inwood, NY 11096-1342

Wayne W. Lapham, U.S. Geological Survey, 12201 Sunrise Valley Dr. MS#413, Reston, VA 20192

Mary L. Laski, OHM Remediation Services Corporation, 16406 U.S. Route 224 East, Findlay, OH 45839-0551

Paul R. Lear, OHM Remediation Services Corporation, 16406 U.S. Route 224 East, Findlay, OH 45839-0551

John T. Lee, Barr Engineering Company, 8300 Norman Center Drive, Suite 300, Minneapolis, MN 55437-1026

Alfred P. Leuschner, Remediation Technologies, Inc., 9 Pond Lane, Suite 3A, Concord, MA 01742-2851

Sharon C. Long, Civil and Environmental Engineering, University of Massachusetts, Amherst, MA 01003-5205

David H. Lucas, Land Tech Remedial, Inc., 569 Main Street, Monroe, CT 06468

Christopher C. Lutes, Acurex Environmental Corporation, 4915 Prospectus Dr., Research Triangle Park, NC 27713

Brian Magee, Ogden Environmental & Energy Services, Westford, MA

Mark T. Mahoney, GEI Consultants, Inc., 1021 Main Street, Winchester, MA 01890

John E. Main, Malden Mills Industries, Inc., 46 Stafford Street, P.O. Box 809, Lawrence, MA 01842

John C. McMillan, ICF Kaiser Engineers, Inc., 1800 Harrison Street, Oakland, CA 94612

Charles A. Menzie, Menzie-Cura & Associates, Inc., 1 Courthouse Lane, Suite 2, Chelmsford, MA 01824

W.P. Miller, Department of Soil and Environmental Science, The University of Georgia, Athens, GA 30605

Mark W. Moeller, Remediation Technologies, Inc., 9 Pond Lane, Concord, MA 01742-2851

Abraham Morduchowitz, Texaco Inc., P.O. Box 509, Beacon, NY 12508

Daniel Morin, D'Aragon, Desbiens, Halde associes Itee, 555 Rene Levesque Boulevard West, 19th Floor, Montreal, Quebec, Canada H2Z 1B1

L.A. Morris, Department of Soil and Environmental Science, The University of Georgia, Athens, GA 30605

Connie E. Morton, Malden Mills Industries, Inc., 46 Stafford Street, P.O. Box 809, Lawrence, MA 01842

Douglas G. Mose, Chemistry Department, George Mason University, Fairfax, VA 22030

J. Mueller, Air Force Center for Environmental Excellence, Loring Air Force Base, ME

James G. Mueller, SBP Technologies, Inc., One Sabine Island Drive, Gulf Breeze, FL 32561

Oliver J. Murphy, Lynntech, Inc., 7610 Eastmark Drive, Suite 105, College Station, TX 77840

George W. Mushrush, Chemistry Department, George Mason University, Fairfax, VA 22030

David Nakles, Remediation Technologies, Inc., One Monroeville Center, Suite 1015, Monroeville, PA 15146-2121

James F. Occhialini, Camp Dresser & McKee Inc., Riverside Technology Center, 840 Memorial Drive, Cambridge, MA 02139

Timothy M. O'Connor, Rhode Island Department of Environmental Management, Office of Waste Management, 235 Promenade Street, 3rd Floor, Providence, RI

Oscar Pancorbo, Massachusetts Department of Environmental Protection, Wall Experiment Station, Lawrence, MA

James F. Pankow, Oregon Graduate Institute of Science and Technology, 20000 NW Walker Rd., Beaverton, OR 97006

Robert A. Paschke, 3M Environmental Technology & Services, P.O. Box 3331, St. Paul, MN 55133-3331

Jeffrey A. Peterson, EMCON, 15055 SW JSequoia Parkway, Suite 140, Portland, OR 97224

Peter Peuron, Orange County Environmental Health, 2009 E. Edinger Avenue, Santa Ana, CA 92705

Ian M. Phillips, GEI Consultants, Inc., 1021 Main Street, Winchester, MA 01890-1970

Eric Powers, Geophex, Ltd., 605 Mercury Street, Raleigh, NC 27603

Nicholas C. Pressly, Pressly Associates, Inc., 15 Ocean Place, Brookhaven, NY 11719

Stephen L. Price, KAI Technologies, Inc., 170 West Road, Suite 4, Portsmouth, NH 03801

Victor S. Prokopenko, Interdisciplinary Scientific and Technical Center "Shelter," Chernobyl City, 255620, Ukraine

Raj V. Rajan, EFX Systems, Inc., 3900 Collins Road, Suite 1011, Lansing, MI 48910

Wanda Ratliff, Bay State Gas Company, Westboro, MA

Mooncalf Reilly, EMCON, 15055 SW Sequoia Parkway, Suite 140, Portland, OR 97224

Chris Renda, Environmental Services Network

Robert J. Roth, Terra Vac-Mid Atlantic Division, 92 N. Main Street, Windsor, NJ 08561

Stephan T. Roy, GZA GeoEnvironmental, Inc., 320 Needham Street, Newton Upper Falls, MA 02164

Eugene Rutz, University of Cincinnati, College of Engineering, Cincinnati, OH 45221

David K. Ryan, Department of Chemistry, University of Massachusetts Lowell, One University Avenue, Lowell, MA 01854

Abigail Salb, School of Engineering, University of Guelph, Ontario, Canada N1G 2W1

John R. Schuring, Department of Civil and Environmental Engineereing, New Jersey Institute of Technology, Newark, NJ 07102

April Seybold, EFX Systems, Inc., 3900 Collins Road, Suite 1011, Lansing, MI 48910

Brett Shamory, Remediation Technologies, Inc., 3040 William Pitt Way, Pittsburgh, PA 15238

Christine J. Shields, Parsons Engineering Science, Inc., 1000 Jorie Boulevard, Suite 250, Oak Brook, IL 60521

Johnette C. Shockley, U.S. Army Corps of Engineers (U.S. Geological Survey)

Gloria A. Skowronski, Pharmacology and Physiology Department, New Jersey Medical School, University of Medicine & Dentistry of New Jersey, Newark, NJ 07103

John R. Smith, Alcoa Technical Center, 100 Technical Drive, Alcoa Center, PA 15069

John C. Snowden, ABB Environmental Services, Inc., 110 Free Street, Portland, ME 04101

Charles K. So, OHM Remediation Services Corporation, 5731 West Las Positas Boulevard, Pleasanton, CA 94588

Warren H. Stiver, School of Engineering, University of Guelph, Ontario, Canada N1G 2W1

D. St. Peter, AFBCA, Loring Air Force Base, ME

Jim Sullivan, Massachusetts Department of Environmental Protection, Wall Experiment Station, Lawrence, MA

M.E. Sumner, Department of Soil and Environmental Science, The University of Georgia, Athens, GA 30605

Lynn Sutton, Geophex, Ltd., 605 Mercury Street, Raleigh, NC 27603

Alexander Taft, Boston Gas Company, Boston, MA

Vladimir V. Tokarevsky, Interdisciplinary Scientific and Technical Center "Shelter," Chernobyl City, 255620, Ukraine

Jason R. Trausch, OHM Remediation Services Corporation, 16406 U.S. Route 224 East, Findlay, OH 45839-0551

Rita M. Turkall, Pharmacology and Physiology Department, New Jersey Medical School, and Clinical Laboratory Sciences Department, School of Health Related Professions, University of Medicine and Dentistry of New Jersey, Newark, NJ 07103

S. Underhill, Bechtel Environmental, Inc., Oak Ridge, TN

Zeynep Ungun, ICF Kaiser Engineers, Inc., 1800 Harrison Street, Oakland, CA 94612

Dennis F. Unites, Atlantic Environmental Services, Inc., 188 Norwich Avenue, P.O. Box 297, Colchester, CT 06415

Sankar N. Venkatraman, Bioremediation Applications Group, McLaren/Hart Inc., 25 Independence Blvd., Warren, NJ 07059

Dan Wagner, EFX Systems, Inc., 3900 Collins Road, Suite 1011, Lansing, MI 48910

Thomas D. Waite, Department of Civil & Architectural Engineering, University of Miami, Coral Gables, FL 33124

Michael P. Walker, GEO Consultants, Inc., 1021 Main Street, Winchester, MA 01890

Evelyn R. Washington, U.S. Environmental Protection Agency, 401 M Street, S.W., Washington, DC 20460

Dan Wiberg, DAHL & Associates, Inc. 4390 McMenemy Road, St. Paul, MN 55127

T.R. Wood, Bechtel Environmental Inc., Oak Ridge, TN

Patrick M. Woodhull, OHM Remediation Services Corporation, 16406 U.S. Route 224 East, Findlay, OH 45839-0551

Michael Wright, Tyree Organization Ltd., 9 Otis Street, Westborough, MA 01581

Dawn A. Zemo, Geomatrix Consultants, Inc., 100 Pine Street, 10th Floor, San Francisco, CA 94111

John F. Zipeto, Bechtel Environmental, Inc., One South Station, Boston, MA 02210

John S. Zogorski, U.S. Geological Survey, 1608 Mountain View Rd., Rapid City, SD 57702

Richard G. Zytner, School of Engineering, University of Guelph, Ontario, Canada N1G 2W1

Glossary of Acronyms

ABB-ES	ABB Environmental Services, Inc.
ABC	asphalt, brick, and concrete
AFB	Air Force Base
AFBCA	Air Force Base Conversion Agency
AFCEE	Air Force Center for Environmental Excellence
AIS	air injection system
AOC	Area of Contamination
APHA	American Public Health Association
APSR	aqueous phase saturation ratio
ARARs	applicable or relevant and appropriate requirements
ARPA	Advanced Research Projects Agency
AS	air sparging
ASA	American Society of Agronomy
AST	aboveground storage tank
ASTM	American Society for Testing and Materials
BAF	bioaccumulation factor
bgs	below ground surface
BOD	biological oxygen demand
BRAC	base realignment and closure
BTEX	benzene, ethylbenzene, toluene, and xylenes
BUD	beneficial use determination
BXSS	Base Exchange Service Station
CA	carbon adsorption
CA/T	Central Artery/Tunnel (Project)
CBs	chlorobenzenes
CCD	charge-coupled device (camera)
CDM	Camp, Dresser, and McKee Consultants, Inc.
CEET	Center for Environmental Engineering and Science Technologies
CERCLA	Comprehensive Environmental Response, Compensation and Liability Act
cfm	cubic feet per minute
cfu	colony-forming unit
CHCs	chlorinated hydrocarbons
CKD	cement kiln dust
CLP	Contract Laboratory Program
CMC	critical micellar concentration
COPC	chemical of potential concern
CPG	critical population groups
CPs	chlorophenols
CQMSP	chemical quality management and sampling plan
CROW™	Contained Recovery of Oil Wastes
CTDEP	Connecticut Department of Environmental Protection
DAR	design analysis report
DCDD	2,7-dichlorodibenzo-p-dioxin
DCE	dichloroethene

DDH	D'Aragon, Desbiens Halde associes Itee
DEP	Department of Environmental Protection
DL	detection limit
DNAPL	dense nonaqueous phase liquid
DO	dissolved oxygen
DOD	Department of Defense
DOE	Department of Energy
DOIT	Demonstration of Innovative Technology (project)
DPE	dual phase extraction
DRO	diesel range organic
DSL	Delhi Loamy Sand
DSS	design support study
DTDs	dielectric track detectors
DTSC	Department of Toxic Substances Control
DW	dry weight
EAEs	environmentally acceptable endpoints
EDD	electronic data delivery
EE/CA	engineering evaluation and cost analysis
EF	enhancement factor
ELISHA	enzyme linked immunosorbent assay
EOD	Explosive Ordnance Demolition
EPA	Environmental Protection Agency
EPC	exposure point concentration
EPH	extractable petroleum hydrocarbons
EPRI	Electric Power Research Institute
ES	entomology shop
ESL	Elora Silt Loam
FBR	fluidized bed reactor
FDTD	finite difference time domain
FFS	Focused Feasibility Study
FHWA	Federal Highway Administration
FID	flame ionization detector
FIVE	fluid injection with vacuum extraction
FJETC	former jet engine test cell
FS	feasibility study
FSSB	former solvent storage building
FTA	fire training area
FTF	fuel tank farm
GAC	granular activated carbon
GAC-FBR	granular activated carbon-fluidized bed reactor
GES	groundwater extraction system
GC	gas chromatography
GC/MS	gas chromatography/mass spectrophotometer
GCW	groundwater circulation well
GIS	Geographical Information System
GLSD	Greater Lawrence Sanitary District

GM	geometric mean
GRI	Gas Research Institute
GROs	gasoline range organics
GS	guide specifications
GSD	geometric standard deviation
GWV	groundwater vent
HAZWRAP	Hazardous Waste Remedial Actions Program
HC	hydrocarbons
HCN	hydrogen cyanide
HDPE	high density polyethylene
HEAST	Health Effects Assessment, Summary Tables
HHRA	human health risk assessment
HI	hazard index
HPLC	high pressure liquid chromatograph
HRLs	health risk limits
HSI	hue-saturation-intensity
HTF	heat transfer fluid
HTRW	hazardous, toxic, and radioactive waste
IR	Interim Response
IRIS	Integrated Risk Information System
I.D.	inside diameter
IRA	Immediate Response Actions
ITA	Innovative Technology Advocate (program)
ITRC	Interstate Technical and Regulatory Cooperation Working Group
KBP	kiln by-product
LDD	least detectable dose
LF	landfill
LHS	Latin hypercube sampling
LNAPL	light nonaqueous-phase liquid
LRT	liquid release test
LSP	Licensed Site Professional
LTTD	Low Temperature Thermal Desorption (task group)
LTU	land treatment unit
MADEP	Massachusetts Department of Environmental Protection
MAHs	monocyclic aromatic hydrocarbons
MASSPORT	Massachusetts Port Authority
MCDD	2-monochlorodibenzo-p-dioxin
MCP	Massachusetts Contingency Plan
MDEP	Massachusetts Department of Environmental Protection
MDL	method detection limit; minimum detection limit
MDPW	Massachusetts Department of Public Works
MDRD	minimum detectable relative difference
MEF	Ministry of the Environment and Wildlife (Quebec)
MEP	master environmental plan
MFX	Model-Free Approach to Low-Dose Extrapolation
MGP	manufactured gas plant

MHD   Massachusetts Highway Department
MODF   mineral oil dielectric fluid
MOU   Memorandum of Understanding
MPN   most probable number
MTBE   methyl-tertiary-butyl-ether
MTS   materials testing site
NAPL   nonaqueous phase liquids
nCHCs   nonchlorinated hydrocarbons
MCRP   National Committee on Radiation Protection
ND   not detectable
NHEERL   National Health and Environmental Effects Research LaboratoryNIBS
            National Institute of Building Sciences
NIOSH   National Institute for Occupational Safety and Health
NOAEL   no observed adverse effect level
NPDES   National Pollution Discharge Elimination System
NPL   National Priority List
NRC   National Research Council
NSF   navy special fuel
NTS   Nevada Test Site
NYSCHWM   New York State Center for Hazardous Waste Management
NYSDEC   New York State Department of Conservation
O.D.   outside diameter
OE*Cert*   Ordnance and Explosives Cost-Effectiveness Risk Tool
OHM   oil and hazardous materials
O&M   operation and maintenance
ONR   Office of Naval Research
OS   oil sump
OSL   oleophilic suction lysimetry
OUs   Operable Units
PA   Preliminary Assessment
PAC   physiologically available cyanide
PAH   polycyclic aromatic hydrocarbons
PCA   pollution control agency
PCBs   polychlorinated biphenyls
PCE   tetrachloroethylene
pcf   pounds per cubic feet
PID/FID   photo ionization/flame ionization detector
PHC   petroleum hydrocarbon
PHDB   petroleum hydrocarbon oxidizing microorganisms
PLC   programmable logic controller
PNAs   polynuclear aromatics
PPDP   power plant drainage pipe
PP&L   Pennsylvania Power and Light Company
PQL   practical quantitative limits
PRGs   preliminary remediation goals
PRP   potentially responsible party

PSA	pressure swing adsorption
PSD	pore size distribution
psf	pounds per square foot
psi	pounds per square inch
QA/QC	quality assurance/quality control
QM	quotient method
RAAs	Remedial Action Alternatives
RALs	Recommended Allowable Limits
RAM	random access memory
RAO	Response Action Outcome
RAP	remedial action plan
RCC	rank correlation coefficient
RCRA	Resource Conservation and Recovery Act
R&D	research and development
REA	radiological exposure assessment
RETEC	Remediation Technologies, Inc.
RfDs	reference doses
RFH	radio frequency heating
RI	remedial investigation
RI/FS	Remedial Investigation/Feasibility Study
RME	reasonable maximally exposed
ROD	Record of Decision
RPD	relative percent difference
RS	Remedial Strategy
SA	study areas
SAB	special Antarctic blend
SAR	specific absorption rate
SCAPS	Site Characterization and Analysis Penetrometer System
SDS	sodium dodecyl sulfate
SI	Site Investigation
SITE	Superfund Innovative Technology Evaluation (program)
SPH	separate phase hydrocarbons
SQC	sediment quality criteria
SRBC	soil risk-based concentration
SRS	simple random sampling
SSI	supplemental site investigation
SSSA	Soil Science Society of America
SVE	soil vapor extraction
SVES	soil vapor extraction system
SVOCs	semivolatile volatile organic compounds
SVP	soil vapor probe
SVV	soil vapor vent
TCD	thermal conductivity detector
TCDD	2,3,7,8-tetrachlorodibenzo-p-dioxin
TCE	trichloroethylene
TCL	target compound list

TCLP	Toxicity Characteristic Leaching Procedure
TCRA	time-critical removal action
TEA	terminal electron acceptors
ThOD	theoretical oxygen demand
TOC	total organic carbon
TOS	total organic solvent
TOV	total organic vapor
TOX	total organic halogen
TPH	total petroleum hydrocarbons
TRI	toxic release inventory
TriCDD	2,3,7-trichlorodibenzo-p-dioxin
TRV	toxicity reference value
TSCA	Toxic Substances Control Act
tsf	ton per square foot
TSTG	Technology Specific Task Group
TTR	Tonopah Test Range
2-D	two-dimensional
UCS	unconfined compressive strength
URI	University Research Initiative (program)
USDA	United States Department of Agriculture
USTs	underground storage tanks
VES	vapor extraction system
VLS	very low strength
VOA	volatile organic analysis
VOC	volatile organic compound
UF	uncertainty factor
URS	URS Consultants, Inc.
USAEC	U.S. Army Environmental Center
USEPA	United States Environmental Protection Agency
USTs	underground storage tanks
UXO	unexploded ordnance
VMB	vehicle maintenance building
WQC	water quality criteria
WRBC	water-risk based concentration
WRI	Western Research Institute
WSF	water-soluble fraction

Index